Political Map of the World

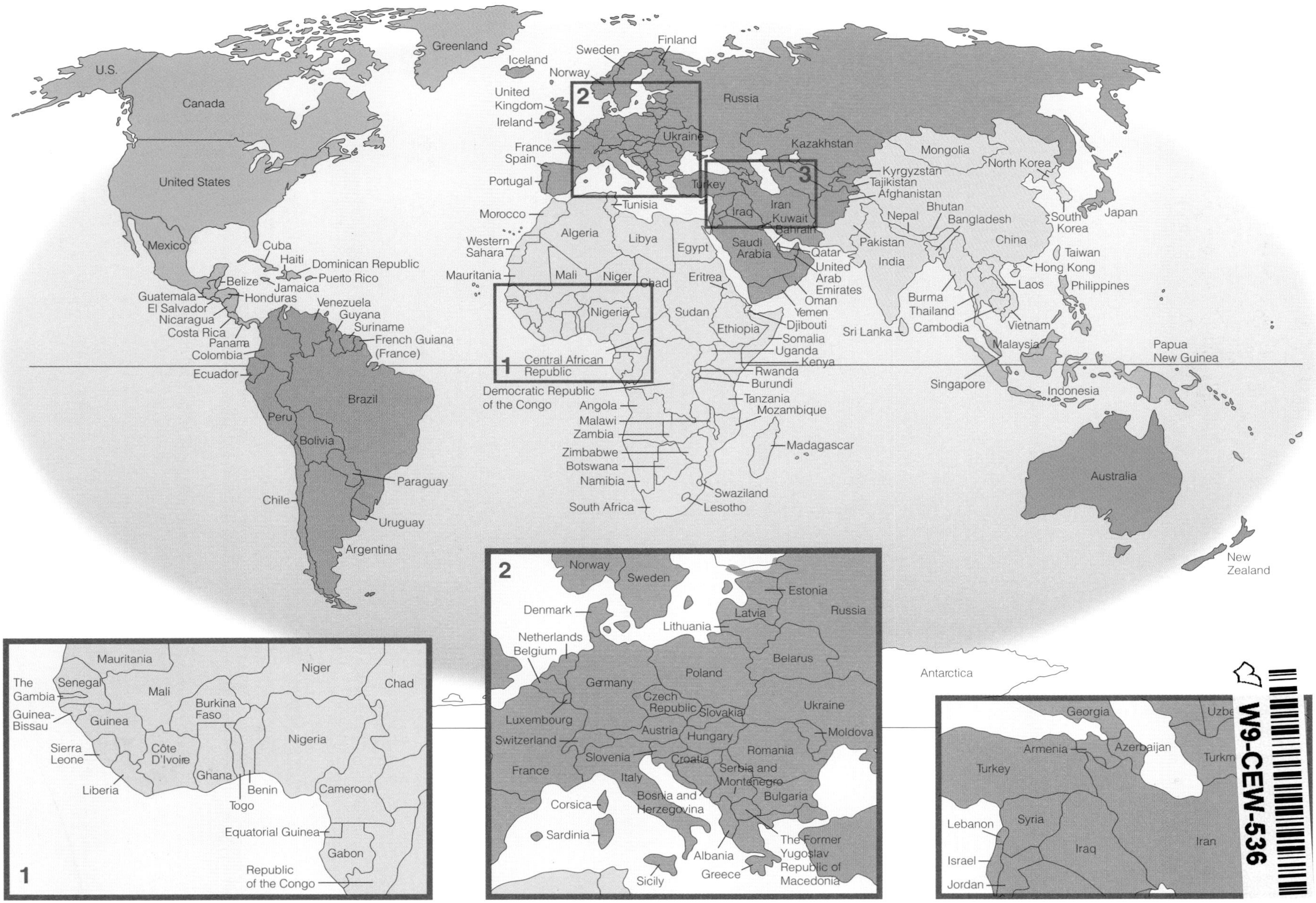

U.S.
Canada
United States
Mexico
Greenland
Iceland
Cuba
Haiti
Dominican Republic
Puerto Rico
Jamaica
Belize
Honduras
Guatemala
El Salvador
Nicaragua
Costa Rica
Panama
Colombia
Venezuela
Guyana
Suriname
French Guiana
(France)
Ecuador
Peru
Brazil
Bolivia
Chile
Paraguay
Uruguay
Argentina
Norway
Sweden
Finland
United
Kingdom
Ireland
France
Spain
Portugal
Ukraine
Russia
Kazakhstan
Mongolia
North Korea
South
Korea
Japan
China
Turkey
Iran
Iraq
Kuwait
Bahrain
Kyrgyzstan
Tajikistan
Afghanistan
Nepal
Bhutan
Bangladesh
Pakistan
India
Saudi
Arabia
Qatar
United
Arab
Emirates
Oman
Yemen
Taiwan
Hong Kong
Philippines
Laos
Burma
Thailand
Vietnam
Cambodia
Sri Lanka
Malaysia
Singapore
Indonesia
Papua
New Guinea
Australia
New
Zealand
Morocco
Western
Sahara
Mauritania
Tunisia
Algeria
Libya
Egypt
Mali
Niger
Chad
Nigeria
Eritrea
Sudan
Ethiopia
Djibouti
Somalia
Uganda
Kenya
Rwanda
Burundi
Tanzania
Mozambique
Madagascar
Central African
Republic
Democratic Republic
of the Congo
Angola
Malawi
Zambia
Zimbabwe
Botswana
Namibia
Swaziland
Lesotho
South Africa
Antarctica

1
Mauritania
Niger
Chad
The
Gambia
Senegal
Mali
Guinea-
Bissau
Guinea
Burkina
Faso
Nigeria
Sierra
Leone
Côte
D'Ivoire
Liberia
Ghana
Benin
Togo
Cameroon
Equatorial Guinea
Gabon
Republic
of the Congo

2
Norway
Sweden
Estonia
Russia
Denmark
Latvia
Lithuania
Netherlands
Belgium
Belarus
Poland
Germany
Czech
Republic
Slovakia
Ukraine
Luxembourg
Austria
Moldova
Switzerland
Hungary
Romania
Slovenia
Croatia
France
Serbia and
Montenegro
Italy
Bosnia and
Herzegovina
Bulgaria
Corsica
Sardinia
Albania
The Former
Yugoslav
Republic of
Macedonia
Sicily
Greece

3
Georgia
Uzbe
Armenia
Azerbaijan
Turkm
Turkey
Syria
Lebanon
Iraq
Iran
Israel
Jordan

THE ENGAGING, APPLIED BEST-SELLER

Kendall's chapter-opening vignettes, timely examples, dynamic visuals, companion videos, and assignable features help students sharpen their sociological skills and connect with sociology!

By introducing chapter topics through first-person accounts of diverse people's experiences and grounding her text's comprehensive coverage of concepts, theories, and research in familiar themes such as homelessness, body image, and global warming, Diana Kendall expertly guides students on an exciting journey that lets them *experience* sociology and see for themselves how it is relevant to their lives. Kendall's personal writing style, emphasis on applications, and compelling everyday examples engage readers and bring sociology to life. The Eighth Edition continues to emphasize diversity and culture and provides updated coverage of topics such as global politics and economics and health care, all along showing students how they can make a difference in their own communities and beyond.

As you peruse the pages of this *Preview*, you'll notice the consistent applied nature of the topic coverage and the repeated connections to students' lives throughout the each chapter . . . just a few of the features that set this book above the rest!

NEW TO THIS EDITION

Kendall expertly illustrates sociology's connections to students' lives

Throughout the Eighth Edition, Diana Kendall uses compelling features to illustrate the various connections between sociology and daily life. With Kendall's fascinating stories and visuals, students quickly learn to see sociology in all aspects of their own lives, and begin to see the relevance of sociology to the world around them.

Kendall's popular chapter-opening first-person commentaries bring a particular social issue to life by presenting diverse peoples' life experiences—and present the chapter topic in concrete terms to which students can relate. The social issue highlighted in the opener is then used as an example throughout the chapter to illustrate chapter concepts. Megan Meier, the teen who committed suicide in response to fake MySpace postings; Supreme Court justice nominee Sonia Sotomayor; and President Barack Obama are among the individuals whose voices are new to this edition.

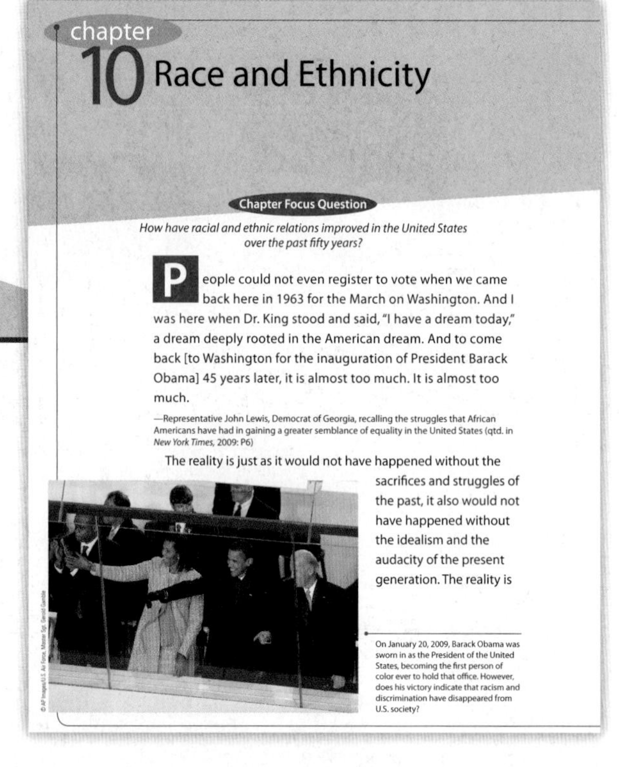

chapter 10 Race and Ethnicity

Chapter Focus Question

How have racial and ethnic relations improved in the United States over the past fifty years?

People could not even register to vote when we came back here in 1963 for the March on Washington. And I was here when Dr. King stood and said, "I have a dream today," a dream deeply rooted in the American dream. And to come back [to Washington for the inauguration of President Barack Obama] 45 years later, it is almost too much. It is almost too much.

—Representative John Lewis, Democrat of Georgia, recalling the struggles that African Americans have had in gaining a greater semblance of equality in the United States (qtd. in *New York Times*, 2009: P6)

The reality is just as it would not have happened without the sacrifices and struggles of the past, it also would not have happened without the idealism and the audacity of the present generation. The reality is

© AP Images/U.S. Air Force, Master Sgt. Gerold Gamble

On January 20, 2009, Barack Obama was sworn in as the President of the United States, becoming the first person of color ever to hold that office. However, does his victory indicate that racism and discrimination have disappeared from U.S. society?

Sociology *Works!*

Why *Place* Matters in Global Poverty

We're deadly poor. We grow just enough food for ourselves to eat, with no surplus grain. We don't have to pay the grain tax anymore, but our lives aren't much better.

—Zhou Zhiwen, a woman who lives in Yangmiao, China, describing what rural poverty is like in her village (qtd. in French, 2008: YT4)

Zhou Zhiwen lives in an area of China that has largely been untouched by the economic boom in her country. Even with the recent abolition of agricultural taxes for people who are impoverished, local villagers such as Zhou continue to live in abject poverty. As more people have risen out of poverty in China's urban centers in recent decades, poverty in the rural areas, mountainous regions, and deserts remains severe. According to some villagers, the central government is "out of touch with rural realities in places like this," and officials have made little effort to take care of the rural poor (French, 2008). China's poverty is widespread, and the income gap between rural and urban residents has widened over the past three decades (IFAD, 2002). Many people live close to, or below, the minimum standard for poverty: Approximately 350 million people in China live below the international poverty line of $1.25 per day (World Bank, 2007, 2009).

For many years, sociologists studying poverty have focused on differences in rural and urban poverty. Throughout the world, they have found that *place* does matter when it comes to finding the deepest pockets of poverty (Rural Policy Research Institute, 2004). Where people live strongly influences how much money they will make, and income inequalities are important indicators of the life chances of entire families. In some developing countries, the rural poor rely primarily on agriculture, fishing, forestry, and sometimes small-scale industries and services for their livelihood (Khan, 2001). When they are unable to derive sufficient economic resources from these endeavors, little else is available to them. Some migrate to urban centers in hopes of finding new opportunities, but many remain behind, living in grinding poverty. When sociologists speak of "place," they are referring to such things as an area's natural environment, which includes its climate, natural resources, and degree of isolation (Rural Policy Research Institute, 2004). Place also involves the economic structure in the area, such as the extent to which adequate amounts of food can be raised to meet people's needs, or whether an individual can earn sufficient money to purchase food.

Can rural poverty be reduced? According to some social policy analysts, broad economic stability, competitive markets, and public investment in *physical* and *social* infrastructure are important prerequisites for a reduction in rural poverty in developing nations (Khan, 2001). From this perspective, a major reduction in rural poverty in China will occur only if people have access to land and credit, education, health care, support services, and entitlements to food through well-designed public works programs and other transfer mechanisms.

Reflect & Analyze

Whether changes that reduce poverty will occur in China's future remains to be seen, but the sociological premise that *place* matters in regard to poverty remains a valid assumption in helping us explain global poverty and inequality. Can you apply this idea to rural and urban areas with which you are familiar in the United States? Why are issues such as this important to each of us even if we do not live in a rural area and have no personal experience with poverty?

A new applications-oriented **Sociology Works!** feature shows how sociological theories and research continue to enhance our understanding of contemporary social issues and our interactions in everyday life. Topics include: "Erving Goffman's Impression Management and Facebook" (Chapter 5); "Why Place Matters in Global Poverty" (Chapter 9); and "Sociology Sheds Light on the Physician-Patient Relationship" (Chapter 18).

Every chapter in the Eighth Edition has been revised to reflect the latest in sociological theory and research. Examples have been updated for timeliness, and all statistics, such as data relating to suicide, crime, demographics, health, and poverty, are the latest available.

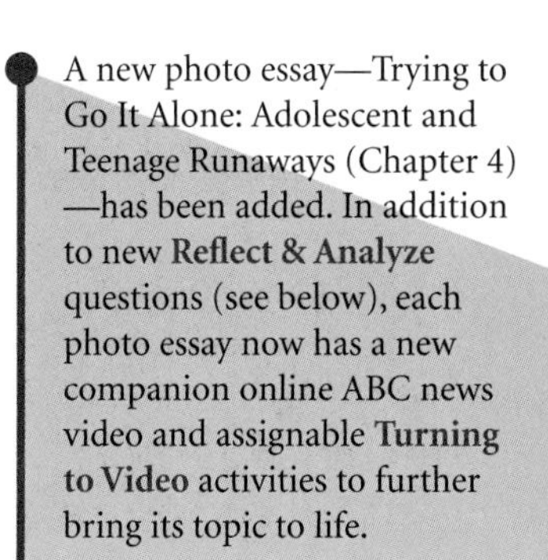

A new photo essay—Trying to Go It Alone: Adolescent and Teenage Runaways (Chapter 4)—has been added. In addition to new **Reflect & Analyze** questions (see below), each photo essay now has a new companion online ABC news video and assignable **Turning to Video** activities to further bring its topic to life.

Photo Essay

Trying to Go It Alone: Runaway Adolescents and Teens

Home. Family. The ideal is that these words evoke a sense of well-being—feelings such as love, understanding, acceptance, and security. However, the reality is that for many of us, especially adolescents and teens, these words trigger negative feelings—fear, anxiety, dread—and an urgent desire to escape. The most recent statistics available indicate that between 1.6 and 2.8 million youths run away each year in the United States. That means one out of every seven children in this country will run away before the age of 18. Included in these numbers are not just runaway but also "throwaway" children, youths who have been forced out of their homes by parents or guardians, or, because they have turned 18, forced out of a foster care system.

What happens to humans' socialization when, as youths, they must fend for themselves to meet their own physical and emotional needs? The images in this essay give you a chance to look at the lives of runaways, from risk factors and precipitating events to means of survival and resources, and the longer-term effects that running away or being thrown away has on individuals and their communities. This essay also provides a glimpse of homeless children in global perspective. As you look at these images, consider the short- and long-term impact that persons age 12 (sometimes younger) to 18 attempting self-sufficiency can have on a society.

Why They Leave

Forty-seven percent of runaway and homeless youths indicate that conflict between them and their parent or guardian is a major problem. Thirty-four percent of runaways (girls and boys) report sexual abuse, and 43 percent report physical abuse before leaving home, abuse that increases youths' risk of being assaulted or raped—or both—on the street. Other problems that runaways report include alcohol and drug use, sexual orientation issues, and mental and physical health issues.

Life on the Street Is Hard

This homeless girl on a sidewalk in Manhattan reflects the reality that youths age 12 to 17 are at higher risk for homelessness than are adults. Available data show that 12 percent of runaway and homeless youths spend at least one night outside in a park, on a street, under a bridge or overhang, or on a rooftop. Seven percent of youths in runaway and homeless shelters and 14 percent of youths on the street surveyed in 1995 reported having traded sex for money, food, shelter, or drugs in the previous 12 months. Other means of survival include shelters and soup kitchens, panhandling, and stealing.

122

ociety's Safety Nets

ales and females run away in equal numbers, lthough females are more likely to seek help rough shelters and hotlines such as the National unaway Switchboard (NRS), where 77 percent of e callers are female. In 2007, NRS handled more an 176,000 calls. NRS data show that the organiation is serving more youths who are contemplatg running away, instead of already having run way, than in the past. Children under age 12 are e fastest-growing group of callers, and in 2007, RS received more crisis calls than it had in the ast.

Home Free Program

In addition to providing its hotline and online services and a professionally developed runaway prevention curriculum titled "Let's Talk," the NRS partners with several organizations, most notably Greyhound Lines, Inc., to offer its Home Free Program, promoted in this poster by the musician Ludacris. More than 10,000 runaways have been reunited with their families, free of charge, since the program was started.

Catching Up

As previously noted, 50 percent of homeless youths age 16 or older reported having dropped out of school, having been expelled, or having been suspended. Even if school had not been a major problem, once adolescents and teens, especially teens, have run away, the disruption to their education may be so significant that they end up dropping out of school. Catching up and moving ahead takes determination—and opportunities such as ones provided by this Skills, Training, Employment, Preparation Services (STEPS) program, where teens can study for a high school General Equivalency Development (GED) test.

123

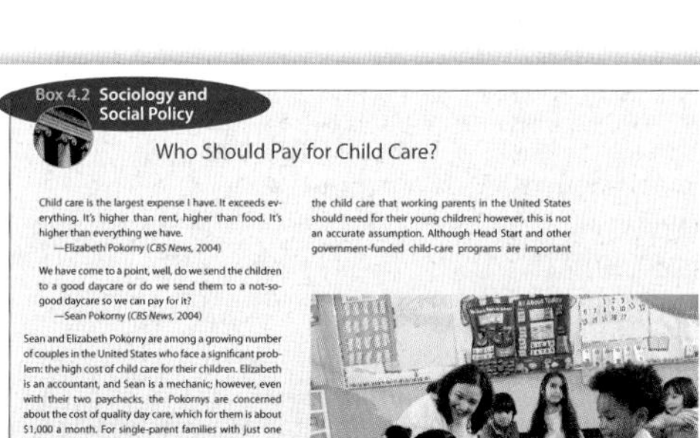

Box 4.2 Sociology and Social Policy

Who Should Pay for Child Care?

> Child care is the largest expense I have. It exceeds everything. It's higher than rent, higher than food. It's higher than everything we have.
> —Elizabeth Pokorny (*CBS News*, 2004)

> We have come to a point, well, do we send the children to a good daycare or do we send them to a not-so-good daycare so we can pay for it?
> —Sean Pokorny (*CBS News*, 2004)

Sean and Elizabeth Pokorny are among a growing number of couples in the United States who face a significant problem: the high cost of child care for their children. Elizabeth is an accountant, and Sean is a mechanic; however, even with their two paychecks, the Pokornys are concerned about the cost of quality day care, which for them is about $1,000 a month. For single-parent families with just one paycheck, the problem may be even more acute.

Although paying for child care is a major issue in the United States, it is not as large a concern in most other high-income countries, where the government typically provides all or almost all of the cost of child care for children above two years of age. By way of example, whereas about 25 to 30 percent of the cost of child care for three- to six-year-olds is provided by the government in the United States, France provides 100 percent, and Denmark, Finland, and Sweden provide about 80 percent (The Future of Children, 2001).

Some people believe that programs such as Head Start and state-funded prekindergarten programs provide all the child care that working parents in the United States should need for their young children; however, this is not an accurate assumption. Although Head Start and other government-funded child-care programs are important to working families, particularly those at or below the poverty line, these programs do not have adequate funding to achieve their primary purpose, which is to prepare children for school, much less to provide quality child care that is free or at least affordable to low-income parents and that is available at the hours around the clock that many of these parents are required to work. Consider, for example, parents who are employed in the food and entertainment industry and must work nights and holidays, or parents in cities such as Las Vegas and Reno, Nevada, where one of the primary sources of jobs is in the gaming industry (such as casinos that remain open 24 hours per day). Their children may need care on a 24/7 basis, rather than during the standard hours of the typical Head Start or state-funded pre-kindergarten program, which typically are available for only a few hours each day.

Day-care centers have become important agents of socialization for increasing numbers of children. Today, about 60 percent of all U.S. preschool children are in day care of one kind or another.

How might parents, especially those in low-income brackets, find affordable child care? Some people believe that it should be the parents' responsibility—that the parents should not have children whom they cannot afford to raise. Other people believe that paying for child care should be the responsibility of the government. However, for the government to contribute more to the cost of child care, we would need a refocused view of this issue, a view which suggests that it is a national concern and that it is critical to meeting two of our nation's priorities: helping families work and making sure that all children enter school ready to learn (Children's Defense Fund, 2003). Today, however, there is currently little hope that political leaders who are working to cut spending and balance budgets will see child care as a top priority.

Still other people believe that if the government is not going to bear the financial responsibility for child care, employers should be the ones to pay that cost because they benefit the most from the hours that their workers spend on the job. Without safe, reliable, affordable care for their children, employees may find it more difficult to be productive workers. Likewise, the cost of quality child care is a concern not only for families in low-incom but also for many middle-class families that are squeeze of high gas prices, food costs, and re mortgage payments.

What are the chances that more employe fund child care? Generally, it appears that too f ers are willing or able to provide real help wit in the twenty-first century. Many employers c sharply rising health care costs for their emp believe that global competition is cutting into ings, thus making funding of child care for th ees a luxury that the employers simply cannot a

Are the *financial constraints* at various levels ment and in corporate America so great that t institutions simply cannot afford to put money care and an investment in the nation's future? *priorities* confused so that we are spending things that are less important to the future of than our children? If we truly mean it when we want to "leave no child behind," perhaps we should rethink our priorities in regard to the funding of child care.

Reflect & Analyze

Do we have a responsibility for the children of this nation? Or are children the sole responsibility of their families, regardless of the parents' ability to pay the high cost of properly caring for and educating children in the United States? What do you think?

Reflect & Analyze

Do we have a responsibility for the children of this nation? Or are children the sole responsibility of their families, regardless of the parents' ability to pay the high cost of properly caring for and educating children in the United States? What do you think?

New **Reflect & Analyze** questions, provided at the end of the **Sociology Works!** feature; photo essays; and **Sociology and Media Framing, Sociology and Social Policy,** and **Sociology in Global Perspectives** boxes are designed to stimulate students' critical-thinking abilities and sociological imagination.

A new **Quick-Start Guide to Using Your Sociological Imagination** is now included at the front of the book, providing students with a visual and question-driven preliminary introduction to this foundational concept.

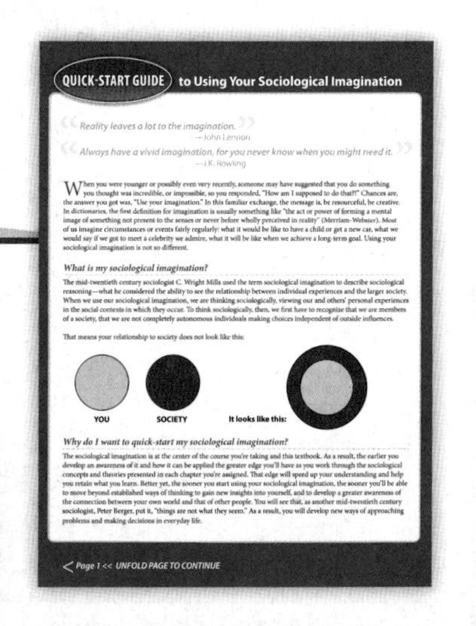

QUICK-START GUIDE to Using Your Sociological Imagination

> *Reality leaves a lot to the imagination.*
> —John Lennon

> *Always have a vivid imagination, for you never know when you might need it.*
> —J.K. Rowling

When you were younger or possibly even very recently, someone may have suggested that you do something you thought was incredible, or impossible, so you responded, "How am I supposed to do that?!" Chances are, the answer you got was, "Use your imagination." In this familiar exchange, the message is, be resourceful, be creative. In dictionaries, the first definition for imagination is usually something like "the act or power of forming a mental image of something not present to the senses or never before wholly perceived in reality" (Merriam-Webster). Most of us imagine circumstances or events fairly regularly: what it would be like to have a child or get a new car, what we would say if we got to meet a celebrity we admire, what it will be like when we achieve a long-term goal. Using your sociological imagination is not so different.

What is my sociological imagination?

The mid-twentieth century sociologist C. Wright Mills used the term sociological imagination to describe sociological reasoning—what he considered the ability to see the relationship between individual experiences and the larger society. When we use our sociological imagination, we are thinking sociologically, viewing our and others' personal experiences in the social contexts in which they occur. To think sociologically, then, we first have to recognize that we are members of a society, that we are not completely autonomous individuals making choices independent of outside influences.

That means your relationship to society does not look like this:

YOU SOCIETY It looks like this:

Why do I want to quick-start my sociological imagination?

The sociological imagination is at the center of the course you're taking and this textbook. As a result, the earlier you develop an awareness of it and how it can be applied the greater edge you'll have as you work through the sociological concepts and theories presented in each chapter you're assigned. That edge will speed up your understanding and help you retain what you learn. Better yet, the sooner you start using your sociological imagination, the sooner you'll be able to move beyond established ways of thinking to gain new insights into yourself, and to develop a greater awareness of the connection between your own world and that of other people. You will see that, as another mid-twentieth century sociologist, Peter Berger, put it, "things are not what they seem." As a result, you will develop new ways of approaching problems and making decisions in everyday life.

Page 1 << UNFOLD PAGE TO CONTINUE

You and your students will enjoy the Eighth Edition's lively and accessible new interior design and its updated and expanded collection of illustrations and photos.

PROVEN FEATURES

The Eighth Edition emphasizes diversity, a global perspective, and personal involvement

Beginning with Kendall's first edition, diversity has been integrated in numerous ways throughout, and this Eighth Edition is no exception. The individuals portrayed and discussed in each chapter accurately mirror the diversity in society itself. As a result, this text speaks to a wide variety of students and captures their interest by taking into account their concerns and perspectives. Moreover, the research used includes the best work of classical and established contemporary sociologists—including women and people of color—and it weaves an inclusive treatment of all people into the examination of sociology throughout. As a result, Diana Kendall helps students consider the significance of the interlocking nature of individuals' class, race, and gender (and, increasingly, age) in all aspects of social life.

Sociology in Global Perspective boxes analyze our interconnected world and reveal how the sociological imagination extends beyond national borders. Global implications of various topics are examined in these captivating boxes, including:

- "Marginal Migration: Moving to a Less Poor Nation" (Chapter 9, *Global Stratification*)
- "Wombs-For-Rent: Outsourcing Births to India (Chapter 15, *Families and Intimate Relationships*)
- "China: A Nation of Environmental Woes and Emergent Social Activism" (Chapter 20, *Behavior, Social Movements, and Social Change*)

Box 9.2 Sociology in Global Perspective

Marginal Migration: Moving to a Less-Poor Nation

We are forced to come back here—not because we like it, but because we are poor. When we cross the border, we are a little better off. We are able to buy shoes and maybe a chicken.

—Anes Moises explaining why he and many other Haitian migrants face continual hardships so that they can live and work in the Dominican Republic (qtd. in DeParle, 2007: A1)

In the Dominican Republic, Anes Moises works in the banana fields and earns six times as much money as he would in his homeland of Haiti. For Anes and the more than 74 million "south-to-south" migrants—people who have moved from one developing country to another—even marginal gains in money and quality of life are important. As one journalist described the living conditions of Haitian migrants in the Dominican Republic, "The scrap-wood shanties on a muddy hillside are a poor man's promised land" (DeParle, 2007: A1). Although we hear much more about the 82 million migrants who have moved "south to north"—from a lower-income nation to a country with a high-income economy—the lived experiences of "south-to-south" migrants are also important in understanding global wealth and poverty in the twenty-first century (Ratha and Shaw, 2007).

What are the typical characteristics of people who move from one poor nation to a slightly less poor one? Do their efforts make a difference in their economic status? Recent studies have found that south-to-south migrants are often poorer than individuals who migrate from lower-income to high-income nations. South-to-south migrants are also more likely to travel without proper documentation and be more vulnerable to unscrupulous people and to apprehension by law enforcement officials than are south-to-north migrants (Ratha and Shaw, 2007). Many south-to-south migrants send money back home to extremely poor family members who reside in remote rural areas. Some analysts estimate that the money sent by these migrants has a significant economic impact on the lives of people in the poorest nations of the world. For example, Haitians residing in the Dominican Republic typically send about $135 million a year to relatives back home (DeParle, 2007: A16). Ironically, numerous jobs are available for Haitians in the Dominican Republic because many Dominicans have migrated to the United States (south-to-north migration) in hopes of finding better jobs and higher wages. According to a recent *New York Times* article, similar patterns exist across many nations as some individuals migrate to high-income economies while others move from a very poor country to a slightly less poor one:

Nicaraguans build Costa Rican buildings. Paraguayans pick Argentine crops. Nepalis dig Indian mines. Indonesians clean Malaysian homes. Farm hands from Burkina Faso tend the fields in Ivory Coast. Some save for the more expensive journeys north, while others find the move from one poor land to another all they will ever afford. With rich countries tightening their borders, migration within the developing world is likely to grow. (DeParle, 2007: A16)

A comparison of the 2007 average per capita income of several nations shows why south-to-south migration will no doubt continue in the future (based on *New York Times*, 2007b: A16):

- Haiti ($480) to the Dominican Republic ($2,850)
- Nicaragua ($1,000) to El Salvador ($2,540)
- Guatemala ($2,640) to Mexico ($7,870)
- Colombia ($2,740) to Panama ($4,890)

Reflect & Analyze

How do these figures compare with what migrants might gain from moving to the United States or other nations with high-income economies? Does this information provide us with new insights on the nature and extent of global stratification and inequality as they affect people living in the United States? Why or why not?

Box 5.2 Framing Homelessness in the Media

Thematic and Episodic Framing

They live—and die—on a traffic island in the middle of a busy downtown street, surviving by panhandling drivers or turning tricks. Everyone in their colony is hooked on drugs or alcohol. They are the harsh face of the homeless in San Francisco.

The traffic island where these homeless people live is a 40-by-75 foot triangle chunk of concrete just west of San Francisco's downtown. . . . The little concrete divider wouldn't get a second glance, or have a name—if not for the colony that lives there in a jumble of shopping carts loaded with everything they own. It's called Homeless Island by the shopkeepers who work near it and the street sweepers who clean it; to the homeless, it is just the Island. The inhabitants live hand-to-mouth, sleep on the cement and abuse booze and drugs, mostly heroin. There are at least 3,000 others like them in San Francisco, social workers say. They are known as the "hard core," the people most visible on the streets, the most difficult to help. . . . (Fagan, 2003)

This news article is an example of typical media framing of stories about homeless people. The full article includes statements about how the homeless of San Francisco use drugs, lack ambition, and present a generally disreputable appearance on the streets. This type of framing of stories about the homeless is not unique. According to the media scholar Eungjun Min (1999: ix), media images typically portray the homeless as "drunk, stoned, crazy, sick, and drug abusers." Such representations of homeless people limit our understanding of the larger issues surrounding the problem of homelessness in the United States.

Most media framing of newspaper articles and television reports about the problem of homelessness can be classified into one of two major categories: thematic framing and episodic framing. *Thematic framing* refers to news stories that focus primarily on statistics about the homeless population and recent trends in homelessness. Examples include stories about changes in the U.S. poverty rate and articles about states and cities that have had the largest increases in poverty. Most articles of this type are abstract and impersonal, primarily presenting data and some expert's interpretation of what those data mean. Media representations of this type convey a message to readers that "the poor and homeless are faceless." According to some analysts, thematic framing of poverty is often dehumanizing because it "ignores the human tragedy of poverty—the suffering, indignities, and misery endured by millions of children and adults" (Mantsios, 2003: 101).

By contrast, *episodic framing* presents public issues such as poverty and homelessness as concrete events, showing them to be specific instances that occur more or less in isolation. For example, a news article may focus on the problems of one homeless family, describing how the parents and kids live in a car and eat meals from a soup kitchen. Often, what is not included is the *big picture of homelessness:* How many people throughout the city or nation are living in their cars or in shelters? What larger structural factors (such as reductions in public and private assistance to the poor, or high rates of unemployment in some regions) contribute to or intensify the problem of homelessness in this country?

For many years, the poor have been a topic of interest to journalists and social commentators. Between 1851 and 1995, the *New York Times* alone printed 4,126 articles that had the word *poverty* in the headline. How stories about the poor and homeless are framed in the media has been and remains an important concern for each of us because these reports influence how we view the less fortunate in our society. If we come to see the problem of homelessness as nothing more than isolated statistical data or as marginal situations that affect only a few people, then we will be unable to make a balanced assessment of the larger social problems involved.

Reflect & Analyze

How are the poor and homeless represented in the news reports and the television entertainment shows that you watch? Are the larger social issues surrounding homelessness discussed within the context of these shows? Should they be?

Kendall gets students thinking about pressing social issues, such as how the media "package" news and entertainment and what effect this framing has on our perception of social relations. The **Media Framing** boxes help foster students' critical-thinking skills. Some of the media framing topics are:

- Framing Culture in the Media: "You Are What You Eat," (Chapter 3, *Culture*).
- Framing Homelessness in the Media: "Thematic and Episodic Framing," (Chapter 5, *Society, Social Structure and Interaction*).
- Framing Health Issues in the Media: "It's Right for You! The Framing of Drug Ads," (Chapter 18, *Health, Health Care, and Disability*).

Kendall's **You Can Make a Difference** boxes—one of the book's signature features—discuss social issues that are related to the chapter topic and provide concrete ideas on how students can address—on a personal level—the issue raised and effect social change. In each **You Can Make a Difference** box, an interesting profile introduces students to individuals and groups and their efforts to find solutions to social problems. Motivating and thought provoking, these boxes serve as models and help generate ideas on how students can become involved.

Box 4.4 **You Can Make a Difference**

Helping a Child Reach Adulthood

After Tina—one of your best friends—moves into a large apartment complex near her university, she keeps hearing a baby cry at all hours of the day and night. Although the crying is coming from the apartment next to Tina's, she never sees anyone come or go from it. On several occasions, she knocks on the door, but no one answers. At first Tina tries to ignore the situation, but eventually she can't sleep or study because the baby keeps crying. Tina decides she must take action and asks you, "What do you think I ought to do?" What advice could you give Tina?

Like Tina, many of us do not know if we should get involved in other people's lives. We also do not know how to report child maltreatment. However, social workers and researchers suggest that bystanders must be willing to get involved in cases of possible abuse or neglect to save a child from harm by others. They also note the importance of people knowing how to report incidents of maltreatment:

- *Report child maltreatment.* Cases of child maltreatment can be reported to any social service or law enforcement agency.
- *Identify yourself to authorities.* Although most agencies are willing to accept anonymous reports, many staff members prefer to know your name, address, telephone number, and other basic information so that they can determine that you are not a self-interested person such as a hostile relative, ex-spouse, or vindictive neighbor.
- *Follow up with authorities.* Once an agency has validated a report of child maltreatment, the agency's first goal is to stop the neglect or abuse of that child, whose health and safety are paramount concerns. However, intervention also has long-term goals. Sometimes, the situation can be improved simply by teaching the parents different values about child rearing or by pointing them to other agencies and organizations that can provide needed help. Other times, it may be necessary to remove the child from the parents' custody and place the child in a foster home, at least temporarily. Either way, the situation for the child will be better than if he or she had been left in an abusive or neglectful home environment.

So the best advice for Tina—or for anyone else who has reason to believe that child maltreatment is occurring—is to report it to the appropriate authorities. In most telephone directories, the number can be located in the government listings section. Online, use keywords such as "helping children" and "child welfare" to search for sources of information and assistance. Here are some other resources for help:

- Childhelp offers a 24-hour crisis hotline, national information, and referral network for support groups and therapists and for reporting suspected abuse: 15757 North 78th Street, Scottsdale, AZ 85260. (800) 422-4453.
  http://www.childhelp.org
- Child Welfare League of America, a Washington, D.C.-area association of nearly 800 public and private nonprofit agencies, serves as an advocacy group for children who have experienced maltreatment: 2345 Crystal Drive, Suite 250, Arlington, VA 22202. (703) 412-2400.
  http://www.cwla.org
- The National Center for Missing and Exploited Children provides brochures about child safety and child protection on request.
  http://www.missingkids.com

© Lisette LeBon/SuperStock

It has been said that it takes a village to raise a child. In contemporary societies, it takes many people pulling together to help a child have a safe and happy childhood and a productive adulthood. Are there ways in which you, like the man in this photo, can help a young person in your community?

CONCEPT QUICK REVIEW

Psychological and Sociological Theories of Human Development

Social Psychological Theories of Human Development	Freud's psychoanalytic perspective	Children first develop the id (drives and needs), then the ego (restrictions on the id), and then the superego (moral and ethical aspects of personality).
	Piaget's cognitive development	Children go through four stages of cognitive (intellectual) development, going from understanding only through sensory contact to engaging in highly abstract thought.
	Kohlberg's stages of moral development	People go through three stages of moral development, from avoidance of unwanted consequences to viewing morality based on human rights.
	Gilligan: gender and moral development	Women go through stages of moral development from personal wants to the greatest good for themselves and others.
Sociological Theories of Human Development	Cooley's looking-glass self	A person's sense of self is derived from his or her perception of how others view him or her.
	Mead's three stages of self-development	In the preparatory stage, children imitate the people around them; in the play stage, children pretend to take the roles of specific people; and in the game stage, children learn the demands and expectations of roles.

The text's highly praised **Concept Quick Reviews** now appear (still in table format) in every chapter and have been redesigned for greater ease of use. These reading and study aids provide concise overviews of key theories and concepts, including "Social Interaction: The Microlevel Perspective" (Chapter 5); "Theoretical Perspectives on Deviance" (Chapter 7); and "Sociological Perspectives on Health and Medicine" (Chapter 18).

Exclusive online and multimedia tools steer students toward success

CengageNOW™ with eBook and InfoTrac®

CENGAGENOW

Instant Access Code 978-0-495-81424-5

This powerful and interactive resource will help students gauge their own unique study needs. Then, it gives them a *Personalized Study plan* that helps them focus study time on the concepts they most need to master. Your students will quickly begin to optimize study time and get one step closer to success.

CengageNOW™ helps students:

- Create a *Personalized Study plan* for each chapter of Kendall's text
- Understand key concepts in the course
- Better prepare for exams and increase their chances of success

CengageNOW™ will help you, as an instructor:

- Plan your curriculum
- Manage your course and communicate with students
- Easily maintain your gradebook—results flow in automatically
- Assign practice or homework to reinforce key concepts
- Assess student performance outcome

How can your students gain access to CengageNOW™?
Access to CengageNOW is web-based and can be packaged with this text at no additional charge. Order new student texts packaged with the access code card. Contact your local Cengage representative for ordering details.

Study Guide

978-0-495-90511-0

Shannon Carter, University of Central Florida

This student study tool contains both brief and detailed chapter outlines, chapter summaries, learning objectives, a list of key terms and key people with page references to the text, questions to guide student reading, Internet and InfoTrac exercises, and practice tests consisting of 20–25 multiple-choice questions, 10–15 true-false questions, 3–5 fill-in-the-blank and short answer questions, and 3–5 essay questions. All multiple-choice, true-false, and fill-in-the-blank and short answer questions include answer explanations and page references to the text.

Telecourse Study Guide

978-0-495-50885-4

This Telecourse Guide for Kendall's text is designed to accompany the "Exploring Society: Introduction to Sociology" telecourse produced by DALLAS TeleLearning of the Dallas County Community College District (DCCCD). This Telecourse Guide provides the essential integration of videos and text, providing students with valuable resources designed to direct their daily study in the "Exploring Society" telecourse.

Study Card

Order the Study Card packaged with the text at no additional cost: 978-1-111-11697-2

See the inside front cover for details.

Careers in Sociology Module

Order the Module packaged with the text at no additional cost: 978-0-538-46336-2

Joan Ferrante, Northern Kentucky University

See the inside front cover for details.

Sociology of Sports Module

978-0-495-59812-1

Jerry M. Lewis, Kent State University

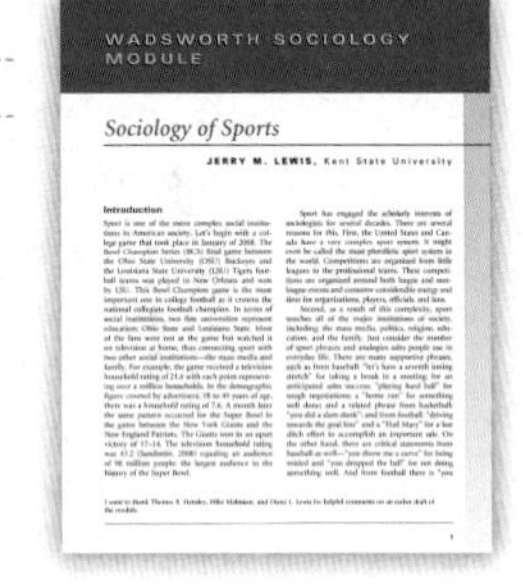

The **Sociology of Sports Module** examines why sociologists are interested in sports, mass media and sports, popular culture and sports (including feature length films on sports), sports and religion, drugs and sports, and violence and sports. As part of Wadsworth's Add-a-Module program, **Sociology of Sports** can be purchased separately, bundled, or customized with Kendall's text.

Companion Website

www.cengage.com/sociology/kendall

When you adopt ***Sociology in Our Times*, Eighth Edition**, you and your students will have access to a rich array of teaching and learning resources that you won't find anywhere else. This outstanding site features chapter-by-chapter online tutorial quizzes, a final exam, chapter outlines, chapter review, chapter-by-chapter web links, flashcards, and more! Instructor resources are password-protected.

Introduction to Sociology Group Activities Workbook

978-0-495-11556-4

Lori Ann Fowler, Tarrant County College

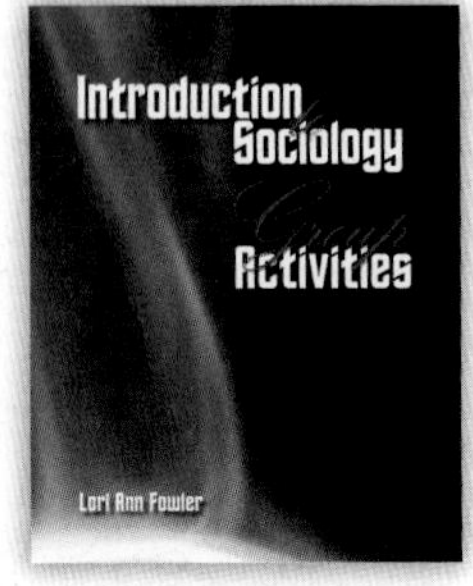

This workbook contains both in and out-of-class group activities (utilizing resources such as MicroCase Online Data exercises) that students can tear out and turn in to the instructor once complete. Also included are ideas for video clips to anchor group discussions, maps, case studies, group quizzes, ethical debates, group questions, group project topics, and ideas for outside readings for students to base group discussions on.

Unparalleled course preparation and management tools increase your efficiency and let you focus on teaching

PowerLecture with JoinIn Student Response System and ExamView®

978-0-495-90512-7

PowerLecture™

ExamView® Assessment Suite

This easy-to-use, one-stop digital library and presentation tool includes preassembled Microsoft® PowerPoint® lecture slides with graphics from the text, making it easy for you to assemble, edit, publish, and present custom lectures for your course. The **PowerLecture** CD-ROM includes book-specific video-based polling and quiz questions that can be used with the JoinIn Student Response System. **PowerLecture** also features ExamView® testing software, which includes all the test items from the printed Test Bank in electronic format, enabling you to create customized tests of up to 250 items that can be delivered in print or online. PowerLecture also includes an electronic version of the Instructor's Resource Manual.

Instructor's Resource Manual

978-0-840-03137-2

Nandi S. Crosby, California State University, Chico

Designed to maximize the effectiveness of your course preparation, this essential manual offers you brief chapter outlines correlated to key ASA guidelines, chapter-specific summaries, key terms, student learning objectives, extensive chapter lecture outlines, chapter review questions, questions for discussion, Internet activities, InfoTrac College Edition discussion, and creative lecture and teaching suggestions. It also includes a Resource Integration Guide with a list of additional print, video, and online resources, including a table of contents for the ABC Video Series for Introductory Sociology and concise user guides for both InfoTrac College Edition and WebTutor.

Test Bank

978-0-495-91135-7

Gary Titchener, Des Moines Area Community College

This test bank consists of revised and updated multiple-choice questions and true/false questions for each chapter of the text, all with answer explanations and page references to the text. Each multiple-choice item has the question type (factual, applied, or conceptual) indicated. Also included are short-answer and essay questions for each chapter. All questions are labeled as new, modified, or pickup so instructors know if the question is new to this edition of the Test Bank, modified but picked up from the previous edition of the Test Bank, or picked up straight from the previous edition of the Test Bank. This edition also lists learning objectives created by the IRM author, creating consistency across supplements. The author of the Test Bank has also keyed each test question to its related learning objective for the chapter.

Tips for Teaching Sociology, Third Edition

978-0-534-54529-1

Jerry M. Lewis, Kent State University

This booklet contains tips on course goals and syllabi, lecture preparation, exams, class exercises, research projects, course evaluations, and more. It is an invaluable tool for first-time instructors of the introductory course, and for veteran instructors in search of new ideas.

Dynamic film and video options spark classroom discussion

ABC Videos

Introductory Sociology Volume I: VHS 978-0-495-00364-9 • DVD 978-0-495-09527-9
Introductory Sociology Volume II: VHS 978-0-495-10081-2 • DVD 978-0-495-10083-6
Introductory Sociology Volume III: VHS 978-0-495-50871-7 • DVD 978-0-495-50872-4
Introductory Sociology Volume IV: VHS 978-0-495-80430-7 • DVD 978-0-495-80437-6

Launch your lectures with exciting video clips from the award-winning news coverage of ABC. Addressing topics covered in a typical course, these videos are divided into short segments—perfect for introducing key concepts in contexts relevant to students' lives.

AIDS in Africa DVD

978-0-495-17183-6

Southern Africa has been overcome by a pandemic of unparalleled proportions. This documentary series focuses on the new democracy of Namibia and the many actions that are being taken to control HIV/AIDS. Included in this series are four documentary films created by the Periclean Scholars at Elon University:

- *Young Struggles, Eternal Faith* focuses on caregivers in the faith community.
- *The Shining Lights of Opuwo* shows how young people share their messages of hope through song and dance.
- *A Measure of Our Humanity* describes HIV/AIDS as an issue related to gender, poverty, stigma, education, and justice.
- *You Wake Me Up* is a story of two HIV+ women and their acts of courage helping other women learn to survive.

Wadsworth's Lecture Launchers for Introductory Sociology

VHS 978-0-534-58839-7 • DVD 978-0-534-58845-8

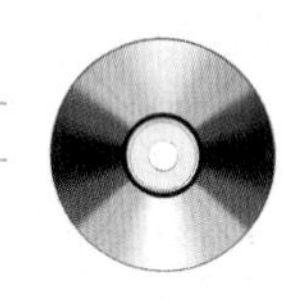

An exclusive offering jointly created by Wadsworth and DALLAS TeleLearning, this video contains a collection of video highlights taken from the *Exploring Society: An Introduction to Sociology* Telecourse (formerly *The Sociological Imagination*). Each video is 3–6 minutes long.

Sociology: Core Concepts Video

VHS 978-0-534-61934-3 • DVD 978-0-534-61935-0

Another exclusive offering jointly created by Wadsworth and DALLAS TeleLearning, this video contains a collection of video highlights taken from the *Exploring Society: An Introduction to Sociology* Telecourse (formerly *The Sociological Imagination*). Each 15–20 minute video segment will enhance student learning of the essential concepts in the introductory course and can be used to initiate class lectures, discussion, and review.

Film Book: Spicing Up Sociology

978-0-495-59900-5

Marisol Clark-Ibanez and Richelle Swan, California State University, San Marcos

Spicing Up Sociology is designed to address the growing interest in using film in the classroom. The authors start the book with the rationale for using film in the classroom, methods for incorporating film into the classroom, and learning outcomes. The authors give a synopsis of a film and a description of what concept in that chapter it gets across. Accompanying each feature film is an activity for students to complete.

edition 8

SOCIOLOGY IN OUR TIMES

DIANA KENDALL
Baylor University

Australia • Brazil • Japan • Korea • Mexico • Singapore • Spain • United Kingdom • United States

Sociology in Our Times, **Eighth Edition**
Diana Kendall

Senior Publisher: Linda Schreiber
Sociology Editor: Erin Mitchell
Developmental Editor: Renee Deljon
Assistant Editor: Rachael Krapf
Editorial Assistant: Rachael Krapf
Media Editor: Melanie Cregger
Marketing Manager: Andrew Keay
Marketing Assistant: Jillian Myers
Marketing Communications Manager: Laura Localio
Content Project Manager: Cheri Palmer
Creative Director: Rob Hugel
Art Director: Caryl Gorska
Print Buyer: Judy Inouye
Rights Acquisitions Account Manager, Text: Roberta Broyer
Rights Acquisitions Account Manager, Image: Leitha Etheridge-Sims
Production Service: Greg Hubit Bookworks
Text Designer: Diane Beasley
Photo Researcher: Terri Wright
Copy Editor: Donald Pharr
Illustrator: Graphic World Illustration Studio
Cover Designer: RHDG
Cover Images (clockwise): Joanna Pecha/Istock; Simon Jarratt/Corbis; SuperStock; Gaby Jalbert, Istock; Allen Simon/Getty
Compositor: Graphic World

Library of Congress Control Number: 2009930836

Student Edition:
ISBN-13: 978-0-495-81391-0
ISBN-10: 0-495-81391-5

Loose-leaf Edition:
ISBN-13: 978-0-495-90510-3
ISBN-10: 0-495-90510-0

Wadsworth
20 Davis Drive
Belmont, CA 94002-3098
USA

Cengage Learning is a leading provider of customized learning solutions with office locations around the globe, including Singapore, the United Kingdom, Australia, Mexico, Brazil, and Japan. Locate your local office at **www.cengage.com/global**.

Cengage Learning products are represented in Canada by Nelson Education, Ltd.

To learn more about Wadsworth, visit **www.cengage.com/wadsworth**
Purchase any of our products at your local college store or at our preferred online store **www.ichapters.com**.

Printed in Canada
1 2 3 4 5 6 7 13 12 11 10 09

Brief Contents

Contents

MICHIGAN
LOTTERY
June 17 2008
LOTTERY WINNER
Fred Topous
$57,000,000
Dollars

Features

17 Religion 550

Features

18 Health, Health Care, and Disability 582

Features

Part 5 Social Dynamics and Social Change

19 Population and Urbanization 618

Photo Essay

Features

20 Collective Behavior, Social Movements, and Social Change 654

Features

Features

Sociology and
Everyday Life

Sociology in
Global Perspective

Sociology and
Social Policy

Media Framing

Sociology *Works!*

You Can Make a Difference

Preface

Welcome to the eighth edition of *Sociology in Our Times*! The twenty-first century offers unprecedented challenges and opportunities for each of us as individuals and for our larger society and world. In the United States, we can no longer take for granted the peace and economic prosperity that many—but far from all—people were able to enjoy in previous decades. However, even as some things change, others remain the same, and among the things that have not changed are the significance of education and the profound importance of understanding how and why people act the way they do. It is also important to analyze how societies grapple with issues such as economic hardship and the threat of terrorist attacks and war, and to gain a better understanding of why many of us seek stability in our social institutions—including family, religion, education, government, and media—even if we believe that some of these institutions might benefit from certain changes.

Like previous editions of this text, which I am gratified has been widely read, the eighth edition of *Sociology in Our Times* highlights the relevance of sociology to help students connect with the subject and the full spectrum of topics and issues it encompasses. It achieves that connection by providing a meaningful, concrete context within which to learn. Specifically, it presents the stories—the *lived experiences*—of real individuals and the social issues they face while discussing a diverse array of classical and contemporary theory and examining interesting and relevant research. The first-person commentaries that open and are revisited throughout each chapter show students how sociology can help them understand the important questions and social issues that not only these other individuals face but that they themselves may face as well.

The individuals presented in each chapter mirror the diversity in society itself not only to accurately represent contemporary society but also to speak to a wide variety of students and capture their interest by taking into account their concerns and perspectives. Moreover, the research used includes the best work of classical and established contemporary sociologists—including many white women and people of color—and it weaves an inclusive treatment of *all* people into the examination of sociology in *all* chapters. By discussing the latest theories and research, *Sociology in Our Times* not only provides students with the most relevant information about sociological thinking but also helps them consider the significance of the interlocking nature of class, race, and gender (and, increasingly, age) in all aspects of social life.

In addition to capturing students' attention, the opening commentaries establish the social themes and issues that each chapter explores to provide additional context for students' learning. Students may not have any prior knowledge of the functionalist perspective, but they are more likely to grasp the concept once it's applied to shopping and consumption, as it is in Chapter 1, which opens with Kay Thayer's comments about her experience with credit cards being marketed to students. Returned to throughout the chapters, the opening personal stories and the themes they introduce function as chapter-length examples that make understanding the concepts, theories, and research presented easier for students.

While helping students learn to appreciate how sociology can help them better understand the world, this text also tries to teach them to see themselves as *members of their communities* and show them what can be done in responding to social issues. As a result, students learn how sociology is not only a collection of concepts and theories but also a field that can make a difference in their lives, their communities, and the world at large.

What's New to the Eighth Edition? For Starters, Sociology Works!

The eighth edition builds on the best of previous editions while offering new insights, learning tools, and opportunities to apply the content of each chapter to relevant sociological issues and major concerns of the twenty-first century. As it is my goal to make each edition better than the previous one, I have revised all the chapters to reflect the latest in sociological theory and research, and have updated examples throughout. Additionally, all statistics, such as data relating to crime, demographics, health, and the economy, are the latest available at the time of this writing.

New chapter openers include the personal stories of Megan Meier, the teen who committed suicide in response to fake MySpace postings; Supreme Court Justice Sonia Sotomayor; and President Barack Obama. Students will also enjoy and benefit from the eighth edition's entirely new—livelier and more functional—**interior design** and its many updated and additional **illustrations, maps, and photos.**

The text's new **Sociology Works!** feature shows how sociological theories and research continue to enhance our understanding of contemporary social issues and our interactions in everyday life. Sociology Works! appears in every chapter and includes "Schools as Laboratories for Getting Along" (Chapter 3), "Erving Goffman's Impression Management and Facebook" (Chapter 5), "Why *Place* Matters in Global Poverty" (Chapter 9), and "Sociology Sheds Light on the Physician–Patient Relationship" (Chapter 18).

To visually capture some of the significant circumstances and issues of our time, this edition again includes **four photo essays,** one of which is new—"Trying to Go It Alone: Adolescent and Teenage Runaways" (Chapter 4). Expanded for improved usability, each essay is now three pages of thought- and conversation-stimulating photos with brief sociological commentary. Each photo essay also has a new companion online **ABC news video** with assignable **Turning to Video** questions to further bring the essay topics to life.

The text's highly praised **Concept Quick Reviews** (previously called Concept Review Tables) appear in every chapter of the eighth edition and have been redesigned for greater ease of use. Still in table format but more prominently displayed than in past editions, these reading and study aids provide concise overviews of key theories and concepts, including Theoretical Perspectives on Deviance (Chapter 7), Sociological Perspectives on Gender Stratification (Chapter 11), and Social Movement Theories (Chapter 20).

Designed to stimulate students' critical thinking ability and sociological imagination, new **Reflect & Analyze** questions are provided at the end of the Sociology Works! feature and photo essays, as well as at the end of the text's media framing, social policy, and global perspectives boxes. This edition offers a strong selection of new boxes, including several updated Sociology and Everyday Life boxes as well as new Sociology in Global Perspective boxes and You Can Make a Difference boxes. A complete listing of the text's boxes and other features, by type and with page references, begins on page xiii.

An innovative new feature appears at the front of the book, the Quick-Start Guide to Using Your Sociological Imagination. Four pages long, this insert orients students with a brief introduction to this foundational concept and then poses Quick-Start questions for each chapter so that students are already thinking sociologically even before they begin reading the first chapter that you assign. *Sociology in Our Times,* eighth edition, also offers instructors a significantly improved **Instructor Resource Manual,** and an **electronic version** of the text is also now available. Full descriptions of the instructor and student supplements begin on page xix.

Overview of the Text's Contents

Sociology in Our Times, eighth edition, comprises 20 carefully written, well-organized chapters to introduce students to the best of sociological thinking. The text's manageable length makes full coverage of the book possible in the time typically allocated to the introductory course. As a result, students are not purchasing a book which contains numerous chapters that the instructor may not have the time or the desire to cover. The availability of *Sociology in Our Times: The Essentials,* seventh edition; electronic versions of this text's chapters for individual sale; and the full spectrum of customization options for this text further ensure that students' textbook investment suits their situation and meets their needs.

Sociology in Our Times is divided into five parts.

Part 1 establishes the foundation for studying society and social life. **Chapter 1** introduces students to the sociological imagination and traces the development of sociological thinking. The chapter sets forth the major theoretical perspectives used by sociologists in analyzing compelling social issues such as the problem of credit card abuse and hyperconsumerism among college students and others. **Chapter 2** focuses on sociological research methods and shows students how sociologists conduct research. This chapter provides a thorough description of both quantitative and qualitative methods of sociological research, and shows how these approaches have been used from the era of Emile Durkheim to the present to study social concerns such as suicide.

In **Chapter 3,** culture is spotlighted as either a stabilizing force or a force that can generate discord, conflict, and even violence in societies. Cultural diversity is discussed as a contemporary issue, and unique coverage is given to popular culture and leisure and to divergent perspectives on popular culture. **Chapter 4** looks at the positive and negative aspects of socialization and presents an innovative analysis of gender and racial–ethnic socialization and issues associated with recent immigration.

Part 2 examines social groups and social control. **Chapter 5** applies the sociological imagination to an examination of society, social structure, and social interaction, using homelessness as a sustained example of the dynamic interplay of structure and interaction in society. Unique to this chapter are discussions of the sociology of emotions and of personal space as viewed through the lenses of race, class, gender, and age.

Chapter 6 analyzes groups and organizations, including innovative forms of social organization and ways in which organizational structures may differentially affect people based on race, class, gender, and age. **Chapter 7** examines how deviance and crime emerge in societies, using diverse theoretical approaches to describe the nature of deviance, crime, and the criminal justice system. Key issues are dramatized for students through an analysis of recent research on peer cliques and gangs.

Part 3 focuses on social differences and social inequality, looking at issues of class, race/ethnicity, sex/gender, and age discrimination. **Chapter 8** focuses on class and stratification in the United States, analyzing the causes and consequences of inequality and poverty, including a discussion of the ideology and accessibility of the American Dream. **Chapter 9** addresses the issue of global stratification and examines differences in wealth and poverty in rich and poor nations around the world. Explanations for these differences are discussed.

The focus of **Chapter 10** is race and ethnicity, which includes an illustration of the historical relationship (or lack of it) between sports and upward mobility by persons from diverse racial–ethnic groups. A thorough analysis of prejudice, discrimination, theoretical perspectives, and the experiences of diverse racial and ethnic groups is presented, along with global racial and ethnic issues. **Chapter 11** examines sex and gender, with special emphasis on gender stratification in historical perspective. Linkages between gender socialization and contemporary gender inequality are described and illustrated by lived experiences and perspectives on body image. **Chapter 12** provides a cutting-edge analysis of aging, including theoretical perspectives and inequalities experienced by people across the life course.

Part 4 offers a systematic discussion of social institutions, making students more aware of the importance of social institutions and showing how a problem in one often has a significant influence on others. The economy and work are explored in **Chapter 13,** including the different types of global economic systems, the social organization of work in the United States, unemployment, and worker resistance and activism. **Chapter 14** discusses the intertwining nature of politics, government, and the media. Political systems are examined in global perspective, and politics and government in the United States are analyzed with attention to governmental bureaucracy and the military–industrial complex.

Chapter 15 focuses on families in global perspective and on the diversity found in U.S. families today. **Chapter 16** investigates education in the United States and contrasts it with systems of education in other nations. In the process, the chapter highlights issues of race, class, and gender inequalities in current U.S. education. In **Chapter 17,** religion is examined in global perspective, including a survey of world religions and an analysis of how religious beliefs affect other aspects of social life. Current trends in U.S. religion are also explored, including various sociological explanations of why people look to religion to find purpose and meaning in life.

Chapter 18 analyzes health, health care, and disability in both U.S. and global perspectives. Among the topics included are social epidemiology, lifestyle factors influencing health and illness, health care organization in the United States and other nations, social implications of advanced medical technology, and holistic and alternative medicine. This chapter is unique in that it contains a thorough discussion of the sociological perspectives on disability and of social inequalities based on disability.

Part 5 surveys social dynamics and social change. **Chapter 19** examines population and urbanization, looking at demography, global population change, and the process and consequences of urbanization. Special attention is given to race- and class-based segregation in urban areas and the crisis in health care in central cities. **Chapter 20** concludes the text with an innovative analysis of collective behavior, social movements, and social change. Environmental activism is used as a sustained example to help students grasp the importance of collective behavior and social movements in producing social change.

Distinctive, Classroom-Tested Features

The following special features are specifically designed to demonstrate the relevance of sociology in our lives, as well as to support students' learning. As the preceding overview of the book's contents shows, these features appear throughout the text, some in every chapter, others in selected chapters.

Unparalleled Coverage of and Attention to Diversity

From its first edition, I have striven to integrate diversity in numerous ways throughout this book. The individuals portrayed and discussed in each chapter accurately mirror the diversity in society itself. As a result, this text speaks to a wide variety of students and captures their interest by taking into account their concerns and perspectives. Moreover, the research used includes the best work of classical and established contemporary sociologists—including many white women and people of color—and it weaves an inclusive treatment of *all* people into the examination of sociology in *all* chapters. Therefore, this text helps students consider the significance of the interlocking nature of individuals' class, race, and gender (and, increasingly, age) in all aspects of social life.

Personal Narratives That Highlight Issues and Serve as Chapter-Length Examples

Authentic first-person commentaries serve as the vignettes that open each chapter and personalize the issue that unifies the chapter's coverage. These lived experiences provide opportunities for students to examine social life beyond their own experiences and for instructors to systematically incorporate into lectures and discussions an array of interesting and relevant topics that help demonstrate to students the value of applying sociology to their everyday lives.

Focus on the Relationship Between Sociology and Everyday Life

Each chapter has a brief quiz that relates the sociological perspective to the pressing social issues presented in the opening vignette. (Answers are provided on a subsequent page.)

Emphasis on the Importance of a Global Perspective

The global implications of all topics are examined throughout each chapter and in the Sociology in Global Perspective boxes, which highlight our interconnected world and reveal how the sociological imagination extends beyond national borders.

Focus on Media Framing

A significant benefit of a sociology course is encouraging critical thinking about such things as how the manner in which the media "package" news and entertainment influences our perception of social issues.

Applying the Sociological Imagination to Social Policy

The Sociology and Social Policy boxes in selected chapters help students understand the connection between sociology and social policy issues in society.

Focus on Making a Difference

Designed to help get students involved in their communities, the You Can Make a Difference boxes look at ways in which students can address, on a personal level, issues raised by the chapter theme.

Census Profiles

This feature highlights current relevant data from the U.S. Census Bureau, providing students with further insight into the United States.

Effective Study Aids

In addition to basic reading and study aids such as chapter outlines, key terms, a running glossary, and our popular online study system CengageNOW™, *Sociology in Our Times* includes the following pedagogical aids to aid students' mastery of the course's content:

- **Concept Quick Review.** These tables categorize and contrast the major theories or perspectives on the specific topics presented in a chapter.
- **Questions for Critical Thinking.** Each chapter concludes with "Questions for Critical Thinking" to encourage students to reflect on important issues, to develop their own critical-thinking skills, and to highlight how ideas presented in one chapter often build on those developed previously.

- **Chapter-opening Focus questions and Sharpening Your Focus questions, and feature-concluding Reflect & Analyze questions.** From activating prior knowledge related to concepts and themes, to highlighting main ideas and reinforcing diverse perspectives, this text's questions consistently contribute to student engagement.
- **End-of-Chapter Summaries in Question-and-Answer Format.** Chapter summaries provide a built-in review for students by reexamining material covered in the chapter in an easy-to-read question-and-answer format to review, highlight, and reinforce the most important concepts and issues discussed in each chapter.

Comprehensive Supplements Package

The eighth edition of *Sociology in Our Times* is accompanied by a wide array of supplements developed to create the best teaching and learning experience inside as well as outside the classroom. All of the continuing supplements have been thoroughly revised and updated, and some new supplements have been added. Cengage Learning prepared the following descriptions, and I invite you to start taking full advantage of the teaching and learning tools available to you by reading this overview.

Supplements for Instructors

Instructor's Edition. The *Instructor's Edition* previews the features that save you time and help students learn, and demonstrates how to integrate our powerful supplements into your curriculum.

Instructor's Resource Manual. Written by Nandi S. Crosby, California State University, Chico, this text's *Instructor's Resource Manual* is designed to maximize the effectiveness of your course preparation. It offers you brief chapter outlines correlated to key ASA guidelines, chapter-specific summaries, key terms, student learning objectives, extensive chapter lecture outlines, chapter review questions, questions for discussion, Internet activities, InfoTrac discussion, and creative lecture and teaching suggestions. It also includes a Resource Integration Guide with a list of additional print, video, and online resources, including a table of contents for the ABC Video Series for Introductory Sociology and concise user guides for both InfoTrac® College Edition and WebTutor™.

Test Bank. Authored by Gary Titchener, Des Moines Area Community College, the eighth edition's test bank consists of revised and updated multiple-choice questions and true/false questions for each chapter of the text, all with answer explanations and page references to the text. Each multiple-choice item has the question type (factual, applied, or conceptual) indicated. Also included are short-answer and essay questions for each chapter. All questions are labeled as new, modified, or pickup so instructors know if the question is new to this edition of the Test Bank, modified but picked up from the previous edition of the Test Bank, or picked up straight from the previous edition of the Test Bank. This edition also lists learning objectives created by the *Instructor's Resource Manual* author, creating consistency across supplements. The author of the Test Bank has also keyed each test question to its related learning objective for the chapter.

Website. When you adopt Kendall's *Sociology in Our Times,* eighth edition, you (and your students!) will have access to a rich array of teaching and learning resources that you won't find anywhere else. This outstanding website features chapter-by-chapter online tutorial quizzes, a final exam, chapter outlines, chapter review, chapter-by-chapter Web links, flashcards, and more! Instructor resources are password protected.

PowerLecture with JoinIn Student Response System™ and ExamView®. This easy-to-use, one-stop digital library and presentation tool includes preassembled Microsoft® PowerPoint® lecture slides with graphics from the text, making it easy for you to assemble, edit, publish, and present custom lectures for your course. The PowerLecture CD-ROM includes book-specific video-based polling and quiz questions that can be used with the JoinIn Student Response System™. PowerLecture also features ExamView® testing software, which includes all the test items from the printed Test Bank in electronic format, enabling you to create customized tests of up to 250 items that can be delivered in print or online. PowerLecture also includes an electronic version of the *Instructor's Resource Manual.*

WebTutor™ on Blackboard® and WebCT®. Creating an engaging e-learning environment is easier than ever with WebTutor Advantage. Save time building or Web-enhancing your course, posting course materials, incorporating multimedia, tracking progress, and more with this customizable, engaging course management tool. WebTutor Advantage saves you time and enhances your students' learning—pairing advanced course management capabilities with text-specific learning tools. View a demo at **http://academic.cengage.com/webtutor.** Printed access cards and instant access codes are available.

CengageNOW® with eBook and InfoTrac. CengageNOW Personalized Study, a diagnostic tool (including a chapter-specific pre-test, individualized study plan, and post-test), helps students master concepts and prepare for exams by creating a study plan based on the students' performance on the pre-test. See full description above, under "Supplements for Instructors." Easily assign Personalized Study for the entire term, and, if you want, results will automatically post to your grade book. CengageNOW also features the most intuitive, easy-to-use online course management and study system on the market. It saves you time through its automatic grading and easy-to-use grade book and provides your students an efficient way to study. Printed access cards and instant access codes are available. Also available: CengageNOW on Blackboard® and WebCT for *Sociology in Our Times,* eighth edition.

Online Activities for Introductory Sociology Courses. Made up of contributions from introductory sociology instructors, this new online supplement is free to adopters of our introductory books and features new classroom activities for professors to use.

***Tips for Teaching Sociology,* Third Edition.** Written by veteran instructor Jerry M. Lewis of Kent State University, this booklet contains tips on course goals and syllabi, lecture preparation, exams, class exercises, research projects, course evaluations, and more. It is an invaluable tool for first-time instructors of the introductory course and for veteran instructors in search of new ideas.

Introduction to Sociology Group Activities Workbook. Prepared by Lori Ann Fowler, Tarrant County College, this workbook contains both in- and out-of-class group activities (using resources such as MicroCase Online Data exercises from Wadsworth's Online Sociology Resource Center) that students can tear out and turn in to the instructor once complete. Also included are ideas for video clips to anchor group discussions, maps, case studies, group quizzes, ethical debates, group questions, group project topics, and ideas for outside readings for students to base group discussions on. This is both a workbook for students and a repository of ideas; instructors can use this guide to get ideas for any introductory sociology class.

ABC Videos: Introductory Sociology. Each of the four volumes of our ABC Videos features short, high-interest clips from current news events as well as historical raw footage going back forty years. Perfect for discussion starters or to enrich your lectures and spark interest in the material in the text, these brief videos provide students with a new lens through which to view the past and present, one that will greatly enhance their knowledge and understanding of significant events and open up to them new dimensions in learning. Clips are drawn from such programs as *World News Tonight, Good Morning America, This Week, PrimeTime Live, 20/20,* and *Nightline,* as well as numerous ABC News specials and material from the Associated Press Television News and British Movietone News collections.

AIDS in Africa DVD. Expand your students' global perspective of HIV/AIDS with this award-winning documentary series focused on controlling HIV/AIDS in southern Africa. Films focus on caregivers in the faith community; how young people share messages of hope through song and dance; the relationship of HIV/AIDS to gender, poverty, stigma, education, and justice; and the story of two HIV-positive women helping others.

Lecture Launchers DVD. An exclusive offering jointly created by Wadsworth Cengage Learning and Dallas TeleLearning, this video contains a collection of video highlights taken from the "Exploring Society: An Introduction to Sociology" telecourse (formerly "The Sociological Imagination"). Each 3- to 6-minute video segment has been specially chosen to enhance and enliven class lectures and discussions of 20 key topics covered in the introduction to sociology course. Accompanying the video is a brief written description of each clip, along with suggested discussion questions to help effectively incorporate the material into the classroom.

Sociology: Core Concepts DVD. An exclusive offering jointly created by Wadsworth Cengage Learning and Dallas TeleLearning, this video contains a collection of video highlights taken from the "Exploring Society: An Introduction to Sociology" telecourse (formerly "The Sociological Imagination"). Each 15- to 20-minute video segment will enhance student learning of the essential concepts in the introductory course and can be used to initiate class lectures, discussion, and review. The video covers topics such as the sociological imagination, stratification, race and ethnic relations, and social change.

Film Book: *Spicing Up Sociology.* Written by Marisol Clark-Ibanez and Richelle Swan of California State University–San Marcos, *Spicing Up Sociology* is designed to address the growing interest in using film in the classroom. The authors start the book with the rationale for using film in the classroom, methods for incorporating film into the classroom, and learning outcomes. The authors give a synopsis of various films and a description of what sociological concept each one demonstrates. Accompanying each feature film is an activity for students to complete.

Extension: Wadsworth's Sociology Reader Database Sampler. Create your own customized reader for your introductory class drawing from dozens of classic and contemporary articles found on the exclusive Wadsworth Cengage Learning Text Choice2 database. Create a customized reader just for your class containing as few as two or three seminal articles or more than a dozen edited selections. With Extension, you can preview articles online, make selections, and add original material of your own to create your printed reader for your class.

Sociology of Careers Module. Written by leading author Joan Ferrante, Northern Kentucky University, the Sociology of Careers module offers the most extensive and useful information on careers that is available. This module provides six career tracks, each of which has a "featured employer," a job description, and a letter of recommendation (written by a professor for a sociology student) or application (written by a sociology student). The module also includes résumé-building tips on how to make the most out of being a sociology major and offers specific course suggestions along with the transferable skills gained by taking them. As part of Wadsworth's Add-a-Module Program, Sociology of Careers can be purchased separately, bundled, or customized with any of our introductory texts. The modules present topics not typically covered in most introductory texts but often requested by instructors.

Sociology of Sports Module. The Sociology of Sports module, authored by Jerry M. Lewis, Kent State University, examines why sociologists are interested in sports, mass media and sports, popular culture and sports (including feature-length films on sports), sports and religion, drugs and sports, and violence and sports. As part of Wadsworth's Add-a-Module Program, Sociology of Sports can be purchased separately, bundled, or customized with any of our introductory texts. The modules present topics not typically covered in most introductory texts but often requested by instructors.

Supplements for Students

Website for Kendall's *Sociology in Our Times,* Eighth Edition. When you adopt Kendall's *Sociology in Our Times,* eighth edition, your students will have access to a rich array of teaching and learning resources that you won't find anywhere else. This outstanding website features chapter-by-chapter online tutorial quizzes, a final exam, chapter outlines, chapter review, chapter-by-chapter Web links, flashcards, and more! (As noted above, instructor resources on this website are password protected.)

Study Card for Intro Sociology. This handy card, created by Matisa Wilbon, Bellarmine University, provides all the important sociological concepts covered in introductory sociology courses, broken down into sections. Providing a large amount of information at a glance, this study card is an invaluable tool for a quick review.

Study Guide. Prepared by Shannon Carter, University of Central Florida, this student study tool contains both brief and detailed chapter outlines, chapter summaries, learning objectives, a list of key terms and key people with page references to the text, questions to guide student reading, Internet and InfoTrac exercises, and practice tests consisting of 20 to 25 multiple-choice questions, 10 to 15 true/false questions, 3 to 5 fill-in-the-blank and short-answer questions, and 3 to 5 essay questions. All multiple-choice, true/false, and fill-in-the-blank and short-answer questions include answer explanations and page references to the text.

Student Course Guide. This Course Guide for *Sociology in Our Times,* eighth edition, is designed to accompany the "Exploring Society: Introduction to Sociology" telecourse produced by DALLAS TeleLearning of the Dallas County Community College District (DCCCD). This course guide provides the essential integration of videos and text, providing students with valuable resources designed to direct their daily study in the "Exploring Society" telecourse. Each chapter of the guide contains a lesson that corresponds to each of the 22 video segments in the telecourse. Each lesson includes the following components: Overview, Lesson Assignment, Lesson

Goal, Lesson Learning Objectives, Review, Lesson Focus Points, Related Activities, Practice Tests, and Answer Key.

CengageNOW with eBook and InfoTrac. CengageNOW Personalized Study, a diagnostic tool (including a chapter-specific pre-test, individualized study plan, and post-test), helps students master concepts and prepare for exams by creating a study plan based on the students' performance on the pre-test. See full description above, under "Supplements for Instructors." Printed access cards and instant access codes are available.

These resources are available to qualified adopters, and ordering options for student supplements are flexible. Please consult your local Cengage Learning sales representative or visit us at **www.cengage.com** for more information, including ISBNs; to receive examination copies of any of these instructor or student resources; or for product demonstrations. Print and e-book versions of this text are available for students to purchase at **www.ichapters.com.**

Acknowledgments

Sociology in Our Times would not have been possible without the insightful critiques of these colleagues, who have reviewed some or all of this book. My profound thanks to each one for engaging in this time-consuming process:

Monifa Brinson-Mulraine, Fairleigh Dickinson University, College at Florham
Toby Buchanan, Dallas Baptist University
Paul Calarco, State University of New York–Albany
Mark J. Gordon, Loyola Marymount University
Allan Hunchuk, Thiel College
John Rinciari, Berkeley College
Laura Toussaint, Green River Community College

I deeply appreciate the energy, creativity, and dedication of the many people responsible for the development and production of this edition of *Sociology in Our Times.* I wish to thank Wadsworth Publishing Company's Michelle Julet, Linda Schreiber, Chris Caldeira, Renee Deljon, Andrew Keay, Laura Localio, Rachael Krapf, and Melanie Cregger for their enthusiasm and insights throughout the development of this text. Many other people worked hard on the production of this eighth edition, especially Cheri Palmer, Greg Hubit, and Donald Pharr. I am extremely grateful to them.

I invite you to send your comments and suggestions about this book to me in care of:

Wadsworth/Cengage Learning
20 Davis Drive
Belmont, CA 94002

About the Author

Diana Kendall is currently Professor of Sociology at Baylor University, where she was recognized as an Outstanding Professor for her research. Dr. Kendall has taught a variety of courses, including Introduction to Sociology; Sociological Theory (undergraduate and graduate); Sociology of Medicine; and Race, Class, and Gender. Previously, she enjoyed many years of teaching sociology and serving as chair of the Social and Behavioral Science Division at Austin Community College.

Diana Kendall received her Ph.D. from the University of Texas at Austin, where she was invited to membership in Phi Kappa Phi Honor Society. Her areas of specialization and primary research interests are sociological theory and the sociology of medicine. In addition to *Sociology in Our Times*, eighth edition, she is the author of *Sociology in Our Times: The Essentials; The Power of Good Deeds: Privileged Women and the Social Reproduction of the Upper Class* (Rowman & Littlefield, 2002); *Framing Class: Media Representations of Wealth and Poverty in America* (Rowman & Littlefield, 2005); and *Members Only: Elite Clubs and the Process of Exclusion* (Rowman & Littlefield, 2008). Professor Kendall is actively involved in national and regional sociological associations, including the American Sociological Association, Sociologists for Women in Society, the Society for the Study of Social Problems, and the Southwestern Sociological Association.

SOCIOLOGY IN OUR TIMES

chapter

1 The Sociological Perspective

Chapter Focus Question

How does sociology add to our knowledge of human societies and of social issues such as consumerism?

I went back to school . . . after losing my job and being told I was over-qualified for most entry-level office positions . . . but not qualified enough for mid-level ones. . . . I thought I could at least make $30,000 per year with my new degree. Once I started looking for work, reality set in. I could barely get jobs paying $24,000. Trying to pay off $30,000 in college loans, about $5,000 in credit card debts, with two young children in daycare, and having a 40-50 mile commute to work has been challenging. As to credit card debt on campus, many college students are in the same boat. They think they will get really good paying jobs once they graduate so all they need to do is get to that point. In reality, that is not going to always be the case. Many students being

According to sociologists, our consumer society continues to grow as more people shop at home via telephone and Internet.

IN THIS CHAPTER

- Putting Social Life into Perspective
- The Importance of a Global Sociological Imagination
- The Origins of Sociological Thinking
- The Development of Modern Sociology
- Contemporary Theoretical Perspectives
- Comparing Sociology with Other Social Sciences

recruited to get credit cards are encouraged to include their college loans and grants as income. I know of friends who were not working but still able to get credit cards probably because there was no income verification. There are errors being made on both sides of the fence. Students need to be more realistic about their future income and learn more about how credit cards fit in personal finances. The credit card companies also need to quit going for the easy dollar by enticing college students who they know are more apt to abuse these cards.

—Kay Thayer (2008) posted this comment on a website, "The Takeaway," after she learned that U.S. lawmakers were attempting to pass rules making it more difficult for credit card companies to bombard college students with applications in light of current financial crises in the United States and other nations.

As we can tell from Kay's statement, she, like millions of college students, learned that it may be difficult to pay off credit card debt accumulated in college and to go out into the world and find a well-paying job, particularly when economic and social conditions throughout the nation are problematic. In the twenty-first century, we live in what is referred to as a "consumer society," and many of us rely on our credit cards when we need to pay for items we want to purchase or for services we need. Many sociologists are interested in studying the *consumer society,* which refers to a society in which discretionary consumption is a mass phenomenon among people across

Sharpening Your Focus

- *What is the sociological imagination?*
- *Why were early thinkers concerned with social order and stability?*
- *Why were later social thinkers concerned with change?*
- *What are the assumptions behind each of the contemporary theoretical perspectives?*

diverse income categories. In the consumer society, for example, purchasing goods and services is not only in the exclusive province of the rich or even individuals in the middle class; people from all but the lowest income categories may spend extensive amounts of time, energy, and money shopping while, at the same time, amassing ever-growing credit card debts (see Baudrillard, 1998/1970; Ritzer, 1995; Schor, 1999). According to sociologists, consumption is a process that extends beyond our individual choices and is rooted in larger structural conditions in the social, political, and economic order in which we live. In recent years, as the United States has experienced major problems in finance and other industries, many people have been hard hit by problems in the larger society such as economic instability, massive job losses, and a declining housing market.

Why have shopping, spending, credit card debt, and bankruptcy become major problems for some people? How are social relations and social meanings shaped by what people in a given society produce and how they consume? What national and worldwide social processes shape the production and consumption of goods, services, and information? In this chapter, we see how the sociological perspective helps us examine complex questions such as these, and we wrestle with some of the difficulties of attempting to study human behavior. Before reading on, take the quiz in Box 1.1, which lists a number of commonsense notions about consumption and credit card debt.

Putting Social Life into Perspective

***Sociology* is the systematic study of human society and social interaction.** It is a *systematic* study because sociologists apply both theoretical perspectives and research methods (or orderly approaches) to examinations of social behavior. Sociologists study human societies and their social interactions to develop theories of how human behavior is shaped by group life and how, in turn, group life is affected by individuals.

Why Study Sociology?

Sociology helps us gain a better understanding of ourselves and our social world. It enables us to see how behavior is largely shaped by the groups to which we belong and the society in which we live.

Most of us take our social world for granted and view our lives in very personal terms. Because of our culture's emphasis on individualism, we often do not consider the complex connections between our own lives and the larger, recurring patterns of the society and world in which we live. Sociology helps us look beyond our personal experiences and gain insights into society and the larger world order. A ***society* is a large social grouping that shares the same geographical territory and is subject to the same political authority and dominant cultural expectations,** such as the United States, Mexico, or Nigeria. Examining the world order helps us understand that each of us is affected by *global interdependence*—a relationship in which the lives of all people are intertwined closely and any one nation's problems are part of a larger global problem.

Individuals can make use of sociology on a more personal level. Sociology enables us to move beyond established ways of thinking, thus allowing us to gain new insights into ourselves and to develop a greater awareness of the connection between our own "world" and that of other people. According to the sociologist Peter Berger (1963: 23), sociological inquiry helps us see that "things are not what they seem." Sociology provides new ways of approaching problems and making decisions in everyday life. Sociology also promotes understanding and tolerance by enabling each of us to look beyond our personal experiences (see ▶ Figure 1.1).

Many of us rely on intuition or common sense gained from personal experience to help us understand our daily lives and other people's behavior. *Commonsense knowledge* guides ordinary conduct in everyday life. We often rely on common sense—or "what everybody knows"—to answer key questions about behavior: Why do people behave the way they do? Who makes the rules? Why do some people break rules and other people follow rules?

Many commonsense notions are actually myths. A *myth* is a popular but false notion that may be used, either intentionally or unintentionally, to perpetuate certain beliefs or "theories" even in the light of conclusive evidence to the contrary. For example, one widely held myth is that "money can buy happiness." By contrast, sociologists strive to use scientific standards, not popular myths or hearsay, in studying society and social interaction. They use systematic research techniques and are accountable to the scientific community for their methods and the presentation of their findings. Although some sociologists argue that sociology must be completely value free—without distorting subjective (personal or emotional) bias—others do not think that total objectivity is an attainable or desirable goal when studying human behavior. However, all sociologists attempt to discover patterns or commonalities in human behavior. For example, when they study shopping behavior or credit card abuse, sociologists look for recurring patterns of behavior and for larger, structural factors that contribute to people's behavior. Women's studies scholar Juliet B. Schor, who wrote *The*

Box 1.1 Sociology and Everyday Life

How Much Do You Know About Consumption and Credit Cards?

True	False	
T	F	1. The average U.S. household owes more than $8,000 in credit card debt.
T	F	2. The average debt owed on undergraduate college students' credit cards is less than $1,000.
T	F	3. Fewer than half of all undergraduate students at four-year colleges have at least one credit card.
T	F	4. In the United States, it is illegal to offer incentives such as free T-shirts and Frisbees to encourage students to apply for credit cards.
T	F	5. Consumer activist groups have been successful in getting Congress to pass a law requiring people under age 21 to get parental approval or show that they have sufficient income prior to obtaining a credit card.
T	F	6. More than one million people in this country file for bankruptcy each year.
T	F	7. If we added up all consumer debt in the United States, we would find that the total amount owed is more than $1.5 trillion.
T	F	8. Overspending is primarily a problem for people in the higher-income brackets in the United States and other affluent nations.

Answers on page 6.

Overspent American (1999: 68), refers to consumption as the "see–want–borrow–buy" process, which she believes is a comparative process in which desire is structured by what we see around us. As sociologists examine patterns such as these, they begin to use the sociological imagination.

The Sociological Imagination

Sociologist C. Wright Mills (1959b) described sociological reasoning as the ***sociological imagination*—the ability to see the relationship between individual experiences and the larger society.** This awareness enables us to understand the link between our personal experiences and the social contexts in which they occur. The sociological imagination helps us distinguish between personal troubles and social (or public) issues. *Personal troubles* are private problems that affect individuals and the networks of people with which they regularly associate. As a result, those problems must be solved by individuals within their immediate social settings. For example, one person being unemployed or running up a high credit card debt could be identified as a personal trouble. *Public issues* are problems that affect large numbers of people and often require solutions at the societal level. Widespread unemployment and massive, nationwide consumer debt are examples of public issues. The sociological imagination helps us place seemingly personal troubles, such as losing one's job or overspending on credit cards, into a larger social context, where we can distinguish whether and how personal troubles may be related to public issues.

Overspending as a Personal Trouble Although the character of the individual can contribute to social problems, some individual experiences are largely beyond the individual's control. They are influenced and in some situations determined by the society as a whole—by its historical development and its organization. In everyday life, we often blame individuals for "creating" their own problems. If a

sociology the systematic study of human society and social interaction.

society a large social grouping that shares the same geographical territory and is subject to the same political authority and dominant cultural expectations.

sociological imagination C. Wright Mills's term for the ability to see the relationship between individual experiences and the larger society.

Box 1.1 Sociology and Everyday Life

Answers to the Sociology Quiz on Consumption and Credit Cards

1. **True.** The credit card debt owed by the average U.S. household in 2007 (the most recent year for which statistics were available) was $9,840.
2. **False.** The average debt on undergraduate college students' credit cards in 2008 was about $2,200, and the typical student amasses almost $20,000 in student loans during his or her undergraduate years.
3. **False.** About 76 percent of undergraduate college students have at least one credit card.
4. **False.** Aggressive marketing of credit cards to college students and others is not illegal; the credit card industry routinely pays colleges and universities fees to rent tables for campus solicitations, and alumni groups offer "affinity" cards linked to the schools.
5. **False.** Although Congress has been encouraged to adopt a measure requiring age or income requirements on the issuance of credit cards, by mid-2009 no laws had been passed in this regard. However, a significant downturn in the economy may have produced a similar effect on the issuance of credit cards and their spending limits.
6. **True.** In the United States, more than one million people filed for bankruptcy in 2006.
7. **True.** The total is about $2.0 trillion when all consumer debt is taken into account.
8. **False.** Recent studies have shown that people in all income brackets have overused credit and not paid off debts in an effort to deal with difficult economic times.

Sources: Based on Americans for Fairness in Lending, 2009; The Motley Fool, 2007; and Weiss, 2008.

person sinks into debt due to overspending or credit card abuse, many people consider it to be the result of his or her own personal failings. However, this approach overlooks debt among people who are in low income brackets, having no way other than debt to gain the basic necessities of life. By contrast, at middle- and upper-income levels, overspending takes on a variety of other meanings.

At the individual level, people may accumulate credit cards and spend more than they can afford, thereby affecting all aspects of their lives, including health, family relationships, and employment stability. Sociologist George Ritzer (1999: 29) suggests that people may overspend through a gradual process in which credit cards "lure people into consumption by easy credit and then entice them into still further consumption by offers

Health and Human Services	Business	Communication	Academia	Law
Counseling Education Medicine Nursing Social Work	Advertising Labor Relations Management Marketing	Broadcasting Public Relations Journalism	Anthropology Economics Geography History Information Studies Media Studies/ Communication Political Science Psychology Sociology	Law Criminal Justice

▶ **Figure 1.1 Fields That Use Social Science Research**
In many careers, including jobs in academia, business, communication, health and human services, and law, the ability to analyze social science research is an important asset.

Source: Based on Katzer, Cook, and Crouch, 1991.

of 'payment holidays,' new cards, and increased credit limits." A classic example of Ritzer's description is Chip H., who describes how he has had problems with overspending on credit cards since his freshman year of college but still carries around seven of them:

> I was pretty good for about a year. Then I bought $200 speakers for my car and that put me over the top. I figured if I'm a little in the hole, then why not spend more. Then if you don't keep track, you end up just paying the interest, and you don't get anywhere. They get you, though. They bump you up another $500 or $600 because of your "outstanding credit." Then you charge more and go into deeper debt. It's like gambling, once you start you can't stop. (qtd. in McDonald, 1997: 4–5)

Chip, like millions of others, stayed heavily in debt, transferring $5,500 of debt from his MasterCard and Visa accounts to an AT&T Universal card that offered him lower interest, at least initially (McDonald, 1997).

Overspending as a Public Issue We can use the sociological imagination to look at the problem of overspending and credit card debt as a public issue—a societal problem. For example, Ritzer (1998) suggests that the relationship between credit card debt and the relatively low *savings rate* in the United States constitutes a public issue. Between 1990 and 2000, credit card debt tripled in the United States while savings diminished. Because savings is money that governments, businesses, and individuals can borrow for expansion, lack of savings may create problems for future economic growth. The rate of bankruptcies in this country is a problem both for financial institutions and the government. As corporations "write off" bad debt from those who declare bankruptcy or simply do not pay their bills, all consumers pay either directly or indirectly for that debt. Finally, poverty is forgotten as a social issue when more-affluent people are having a spending holiday and consuming all, or more than, they can afford to purchase.

Courtesy of Charlene Sullivan

According to management professor Charlene Sullivan, credit cards are available today to just about anyone; however, she advises consumers to limit the number of cards they carry and to think carefully about the interest rates, services, and flexibility of each card. She also reminds us that the responsibility falls on credit card holders to understand and control their credit behavior.

Some practices of the credit card industry are also a public issue (Ritzer, 1998). In a study of credit card use among college students, the sociologist Robert D. Manning (1999) found that students are aggressively targeted through marketing campaigns by credit card companies even though it is an accepted fact that some of the students will ruin their credit while still in college. Taking a walking tour of most large university campuses at registration time provides ample evidence of aggressive credit card marketing to students (and sometimes to faculty and staff). Card offers are on or near campus, tucked into school newspapers, and distributed by area bookstores that sell texts and school supplies. Solicitation letters begin arriving in students' mailboxes early in the freshman year. As graduation time approaches, seniors are regaled with letters of "congratulations" from the credit card industry and one last appeal to get a "lifestyle" credit card that benefits the school's alumni association with a small contribution when purchases are charged on the card. The problem has grown so great that some in the credit card industry have acknowledged it is a problem. As one bank vice president stated, "I've seen customers who've had as many as 13 Visas and MasterCards. . . . Banks are guilty in that they make credit too easily available" (*ABA Banking Journal*, 1990: 42).

As these examples show, Mills's *The Sociological Imagination* (1959b) remains useful for examining issues in the twenty-first century because it helps integrate microlevel (individual and small group) troubles with compelling public issues of our day. Recently, his ideas have been applied at the global level as well.

The Importance of a Global Sociological Imagination

Although existing sociological theory and research provide the foundation for sociological thinking, we must reach beyond past studies that have focused primarily on the United States to develop a more comprehensive *global* approach for the future. In the twenty-first century, we face important challenges in a rapidly changing nation and world. The world's ***high-income countries*** **are nations with highly industrialized economies; technologically advanced industrial, administrative, and service occupations; and relatively high levels of national and personal income.** Examples include the United States, Canada, Australia, New Zealand, Japan, and the countries of Western Europe (see ▶ Map 1.1).

As compared with other nations of the world, many high-income nations have a high standard of living and a lower death rate due to advances in nutrition and medical technology. However, everyone living in a so-called high-income country does not necessarily have a high income or an outstanding quality of life. Even among middle- and upper-income people, problems such as personal debt may threaten economic and social stability. For example, more than 1.1 million people in this country filed for bankruptcy in 2006, and more than 97 percent of all U.S. bankruptcies were filed by consumers (U.S. Courts, 2006).

In contrast, ***middle-income countries*** **are nations with industrializing economies, particularly in urban areas, and moderate levels of national and personal income.** Examples of middle-income countries include the nations of Eastern Europe and many Latin American countries, where nations such as Brazil and Mexico are industrializing rapidly. ***Low-income countries*** **are primarily agrarian nations with little industrialization and low levels of national and personal income.** Examples of low-income countries include many of the nations of Africa and Asia, particularly the People's Republic of China and India, where people typically work the land and are among the poorest in the world. However, generalizations are difficult to make because there are wide differences in income and

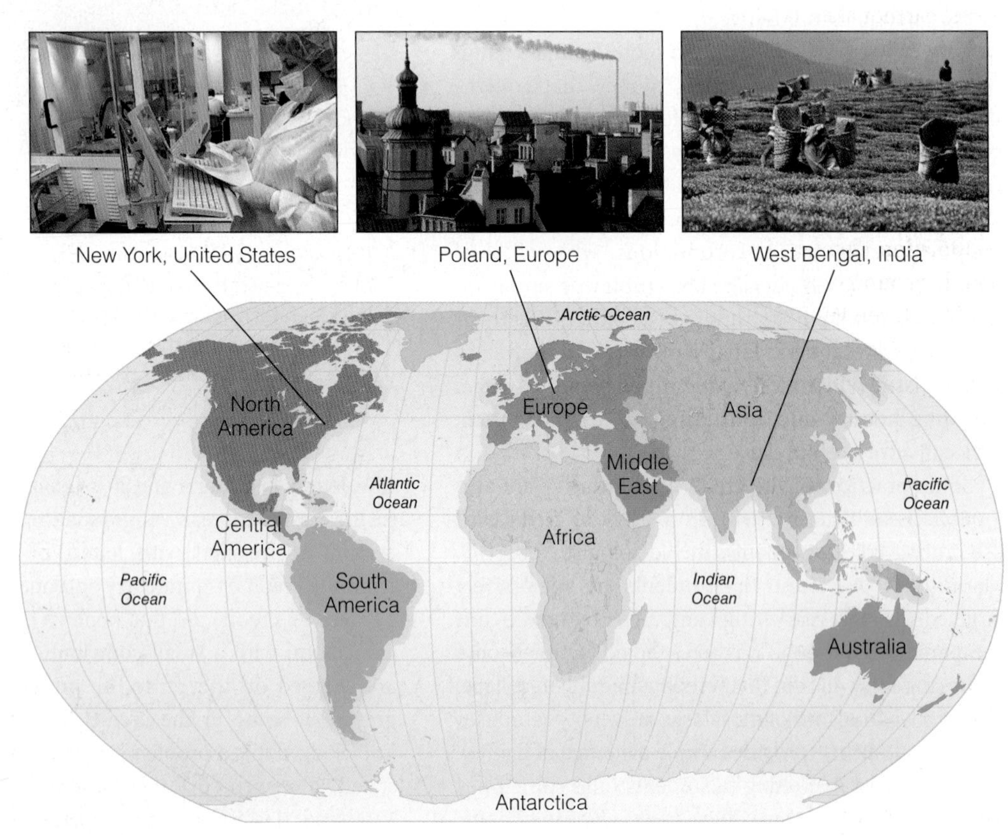

▶ **Map 1.1 The World's Economies in the Early Twenty-First Century**
High-income, middle-income, and low-income countries.

standards of living within many nations (see Chapter 9, "Global Stratification").

The global expansion of credit cards and other forms of consumerism, including the proliferation of "big-box" retail establishments such as Wal-Mart, shows the influence of U.S.-based megacorporations on other nations of the world. Consider Wal-Mart, for example. Sam Walton opened his first Wal-Mart store in Rogers, Arkansas, in 1962, and the company's home office was established in Bentonville, Arkansas, in the early 1970s. From a small-scale, regional operation in Arkansas, the Wal-Mart chain has now built a worldwide empire. Although the global expansion of credit cards and Wal-Mart Superstores has produced benefits for some people, it has also affected the everyday lives of many individuals around the world (see Box 1.2).

Throughout this text, we will continue to develop our sociological imaginations by examining social life in the United States and other nations. The future of our nation is deeply intertwined with the future of all other nations of the world on economic, political, environmental, and humanitarian levels. We buy many goods and services that were produced in other nations, and we sell much of what we produce to the people of other nations. Peace in other nations is important if we are to ensure peace within our borders. Famine, unrest, and brutality in other regions of the world must be of concern to people in the United States. Moreover, fires, earthquakes, famine, or environmental pollution in one nation typically has an adverse influence on other nations as well. Global problems contribute to the large influx of immigrants who arrive in the United States annually. These immigrants bring with them a rich diversity of language, customs, religions, and previous life experiences; they also contribute to dramatic population changes that will have a long-term effect on this country.

Whatever your race/ethnicity, class, sex, or age, are you able to include in your thinking the perspectives of people who are quite different from you in experiences and points of view? Before you answer this question, a few definitions are in order. *Race* is a term used by many people to specify groups of people distinguished by physical characteristics such as skin color; in fact, there are no "pure" racial types, and the concept of race is considered by most sociologists to be a social construction that people use to justify existing social inequalities. *Ethnicity* refers to the cultural heritage or identity of a group and is based on factors such as language or country of origin. *Class* is the relative location of a person or group within the larger society, based on wealth, power, prestige, or other valued resources. *Sex* refers to the biological and anatomical differences between females and males. By contrast, *gender* refers to the meanings, beliefs, and practices associated with sex differences, referred to as *femininity* and *masculinity*.

In forming your own global sociological imagination and in seeing the possibilities for sociology in the twenty-first century, it will be helpful for you to understand the development of the discipline.

The Origins of Sociological Thinking

Throughout history, social philosophers and religious authorities have made countless observations about human behavior, but the first systematic analysis of society is found in the philosophies of early Greek philosophers such as Plato (c. 427–347 B.C.E.) and Aristotle (384–322 B.C.E.). For example, Aristotle was concerned with developing a system of knowledge, and he engaged in theorizing and the empirical analysis of data collected from people in Greek cities regarding their views about social life when ruled by kings or aristocracies or when living in democracies (Collins, 1994). However, early thinkers such as Plato and Aristotle provided thoughts on what they believed society *ought* to be like, rather than describing how society actually *was*.

Social thought began to change rapidly in the seventeenth century with the scientific revolution. Like their predecessors in the natural sciences, social thinkers sought to develop a scientific understanding of social life, believing that their work might enable people to reach their full potential. The contributions of Isaac Newton (1642–1727) to modern science, including the discovery of the laws of gravity and motion and the development of calculus, inspired social thinkers to believe that similar advances could be made in the systematic study of human behavior. As Newton advanced the

high-income countries (sometimes referred to as **industrial countries**) nations with highly industrialized economies; technologically advanced industrial, administrative, and service occupations; and relatively high levels of national and personal income.

middle-income countries (sometimes referred to as **developing countries**) nations with industrializing economies and moderate levels of national and personal income.

low-income countries (sometimes referred to as **underdeveloped countries**) nations with little industrialization and low levels of national and personal income.

Box 1.2 Sociology in Global Perspective

The Global Wal-Mart Effect? Big-Box Stores and Credit Cards

Did you know that:

- More than half of all Wal-Mart stores worldwide are located outside the United States?
- Wal-Mart operates more than 100 stores, including supercenters, neighborhood markets, and Sam's Clubs, in China?
- Wal-Mart is a major player in the credit card business in China, where people traditionally have been opposed to purchasing products on credit?

Although most of us are aware that Wal-Mart stores are visible in virtually every community in the United States, we are less aware of the extent to which Wal-Mart and other "big-box" stores are changing the face of the world economy as the megacorporations that own them expand their operations into other nations—and into the credit card business.

The strategic placement of Wal-Mart stores both here and abroad accounts for part of the financial success of this retailing giant, but another American export—credit cards—is also part of the company's business plan. Credit cards are changing the way that people shop and how they think about spending money in emerging nations such as China. For example, Wal-Mart China is aggressively seeking both shoppers and credit card holders. By encouraging people to spend money now rather than save it for later, corporations such as Wal-Mart that issue "co-branded" credit cards gain in two ways: (1) people buy more goods than they would otherwise acquire, thus increasing sales; and (2) the corporation whose "brand" is on the credit card increases its earnings as a result of the interest the cardholder pays on credit card debt.

Co-branded credit cards are issued jointly by a bank (which provides the credit) and a business (such as an airline or a retailer) that offers some sort of reward for using its credit card. The credit card is a form of *revolving credit*—an agreement under which a person may make a minimum payment on the total balance each month, pay interest on the unpaid balance, and purchase additional items within a preestablished credit limit. As shoppers pay interest on their credit card debt, corporations such as Wal-Mart win twice because they make profits from the interest (which is shared between the bank and the business) as well as the initial sale of the goods that were charged on the card.

The motto for the Wal-Mart credit card in China is "maximizing value, enjoying life," and this idea encourages a change in attitude from the past, when—regardless of income level—most residents of that country did not possess credit cards. With the recent economic depression in China, its citizens now use credit cards for more than $300 billion worth of transactions annually (Kurlantzick, 2003). This has brought a corresponding surge in credit card debt, which can be partly attributed to aggressive marketing by transnational retailers but also to credit card companies encouraging consumers to buy now, pay later.

Reflect & Analyze

If some of the social theorists discussed in Chapter 1 (such as Karl Marx) were alive today, how might they describe what is happening in regard to consumerism and debt in the twenty-first century? What new insights might we gain regarding the global economy if we use our sociological imagination to think about this issue? What do you think?

Sources: Based on *The Economist*, 2006; Wal-Mart China, 2006; and Wal-Mart, 2008.

An exciting aspect of studying sociology is comparing our own lives with those of people around the world. Global consumerism, as evidenced by the opening of the new Wal-Mart Supercenter in Shanghai, China, provides a window through which we can observe how issues such as shopping and credit affect all of us. Which aspects of this photo reflect local culture? Which aspects reflect a global cultural phenomenon?

cause of physics and the natural sciences, he was viewed by many as the model of a true scientist. Moreover, his belief that the universe is an orderly, self-regulating system strongly influenced the thinking of early social theorists.

Sociology and the Age of Enlightenment

The origins of sociological thinking as we know it today can be traced to the scientific revolution in the late seventeenth and mid-eighteenth centuries and to the Age of Enlightenment. In this period of European thought, emphasis was placed on the individual's possession of critical reasoning and experience. There was also widespread skepticism regarding the primacy of religion as a source of knowledge and heartfelt opposition to traditional authority. A basic assumption of the Enlightenment was that scientific laws had been designed with a view to human happiness and that the "invisible hand" of either Providence or the emerging economic system of capitalism would ensure that the individual's pursuit of enlightened self-interest would always be conducive to the welfare of society as a whole.

In France, the Enlightenment (also referred to as the *Age of Reason*) was dominated by a group of thinkers referred to collectively as the *philosophes*. The philosophes included such well-known intellectuals as Charles Montesquieu (1689–1755), Jean-Jacques Rousseau (1712–1778), and Jacques Turgot (1727–1781). They defined a *philosophe* as one who, trampling on prejudice, tradition, universal consent, and authority—in a word, all that enslaves most minds—dares to think for himself, to go back and search for the clearest general principles, and to admit nothing except on the testimony of his experience and reason (Kramnick, 1995). For the most part, these men were optimistic about the future, believing that human society could be improved through scientific discoveries. In this view, if people were free from the ignorance and superstition of the past, they could create new forms of political and economic organization such as democracy and capitalism, which would eventually produce wealth and destroy aristocracy and other oppressive forms of political leadership.

Although women were categorically excluded from much of public life in France because of the sexism of the day, some women strongly influenced the philosophes and their thinking through their participation in the *salon*—an open house held to stimulate discussion and intellectual debate. Salons provided a place for intellectuals and authors to discuss ideas and opinions and for women and men to engage in witty repartee regarding the issues of the day, but the "brotherhood" of philosophes typically viewed the women primarily as good listeners or mistresses more than as intellectual equals, even though the men sometimes later adopted the women's ideas as if they were their own. However, the writings of Mary Wollstonecraft (1759–1797) reflect the Enlightenment spirit, and her works have recently received recognition for influencing people's thoughts on the idea of human equality, particularly as it relates to social equality and women's right to education.

For women and men alike, the idea of observing how people lived in order to find out what they thought, and doing so in a systematic manner that could be verified, did not take hold until sweeping political and economic changes in the late eighteenth and early nineteenth centuries caused many people to realize that several of the answers provided by philosophers and theologians to some very pressing questions no longer seemed relevant. Many of these questions concerned the social upheaval brought about by the age of revolution, particularly the American Revolution of 1776 and the French Revolution of 1789, and the rapid industrialization and urbanization that occurred first in Britain, then in Western Europe, and later in the United States.

Sociology and the Age of Revolution, Industrialization, and Urbanization

Several types of revolution that took place in the eighteenth century had a profound influence on the origins of sociology. The Enlightenment produced an *intellectual revolution* in how people thought about social change, progress, and critical thinking. The optimistic views of the philosophes and other social thinkers regarding progress and equal opportunity (at least for some people) became part of the impetus for *political revolutions* and *economic revolutions*, first in America and then in France. The Enlightenment thinkers had emphasized a sense of common purpose and hope for human progress; the French Revolution and its aftermath replaced these ideals with discord and overt conflict (see Schama, 1989; Arendt, 1973).

During the nineteenth and early twentieth centuries, another form of revolution occurred: the *Industrial Revolution*. ***Industrialization* is the process by which societies are transformed from dependence on agriculture and handmade products to an emphasis on manufacturing and related industries.** This process first

industrialization the process by which societies are transformed from dependence on agriculture and handmade products to an emphasis on manufacturing and related industries.

occurred during the Industrial Revolution in Britain between 1760 and 1850, and was soon repeated throughout Western Europe. By the mid-nineteenth century, industrialization was well under way in the United States. Massive economic, technological, and social changes occurred as machine technology and the factory system shifted the economic base of these nations from agriculture to manufacturing. A new social class of industrialists emerged in textiles, iron smelting, and related industries. Many people who had labored on the land were forced to leave their tightly knit rural communities and sacrifice well-defined social relationships to seek employment as factory workers in the emerging cities, which became the centers of industrial work.

Urbanization accompanied modernization and the rapid process of industrialization. ***Urbanization* is the process by which an increasing proportion of a population lives in cities rather than in rural areas.** Although cities existed long before the Industrial Revolution, the development of the factory system led to a rapid increase in both the number of cities and the size of their populations. People from very diverse backgrounds worked together in the same factory. At the same time, many people shifted from being *producers* to being *consumers*. For example, families living in the cities had to buy food with their wages because they could no longer grow their own crops to consume or to barter for other resources. Similarly, people had to pay rent for their lodging because they could no longer exchange their services for shelter.

These living and working conditions led to the development of new social problems: inadequate housing, crowding, unsanitary conditions, poverty, pollution, and crime. Wages were so low that entire families—including very young children—were forced to work, often under hazardous conditions and with no job security. As these conditions became more visible, a new breed of social thinkers turned its attention to trying to understand why and how society was changing.

The Development of Modern Sociology

At the same time that urban problems were growing worse, natural scientists had been using reason, or rational thinking, to discover the laws of physics and the movement of the planets. Social thinkers started to believe that by applying the methods developed by the natural sciences, they might discover the laws of human behavior and apply these laws to solve social problems. Historically, the time was ripe for such thoughts because the Age of Enlightenment had produced a belief in reason and humanity's ability to perfect itself.

Early Thinkers: A Concern with Social Order and Stability

Early social thinkers—such as Auguste Comte, Harriet Martineau, Herbert Spencer, and Emile Durkheim—were interested in analyzing social order and stability, and many of their ideas had a dramatic influence on modern sociology.

Auguste Comte The French philosopher Auguste Comte (1798–1857) coined the term *sociology* from the Latin *socius* ("social, being with others") and the Greek *logos* ("study of") to describe a new science that would engage in the study of society. Even though he never actually conducted sociological research, Comte is considered by some to be the "founder of sociology." Comte's theory that societies contain *social statics* (forces for social order and stability) and *social dynamics* (forces for conflict and change) contin-

© Hulton Archive/Getty Images

As the Industrial Revolution swept through the United States beginning in the nineteenth century, sights like this became increasingly common. This early automobile assembly line is symbolic of the factory system that shifted the base of the U.S. economy from agriculture to manufacturing. What new technologies are transforming the U.S. economy in the twenty-first century?

Auguste Comte

ues to be used, although not in these exact terms, in contemporary sociology.

Drawing heavily on the ideas of his mentor, Count Henri de Saint-Simon, Comte stressed that the methods of the natural sciences should be applied to the objective study of society. Saint-Simon's primary interest in studying society was social reform, but Comte sought to unlock the secrets of society so that intellectuals like himself could become the new secular (as contrasted with religious) "high priests" of society (Nisbet, 1979). For Comte, the best policies involved order and authority. He envisioned that a new consensus would emerge on social issues and that the new science of sociology would play a significant part in the reorganization of society (Lenzer, 1998).

Comte's philosophy became known as ***positivism*—a belief that the world can best be understood through scientific inquiry.** Comte believed that objective, bias-free knowledge was attainable only through the use of science rather than religion. However, scientific knowledge was "relative knowledge," not absolute and final. Comte's positivism had two dimensions: (1) methodological—the application of scientific knowledge to both physical and social phenomena—and (2) social and political—the use of such knowledge to predict the likely results of different policies so that the best one could be chosen.

The ideas of Saint-Simon and Comte regarding the objective, scientific study of society are deeply embedded in the discipline of sociology. Of particular importance is Comte's idea that the nature of human thinking and knowledge passed through several stages as societies evolved from simple to more complex. Comte described how the idea systems and their corresponding social structural arrangements changed in what he termed the *law of the three stages:* the theological, metaphysical, and scientific (or positivistic) stages. Comte believed that knowledge began in the *theological stage*—explanations were based on religion and the supernatural. Next, knowledge moved to the *metaphysical stage*—explanations were based on abstract philosophical speculation. Finally, knowledge would reach the *scientific* or *positive stage*—explanations are based on systematic observation, experimentation, comparison, and historical analysis. Shifts in the forms of knowledge in societies were linked to changes in the structural systems of society. In the theological stage, kinship was the most prominent unit of society; however, in the metaphysical stage, the state became the prominent unit, and control shifted from small groups to the state, military, and law. In the scientific or positive stage, industry became the prominent structural unit in society, and scientists became the spiritual leaders, replacing in importance the priests and philosophers of the previous stages of knowledge. For Comte, this progression through the three stages constituted the basic law of social dynamics, and, when coupled with the laws of statics (which emphasized social order and stability), the new science of sociology could bring about positive social change.

Harriet Martineau Comte's works were made more accessible for a wide variety of scholars through the efforts of the British sociologist Harriet Martineau (1802–1876). Until recently, Martineau received no recognition in the field of sociology, partly because she was a woman in a male-dominated discipline and society. Not only did she translate and condense Comte's work, but she was also an active sociologist in her own right. Martineau studied the social customs of Britain and the United States and analyzed the consequences of industrialization and capitalism. In *Society in America* (1962/1837), she examined religion, politics, child rearing, slavery, and immigration in the United States, paying special attention to social distinctions based on class, race, and gender. Her works explore the status of women, children, and "sufferers" (persons who are

urbanization the process by which an increasing proportion of a population lives in cities rather than in rural areas.

positivism a term describing Auguste Comte's belief that the world can best be understood through scientific inquiry.

© Spencer Arnold/Getty Images

Harriet Martineau

considered to be criminal, mentally ill, handicapped, poor, or alcoholic).

Based on her reading of Mary Wollstonecraft's *A Vindication of the Rights of Women* (1974/1797), Martineau advocated racial and gender equality. She was also committed to creating a science of society that would be grounded in empirical observations and widely accessible to people. She argued that sociologists should be impartial in their assessment of society but that it is entirely appropriate to compare the existing state of society with the principles on which it was founded (Lengermann and Niebrugge-Brantley, 1998).

Some scholars believe that Martineau's place in the history of sociology should be as a founding member of this field of study, not just as the translator of Auguste Comte's work (Hoecker-Drysdale, 1992; Lengermann and Niebrugge-Brantley, 1998). Others have highlighted her influence in spreading the idea that societal progress could be brought about by the spread of democracy and the growth of industrial capitalism (Polanyi, 1944). Martineau believed that a better society would emerge if women and men were treated equally, enlightened reform occurred, and cooperation existed among people in all social classes (but led by the middle class).

In keeping with the sociological imagination, Martineau not only analyzed large-scale social structures in society, but she also explored how these factors influenced the lives of people, particularly women, children, and those who were marginalized by virtue of being criminal, mentally ill, disabled, poor, or alcoholic (Lengermann and Niebrugge-Brantley, 1998). She remained convinced that sociology, the "true science of human nature," could bring about new knowledge and understanding, enlarging people's capacity to create a just society and live heroic lives (Hoecker-Drysdale, 1992).

Herbert Spencer Unlike Comte, who was strongly influenced by the upheavals of the French Revolution, the British social theorist Herbert Spencer (1820–1903) was born in a more peaceful and optimistic period in his country's history. Spencer's major contribution to sociology was an evolutionary perspective on social order and social change. Although the term *evolution* has various meanings, evolutionary theory should be taken to mean "a theory to explain the mechanisms of organic/social change" (Haines, 1997: 81). According to Spencer's Theory of General Evolution, society, like a biological organism, has various interdependent parts (such as the family, the economy, and the government) that work to ensure the stability and survival of the entire society.

Spencer believed that societies developed through a process of "struggle" (for existence) and "fitness" (for survival), which he referred to as the "survival of the fittest." Because this phrase is often attributed to Charles Darwin, Spencer's view of society is known as ***social Darwinism*—the belief that those species of animals, including human beings, best adapted to their environment survive and prosper, whereas those poorly adapted die out.** Spencer equated this process of *natural selection* with progress because only the "fittest" members of society would survive the competition, and the "unfit" would be filtered out of society. Based on this belief, he strongly opposed any social reform that might interfere with the natural selection process and, thus, damage society by favoring its least-worthy members.

Critics have suggested that many of Spencer's ideas contain serious flaws. For one thing, societies are not the same as biological systems; people are able to cre-

© Hulton Archive/Getty Images

Herbert Spencer

ate and transform the environment in which they live. Moreover, the notion of the survival of the fittest can easily be used to justify class, racial–ethnic, and gender inequalities and to rationalize the lack of action to eliminate harmful practices that contribute to such inequalities. Not surprisingly, Spencer's "hands-off" view was applauded by many wealthy industrialists of his day. John D. Rockefeller, who gained monopolistic control of much of the U.S. oil industry early in the twentieth century, maintained that the growth of giant businesses was merely the "survival of the fittest" (Feagin, Baker, and Feagin, 2006).

Social Darwinism served as a rationalization for some people's assertion of the superiority of the white race. After the Civil War, it was used to justify the repression and neglect of African Americans as well as the policies that resulted in the annihilation of Native American populations. Although some social reformers spoke out against these justifications, "scientific" racism continued to exist (Turner, Singleton, and Musick, 1984). In both positive and negative ways, many of Spencer's ideas and concepts have been deeply embedded in social thinking and public policy for over a century.

Emile Durkheim French sociologist Emile Durkheim (1858–1917) was an avowed critic of some of Spencer's views while incorporating others into his own writing. Durkheim stressed that people are the product of their social environment and that behavior cannot be fully understood in terms of *individual* biological and psychological traits. He believed that the limits of human potential are *socially* based, not *biologically* based. As Durkheim saw religious traditions evaporating in his society, he searched for a scientific, rational way to provide for societal integration and stability (Hadden, 1997).

Emile Durkheim

In *The Rules of Sociological Method* (1964a/1895), Durkheim set forth one of his most important contributions to sociology: the idea that societies are built on social facts. ***Social facts* are patterned ways of acting, thinking, and feeling that exist *outside* any one individual but that exert social control over each person.** Durkheim believed that social facts must be explained by other social facts—by reference to the social structure rather than to individual attributes.

Durkheim was concerned with social order and social stability because he lived during the period of rapid social changes in Europe resulting from industrialization and urbanization. His recurring question was this: How do societies manage to hold together? In *The Division of Labor in Society* (1933/1893), Durkheim concluded that preindustrial societies were held together by strong traditions and by members' shared moral beliefs and values. As societies industrialized, more specialized economic activity became the basis of the social bond because people became dependent on one another.

Durkheim observed that rapid social change and a more specialized division of labor produce *strains* in society. These strains lead to a breakdown in traditional organization, values, and authority and to a dramatic increase in ***anomie*—a condition in which social control becomes ineffective as a result of the loss of shared values and of a sense of purpose in society.** According to Durkheim, anomie is most likely to occur during a period of rapid social change. In *Suicide* (1964b/1897), he explored the relationship between anomic social conditions and suicide, as discussed in Chapter 2.

Durkheim's contributions to sociology are so significant that he has been referred to as "*the* crucial figure in the development of sociology as an academic discipline [and as] one of the deepest roots of the sociological imagination" (Tiryakian, 1978: 187). He has long been viewed as a proponent of the scientific approach

social Darwinism Herbert Spencer's belief that those species of animals, including human beings, best adapted to their environment survive and prosper, whereas those poorly adapted die out.

social facts Emile Durkheim's term for patterned ways of acting, thinking, and feeling that exist *outside* any one individual but that exert social control over each person.

anomie Emile Durkheim's designation for a condition in which social control becomes ineffective as a result of the loss of shared values and of a sense of purpose in society.

to examining social facts that lie outside individuals. He is also described as the founding figure of the functionalist theoretical tradition. Recently, scholars have acknowledged Durkheim's influence on contemporary social theory, including the structuralist and postmodernist schools of thought. Like Comte, Martineau, and Spencer, Durkheim emphasized that sociology should be a science based on observation and the systematic study of social facts rather than on individual characteristics or traits.

Can Durkheim's ideas be applied to our ongoing analysis of credit cards? Durkheim was interested in examining the "social glue" that could hold contemporary societies together and provide people with a "sense of belonging." Ironically, the credit card industry has created what we might call a "pseudo-sense of belonging" through the creation of "affinity cards" designed to encourage members of an organization (such as a university alumni association) or people who share interests and activities (such as dog owners and skydiving enthusiasts) to possess a particular card. In later chapters, we examine Durkheim's theoretical contributions to diverse subjects ranging from suicide and deviance to education and religion.

Differing Views on the Status Quo: Stability Versus Change

Together with Karl Marx, Max Weber, and Georg Simmel, Durkheim established the course for modern sociology. We will look first at Marx's and Weber's divergent thoughts about conflict and social change in societies, and then at Georg Simmel's analysis of society.

Karl Marx In sharp contrast to Durkheim's focus on the stability of society, German economist and philosopher Karl Marx (1818–1883) stressed that history is a continuous clash between conflicting ideas and forces. He believed that conflict—especially class conflict—is necessary in order to produce social change and a better society. For Marx, the most important changes were economic. He concluded that the capitalist economic system was responsible for the overwhelming poverty that he observed in London at the beginning of the Industrial Revolution (Marx and Engels, 1967/1848).

In the Marxian framework, *class conflict* is the struggle between the capitalist class and the working class. The capitalist class, or *bourgeoisie,* comprises those who own and control the means of production—the tools, land, factories, and money for investment that form the economic basis of a society. The working class, or *proletariat,* is composed of those who must sell their labor because they have no other means to earn a livelihood. From Marx's viewpoint, the capitalist class controls and exploits the masses of struggling work-

Karl Marx

ers by paying less than the value of their labor. This exploitation results in workers' *alienation*—a feeling of powerlessness and estrangement from other people and from themselves (see a contemporary discussion of alienation based of Marx's perspective in Sociology Works!). Marx predicted that the working class would become aware of its exploitation, overthrow the capitalists, and establish a free and classless society.

Can Marx provide useful insights on the means of consumption? Although Marx primarily analyzed the process of production, he linked production and consumption in his definition of *commodities* as products that workers produce. Marx believed that commodities have a use value and an exchange value. *Use value* refers to objects that people produce to meet their personal needs or the needs of those in their immediate surroundings. By contrast, *exchange value* refers to the value that a commodity has when it is exchanged for money in the open market. In turn, this money is used to acquire other use values, and the cycle continues. According to Marx, commodities play a central role in capitalism, but the workers who give value to the commodities eventually fail to see this fact. Marx coined the phrase the *fetishism of commodities* to describe the situation in which workers fail to recognize that their labor gives the commodity its value and instead come to believe that a commodity's value is based on the natural properties of the thing itself. By extending Marx's idea in this regard, we might conclude that the workers did not rebel against capitalism for several reasons: (1) they falsely believed that what capitalists did was in their own best interests as well, (2) they believed that the products they produced had a value in the marketplace that was independent of anything the workers

Sociology *Works!*

Marx's Preview of Contemporary Alienation

The texture of discontent (or lack thereof) can say a lot about a nation, and that Americans today are less likely to rebel may not be an entirely positive sign.

It certainly doesn't mean we have more love, patience or tolerance for one another. Indeed, it may mean just the opposite, that we tend not to trust one another and that we are more alienated from our neighbors than ever before. The lack of direct action could signal the weakening of a social contract that keeps people meaningfully invested in the fate of our country—which may, in turn, be hindering our ability to resolve this [financial] crisis.

—Sudhir Venkatesh (2009: WK10), a contemporary sociologist who sees alienation as one possible explanation of why many people are not more actively protesting against the actions of high-powered individuals and organizations that have contributed to our current financial plight

Social scientists have long been fascinated by alienation. This concept is often attributed to the economist and philosopher Karl Marx. As further discussed in Chapter 8, *alienation* is a term used to refer to an individual's feeling of *powerlessness* and *estrangement* from other people and from oneself. Marx specifically linked alienation to social relations that are inherent in capitalism; however, more recent social thinkers have expanded his ideas to include social psychological feelings of powerlessness, meaninglessness, and isolation. These may be present because people experience social injustice and vast economic inequalities in contemporary societies.

In the sociologist Sudhir Venkatesh's explanation of why people are not more rebellious when they are grappling with issues of corporate malfeasance and their own economic hardship, Venkatesh states that we hear a lot about "populist rage" but actually see very few protests or acts of outright rebellion: People may feel alienated, but they do not sense that there is much that they can do about the problem. They also do not feel strong social ties with other individuals that might lead them to bond together for joint action. According to Venkatesh (2009: WK10), rather than coming together for social action, we often express our individual frustration in a blog or on a radio call-in show. Or we may entertain ourselves with technology, such as our computers or cell phones, which actually separates us from other people and turns our communications into something that is "indirect, impersonal and emotionally flat." According to Venkatesh, "With headsets on and our hands busily texting, we are less aware of one another's behavior in public space." And this lack of awareness contributes to, rather than reduces, our alienation from one another and from the larger society of which we are a part.

If we apply the earlier theorizing of Marx regarding alienation to the contemporary views of Venkatesh on the weakening social contract, we can gain new insights on how sociology works. Today, it is necessary for us to view pressing social issues as important problems that we must all work together to solve. This viewpoint requires us to come together and talk about how we might solve problems rather than continuing to live in our own isolated social worlds, where many people feel alienated from other individuals.

Reflect & Analyze

Why are we often more concerned about trivial matters, such as who will be the big winner in a sporting event or on a TV reality show competition, than we are about how to reduce or eliminate some of our most pressing social and economic concerns? What might we learn from Marx's concept of alienation that would help us show that sociology works today?

did, and (3) they came to view ownership of the commodities as a desirable end in itself and to work longer hours so that they could afford to purchase more goods and services.

Although Marx's ideas on exploitation of workers cannot be fully developed into a theory of consumer exploitation, it has been argued that a form of exploitation does occur when capitalists "devote increasing attention to getting consumers to buy more goods and services" (Ritzer, 1995: 19). The primary ways by which capitalists can increase their profits are cutting costs and selling more products. To encourage continual increases in spending (and thus profits), capitalists have created mega-shopping malls, cable television shopping networks, and online shopping in order to provide consumers with greater opportunities to purchase more goods, increasing the consumers' credit card debt and forcing them to continue to work in order to pay their bills, but also raising privacy issues (see Box 1.3). Perhaps the ultimate agents of consumption are the industries that produce *desire* and make it possible for people to consume beyond their means.

Box 1.3 Sociology and Social Policy

Online Shopping and Your Privacy

Motorcycle jacket for kid brother on the Internet—$300
Monogrammed golf balls for dad on the Internet—$50
Vintage smoking robe for husband on the Internet—$80
Not having to hear "attention shoppers"—not even once—priceless.

The way to pay on the Internet and everywhere else you see the MasterCard logo: MasterCard.

—MasterCard advertisement (qtd. in Manning, 2000: 114)

This advertisement taps into a vital source of revenue for companies that issue credit cards: Online customers are an increasing percentage of those persons who use credit cards to make daily purchases. Some analysts estimate that online shopping generates more than $75 billion per year in revenues, and much of that $75 billion is based on credit (or debit) card purchases.

Earlier in this chapter, we mentioned that industrialization and urbanization were important historical factors that brought about significant changes in social life. Today, however, social life continues to change rapidly as the Internet increasingly becomes an integral part of our daily lives, including how we gather information, how we go about shopping, and how we view our privacy.

The Internet raises some important questions: Who is watching your online activity? How far are companies willing to go in "snooping" on those who visit their websites? At the time of this writing, companies that sell products or services on the Internet are not required to respect the privacy of their shoppers. According to the American Bar Association (2003), "This means the seller may collect data on which site pages you visit, which products you buy, when you buy them, and where you ship them. Then, the seller may share the information with other companies or sell it to them." Some websites have privacy policies posted but still insert "cookies" onto the hard drive of your computer. These cookies help the site's owner know where you go and what you do on the site. In some cases, the site owner records your e-mail address and begins sending you e-mail messages ("spam") about that company's products, whether you want to receive them or not.

To offset people's fears of invasion of privacy or abuse of their credit card information, corporations in the online sales business have sought to reassure customers that they are not being tracked and that it is safe to give out personal information online. However, the American Bar Association (ABA) advises caution in Internet interactions. According to the ABA, consumers using a credit card for an online purchase should ask whether or not their credit card number will be kept on file by the seller for automatic use in future orders. Online shoppers should also find out what information the seller is gathering about you, how the seller will use this information, and whether you can "opt out" of having this information gathered on you (American Bar Association, 2003).

Federal law requires banks, insurance companies, and other *financial services companies* with which you do business to notify you annually in writing with regard to what personal information they collect about you and what they do with that information. In response to that notice, you have the right to place certain restrictions on how the company uses that information.

Reflect & Analyze

Should federal law require all companies that obtain information about you on the Internet to give you that same notice and that same right, or would the companies that are most likely to misuse information simply ignore such a law? How can our study of sociology make us more aware of key social policy issues—such as this—that affect our everyday life?

Sources: Based on American Bar Association, 2003; and Ritzer, 1999.

Do you feel comfortable shopping online? Do you care if retailers use your private information for their own purposes? Why or why not?

© Colin Young-Wolff/PhotoEdit

These include the credit card, advertising, and marketing industries, which encourage consumers to spend more money, in many cases far beyond their available cash, on goods and services (Ritzer, 1995, 1999).

Marx's theories provide a springboard for neo-Marxist analysts and other scholars to examine the economic, political, and social relations embedded in production and consumption in historical and contemporary societies. But what is Marx's place in the history of sociology? Marx is regarded as one of the most profound sociological thinkers, one who combined ideas derived from philosophy, history, and the social sciences into a new theoretical configuration. However, his social and economic analyses have also inspired heated debates among generations of social scientists. Central to his view was the belief that society should not just be studied but should also be changed, because the *status quo* (the existing state of society) involved the oppression of most of the population by a small group of wealthy people. Those who believe that sociology should be value free are uncomfortable with Marx's advocacy of what some perceive to be radical social change. Scholars who examine society through the lens of race, gender, and class believe that his analysis places too much emphasis on class relations, often to the exclusion of issues regarding race/ethnicity and gender.

In recent decades, scholars have shown renewed interest in Marx's *social theory,* as opposed to his radical ideology (see Postone, 1997; Lewis, 1998). Throughout this text, we will continue to explore Marx's various contributions to sociological thinking.

Max Weber German social scientist Max Weber (pronounced VAY-ber) (1864–1920) was also concerned about the changes brought about by the Industrial Revolution. Although he disagreed with Marx's idea that economics is *the* central force in social change, Weber acknowledged that economic interests are important in shaping human action. Even so, he thought that economic systems are heavily influenced by other factors in a society. As we will see in Chapter 17 ("Religion"), one of Weber's most important works, *The Protestant Ethic and the Spirit of Capitalism* (1976/1904–1905), evaluated the role of the Protestant Reformation in producing a social climate in which capitalism could exist and flourish.

Unlike many early analysts, who believed that values could not be separated from the research process, Weber emphasized that sociology should be *value free*—research should be conducted in a scientific manner and should exclude the researcher's personal values and economic interests (Turner, Beeghley, and Powers, 2002). However, Weber realized that social behavior cannot be analyzed by the objective criteria

Max Weber

that we use to measure such things as temperature or weight. Although he recognized that sociologists cannot be totally value free, Weber stressed that they should employ *verstehen* (German for "understanding" or "insight") to gain the ability to see the world as others see it. In contemporary sociology, Weber's idea has been incorporated into the concept of the sociological imagination (discussed earlier in this chapter).

Weber was also concerned that large-scale organizations (bureaucracies) were becoming increasingly oriented toward routine administration and a specialized division of labor, which he believed were destructive to human vitality and freedom. According to Weber, rational bureaucracy, rather than class struggle, was the most significant factor in determining the social relations among people in industrial societies. In this view, bureaucratic domination can be used to maintain powerful (capitalist) interests in society. As discussed in Chapter 6 ("Groups and Organizations"), Weber's work on bureaucracy has had a far-reaching impact.

What might Weber's work contribute to a contemporary study of consumerism and the credit card industry? One of Weber's most useful concepts in this regard is *rationalization:* "the process by which the modern world has come to be increasingly dominated by structures devoted to efficiency, calculability, predictability, and technological control" (Ritzer, 1995: 21). According to Ritzer, the credit card industry has contributed to the rationalization process by the *efficiency* with which it makes loans and deals with consumers. For example, prior to the introduction of credit cards, the process of obtaining a loan was slow and cumbersome. Today, the process of obtaining a credit card is highly efficient. It may take only minutes from the time a brief questionnaire is filled out until credit records are checked by computer and the application is approved or disapproved. *Calculability* is demonstrated by scorecards that allow lenders to score potential borrowers based on prior statistics of other people's performance in paying their bills. Factors that are typically calculated include home ownership versus renting, length of time with present employer, and current bank and/or credit card references (see ▶ Figure 1.2).

The *predictability* of credit cards is easy to see. If the cardholder is current on paying bills and the merchant accepts that particular kind of card, the person knows that he or she will not be turned down on a purchase. Even the general appearance of the cards is highly predictable. Finally, the use of *technological control* in the contemporary rationalization process is apparent in the credit card industry. These technologies range from the computerized system that determines whether or not a new credit card will be issued to cards embedded with computer chips, ATM machines, and online systems that permit instantaneous transfers of funds. As Ritzer's application of the concept of rationalization to the credit card industry shows, many of Weber's ideas have served as the springboard for contemporary sociological theories and research.

Weber made significant contributions to modern sociology by emphasizing the goal of value-free inquiry and the necessity of understanding how others see the world. He also provided important insights on the process of rationalization, bureaucracy, religion, and many other topics. In his writings, Weber was more aware of women's issues than were many of the scholars of his day. Perhaps his awareness at least partially resulted from the fact that his wife, Marianne Weber, was an important figure in the women's movement in Germany in the early twentieth century (Roth, 1988).

Georg Simmel At about the same time that Durkheim was developing the field of sociology in France, the German sociologist Georg Simmel (pronounced ZIM-mel) (1858–1918) was theorizing about society as a web of patterned interactions among people. The main purpose of sociology, according to Simmel, should be to examine these social interaction processes within groups. In *The Sociology of Georg*

The more points you score, the more likely you are to receive credit, and a higher score may also result in your being charged lower interest rates on loans. The total score runs from 300 to 850, and most people will have a score between 600 and 800. Fifteen percent of the population will score below that range, and thirteen percent will score 800 or above. Here are the factors that go into your score:

	Score
☐ Bill payment history (35 percent of the total score), especially recently. Consistently paying your bills on time raises your score; being behind lowers it. Having an account sent to collections is bad. Filing banckruptcy *really* lowers your score.	
☐ Amount you owe and available credit (30 percent of total score). This includes money owed on credit cards, car loans, mortgages, and other debt as compared with the total amount of credit you have available.	
☐ Length of credit history (15 percent of total score). The longer you've had credit, the higher your score in this category.	
☐ Mix of credit (10 percent of total score). A person with both revolving credit (such as credit cards) and installment credit (such as car loans and mortgages) will have a higher score in this category.	
☐ Recent applications for credit (10 percent of total score). It doesn't hurt to shop several sources for the best interest rates, but if you are delinquent in paying your bills, applying for new credit from other sources may look like you are fighting off bankruptcy.	
Total Score:	

▶ **Figure 1.2**
Typical Credit Report "Scorecard"

Source: Based on Bankrate.com, 2003.

© The Granger Collection, New York

Georg Simmel

Simmel (1950/1902–1917), he analyzed how social interactions vary depending on the size of the social group. He concluded that interaction patterns differed between a *dyad,* a social group with two members, and a *triad,* a social group with three members. He developed *formal sociology,* an approach that focuses attention on the universal recurring social forms that underlie the varying content of social interaction. Simmel referred to these forms as the "geometry of social life." He also distinguished between the *forms* of social interaction (such as cooperation or conflict) and the *content* of social interaction in different contexts (for example, between leaders and followers).

Like the other social thinkers of his day, Simmel analyzed the impact of industrialization and urbanization on people's lives. He concluded that class conflict was becoming more pronounced in modern industrial societies. He also linked the increase in individualism, as opposed to concern for the group, to the fact that people now had many cross-cutting "social spheres"—membership in a number of organizations and voluntary associations—rather than having the singular community ties of the past. Simmel also assessed the costs of "progress" on the upper-class city dweller, who, he believed, had to develop certain techniques to survive the overwhelming stimulation of the city. Simmel's ultimate concern was to protect the autonomy of the individual in society.

The Philosophy of Money (1990/1907), one of Simmel's most insightful studies, sheds light on the issue of consumerism. According to Simmel, money takes on a life of its own as people come to see money and the things that it can purchase as an end in themselves. Eventually, everything (and everybody) is seen as having a price, and people become blasé, losing the ability to differentiate between what is really of value and what is not. If money increases imprudence in consumption, credit cards afford even greater opportunities for people to spend money they do not have for things they do not need and, in the process, to sink deeper into debt (Ritzer, 1995). An example is Diane Curran, a teacher in Syracuse, New York, who accumulated $27,452 of debt on a dozen credit cards and other loans (Frank, 1999). Even when her monthly credit card payments equaled her take-home pay, she was still acquiring new credit cards, which she used to keep up with the other credit card payments. After Curran was forced to file for bankruptcy, she stated that "I wish somebody had cut me off 10 years earlier" (qtd. in Hays, 1996: B1, B6).

Simmel's perspective on money is only one of many possible examples of how his writings provide insights into social life. Simmel's contributions to sociology are significant. He wrote more than thirty books and numerous essays on diverse topics, leading some critics to state that his work was fragmentary and piecemeal. However, his thinking has influenced a wide array of

© Simon Jarratt/Corbis

According to the sociologist Georg Simmel, society is a web of patterned interactions among people. If we focus on the behavior of individuals only, we may miss the underlying forms that make up the "geometry of social life."

sociologists, including the members of the "Chicago School" in the United States.

The Beginnings of Sociology in the United States

From Western Europe, sociology spread in the 1890s to the United States, where it thrived as a result of the intellectual climate and the rapid rate of social change. The first departments of sociology in the United States were located at the University of Chicago and at Atlanta University, then an African American school.

The Chicago School The first department of sociology in the United States was established at the University of Chicago, where the faculty was instrumental in starting the American Sociological Society (now known as the American Sociological Association). Robert E. Park (1864–1944), a member of the Chicago faculty, asserted that urbanization had a disintegrating influence on social life by producing an increase in the crime rate and in racial and class antagonisms that contributed to the segregation and isolation of neighborhoods (Ross, 1991). George Herbert Mead (1863–1931), another member of the faculty at Chicago, founded the symbolic interaction perspective, which is discussed later in this chapter. Mead made many significant contributions to sociology. Among these were his emphasis on the importance of studying the group ("the social") rather than starting with separate individuals. Mead also called our attention to the importance of shared communication among people based on language and gestures. As discussed in Chapter 4 ("Socialization"), Mead gave us important insights on how we develop our self-concept through interaction with those persons who are the most significant influences in our lives.

Jane Addams Jane Addams (1860–1935) is one of the best-known early women sociologists in the United States because she founded Hull House, one of the most famous settlement houses, in an impoverished area of Chicago. Throughout her career, she was actively engaged in sociological endeavors: She lectured at numerous colleges, was a charter member of the American Sociological Society, and published a number of articles and books. Addams was one of the authors of *Hull-House Maps and Papers,* a groundbreaking book that used a methodological technique employed by sociologists for the next forty years (Deegan, 1988). She was also awarded a Nobel Prize for her assistance to the underprivileged.

W. E. B. Du Bois and Atlanta University The second department of sociology in the United States was founded by W. E. B. Du Bois (1868–1963) at Atlanta University. He created a laboratory of sociology, instituted a program of systematic research, founded and conducted regular sociological conferences on research, founded two journals, and established a record of valuable publications. His classic work, *The Philadelphia Negro: A Social Study* (1967/1899), was based on his research into Philadelphia's African American community and stressed the strengths and weaknesses of a community wrestling with overwhelming social problems. Du Bois was one of the first scholars to note that a dual heritage creates conflict for people of color.

Jane Addams

W. E. B. Du Bois

He called this duality *double-consciousness*—the identity conflict of being a black and an American. Du Bois pointed out that although people in this country espouse such values as democracy, freedom, and equality, they also accept racism and group discrimination. African Americans are the victims of these conflicting values and the actions that result from them (Benjamin, 1991).

Contemporary Theoretical Perspectives

Given the many and varied ideas and trends that influenced the development of sociology, how do contemporary sociologists view society? Some see it as basically a stable and ongoing entity; others view it in terms of many groups competing for scarce resources; still others describe it based on the everyday, routine interactions among individuals. Each of these views represents a method of examining the same phenomena. Each is based on general ideas as to how social life is organized and represents an effort to link specific observations in a meaningful way. Each uses a ***theory*—a set of logically interrelated statements that attempts to describe, explain, and (occasionally) predict social events.** Each theory helps interpret reality in a distinct way by providing a framework in which observations may be logically ordered. Sociologists refer to this theoretical framework as a *perspective*—an overall approach to or viewpoint on some subject. Three major theoretical perspectives have emerged in sociology: the functionalist, conflict, and symbolic interactionist perspectives. Other perspectives, such as postmodernism, have emerged and gained acceptance among some social thinkers more recently. Before turning to the specifics of these perspectives, however, we should note that some theorists and theories do not neatly fit into any of these perspectives.

Functionalist Perspectives

Also known as *functionalism* and *structural functionalism*, ***functionalist perspectives* are based on the assumption that society is a stable, orderly system.** This stable system is characterized by *societal consensus*, whereby the majority of members share a common set of values, beliefs, and behavioral expectations. According to this perspective, a society is composed of interrelated parts, each of which serves a function and (ideally) contributes to the overall stability of the society. Societies develop social structures, or institutions, that persist because they play a part in helping society survive. These institutions include the family, education, government, religion, and the economy. If anything adverse happens to one of these institutions or parts, all other parts are affected and the system no longer functions properly. As Durkheim noted, rapid social change and a more specialized division of labor produce *strains* in society that lead to a breakdown in these traditional institutions and may result in social problems such as an increase in crime and suicide rates.

Talcott Parsons and Robert Merton Talcott Parsons (1902–1979), perhaps the most influential contemporary advocate of the functionalist perspective, stressed that all societies must provide for meeting social needs in order to survive. Parsons (1955) suggested, for example, that a division of labor (distinct, specialized functions) between husband and wife is essential for family stability and social order. The husband/father performs the *instrumental tasks*, which involve leadership and decision-making responsibilities in the home and employment outside the home to support the family. The wife/mother is responsible for the *expressive tasks*, including housework, caring for the children, and providing emotional support for the entire family. Parsons believed that other institutions, including school, church, and government, must function to assist the family and that all institutions must work together to preserve the system over time (Parsons, 1955).

Functionalism was refined further by Robert K. Merton (1910–2003), who distinguished between manifest and latent functions of social institutions. ***Manifest functions* are intended and/or overtly recognized by the participants in a social unit.** In contrast, ***latent functions* are unintended functions that are hidden and remain unacknowledged by participants.** For example, a manifest function of education is the transmission of knowledge and skills from one generation to the

theory a set of logically interrelated statements that attempts to describe, explain, and (occasionally) predict social events.

functionalist perspectives the sociological approach that views society as a stable, orderly system.

manifest functions functions that are intended and/or overtly recognized by the participants in a social unit.

latent functions unintended functions that are hidden and remain unacknowledged by participants.

next; a latent function is the establishment of social relations and networks. Merton noted that all features of a social system may not be functional at all times; *dysfunctions* are the undesirable consequences of any element of a society. A dysfunction of education in the United States is the perpetuation of gender, racial–ethnic, and class inequalities. Such dysfunctions may threaten the capacity of a society to adapt and survive (Merton, 1968).

Talcott Parsons

Robert Merton

Applying a Functional Perspective to Shopping and Consumption How might functionalists analyze shopping and consumption? When we examine the part-to-whole relationships of contemporary society in high-income nations, it immediately becomes apparent that each social institution depends on the others for its well-being. For example, a booming economy benefits other social institutions, including the family (members are gainfully employed), religion (churches, mosques, synagogues, and temples receive larger contributions), and education (school taxes are higher when property values are higher). A strong economy also makes it possible for more people to purchase more goods and services. Due to the significance of the strength of the economy, the U.S. Census Bureau conducts surveys (for the Bureau of Labor Statistics) to determine how people are spending their money (see the "Census Profiles" feature). If people have "extra" money to spend and can afford leisure time away from work, they are more likely to dine out, take trips, and purchase things they might otherwise forgo.

Clearly, the manifest functions of shopping and consumption include purchasing necessary items such as food, clothing, household items, and some-

Shopping malls are a reflection of a consumer society. A manifest function of a shopping mall is to sell goods and services to shoppers; however, a latent function may be to provide a communal area in which people can visit friends and eat. For this reason, food courts have proven to be a boon in shopping malls around the globe.

times transportation. In contemporary societies, purchasing entertainment and information is another function of shopping, both in actual stores and in virtual stores online. But what are the latent functions of shopping malls, for example? Many teens go to the mall to "hang out," visit with friends, maybe buy a T-shirt, and eat lunch at the food court. People of all ages go shopping for pleasure, relaxation, and perhaps to enhance their feelings of self-worth. ("If I buy this product, I'll look younger/beautiful/handsome/sexy, etc.!") As one scholar noted, "Shopping entails the joy of going into a safe spot filled with things to look at where [shoppers] are treated deferentially. Although no one has hooked up a turbo lie detector to a shopper out for fun, if they did, the machine would register increased arousal, heightened involvement, perceived freedom, and fantasy fulfillment" (Twitchell, 1999: 243).

However, shopping and consuming may also produce problems or dysfunctions. Some people are "shopaholics" or "credit card junkies" who cannot stop spending money; others are kleptomaniacs, who steal products rather than pay for them. In the end, however, the typical functionalist approach to consumerism is shown in this comment by one scholar: "Let's face it, the idea that consumerism creates artificial desires rests on a wistful ignorance of history and human nature, on the hazy, romantic feeling that there existed some halcyon era of noble savages with purely natural needs. Once fed and sheltered, our needs have always been cultural, not natural. Until there is some other system to codify and satisfy those needs and yearnings, capitalism—and the culture it carries with it—will continue not just to thrive but to triumph" (Twitchell, 1999: 283).

Conflict Perspectives

According to ***conflict perspectives*, groups in society are engaged in a continuous power struggle for control of scarce resources.** Conflict may take the form of politics, litigation, negotiations, or family discussions about financial matters. Simmel, Marx, and Weber contributed significantly to this perspective by focusing on the inevitability of clashes between social groups. Today, advocates of the conflict perspective view social life as a continuous power struggle among competing social groups.

Max Weber and C. Wright Mills As previously discussed, Karl Marx focused on the exploitation and oppression of the proletariat (the workers) by the bourgeoisie (the owners or capitalist class). Max Weber recognized the importance of economic conditions in producing inequality and conflict in society but added

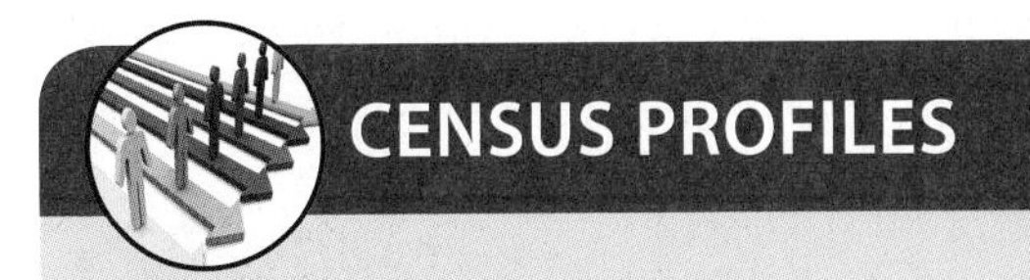

CENSUS PROFILES

Consumer Spending

The U.S. Census Bureau provides a wealth of data that helps sociologists and other researchers answer questions about the characteristics of the U.S. population. Although the decennial census occurs only once every ten years, the Census Bureau conducts surveys and produces reports on many topics throughout every year. For example, the Census Bureau conducts surveys that the Bureau of Labor Statistics uses to compute the Consumer Price Index, which is a measure of the average change over time in the prices paid by urban consumers for consumer goods and services. This index is used to automatically provide cost-of-living wage adjustments to millions of U.S. workers and is a measure of how the nation's economy is performing.

The Consumer Price Index is based on a survey of 7,500 randomly selected households in which people keep a diary of all expenditures they make over a period of time and record whether those expenditures occur on a regular basis (such as for food or rent) or involve relatively large purchases (such as a house or a car). The Census Bureau also conducts interviews to obtain additional data about people's expenditures.

According to the most recent report that has been published based on those surveys, here is the distribution of how we spend our money on an annual basis:

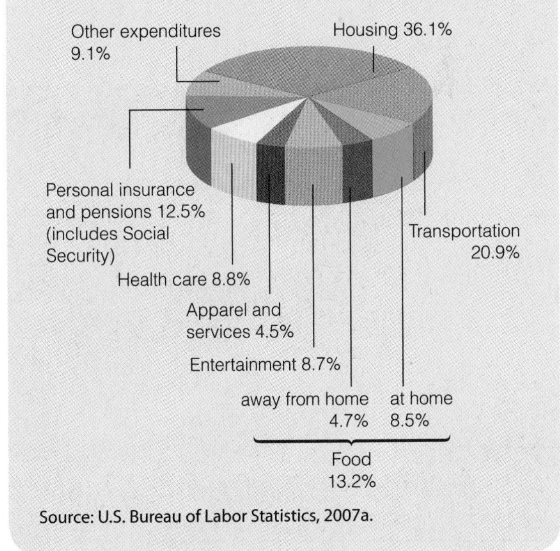

Source: U.S. Bureau of Labor Statistics, 2007a.

conflict perspectives the sociological approach that views groups in society as engaged in a continuous power struggle for control of scarce resources.

© Hulton Archive/Getty Images

C. Wright Mills

power and *prestige* as other sources of inequality. Weber (1968/1922) defined *power* as the ability of a person within a social relationship to carry out his or her own will despite resistance from others, and *prestige* as a positive or negative social estimation of honor (Weber, 1968/1922).

C. Wright Mills (1916–1962), a key figure in the development of contemporary conflict theory, encouraged sociologists to get involved in social reform. He contended that value-free sociology was impossible because social scientists must make value-related choices—including the topics they investigate and the theoretical approaches they adopt. Mills encouraged everyone to look beneath everyday events in order to observe the major resource and power inequalities that exist in society. He believed that the most important decisions in the United States are made largely behind the scenes by the *power elite*—a small clique composed of the top corporate, political, and military officials. Mills's power elite theory is discussed in Chapter 14 ("Politics and Government in Global Perspective").

The conflict perspective is not one unified theory but rather encompasses several branches. One branch is the neo-Marxist approach, which views struggle between the classes as inevitable and as a prime source of social change. A second branch focuses on racial–ethnic inequalities and the continued exploitation of members of some racial–ethnic groups. A third branch is the feminist perspective, which focuses on gender issues.

The Feminist Approach A feminist approach (or "feminism") directs attention to women's experiences and the importance of gender as an element of social structure. This approach is based on the belief that "women and men are equal and should be equally valued as well as have equal rights" (Basow, 1992). According to feminists (including many men as well as women), we live in a patriarchy, a system in which men dominate women and in which things that are considered to be "male" or "masculine" are more highly valued than those considered to be "female" or

© AP Images/Denis Farrell

As one of the wealthiest and most-beloved entertainers in the world, Oprah Winfrey is an example of Max Weber's concept of prestige—a positive estimate of honor. Ms. Winfrey has used her success and prestige to do good works for many causes, including starting the Oprah Winfrey Leadership Academy for Girls in South Africa.

"feminine." The feminist perspective assumes that gender is socially created, rather than determined by one's biological inheritance, and that change is essential in order for people to achieve their human potential without limits based on gender. It also assumes that society reinforces social expectations through social learning, which is acquired through social institutions such as education, religion, and the political and economic structure of society. Some feminists argue that women's subordination can end only after the patriarchal system becomes obsolete. However, note that feminism is not one single, unified approach. Rather, there are several feminist perspectives, which are discussed in Chapter 11 ("Sex and Gender").

Applying Conflict Perspectives to Shopping and Consumption How might advocates of a conflict approach analyze the process of shopping and consumption? A contemporary conflict analysis of consumption might look at how inequalities based on racism, sexism, and income differentials affect people's ability to acquire the things they need and want. It might also look at inequalities regarding the issuance of credit cards and access to "cathedrals of consumption" such as mega-shopping malls and tourist resorts (see Ritzer, 1999: 197–214). However, one of the earliest social theorists to discuss the relationship between social class and consumption patterns was the U.S. social scientist Thorstein Veblen (1857–1929). In *The Theory of the Leisure Class* (1967/1899), Veblen described early wealthy U.S. industrialists as engaging in *conspicuous consumption*—the continuous public display of one's wealth and status through purchases such as expensive houses, clothing, motor vehicles, and other consumer goods. According to Veblen, the leisurely lifestyle of the upper classes typically does not provide them with adequate opportunities to show off their wealth and status. In order to attract public admiration, the wealthy often engage in consumption and leisure activities that are both highly visible and highly wasteful. Examples of conspicuous consumption range from Cornelius Vanderbilt's 8 lavish mansions (including one with 137 rooms) and 10 major summer estates in the Gilded Age (about 1890 to the beginning of World War I) to the 2,400 pairs of shoes owned by Imelda Marcos, wife of the late President Ferdinand Marcos of the Philippines (Frank, 1999; Twitchell, 1999). However, as Ritzer (1999) points out, some of today's wealthiest people engage in *inconspicuous consumption,* perhaps to maintain a low public profile or out of fear for their own safety.

According to contemporary social analysts, conspicuous consumption has become more widely acceptable at all income levels, and some middle- and lower-income individuals and families now use as their frame of reference the lifestyles of the more affluent in their communities. As a result, many families live on credit in order to purchase the goods and services that they would like to have or that keep them on the competitive edge with their friends, neighbors, and co-workers (Schor, 1999). However, others may decide not to overspend, instead seeking to make changes in their lives and encouraging others to do likewise (see Box 1.4).

Living in a society that overemphasizes consumption is particularly difficult for people in low-income categories, as women's studies scholar Juliet B. Schor (1999: 39) states: "For many low-income individuals, the lure of consumerism is hard to resist. When the money isn't there, however, feelings of deprivation, personal failure, and deep psychic pain result. In a culture where consuming means so much, not having money is a profound social disability. For parents, faced with the desires of their children, the failure can feel overwhelming." According to conflict theorists, the economic gains of the upper classes are often at the expense of those in the lower classes, who may have had to struggle (sometimes unsuccessfully) to have adequate food, clothing, and shelter for themselves and their children. Chapter 8 ("Class and Stratification in the United States") and Chapter 9 ("Global Stratification") discuss contemporary conflict perspectives on class-based inequalities.

Symbolic Interactionist Perspectives

The conflict and functionalist perspectives have been criticized for focusing primarily on macrolevel analysis. A ***macrolevel analysis* examines whole societies, large-scale social structures, and social systems** instead of looking at important social dynamics in individuals' lives. Our third perspective, symbolic interactionism, fills this void by examining people's day-to-day interactions and their behavior in groups. Thus, symbolic interactionist approaches are based on a ***microlevel analysis,* which focuses on small groups rather than on large-scale social structures.**

We can trace the origins of this perspective to the Chicago School, especially George Herbert Mead and Herbert Blumer (1900–1986), who is credited with coining the term *symbolic interactionism.* According

macrolevel analysis an approach that examines whole societies, large-scale social structures, and social systems.

microlevel analysis sociological theory and research that focus on small groups rather than on large-scale social structures.

Box 1.4 You Can Make a Difference

Dealing with Money Matters in a Material World

"Retail therapy always revives her."

—Lisa Tisdale, mother of *High School Musical* star Ashley Tisdale, describing how her daughter, a self-confessed "shopaholic," energizes herself after a busy day promoting her music (*People,* 2007: 123)

Are celebrities such as Ashley alone in their desire to shop continually? Increasingly, people of all ages and social classes find that shopping is a central part of their lives. Consider these comments from college students:

- "I didn't really need it, but I bought it anyway."
- "With my rent due, I shouldn't have bought this skirt."
- "I felt so much better after I went shopping, even if my credit card feels sick."

These comments are a reflection of how many people feel after shopping. Madonna's song about being "a material girl in a material world" resonates with many of us because we are continually surrounded by advertisements and shopping malls that set in front of us a veritable banquet of merchandise to buy. However, shopping that gets out of hand is a serious habit that may have lasting psychological and economic consequences. If we are aware of these problems, we may be able to prevent them by helping ourselves or other people avoid hyperconsumerism.

Do you know the symptoms of compulsive overspending and debt dependency? Consider these questions:

- Do you or someone you know spend large amounts of time shopping or thinking about going shopping?
- Do you or someone you know rush to the store or to the computer for online shopping when feeling frustrated, depressed, or otherwise "out of sorts"?
- Do you or someone you know routinely argue with parents, friends, or partners about spending too much money or overcharging on credit cards?
- Do you or someone you know hide purchases or make dishonest statements—such as "It was a gift from a friend"—to explain where new merchandise came from?

Economist Juliet Schor (1999), who has extensively studied the problems associated with excessive spending and credit card debt, believes that each of us can empower ourselves and help others as well if we follow simple steps in our consumer behavior. Among these steps are *controlling desire* by gaining knowledge of the process of consumption and its effect on people, *helping to make exclusivity uncool* by demystifying the belief that people are "better" simply because they own excessively expensive items, and *discouraging competitive consumption* by encouraging our friends and acquaintances to spend less on presents and other purchases. Finally, Schor suggests that we should *become educated consumers* and *avoid use of shopping as a form of therapy.* By following Schor's simple steps and encouraging our friends and relatives to do likewise, we may be able to free ourselves from the demands of a hyperconsumer society that continually bombards us with messages indicating that we should spend more and go deeper in debt on our credit cards.

Can an individual make a difference using these suggestions? Certainly, these ideas may change an individ-

to ***symbolic interactionist perspectives,* society is the sum of the interactions of individuals and groups.** Theorists using this perspective focus on the process of *interaction*—defined as immediate reciprocally oriented communication between two or more people—and the part that *symbols* play in giving meaning to human communication. A *symbol* is anything that meaningfully represents something else. Examples of symbols include signs, gestures, written language, and shared values. Symbolic interaction occurs when people communicate through the use of symbols; for example, a gift of food—a cake or a casserole—to a newcomer in a neighborhood is a symbol of welcome and friendship. But symbolic communication occurs in a variety of forms, including facial gestures, posture, tone of voice, and other symbolic gestures (such as a handshake or a clenched fist).

Symbols are instrumental in helping people derive meanings from social situations. In social encounters, each person's interpretation or definition of a given situation becomes a *subjective reality* from that person's viewpoint. We often assume that what we consider to be "reality" is shared by others; however, this assumption is often incorrect. Subjective reality is acquired and shared through agreed-upon symbols, especially language. If a person shouts "Fire!" in a crowded movie theater, for example, that language produces the same response (attempting to escape) in all of those who hear and understand it. When people in a group do not share the same meaning for a given symbol, however, confusion results; for example, people who did not know the meaning of the word *fire* would not know what the commotion was about. How people *interpret* the messages they receive and the situations

© Michael Bezjian/WireImage/ Getty Images

Ashley Tisdale, star of *High School Musical,* is often photographed while shopping or while walking down the street with numerous shopping bags on her arms. One magazine even offered readers the opportunity to see what Tisdale purchased one day by logging on to its website. Does continual media coverage of the behavior of celebrities encourage excessive consumerism? What do you think?

ual's shopping and overspending habits. However, if we apply C. Wright Mills's (1959b) *sociological imagination,* including the distinction between personal troubles and public issues (as discussed in this chapter), we see that these suggestions focus exclusively on what *individuals* can do to change their own behavior. As a result, this approach may be somewhat useful but may still overlook the larger structural factors that contribute to people's overspending. According to the sociologist George Ritzer (1999), individual actions in this regard are likely to fail as long as there are no changes in the larger society, particularly in the cathedrals of consumption, advertisers, credit card companies, and other businesses that have a vested interest in promoting hyperconsumption. Some organizations suggest that the only way to reduce overconsumption and credit card debt is through activism, such as getting the age limit raised at which people can be issued their first credit card. In this view, those who want to make a difference could also become involved in advocating social change. If you would like to know more about the simplicity or downshifting movements, here are several organizations to contact:

- New American Dream, 6930 Carroll Avenue, Suite 900, Takoma Park, MD 20912. Online: **http://www.newdream.org**
- The Media Foundation, 1243 W. 7th Avenue, Vancouver, BC V6H 1B7, Canada. Online: **http://www.adbusters.org**
- Green America, 1612 K Street, Washington, DC 20006. Online: **http://www.coopamerica.org**

Sources: Based on Durling, 2005; Ritzer, 1999; and Schor, 1999.

they encounter becomes their subjective reality and may strongly influence their behavior.

Symbolic interactionists attempt to study how people make sense of their life situations and the way they go about their activities, in conjunction with others, on a day-to-day basis (Prus, 1996). How do people develop the capacity to think and act in socially prescribed ways? According to symbolic interactionists, our thoughts and behavior are shaped by our social interactions with others. Early theorists such as Charles H. Cooley and George Herbert Mead explored how individual personalities are developed from social experience and concluded that we would not have an identity, a "self," without communication with other people. This idea is developed in Cooley's notion of the "looking-glass self" and Mead's "generalized other," as discussed in Chapter 4 ("Socialization"). From this perspective, the attainment of language is essential not only for the development of a "self" but also for establishing common understandings about social life.

How do symbolic interactionists view social organization and the larger society? According to symbolic interactionists, social organization and society are possible only through people's everyday interactions. In other words, group life takes its shape as people interact with one another (Blumer, 1986/1969). Although macrolevel factors such as economic and political institutions constrain and define the forms of

symbolic interactionist perspectives the sociological approach that views society as the sum of the interactions of individuals and groups.

© Bill Aron/PhotoEdit

Sporting events are a prime location for seeing how college students use symbols to convey shared meanings. From the colors of clothing to hand gestures, students show pride in their school.

interaction that we have with others, the social world is dynamic and always changing. Chapter 5 ("Society, Social Structure, and Interaction") explores two similar approaches—rational choice and exchange theories—that focus specifically on how people rationally try to get what they need by exchanging valued resources with others.

As we attempt to present ourselves to others in a particular way, we engage in behavior that the sociologist Erving Goffman (1959) referred to as "impression management." Chapter 5 also presents some of Goffman's ideas, including *dramaturgical analysis,* which envisions that individuals go through their life somewhat like actors performing on a stage, playing out their roles before other people. Symbolic interactionism involves both a theoretical perspective and specific research methods, such as observation, participant observation, and interviews, that focus on individual and small-group behavior (see Chapter 2, "Sociological Research Methods").

Applying Symbolic Interactionist Perspectives to Shopping and Consumption Sociologists applying a symbolic interactionist framework to the study of shopping and consumption would primarily focus on a microlevel analysis of people's face-to-face interactions and the roles that people play in society. In our efforts to interact with others, we define any situation according to our own subjective reality. This theoretical viewpoint applies to shopping and consumption just as it does to other types of conduct. For example, when a customer goes into a store to make a purchase and offers a credit card to the cashier, what meanings are embedded in the interaction process that takes place between the two of them? The roles that the two people play are based on their histories of interaction in previous situations. They bring to the present encounter symbolically charged ideas, based on previous experiences. Each person also has a certain level of emotional energy available for each interaction. When we are feeling positive, we have a high level of emotional energy, and the opposite is also true. Each time we engage in a new interaction, the situation has to be negotiated all over again, and the outcome cannot be known beforehand (Collins, 1987).

In the case of the shopper–cashier interaction, how successful will the interaction be for each of them? The answer to this question depends on a kind of social marketplace in which such interactions can either raise or lower one's emotional energy (Collins, 1987). If the customer's credit card is rejected, he or she may come away with lower emotional energy. If the customer is angry at the cashier, he or she may attempt to "save face" by reacting in a haughty manner regarding the rejection of the card. ("What's wrong with you? Can't you do anything right? I'll never shop here again!") If this type of encounter occurs, the cashier may also come out of the interaction with a lower level of emotional energy, which may affect the cashier's interactions with subsequent customers. Likewise, the next time the customer uses a credit card, he or she may say something like "I hope this card isn't over its limit. Sometimes I lose track," even if the person knows that the card's credit limit has not been exceeded. This is only one of many ways in which the rich tradition of symbolic interactionism might be used to examine shopping and consumption. Other areas of interest might include the social nature of the shopping experience, social interaction patterns in families regarding credit card debts, and why we might spend money to impress others.

Postmodern Perspectives

According to ***postmodern perspectives,*** **existing theories have been unsuccessful in explaining social life in contemporary societies that are characterized by postindustrialization, consumerism, and global communications.** Postmodern social theorists reject the theoretical perspectives we have previously discussed, as well as how those thinkers created the theories (Ritzer, 1996). These theorists oppose the grand narratives that characterize modern thinking and believe that boundaries should not be placed on academic disciplines—such as philosophy, literature, art, and the social sciences—when much could be learned by sharing ideas.

Just as functionalist, conflict, and symbolic interactionist perspectives emerged in the aftermath of the Industrial Revolution, postmodern theories emerged after World War II (in the late 1940s) and reflected the belief that some nations were entering a period of postindustrialization. Postmodern (or "postindustrial") societies are characterized by an *information explosion* and an economy in which large numbers of people either provide or apply information, or they are employed in professional occupations (such as lawyers and physicians) or service jobs (such as fast-food servers and health care workers). There is a corresponding *rise of a consumer society* and the emergence of a *global village* in which people around the world communicate with one another by electronic technologies such as television, telephone, fax, e-mail, and the Internet.

Jean Baudrillard, a well-known French social theorist, is one of the key figures in postmodern theory, even though he would dispute this label. Baudrillard has extensively explored how the shift from production of goods (such as in the era of Marx and Weber) to consumption of information, services, and products in contemporary societies has created a new form of social control. According to Baudrillard's approach, capitalists strive to control people's shopping habits, much like the output of factory workers in industrial economies, to enhance their profits and to keep everyday people from rebelling against social inequality (1998/1970). How does this work? When consumers are encouraged to purchase more than they need or can afford, they often sink deeper in debt and must keep working to meet their monthly payments. Instead of consumption being related to our needs, it is based on factors such as our "wants" and the need we feel to distinguish ourselves from others. We will look at this idea in more detail in the next section, where we apply a postmodern perspective to shopping and consumption. We will also return to Baudrillard's general ideas on postmodern societies in Chapter 3 ("Culture").

Today, postmodern theory remains an emerging perspective in the social sciences. How influential will this approach be? It remains to be seen what influence postmodern thinkers will have on the social sciences. Although this approach opens up broad new avenues of inquiry by challenging existing perspectives and questioning current belief systems, it also tends to ignore many of the central social problems of our time—such as inequalities based on race, class, and gender, and global political and economic oppression (Ritzer, 1996).

Applying Postmodern Perspectives to Shopping and Consumption According to some social theorists, the postmodern society is a consumer society. The focus of the capitalist economy has shifted from production to consumption. Today, the emphasis is on getting people to consume more and to own a greater variety of things. As previously discussed, credit cards may encourage people to spend more money than they should, and often more than they can afford (Ritzer, 1998). Television shopping networks and cybermalls make it possible for people to shop around the clock without having to leave home or encounter "real" people. As Ritzer (1998: 121) explains, "So many of our interactions in these settings . . . are simulated, and we become so accustomed to them, that in the end all we have are simulated interactions; there are no more 'real' interactions. The entire distinction between the simulated and the real is lost; simulated interaction *is* the reality" (see also Baudrillard, 1983). Similarly, Ritzer (1998: 121) points out that a credit card is a simulation:

> Any given credit card is a simulation of all other cards of the same brand; there was no "original" card from which all others are copied; there is no "real" credit card. Furthermore, credit cards can be seen as simulations of simulations. That is, they simulate currency, but each bill is a simulation, a copy, of every other bill and, again, there was never an original bill from which all others have been copied. But currencies, in turn, can be seen as simulations of material wealth, or of the faith one has in the Treasury, or whatever one imagines to be the "real" basis of wealth. Thus, the credit card shows how we live in a world characterized by a never-ending spiral of simulation built upon simulation.

As this example suggests, postmodern theorists do not focus on actors (human agents) as they go about their everyday lives, but instead offer more-abstract conceptions of what constitutes "reality." For postmodernists, social life is not an objective reality waiting for us to discover how it works. Rather, what we experience as social life is actually nothing more or less than how we think about it, and there are many diverse ways of

postmodern perspectives the sociological approach that attempts to explain social life in modern societies that are characterized by postindustrialization, consumerism, and global communications.

CONCEPT QUICK REVIEW

The Major Theoretical Perspectives

Perspective	Analysis Level	View of Society
Functionalist	Macrolevel	Society is composed of interrelated parts that work together to maintain stability within society. This stability is threatened by dysfunctional acts and institutions.
Conflict	Macrolevel	Society is characterized by social inequality; social life is a struggle for scarce resources. Social arrangements benefit some groups at the expense of others.
Symbolic Interactionist	Microlevel	Society is the sum of the interactions of people and groups. Behavior is learned in interaction with other people; how people define a situation becomes the foundation for how they behave.
Postmodernist	Macrolevel/Microlevel	Societies characterized by postindustrialization, consumerism, and global communications bring into question existing assumptions about social life and the nature of reality.

doing that. According to a postmodernist perspective, the Enlightenment goal of intentionally creating a better world out of some knowable truth is an illusion. Although some might choose to dismiss postmodernist approaches, they do give us new and important questions to think about regarding the nature of social life.

The Concept Quick Review reviews all four of these perspectives. Throughout this book, we will be using these perspectives as lenses through which to view our social world.

Comparing Sociology with Other Social Sciences

In this chapter, we have discussed how sociologists examine social life. We have focused on shopping and consumption as an example of the many topics studied by sociologists and other social scientists. Let's briefly examine how other social sciences investigate human relationships and then compare each discipline to sociology.

Anthropology

Anthropologists and sociologists are interested in studying human behavior; however, there are differences between the two disciplines. Anthropology seeks to understand human existence over geographic space and evolutionary time (American Anthropological Association, 2001), whereas sociology seeks to understand contemporary social organization, relations, and change. Some anthropologists focus on the beginnings of human history, millions of years ago, whereas others primarily study contemporary societies. Anthropology is divided into four main subfields: sociocultural, linguistic, archaeological, and biological anthropology. Cultural anthropologists focus on culture and its many manifestations, including art, religion, and politics. Linguistic anthropologists primarily study language because culture itself depends on language. Archaeologists are interested in discovering and analyzing material artifacts—such as cave paintings, discarded stone tools, and abandoned baskets—from which they piece together a record of social life in earlier societies and assess what this information adds to our knowledge of contemporary cultures. Biological (or physical) anthropologists study the biological origins, evolutionary development, and genetic diversity of primates, including Homo sapiens (human beings). Clearly, some cultural anthropologists would be interested in the issue we have looked at in this chapter—why people consume items of material culture and how their behavior as consumers is interwoven with all other aspects of social life in this society.

Psychology

Psychology is the systematic study of behavior and mental processes—what occurs in the mind. Psychologists focus not only on behavior that is directly observable, such as talking, laughing, and eating, but also on mental processes that cannot be directly observed, such as thinking and dreaming. For psychologists, behavior and mental processes are interwoven; therefore, to understand behavior, they examine the emotions that underlie people's actions. For example, a psychologist interested in studying why some individuals have excessive credit card debt might identify the specific emotions that a person has when purchasing an expensive item that is well beyond his or her budget.

Psychology is a diverse field. Some psychologists work in clinical settings, where they diagnose and treat psychological disorders; others practice in schools, where they are concerned with the intellectual, social, and emotional development of schoolchildren. Still other psychologists work in business, industry, and other work-related settings. Another branch of psychology—social psychology—is similar to sociology in that it emphasizes how social conditions affect individual behavior. Social psychological perspectives on human development are useful to sociologists who study the process of socialization (see Chapter 4, "Socialization"). However, a distinction between psychology and sociology is the extent to which most psychological studies focus on internal factors relating to the individual in their explanations of human behavior, whereas sociological research examines the effects of groups, organizations, and social institutions on social life.

Economics

Unlike the other social sciences we have discussed, economics concentrates primarily on a single institution in society—the economy (see Chapter 13, "The Economy and Work in Global Perspective"). Economists attempt to explain how the limited resources of a society are allocated among competing demands. Economics is divided into two different branches. Macroeconomics looks at such issues as the total amount of goods and services produced by a society; microeconomics studies such issues as decisions made by individual businesses. Thus, consumerism and credit card debt would be of interest to economists because such topics can be analyzed at global, national, and individual levels. However, a distinction between economics and sociology is that economists focus on the complex workings of economic systems (such as monetary policy, inflation, and the national debt), whereas sociologists focus on a number of social institutions, one of which is the economy.

Political Science

Political science is the academic discipline that studies political institutions such as the state, government, and political parties (see Chapter 14, "Politics and Government in Global Perspective"). Political scientists study power relations and seek to determine how power is distributed in various types of political systems. Some political scientists focus primarily on international relations and similarities and differences in the political institutions across nations. Other political scientists look at the political institutions in a particular country. An example is a political scientist interested in studying consumerism in the United States who systematically examines how the political process—such as the efforts of lobbyists and interest groups to influence governmental policies—affects credit card interest rates and consumer spending in this country. Political scientists concentrate on political institutions, whereas sociologists study these institutions within the larger context of other social institutions such as families, religion, education, and the media.

In Sum

Clearly, the areas of interest and research in the social sciences overlap in that the goal of scholars, teachers, and students is to learn more about human behavior, including its causes and consequences. In applied sociology, there is increasing collaboration among researchers across disciplines to develop a more holistic, integrated view of how human behavior and social life take place in societies. Join me as, throughout this book, we examine specific ways in which sociology creates its own realm of knowledge yet benefits from the other social sciences in order to provide new information and personal insights that we can use in our everyday lives.

Chapter Review

• What is sociology, and how can it help us to understand ourselves and others?

Sociology is the systematic study of human society and social interaction. We study sociology to understand how human behavior is shaped by group life and, in turn, how group life is affected by individuals. Our culture tends to emphasize individualism, and sociology pushes us to consider more-complex connections between our personal lives and the larger world.

• What is the sociological imagination, and why is it important to have a global sociological imagination?

According to C. Wright Mills, the sociological imagination helps us understand how seemingly personal troubles, such as suicide, are actually related to larger social forces. It is the ability to see the relationship between individual experiences and the larger society. It is important to have a global sociological imagination because the future of this nation is deeply intertwined with the future of all nations of the world on economic, political, and humanitarian levels.

• What factors contributed to the emergence of sociology as a discipline?

Industrialization and urbanization increased rapidly in the late eighteenth century, and social thinkers began to examine the consequences of these powerful forces. Auguste Comte coined the term *sociology* to describe a new science that would engage in the study of society.

• What are the major contributions of early sociologists such as Durkheim, Marx, and Weber?

The ideas of Emile Durkheim, Karl Marx, and Max Weber helped lead the way to contemporary sociology. Durkheim argued that societies are built on social facts, that rapid social change produces strains in society, and that the loss of shared values and purpose can lead to a condition of anomie. Marx stressed that within society there is a continuous clash between the owners of the means of production and the workers who have no choice but to sell their labor to others. According to Weber, sociology should be value free, and people should become more aware of the role that bureaucracies play in daily life.

• How did Simmel's perspective differ from that of other early sociologists?

Whereas other sociologists primarily focused on society as a whole, Simmel explored small social groups and argued that society was best seen as a web of patterned interactions among people.

• What are the major contemporary sociological perspectives?

Functionalist perspectives assume that society is a stable, orderly system characterized by societal consensus. Conflict perspectives argue that society is a continuous power struggle among competing groups, often based on class, race, ethnicity, or gender. Interactionist perspectives focus on how people make sense of their everyday social interactions, which are made possible by the use of mutually understood symbols. From an alternative perspective, postmodern theorists believe that entirely new ways of examining social life are needed and that it is time to move beyond functionalist, conflict, and interactionist approaches.

www.cengage.com/login

Want to maximize your online study time? Take this easy-to-use study system's diagnostic pre-test, and it will create a personalized study plan for you. By helping you to identify the topics that you need to understand better and then directing you to valuable online resources, it can speed up your chapter review. CengageNOW even provides a post-test so you can confirm that you are ready for an exam.

Key Terms

anomie 15
conflict perspectives 25
functionalist perspectives 23
high-income countries 8
industrialization 11
latent functions 23
low-income countries 8
macrolevel analysis 27
manifest functions 23
microlevel analysis 27
middle-income countries 8
positivism 13
postmodern perspectives 30
social Darwinism 14
social facts 15
society 4
sociological imagination 5
sociology 4
symbolic interactionist perspectives 27
theory 23
urbanization 12

Questions for Critical Thinking

1. What does C. Wright Mills mean when he says the sociological imagination helps us "to grasp history and biography and the relations between the two within society?" (Mills, 1959b: 6). How might this idea be applied to today's consumer society?
2. As a sociologist, how would you remain objective yet see the world as others see it? Would you make subjective decisions when trying to understand the perspectives of others?
3. Early social thinkers were concerned about stability in times of rapid change. In our more global world, is stability still a primary goal? Or is constant conflict important for the well-being of all humans? Use the conflict and functionalist perspectives to bolster your analysis.
4. Some social analysts believe that college students relate better to commercials and advertising culture than they do to history, literature, or probably anything else (Twitchell, 1996). How would you use the various sociological perspectives to explore the validity of this assertion in regard to students on your college campus?

The Kendall Companion Website

www.cengage.com/sociology/kendall

Visit this book's companion website, where you'll find more resources to help you study and successfully complete course projects. Resources include quizzes and flash cards, as well as special features such as an interactive sociology timeline, maps, General Social Survey (GSS) data, and Census 2000 data. The site also provides links to useful websites that have been selected for their relevance to the topics in this chapter and include those listed below. (*Note:* Visit the book's website for updated URLs.)

Dead Sociologists Index
http://media.pfeiffer.edu/lridener/DSS

Visit this website to learn more about the sociologists discussed in this chapter. Click on the name of an individual you wish to study, and access biographical information, summaries of key ideas, and selections from original works.

A Sociological Tour Through Cyberspace
http://www.trinity.edu/mkearl

Professor Michael C. Kearl of Trinity University in San Antonio has developed a comprehensive gateway site to a number of Internet resources in the field of sociology. For Chapter 1, click on "General Sociological Resources" and "Sociological Theory."

American Sociological Society (ASA)
http://www.asanet.org

The American Sociological Association (ASA) is a national organization for sociologists. Its home page provides information on annual meetings, resources available for sociological research, information on careers in sociology, special reports on current activities in social policy, and information on how students can get involved in this sociological organization, including the Minority Fellowship Program.

chapter

2 Sociological Research Methods

Chapter Focus Question

How do sociological theory and research add to our knowledge of human societies and social issues such as suicide?

His name was Josh Evans. He was 16 years old. And he was hot.

"Mom! Mom! Mom! Look at him!" Tina Meier recalls her daughter saying.

Josh had contacted Megan Meier through her MySpace page and wanted to be added as a friend.

"Yes, he's cute," Tina Meier told her daughter. "Do you know who he is?"

"No, but look at him! He's hot! Please, please, can I add him?"

Mom said yes. And for six weeks Megan and Josh—under Tina's watchful eye—became acquainted in the virtual world of MySpace.

Josh said he was born in Florida and recently had moved to

Tina Meier, the mother of Megan Meier, displays photographs of her daughter, who committed suicide at the age of thirteen after reacting to fake MySpace postings by the mother of one of her former friends. Studying sociology provides us with new insights on problems such as suicide by making us aware that much more goes on in social life than we initially observe.

IN THIS CHAPTER

- Why Is Sociological Research Necessary?
- The Sociological Research Process
- Research Methods
- Ethical Issues in Sociological Research

O'Fallon [Missouri]. He was homeschooled. He played the guitar and drums. . . .

As for 13-year-old Megan . . . [she] loved swimming, boating, fishing, dogs, rap music and boys.

But her life had not always been easy, her mother says. She was heavy and for years had tried to lose weight. She had attention deficit disorder and battled depression. . . . But things were going exceptionally well. She had shed 20 pounds, getting down to 175. She was 5 foot 5-1/2 inches tall. . . .

Part of the reason for Megan's rosy outlook was Josh, Tina says. After school Megan would rush to the computer. . . . It did seem odd, Tina says, that Josh never asked for Megan's phone number. And when Megan asked for his, she says, Josh said he didn't have a cell and his mother did not yet have a landline.

And then on Sunday, Oct. 15, 2006, Megan received a puzzling and disturbing message from Josh. Tina recalls that it said, "I don't know if I want to be friends with you anymore because I've heard that you are not very nice to your friends."

(Continued)

Sharpening Your Focus

- *What is the relationship between theory and research?*
- *What are the steps in the conventional research process?*
- *What can qualitative methods add to our understanding of human behavior?*
- *Why is it important to have a variety of research methods available?*
- *What has research contributed to our understanding of suicide?*
- *Why is a code of ethics necessary for sociological research?*

Frantic, Megan shot back: "What are you talking about?" (Pokin, 2007)

This and other hostile instant message exchanges set into motion the final, disturbing episode in the life of Megan Meier, as she was suddenly confronted with not only the anger and cynicism of a young man she thought she knew and trusted but also the bullying of other young people gathered on the social networking site MySpace who also sent a barrage of hate-filled messages that called Megan a liar and much worse.

"Mom, they're being Horrible!" Megan said, sobbing into the phone when her mother called. After an hour, Megan ran into her bedroom and hanged herself with a belt.

"She felt there was no way out," Ms. Meier said. (Maag, 2007)

—the parents of Megan Meier recalling the events leading up to her suicide at only thirteen years of age

Clearly, the suicide of Megan deeply touched her parents and friends while raising many issues about the problem of cyber-bullying. In the aftermath of Megan's tragic death, her parents tried to send an instant message to Josh Evans to inform him about the destructive nature of his actions, only to learn that his MySpace account had been deleted. Six weeks after Megan's death, her parents learned that Josh Evans never existed: His fake persona allegedly had been created by a mother whose daughter was once Megan's friend. According to some media reports, this parent created a fake MySpace account for "Josh Evans" so that she could find out what Megan would say about her daughter and other people. Subsequently, other members gathered on MySpace and—not knowing that Josh Evans did not exist—jumped into the fray and began hurling accusations at Megan and bullying her. A local ordinance in Megan's hometown now prohibits any harassment that uses the Internet, text messaging services, or any other electronic medium.

Although we will never know the full story of Megan's life, this tragic occurrence brings us to a larger sociological question: Why does anyone commit suicide? Is suicide purely an individual phenomenon, or is it related to our social interactions and the social environment and society in which we live?

In this chapter, we examine how sociological theories and research can help us understand the seemingly individualistic act of taking one's own life. We will see how sociological theory and research methods might be used to answer complex questions, and we will wrestle with some of the difficulties that sociologists experience as they study human behavior.

Why Is Sociological Research Necessary?

Sociologists obtain their knowledge of human behavior through research, which results in a body of information that helps us move beyond guesswork and common sense in understanding society. The sociological perspective incorporates theory and research to arrive at a more accurate understanding of the "hows" and "whys" of human social interaction. Once we have an informed perspective about social issues, such as who commits suicide and why, we are in a better position to find solutions and make changes. Social research, then, is a key part of sociology.

Common Sense and Sociological Research

Most of us have commonsense ideas about suicide. Common sense, for example, may tell us that people who threaten suicide will not commit suicide. Socio-

Box 2.1 Sociology and Everyday Life

How Much Do You Know About Suicide?

True	False	
T	F	1. For people thinking of suicide, it is difficult, if not impossible, to see the bright side of life.
T	F	2. People who talk about suicide don't do it.
T	F	3. Once people contemplate or attempt suicide, they must be considered suicidal for the rest of their lives.
T	F	4. In the United States, suicide occurs on the average of one every sixteen minutes.
T	F	5. Accidents and injuries sustained by teenagers and young adults may indicate suicidal inclinations.
T	F	6. Alcohol and drugs are outlets for anger and thus reduce the risk of suicide.
T	F	7. Suicide rates for African Americans are higher than for white Americans.
T	F	8. Suicidal people are fully intent upon dying.

Answers on page 40.

logical research indicates that this assumption is frequently incorrect: People who threaten to kill themselves are often sending messages to others and may indeed attempt suicide. Common sense may also tell us that suicide is caused by despair or depression. However, research suggests that suicide is sometimes used as a means of lashing out at friends and relatives because of real or imagined wrongs. Before reading on, take the quiz in Box 2.1, which lists a number of commonsense notions about suicide.

Historically, the commonsense view of suicide was that it was a sin, a crime, and a mental illness (Evans and Farberow, 1988). Emile Durkheim refused to accept these explanations. In what is probably the first sociological study to use scientific research methods, he related suicide to the issue of cohesiveness (or lack of cohesiveness) in society instead of viewing suicide as an isolated act that could be understood only by studying individual personalities or inherited tendencies. In *Suicide* (1964b/1897), Durkheim documented his contention that a high suicide rate was symptomatic of large-scale societal problems. In the process, he developed an approach to research that influences researchers to this day (see "Sociology Works!"). As we discuss sociological research, we will use the problem of suicide to demonstrate the research process.

Because much of sociology deals with everyday life, we might think that common sense, our own personal experiences, and the media are the best sources of information. However, our personal experiences are subjective, and much of the information provided by the media comes from sources seeking support for a particular point of view. The content of the media is also influenced by the continual need for audience ratings.

We need to be able to evaluate the information we receive. This is especially true because the quantity—but, in some instances,

Although scientific studies about human behavior are readily available to most of us, and trained professionals can help us with our personal and social problems, many individuals rely—at least in part—on the advice of psychics such as the tarot card reader shown here. What problems might occur as a result of relying on psychics for counseling and advice?

Box 2.1 Sociology and Everyday Life

Answers to the Sociology Quiz on Suicide

1. True. To people thinking of suicide, an acknowledgment that there is a bright side only confirms and conveys the message that they have failed; otherwise, they, too, could see the bright side of life.

2. False. Some people who talk about suicide do kill themselves. Warning signals of possible suicide attempts include talk of suicide, the desire not to exist anymore, despair, and hopelessness.

3. False. Most people think of suicide for only a limited amount of time. When the crisis is over and the problems leading to suicidal thoughts are resolved, people usually cease to think of suicide as an option.

4. True. A suicide occurs on the average of every sixteen minutes in the United States; however, this rate differs with respect to the sex, race/ethnicity, and age of the individual. For example, men are three times more likely to kill themselves than are women.

5. True. Accidents, injuries, and other types of life-threatening behavior may be signs that a person is on a course of self-destruction. One study concluded that the incidence of suicide was twelve times higher among adolescents and young adults who had been previously hospitalized because of an injury.

6. False. Excessive use of alcohol or drugs may enhance a person's feelings of anger and frustration, making suicide a greater possibility. This risk appears to be especially high for men who abuse alcohol or drugs.

7. False. Suicide rates are much higher among white Americans than African Americans. For example, in 2004 the overall U.S. suicide rate was 11.0 per 100,000 population. For white males, the rate was 19.6; for African American males, it was 9.0. For white females, it was 5.1; for African American females, the rate was 1.8.

8. False. Suicidal people often have an ambivalence about dying—they want to live and to die at the same time. They want to end the pain or problems they are experiencing, but they also wish that something or someone would remove the pain or problem so that life could continue.

Sources: Based on American Association of Suicidology, 2006; Leenaars, 1991; Levy and Deykin, 1989; and National Center for Injury Prevention and Control, 2006.

not the quality—of information available has grown dramatically as a result of the information explosion brought about by computers and by the telecommunications industry.

Sociology and Scientific Evidence

In taking this course, you will be studying social science research and may be asked to write research reports or read and evaluate journal articles. If you attend graduate or professional school in fields that use sociological research, you will be expected to evaluate existing research and perhaps do your own. Hopefully, you will find that social research is relevant to the practical, everyday concerns of the real world.

Sociology involves *debunking*—the unmasking of fallacies (false or mistaken ideas or opinions) in the everyday and official interpretations of society (Mills, 1959b). Because problems such as suicide involve threats to existing societal values, we cannot analyze these problems without acknowledging the values involved. For example, should assisted suicide for terminally ill patients who wish to die be legal? We often answer questions like this by using either the normative or the empirical approach. The *normative approach* uses religion, customs, habits, traditions, and law to answer important questions. It is based on strong beliefs about what is right and wrong and what "ought to be" in society. Issues such as assisted suicide are often answered by the normative approach. From a legal standpoint, the consequences of assisting in another person's suicide may be severe.

Although these issues are immediate and profound, some sociologists discourage the use of the normative approach in their field and advocate the use of the empirical approach instead. The *empirical approach* attempts to answer questions through systematic collection and analysis of data. This approach is referred to

Sociology *Works!*

Durkheim's Sociology of Suicide and Twenty-First-Century India

The bond attaching [people] to life slackens because the bond which attaches [them] to society is itself slack.
—Emile Durkheim, *Suicide* (1964b/1897)

Although this statement described social conditions accompanying the high rates of suicide found in late-nineteenth-century France, Durkheim's words ring true today as we look at contemporary suicide rates for cities such as Bangalore, which some refer to as "India's Suicide City" (Guha, 2004).

At first glance, we might think that the outsourcing of jobs in the technology sector—from high-income nations such as the United States to India—would provide happiness and job satisfaction for individuals in cities such as Bangalore and New Delhi who have gained new opportunities and higher salaries in recent years as a result of outsourcing. News stories have focused on the wealth of opportunities that these outsourced jobs have brought to millions of men and women in India, most of whom are in their twenties and thirties and who now earn larger incomes than do their parents and many of their contemporaries. However, the underlying story of what is really going on in these cities stands in stark contrast: Rapid urbanization and fast-paced changes in the economy and society are weakening social ties that have been so important to individuals. Social bonds have been weakened or dissolved as people move away from their families and their community. Life in the cities moves at a much faster pace than in the rural areas, and many individuals experience loneliness, sleep disorders, family discord, and major health risks such as heart disease and depression (Mahapatra, 2007). In the words of Ramachandra Guha (2004), a historian residing in India, Durkheim's sociology of suicide remains highly relevant to finding new answers to this challenging problem: "The rash of suicides in city and village is a qualitatively new development in our history. We sense that tragedies are as much social as they are individual. But we know very little of what lies behind them. What we now await, in sum, is an Indian Durkheim."

Reflect & Analyze

How does sociology help us to examine seemingly private acts such as suicide within a larger social context? Why are some people more inclined to commit suicide if they are not part of a strong social fabric?

as the conventional model, or the "scientific method," and is based on the assumption that knowledge is best gained by direct, systematic observation. Many sociologists believe that two basic scientific standards must be met: (1) scientific beliefs should be supported by good evidence or information, and (2) these beliefs should be open to public debate and critiques from other scholars, with alternative interpretations being considered (Cancian, 1992).

Sociologists typically use two types of empirical studies: descriptive and explanatory. *Descriptive studies* attempt to describe social reality or provide facts about some group, practice, or event. Studies of this type are designed to find out what is happening to whom, where, and when. For example, a descriptive study of suicide might attempt to determine the number of people who recently thought about committing suicide. On other topics, well-known descriptive studies include the U.S. Census and the FBI's Uniform Crime Reports. However, it is important to note that even studies which are considered to be "objective" have certain biases because of the limitations inherent in doing certain types of research, as discussed in this chapter and in Chapter 7 ("Deviance and Crime"). By contrast, *explanatory studies* attempt to explain cause-and-effect relationships and to provide information on why certain events do or do not occur. In an explanatory study of suicide, we might ask questions such as these: Why do African American men over age sixty-five have a significantly lower suicide rate than white males in the same age bracket? Why are women more likely to attempt suicide than men? Sociologists engage in theorizing and conducting research in order to describe, explain, and sometimes predict how and why people will act in certain situations.

The Theory and Research Cycle

The relationship between theory and research has been referred to as a continuous cycle, as shown in ▶ Figure 2.1 (Wallace, 1971). You will recall that a *theory* is a set of logically interrelated statements that attempts to describe, explain, and (occasionally) predict social events. A theory attempts to explain why something is the way it is. *Research* is the process of systematically collecting information for the purpose of testing

Why do older African American men have a lower rate of suicide than white males of similar ages? Questions such as this often serve as the foundation for an explanatory study as sociologists attempt to understand and describe certain cause-and-effect relationships.

an existing theory or generating a new one. The theory and research cycle consists of deductive and inductive approaches. In the *deductive approach,* the researcher begins with a theory and uses research to test the theory. This approach proceeds as follows: (1) theories generate hypotheses, (2) hypotheses lead to observations (data gathering), (3) observations lead to the formation of generalizations, and (4) generalizations are used to support the theory, to suggest modifications to it, or to refute it. To illustrate, if we use the deductive method to determine why people commit suicide, we start by formulating a theory about the "causes" of suicide and then test our theory by collecting and analyzing data (for example, vital statistics on suicides or surveys to determine whether adult church members view suicide differently from nonmembers).

DEDUCTIVE SCIENCE

Theories

Hypotheses

Observations

Generalizations

INDUCTIVE SCIENCE

▶ Figure 2.1 The Theory and Research Cycle
The theory and research cycle can be compared to a relay race; although all participants do not necessarily start or stop at the same point, they share a common goal—to examine all levels of social life.

Source: Adapted from Walter Wallace, *The Logic of Science in Sociology.* New York: Aldine de Gruyter, 1971.

In the *inductive approach,* the researcher collects information or data (facts or evidence) and then generates theories from the analysis of that data. Under the inductive approach, we would proceed as follows: (1) specific observations suggest generalizations, (2) generalizations produce a tentative theory, (3) the theory is tested through the formation of hypotheses, and (4) hypotheses may provide suggestions for additional observations. Using the inductive approach to study suicide, we might start by simultaneously collecting and analyzing data related to suicidal behavior and then generate a theory (see Glaser and Strauss, 1967; Reinharz, 1992). Researchers may break into the cycle at different points depending on what they want to know and what information is available.

Theory gives meaning to research; research helps support theory. For example, data collected from interviews with 25 women aged 15 to 24 who recently attempted suicide will not give us an explanation of *why* women are more likely than men to *attempt* to take their own lives. Similarly, theories unsupported by data are meaningless. Suppose, for instance, that we made the following assertions: Women are more likely to attempt suicide because of problems in their personal relationships, whereas men are more likely to be suicidal when they have economic difficulties, are unemployed, or experience a severe physical illness (Canetto, 1992). Our assertions are unsupported because we have not tested their validity.

Research helps us question such assumptions about suicide and other social concerns. Sociologists suggest that a healthy skepticism (a feature of science) is important in research because it keeps us open to the possibility of alternative explanations. Some degree of skepticism is built into each step of the research process. With that in mind, let's explore the steps in the sociological research process.

The Sociological Research Process

Not all sociologists conduct research in the same manner. Some researchers primarily engage in quantitative research, whereas others engage in qualitative research. With *quantitative research,* the goal is scientific objectivity, and the focus is on data that can be measured numerically. Quantitative research typically emphasizes complex statistical techniques. Most sociological studies on suicide have used quantitative research. They have compared rates of suicide with almost every conceivable variable, including age, sex, race/ethnicity, education, and even sports participation (see Lester, 1992). For example, researchers in one study examined the effects of church membership, divorce, and migration on suicide rates in the United States and concluded that suicide rates are typically higher where divorce and migration rates are higher and church membership is lower (Breault, 1986). (The "Understanding Statistical Data Presentations" box explains how to read numerical tables, how to interpret the data and draw conclusions, and how to calculate ratios and rates.)

With *qualitative research,* interpretive description (words) rather than statistics (numbers) is used to analyze underlying meanings and patterns of social relationships. An example of qualitative research is a study in which the researcher systematically analyzed the contents of the notes of suicide victims to determine recurring themes, such as a feeling of despair or failure. Through this study, the researcher hoped to determine if any patterns could be found that would help in understanding why people might kill themselves (Leenaars, 1988).

The "Conventional" Research Model

Research models are tailored to the specific problem being investigated and the focus of the researcher. Both quantitative research and qualitative research contribute to our knowledge of society and human social interaction, and both involve a series of steps, as shown in ▶ Figure 2.2. We will now trace the steps in the "conventional" research model, which focuses on

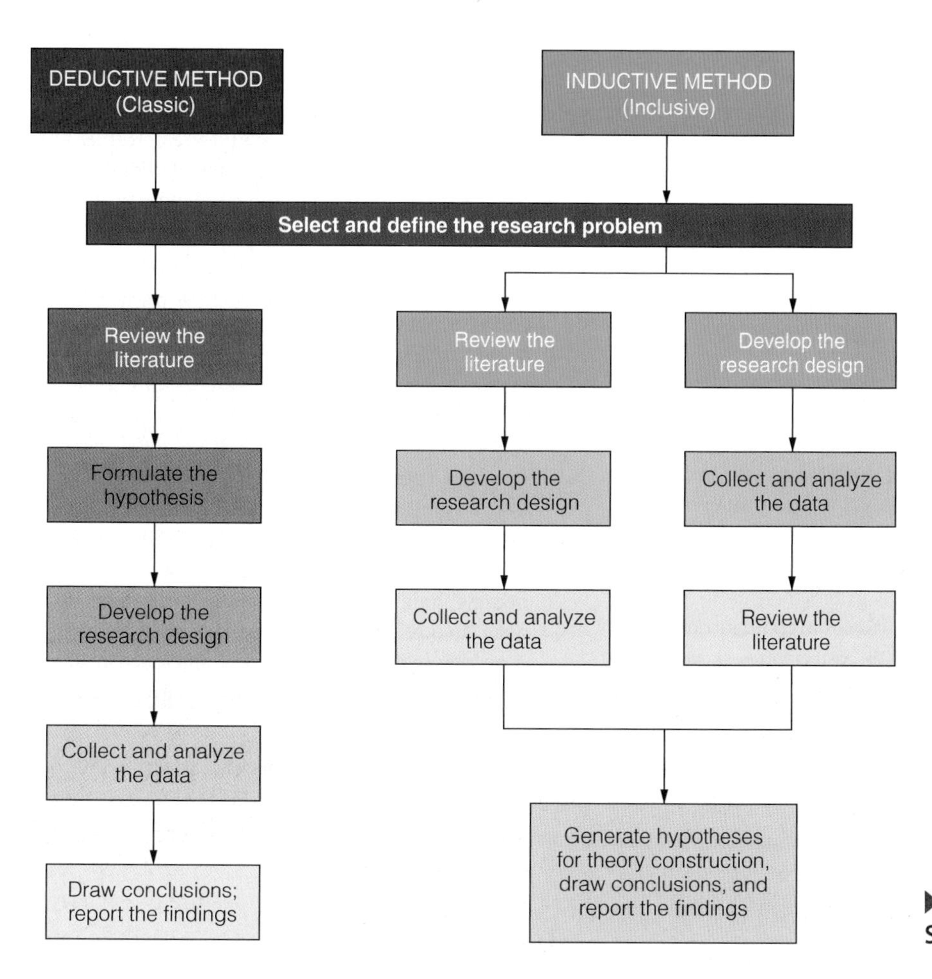

▶ **Figure 2.2 Steps in Sociological Research**

Understanding Statistical Data Presentations

Are men or women more likely to commit suicide? Are suicide rates increasing or decreasing? These questions can be answered in numerical terms. Sociologists often use statistical tables as a concise way to present data because such tables convey a large amount of information in a relatively small space; Table 1 gives an example. To understand a table, follow these steps:

1. *Read the title.* The title indicates the topic. From the title, "U.S. Suicides, by Sex and Method Used, 1984 and 2005," we learn that the table shows relationships between two variables: sex and method of suicide used. It also indicates that the table contains data for two different time periods: 1984 and 2005.
2. *Check the source and other explanatory notes.* In this case, the source is *National Center for Injury Prevention and Control, 2006.* Checking the source helps determine its reliability and timeliness. The first footnote indicates that the table includes only people who reside in the United States. The next footnote reflects that, due to rounding, the percentages in a column may not total 100.0%. The final two footnotes provide more information about exactly what is included in each category.
3. *Read the headings for each column and each row.* The main column headings in Table 1 are "Method," "Males," and "Females." These latter two column headings are divided into two groups: 1984 and 2005. The columns present information (usually numbers) arranged vertically. The rows present information horizontally. Here, the row headings indicate suicide methods.
4. *Examine and compare the data.* To examine the data, determine what units of measurement have been used. In ◆ Table 1 the figures are numerical counts (for example, the total number of reported female suicides by poisoning in 2005 was 2,632) and percentages (for example, in 2005, poisoning accounted for 39.1 percent of all female suicides reported). A *percentage,* or proportion, shows how many of a given item there are in every one hundred. Percentages allow us to compare groups of different sizes. For example, percentages show the proportion of people who used each method, thus giving a more meaningful comparison.
5. *Draw conclusions.* By looking for patterns, some conclusions can be drawn from Table 1.
   a. *Determining the increase or decrease.* Between 1984 and 2005, reported male suicides by firearms increased from 14,504 to 14,916—an increase of 412—while female suicides by firearms decreased by 523. This represents a *total* increase (for males) and decrease (for females) in suicides by firearms for the two years being compared. The *amount* of increase or decrease can be stated as a percentage: Total male suicides by firearms were about 2.8 percent higher in 2005, calculated by dividing the total increase (412) by the earlier (lower) number. Total female suicides by firearms were about 20 percent lower in 2005, calculated by dividing the total decrease (523) by the earlier (higher) number.
   b. *Drawing appropriate conclusions.* The number of female suicides by firearms decreased about 20 percent between 1984 and 2005; the number for poisoning increased by about 9.4 percent. We might conclude that more women preferred poisoning over firearms as a means of killing themselves in 2005 than in 1984. Does that mean fewer women had access to guns in 2005? That poisoning oneself became more acceptable? Such generalizations do not take into account that we are only looking at data from two years and that the difference in statistics for those two years may not really represent a trend.

Source: National Center for Injury Prevention and Control, 2006.

◆ Table 1 U.S. Suicides, by Sex and Method Used, 1984 and 2005[a]

	MALES		FEMALES	
METHOD	**1984**	**2005**	**1984**	**2005**
Total	22,689	25,907	6,597	6,730
Firearms (% of total)[b]	14,504 (64.0)	14,916 (57.6.)	2,609 (39.5)	2,086 (30.9)
Poisoning[c] (% of total)[b]	3,203 (14.1)	3,112 (12.0)	2,406 (36.5)	2,632 (39.1)
Suffocation[d] (% of total)[b]	3,478 (15.3)	5,887 (22.7)	863 (13.0)	1,361 (20.2)
Other (% of total)[b]	1,504 (6.6)	1,992 (7.7)	719 (10.9)	651 (9.8)

[a]Excludes deaths of nonresidents of the United States.
[b]Due to rounding, the percentages in a column may not add up to 100.0%.
[c]Includes solids, liquids, and gases.
[d]Includes hanging and strangulation.

quantitative research. Then we will describe an alternative model that emphasizes qualitative research.

1. *Select and define the research problem.* When you engage in research, the first step is to select and clearly define the research topic. Sometimes, a specific experience such as having known someone who committed suicide can trigger your interest in a topic. Other times, you might select topics to fill gaps or challenge misconceptions in existing research or to test a specific theory (Babbie, 2004). Emile Durkheim selected suicide because he wanted to demonstrate the importance of *society* in situations that might appear to be arbitrary acts by individuals. Suicide was a suitable topic because it was widely believed that suicide was a uniquely individualistic act. However, Durkheim emphasized that *suicide rates* provide better explanations for suicide than do *individual acts* of suicide. He reasoned that if suicide were purely an individual act, then the rate of suicide (the relative number of people who kill themselves each year) should be the same for every group regardless of culture and social structure (see Box 2.2 on page 48 for a current example). Moreover, Durkheim wanted to know why there were different rates of suicide—whether factors such as religion, marital status, sex, and age had an effect on social cohesion.
2. *Review previous research.* Before you begin your research, it is important to review the literature to see what others have written about the topic. Analyzing what previous researchers have found helps to clarify issues and focus the direction of your own research. But when Durkheim began his study, very little sociological literature existed for him to review other than the works of Henry Morselli (1975/1881), who concluded that suicide was a part of an evolutionary process whereby "weak-brained" individuals were sorted out by insanity and voluntary death.
3. *Formulate the hypothesis (if applicable).* You may formulate a ***hypothesis*—a statement of the relationship between two or more concepts.** Concepts are the abstract elements representing some aspect of the world in simplified form (such as "social integration" or "loneliness"). As you formulate your hypothesis about suicide, you may need to convert concepts to variables. A *variable* is any concept with measurable traits or characteristics that can change or vary from one person, time, situation, or society to another. Variables are the observable and/or measurable counterparts of concepts. For example, "suicide" is a concept; the "rate of suicide" is a variable.

   The most fundamental relationship in a hypothesis is between a dependent variable and one or more independent variables (see ▶ Figure 2.3). The ***independent variable* is presumed to cause or determine a dependent variable.** Age, sex, race, and ethnicity are often used as independent variables. The ***dependent variable* is assumed to depend on or be caused by the independent variable(s)** (Babbie, 2004). Durkheim used the degree of social integration in society as the independent variable to determine its influence on the dependent variable, the rate of suicide.

   Whether a variable is dependent or independent depends on the context in which it is used. To use variables in the contemporary research process, sociologists create operational definitions. An *operational definition* is an explanation of an abstract concept in terms of observable features that are specific enough to measure the variable. For example, suppose that your goal is to earn an *A* in this course. Your professor may have created an operational definition by defining an *A* as earning an exam average of 90 percent or above (Babbie, 2004).

   Events such as suicide are too complex to be caused by any one variable. Therefore, they must be explained in terms of *multiple causation*—that is, an event occurs as a result of many factors operating in combination. What *does* cause suicide? Social scientists cite multiple causes, including rapid social change, economic conditions, hopeless poverty, and lack of religiosity (the degree to which an individual or group feels committed to a particular system of religious beliefs). Usually, no one factor will cause a person to commit suicide. Rather, other factors must combine with a factor such as poverty to cause a person to commit suicide. Sociologists cannot produce an equation (such as poverty + homelessness = suicide) to predict a social occurrence. Not all social research makes use of hypotheses.
4. *Develop the research design.* In developing the research design, you must first consider the units of analysis and the time frame of the study. A *unit of analysis* is *what* or *whom* is being studied (Babbie, 2004). In social science research, individuals

hypothesis in research studies, a tentative statement of the relationship between two or more concepts.

independent variable a variable that is presumed to cause or determine a dependent variable.

dependent variable a variable that is assumed to depend on or be caused by one or more other (independent) variables.

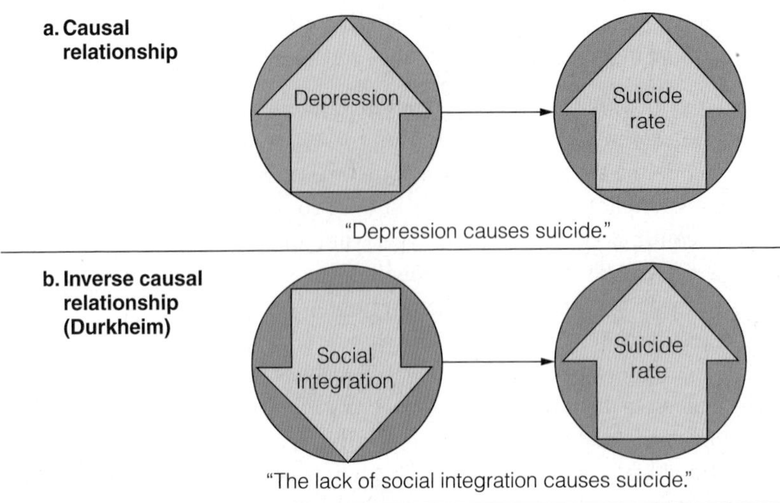

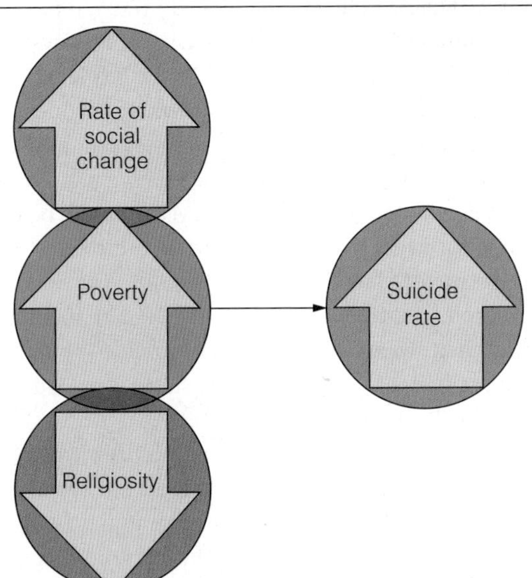

▶ **Figure 2.3 Hypothesized Relationships Between Variables**
A causal hypothesis connects one or more independent (causal) variables with a dependent (affected) variable. The diagram illustrates three hypotheses about the causes of suicide. To test these hypotheses, social scientists would need to operationalize the variables (define them in measurable terms) and then investigate whether the data support the proposed explanation.

are the most typical unit of analysis. Social groups (such as families, cities, or geographic regions), organizations (such as clubs, labor unions, or political parties), and social artifacts (such as books, paintings, or weddings) may also be units of analysis. Durkheim's unit of analysis was social groups, not individuals, because he believed that the study of individual cases of suicide would not explain the rates of suicide.

After determining the unit of analysis for your study, you must select a time frame for study: cross-sectional or longitudinal. *Cross-sectional studies* are based on observations that take place at a single point in time; these studies focus on behavior or responses at a specific moment. *Longitudinal studies* are concerned with what is happening over a period of time or at several different points in time; they focus on processes and social change. Some longitudinal studies are designed to examine the same set of people each time, whereas others look at trends within a general population. Using longitudinal data, Durkheim was able to compare suicide rates over a period of time in France and other European nations.

5. *Collect and analyze the data.* Your next step is to collect and analyze data. You must decide which population—persons about whom we want to be able to draw conclusions—will be observed or questioned. Then it is necessary to select a sample of people from the larger population to be studied. It is important that the sample accurately represent the larger population. For example, if you arbitrarily selected five students from your sociology class to interview, they probably would not be representative of your school's total student body. However, if you selected five students from the total student body by a random sample, they might be closer to being representative (although a random sample of five students would be too small to yield much useful data). In ***random sampling,*** **every member of an entire population being studied has the same**

An operational definition is an explanation of an abstract concept in terms of observable features that are specific enough to measure the variable. For example, the operational definition of an *A* may be an exam average of 90 percent or above. After college professors have established the grading requirements for a course, students seek to meet those expectations by performing well on examinations.

chance of being selected. You would have a more representative sample of the total student body, for example, if you placed all the students' names in a rotating drum and conducted a drawing. By contrast, in ***probability sampling,*** **participants are deliberately chosen because they have specific characteristics,** possibly including such factors as age, sex, race/ethnicity, and educational attainment.

In addition to problems with sampling, sociologists must maintain the validity and reliability of the data they collect. ***Validity*** **is the extent to which a study or research instrument accurately measures what it is supposed to measure.** For example, sociologists who analyze the relationship between religious beliefs and suicide must determine whether "church membership" is an accurate indicator of a person's religious beliefs. In fact, one person may be very religious but not belong to a specific church, whereas another person may be a member of a church yet not hold any deep religious convictions. To maintain validity, some sociologists study the relationship between suicide and religion not only in terms of people's specific behaviors (e.g., frequency of attendance at church services) but also as a set of values, beliefs, or attitudes (Breault, 1986). ***Reliability*** **is the extent to which a study or research instrument yields consistent results** when applied to different individuals at one time or to the same individuals over time. An important issue in reliability is the fact that sociologists have found that the characteristics of interviewers and how they ask questions may produce different answers from the people being interviewed. As a result, different studies of college students who have contemplated suicide may arrive at different conclusions. Problems of validity are also linked to how data is analyzed. *Analysis* is the process through which data are organized so that comparisons can be made and conclusions drawn. Sociologists use many techniques to analyze data. The process for each type of research method is discussed later in this chapter.

In Durkheim's study, he collected data from vital statistics for approximately 26,000 suicides. He classified them separately according to age, sex, marital status, presence or absence of children in the family, religion, geographic location, calendar date, method of suicide, and a number of other variables. As Durkheim analyzed his data, four distinct categories of suicide emerged: egoistic, altruistic, anomic, and fatalistic. *Egoistic suicide* occurs among people who are isolated from any social group. For example, Durkheim concluded that suicide rates were relatively high in Protestant countries in Europe because Protestants believed in individualism and were more loosely tied to the church than were Catholics. Single people had proportionately higher suicide rates than married persons because they had a low degree of social integration, which contributed to their loneliness. In contrast, *altruistic suicide* occurs among individuals who are excessively integrated into society. An example is military leaders who kill themselves after

random sampling a study approach in which every member of an entire population being studied has the same chance of being selected.

probability sampling choosing participants for a study on the basis of specific characteristics, possibly including such factors as age, sex, race/ethnicity, and educational attainment.

validity in sociological research, the extent to which a study or research instrument accurately measures what it is supposed to measure.

reliability in sociological research, the extent to which a study or research instrument yields consistent results when applied to different individuals at one time or to the same individuals over time.

Box 2.2 Sociology in Global Perspective

Comparing Suicide Statistics from Different Nations

We're often told that this is the age of convergence. As global optimists see it, free-market capitalism, democratic norms, and maybe even a better appreciation for the sanctity of life are gaining ascendancy across the borders. In the rich world at least, you get the sense that a country's unique way of looking at the world will eventually be submerged by these big global trends.

There's some truth to that. But every once in a while here in Japan, I'm abruptly reminded that some things about this remarkable culture I'll never begin to fathom. To my mind the most fundamental one is the prevalence of suicide in a nation that boasts some of the highest living standards and longest life expectancies in the world.

—Brian Bremner (2000), Tokyo bureau chief for *Business Week,* stating his concern about the "suicide epidemic" in Japan

When sociologists select and define a research problem, they may look at a current social phenomenon such as statistical trends in suicides or how the rates of suicide compare across nations. If we look at the rates of suicide per 100,000 people in various countries, will these rates differ? In fact, the answer to this question is "yes." There is a wide disparity among suicide rates in various nations. For example, Lithuania has a suicide rate of 39 per 100,000 people, as compared with the United States, which has a suicide rate of 11 per 100,000 people.

Can any patterns be identified in regard to Lithuanian suicides? One that has been identified is that the suicide rate has risen in Lithuania over the past three decades as that nation has gone through a lengthy period of economic, political, and social upheaval.

© Junko Kimura/Getty Images

Global economic woes and worries about the health of banks in Japan have contributed to heightening concern among many Japanese workers. Sociologists continue to explore the relationship between economic problems and suicide rates in nations such as Japan, where the recent suicide rate has been twice that of the United States.

defeat in battle because they have so strongly identified themselves with their cause that they believe they cannot live with defeat. According to Durkheim, people are more likely to kill themselves when social cohesion is either very weak or very strong.

Durkheim further observed that degree of social integration is not the only variable that influences suicide rates. Rapid social change and shifts in moral values make it difficult for people to know what is right and wrong. *Anomic suicide* results from a lack of shared values or purpose and from the absence of social regulation. By contrast, excessive regulation and oppressive discipline may contribute to *fatalistic suicide,* as in the suicides of slaves.

6. *Draw conclusions and report the findings.* After analyzing the data, your first step in drawing conclusions is to return to your hypothesis or research objective to clarify how the data relate both to the hypothesis and to the larger issues being addressed. At this stage, you note the limitations of the study, such as problems with the sample, the influence of variables over which you had no control, or variables that your study was unable to measure.

At the end of your research, it is important to report your findings. This report usually includes a review of each step you took so that others can replicate your work in substantially the same way that it was originally conducted. Social scientists generally present their findings in papers at professional meetings and publish them in academic journals and scholarly books. Durkheim reported his findings in his book

Comparative Suicide Rates

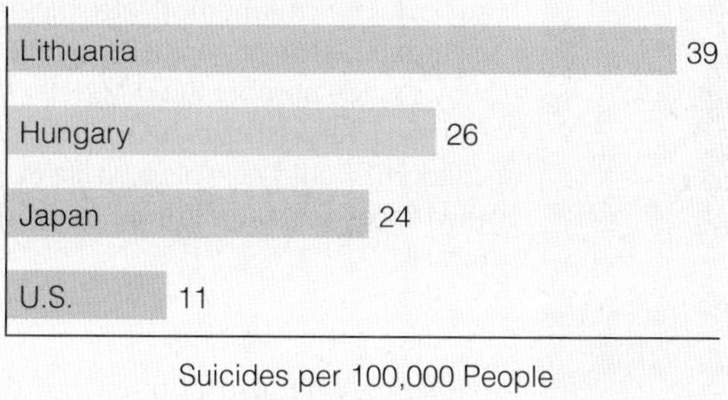

Country	Suicides per 100,000 People
Lithuania	39
Hungary	26
Japan	24
U.S.	11

Suicides per 100,000 People

Source: World Health Organization, 2004b.

We might ask this: Is Lithuania's suicide rate increase due to that social upheaval? Japan has also had an upswing in the number of reported suicides, culminating in the late 1990s, when there was an unprecedented 35 percent surge in suicides. The suicide rate in Japan (24.2 per 100,000 people) is less than that of Lithuania but twice that of the United States. To put the suicide rate in Japan in perspective, that country has three times as many reported suicide victims as it does traffic fatalities. In 2005, the rate of suicide for males in Japan was 36.1 per 100,000, with many of the men being in their forties and fifties, and many of them having recently experienced major financial difficulties. By contrast, the female rate of suicide in Japan for the same year was 12.9 per 100,000.

Using theory as the foundation for our research on Japanese and Lithuanian suicides, we might reflect on Emile Durkheim's types of suicides and test whether or not some of these suicides might be best described as *anomic suicides,* which may be brought about by rapid social change. Japan has experienced a prolonged economic slump that has particularly affected the employment rates of middle-aged men. In a society where people are accustomed to full employment (meaning that most people who wanted a job could have one), many people have found themselves unemployed for the first time. Recent statistics showed that 47 percent of those who killed themselves in Japan were unemployed. Business failures and inability to meet basic living costs were two of the major reasons cited for the upswing in deaths among Japanese men. An example was Masaaki Kobayashi, age fifty-one, who left a suicide note asking his firm's accountant to use Kobayashi's $3.14 million in life insurance to raise money to head off the bankruptcy of his company.

Some social analysts have suggested that additional factors come into play in explaining the high rates of suicide in any nation. Although there is no consensus on the factors associated with suicide in Japan, some analysts attribute the problem to "cultural factors" such as a belief that the group is more important than the individual and that, under certain conditions, suicide is an honorable deed. Examples include the samurais who committed hara-kiri and the kamikaze pilots of World War II. Other analysts believe that the two main Japanese religions—Shintoism and Buddhism—are related to the high rate of suicides because these religions have no moral prohibition on self-killing. However, it is important for us to realize that these views about cultural differences as factors remain nothing more than assumptions until more social scientists conduct systematic research on these issues and are able to more conclusively demonstrate that cause-and-effect relationships exist.

Reflect & Analyze

Which of Durkheim's types of suicide do you think might be applicable in a study of suicide in Japan? Would the same types of suicide be applicable to a study of suicide in the United States? Why or why not?

Sources: Based on Bremner, 2000; Lamar, 2000; Lev, 1998; and World Health Organization, 2004b.

Suicide (1964b/1897), in which he concluded that suicide is an indicator of the moral condition of a society and that the suicide rate reflects such factors as the presence or absence of strong social bonds among people, and shifting standards of behavior.

We have traced the steps in the "conventional" research process (based on quantitative research). But what steps might be taken in an alternative approach based on qualitative research?

A Qualitative Research Model

Although the same underlying logic is involved in both quantitative and qualitative sociological research, the *styles* of these two models are very different (King, Keohane, and Verba, 1994). As previously stated, qualitative research is more likely to be used when the research question does not easily lend itself to numbers and statistical methods. As compared to a quantitative model, a qualitative approach often involves a different type of research question and a smaller number of cases. As a result, the outcome of a qualitative study is a complex, more holistic picture of some particular social phenomenon or human problem (King, Keohane, and Verba, 1994; Creswell, 1998).

How might qualitative research be used to study suicidal behavior? In studying different rates of suicide among women and men, for example, the social psychologist Silvia Canetto (1992) questioned whether existing theories and quantitative research provided an adequate explanation for gender differences in suicidal behavior and decided that she would explore

Sociological research on suicide has begun to look at issues such as what social factors might motivate suicide bombers. Some researchers might ask why suicide bomber Raed Abdel-Hameed Mesk (shown here with his children) would take his own life in the process of committing a terrorist attack.

alternate explanations. As a result, Canetto redefined the concept of suicidal behavior to focus on outcome ("fatal" versus "nonfatal") rather than in terms of intent ("completed" versus "attempted"). Analyzing previous research, Canetto learned that most studies linked suicidal behavior in women to problems in their personal relationships, particularly with members of the opposite sex, whereas men's suicides were most often linked to performance pressure, especially when their self-esteem and independence were threatened. However, from her analysis of existing research, Canetto believed that gender differences in suicidal behavior are more closely associated with beliefs about and expectations for men and women in a particular culture rather than purely interpersonal crises (Canetto, 1992).

As in Canetto's case, researchers using a qualitative approach may engage in *problem formulation* to clarify the research question and to develop questions of concern and interest to the research participants (Reinharz, 1992). To create a research design for Canetto's study, we might start with the proposition that most studies may have attributed women's and men's suicidal behavior to the wrong causes. Next, we might decide to interview people who have attempted suicide by using a collaborative approach in which the participants suggest avenues of inquiry that the researcher should explore (Reinharz, 1992).

Although Canetto did not gather data in her study, she made an important contribution to our knowledge about gender differences in suicidal behavior by suggesting that there may be a relationship between suicide and feelings of fear, especially in cases of domestic violence. She also pointed out that cultural norms often encourage nonfatal suicide in women and fatal suicide in men (e.g., "real men" don't fail when they take their own life). Canetto concluded that most researchers do not explore social structure factors such as the effect of low income or restricted job mobility on women's suicidal behavior. Similarly, men's suicidal behavior tends to be linked to the lack of relationships with other people and the loss of social privilege (such as might occur at retirement).

In a qualitative approach, the next step is to collect and analyze data to assess the validity of the starting proposition. Qualitative researchers typically gather data in natural settings, such as where people live or work, rather than in a laboratory or other research setting. In this environment, the researcher can play a background rather than a foreground role, and the data analysis frequently uses the language of the people being studied, not the researcher. Often, this approach generates new theories and innovative research that incorporate the perspectives of people previously excluded on the basis of race, class, gender, sexual orientation, or other attributes.

Although the qualitative approach follows the conventional research approach in presenting a problem, asking a question, collecting and analyzing data, and seeking to answer the question, it also has several unique features (based on Creswell, 1998, and Kvale, 1996):

1. *The researcher begins with a general approach rather than a highly detailed plan.* Flexibility is necessary because of the nature of the research question. The topic needs to be explored so that we can know "how" or "what" is going on, but we may not be able to explain "why" a particular social phenomenon is occurring.
2. *The researcher has to decide when the literature review and theory application should take place.* Initial work may involve redefining existing concepts or reconceptualizing how existing studies have been

conducted. The literature review may take place at an early stage, before the research design is fully developed, or it may occur after the development of the research design, and after the data collection has already occurred. Many of us who teach sociological theory would like to see greater use of theory to inform both qualitative and quantitative studies because this approach provides a framework for interpreting the data collected (see also Kvale, 1996).

3. *The study presents a detailed view of the topic.* Qualitative research usually involves a smaller number of cases and many variables, whereas quantitative researchers typically work with a few variables and many cases (Creswell, 1998).
4. *Access to people or other resources that can provide the necessary data is crucial.* Unlike the quantitative researcher, who often uses existing databases, many qualitative researchers generate their own data. As a result, it is necessary to have access to people and build rapport with them.
5. *Appropriate research method(s) are important for acquiring useful qualitative data.* Qualitative studies are often based on field research such as observation, participant observation, case studies, ethnography, and unstructured interviews, as discussed in the next section.

Conducting surveys and polls is an important means of gathering data from respondents. Some surveys take place on street corners; increasingly, however, such surveys are done by telephone, Internet, or other means.

Research Methods

How do sociologists know which research method to use? Are some approaches better than others? Which method is best for a particular problem? ***Research methods* are specific strategies or techniques for systematically conducting research.** The methods should be acceptable to a larger community of scholars and nonacademic researchers who routinely engage in research endeavors. Qualitative researchers frequently attempt to study the social world from the point of view of the people they are studying. By contrast, quantitative researchers generally use surveys, secondary analyses of existing statistical data, and experimental designs. We will now look at these research methods.

Survey Research

A *survey* is a poll in which the researcher gathers facts or attempts to determine the relationships among facts. Surveys are often done when the researcher wants to describe, compare, and predict knowledge, attitudes, and behavior. For example, a community survey might describe and compare such things as income, educational level, and type of employment in regard to people's attitudes about a juvenile curfew ordinance that prohibits adolescents from being out on the streets at certain nighttime hours.

Researchers frequently select a representative sample (a small group of respondents) from a larger population (the total group of people) to answer questions about their attitudes, opinions, or behavior. For example, if the larger population consists of 10,000 people, 51 percent of whom are female, with 30 percent over age 35, a representative sample will have fewer people (perhaps 1,000) but must still consist of 51 percent females, with 30 percent over age 35

research methods specific strategies or techniques for systematically conducting research.

survey a poll in which the researcher gathers facts or attempts to determine the relationships among facts.

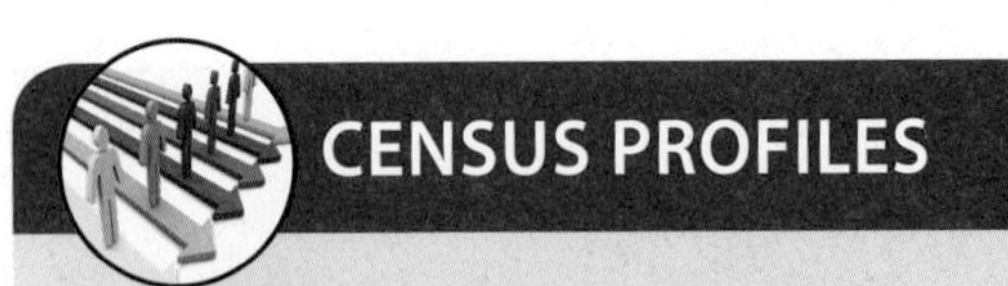

CENSUS PROFILES

How People in the United States Self-Identify Regarding Race

Beginning with Census 2000, the U.S. Census Bureau has made it possible for people responding to census questions regarding their race to mark more than one racial category. Although the vast majority of respondents select only one category (see below), the Census Bureau reports that in 2003 approximately 4.3 million people (1.48 percent of the population) in the United States self-identified as being of more than one race. As a result, if you look at the figures set forth below, they total more than 100 percent of the total population. How can this be? Simply stated, some individuals are counted at least twice, based on the number of racial categories they listed.

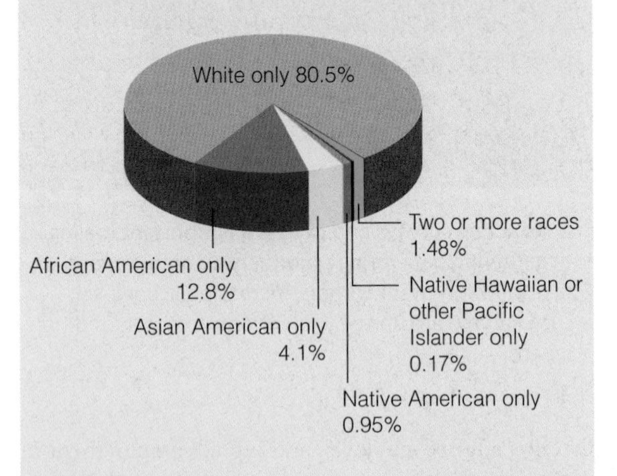

Race	Percentage of Total Population
White alone or in combination with one or more other races	81.8
African American alone or in combination with one or more other races	13.3
Asian American alone or in combination with one or more other races	4.6
Native American alone or in combination with one or more other races	1.5
Native Hawaiian or other Pacific Islander alone or in combination with one or more other races	0.3
Total	101.5

Source: U.S. Census Bureau, 2007.

(Fink, 1995). ***Respondents* are persons who provide data for analysis through interviews or questionnaires.** The Gallup and Harris polls are among the most widely known large-scale surveys; however, government agencies such as the U.S. Census Bureau conduct a variety of surveys as well. Unlike many polls that use various methods of gaining a representative sample of the larger population, the Census Bureau attempts to gain information from all persons in the United States. The decennial census occurs every 10 years, in the years ending in "0." The purpose of this census is to count the population and housing units of the entire United States. The population count determines how seats in the U.S. House of Representatives are apportioned; however, census figures are also used in formulating public policy and in planning and decision making in the private sector. The Census Bureau attempts to survey the *entire* U.S. population by using two forms—a "short form" of questions asked of *all* respondents, and a "long form" that contains additional questions asked of a *representative sample* of about one in six respondents. Statistics from the Census Bureau provide information that sociologists use in their research. An example is shown in the Census Profiles feature: "How People in the United States Self-Identify Regarding Race." Note that because of recent changes in the methods used to collect data by the Census Bureau, information on race from the 2000 census is not directly comparable with data from earlier censuses.

Surveys are the most widely used research method in the social sciences because they make it possible to study things that are not directly observable—such as people's attitudes and beliefs—and to describe a population too large to observe directly (Babbie, 2004). Let's take a brief look at the most frequently used types of surveys.

Types of Surveys Survey data are collected by using self-administered questionnaires, face-to-face interviews, and/or telephone interviews. A ***questionnaire* is a printed research instrument containing a series of items to which subjects respond.** Items are often in the form of statements with which the respondent is asked to "agree" or "disagree." Questionnaires may be administered by interviewers in face-to-face encounters or by telephone, but the most commonly used technique is the *self-administered questionnaire.* The questionnaires are typically mailed or delivered to the respondents' homes; however, they may also be administered to groups of respondents gathered at the same place at the same time.

Sociologist Kevin E. Early (1992), for example, conducted a survey regarding the lower rates of suicide among African Americans than whites in the United

States. Early collected and analyzed survey data to test his hypothesis that "the black church's influence is an essential factor in ameliorating and buffering social forces that otherwise would lead to suicide." A self-administered questionnaire was completed by congregation members in conjunction with services at six black churches considered representative of the thirty-seven black churches in Gainesville, Florida.

Self-administered questionnaires have certain strengths. They are relatively simple and inexpensive to administer, they allow for rapid data collection and analysis, and they permit respondents to remain anonymous (an important consideration when the questions are of a personal nature). A major disadvantage is the low response rate. Mailed surveys sometimes have a response rate as low as 10 percent—and a 50 percent response rate is considered by some to be minimally adequate (Babbie, 2004). The response rate is usually somewhat higher if the survey is handed out to a group that is asked to fill it out on the spot. Moreover, for surveys involving ethnically diverse or international respondents, the questionnaire must be available in languages other than English (Fink, 1995).

Survey data may also be collected by interviews. An ***interview*** **is a data-collection encounter in which an interviewer asks the respondent questions and records the answers.** Survey research often uses *structured interviews,* in which the interviewer asks questions from a standardized questionnaire. Structured interviews tend to produce uniform or replicable data that can be elicited time after time by different interviews. For example, in addition to surveying congregation members, Early (1992) conducted interviews with pastors of African American churches, using a series of open-ended questions. Next, he read four vignettes (stories about people) relating to suicide to the pastors and then asked questions designed to determine the pastors' opinions and attitudes concerning the behavior displayed in the vignettes. His goal was to learn the extent to which the African American church reinforces attitudes, values, beliefs, and norms that discourage suicide.

Unlike the open-ended questions used in Early's more qualitative approach, closed-ended questions may be used when researchers want to have a large number of respondents and to generate standardized answers to questions. For example, in a study involving forty-nine hospital accident and emergency departments and psychiatric services in the United Kingdom, researchers developed closed-ended questions to be asked of patients who had attempted suicide by poisoning. The purpose of the study was to determine if media representation of suicide and deliberate self-harm encouraged suicidal behavior in vulnerable individuals. Specifically, the researchers wanted to know if the rate of self-poisoning and the choice of overdose drugs were influenced by a television drama, *Casualty,* which portrayed an acetaminophen overdose in one of its episodes. Questionnaires were completed by more than 1,000 self-poisoning patients during the three-week periods before and after the program was broadcast. Was there a direct link between viewing the episode and the person's decision to take an overdose, choice of drug, and speed with which he or she arrived at the hospital? According to the researchers, there was a 17 percent increase in the number of hospital patients who reported that they engaged in self-poisoning in the week after the broadcast, and a 9 percent increase in the second week. Moreover, the rate of poisonings by acetaminophen increased more than that of any other drug. In fact, 20 percent of the patients interviewed indicated that the program had influenced their decision to overdose, and 17 percent said it had influenced their drug choice (Hawton, Simkin, Deeks, et al., 1999).

Interviews have specific advantages. They are usually more effective in dealing with complicated issues and provide an opportunity for face-to-face communication between the interviewer and the respondent. When open-ended questions are used, the researcher may gain new perspectives. The pastors interviewed in Early's study distinguished between suicide (which is "unthinkable for black people" because it is a "white thing, not a black thing") and alcohol abuse, drug addiction, and homicide (which are wrong and "sinful" but are "an understandable response to the socioeconomic and political conditions of blacks in the United States") (Early, 1992: 79). When closed-ended questions are used, it is easier for interviewers to code responses and for researchers to compare individuals' responses across categories of interest. For example, researchers in the overdose study were able to compare such variables as sex, age, choice of overdose drug, history of taking overdoses, and whether the choice of substance was influenced by television programs. Based on their findings, the researchers argued that media portrayals of self-poisoning or self-injury on popular television shows may contribute to self-harming behavior and choice of method used,

respondents persons who provide data for analysis through interviews or questionnaires.

questionnaire a printed research instrument containing a series of items to which subjects respond.

interview a research method using a data-collection encounter in which an interviewer asks the respondent questions and records the answers.

and thus should be of concern to the general public and to media producers as well (Hawton, Simkin, Deeks, et al., 1999). As this and other research studies show, interviews provide a wide variety of useful information; however, a major disadvantage is the cost and time involved in conducting the interviews and analyzing the results. Also, one weakness of interviews is that in responding to the questions asked, people may be influenced by the interviewer's race, age, sex, size, or other attributes.

A quicker method of administering questionnaires is the *telephone survey,* which is becoming an increasingly popular way to collect data. Telephone surveys save time and money as compared to self-administered questionnaires or face-to-face interviews. Some respondents may be more honest than when they are facing an interviewer. Telephone surveys also give greater control over data collection and provide greater personal safety for respondents and researchers than do personal encounters. In *computer-assisted telephone interviewing* (sometimes called CATI), the interviewer uses a computer to dial random telephone numbers, reads the questions shown on the video monitor to the respondent, and then types the responses into the computer terminal. The answers are immediately stored in the central computer, which automatically prepares them for data analysis. Although use of the CATI system overcomes the problem of unlisted telephone numbers by randomly dialing numbers, it is limited by people's widespread use of answering machines, voice mail, and caller ID to filter their incoming telephone calls.

Computer-assisted telephone interviewing is an easy and cost-efficient method of conducting research. The widespread use of answering machines, voice mail, and caller ID may make this form of research more difficult in the twenty-first century.

Strengths and Weaknesses of Surveys Survey research has several important strengths. First, it is useful in describing the characteristics of a large population without having to interview each person in that population. Second, survey research enables the researcher to search for causes and effects and to assess the relative importance of a number of variables. In recent years, computer technology has enhanced our ability to do *multivariate analysis*—research involving more than two independent variables. For example, to assess the influence of religion on suicidal behavior among African Americans, a researcher might look at the effects of age, sex, income level, and other variables all at once to determine which of these independent variables influences suicide the most or least and how influential each variable is relative to the others. Third, survey research can be useful in analyzing social change or in documenting the existence of a social problem. Contemporary scholars have used survey research to provide information about such problems as racial discrimination, sexual harassment, and sex-based inequality in employment by documenting the fact that they are more widespread than previously thought (Reinharz, 1992).

Survey research also has weaknesses. One is that the use of standardized questions tends to force respondents into categories in which they may or may not belong. Another weakness concerns validity. People's opinions on issues seldom take the form of a standard response ranging from "strongly agree" to "strongly disagree." Moreover, as in other types of research, people may be less than truthful, especially on emotionally charged issues such as suicide, thus making reliance on self-reported attitudes problematic. Some scholars have also criticized the way survey data are used. They believe that survey data do not always constitute the "hard facts" that other analysts may use to justify changes in public policy or law. For example, survey statistics may over- or underestimate the extent of a problem and work against some categories of people more than others, as shown in ◆ Table 2.1.

Secondary Analysis of Existing Data

In ***secondary analysis,*** **researchers use existing material and analyze data that were originally collected by others.** Existing data sources include public re-

◆ **Table 2.1 Statistics: What We Know (and Don't Know)**

	TOPIC	
	Homelessness in the United States	**Suicide in the United States**
Research Finding	At least 250,000 people in this country are homeless.	At least 32,439 Americans committed suicide in 2004.
Possible Problem	Does that badly underestimate the total number of homeless people?	Are suicide rates different for some categories of U.S. citizens?
Explanation	The homeless are difficult to count, frequently attempting to avoid interviews with census takers. Critics of the census figures assert the actual number may be 3 million and that the government intentionally undercounts the homeless.	U.S. census data place Latinos/as in the category of whites. Other than African Americans, all other people of color are listed as "nonwhite—other." Thus, census data on specific categories are not available.

As the examples in this table show, statistics provide certain insights on the prevalence of social issues such as homelessness and suicide but do not always provide the *answer* regarding the nature and extent of the problem. What difficulties do researchers encounter when gathering data on people?

cords, official reports of organizations and government agencies, and surveys conducted by researchers in universities and private corporations. Research data gathered from studies are available in data banks, such as the Inter-University Consortium for Political and Social Research, the National Opinion Research Center (NORC), and the Roper Public Opinion Research Center. Other sources of data for secondary analysis are books, magazines, newspapers, radio and television programs, and personal documents. Secondary analysis is referred to as *unobtrusive research* because it has no impact on the people being studied. In Durkheim's study of suicide, for example, his analysis of existing statistics on suicide did nothing to increase or decrease the number of people who *actually* committed suicide.

Analyzing Existing Statistics Secondary analysis may involve obtaining *raw data* collected by other researchers and undertaking a statistical analysis of the data, or it may involve the use of other researchers' existing statistical analyses. In analysis of existing statistics, the unit of analysis is often *not* the individual. Most existing statistics are *aggregated:* They describe a group. Durkheim wanted to determine whether Protestants or Catholics were more likely to commit suicide; however, none of the available records indicated the religion of those who committed suicide. Although Durkheim suggested that Protestants were more likely to commit suicide than Catholics, it was impossible for him to determine that from the existing data.

In a contemporary study of suicide, K. D. Breault (1986) analyzed secondary data collected by government agencies to test Durkheim's hypothesis that religion and social integration provide protection from suicide. Using suicide as the dependent variable and church membership, divorce, unemployment, and female labor force participation as several of his independent variables, Breault performed a series of sophisticated statistical analyses and concluded that the data supported Durkheim's views on social integration and his theory of egoistic suicide. He also found support for Durkheim's proposition that Catholics are less likely to commit suicide than are Protestants. However, it should be noted that Durkheim did not attribute lower rates of suicide among Catholics to the role of church beliefs as much as to the tendency of Catholicism to promote social integration through rituals and regulation of standards of faith and moral conduct (Ellison, Burr, and McCall, 1997).

Numerous other studies have used secondary data to examine the relationship between religious factors and rates of suicide. For example, in a recent study, researchers used data from sources including the National Center for Health Statistics and a large survey of religious denominations to examine the extent to which religious homogeneity—how well community residents adhere to a single religion or a small number of faiths—is associated with lower suicide rates (Ellison, Burr, and McCall, 1997). Using larger categories such as conservative Protestant, moderate Protestant, Catholic, Mormon, Orthodox, and Jewish, the researchers concluded that religious homogeneity was linked with lower suicide rates, particularly in the northeastern and southern United States (Ellison, Burr, and McCall, 1997).

secondary analysis a research method in which researchers use existing material and analyze data that were originally collected by others.

Box 2.3 Framing Suicide in the Media

Sociology Versus Sensationalism

Front Page: *New York Post,* March 10, 2004:

> The daughter of a Silicon Valley executive has become the fourth New York University student to die in a plunge this academic year. [Name of student,] 19, jumped . . . from the roof of her boyfriend's 24-story apartment building Saturday after having a fight with him. (*New York Post,* 2004: 1)

***New York Times,* March 10, 2004:**

> The suicide of a New York University student who fell to her death from a rooftop off campus on Saturday chilled students, who learned about it in an e-mail message from university officials on Monday. It was the fourth N.Y.U. student death this year. One N.Y.U. student committed suicide last semester by jumping from a high floor of the Elmer Holmes Bobst Library. A second student also jumped from a high floor in the library, but the city medical examiner's office found that he had been on drugs and ruled his death an accident. They have not yet ruled on the death of a third student, a young woman who fell to her death from the window of a friend's apartment near Washington Square Park. . . . (Arenson, 2004)

Compare these two news accounts of a college student's suicide. One source is a tabloid; the other is a so-called mainstream national newspaper. Tabloids have a newspaper format but are smaller in size. They typically provide readers with a condensed version of the news and contain illustrated, often sensational material that they hope will encourage people to buy that day's paper at the newsstand or vending machine. It is no surprise that in its effort to attract readers, the tabloid placed a large, four-color photo of the student jumping to her death on the front page and suggested a cause of suicide (a fight with her boyfriend) that might entice readers. By contrast, the *New York Times* article begins with a description of how this student's death might affect other students and explains how school officials notified them of the tragedy. No picture was included with the coverage.

As these examples show, the media offer us different vantage points from which to view a given social event based on how they *frame* the information they provide to their audience. The term *media framing* refers to the process by which information and entertainment are packaged by the mass media (newspapers, magazines, radio and television networks and stations, and the Internet) before being presented to an audience. This process includes factors such as the amount of exposure given to a story, where it is placed, the positive or negative tone it conveys, and its accompanying headlines, photographs, or other visual or auditory effects (if any). Through framing, the media emphasize some beliefs and values over others and manipulate salience by directing people's attention to

Analyzing Content ***Content analysis* is the systematic examination of cultural artifacts or various forms of communication to extract thematic data and draw conclusions about social life.** *Cultural artifacts* are products of individual activity, social organizations, technology, and cultural patterns (Reinharz, 1992). Among the materials studied are *written records,* such as diaries, love letters, poems, books, and graffiti, and *narratives and visual texts,* such as movies, television programs, advertisements, and greeting cards. Also studied are *material culture,* such as music, art, and even garbage, and *behavioral residues,* such as patterns of wear and tear on the floors in front of various exhibits at museums to determine which exhibits are the most popular (see Webb et al., 1966). Harriet Martineau stated that more could be learned about a society in a day by studying "things" than by talking with individuals for a year (Martineau, 1988/1838). Researchers may look for regular patterns, such as frequency of suicide as a topic on television talk shows. They may also examine subject matter to determine how it has been handled, such as how the mass media handle the subject of suicide (see Box 2.3 for an example).

Content analysis provides objective coding procedures for analyzing written material (see Berg, 1998; Manning and Cullum-Swan, 1994). It also allows for the counting and arranging of data into clearly identifiable categories (manifest coding) and provides for the creation of analytically developed categories (latent or open coding). Using latent or open coding, it is possible to identify general themes, create generalizations, and develop "grounded theoretical" explanations (Glaser and Strauss, 1967). As this explanation suggests, researchers use both qualitative and quantitative procedures in content analysis.

How might a social scientist use content analysis in research on why people commit suicide? Suicide notes and diaries are useful forms of cultural artifacts. Suicide notes have been subjected to extensive analysis because

some ideas while ignoring others. As such, a frame constitutes a story line or an unfolding narrative about an issue. These narratives are organizations of experience that bring order to events. Consequently, such narratives wield power because they influence how we make sense of the world (Kendall, 2005).

Thinking sociologically, what problems exist in how the media frame stories about a social issue such as suicide? As discussed in this chapter, the early sociologist Emile Durkheim believed that we should view even seemingly individual actions—such as suicide—from a *sociological perspective* that focuses on the part that social groups and societies play in patterns of behavior rather than focusing on the *individual* attributes of people who commit such acts. If we use a sociological approach to analyzing how the media frame stories about suicide, here are three issues to consider:

- Do television and newspaper reports of suicides simplify the *reasons* for the suicide? Many factors interact in a complex manner to contribute to a person's decision to commit suicide, but one or more factors do not necessarily *cause* a suicide to occur. The media often report a final precipitating situation (such as a fight with a boyfriend or girlfriend, losing one's job, or getting a divorce) that distressed the individual prior to suicide but do not inform audiences that this was not the *only* cause of this suicide.
- Are readers and viewers provided with *repetitive, ongoing, and excessive reporting* on a suicide? Repeated sensationalistic framing of stories about suicide may contribute to *suicide contagion*. Some people are more at risk because of age, stress, and/or other personal problems. For example, *suicide clusters* (suicides occurring close together in time and location) are most common among people fifteen to twenty-four years of age who may not have known the original suicide victim and learned of the details from the media only.
- Do the media use dramatic photographs (such as of the person committing suicide or friends and family weeping and showing great emotion at the victim's funeral) primarily to sell papers or increase viewer ratings? Photographs and other sensational material are potentially most damaging for people who have long-standing mental health problems or who have limited social networks to provide them with hope, encouragement, and guidance (American Association of Suicidology, 2006).

In our mass-mediated culture, many sociologists agree that the media do much more than simply *mirror* society: The media help shape our society and our cultural perceptions on many issues, including seemingly individual behavior such as suicide.

Reflect & Analyze

Have you ever read two media accounts of the same event that were examples of very different approaches to framing? Are we influenced by how the media frame news and entertainment stories even when we think we are unaffected by such coverage? We will return to this issue in some later chapters.

they are "ultrapersonal documents" that are not solicited by others and frequently are written just before the person's death (Leenaars, 1988: 34). Many notes provide new levels of meaning regarding the *individuality* of the person who committed or attempted suicide. Suicide notes and diaries often reveal that people committing suicide consider their death as a "passing on to another world" or simply "escaping this world." Some notes indicate that people may want to get revenge and make other people feel guilty or responsible for their suicide: "Now you'll be sorry for what you did" or "It's all your fault!" Thus, suicide notes may be a valuable starting point for finding patterns of suicidal behavior and determining the characteristics of people who are most likely to commit suicide (Leenaars, 1988). Today, researchers analyze the suicide notes of both women and men. However, earlier studies of suicide notes primarily focused only on those written by men, even though women have been found to leave notes more often than do men (Lester, 1988, 1992).

Strengths and Weaknesses of Secondary Analysis

One strength of secondary analysis is that data are readily available and inexpensive. Another is that, because the researcher often does not collect the data personally, the chances of bias may be reduced. In addition, the use of existing sources makes it possible to analyze longitudinal data to provide a historical context within which to locate original research. However, secondary analysis has inherent problems. For one thing, the data may be incomplete, unauthentic, or inaccurate. A second issue is that the various data from which content analysis is done may not be strictly comparable with one another (Reinharz, 1992), and *coding* this data—

content analysis the systematic examination of cultural artifacts or various forms of communication to extract thematic data and draw conclusions about social life.

sorting, categorizing, and organizing them into conceptual categories—may be difficult (Babbie, 2004).

Field Research

***Field research* is the study of social life in its natural setting: observing and interviewing people where they live, work, and play.** Some kinds of behavior can be studied best by "being there"; a fuller understanding can be developed through observations, face-to-face discussions, and participation in events. Researchers use these methods to generate *qualitative* data: observations that are best described verbally rather than numerically. Although field research is less structured and more flexible than the other methods we have discussed, it still places many demands on the researcher. To engage in field research, sociologists must select the method or combination of methods that will best reveal what they want to know. For example, they must decide how to approach the target group, whether to identify themselves as researchers, and whether to participate in the events they are observing.

Participant Observation Sociologists who are interested in observing social interaction as it occurs may use participant observation. ***Participant observation* refers to the process of collecting data while being part of the activities of the group that the researcher is studying.** As this definition states, the researcher gains insight into some aspect of social life by participating in what is going on while observing what is taking place.

Let's assume that you wanted to study how volunteers at a suicide prevention center learned how to counsel people by telephone. You might become a volunteer-in-training and attend the orientation sessions for volunteers, taking notes on how others responded to the information being provided and how various volunteers interacted with one another. Then you might serve as a "hotline" volunteer, not only seeking to help the people who called but also observing how the volunteers interacted with callers. Throughout this process, of course, you would have to be aware of issues relating to ethics and with conducting research with human subjects, as discussed later in this chapter.

What are the strengths of participant observation research? Participant observation generates more "inside" information than simply asking questions or observing from the outside. For example, to learn more about how coroners make a ruling of "suicide" in connection with a death and to analyze what, if any, effect such a ruling has on the accuracy of "official" suicide statistics, sociologist Steve Taylor (1982) engaged in participant observation at a coroner's office over a six-month period. As he followed a number of cases from the initial report of death through the various stages of investigation, Taylor found that it was important to "be around" so that he could listen to the discussion of particular cases and ask the coroners questions. According to Taylor, intuition and guesswork play a much larger part in coroners' decisions than they are willing to acknowledge.

What are the limitations of participant observation research? This type of research requires time and expertise on the part of the researcher, who must be able to become a participant while still maintaining some distance from those being observed. Interpreting the results of this type of research also requires that the researchers distance themselves from people with whom they may have spent a great deal of time and with whom they developed personal relationships during the research project.

Case Studies Most participant observation research takes the form of a *case study,* which is often an in-depth, multifaceted investigation of a single event, person, or social grouping (Feagin, Orum, and Sjoberg, 1991). However, a case study may also involve multiple cases and is then referred to as a *collective case study* (Stake, 1995). Whether the case is single or collective, most case studies require detailed, in-depth data collection involving multiple sources of rich information such as documents and records and the use of methods such as participant observation, unstructured or in-depth interviews, and life histories (Creswell, 1998). As they collect extensive amounts of data, the researchers seek to develop a detailed description of the case, to analyze the themes or issues that emerge, and to interpret or create their own assertions about the case (Stake, 1995).

When do social scientists decide to do case studies? Initially, some researchers have only a general idea of what they wish to investigate. In other cases, they literally "back into" the research. They may find themselves close to interesting people or situations. For example, the anthropologist Elliot Liebow "backed into" his study of single, homeless women living in emergency shelters by becoming a volunteer at a shelter. As he got to know the women, Liebow became fascinated with their lives and survival strategies. Prior to Liebow's research, most studies of the homeless focused primarily on men. These studies typically asked questions such as "How many homeless are there?" and "What proportion of the homeless are chronically mentally ill?" By contrast, Liebow wanted to know more about the homeless women themselves, wondering such things as "What are they carrying in those [shopping] bags?" (Coughlin, 1993: A8). Liebow spent the next four years engaged in participant observation research that culminated in his book *Tell Them Who I Am* (1993).

© Ariel Skelley/Getty Images

Field research takes place in a wide variety of settings. For example, how might sociologists study the ways in which parents and their college-age children cope with change when the students first leave home and move into college housing?

In participant observation studies, the researcher must decide whether to let people know they are being studied. After Liebow decided that he would like to take notes on informal conversations and conduct interviews with the women, he asked the shelter director and the women for permission and told them that he would like to write about them. Liebow's findings are discussed in Chapter 5 ("Society, Social Structure, and Interaction"). Although some social scientists gain permission from their subjects, others fear that people will refuse to participate or will change their behavior if they know they are being observed. On the one hand, researchers who do not obtain consent from their subjects may be acting unethically. On the other hand, when subjects know they are being observed, they risk succumbing to the Hawthorne effect (discussed later in this chapter).

The next step is to gain the trust of participants. Liebow had previous experience in blending in with individuals he wanted to observe when he gained the trust of young, lower-class African American men who talked and passed time on an inner-city street corner in Washington, D.C., in the 1960s. In his classic study *Tally's Corner* (1967), Liebow described how he (as a thirty-seven-year-old white anthropology graduate student) played pool and drank beer with his subjects. While interacting with the men, Liebow gathered a large volume of data which led him to conclude that his subjects had created their own "society" after being unable to find a place in the existing one. Liebow found "insiders" to help him gain the trust of other participants in his research. In a participant observation study, you may wish to identify possible *informants*—individuals who introduce you to others, give you suggestions about how to "get around" in the natural setting, and provide you with essential insider information on what you are observing. Informants are especially useful in the community study/ethnography fields.

Ethnography An ***ethnography*** **is a detailed study of the life and activities of a group of people by researchers who may live with that group over a period of years** (Feagin, Orum, and Sjoberg, 1991). Although this approach is similar in some ways to participant observation, these studies typically take place over much longer periods of time. In fact, ethnography has been referred to as "the study of the way of life of a group of people" (Prus, 1996). For example, *Middletown* and *Middletown in Transition* describe the sociologists Robert Lynd and Helen Lynd's (1929, 1937) study in Muncie, Indiana. The Lynds, who lived in this midwestern town for a number of years, applied ethnographic research to the daily lives of residents, conducting interviews and reading newspaper files in order to build a historical base for their own research. The Lynds showed how a dominant family "ruled" the city and how the working class developed as a result of industry moving into Muncie. They concluded that the people had strong beliefs about the importance of religion, hard work, self-reliance, and civic pride. When a team of sociologists returned to Muncie in the late 1970s, they found that the people there still held these views (Bahr and Caplow, 1991).

field research the study of social life in its natural setting: observing and interviewing people where they live, work, and play.

participant observation a research method in which researchers collect data while being part of the activities of the group being studied.

ethnography a detailed study of the life and activities of a group of people by researchers who may live with that group over a period of years.

In another classic study, *Street Corner Society,* the sociologist William F. Whyte (1988/1943) conducted long-term participant observation studies in Boston's low-income Italian neighborhoods. Whereas "outsiders" generally regarded these neighborhoods as disorganized slums with high crime rates, Whyte found the residents to be hardworking people who tried to take care of one another. More recently, the sociologist Elijah Anderson (1990) conducted a study in two Philadelphia neighborhoods—one populated by low-income African Americans, the other racially mixed but becoming increasingly middle- to upper-income and white. Over the course of fourteen years, Anderson spent numerous hours on the streets, talking and listening to the people (Anderson, 1990: ix). In this longitudinal study, Anderson was able to document the changes brought about by drug abuse, loss of jobs, decreases in city services despite increases in taxes, and the eventual exodus of middle-income people. As these examples show, ethnographic work involves not only immersing oneself into the group or community that the researcher studies but also engaging in dialogue to learn more about social life through ongoing interaction with others (Burawoy, 1991).

Unstructured Interviews An ***unstructured interview* is an extended, open-ended interaction between an interviewer and an interviewee.** This type of interview is referred to as an *unstructured,* or *nonstandardized, interview* because few predetermined or standardized procedures are established for conducting it. Because many decisions have to be made during the interview, this approach requires that the researcher have a high level of skill in interviewing and extensive knowledge regarding the interview topic (Kvale, 1996). Here, the interviewer has a general plan of inquiry but not a specific set of questions that must be asked, as is often the case with surveys. Unstructured interviews are essentially conversations in which interviewers establish the general direction by asking open-ended questions, to which interviewees may respond flexibly. Interviewers have the ability to "shift gears" to pursue specific topics raised by interviewees because answers to one question are used to suggest the next question or new areas of inquiry.

Sociologist Joe Feagin's (1991) study of middle-class African Americans is an example of research that used in-depth interviews to examine public discrimination and victims' coping strategies. No specific questions were asked regarding discrimination in public accommodations or other public places. Rather, discussion of discrimination was generated by answers to general questions about barriers to personal goals or in digressions in answers to specific questions about employment, education, and housing (Feagin, 1991).

Even in unstructured interviews, researchers must prepare a few general or "lead-in" questions to get the interview started. Following the interviewee's initial responses, the interviewer may wish to ask additional questions on the same topic, probe for more information (by using questions such as "In what ways?" or "Anything else?"), or introduce a new line of inquiry. At all points in the interview, *careful listening* is essential. It provides the opportunity to introduce new questions as the interview proceeds while simultaneously keeping the interview focused on the research topic. It also enables the interviewer to envision the interviewees' experiences and to glean multiple levels of meaning.

The Interview and Sampling Process Before conducting in-depth interviews, researchers must make a number of decisions, including how the people to be interviewed will be selected. Respondents for unstruc-

© Addison Geary

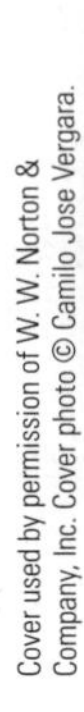

Elijah Anderson (at left in photo) conducted an ethnographic study of two very different Philadelphia neighborhoods that became the basis for his landmark study *Code of the Street.* What can researchers learn from ethnographic research that might be less apparent if they used other methods to study human behavior?

tured interviews are often chosen by "snowball sampling." In *snowball sampling,* the researcher interviews a few individuals who possess a certain characteristic; these interviewees are then asked to supply the names of others with the same characteristic. The process continues until the sample has "snowballed" into an acceptable size and no new information of any significance is being gained.

Researchers must make other key decisions. Will people be interviewed more than once? If so, how long will the interviews be? Are there a specific number and order of questions to be followed? Will the interviewees have an opportunity to question the interviewer? Where will the interview take place? How will information be recorded? Who should do the interviewing? Who should be present at the interview? (Reinharz, 1992; Kvale, 1996). Unstructured, open-ended interviews do not mean that the researcher simply walks into a room, has a conversation with someone, and the research is complete. Planning and preparation are essential. Similarly, the follow-up, analysis of data, and write-up of the study must be carefully designed and carried out.

Interviews and Theory Construction In-depth interviews, along with participant observation and case studies, are frequently used to develop theories through observation. The term *grounded theory* was developed by sociologists Barney Glaser and Anselm Strauss (1967) to describe this inductive method of theory construction. Researchers who use grounded theory collect and analyze data simultaneously. For example, after in-depth interviews with 106 suicide attempters, researchers in one study concluded that half of the individuals who attempted suicide wanted *both* to live *and* to die at the time of their attempt. From these unstructured interviews, it became obvious that ambivalence led about half of "serious" suicidal attempters to "literally gamble with death" (Kovacs and Beck, 1977, qtd. in Taylor, 1982: 144). After asking their initial unstructured questions of the interviewees, Kovacs and Taylor decided to widen the research question from "Why do people kill themselves?" to a broader question: "Why do people engage in acts of self damage which may result in death?" In other words, uncertainty of outcome is a common feature of most suicidal acts. In previous studies, researchers had simply assumed that in "dangerous attempts" the individual really wanted to die whereas in "moderate" attempts the person was ambivalent (Taylor, 1982: 160).

Strengths and Weaknesses of Field Research Participant observation research, case studies, ethnography, and unstructured interviews provide opportunities for researchers to view from the inside what may not be obvious to an outside observer. They are useful when attitudes and behaviors can be understood best within their natural setting or when the researcher wants to study social processes and change over a period of time. They provide a wealth of information about the reactions of people and give us an opportunity to generate theories from the data collected (Whyte, 1989). For example, through unstructured interviews, researchers gain access to "people's ideas, thoughts, and memories in their own words rather than in the words of the researcher" (Reinharz, 1992: 19). Research of this type is important for the study of race, ethnicity, and gender because it often includes those who have been previously excluded from studies and provides information about them.

Social scientists who believe that quantitative research methods (such as survey research) provide the most scientific and accurate means of measuring attitudes, beliefs, and behavior are often critical of data obtained through field research. They argue that what is learned from a specific group or community cannot be generalized to a larger population. They also suggest that the data collected in natural settings are descriptive and do not lend themselves to precise measurement. Researchers who want to determine cause and effect or to test a theory emphasize that it is impossible to demonstrate such relationships from participant observation studies. For these reasons and others, some qualitative researchers (particularly ethnographers) use computer-assisted qualitative data analysis (CAQDA) programs. Such programs make it easier for researchers to enter, organize, annotate, code, retrieve, count, and analyze data (Dohan and Sanchez-Jankowski, 1998). However, other ethnographers and field researchers do not use CAQDA programs in their research (see Charmaz and Olesen, 1997; Horowitz, 1997; and Morrill and Fine, 1997).

Experiments

An ***experiment*** **is a carefully designed situation in which the researcher studies the impact of certain variables on subjects' attitudes or behavior.** Experiments are designed to create "real-life" situations, ideally

unstructured interview an extended, open-ended interaction between an interviewer and an interviewee.

experiment a research method involving a carefully designed situation in which the researcher studies the impact of certain variables on subjects' attitudes or behavior.

under controlled circumstances, in which the influence of different variables can be modified and measured.

Types of Experiments Conventional experiments require that subjects be divided into two groups: an experimental group and a control group. The ***experimental group* contains the subjects who are exposed to an independent variable** (the experimental condition) to study its effect on them. The ***control group* contains the subjects who are not exposed to the independent variable.** The members of the two groups are matched for similar characteristics so that comparisons may be made between the groups. In the simplest experimental design, subjects are (1) pretested (measured) in terms of the dependent variable in the hypothesis, (2) exposed to a stimulus representing an independent variable, and (3) post-tested (remeasured) in terms of the dependent variable. The experimental and control groups are then compared to see if they differ in relation to the dependent variable, and the hypothesis stating the relationship of the two variables is confirmed or rejected.

In a *laboratory experiment,* subjects are studied in a closed setting so that researchers can maintain as much control as possible over the research. For example, if you wanted to examine the influence of the media on attitudes regarding suicide, you might decide to use a laboratory experiment. Sociologist Arturo Biblarz and colleagues (1991) designed a laboratory study to investigate the effects of the media on people's attitudes toward suicide. Researchers showed one group of subjects a film about suicide, showed a second group a film about violence, and showed a third a film containing neither suicide nor violence. Some evidence was found that media exposure to suicidal acts or violence may arouse an emotional state favorable to suicidal behavior, especially in those persons already "at risk" for suicide.

Not all experiments occur in laboratory settings. *Natural experiments* are real-life occurrences such as floods and other disasters that provide researchers with "living laboratories." Sociologist Kai Erikson (1976) studied the consequences of a deadly 1972 flood in Buffalo Creek, West Virginia, and found that extensive disruption of community ties occurred. Natural experiments cannot be replicated because it is impossible to re-create the exact conditions, nor would we want to do so.

Demonstrating Cause-and-Effect Relationships Researchers may use experiments when they want to demonstrate that a cause-and-effect relationship exists between variables. In order to show that a change in one variable causes a change in another, these three conditions must be fulfilled:

1. *You must show that a correlation exists between the two variables.* ***Correlation* exists when two variables are associated more frequently than could be expected by chance** (Hoover, 1992). For example, suppose that you wanted to test the hypothesis that the availability of a crisis intervention center with a twenty-four-hour counseling "hotline" on your campus causes a change in students' attitudes toward suicide (see ▶ Figure 2.4). To demonstrate correlation, you would need to show that the students had different attitudes toward committing suicide depending on whether they had any experience with the crisis intervention center.
2. *You must ensure that the independent variable preceded the dependent variable.* If differences in students' attitudes toward suicide were evident before the students were exposed to the intervention center, exposure to the center could not be the cause of these differences.
3. *You must make sure that any change in the dependent variable was not due to an extraneous variable*—one outside the stated hypothesis. If some of the students receive counseling from off-campus psychiatrists, any change in attitude that they experience could be due to this third variable and not to the hotline. This is referred to as a *spurious correlation*—the association of two variables that is actually caused by a third variable and does not demonstrate a cause-and-effect relationship.

Do extremely violent video games cause an increase in violent tendencies in their users? Experiments are one way to test this hypothesis.

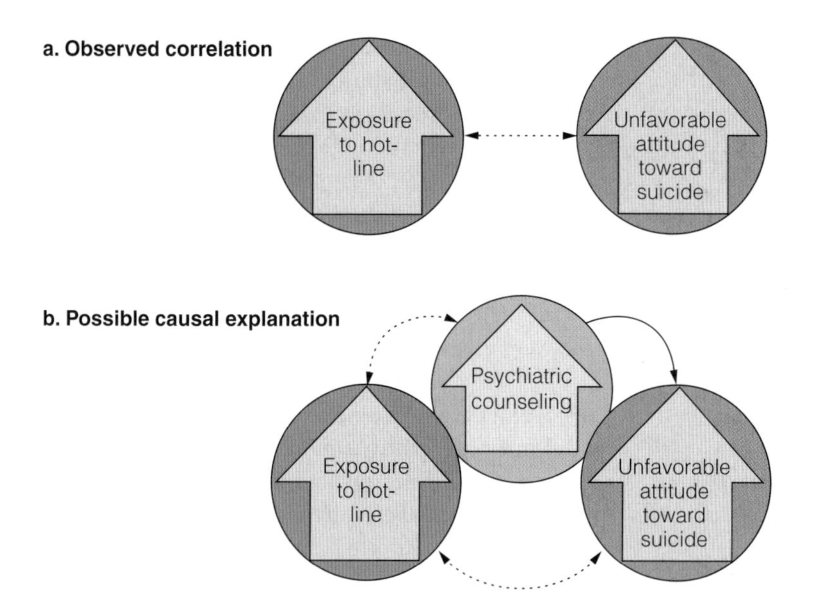

▶ **Figure 2.4 Correlation Versus Causation**
A study might find that exposure to a suicide hotline is associated (correlated) with a change in attitude toward suicide. But if some of the people who were exposed to the hotline also received psychiatric counseling, the counseling may be the "hidden" cause of the observed change in attitude. In general, correlations alone do not prove causation.

Strengths and Weaknesses of Experiments The major advantage of the controlled experiment is the researcher's control over the environment and the ability to isolate the experimental variable. Because many experiments require relatively little time and money and can be conducted with limited numbers of subjects, it is possible for researchers to replicate an experiment several times by using different groups of subjects. Replication strengthens claims about the validity and generalizability of the original research findings (Babbie, 2004).

Perhaps the greatest limitation of experiments is that they are artificial. Social processes that occur in a laboratory setting often do not occur in the same way in real-life settings. For example, social scientists frequently rely on volunteers or captive audiences. As a result, the subjects of most experiments may not be representative of a larger population, and the findings cannot be generalized to other groups.

Experiments have several other limitations. First, the rigid control and manipulation of variables demanded by experiments do not allow for a more communal approach to data gathering. Second, biases can influence each of the stages in an experiment, and research subjects may become the objects of sex/class/race biases. Third, the unnatural characteristics of laboratory experiments and of group competition in such settings have a negative effect on subjects (Reinharz, 1992).

Researchers acknowledge that experiments have the additional problem of *reactivity*—the tendency of subjects to change their behavior in response to the researcher or to the fact that they know they are being studied. This problem was first noted in a study conducted between 1927 and 1932 by the social psychologist Elton Mayo, who used a series of experiments to determine how worker productivity and morale might be improved at Western Electric's Hawthorne plant. To identify variables that tend to increase worker productivity, Mayo separated one group of women (the experimental group) from the other workers and then systematically varied factors in that group's work environment while closely observing them. Meanwhile, the working conditions of the other workers (the control group) were not changed. The researchers tested a number of hypotheses, including one stating that an increase in the amount of lighting would raise the workers' productivity. Much to the researchers' surprise, the level of productivity rose not only when the lighting was brightened but also when it was dimmed.

Indeed, all of the changes increased productivity. Mayo concluded that the subjects were trying to please the researchers because of the interest being shown in the subjects (Roethlisberger and Dickson, 1939). Thus, the ***Hawthorne effect*** **refers to changes in the subject's behavior caused by the researcher's presence or by the subject's awareness of being studied.**

experimental group in an experiment, the group that contains the subjects who are exposed to an independent variable (the experimental condition) to study its effect on them.

control group in an experiment, the group containing the subjects who are not exposed to the independent variable.

correlation a relationship that exists when two variables are associated more frequently than could be expected by chance.

Hawthorne effect a phenomenon in which changes in a subject's behavior are caused by the researcher's presence or by the subject's awareness of being studied.

Other aspects of this study are discussed in Chapter 6 ("Groups and Organizations").

Multiple Methods: Triangulation

What is the best method for studying a particular topic? How can we get accurate answers to questions about suicide and other important social concerns? The Concept Quick Review compares the various social research methods. There is no one best research method because of the "complexity of social reality and the limitations of all research methodologies" (Snow and Anderson, 1991: 158).

Many sociologists believe that it is best to combine multiple methods in a given study. *Triangulation* is the term used to describe this approach (Denzin, 1989). Triangulation refers not only to research methods but also to multiple data sources, investigators, and theoretical perspectives in a study. Multiple data sources include persons, situations, contexts, and time (Snow and Anderson, 1991). For example, in a study of "unattached homeless men and women living in and passing through Austin, Texas, in the mid-1980s," sociologists David Snow and Leon Anderson (1991: 158) used as their primary data sources "the homeless themselves and the array of settings, agency personnel, business proprietors, city officials, and neighborhood activities relevant to the routines of the homeless." Snow and Anderson gained a detailed portrait of the homeless and their experiences and institutional contacts by tracking more than seven hundred homeless individuals through a network of seven institutions with which they had varying degrees of contact.

The study also tracked a number of the individuals over a period of time and used a variety of methods, including "participant observation and informal, conversational interviewing with the homeless; participant and nonparticipation observation, coupled with formal and informal interviewing in street agencies and settings; and a systematic survey of agency records" (Snow and Anderson, 1991: 158–169). This study is discussed in depth in Chapter 5 ("Society, Social Structure, and Interaction").

Multiple methods and approaches provide a wider scope of information and enhance our understanding of critical issues. Many researchers also use multiple

CONCEPT QUICK REVIEW

Strengths and Weaknesses of Social Research Methods

Research Method	Strengths	Weaknesses
Experiments (Laboratory, Field, Natural)	Control over research Ability to isolate experimental factors Relatively little time and money required Replication possible, except for natural experiments	Artificial by nature Frequent reliance on volunteers or captive audiences Ethical questions of deception
Survey Research (Questionnaire, Interview, Telephone Survey)	Useful in describing features of a large population without interviewing everyone Relatively large samples possible Multivariate analysis possible	Potentially forced answers Respondent untruthfulness on emotional issues Data that are not always "hard facts" presented as such in statistical analyses
Secondary Analysis of Existing Data (Existing Statistics, Content Analysis)	Data often readily available, inexpensive to collect Longitudinal and comparative studies possible Replication possible	Difficulty in determining accuracy of some of the data Failure of data gathered by others to meet goals of current research Questions of privacy when using diaries, other personal documents
Field Research (Participant Observation, Case Study, Ethnography, Unstructured Interview)	Opportunity to gain insider's view Useful for studying attitudes and behavior in natural settings Longitudinal/comparative studies possible Documentation of important social problems of excluded groups possible Access to people's ideas in their words Forum for previously excluded groups Documentation of need for social reform	Problems in generalizing results to a larger population Nonprecise data measurements Inability to demonstrate cause/effect relationship or test theories Difficult to make comparisons because of lack of structure Not a representative sample

Multiple research methods are often used to gain information about important social concerns. Which methods might be most effective in learning more about the problems of the homeless, such as these street people warming themselves on a warm grate in Moscow, Russia?

methods to validate or refine one type of data by use of another type.

Ethical Issues in Sociological Research

The study of people ("human subjects") raises vital questions about ethical concerns in sociological research. Beginning in the 1960s, the U.S. government set up regulations for "the protection of human subjects." Because of scientific abuses in the past, researchers are now mandated to weigh the societal benefits of research against the potential physical and emotional costs to participants. Researchers are required to obtain written "informed consent" statements from the persons they study. However, these guidelines have produced many new questions. What constitutes "informed consent"? What constitutes harm to a person? How do researchers protect the identity and confidentiality of their sources?

The ASA *Code of Ethics*

The American Sociological Association (ASA) *Code of Ethics* (1997) sets forth certain basic standards that sociologists must follow in conducting research:

1. Researchers must endeavor to maintain objectivity and integrity in their research by disclosing their research findings in full and including all possible interpretations of the data (even those interpretations that do not support their own viewpoints).
2. Researchers must safeguard the participants' right to privacy and dignity while protecting them from harm.
3. Researchers must protect confidential information provided by participants, even when this information is not considered to be "privileged" (legally protected, as is the case between doctor and patient and between attorney and client) and legal pressure is applied to reveal this information.
4. Researchers must acknowledge research collaboration and assistance they receive from others and disclose all sources of financial support.

Sociologists are obligated to adhere to this code and to protect research participants; however, many ethical issues arise that cannot be easily resolved. Ethics in sociological research is a difficult and often ambiguous topic. But ethical issues cannot be ignored by researchers, whether they are sociology professors, graduate students conducting investigations for their dissertations, or undergraduates conducting a class research project. Sociologists have a burden of "self-reflection" —of seeking to understand the role they play in contemporary social processes while at the same time assessing how these social processes affect their findings (Gouldner, 1970).

How honest do researchers have to be with potential participants? Let's look at two specific cases in point. Where does the "right to know" end and the "right to privacy" begin in these situations?

The Zellner Research

Sociologist William Zellner (1978) sought to interview the family, friends, and acquaintances of persons killed in single-car crashes that he thought might have been "autocides." Zellner wondered if some automobile "accidents" were actually suicides—instances in which the individual wished to protect other people and perhaps make it easier for them to collect insurance benefits that might not be paid if the death was a suicide. By interviewing people who knew the victims, Zellner hoped to obtain information that would help determine if the deaths were accidental or intentional. To recruit respondents, he suggested that their

Box 2.4 You Can Make a Difference

Responding to a Cry for Help

Chad felt that he knew Frank quite well. After all, they had been roommates for two years at State U. As a result, Chad was taken aback when Frank became very withdrawn, sleeping most of the day and mumbling about how unhappy he was. One evening, Chad began to wonder whether he needed to do something because Frank had begun to talk about "ending it all" and saying things like "the world will be better off without me." If you were in Chad's place, would you know the warning signs that you should look for? Do you know what you might do to help someone like Frank?

The American Foundation for Suicide Prevention, a national nonprofit organization dedicated to funding research, education, and treatment programs for depression and suicide prevention, suggests that each of us should be aware of these warning signs of suicide:

- *Talking about death or suicide.* Be alert to such statements as "Everyone would be better off without me." Sometimes, individuals who are thinking about suicide speak as if they are saying good-bye.
- *Making plans.* The person may do such things as giving away valuable items, paying off debts, and otherwise "putting things in order."
- *Showing signs of depression.* Although most depressed people are not suicidal, most suicidal people are depressed. Serious depression tends to be expressed as a loss of pleasure or withdrawal from activities that a person has previously enjoyed. It is especially important to note whether five of the following symptoms are present almost every day for several weeks: change in appetite or weight, change in sleeping patterns, speaking or moving with unusual speed or slowness, loss of interest in usual activities, decrease in sexual drive, fatigue, feelings of guilt or worthlessness, and indecisiveness or inability to concentrate.

The possibility of suicide must be taken seriously: Most people who commit suicide give some warning to family members or friends. Instead of attempting to argue the person out of suicide or saying "You have so much to live for," let the person know that you care and understand, and that

participation in his study might reduce the number of accidents in the future; however, he did not mention that he suspected autocide. In each interview, he asked if the deceased had recently talked about suicide or about himself or herself in a negative manner.

From the data he collected, Zellner concluded that at least 12 percent of the fatal single-occupant crashes were suicides. He also learned that in a number of the crashes, other people (innocent bystanders) were killed or critically injured. Was Zellner's research unethical because he misrepresented the reasons for his study? In this situation, does the right to know outweigh the right to privacy?

The Humphreys Research

Laud Humphreys (1970), then a sociology graduate student, decided to study homosexual conduct as a topic for his doctoral dissertation. His research focused on homosexual acts between strangers meeting in "tearooms," public restrooms in parks. He did *not* ask permission of his subjects, nor did he inform them that they were being studied. Instead, Humphreys showed up at public restrooms that were known to be tearooms and offered to be the lookout while others engaged in homosexual acts. Then he systematically recorded the encounters that took place.

Humphreys was interested in the fact that the tearoom participants seemed to live "normal" lives apart from these encounters, and he decided to learn more about their everyday lives. To determine who they were, he wrote down their auto license numbers and tracked down their names and addresses. Later, he arranged for these men to be included in a medical survey so that he could go out and interview them personally. He wore different disguises and drove a different car so that they would not recognize him (Henslin, 1997). From these interviews, he collected personal information and determined that most of the men were married and lived very conventional lives.

Would Humphreys have gained access to these subjects if he had identified himself as a researcher? Probably not—nevertheless, the fact that he did not do so produced widespread criticism from sociologists and journalists. Despite the fact that his study, *Tearoom Trade* (1970), won an award for its scholarship, the controversy surrounding his project was never fully resolved.

In this chapter, we have looked at the research process and the methods used to pursue sociological knowledge. We have also critiqued many of the existing approaches and suggested alternate ways of pursuing research. The important thing to realize is that re-his or her problems can be solved. Urge the person to see a school counselor, a physician, or a mental health professional immediately. If you think the person is in imminent danger of committing suicide, you should take the person to an emergency room or a walk-in clinic at a psychiatric hospital. It is best to remain with the person until help is available.

© Geri Engberg/The Image Works

Can a suicide crisis center prevent a person from committing suicide? People who understand factors that contribute to suicide may be able to better counsel those who call for help.

For more information about suicide prevention, contact the following organizations:

- American Foundation for Suicide Prevention, 120 Wall Street, 22nd Floor, New York, NY 10005 (888-333-2377). AFSP is a leading not-for-profit organization dedicated to understanding and preventing suicide through research and education.
- Suicide Awareness Voices of Education (**http://www.save.org**) is a resource index with links to other valuable resources, such as "Questions Most Frequently Asked on Suicide," "Symptoms of Depression and Danger Signs of Suicide," and "What to Do If Someone You Love Is Suicidal."
- Befrienders Worldwide (**http://www.befrienders.org**) is a website providing information for anyone feeling depressed or suicidal or who is worried about a friend or relative who feels that way. Includes a directory of suicide and crisis helplines.

search is the "lifeblood" of sociology. Theory provides the framework for an analysis, and research takes us beyond common sense and provides opportunities for us to use our sociological imagination to generate new knowledge. For example, as we have seen in this chapter, suicide cannot be explained by common sense or a few isolated variables. In answering questions such as "Why do people commit suicide?" we have to take into account many aspects of personal choice and social structure that are related to one another in extremely complex ways. Research can help us unravel the complexities of social life if sociologists observe, talk to, and interact with people in real-life situations (Feagin, Orum, and Sjoberg, 1991).

Our challenge today is to find new ways to integrate knowledge and action and to include all people in the research process in order to help fill the gaps in our existing knowledge about social life and how it is shaped by gender, race, class, age, and the broader social and cultural contexts in which everyday life occurs (Cancian, 1992). Each of us can and should find new ways to integrate knowledge and action into our daily lives (see Box 2.4).

Chapter Review

• How does sociological research differ from commonsense knowledge?

Sociological research provides a factual and objective counterpoint to commonsense knowledge and ill-informed sources of information. It is based on an empirical approach that answers questions through a direct, systematic collection and analysis of data.

• What is the relationship between theory and research?

Theory and research form a continuous cycle that encompasses both deductive and inductive approaches. With the deductive approach, the researcher begins with a theory and then collects and analyzes research to test it. With the inductive approach, the researcher collects and analyzes data and then generates a theory based on that analysis.

- **How does quantitative research differ from qualitative research?**

Quantitative research focuses on data that can be measured numerically (comparing rates of suicide, for example). Qualitative research focuses on interpretive description (words) rather than statistics to analyze underlying meanings and patterns of social relationships.

- **What are the key steps in the conventional research process?**

A conventional research process based on deduction and the quantitative approach has these key steps: (1) selecting and defining the research problem; (2) reviewing previous research; (3) formulating the hypothesis, which involves constructing variables; (4) developing the research design; (5) collecting and analyzing the data; and (6) drawing conclusions and reporting the findings.

- **What steps are often taken by researchers using the qualitative approach?**

A researcher taking the qualitative approach might (1) formulate the problem to be studied instead of creating a hypothesis, (2) collect and analyze the data, and (3) report the results.

- **What are the major types of research methods?**

The main types of research methods are surveys, secondary analysis of existing data, field research, and experiments. Surveys are polls used to gather facts about people's attitudes, opinions, or behaviors; a representative sample of respondents provides data through questionnaires or interviews. In secondary analysis, researchers analyze existing data, such as a government census, or cultural artifacts, such as a diary. In field research, sociologists study social life in its natural setting through participant observation, case studies, unstructured interviews, and ethnography. Through experiments, researchers study the impact of certain variables on their subjects.

- **What ethical issues are involved in sociological research?**

Because sociology involves the study of people ("human subjects"), researchers are required to obtain the informed consent of the people they study; however, in some instances what constitutes "informed consent" may be difficult to determine.

www.cengage.com/login

Want to maximize your online study time? Take this easy-to-use study system's diagnostic pre-test, and it will create a personalized study plan for you. By helping you to identify the topics that you need to understand better and then directing you to valuable online resources, it can speed up your chapter review. CengageNOW even provides a post-test so you can confirm that you are ready for an exam.

Key Terms

content analysis 56
control group 62
correlation 62
dependent variable 45
ethnography 59
experiment 61
experimental group 62
field research 58
Hawthorne effect 63
hypothesis 45
independent variable 45
interview 53
participant observation 58
probability sampling 47
questionnaire 52
random sampling 46
reliability 47
research methods 51
respondents 52
secondary analysis 54
survey 51
unstructured interview 60
validity 47

Questions for Critical Thinking

1. The agency that funds the local suicide clinic has asked you to study the clinic's effectiveness in preventing suicide. What would you need to measure? What can you measure? What research method(s) would provide the best data for analysis?
2. Recent studies have suggested that groups with high levels of *suicide acceptability* (holding the belief that suicide is an acceptable way to end one's life under certain circumstances) tend to have a higher than average suicide risk (Stack and Wasserman, 1995; Stack, 1998). What implications might such findings have on public policy issues such as the legalization of physician-assisted suicide and euthanasia? What implications might the findings have on an individual who is thinking about committing suicide? Analyze your responses using a sociological perspective.
3. In high-income nations, computers have changed many aspects of people's lives. Thinking about the various research methods discussed in this chapter, which approaches do you believe would be most affected by greater reliance on computers for collecting, organizing, and analyzing data? What are the advantages and limitations of conducting sociological research via the Internet?

The Kendall Companion Website

www.cengage.com/sociology/kendall

Visit this book's companion website, where you'll find more resources to help you study and successfully complete course projects. Resources include quizzes and flash cards, as well as special features such as an interactive sociology timeline, maps, General Social Survey (GSS) data, and Census 2000 data. The site also provides links to useful websites that have been selected for their relevance to the topics in this chapter and include those listed below. (*Note:* Visit the book's website for updated URLs.)

The Gallup Organization

http://www.gallup.com

Perhaps best known for its Gallup poll, the Gallup Organization specializes in gauging and understanding human attitudes and behaviors. Visit this site and click on "Gallup Poll News Service" to access public opinion surveys on a wide variety of topics. You can look through an alphabetical index of poll topics or conduct a keyword search.

Research Methods and Statistics

http://www.trinity.edu/mkearl/methods.html#ms

Visit Professor Michael C. Kearl's research methods and statistics website to access resources on a wide range of topics, including strategies for data collection, tactics for improving statistical and methodological skills, and information on how to write a research paper.

ASA Code of Ethics

http://www.asanet.org/cs/root/leftnav/ethics/ethics

Learn more about the principles and ethical standards that sociologists must follow in conducting research by visiting the complete online version of the American Sociological Association's *Code of Ethics* (1997).

chapter

3 Culture

Chapter Focus Question

What part does culture play in shaping people and the social relations in which they participate?

At home, I kept opening the refrigerator and cupboards, wishing for American foods to magically appear. I wanted what the other kids had: Bundt cakes and casseroles, Cheetos and Doritos. . . . The more American foods I ate, the more my desires multiplied, outpacing my interest in Vietnamese food. I had memorized the menu at Dairy Cone, the sugary options in the cereal aisle at Meijer's [grocery], and every inch of the candy display at Gas City: the rows of gum, the rows of chocolate, the rows without chocolate. . . . I knew Reese's peanut butter cups, Twix, Heath Crunch, Nestlé Crunch, Baby Ruth, Bar None, Oh Henry!, Mounds and Almond Joy, Snickers, Mr. Goodbar[,] . . . Milk Duds, [and] Junior Mints. I dreamed of taking it all, plus the freezer full of Popsicles and nutty,

© Bob Daemmrich/The Image Works

How is the food that we consume linked to our identity and the larger culture of which we are a part? Do people who identify with more than one culture face more-complex issues when it comes to food preferences?

IN THIS CHAPTER

- Culture and Society in a Changing World
- Components of Culture
- Technology, Cultural Change, and Diversity
- A Global Popular Culture?
- Sociological Analysis of Culture
- Culture in the Future

chocolate-coated ice cream drumsticks. I dreamed of Little Debbie, Dolly Madison, Swiss Miss, all the bakeries presided over by prim and proper girls.

—Bich Minh Nguyen (2007: 50–51), an English professor at Purdue University, describing how food served as a powerful cultural symbol in her childhood as a Vietnamese American

Growing up in Oakland . . . I came to dislike Chinese food. That may have been, in part, because I was Chinese and desperately wanted to be American. I *was* American, of course, but being born and raised in Chinatown—in a restaurant my parents operated, in fact—I didn't feel much like the people I saw outside Chinatown, or in books and movies.

It didn't help that for lunch at school, my mother would pack—*Ai ya!*—Chinese food. Barbecued pork sandwiches, not ham and cheese; Chinese pears, not apples. At home—that is, at the New Eastern Café—it was Chinese food night after night. No wonder I would sneak off, on the way to Chinese school, to Hamburger Gus for a helping of thick-cut French fries.

—author Ben Fong-Torres (2007: 11) describing his experiences as a Chinese American who desired to "Americanize" his eating habits

Sharpening Your Focus

- *What are the essential components of culture?*
- *To what degree are we shaped by popular culture?*
- *How do subcultures and countercultures reflect diversity within a society?*
- *How do the various sociological perspectives view culture?*

Why are these authors concerned about the food they ate as children? For all of us, the food we consume is linked to our identity and to the larger culture of which we are a part. For people who identify with more than one culture, food and eating patterns may become a very complex issue. To some people, food consumption is nothing more than how we meet a basic biological need; however, many sociologists are interested in the sociology of food and eating because of their cultural significance in our lives (see Mennell, 1996; Mennell, Murcott, and van Otterloo, 1993).

What is culture? ***Culture* is the knowledge, language, values, customs, and material objects that are passed from person to person and from one generation to the next in a human group or society.** As previously defined, a *society* is a large social grouping that occupies the same geographic territory and is subject to the same political authority and dominant cultural expectations. Whereas a society is composed of people, a culture is composed of ideas, behavior, and material possessions. Society and culture are interdependent; neither could exist without the other.

In this chapter, we examine society and culture, with special attention to how our material culture, including the food we eat, is related to our beliefs, values, and actions. We also analyze culture from functionalist, conflict, symbolic interactionist, and postmodern perspectives. Before reading on, test your knowledge of food and culture by answering the questions in Box 3.1.

a. HORNS: "Hook 'em Horns" or "your spouse is unfaithful"

b. CIRCLE: "OK (absolutely fine)" or "I'll kill you"

c. THUMBS UP: "Great," or an obscenity

▶ **Figure 3.1 Hand Gestures with Different Meanings**

As international travelers and businesspeople have learned, hand gestures may have very different meanings in different cultures.

Culture and Society in a Changing World

Understanding how culture affects our lives helps us develop a sociological imagination. When we meet someone from a culture vastly different from our own, or when we travel in another country, it may be easier to perceive the enormous influence of culture on people's lives. However, as our society has become more diverse, and communication among members of international cultures more frequent, the need to appreciate diversity and to understand how people in other cultures view their world has also increased (Samovar and Porter, 1991b). For example, many international travelers and businesspeople have learned the importance of knowing what gestures mean in various nations (see ▶ Figure 3.1). Although the "hook 'em Horns" sign—the pinky and index finger raised up and the middle two fingers folded down—is used by fans to express their support for University of Texas at Austin sports teams, for millions of Italians the same gesture means "Your spouse is being unfaithful." In Argentina, rotating one's index finger around the front of the ear means "You have a telephone call," but in the United States it usually suggests that a person is "crazy" (Axtell, 1991). Similarly, making a circle with

Box 3.1 Sociology and Everyday Life

How Much Do You Know About Global Food and Culture?

True	False	
T	F	1. Cheese is a universal food enjoyed by people of all nations and cultures.
T	F	2. Giving round-shaped foods to the parents of new babies is considered to be lucky in some cultures.
T	F	3. Wedding cakes are made of similar ingredients in all countries, regardless of culture or religion.
T	F	4. Food is an important part of religious observance for many different faiths.
T	F	5. In authentic Chinese cuisine, cooking methods are divided into "yin" and "yang" qualities.
T	F	6. Because of the fast pace of life in the United States, virtually everyone relies on mixes and instant foods at home and fast foods when eating out.
T	F	7. Potatoes are the most popular mainstay in the diet of first- and second-generation immigrants who have arrived in the United States over the past forty years.
T	F	8. According to sociologists, individuals may be offended when a person from another culture does not understand local food preferences or the cultural traditions associated with eating, even if the person is obviously an "outsider" or a "tourist."

Answers on page 74.

your thumb and index finger indicates "OK" in the United States, but in Tunisia it means "I'll kill you!" (Samovar and Porter, 1991a).

The Importance of Culture

How important is culture in determining how people think and act on a daily basis? Simply stated, culture is essential for our individual survival and for our communication with other people. We rely on culture because we are not born with the information we need to survive. We do not know how to take care of ourselves, how to behave, how to dress, what to eat, which gods to worship, or how to make or spend money. We must learn about culture through interaction, observation, and imitation in order to participate as members of the group. Sharing a common culture with others simplifies day-to-day interactions. However, we must also understand other cultures and the world views therein.

Just as culture is essential for individuals, it is also fundamental for the survival of societies. Culture has been described as "the common denominator that makes the actions of individuals intelligible to the group" (Haviland, 1993: 30). Some system of rule making and enforcing necessarily exists in all societies. What would happen, for example, if *all* rules and laws in the United States suddenly disappeared? At a basic level, we need rules in order to navigate our bicycles and cars through traffic. At a more abstract level, we need laws to establish and protect our rights.

In order to survive, societies need rules about civility and tolerance toward others. We are not born knowing how to express kindness or hatred toward others, although some people may say "Well, that's just human nature" when explaining someone's behavior. Such a statement is built on the assumption that what we do as human beings is determined by *nature* (our biological and genetic makeup) rather than *nurture* (our social environment)—in other words, that our behavior is instinctive. An *instinct* is an unlearned, biologically determined behavior pattern common to all members of a species that predictably occurs whenever certain environmental conditions exist. For example, spiders do not learn to build webs. They build webs because of instincts that are triggered by basic biological needs such as protection and reproduction.

Humans do not have instincts. What we most often think of as instinctive behavior can actually be attributed to reflexes and drives. A *reflex* is an unlearned, biologically determined involuntary response to some physical stimuli (such as a sneeze after breathing

culture the knowledge, language, values, customs, and material objects that are passed from person to person and from one generation to the next in a human group or society.

Box 3.1 Sociology and Everyday Life

Answers to the Sociology Quiz on Global Food and Culture

1. False. Although cheese is a popular food in many cultures, most of the people living in China find cheese very distasteful and prefer delicacies such as duck's feet.

2. True. Round foods such as pears, grapes, and moon cakes are given to celebrate the birth of babies because the shape of the food is believed to symbolize family unity.

3. False. Although wedding cakes are a tradition in virtually all nations and cultures, the ingredients of the cake—as well as other foods served at the celebration—vary widely at this important family celebration. The traditional wedding cake in Italy is made from biscuits, for example, whereas in Norway the wedding cake is made from bread topped with cream, cheese, and syrup.

4. True. Many faiths, including Christianity, Judaism, Islam, Hinduism, and Buddhism, have dietary rules and rituals that involve food; however, these practices and beliefs vary widely among individuals and communities. For some people, food forms an integral part of religion in their life; for others, food is less relevant.

5. True. Just as foods are divided into yin foods (e.g., bean sprouts, cabbage, and carrots) and yang foods (beef, chicken, eggs, and mushrooms), cooking methods are also referred to as having yin qualities (e.g., boiling, poaching, and steaming) or yang qualities (deep-frying, roasting, and stir-frying). For many Chinese Americans, yin and yang are complementary pairs that should be incorporated into all aspects of social life, including the ingredients and preparation of foods.

6. False. Although more people now rely on fast foods, there is a "slow food" movement afoot to encourage people to prepare their food from scratch for a healthier lifestyle. Also, some cultural and religious communities—such as the Amish of Ohio, Pennsylvania, and Indiana—encourage families to prepare their food from scratch and to preserve their own fruits, vegetables, and meats. Rural families are more likely to grow their own food or prepare it from scratch than are families residing in urban areas.

7. False. Rice is a popular mainstay in the diets of people from diverse cultural backgrounds who have arrived in the United States over the past four decades. Groups ranging from the Hmong and Vietnamese to Puerto Ricans and Mexican Americans use rice as a central ingredient in their diets. Among some in the younger generations, however, food choices have become increasingly Americanized, and items such as french fries and pizza have become very popular.

8. True. Cultural diversity is a major issue in eating, and people in some cultures, religions, and nations expect that even an "outsider" will have a basic familiarity with, and respect for, their traditions and practices. However, social analysts also suggest that we should not generalize or imply that certain characteristics apply to *all* people in a cultural group or nation.

Sources: Based on Better Health Channel, 2007; Ohio State University, 2007; and PBS, 2005a.

some pepper in through the nose or the blinking of an eye when a speck of dust gets in it). *Drives* are unlearned, biologically determined impulses common to all members of a species that satisfy needs such as sleep, food, water, and sexual gratification. Reflexes and drives do not determine how people will behave in human societies; even the expression of these biological characteristics is channeled by culture. For example, we may be taught that the "appropriate" way to sneeze (an involuntary response) is to use a tissue or turn our head away from others (a learned response). Similarly, we may learn to sleep on mats or in beds. Most contemporary sociologists agree that culture and social learning, not nature, account for virtually all of our behavior patterns.

Because humans cannot rely on instincts in order to survive, culture is a "tool kit" for survival. According to the sociologist Ann Swidler (1986: 273), culture is a "tool kit of symbols, stories, rituals, and world views, which people may use in varying configurations to solve different kinds of problems." The tools we choose will vary according to our own personality and the situations we face. We are not puppets on a string; we make choices from among the items in our own "tool box."

© Celia Peterson/Getty Images

© Frans Lemmens/Getty Images

© Mark Henley/Impact/HIP/The Image Works

Food is a universal type of material culture, but what people eat and how they eat it vary widely, as shown in these cross-cultural examples from the United Arab Emirates (upper left), Holland (upper right), and China (bottom photo). What might be some reasons for the similarities and differences that you see in these photos?

Material Culture and Nonmaterial Culture

Our cultural tool box is divided into two major parts: material culture and nonmaterial culture (Ogburn, 1966/1922). ***Material culture* consists of the physical or tangible creations that members of a society make, use, and share.** Initially, items of material culture begin as raw materials or resources such as ore, trees, and oil. Through technology, these raw materials are transformed into usable items (ranging from books and computers to guns and tanks). Sociologists define ***technology*** as **the knowledge, techniques, and tools that make it possible for people to transform resources into usable forms, and the knowledge and skills required to use them after they are developed.** From this standpoint, technology is both concrete and abstract. For example, technology includes a pair of scissors and the knowledge and skill necessary to make them from iron, carbon, and chromium (Westrum, 1991). At the most basic level, material culture is important because it is our buffer against the environment. For example, we create shelter to protect ourselves from the weather and to provide ourselves with privacy. Beyond the survival level, we make, use, and share objects that are both interesting and important to us. Why are you wearing the particular clothes that you have on today? Perhaps you're communicating something about yourself, such as where you attend school, what kind of music you like, or where you went on vacation.

***Nonmaterial culture* consists of the abstract or intangible human creations of society that influence**

material culture a component of culture that consists of the physical or tangible creations (such as clothing, shelter, and art) that members of a society make, use, and share.

technology the knowledge, techniques, and tools that allow people to transform resources into a usable form and the knowledge and skills required to use what is developed.

nonmaterial culture a component of culture that consists of the abstract or intangible human creations of society (such as attitudes, beliefs, and values) that influence people's behavior.

© Spencer Grant/PhotoEdit

© Mark Richards/PhotoEdit

© Michael Greenlar/The Image Works

The customs and rituals associated with weddings are one example of nonmaterial culture. What can you infer about beliefs and attitudes concerning marriage in the societies represented by these photographs?

people's behavior. Language, beliefs, values, rules of behavior, family patterns, and political systems are examples of nonmaterial culture. A central component of nonmaterial culture is *beliefs*—the mental acceptance or conviction that certain things are true or real. Beliefs may be based on tradition, faith, experience, scientific research, or some combination of these. Faith in a supreme being and trust in another person are examples of beliefs. We may also have a belief in items of material culture. When we travel by airplane, for instance, we believe that it is possible to fly at 33,000 feet and to arrive at our destination even though we know that we could not do this without the airplane itself.

Cultural Universals

Because all humans face the same basic needs (such as for food, clothing, and shelter), we engage in similar activities that contribute to our survival. Anthropologist George Murdock (1945: 124) compiled a list of over seventy ***cultural universals*—customs and practices that occur across all societies.** His categories included appearance (such as bodily adornment and hairstyles), activities (such as sports, dancing, games, joking, and visiting), social institutions (such as family, law, and religion), and customary practices (such as cooking, folklore, gift giving, and hospitality). These general customs and practices may be present in all cultures, but their specific forms vary from one group to another and from one time to another within the same group. For example, although telling jokes may be a universal practice, what is considered to be a joke in one society may be an insult in another.

How do sociologists view cultural universals? In terms of their functions, cultural universals are useful because they ensure the smooth and continual operation of society (Radcliffe-Brown, 1952). A society must meet basic human needs by providing food, shelter, and some degree of safety for its members so that they will survive. Children and other new members (such as immigrants) must be taught the ways of the group. A society must also settle disputes and deal with people's emotions. All the while, the self-interest of individuals must be balanced with the needs of society as a whole. Cultural universals help fulfill these important functions of society.

From another perspective, however, cultural universals are not the result of functional necessity; these practices may have been *imposed* by members of one society on members of another. Similar customs and practices do not necessarily constitute cultural universals. They may be an indication that a conquering nation used its power to enforce certain types of behavior on those who were defeated (Sargent, 1987). Sociologists might ask questions such as "Who determines the dominant cultural patterns?" For example, although religion is a cultural universal, the traditional religious practices of indigenous peoples (those who first live in an area) have often been repressed and even stamped out by subsequent settlers or conquerors who have gained political and economic power over them. However, many people believe there is cause for optimism in the United States because the democratic ideas of this nation provide more guarantees of religious freedom than might be found in some other nations.

Components of Culture

Even though the specifics of individual cultures vary widely, all cultures have four common nonmaterial cultural components: symbols, language, values, and norms. These components contribute to both harmony and strife in a society.

Symbols

A ***symbol*** **is anything that meaningfully represents something else.** Culture could not exist without symbols because there would be no shared meanings among people. Symbols can simultaneously produce loyalty and animosity, and love and hate. They help us communicate ideas because they express abstract concepts with visible objects. For example, flags can stand for patriotism, nationalism, school spirit, or religious beliefs held by members of a group or society. Symbols can stand for love (a heart on a valentine), peace (a dove), or hate (a Nazi swastika), just as words can be used to convey these meanings. Symbols can also transmit other types of ideas. A siren is a symbol that denotes an emergency situation and sends the message to clear the way immediately. Gestures are also a symbolic form of communication—a movement of the head, body, or hands can express our ideas or feelings to others. For example, in the United States, pointing toward your chest with your thumb or finger is a symbol for "me."

Symbols affect our thoughts about class. For example, how a person is dressed or the kind of car that he or she drives is often at least subconsciously used as a measure of that individual's economic standing or position. With regard to clothing, although many people wear casual clothes on a daily basis, where the clothing was purchased is sometimes used as a symbol of social status. Were the items purchased at Wal-Mart, Old Navy, Abercrombie & Fitch, or Saks Fifth Avenue? What indicators are there on the items of clothing—such as the Nike *swoosh,* some other logo, or a brand name—that say something about the status of the product? Automobiles and their logos are also symbols that have cultural meaning beyond the shopping environment in which they originate.

Finally, symbols may be specific to a given culture and have special meaning to individuals who share that culture but not necessarily to other people. Consider, for example, the use of certain foods to celebrate the Chinese New Year: Bamboo shoots and black moss seaweed both represent wealth, peanuts and noodles symbolize a long life, and tangerines represent good luck. What foods in other cultures represent "good luck" or prosperity?

Language

Language **is a set of symbols that expresses ideas and enables people to think and communicate with one another.** Verbal (spoken) language and nonverbal (written or gestured) language help us describe reality. One of our most important human attributes is the ability to use language to share our experiences, feelings, and knowledge with others. Language can create visual images in our head, such as "the kittens look like little cotton balls" (Samovar and Porter, 1991a). Language also allows people to distinguish themselves from outsiders and to maintain group boundaries and solidarity (Farb, 1973).

Language is not solely a human characteristic. Other animals use sounds, gestures, touch, and smell to communicate with one another, but they use signals with fixed meanings that are limited to the immediate situation (the present) and cannot encompass past or future situations. For example, chimpanzees can use elements of standard American Sign Language and manipulate physical objects to make "sentences," but they are not physically endowed with the vocal apparatus needed to form the consonants required for

cultural universals customs and practices that occur across all societies.

symbol anything that meaningfully represents something else.

language a set of symbols that expresses ideas and enables people to think and communicate with one another.

oral language. As a result, nonhuman animals cannot transmit the more complex aspects of culture to their offspring. Humans have a unique ability to manipulate symbols to express abstract concepts and rules, and thus to create and transmit culture from one generation to the next.

Language and Social Reality Does language *create* or simply *communicate* reality? Anthropological linguists Edward Sapir and Benjamin Whorf have suggested that language not only expresses our thoughts and perceptions but also influences our perception of reality. According to the ***Sapir–Whorf hypothesis,* language shapes the view of reality of its speakers** (Whorf, 1956; Sapir, 1961). If people are able to think only through language, then language must precede thought. If language actually shapes the reality we perceive and experience, then some aspects of the world are viewed as important and others are virtually neglected because people know the world only in terms of the vocabulary and grammar of their own language.

If language does create reality, are we trapped by our language? Many social scientists agree that the Sapir–Whorf hypothesis overstates the relationship between language and our thoughts and behavior patterns. Although they acknowledge that language has many subtle meanings and that words used by people reflect their central concerns, most sociologists contend that language may *influence* our behavior and interpretation of social reality but not *determine* them.

Language and Gender What is the relationship between language and gender? What cultural assumptions about women and men does language reflect? Scholars have suggested several ways in which language and gender are intertwined:

- The English language ignores women by using the masculine form to refer to human beings in general. For example, the word *man* is used generically in words such as *chairman* and *mankind,* which allegedly include both men and women.
- Use of the pronouns *he* and *she* affects our thinking about gender. Pronouns show the gender of the *person* we expect to be in a particular occupation. For instance, nurses, secretaries, and schoolteachers are usually referred to as *she,* but doctors, engineers, electricians, and presidents are referred to as *he.*
- Words have positive connotations when relating to male power, prestige, and leadership; when relating to women, they carry negative overtones of weakness, inferiority, and immaturity (Epstein, 1988: 224). ◆ Table 3.1 shows how gender-based language reflects the traditional acceptance of men and women in certain positions, implying that the jobs are different when filled by women rather than men.
- A language-based predisposition to think about women in sexual terms reinforces the notion that women are sexual objects. Women are often described by terms such as *fox, broad, bitch, babe,* or

◆ **Table 3.1 Language and Gender**

Male Term	Female Term	Neutral Term
Teacher	Teacher	Teacher
Chairman	Chairwoman	Chair, chairperson
Congressman	Congresswoman	Representative
Policeman	Policewoman	Police officer
Fireman	Lady fireman	Firefighter
Airline steward	Airline stewardess	Flight attendant
Race car driver	Woman race car driver	Race car driver
Wrestler	Lady/woman wrestler	Wrestler
Professor	Female/woman professor	Professor
Doctor	Lady/woman doctor	Doctor
Bachelor	Spinster/old maid	Single person
Male prostitute	Prostitute	Prostitute
Welfare recipient	Welfare mother	Welfare recipient
Worker/employee	Working mother	Worker/employee
Janitor/maintenance man	Maid/cleaning lady	Custodial attendant

Sources: Adapted from Korsmeyer, 1981: 122; and Miller and Swift, 1991.

doll, which ascribe childlike or even petlike characteristics to them. By contrast, men have performance pressures placed on them by being defined in terms of their sexual prowess, such as *dude, stud,* and *hunk* (Baker, 1993).

Gender in language has been debated and studied extensively in recent years, and some changes have occurred. The preference of many women to be called *Ms.* (rather than *Miss* or *Mrs.* in reference to their marital status) has received a substantial degree of acceptance in public life and the media. Many organizations and publications have established guidelines for the use of nonsexist language and have changed titles such as *chairman* to *chair* or *chairperson.* "Men Working" signs in many areas have been replaced with "People Working." Some occupations have been given "genderless" titles, such as *firefighter* or *flight attendant.* To develop a more inclusive and equitable society, many scholars suggest that a more inclusive language is needed (see Basow, 1992). Yet many people resist change, arguing that the English language is being ruined (Epstein, 1988).

Language, Race, and Ethnicity Language may create and reinforce our perceptions about race and ethnicity by transmitting preconceived ideas about the superiority of one category of people over another. Let's look at a few images conveyed by words in the English language in regard to race/ethnicity:

- Words may have more than one meaning and create and/or reinforce negative images. Terms such as *blackhearted* (malevolent) and expressions such as *a black mark* (a detrimental fact) and *Chinaman's chance of success* (unlikely to succeed) associate the words *black* or *Chinaman* with negative associations and derogatory imagery. By contrast, expressions such as *that's white of you* and *the good guys wear white hats* reinforce positive associations with the color white.
- Overtly derogatory terms such as *nigger, kike, gook, honkey, chink, spic,* and other racial–ethnic slurs have been "popularized" in movies, music, comic routines, and so on. Such derogatory terms are often used in conjunction with physical threats against persons and are increasingly viewed as words that should not be used even in a supposedly "joking" manner.
- Words are frequently used to create or reinforce perceptions about a group. For example, Native Americans have been referred to as "savages" and "primitive," and African Americans have been described as "uncivilized," "cannibalistic," and "pagan."
- The "voice" of verbs may minimize or incorrectly identify the activities or achievements of people of color. For example, the use of the passive voice in the statement "African Americans *were given* the right to vote" ignores how African Americans *fought* for that right. Active-voice verbs may also inaccurately attribute achievements to people or groups. Some historians argue that cultural bias is shown by the very notion that "Columbus discovered America"—given that America was already inhabited by people who later became known as Native Americans (see Stannard, 1992; Takaki, 1993).
- Adjectives that typically have positive connotations can have entirely different meanings when used in certain contexts. Regarding employment, someone may say that a person of color is "qualified" for a position when it is taken for granted that whites in the same position *are* qualified (see Moore, 1992).

In addition to these concerns about the English language, problems also arise when more than one language is involved. Across the nation, the question of whether or not the United States should have an "official" language continues to arise. Some people believe that there is no need to designate an official language; other people believe that English should be designated as the official language and that the use of any other language should be discouraged or negatively sanctioned.

Certain jobs are stereotypically considered to be "men's jobs"; others are "women's jobs." Is your perception of a male flight attendant the same as your perception of a female flight attendant?

Sapir–Whorf hypothesis the proposition that language shapes the view of reality of its speakers.

Rapid changes in language and culture in the United States are reflected in this sign at a shopping center. How do functionalist and conflict theorists' views regarding language differ?

Recently, the city council in Farmers Branch—a suburb of Dallas, Texas—adopted a resolution declaring English as the official language of that city. According to the resolution, the use of a common language "removes barriers of misunderstanding and helps to unify the people of Farmers Branch, [the state of Texas,] and the United States and helps to enable the full economic and civic participation of all of its citizens . . ." (City of Farmers Branch, 2006). This resolution was passed at the same time as a local law that banned "illegal immigrants" from renting apartments in Farmers Branch. Are deep-seated social and cultural issues embedded in social policy decisions such as these? Although the United States has always been a nation of immigrants, in recent decades this country has experienced rapid changes in population that have brought about greater diversity in languages and cultures. Recent data gathered by the U.S. Census Bureau (see "Census Profiles: Languages Spoken in U.S. Households") indicate that although more than 80 percent of the people in this country speak only English at home, almost 20 percent speak a language other than English. The largest portion (over 10 percent of the U.S. population) of non-English speakers speak Spanish at home.

If we think about language from a functionalist perspective, we see that a shared language is essential to maintaining a common culture. From this approach, language is a stabilizing force in society and an important means of cultural transmission. Through language, children learn about their cultural heritage and develop a sense of personal identity in relationship to their group. For example, Latinos/as in New Mexico and south Texas use *dichos*—proverbs or sayings that are unique to the Spanish language—as a means of expressing themselves and as a reflection of their cultural heritage. Examples of *dichos* include *Anda tu camino sin ayuda de vecino* ("Walk your own road without the help of a neighbor") and *Amor de lejos es para pendejos* ("A long-distance romance is for fools"). *Dichos* are passed from generation to generation as a priceless verbal tradition whereby people can give advice or teach a lesson (Gandara, 1995).

On the other hand, if we look at language from a conflict approach, language is a source of power and a means of social control. Language may be used to perpetuate inequalities between people and between groups because words can be used (whether or not intentionally) to "keep people in their place." As the linguist Deborah Tannen (1993: B5) has suggested, "The devastating group hatreds that result in so much suffering in our own country and around the world are related in origin to the small intolerances in our everyday conversations—our readiness to attribute good intentions to ourselves and bad intentions to others." Language, then, is a reflection of our feelings and values.

Values

***Values* are collective ideas about what is right or wrong, good or bad, and desirable or undesirable in a particular culture** (Williams, 1970). Values do not dictate which behaviors are appropriate and which ones are not, but they provide us with the criteria by which we evaluate people, objects, and events. Values typically come in pairs of positive and negative values, such as being brave or cowardly, hardworking or lazy. Because we use values to justify our behavior, we tend to defend them staunchly (Kluckhohn, 1961).

Core American Values Do we have shared values in the United States? Sociologists disagree about the extent to which all people in this country share a core set of values. Functionalists tend to believe that shared values are essential for the maintenance of a society, and scholars using a functionalist approach have conducted most of the research on core values. Analysts who focus on the importance of core values maintain that the following ten values, identified forty years ago by sociologist Robin M. Williams, Jr. (1970), are still very important to people in the United States:

1. *Individualism.* People are responsible for their own success or failure. Individual ability and hard work are the keys to success. Those who do not

CENSUS PROFILES

Languages Spoken in U.S. Households

Among the categories of information gathered by the U.S. Census Bureau is data on the languages spoken in U.S. households. As shown below, English is the only language spoken at home in more than 80 percent of U.S. households; however, in almost 20 percent of U.S. households, some other language is the primary language spoken at home.

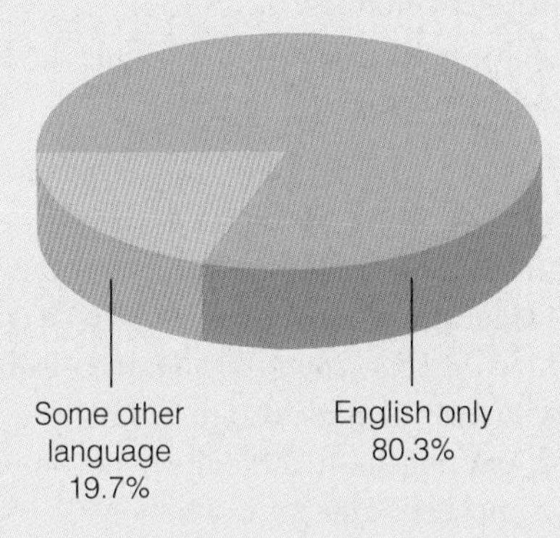

People who speak a language other than English at home are asked not only to indicate which other languages they speak but also how well they speak English. Approximately 44 percent of people who speak a language other than English at home report that they speak English "less than well." The principal languages other than English that are most frequently spoken at home are shown in the following chart. Do you think that changes in the languages spoken in this country will bring about other significant changes in U.S. culture? Why or why not?

Languages Spoken at Home Other Than English, by Percentage

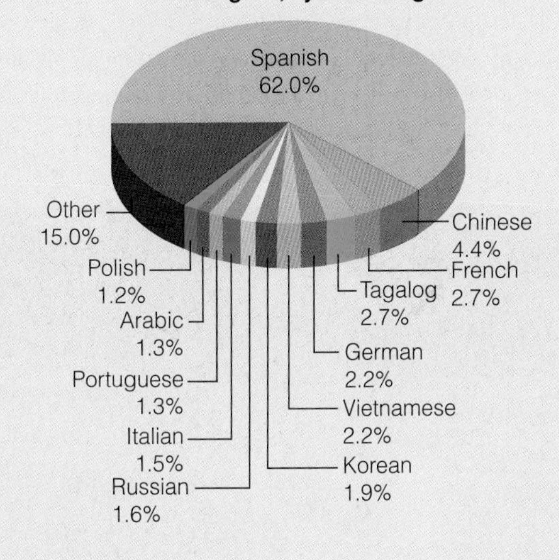

Source: U.S. Census Bureau, 2008.

succeed have only themselves to blame because of their lack of ability, laziness, immorality, or other character defects.

2. *Achievement and success.* Personal achievement results from successful competition with others. Individuals are encouraged to do better than others in school and to work in order to gain wealth, power, and prestige. Material possessions are seen as a sign of personal achievement.
3. *Activity and work.* People who are industrious are praised for their achievement; those perceived as lazy are ridiculed. From the time of the early Puritans, work has been viewed as important. Even during their leisure time, many people "work" in their play. Think, for example, of all the individuals who take exercise classes, run in marathons, garden, repair or restore cars, and so on in their spare time.
4. *Science and technology.* People in the United States have a great deal of faith in science and technology. They expect scientific and technological advances ultimately to control nature, the aging process, and even death.
5. *Progress and material comfort.* The material comforts of life include not only basic necessities (such as adequate shelter, nutrition, and medical care) but also the goods and services that make life easier and more pleasant.
6. *Efficiency and practicality.* People want things to be bigger, better, and faster. As a result, great value is placed on efficiency ("How well does it work?") and practicality ("Is this a realistic thing to do?").
7. *Equality.* Since colonial times, overt class distinctions have been rejected in the United States. However, "equality" has been defined as "equality of *opportunity*"—an assumed equal chance to achieve success—not as "equality of *outcome.*"
8. *Morality and humanitarianism.* Aiding others, especially following natural disasters (such as floods

values collective ideas about what is right or wrong, good or bad, and desirable or undesirable in a particular culture.

or hurricanes), is seen as a value. The notion of helping others was originally a part of religious teachings and tied to the idea of morality. Today, people engage in humanitarian acts without necessarily perceiving that it is the "moral" thing to do.

9. *Freedom and liberty.* Individual freedom is highly valued in the United States. The idea of freedom includes the right to private ownership of property, the ability to engage in private enterprise, freedom of the press, and other freedoms that are considered to be "basic" rights.
10. *Racism and group superiority.* People value their own racial or ethnic group above all others. Such feelings of superiority may lead to discrimination; slavery and segregation laws are classic examples. Many people also believe in the superiority of their country and that "the American way of life" is best.

© Gregory Shamus/Getty Images

Basketball star LeBron James is widely considered to be one of the greatest players in the history of the NBA. Which core American values are reflected in sports such as basketball?

Do you think that these values are still important today? Are there core values that you believe should be added to this list? Although sociologists have not agreed upon a specific list of emerging core values, various social analysts have suggested that some additional shared values in the United States today include the following:

- Ecological sensitivity, with an increased awareness of global problems such as overpopulation and global warming.
- Emphasis on developing and maintaining relationships through honesty and with openness, fairness, and tolerance of others.
- Spirituality and a need for meaning in life that reaches beyond oneself.

Value Contradictions Is it possible for there to be a contradiction between values in a society? Yes, all societies—including the United States—have value contradictions. *Value contradictions* are values that conflict with one another or are mutually exclusive (achieving one makes it difficult, if not impossible, to achieve another). There are situations in which the core values of morality and humanitarianism may conflict with values of individual achievement and success. For example, humanitarian values reflected in welfare and other government aid programs continue to come into conflict with values that emphasize hard work and personal achievement. Today, some people are more ambivalent about helping people who are chronically poor or homeless than they are about helping recent victims of major natural disasters such as floods, hurricanes, or earthquakes. In the aftermath of Hurricane Katrina in 2005, for instance, many people in the United States were more willing to make generous contributions to help the survivors of this natural disaster than they were to help the long-term homeless and disadvantaged throughout the nation.

Ideal Versus Real Culture What is the relationship between values and human behavior? According to sociologists, we do not always act in accord with our stated values. Sociologists refer to this contradiction as a gap between ideal culture and real culture. *Ideal culture* refers to the values and standards of behavior that people in a society profess to hold. *Real culture* refers to the values and standards of behavior that people actually follow. For example, we may claim to be law-abiding (ideal cultural value) but smoke marijuana (real cultural behavior), or we may regularly drive over the speed limit but think of ourselves as "good citizens."

Numerous studies have shown a discrepancy between ideal cultural values and people's actual behavior. For example, a University of Arizona study known as the "Garbage Project" analyzed household waste to determine the rate of alcohol consumption in Tucson, Arizona. When people were asked about their level of alcohol consumption, individuals who lived in some areas of the city reported very low levels of alcohol use. However, when researchers analyzed their garbage, the researchers found that over 80 percent of those households consumed some beer, and more than half discarded eight or more empty beer cans a week (Haviland, 1993). This is only one of many examples of how people's self-reporting of their beliefs or values may differ from their actual behavior. For this reason, societies have specific norms that govern human behavior.

Norms

Values provide ideals or beliefs about behavior but do not state explicitly how we should behave. Norms, on the other hand, do have specific behavioral expectations. ***Norms* are established rules of behavior or standards of conduct.** *Prescriptive norms* state what behavior is appropriate or acceptable. For example, persons making a certain amount of money are expected to file a tax return and pay any taxes they owe. Norms based on custom direct us to open a door for a person carrying a heavy load. By contrast, *proscriptive norms* state what behavior is inappropriate or unacceptable. Laws that prohibit us from driving over the speed limit and "good manners" that preclude you from talking on your cell phone during class are examples. Prescriptive and proscriptive norms operate at all levels of society, from our everyday actions to the formulation of laws.

Formal and Informal Norms Not all norms are of equal importance; those that are most crucial are formalized. *Formal norms* are written down and involve specific punishments for violators. Laws are the most common type of formal norms; they have been codified and may be enforced by sanctions. ***Sanctions* are rewards for appropriate behavior or penalties for inappropriate behavior.** Examples of *positive sanctions* include praise, honors, or medals for conformity to specific norms. *Negative sanctions* range from mild disapproval to the death penalty. In the case of law, formal sanctions are clearly defined and can be administered only by persons in certain official positions (such as police officers and judges), who are given the authority to impose the sanctions.

Norms considered to be less important are referred to as *informal norms*—unwritten standards of behavior understood by people who share a common identity. When individuals violate informal norms, other people may apply informal sanctions. *Informal sanctions* are not clearly defined and can be applied by any member of a group (such as frowning at someone or making a negative comment or gesture).

Folkways Norms are also classified according to their relative social importance. ***Folkways* are informal norms or everyday customs that may be violated without serious consequences within a particular culture** (Sumner, 1959/1906). They provide rules for conduct but are not considered to be essential to society's survival. In the United States, folkways include using underarm deodorant, brushing our teeth, and wearing appropriate clothing for a specific occasion. Often, folkways are not enforced; when they are enforced, the resulting sanctions tend to be informal and relatively mild.

Folkways are culture specific; they are learned patterns of behavior that can vary markedly from one society to another. In Japan, for example, where the walls of restroom stalls reach to the floor, folkways dictate that a person should knock on the door before entering a stall (you cannot tell if anyone is there without knocking). However, people in the United States find it disconcerting when someone knocks on the door of the stall (A. Collins, 1991).

Mores Other norms are considered to be highly essential to the stability of society. ***Mores* are a particular culture's strongly held norms with moral and ethical connotations that may not be violated without serious consequences.** Because mores (pronounced MOR-ays) are based on cultural values and are considered to be crucial for the well-being of the group, violators are subject to more severe negative sanctions (such as ridicule, loss of employment, or imprisonment) than are those who fail to adhere to folkways. The strongest mores are referred to as taboos. ***Taboos* are mores so strong that their violation is considered to be extremely offensive and even unmentionable.** Violation of taboos is punishable by the group or even, according to certain belief systems, by a supernatural force. The

norms established rules of behavior or standards of conduct.

sanctions rewards for appropriate behavior or penalties for inappropriate behavior.

folkways informal norms or everyday customs that may be violated without serious consequences within a particular culture.

mores strongly held norms with moral and ethical connotations that may not be violated without serious consequences in a particular culture.

taboos mores so strong that their violation is considered to be extremely offensive and even unmentionable.

chapter

4 Socialization

Chapter Focus Question

What happens when children do not have an environment that supports positive socialization?

I don't want to be crippled by things that happened in the past. [My father] was a free bird, you know? He couldn't handle being a father. [When I learned that my father was terminally ill with cancer,] I thought, even if it's seven months, I want to get to know this person. Sometimes I'd just sit there, and he'd say, "Stop staring at me." Then sometimes we'd talk about the past. One day he said, "Perfect. You were made perfect." I just started crying. I was like, "Thank you, God, for letting me have this moment."

—In an interview, actor Drew Barrymore explains how she reconciled with her father, from whom she had been estranged for many years, a short time before his death (Lynch and Gold, 2005: 96, 98). As her previous autobiography, *Little Girl Lost,* described, Barrymore and her father had experienced problems in the past:

I think my mom and dad were boyfriend and girlfriend for a couple of years, but they were apart by the time I was born. . . .

Actress Drew Barrymore's description of childhood maltreatment in her family makes us aware of the importance of early socialization in all our lives. As an adult, she has experienced many happier times, including the honor of receiving a star on Hollywood's Walk of Fame.

IN THIS CHAPTER

- Why Is Socialization Important Around the Globe?
- Social Psychological Theories of Human Development
- Sociological Theories of Human Development
- Agents of Socialization
- Gender and Racial–Ethnic Socialization
- Socialization Through the Life Course
- Resocialization
- Socialization in the Future

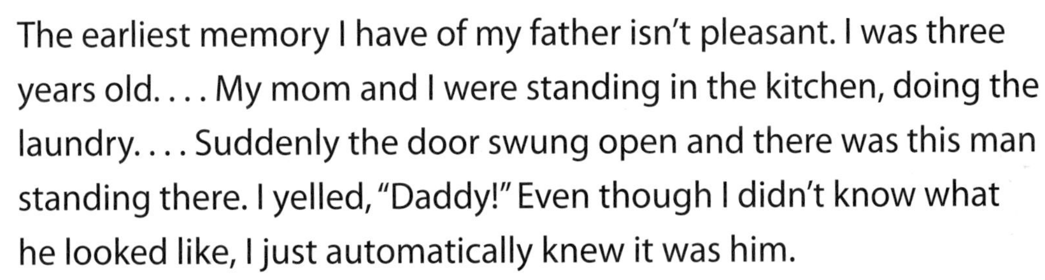

The earliest memory I have of my father isn't pleasant. I was three years old. . . . My mom and I were standing in the kitchen, doing the laundry. . . . Suddenly the door swung open and there was this man standing there. I yelled, "Daddy!" Even though I didn't know what he looked like, I just automatically knew it was him.

He paused in the doorway, like he was making a dramatic entrance, and I think he said something, but he was so drunk, it was unintelligible. It sounded more like a growl. We stood there, staring at him. I was so excited to see him. . . . I didn't really know what my dad was like, but I learned real fast. In a blur of anger he roared into the room and threw my mom down on the ground. Then he turned on me. I didn't know what was happening. I was still excited to see him, still hearing the echo of my gleeful yell, "Daddy!" when he picked me up and threw me into the wall. Luckily, half of my body landed on a big sack of laundry, and I wasn't hurt. But my dad didn't even look back at me. He turned and grabbed a bottle of tequila, shattered a bunch of glasses all over the floor, and then stormed out of the house. . . . And that was it. That was the first time I remember seeing my dad. (Barrymore, 1994: 185–187)

Sharpening Your Focus

- *What purpose does socialization serve?*
- *How do individuals develop a sense of self?*
- *How does socialization occur?*
- *Who experiences resocialization?*

For Barrymore and many other people, early interactions with their parents have had a profound influence on their later lives. Clearly, the parent–child relationship is a significant factor in the process of socialization, which is of interest to sociologists. Although most children are nurtured, trusted, and loved by their parents, Barrymore's experience is not an isolated incident: Large numbers of children experience maltreatment at the hands of family members or other caregivers such as babysitters or child-care workers. Child maltreatment includes physical abuse, sexual abuse, physical neglect, and emotional mistreatment of children and young adolescents. Such maltreatment is of interest to sociologists because it has a serious impact on a child's social growth, behavior, and self-image—all of which develop within the process of socialization. By contrast, children who are treated with respect by their parents are more likely to develop a positive self-image and learn healthy conduct because their parents provide appropriate models of behavior.

In this chapter, we examine why socialization is so crucial, and we discuss both sociological and social psychological theories of human development. We look at the dynamics of socialization—how it occurs and what shapes it. Throughout the chapter, we focus on positive and negative aspects of the socialization process. Before reading on, test your knowledge of socialization and child care by taking the quiz in Box 4.1.

Why Is Socialization Important Around the Globe?

***Socialization* is the lifelong process of social interaction through which individuals acquire a self-identity and the physical, mental, and social skills needed for survival in society.** It is the essential link between the individual and society. Socialization enables each of us to develop our human potential and to learn the ways of thinking, talking, and acting that are necessary for social living.

Socialization is essential for the individual's survival and for human development. The many people who met the early material and social needs of each of us were central to our establishing our own identity. During the first three years of our life, we begin to develop both a unique identity and the ability to manipulate things and to walk. We acquire sophisticated cognitive tools for thinking and for analyzing a wide variety of situations, and we learn effective communication skills. In the process, we begin a relatively long socialization process that culminates in our integration into a complex social and cultural system (Garcia Coll, 1990).

Socialization is also essential for the survival and stability of society. Members of a society must be socialized to support and maintain the existing social structure. From a functionalist perspective, individual conformity to existing norms is not taken for granted; rather, basic individual needs and desires must be balanced against the needs of the social structure. The socialization process is most effective when people conform to the norms of society because they believe that this is the best course of action. Socialization enables a society to "reproduce" itself by passing on its culture from one generation to the next.

Although the techniques used to teach newcomers the beliefs, values, and rules of behavior are somewhat similar in many nations, the *content* of socialization differs greatly from society to society. How people walk, talk, eat, make love, and wage war are all functions of the culture in which they are raised. At the same time, we are also influenced by our exposure to subcultures of class, race, ethnicity, religion, and gender. In addition, each of us has unique experiences in our families and friendship groupings. The kind of human being that we become depends greatly on the particular society and social groups that surround us at birth and during early childhood. What we believe about ourselves, our society, and the world does not spring full-blown from inside ourselves; rather, we learn these things from our interactions with others.

Human Development: Biology and Society

What does it mean to be "human"? To be human includes being conscious of ourselves as individuals with unique identities, personalities, and relationships with others. As humans, we have ideas, emotions, and values. We have the capacity to think and to make rational decisions. But what is the source of "humanness"? Are we born with these human characteristics, or do we develop them through our interactions with others?

When we are born, we are totally dependent on others for our survival. We cannot turn ourselves over, speak, reason, plan, or do many of the things that are associated with being human. Although we can nurse, wet, and cry, most small mammals can also do those things. As discussed in Chapter 3, we humans differ from nonhuman animals because we lack instincts and must rely on learning for our survival. Human infants have the potential for developing human characteristics if they are exposed to an adequate socialization process.

Every human being is a product of biology, society, and personal experiences—that is, of heredity and

Box 4.1 Sociology and Everyday Life

How Much Do You Know About Early Socialization and Child Care?

True	False	
T	F	1. In the United States, full-day child care often costs as much per year as college tuition at a public college or university.
T	F	2. The cost of child care is a major problem for many U.S. families.
T	F	3. After-school programs have greatly reduced the number of children who are home alone after school.
T	F	4. The average annual salary of a child-care worker is less than the average yearly salaries for funeral attendants or garbage collectors.
T	F	5. All states require teachers in child-care centers to have training in their field and to pass a licensing examination.
T	F	6. In a family in which child abuse occurs, all the children are likely to be victims.
T	F	7. It is against the law to fail to report child abuse.
T	F	8. Some people are "born" child abusers, whereas others learn abusive behavior from their family and friends.

Answers on page 106.

environment or, in even more basic terms, "nature" and "nurture." How much of our development can be explained by socialization? How much by our genetic heritage? Sociologists focus on how humans design their own culture and transmit it from generation to generation through socialization. By contrast, sociobiologists assert that nature, in the form of our genetic makeup, is a major factor in shaping human behavior. ***Sociobiology* is the systematic study of how biology affects social behavior** (Wilson, 1975). According to the zoologist Edward O. Wilson, who pioneered sociobiology, genetic inheritance underlies many forms of social behavior such as war and peace, envy and concern for others, and competition and cooperation. Most sociologists disagree with the notion that biological principles can be used to explain all human behavior. Obviously, however, some aspects of our physical makeup—such as eye color, hair color, height, and weight—are largely determined by our heredity.

How important is social influence ("nurture") in human development? There is hardly a single behavior that is not influenced socially. Except for simple reflexes, most human actions are social, either in their causes or in their consequences. Even solitary actions such as crying and brushing our teeth are ultimately social. We cry because someone has hurt us. We brush our teeth because our parents (or dentist) told us it was important. Social environment probably has a greater effect than heredity on the way we develop and the way we act. However, heredity does provide the basic material from which other people help to mold an individual's human characteristics.

Our biological needs and emotional needs are related in a complex equation. Children whose needs are met in settings characterized by affection, warmth, and closeness see the world as a safe and comfortable place and see other people as trustworthy and helpful. By contrast, infants and children who receive less-than-adequate care or who are emotionally rejected or abused often view the world as hostile and have feelings of suspicion and fear.

Problems Associated with Social Isolation and Maltreatment

Social environment, then, is a crucial part of an individual's socialization. Even nonhuman primates such as monkeys and chimpanzees need social contact with others of their species in order to develop properly. As we will see, appropriate social contact is even more important for humans.

socialization the lifelong process of social interaction through which individuals acquire a self-identity and the physical, mental, and social skills needed for survival in society.

sociobiology the systematic study of how biology affects social behavior.

Box 4.1 Sociology and Everyday Life

Answers to the Sociology Quiz on Early Socialization and Child Care

1. True. Full-day child care typically costs between $4,000 and $10,000 per child per year, which is as much or more than tuition at many public colleges and universities.

2. True. Child care outside the home is a major financial burden, particularly for the one out of every four families with young children but with an income of less than $25,000 a year.

3. False. Although after-school programs have slightly reduced the number of children at home alone after school, nearly seven million school-age children are alone each week while their parents work.

4. True. The average salary for a child-care worker is only $15,430 per year, which is less than the yearly salaries for people in many other employment categories.

5. False. Although all states require hairdressers and manicurists to have about 1,500 hours of training at an accredited school, only 11 states require child-care providers to have any early childhood training prior to taking care of children.

6. False. In some families, one child may be the victim of repeated abuse, whereas others are not.

7. True. In the United States, all states have reporting requirements for child maltreatment; however, there has been inconsistent compliance with these legal mandates. Some states have requirements that everyone who suspects abuse or neglect must report it. Other states mandate reporting only by certain persons, such as medical personnel and child-care providers.

8. False. No one is "born" to be an abuser. People learn abusive behavior from their family and friends.

Source: Based on Children's Defense Fund, 2002.

Isolation and Nonhuman Primates Researchers have attempted to demonstrate the effects of social isolation on nonhuman primates raised without contact with others of their own species. In a series of laboratory experiments, the psychologists Harry and Margaret Harlow (1962, 1977) took infant rhesus monkeys from their mothers and isolated them in separate cages. Each cage contained two nonliving "mother substitutes" made of wire, one with a feeding bottle attached and the other covered with soft terry cloth but without a bottle. The infant monkeys instinctively clung to the cloth "mother" and would not abandon it until hunger drove them to the bottle attached to the wire "mother." As soon as they were full, they went back to the cloth "mother" seeking warmth, affection, and physical comfort.

The Harlows' experiments show the detrimental effects of isolation on nonhuman primates. When the young monkeys were later introduced to other members of their species, they cringed in the corner. Having been deprived of social contact with other monkeys during their first six months of life, they never learned

As Harry and Margaret Harlow discovered, humans are not the only primates that need contact with others. Deprived of its mother, this infant monkey found a substitute.

how to relate to other monkeys or to become well-adjusted adults—they were fearful of or hostile toward other monkeys (Harlow and Harlow, 1962, 1977).

Because humans rely more heavily on social learning than do monkeys, the process of socialization is even more important for us.

Isolated Children Of course, sociologists would never place children in isolated circumstances so that they could observe what happened to them. However, some cases have arisen in which parents or other caregivers failed to fulfill their responsibilities, leaving children alone or placing them in isolated circumstances. From analysis of these situations, social scientists have documented cases in which children were deliberately raised in isolation. A look at the lives of two children who suffered such emotional abuse provides important insights into the importance of a positive socialization process and the negative effects of social isolation.

Anna Born in 1932 to an unmarried, mentally impaired woman, Anna was an unwanted child. She was kept in an attic-like room in her grandfather's house. Her mother, who worked on the farm all day and often went out at night, gave Anna just enough care to keep her alive; she received no other care. Sociologist Kingsley Davis (1940) described Anna's condition when she was found in 1938:

> [Anna] had no glimmering of speech, absolutely no ability to walk, no sense of gesture, not the least capacity to feed herself even when the food was put in front of her, and no comprehension of cleanliness. She was so apathetic that it was hard to tell whether or not she could hear. And all of this at the age of nearly six years.

When she was placed in a special school and given the necessary care, Anna slowly learned to walk, talk, and care for herself. Just before her death at the age of ten, Anna reportedly could follow directions, talk in phrases, wash her hands, brush her teeth, and try to help other children (Davis, 1940).

Genie Almost four decades later, Genie was found in 1970 at the age of thirteen. She had been locked in a bedroom alone, alternately strapped down to a child's potty chair or straitjacketed into a sleeping bag, since she was twenty months old. She had been fed baby food and beaten with a wooden paddle when she whimpered. She had not heard the sounds of human speech because no one talked to her and there was no television or radio in her room (Curtiss, 1977; Pines, 1981). Genie was placed in a pediatric hospital, where one of the psychologists described her condition:

> At the time of her admission she was virtually unsocialized. She could not stand erect, salivated continuously, had never been toilet-trained and had no control over her urinary or bowel functions. She was unable to chew solid food and had the weight, height and appearance of a child half her age. (Rigler, 1993: 35)

© Bill Aron/PhotoEdit

What are the consequences to children of isolation and physical abuse, as contrasted with social interaction and parental affection? Sociologists emphasize that social environment is a crucial part of an individual's socialization.

© Jose Luis Pelaez, Inc./Getty Images

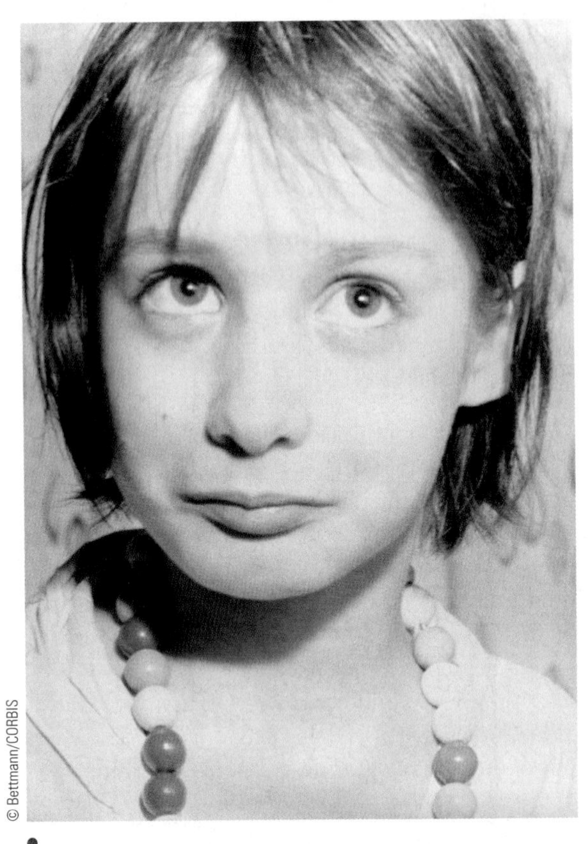

A victim of extreme child abuse, Genie was isolated from human contact and tortured until she was rescued at the age of thirteen. Subsequent attempts to socialize her were largely unsuccessful.

In addition to her physical condition, Genie showed psychological traits associated with neglect, as described by one of her psychiatrists:

> If you gave [Genie] a toy, she would reach out and touch it, hold it, caress it with her fingertips, as though she didn't trust her eyes. She would rub it against her cheek to feel it. So when I met her and she began to notice me standing beside her bed, I held my hand out and she reached out and took my hand and carefully felt my thumb and fingers individually, and then put my hand against her cheek. She was exactly like a blind child. (Rymer, 1993: 45)

Extensive therapy was used in an attempt to socialize Genie and develop her language abilities (Curtiss, 1977; Pines, 1981). These efforts met with limited success: In the 1990s, Genie was living in a board-and-care home for retarded adults (see Angier, 1993; Rigler, 1993; Rymer, 1993).

Why do we discuss children who have been the victims of maltreatment in a chapter that looks at the socialization process? The answer lies in the fact that such cases are important to our understanding of the socialization process because they show the importance of this process and reflect how detrimental that social isolation and neglect can be to the well-being of people.

Child Maltreatment What do the terms *child maltreatment* and *child abuse* mean to you? When asked what constitutes child maltreatment, many people first think of cases that involve severe physical injuries or sexual abuse. However, neglect is the most frequent form of child maltreatment (Dubowitz et al., 1993). Child neglect occurs when children's basic needs—including emotional warmth and security, adequate shelter, food, health care, education, clothing, and protection—are not met, regardless of cause (Dubowitz et al., 1993: 12). Neglect often involves acts of omission (where parents or caregivers fail to provide adequate physical or emotional care for children) rather than acts of commission (such as physical or sexual abuse). Of course, what constitutes child maltreatment differs from society to society.

Social Psychological Theories of Human Development

Over the past hundred years, a variety of psychological and sociological theories have been developed not only to explain child abuse but also to describe how a positive process of socialization occurs. Let's look first at several psychological theories that focus primarily on how the individual personality develops.

Freud and the Psychoanalytic Perspective

The basic assumption in Sigmund Freud's (1924) psychoanalytic approach is that human behavior and personality originate from unconscious forces within individuals. Sigmund Freud (1856–1939), who is known as the founder of psychoanalytic theory, developed his major theories in the Victorian era, when biological explanations of human behavior were prevalent. It was also an era of extreme sexual repression and male dominance when compared to contemporary U.S. standards. Freud's theory was greatly influenced by these cultural factors, as reflected in the importance he assigned to sexual motives in explaining behavior. For example, Freud based his ideas on the belief that people have two basic tendencies: the urge to survive and the urge to procreate.

According to Freud (1924), human development occurs in three states that reflect different levels of the personality, which he referred to as the *id*, *ego*, and *su-*

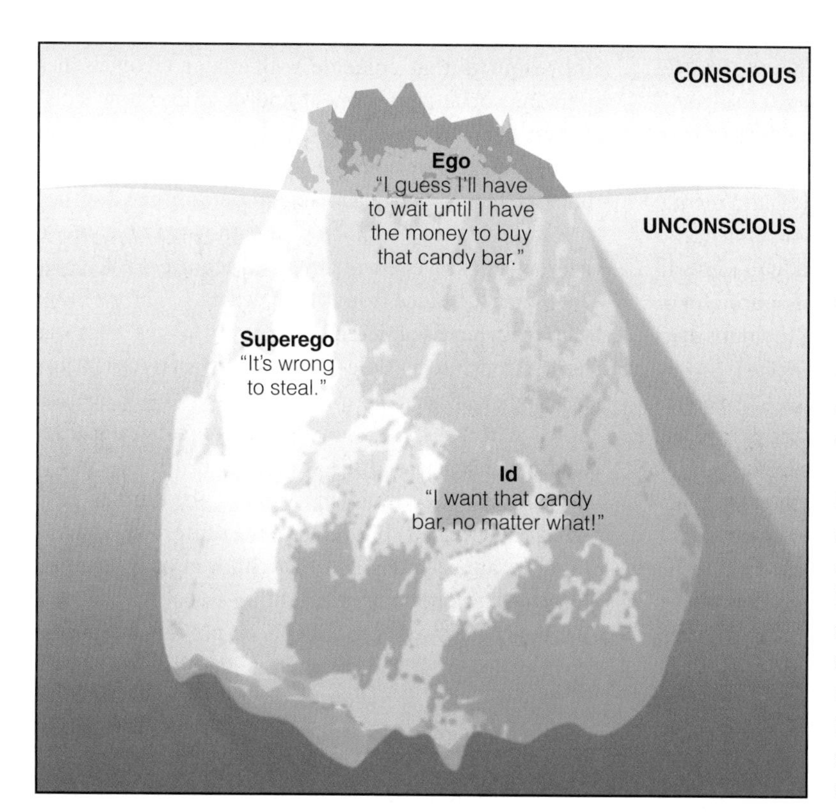

▶ **Figure 4.1 Freud's Theory of Personality**
This illustration shows how Freud might picture a person's internal conflict over whether to commit an antisocial act such as stealing a candy bar. In addition to dividing personality into three components, Freud theorized that our personalities are largely unconscious—hidden from our normal awareness. To dramatize his point, Freud compared conscious awareness (portions of the ego and superego) to the visible tip of an iceberg. Most of personality—including the id, with its raw desires and impulses—lies submerged in our subconscious.

perego. The ***id* is the component of personality that includes all of the individual's basic biological drives and needs that demand immediate gratification.** For Freud, the newborn child's personality is all id, and from birth the child finds that urges for self-gratification—such as wanting to be held, fed, or changed—are not going to be satisfied immediately. However, id remains with people throughout their life in the form of *psychic energy,* the urges and desires that account for behavior. By contrast, the second level of personality—the ego—develops as infants discover that their most basic desires are not always going to be met by others. The ***ego* is the rational, reality-oriented component of personality that imposes restrictions on the innate, pleasure-seeking drives of the id.** The ego channels the desire of the id for immediate gratification into the most advantageous direction for the individual. The third level of personality—the superego—is in opposition to both the id and the ego. The ***superego,* or conscience, consists of the moral and ethical aspects of personality.** It is first expressed as the recognition of parental control and eventually matures as the child learns that parental control is a reflection of the values and moral demands of the larger society. When a person is well adjusted, the ego successfully manages the opposing forces of the id and the superego. ▶ Figure 4.1 illustrates Freud's theory of personality.

Although subject to harsh criticism, Freud's theory made people aware of the importance of early childhood experiences, including abuse and neglect. His theories have also had a profound influence on contemporary mental health practitioners and on other human development theories.

Erikson and Psychosocial Development

Erik H. Erikson (1902–1994) drew from Freud's theory and identified eight psychosocial stages of development. According to Erikson (1980/1959), each stage is accompanied by a crisis or potential crisis that involves transitions in social relationships:

1. *Trust versus mistrust* (birth to age one). If infants receive good care and nurturing (characterized by emotional warmth, security, and love) from their

id Sigmund Freud's term for the component of personality that includes all of the individual's basic biological drives and needs that demand immediate gratification.

ego according to Sigmund Freud, the rational, reality-oriented component of personality that imposes restrictions on the innate pleasure-seeking drives of the id.

superego Sigmund Freud's term for the conscience, consisting of the moral and ethical aspects of personality.

parents, they will develop a sense of trust. If they do not receive such care, they will become mistrustful and anxious about their surroundings.

2. *Autonomy versus shame and doubt* (age one to three). As children gain a feeling of control over their behavior and develop a variety of physical and mental abilities, they begin to assert their independence. If allowed to explore their environment, children will grow more autonomous. If parents disapprove of or discourage them, children will begin to doubt their abilities.
3. *Initiative versus guilt* (age three to five). If parents encourage initiative during this stage, children will develop a sense of initiative. If parents make children feel that their actions are bad or that they are a nuisance, children may develop a strong sense of guilt.
4. *Industry versus inferiority* (age six to eleven). At this stage, children want to manipulate objects and learn how things work. Adults who encourage children's efforts and praise the results—both at home and at school—produce a feeling of industry in children. Feelings of inferiority result when parents or teachers appear to view children's efforts as silly or as a nuisance.
5. *Identity versus role confusion* (age twelve to eighteen). During this stage, adolescents attempt to develop a sense of identity. As young people take on new roles, the new roles must be combined with the old ones to create a strong self-identity. Role confusion results when individuals fail to acquire an accurate sense of personal identity.
6. *Intimacy versus isolation* (age eighteen to thirty-five). The challenge of this stage (which covers courtship and early family life) is to develop close and meaningful relationships. If individuals establish successful relationships, intimacy ensues. If they fail to do so, they may feel isolated.
7. *Generativity versus self-absorption* (age thirty-five to fifty-five). Generativity means looking beyond oneself and being concerned about the next generation and the future of the world in general. Self-absorbed people may be preoccupied with their own well-being and material gains or be overwhelmed by stagnation, boredom, and interpersonal impoverishment.
8. *Integrity versus despair* (maturity and old age). Integrity results when individuals have resolved previous psychosocial crises and are able to look back at their life as having been meaningful and personally fulfilling. Despair results when previous crises remain unresolved and individuals view their life as a series of disappointments, failures, and misfortunes.

Erikson's psychosocial stages broaden the framework of Freud's theory by focusing on social and cultural forces and by examining development throughout the life course. The psychosocial approach encompasses the conflicts that coincide with major changes in a person's social environment and describes how satisfactory resolution of these conflicts results in positive development. For example, if adolescents who experience an identity crisis are able to determine who they are and what they want from life, they may be able to achieve a positive self-identity and acquire greater psychological distance from their parents.

Critics have pointed out that Erikson's research was limited to white, middle-class respondents from industrial societies (Slugoski and Ginsburg, 1989). However, other scholars have used his theoretical framework to examine racial–ethnic variations in the process of psychosocial development. Most of the studies have concluded that all children face the same developmental tasks at each stage but that children of color often have greater difficulty in obtaining a positive outcome because of experiences with racial prejudice and discrimination in society (Rotheram and Phinney, 1987). Although establishing an identity is difficult for most adolescents, one study found that it was especially problematic for children of recent Asian American immigrants who had experienced high levels of stress related to immigration (Huang and Ying, 1989).

Piaget and Cognitive Development

Unlike psychoanalytic approaches, which focus primarily on personality development, cognitive approaches emphasize the intellectual (cognitive) development of children. The Swiss psychologist Jean Piaget (1896–1980) was a pioneer in the field of cognitive development. Cognitive theorists are interested in how people obtain, process, and use information—that is, in how we think. Cognitive development relates to changes over time in how we think.

According to Piaget (1954), in each stage of human development (from birth through adolescence), children's activities are governed by their perception of the world around them. His four stages of cognitive development are organized around specific tasks that, when mastered, lead to the acquisition of new mental capacities, which then serve as the basis for the next level of development. Thus, development is a continuous process of successive changes in which the child must go through each stage in the sequence before moving on to the next one. However, Piaget believed that the length of time each child remained in a specific stage would vary based on the child's individual attributes and the cultural context in which the development process occurred.

1. *Sensorimotor stage* (birth to age two). Children understand the world only through sensory contact

© Tony Freeman/PhotoEdit

© Tony Freeman/PhotoEdit

Psychologist Jean Piaget identified four stages of cognitive development, including the preoperative stage, in which children have limited ability to realize that physical objects may change in shape or appearance. Piaget poured liquid from one beaker into a taller, narrower beaker and then asked children about the amounts of liquid in each beaker.

and immediate action; they cannot engage in symbolic thought or use language. Children gradually comprehend *object permanence*—the realization that objects exist even when the items are placed out of their sight.

2. *Preoperational stage* (age two to seven). Children begin to use words as mental symbols and to form mental images. However, they have limited ability to use logic to solve problems or to realize that physical objects may change in shape or appearance but still retain their physical properties.
3. *Concrete operational stage* (age seven to eleven). Children think in terms of tangible objects and actual events. They can draw conclusions about the likely physical consequences of an action without always having to try the action out. Children begin to take the role of others and start to empathize with the viewpoints of others.
4. *Formal operational stage* (age twelve through adolescence). Adolescents have the potential to engage in highly abstract thought and understand places, things, and events they have never seen. They can think about the future and evaluate different options or courses of action.

Using this cognitive model, Piaget (1932) also investigated moral development. In one study, he told stories and asked children to judge how "good" or "bad" the characters were. One story involved a child who *accidentally* broke fifteen cups while another *deliberately* broke one cup. Piaget asked the children in his study if they thought one child's behavior was worse than the other's. From his research, Piaget concluded that younger children (lasting until about age eight or ten) believe that it is more evil to break a large number of cups (or steal large sums of money) than to break one cup (or steal small sums of money) for whatever reason. In contrast, older children (beginning at about age eleven) are more likely to consider principles, including the intentions and motives behind people's behavior.

Piaget's stages of cognitive development provide us with useful insights on children's logical thinking and how children invent or construct the rules that govern their understanding of the world. His views on moral development show that children move from greater external influence, such as parental and other forms of moral authority, to being more autonomous, based on their own moral judgments about behavior. However, critics have pointed out that his theory says little about children's individual differences, including how gender or culture may influence children's beliefs and actions.

Kohlberg and the Stages of Moral Development

Lawrence Kohlberg (1927–1987) elaborated on Piaget's theories of cognitive reasoning by conducting a series of studies in which children, adolescents, and adults were presented with moral dilemmas that took the form of stories. Based on his findings, Kohlberg (1969, 1981) classified moral reasoning into three sequential levels:

1. *Preconventional level* (age seven to ten). Children's perceptions are based on punishment and obedience.

© PhotoAlto/Alamy

How do these teenagers' perceptions of the world differ from their perceptions ten years earlier, according to Piaget?

Evil behavior is that which is likely to be punished; good conduct is based on obedience and avoidance of unwanted consequences.

2. *Conventional level* (age ten through adulthood). People are most concerned with how they are perceived by their peers and with how one conforms to rules.
3. *Postconventional level* (few adults reach this stage). People view morality in terms of individual rights; "moral conduct" is judged by principles based on human rights that transcend government and laws.

Although Kohlberg presents interesting ideas about the moral judgments of children, some critics have challenged the universality of his stages of moral development. They have also suggested that the elaborate "moral dilemmas" he used are too abstract for children. In one story, for example, a husband contemplates stealing for his critically ill wife medicine that he cannot afford. When questions are made simpler, or when children and adolescents are observed in natural (as opposed to laboratory) settings, they often demonstrate sophisticated levels of moral reasoning (Darley and Shultz, 1990; Lapsley, 1990).

Gilligan's View on Gender and Moral Development

Psychologist Carol Gilligan (b. 1936) is one of the major critics of Kohlberg's theory of moral development. According to Gilligan (1982), Kohlberg's model was developed solely on the basis of research with male respondents, and women and men often have divergent views on morality based on differences in socialization and life experiences. Gilligan believes that men become more concerned with law and order but that women analyze social relationships and the social consequences of behavior. For example, in Kohlberg's story about the man who is thinking about stealing medicine for his wife, Gilligan argues that male respondents are more likely to use *abstract standards* of right and wrong, whereas female respondents are more likely to be concerned about what *consequences* his stealing the drug might have for the man and his family. Does this constitute a "moral deficiency" on the part of either women or men? Not according to Gilligan.

To correct what she perceived to be a male bias in Kohlberg's research, Gilligan (1982) examined morality in women by interviewing twenty-nine pregnant women who were contemplating having an abortion. Based on her research, Gilligan concluded that Kohlberg's stages do not reflect the ways that many women think about moral problems. As a result, Gilligan identified three stages in female moral development. In stage 1, the woman is motivated primarily by selfish concerns ("This is what I want . . . this is what I need"). In stage 2, she increasingly recognizes her responsibility to others. In stage 3, she makes a decision based on her desire to do the greatest good for both herself and for others. Gilligan argued that men are socialized to make moral decisions based on a justice perspective ("What is the fairest thing to do?"), whereas women are socialized to make such decisions on a care and responsibility perspective ("Who will be hurt least?").

Subsequent research that directly compared women's and men's reasoning about moral dilemmas has supported some of Gilligan's assertions but not others. For example, some researchers have not found that women are more compassionate than men (Tavris, 1993). Overall, however, Gilligan's argument that people make moral decisions according to both abstract principles of justice and principles of compassion and care is an important contribution to our knowledge about moral reasoning. Her book *In a Different Voice* (1982) also made social scientists more aware that the same situation may be viewed quite differently by men and by women.

Sociological Theories of Human Development

Although social scientists acknowledge the contributions of psychoanalytic and psychologically based explanations of human development, sociologists believe that it is important to bring a sociological perspective to bear on how people develop an awareness of self and learn about the culture in which they live. According to a sociological perspective, we cannot form a sense of self or personal identity without intense social contact with others. The self represents the sum total of perceptions and feelings that an individual has of being a distinct, unique person—a sense of who and what one is. When we speak of the "self," we typically use words such as *I, me, my, mine,* and *myself* (Cooley, 1998/1902). This sense of self (also referred to *self-concept*) is not present at birth; it arises in the process of social experience. ***Self-concept* is the totality of our beliefs and feelings about ourselves.** Four components make up our self-concept: (1) the physical self ("I am tall"), (2) the active self ("I am good at soccer"), (3) the social self ("I am nice to others"), and (4) the psychological self ("I believe in world peace"). Between early and late childhood, a child's focus tends to shift from the physical and active dimensions of self toward the social and psychological aspects. Self-concept is the foundation for communication with others; it continues to develop and change throughout our lives.

Our *self-identity* is our perception about what kind of person we are. As we have seen, socially isolated children do not have typical self-identities; they have had no experience of "humanness." According to symbolic interactionists, we do not know who we are until we see ourselves as we believe that others see us. We gain information about the self largely through language, symbols, and interaction with others. Our interpretation and evaluation of these messages are central to the social construction of our identity. However, we are not just passive reactors to situations, programmed by society to respond in fixed ways. Instead, we are active agents who develop plans out of the pieces supplied by culture and attempt to execute these plans in social encounters (McCall and Simmons, 1978).

Cooley, Mead, and Symbolic Interactionist Perspectives

Social constructionism is a term that is applied to theories that emphasize the socially created nature of social life. This perspective is linked to symbolic interactionist theory, and its roots can be traced to the Chicago School and early theorists such as Charles Horton Cooley and George Herbert Mead.

Cooley and the Looking-Glass Self According to the sociologist Charles Horton Cooley (1864–1929), the ***looking-glass self* refers to the way in which a person's sense of self is derived from the perceptions of others.** Our looking-glass self is not who we actually are or what people actually think about us; rather, it is based on our perception of *how* other people think of us (Cooley, 1998/1902). Cooley asserted that we base our perception of who we are on how we think other people see us and on whether this opinion seems good or bad to us.

As ▶ Figure 4.2 on the next page shows, the looking-glass self is a self-concept derived from a three-step process:

1. We imagine how our personality and appearance will look to other people. We may imagine that we are attractive or unattractive, heavy or slim, friendly or unfriendly, and so on.
2. We imagine how other people judge the appearance and personality that we think we present. This step involves our perception of how we think they are judging us. We may be correct or incorrect!
3. We develop a self-concept. If we think the evaluation of others is favorable, our self-concept is enhanced. If we think the evaluation is unfavorable, our self-concept is diminished. (Cooley, 1998/1902)

According to Cooley, we use our interactions with others as a mirror for our own thoughts and actions; our sense of self depends on how we interpret what others do and say. Consequently, our sense of self is not permanently fixed; it is always developing as we interact with others in the larger society. For Cooley, self and society are merely two sides of the same coin: "Self and society go together, as phases of a common whole. I am aware of the social groups in which I live as immediately and authentically as I am aware of myself" (Cooley, 1963/1909: 8–9). Accordingly, the self develops only through contact with others, just as social institutions and societies do not exist independently of the interaction of acting individuals (Schubert, 1998). By developing the idea of the looking-glass self, Cooley made us aware of the mutual interrelationship between the individual and society—namely, that society shapes people and people shape society.

Mead and Role-Taking George Herbert Mead (1863–1931) extended Cooley's insights by linking the idea of self-concept to ***role-taking*—the process by which a person mentally assumes the role of another person or group in order to understand the world from that person's or group's point of view.** Role-taking often

self-concept the totality of our beliefs and feelings about ourselves.

looking-glass self Charles Horton Cooley's term for the way in which a person's sense of self is derived from the perceptions of others.

role-taking the process by which a person mentally assumes the role of another person in order to understand the world from that person's point of view.

▶ **Figure 4.2 How the Looking-Glass Self Works**

Source: Based on Katzer, Cook, and Crouch, 1991.

occurs through play and games, as children try out different roles (such as being mommy, daddy, doctor, or teacher) and gain an appreciation of them. First, people come to take the role of the other (role-taking). By taking the roles of others, the individual hopes to ascertain the intention or direction of the acts of others. Then the person begins to construct his or her own roles (role-making) and to anticipate other individuals' responses. Finally, the person plays at her or his particular role (role-playing).

According to Mead (1934), in the early months of life, children do not realize that they are separate from others. However, they do begin early on to see a mirrored image of themselves in others. Shortly after birth, infants start to notice the faces of those around them, especially the significant others, whose faces start to have meaning because they are associated with experiences such as feeding and cuddling. ***Significant others*** **are those persons whose care, affection, and approval are especially desired and who are most important in the development of the self.** Gradually, we distinguish ourselves from our caregivers and begin to perceive ourselves in contrast to them. As we develop language skills and learn to understand symbols, we begin to develop a self-concept. When we can represent ourselves in our minds as objects distinct from everything else, our self has been formed.

Mead (1934) divided the self into the "I" and the "me." The "I" is the subjective element of the self and represents the spontaneous and unique traits of each person. The "me" is the objective element of the self, which is composed of the internalized attitudes and demands of other members of society and the individual's awareness of those demands. Both the "I" and the "me" are needed to form the social self. The unity of the two constitutes the full development of the individual. According to Mead, the "I" develops first, and the "me" takes form during the three stages of self development:

1. During the *preparatory stage*, up to about age three, interactions lack meaning, and children largely imitate the people around them. At this stage, children are preparing for role-taking.
2. In the *play stage*, from about age three to five, children learn to use language and other symbols, thus enabling them to pretend to take the roles of specific people. At this stage, they begin to see themselves in relation to others, but they do not see role-taking as something they have to do.
3. During the *game stage*, which begins in the early school years, children understand not only their own social position but also the positions of others around them. In contrast to play, games are structured by rules, are often competitive, and involve a number of other "players." At this time, children become concerned about the demands and expectations of others and of the larger society. Mead used the example of a baseball game to describe this stage because children, like baseball players, must take into account the roles of all the other players at the same time. Mead's concept of the ***generalized other*** **refers to the child's awareness of the demands and expectations of the society as a whole or of the child's subculture.**

© Digital Vision/Alamay

© Daniel Bosler/Getty Images

© Stella/Getty Images

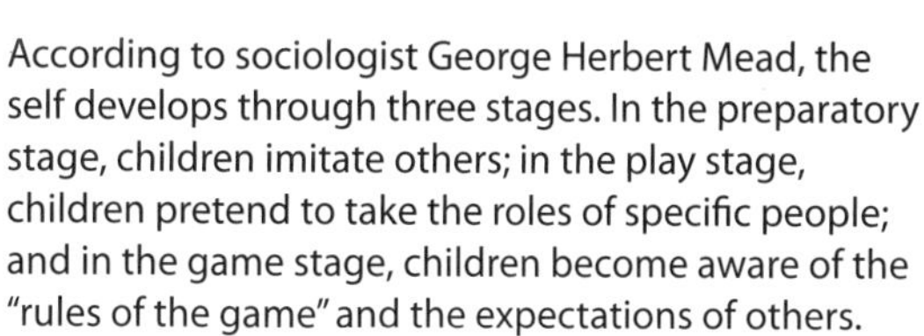

According to sociologist George Herbert Mead, the self develops through three stages. In the preparatory stage, children imitate others; in the play stage, children pretend to take the roles of specific people; and in the game stage, children become aware of the "rules of the game" and the expectations of others.

Is socialization a one-way process? No, according to Mead. Socialization is a two-way process between society and the individual. Just as the society in which we live helps determine what kind of individuals we will become, we have the ability to shape certain aspects of our social environment and perhaps even the larger society.

How useful are symbolic interactionist perspectives such as Cooley's and Mead's in enhancing our understanding of the socialization process? Certainly, this approach contributes to our understanding of how the self develops. Cooley's idea of the looking-glass self makes us aware that our perception of how we think others see us is not always correct. Mead extended Cooley's ideas by emphasizing the cognitive skills acquired through role-taking. His concept of the generalized other helps us see that the self is a social creation. According to Mead (1934: 196), "Selves can only exist in definite relations to other selves. No hard-and-fast line can be drawn between our own selves and the selves of others." As shown in "Sociology Works!," some of Mead's ideas have important current applications.

However, the viewpoints of symbolic interactionists such as Cooley and Mead have certain limitations. Sociologist Anne Kaspar (1986) suggests that Mead's ideas about the social self may be more applicable to men than to women because women are more likely to experience inherent conflicts between the meanings they derive from their personal experiences and those they take from the culture, particularly in regard to balancing the responsibilities of family life and paid employment. (This chapter's Concept Quick Review on page 117 summarizes the major theories of human development.)

Recent Symbolic Interactionist Perspectives

The symbolic interactionist approach emphasizes that socialization is a collective process in which children are active and creative agents, not just passive recipients of

significant others those persons whose care, affection, and approval are especially desired and who are most important in the development of the self.

generalized other George Herbert Mead's term for the child's awareness of the demands and expectations of the society as a whole or of the child's subculture.

Sociology *Works!*

"Good Job!": Mead's Generalized Other and the Issue of Excessive Praise

> Hang out at a playground, visit a school, or show up at a child's birthday party, and there's one phrase you can count on hearing repeatedly: "Good job!" Even tiny infants are praised for smacking their hands together ("Good clapping!"). Many of us blurt out these judgments of our children to the point that it has become almost a verbal tic. (Kohn, 2001)

Educational analyst Alfie Kohn describes the common practice of praising children for practically everything they say or do. According to Kohn, excessive praise or unearned compliments may be problematic for children because, rather than bolstering their self-esteem, such praise may increase a child's dependence on adults. As children increasingly rely on constant praise and on significant others to identify what is good or bad about their performance, they may not develop the ability to make meaningful judgments about what they have done. As Kohn suggests (2001), "Sadly, some of these kids will grow into adults who continue to need someone to pat them on the head and tell them whether what they did was OK."

Kohn's ideas remind us of the earlier sociological insights of George Herbert Mead, who described how children learn to take into account the expectations of the larger society and to balance the "I" (the subjective element of the self: the spontaneous and unique traits of each person) with the "me" (the objective element of the self: the internalized attitudes and demands of other members of society and the individual's awareness of those demands). As Mead (1934: 160) stated, "What goes on in the game goes on in the life of the child at all times. He is continually taking the attitudes of those about him, especially the roles of those who in some sense control him and on whom he depends." According to Mead, role-taking is vital to the formation of a mature sense of self as each individual learns to visualize the intentions and expectations of other people and groups. Excessively praising children may make it more difficult for them to develop a positive self-concept and visualize an accurate picture of what is expected of them as they grow into young adulthood.

Does this mean that children should not be praised? Definitely not! It means that we should think about when and how to praise children. What children may need sometimes is not praise, but encouragement. As child development specialist Docia Zavitkovsky has stated,

> I sometimes say that praise is fine "when praise is due." We get into the habit of praising when it isn't praise that is appropriate but encouragement. For example, we're always saying to young children: "Oh, what a beautiful picture," even when their pictures aren't necessarily beautiful. So why not really look at each picture? Maybe a child has painted a picture with many wonderful colors. Why don't we comment on that—on the reality of the picture? (qtd. in Scholastic Parent & Child, 2007)

From this perspective, positive feedback can have a very important influence on a child's self-esteem because he or she can learn how to do a "good job" when engaging in a specific activity or accomplishing a task rather than simply being praised for any effort expended. Mead's concept of the *generalized other* makes us aware of the importance of other people's actions in how self-concept develops.

Reflect & Analyze

To what extent are parents responsible for how self-concept develops in their child? Also, when we are dealing with peers, how might we thoughtfully use the phrase "Good job!" without making it into an overworked expression?

the socialization process. From this view, childhood is a *socially constructed* category (Adler and Adler, 1998). Children are capable of actively constructing their own shared meanings as they acquire language skills and accumulate interactive experiences (Qvortrup, 1990). According to the sociologist William A. Corsaro's (1985, 1997) "orb web model," children's cultural knowledge reflects not only the beliefs of the adult world but also the unique interpretations and aspects of the children's own peer culture. Corsaro (1992: 162) states that *peer culture* is "a stable set of activities or routines, artifacts, values, and concerns that children produce and share." This peer culture emerges through interactions as children "borrow" from the adult culture but transform it so that it fits their own situation. Based on ethnographic studies of U.S. and Italian preschoolers, Corsaro found that very young children engage in predictable patterns of interaction. For example, when playing together, children often permit some children to gain access to their group and play area while preventing others from becoming a part of their group. Children also play "approach–avoidance" games in which they alternate between approaching a threatening person or group and then running away. In fact, Corsaro (1992) believes that the peer group is the most significant public realm for children. (Peer groups as agents of socialization are dis-

CONCEPT QUICK REVIEW

Psychological and Sociological Theories of Human Development

Social Psychological Theories of Human Development	Freud's psychoanalytic perspective	Children first develop the id (drives and needs), then the ego (restrictions on the id), and then the superego (moral and ethical aspects of personality).
	Piaget's cognitive development	Children go through four stages of cognitive (intellectual) development, going from understanding only through sensory contact to engaging in highly abstract thought.
	Kohlberg's stages of moral development	People go through three stages of moral development, from avoidance of unwanted consequences to viewing morality based on human rights.
	Gilligan: gender and moral development	Women go through stages of moral development from personal wants to the greatest good for themselves and others.
Sociological Theories of Human Development	Cooley's looking-glass self	A person's sense of self is derived from his or her perception of how others view him or her.
	Mead's three stages of self-development	In the preparatory stage, children imitate the people around them; in the play stage, children pretend to take the roles of specific people; and in the game stage, children learn the demands and expectations of roles.

cussed later in the chapter.) This approach contributes to our knowledge about human development because it focuses on group life rather than individuals. Researchers using this approach "look at social relations, the organization and meanings of social situations, and the collective practices through which children create and recreate key constructs in their daily interactions" (Adler and Adler, 1998: 10; see also Thorne, 1993; Eder, 1995).

Ecological Perspectives

Another approach that emphasizes cultural or environmental influences on human development is the ecological perspective. One of the best-known ecological approaches is developmental psychologist Urie Bronfenbrenner's (1989) *ecological systems theory.* The ecological systems in this theory are the interactions a child has with other people, and how those interactions are affected by other people and situations. The four ecological systems are as follows, starting with the one closest to the child: the *microsystem,* the *mesosystem,* the *exosystem,* and the *macrosystem.* In the microsystem, a child is engaged in immediate face-to-face interaction with the child's parents, siblings, and other immediate family members. By contrast, in the mesosystem, the child's interactions with family members are influenced by the interactions of those family members. For example, how the mother reacts to her son is influenced by how she is getting along with the father. The exosystem relates to how the immediate family members are influenced by another setting, such as the mother's job. Finally, the macrosystem involves how interaction with the child is affected by all the components of the larger society, including public policy, such as child-care legislation.

Bronfenbrenner's ecological perspective provides interesting insights on the overall context in which child development occurs. However, research using this approach is somewhat difficult because of the complex nature of the systems approach that he suggests. As a result, many sociological studies have focused on specific agents of socialization rather than the larger societal context in which child development occurs.

Agents of Socialization

***Agents of socialization* are the persons, groups, or institutions that teach us what we need to know in order to participate in society.** We are exposed to many agents of socialization throughout our lifetime;

agents of socialization the persons, groups, or institutions that teach us what we need to know in order to participate in society.

in turn, we have an influence on those socializing agents and organizations. Here, we look at the most pervasive ones in childhood—the family, the school, peer groups, and the mass media.

The Family

The family is the most important agent of socialization in all societies. From our infancy onward, our families transmit cultural and social values to us. As discussed later in this book, families vary in size and structure. Some families consist of two parents and their biological children, whereas others consist of a single parent and one or more children. Still other families reflect changing patterns of divorce and remarriage, and an increasing number are made up of same-sex partners and their children. Over time, patterns have changed in some two-parent families so that fathers, rather than mothers, are the primary daytime agents of socialization for their young children.

Theorists using a functionalist perspective emphasize that families serve important functions in society because they are the primary locus for the procreation and socialization of children. Most of us form an emerging sense of self and acquire most of our beliefs and values within the family context. We also learn about the larger dominant culture (including language, attitudes, beliefs, values, and norms) and the primary subcultures to which our parents and other relatives belong.

Families are also the primary source of emotional support. Ideally, people receive love, understanding, security, acceptance, intimacy, and companionship within families. The role of the family is especially significant because young children have little social experience beyond the family's boundaries; they have no basis for comparing or evaluating how they are treated by their own family.

To a large extent, the family is where we acquire our specific social position in society. From birth, we are a part of the specific racial, ethnic, class, religious, and regional subcultural grouping of our family. Studies show that families socialize their children somewhat differently based on race, ethnicity, and class (Kohn, 1977; Kohn et al., 1990; Harrison et al., 1990). For example, sociologist Melvin Kohn (1977; Kohn et al., 1990) has suggested that social class (as measured by parental occupation) is one of the strongest influences on what and how parents teach their children. On the one hand, working-class parents, who are closely supervised and expected to follow orders at work, typically emphasize to their children the importance of obedience and conformity. On the other hand, parents from the middle and professional classes, who have more freedom and flexibility at work, tend to give their children more freedom to make their own decisions and to be creative. Kohn concluded that differences in parents' occupations were a better predictor of child-rearing practices than was social class itself.

Whether or not Kohn's findings are valid today, the issues he examined make us aware that not everyone has the same family experiences. Many factors—including our cultural background, nation of origin, religion, and gender—are important in determining how we are socialized by family members and others who are a part of our daily life.

Conflict theorists stress that socialization contributes to false consciousness—a lack of awareness and a distorted perception of the reality of class as it affects all aspects of social life. As a result, socialization reaffirms and reproduces the class structure in the next generation rather than challenging the conditions that presently exist. For example, children in low-income families may be unintentionally socialized to believe that acquiring an education and aspiring to lofty ambitions are pointless because of existing economic conditions in the family (Ballantine, 2001). By contrast, middle- and upper-income families typically instill ideas of monetary and social success in children while encouraging them to think and behave in "socially acceptable" ways.

The social constructionist/symbolic interactionist perspective helps us recognize that children affect their parents' lives and change the overall household environment. When we examine the context in which family life takes place, we also see that grandparents and other relatives have a strong influence on how parents socialize their children. In turn, the children's be-

As this celebration attended by three generations of family members illustrates, socialization enables society to "reproduce" itself.

havior may have an effect on how parents, siblings, and grandparents get along with one another. For example, in families where there is already intense personal conflict, the birth of an infant may intensify the stress and discord, sometimes resulting in child maltreatment, spousal battering, or elder abuse. By contrast, in families where partners feel happiness and personal satisfaction, the birth of an infant may contribute to the success of the marriage and bring about positive interpersonal communications among relatives.

The School

As the amount of specialized technical and scientific knowledge has expanded rapidly and as the amount of time that children are in educational settings has increased, schools continue to play an enormous role in the socialization of young people. For many people, the formal education process is an undertaking that lasts up to twenty years.

As the number of one-parent families and families in which both parents work outside the home has increased dramatically, the number of children in day-care and preschool programs has also grown rapidly. Currently, about 60 percent of all U.S. preschool children are in day care, either in private homes or institutional settings, and this percentage continues to climb (Children's Defense Fund, 2002). Generally, studies have found that quality day-care and preschool programs have a positive effect on the overall socialization of children. These programs provide children with the opportunity to have frequent interactions with teachers and to learn how to build their language and literacy skills. High-quality programs also have a positive effect on the academic performance of children, particularly those from low-income families. For example, several states with pre-kindergarten programs reported an increase in children's math and reading scores, school attendance records, and parents' involvement in their children's education (Children's Defense Fund, 2002). Today, however, the cost of child-care programs has become a major concern for many families (see Box 4.2).

Although schools teach specific knowledge and skills, they also have a profound effect on children's self-image, beliefs, and values. As children enter school for the first time, they are evaluated and systematically compared with one another by the teacher. A permanent, official record is kept of each child's personal behavior and academic activities. From a functionalist perspective, schools are responsible for (1) socialization, or teaching students to be productive members of society; (2) transmission of culture; (3) social control and personal development; and (4) the selection, training, and placement of individuals on different rungs in the society (Ballantine, 2001).

In contrast, conflict theorists assert that students have different experiences in the school system depending on their social class, their racial–ethnic background, the neighborhood in which they live, their gender, and other factors. According to the sociologists Samuel Bowles and Herbert Gintis (1976), much of what happens in school amounts to teaching a hidden curriculum in which children learn to be neat, to be on time, to be quiet, to wait their turn, and to remain attentive to their work. Thus, schools do not socialize children for their own well-being but rather for their later roles in the work force, where it is important to be punctual and to show deference to supervisors. Students who are destined for leadership or elite positions acquire different skills and knowledge than those who will enter working-class and middle-class occupations (see Cookson and Persell, 1985).

Symbolic interactionists examining socialization in the school environment might focus on how daily interactions and practices in schools affect the construction of students' beliefs regarding such things as patriotism, feelings of aggression or cooperation, and gender practices as they influence girls and boys. For example, some studies have shown that the school environment often fosters a high degree of gender segregation, including having boys and girls line up separately to participate in different types of extracurricular activities in middle schools and high schools (Eder, 1995; Thorne, 1993).

Peer Groups

As soon as we are old enough to have acquaintances outside the home, most of us begin to rely heavily on peer groups as a source of information and approval about social behavior. A ***peer group* is a group of people who are linked by common interests, equal social position, and (usually) similar age.** In early childhood, peer groups are often composed of classmates in day care, preschool, and elementary school. Recent studies have found that preadolescence—the latter part of the elementary school years—is an age period in which children's peer culture has an important effect on how children perceive themselves and how they internalize society's expectations (Adler and Adler, 1998). In adolescence, peer groups are typically made up of people with similar interests and social activities. As adults, we continue to participate in peer

peer group a group of people who are linked by common interests, equal social position, and (usually) similar age.

Box 4.2 Sociology and Social Policy

Who Should Pay for Child Care?

> Child care is the largest expense I have. It exceeds everything. It's higher than rent, higher than food. It's higher than everything we have.
>
> —Elizabeth Pokorny (*CBS News,* 2004)

> We have come to a point, well, do we send the children to a good daycare or do we send them to a not-so-good daycare so we can pay for it?
>
> —Sean Pokorny (*CBS News,* 2004)

Sean and Elizabeth Pokorny are among a growing number of couples in the United States who face a significant problem: the high cost of child care for their children. Elizabeth is an accountant, and Sean is a mechanic; however, even with their two paychecks, the Pokornys are concerned about the cost of quality day care, which for them is about $1,000 a month. For single-parent families with just one paycheck, the problem may be even more acute.

Although paying for child care is a major issue in the United States, it is not as large a concern in most other high-income countries, where the government typically provides all or almost all of the cost of child care for children above two years of age. By way of example, whereas about 25 to 30 percent of the cost of child care for three- to six-year-olds is provided by the government in the United States, France provides 100 percent, and Denmark, Finland, and Sweden provide about 80 percent (The Future of Children, 2001).

Some people believe that programs such as Head Start and state-funded prekindergarten programs provide all the child care that working parents in the United States should need for their young children; however, this is not an accurate assumption. Although Head Start and other government-funded child-care programs are important

© Yellow Dog Productions/Getty Images

Day-care centers have become important agents of socialization for increasing numbers of children. Today, about 60 percent of all U.S. preschool children are in day care of one kind or another.

groups of people with whom we share common interests and comparable occupations, income, and/or social position.

Peer groups function as agents of socialization by contributing to our sense of "belonging" and our feelings of self-worth. As early as the preschool years, peer groups provide children with an opportunity for successful adaptation to situations such as gaining access to ongoing play, protecting shared activities from intruders, and building solidarity and mutual trust during ongoing activities (Corsaro, 1985; Rizzo and Corsaro, 1995). Unlike families and schools, peer groups provide children and adolescents with some degree of freedom from parents and other authority figures (Corsaro, 1992). Although peer groups afford children some degree of freedom, they also teach cultural norms such as what constitutes "acceptable" behavior in a specific situation. Peer groups simultaneously reflect the larger culture and serve as a conduit for passing on culture to young people. As a result, the peer group is both a product of culture and one of its major transmitters (Elkin and Handel, 1989).

Is there such a thing as "peer pressure"? Individuals must earn their acceptance with their peers by conforming to a given group's norms, attitudes, speech patterns, and dress codes. When we conform to our peer group's expectations, we are rewarded; if we do not conform, we may be ridiculed or even expelled from the group. Conforming to the demands of peers frequently places children and adolescents at cross-purposes with their parents. For example, young people are frequently under pressure to obtain certain valued material possessions (such as toys, clothing, athletic shoes, or cell phones); they then pass the pressure on to their parents through emotional pleas to purchase the desired items. Peer pressure and the adult tensions that often accompany this kind of pressure are not unique to families in the United States (see Box 4.3 on page 126).

to working families, particularly those at or below the poverty line, these programs do not have adequate funding to achieve their primary purpose, which is to prepare children for school, much less to provide quality child care that is free or at least affordable to low-income parents and that is available at the hours around the clock that many of these parents are required to work. Consider, for example, parents who are employed in the food and entertainment industry and must work nights and holidays, or parents in cities such as Las Vegas and Reno, Nevada, where one of the primary sources of jobs is in the gaming industry (such as casinos that remain open 24 hours per day). Their children may need care on a 24/7 basis, rather than during the standard hours of the typical Head Start or state-funded pre-kindergarten program, which typically are available for only a few hours each day.

How might parents, especially those in low-income brackets, find affordable child care? Some people believe that it should be the parents' responsibility—that the parents should not have children whom they cannot afford to raise. Other people believe that paying for child care should be the responsibility of the government. However, for the government to contribute more to the cost of child care, we would need a refocused view of this issue, a view which suggests that it is a national concern and that it is critical to meeting two of our nation's priorities: helping families work and making sure that all children enter school ready to learn (Children's Defense Fund, 2003). Today, however, there is currently little hope that political leaders who are working to cut spending and balance budgets will see child care as a top priority.

Still other people believe that if the government is not going to bear the financial responsibility for child care, employers should be the ones to pay that cost because they benefit the most from the hours that their workers spend on the job. Without safe, reliable, affordable care for their children, employees may find it more difficult to be productive workers. Likewise, the cost of quality child care is a concern not only for families in low-income brackets, but also for many middle-class families that are feeling the squeeze of high gas prices, food costs, and rent or home-mortgage payments.

What are the chances that more employers will help fund child care? Generally, it appears that too few employers are willing or able to provide real help with child care in the twenty-first century. Many employers complain of sharply rising health care costs for their employees and believe that global competition is cutting into their earnings, thus making funding of child care for their employees a luxury that the employers simply cannot afford.

Are the *financial constraints* at various levels of government and in corporate America so great that these social institutions simply cannot afford to put money into child care and an investment in the nation's future? Or are our *priorities* confused so that we are spending money for things that are less important to the future of the nation than our children? If we truly mean it when we say that we want to "leave no child behind," perhaps we should rethink our priorities in regard to the funding of child care.

Reflect & Analyze

Do we have a responsibility for the children of this nation? Or are children the sole responsibility of their families, regardless of the parents' ability to pay the high cost of properly caring for and educating children in the United States? What do you think?

Mass Media

An agent of socialization that has a profound impact on both children and adults is the *mass media*, composed of large-scale organizations that use print or electronic means (such as radio, television, film, and the Internet) to communicate with large numbers of people. The media function as socializing agents in several ways: (1) they inform us about events; (2) they introduce us to a wide variety of people; (3) they provide an array of viewpoints on current issues; (4) they make us aware of products and services that, if we purchase them, will supposedly help us to be accepted by others; and (5) they entertain us by providing the opportunity to live vicariously (through other people's experiences). Although most of us take for granted that the media play an important part in contemporary socialization, we frequently underestimate the enormous influence this agent of socialization may have on children's attitudes and behavior.

The pleasure of participating in activities with friends is one of the many attractions of adolescent peer groups. What groups have contributed the most to your sense of belonging and self-worth?

Trying to Go It Alone: Runaway Adolescents and Teens

Home. Family. The ideal is that these words evoke a sense of well-being—feelings such as love, understanding, acceptance, and security. However, the reality is that for many of us, especially adolescents and teens, these words trigger negative feelings—fear, anxiety, dread—and an urgent desire to escape. The most recent statistics available indicate that between 1.6 and 2.8 million youths run away each year in the United States. That means one out of every seven children in this country will run away before the age of 18. Included in these numbers are not just runaway but also "throwaway" children, youths who have been forced out of their homes by parents or guardians, or, because they have turned 18, forced out of a foster care system.

What happens to humans' socialization when, as youths, they must fend for themselves to meet their own physical and emotional needs? The images in this essay give you a chance to look at the lives of runaways, from risk factors and precipitating events to means of survival and resources, and the longer-term effects that running away or being thrown away has on individuals and their communities. This essay also provides a glimpse of homeless children in global perspective. As you look at these images, consider the short- and long-term impact that persons age 12 (sometimes younger) to 18 attempting self-sufficiency can have on a society.

Why They Leave

Forty-seven percent of runaway and homeless youths indicate that conflict between them and their parent or guardian is a major problem. Thirty-four percent of runaways (girls and boys) report sexual abuse, and 43 percent report physical abuse before leaving home, abuse that increases youths' risk of being assaulted or raped—or both—on the street. Other problems that runaways report include alcohol and drug use, sexual orientation issues, and mental and physical health issues.

Life on the Street Is Hard

This homeless girl on a sidewalk in Manhattan reflects the reality that youths age 12 to 17 are at higher risk for homelessness than are adults. Available data show that 12 percent of runaway and homeless youths spend at least one night outside in a park, on a street, under a bridge or overhang, or on a rooftop. Seven percent of youths in runaway and homeless shelters and 14 percent of youths on the street surveyed in 1995 reported having traded sex for money, food, shelter, or drugs in the previous 12 months. Other means of survival include shelters and soup kitchens, panhandling, and stealing.

Society's Safety Nets

Males and females run away in equal numbers, although females are more likely to seek help through shelters and hotlines such as the National Runaway Switchboard (NRS), where 77 percent of the callers are female. In 2007, NRS handled more than 176,000 calls. NRS data show that the organization is serving more youths who are contemplating running away, instead of already having run away, than in the past. Children under age 12 are the fastest-growing group of callers, and in 2007, NRS received more crisis calls than it had in the past.

© Tim Boyle/Getty Images

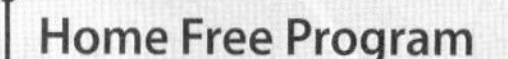

Courtesy of the National Runaway Switchboard

Home Free Program

In addition to providing its hotline and online services and a professionally developed runaway prevention curriculum titled "Let's Talk," the NRS partners with several organizations, most notably Greyhound Lines, Inc., to offer its Home Free Program, promoted in this poster by the musician Ludacris. More than 10,000 runaways have been reunited with their families, free of charge, since the program was started.

© DEAN J. KOEPFLER/Tacoma News Tribune/MCT/Landov

Catching Up

As previously noted, 50 percent of homeless youths age 16 or older reported having dropped out of school, having been expelled, or having been suspended. Even if school had not been a major problem, once adolescents and teens, especially teens, have run away, the disruption to their education may be so significant that they end up dropping out of school. Catching up and moving ahead takes determination—and opportunities such as ones provided by this Skills, Training, Employment, Preparation Services (STEPS) program, where teens can study for a high school General Equivalency Development (GED) test.

© AP Images

Do you believe that what this child is learning here will have an influence on her actions in the future? What other childhood experiences might offset early negative racial socialization?

directly or intentionally. Scholars have found that ethnic values and attitudes begin to crystallize among children as young as age four (Goodman, 1964; Porter, 1971). By this age, the society's ethnic hierarchy has become apparent to the child (Marger, 2003). Some minority parents believe that racial socialization is essential because it provides children with the skills and abilities they will need to survive in the larger society (Hale-Benson, 1986). For example, Chuck Hayashi, a Japanese American, describes how his parents helped him:

> [Racism] never tore me up, but what helped me was just little things my parents used to tell me. Like, yes, you are different, but the people that really count will overlook things like that. My dad would say things like that. But he would also say, look, you are a minority, and you will have to compete for jobs and things with the majority. And he would tell me. I know it was to make me work harder because he would say we had to do twice as good to get the jobs. He told me that several times. (qtd. in Tuan, 1998: 69)

Scholars may be hesitant to point out differences in socialization practices among diverse racial–ethnic and social class groupings because such differences have typically been interpreted by others to be a sign of inadequate (or inferior) socialization practices.

Socialization Through the Life Course

Why is socialization a lifelong process? Throughout our lives, we continue to learn. Each time we experience a change in status (such as becoming a college student or getting married), we learn a new set of rules, roles, and relationships. Even before we achieve a new status, we often participate in ***anticipatory socialization*—the process by which knowledge and skills are learned for future roles.** Many societies organize social activities according to age and gather data regarding the age composition of the people who live in that society. For example, the U.S. Census Bureau gathers and maintains those data in the United States (see "Census Profiles: Age of the U.S. Population"). Some societies have distinct *rites of passage,* based on age or other factors, that publicly dramatize and validate changes in a person's status. In the United States and other industrialized societies, the most common categories of age are infancy, childhood, adolescence, and adulthood (often subdivided into young adulthood, middle adulthood, and older adulthood).

Infancy and Childhood

Some social scientists believe that a child's sense of self is formed at a very early age and that it is difficult to change this self-perception later in life. Symbolic interactionists emphasize that during infancy and early childhood, family support and guidance are crucial to a child's developing self-concept. In some families, children are provided with emotional warmth, feelings of mutual trust, and a sense of security. These families come closer to our ideal cultural belief that childhood should be a time of carefree play, safety, and freedom from economic, political, and sexual responsibilities. However, other families reflect the discrepancy between cultural ideals and reality—children grow up in a setting characterized by fear, danger, and risks that are created by parental neglect, emotional maltreatment, or premature economic and sexual demands (Knudsen, 1992).

Abused children often experience low self-esteem, an inability to trust others, feelings of isolationism and powerlessness, and denial of their feelings. However, the manner in which parental abuse affects children's ongoing development is subject to much debate and uncertainty. For example, some scholars and therapists assert that the intergenerational hypothesis—the idea that abused children will become abusive parents—is valid, but others have found little support for this hypothesis (Knudsen, 1992).

According to the developmental psychologist Urie Bronfenbrenner (1990), mutual interaction with a caring adult—and preferably a number of nurturing adults—is essential for the child's emotional, physical, intellectual, and social growth. However, Bronfenbrenner also states that at the macrosystem level, it is necessary for communities and the major economic, social, and political institutions of the entire society to provide the public policies and practices that support positive child-rearing activities on the part of families.

Adolescence

In industrialized societies, the adolescent (or teenage) years represent a buffer between childhood and adulthood. In the United States, no specific rites of passage exist to mark children's move into adulthood; therefore, young people have to pursue their own routes to self-identity and adulthood (Gilmore, 1990). Anticipatory socialization is often associated with adolescence, during which many young people spend much of their time planning or being educated for future roles they hope to occupy. However, other adolescents (such as eleven- and twelve-year-old mothers) may have to plunge into adult responsibilities at this time. Adolescence is often characterized by emotional and social unrest. In the process of developing their own identities, some young people come into conflict with parents, teachers, and other authority figures who attempt to restrict their freedom. Adolescents may also find themselves caught between the demands of adulthood and their own lack of financial independence and experience in the job market. The experiences of individuals during adolescence vary according to race, class, and gender. Based on their family's economic situation, some young people move directly into the adult world of work. However, those from upper-middle-class and upper-class families may extend adolescence into their late twenties or early thirties by attending graduate or professional school and then receiving additional advice and financial support from their parents as they start their own families, careers, or businesses.

CENSUS PROFILES

Age of the U.S. Population

Just as age is a crucial variable in the socialization process, the U.S. Census Bureau gathers data about people's age so that the government and other interested parties will know how many individuals residing in this country are in different age categories. This chapter examines how a person's age is related to socialization and one's life experiences. Shown below is a depiction of the nation's population in the year 2007, separated into three broad age categories.

Below age 25

Age 25 to 54

Age 55 and above

Can age be a source of social cohesion among people? Why might age differences produce conflict among individuals in different age groups? What do you think?

Source: U.S. Census Bureau, 2007.

© David Young-Wolff/PhotoEdit

An important rite of passage for many Latinas is the *quinceañera*—a celebration of their fifteenth birthday and their passage into womanhood. Can you see how this occasion might also be a form of anticipatory socialization?

anticipatory socialization the process by which knowledge and skills are learned for future roles.

which makes it possible for children to experience many things outside their own homes and schools and to communicate regularly with people around the world. It is very likely that socialization in the future will be vastly different in the world of global instant communication than it has been in the past.

Chapter Review

- **What is socialization, and why is it important for human beings?**

Socialization is the lifelong process through which individuals acquire their self-identity and learn the physical, mental, and social skills needed for survival in society. The kind of person we become depends greatly on what we learn during our formative years from our surrounding social groups and social environment.

- **How much of our unique human characteristics comes from heredity and how much from our social environment?**

As individual human beings, we have unique identities, personalities, and relationships with others. Each of us is a product of two forces: (1) heredity, referred to as "nature," and (2) the social environment, referred to as "nurture." Whereas biology dictates our physical makeup, the social environment largely determines how we develop and behave.

- **Why is social contact essential for human beings?**

Social contact is essential in developing a self, or self-concept, which represents an individual's perceptions and feelings of being a distinct or separate person. Much of what we think about ourselves is gained from our interactions with others and from what we perceive that others think of us.

- **What are the main social psychological theories on human development?**

According to Sigmund Freud, the self emerges from three interrelated forces: the id, the ego, and the superego. When a person is well adjusted, the three forces act in balance. Jean Piaget identified four cognitive stages of development; each child must go through each stage in sequence before moving on to the next one, although some children move through them faster than others.

- **How do sociologists believe that we develop a self-concept?**

According to Charles Horton Cooley's concept of the looking-glass self, we develop a self-concept as we see ourselves through the perceptions of others. Our initial sense of self is typically based on how families perceive and treat us. George Herbert Mead suggested that we develop a self-concept through role-taking and learning the rules of social interaction. According to Mead, the self is divided into the "I" and the "me." The "I" represents the spontaneous and unique traits of each person. The "me" represents the internalized attitudes and demands of other members of society.

- **What are the primary agents of socialization?**

The agents of socialization include the family, schools, peer groups, and the media. Our families, which transmit cultural and social values to us, are the most important agents of socialization in all societies, serving these functions: (1) procreating and socializing children, (2) providing emotional support, and (3) assigning social position. Schools primarily teach knowledge and skills but also have a profound influence on the self-image, beliefs, and values of children. Peer groups contribute to our sense of belonging and self-worth, and are a key source of information about acceptable behavior. The media function as socializing agents by (1) informing us about world events, (2) introducing us to a wide variety of people, and (3) providing an opportunity to live vicariously through other people's experiences.

- **When does socialization end?**

Socialization is ongoing throughout the life course. We learn knowledge and skills for future roles through anticipatory socialization. Parents are socialized by their own children, and adults learn through workplace socialization. Resocialization is the process of learning new attitudes, values, and behaviors, either voluntarily or involuntarily.

www.cengage.com/login

Want to maximize your online study time? Take this easy-to-use study system's diagnostic pre-test, and it will create a personalized study plan for you. By helping you to identify the topics that you need to understand better and then directing you to valuable online resources, it can speed up your chapter review. CengageNOW even provides a post-test so you can confirm that you are ready for an exam.

Key Terms

ageism 130
agents of socialization 117
anticipatory socialization 128
ego 109
gender socialization 125
generalized other 114
id 109
looking-glass self 113
peer group 119
racial socialization 127
resocialization 131
role-taking 113
self-concept 113
significant others 114
social devaluation 130
socialization 104
sociobiology 105
superego 109
total institution 131

Questions for Critical Thinking

1. Consider the concept of the looking-glass self. How do you think others perceive you? Do you think most people perceive you correctly?
2. What are your "I" traits? What are your "me" traits? Which ones are stronger?
3. What are some different ways that you might study the effect of toys on the socialization of children? How could you isolate the toy variable from other variables that influence children's socialization?
4. Is the attempted rehabilitation of criminal offenders—through boot camp programs, for example—a form of socialization or resocialization?

The Kendall Companion Website

www.cengage.com/sociology/kendall

Visit this book's companion website, where you'll find more resources to help you study and successfully complete course projects. Resources include quizzes and flash cards, as well as special features such as an interactive sociology timeline, maps, General Social Survey (GSS) data, and Census 2000 data. The site also provides links to useful websites that have been selected for their relevance to the topics in this chapter and include those listed below. (*Note:* Visit the book's website for updated URLs.)

Social Psychology Index

http://www.trinity.edu/mkearl/socpsy.html#in

From Michael Kearl's award-winning sociology Web gateway, this page will link you to a number of resources on topics discussed in this chapter, including the nature-versus-nurture debate, the looking-glass self, agents of socialization, and much more.

Rob Kling Center for Social Informatics

http://rkcsi.indiana.edu

The Rob Kling Center for Social Informatics at Indiana University conducts research on the influence of technology and computerization on society. Its website will link you to journals, papers, workshops, and other resources that examine the social aspects of computerization.

chapter

5 Society, Social Structure, and Interaction

Chapter Focus Question

How is homelessness related to the social structure of a society?

I began Dumpster diving [scavenging in a large garbage bin] about a year before I became homeless. . . . The area I frequent is inhabited by many affluent college students. I am not here by chance; the Dumpsters in this area are very rich. Students throw out many good things, including food. In particular they tend to throw everything out when they move at the end of a semester, before and after breaks, and around midterm, when many of them despair of college. So I find it advantageous to keep an eye on the academic calendar.

I learned to scavenge gradually, on my own. Since then I have initiated several companions into the trade. I have learned that there is a predictable series of stages a person goes through in learning to scavenge.

At first the new scavenger is filled with disgust and self-loathing. He is ashamed of being seen and may

All activities in life—including scavenging in garbage bins and living "on the streets"—are social in nature.

IN THIS CHAPTER

lurk around, trying to duck behind things, or he may dive at night. (In fact, most people instinctively look away from a scavenger. By skulking around, the novice calls attention to himself and arouses suspicion. Diving at night is ineffective and needlessly messy.) . . . That stage passes with experience. The scavenger finds a pair of running shoes that fit and look and smell brand-new. . . . He begins to understand: People throw away perfectly good stuff, a lot of perfectly good stuff.

At this stage, Dumpster shyness begins to dissipate. The diver, after all, has the last laugh. He is finding all manner of good things that are his for the taking. Those who disparage his profession are the fools, not he.

—Author Lars Eighner recalls his experiences as a Dumpster diver while living under a shower curtain in a stand of bamboo in a public park. Eighner became homeless when he was evicted from his "shack" after being unemployed for about a year. (Eighner, 1993: 111–119)

Eighner's "diving" activities reflect a specific pattern of social behavior. All activities in life—including scavenging in garbage bins and living "on the streets"—are social in nature. Homeless persons and domiciled persons (those with homes) live in social worlds that have predictable patterns of social interaction. ***Social interaction*** **is the process by which people act toward or respond to other people** and is the foundation for all relationships and groups in society. In this chapter, we look at the relationship between social structure and

Sharpening Your Focus

- *How do societies change over time?*
- *What are the components of social structure?*
- *Why do societies have shared patterns of social interaction?*
- *How are daily interactions similar to being onstage?*
- *Do positive changes in society occur through individual efforts or institutional efforts?*

social interaction. In the process, homelessness is used as an example of how social problems occur and how they may be perpetuated within social structures and patterns of interaction.

***Social structure* is the complex framework of societal institutions (such as the economy, politics, and religion) and the social practices (such as rules and social roles) that make up a society and that organize and establish limits on people's behavior.** This structure is essential for the survival of society and for the well-being of individuals because it provides a social web of familial support and social relationships that connects each of us to the larger society. Many homeless people have lost this vital linkage. As a result, they often experience a loss of personal dignity and a sense of moral worth because of their "homeless" condition (Snow and Anderson, 1993).

Who are the homeless? Before reading on, take the quiz on homelessness in Box 5.1. The characteristics of the homeless population in the United States vary widely. Among the homeless are single men, single women, and families. In recent years, families with children have accounted for 40 percent of the homeless population (U.S. Conference of Mayors, 2005). Further, people of color are overrepresented among the homeless. In 2004, African Americans made up 49 percent of the homeless population, whites (Caucasians) 35 percent, Latinas/os (Hispanics) 13 percent, Native Americans 2 percent, and Asian Americans 1 percent (U.S. Conference of Mayors, 2005). These percentages obviously vary across communities and different areas of the country.

Homeless persons come from all walks of life. They include undocumented workers, parolees, runaway youths and children, Vietnam veterans, and the elderly. They live in cities, suburbs, and rural areas. Contrary to popular myths, most of the homeless are not on the streets by choice or because they were deinstitutionalized by mental hospitals. Not all of the homeless are unemployed. About 22 percent of homeless people hold full- or part-time jobs but earn too little to find an affordable place to live (U.S. Conference of Mayors, 2005).

Social Structure: The Macrolevel Perspective

Social structure provides the framework within which we interact with others. This framework is an orderly, fixed arrangement of parts that together make up the whole group or society (see ▶ Figure 5.1). As defined in Chapter 1, a *society* is a large social grouping that shares the same geographical territory and is subject to the same political authority and dominant cultural expectations. At the macrolevel, the social structure of a society has several essential elements: social institutions, groups, statuses, roles, and norms.

Functional theorists emphasize that social structure is essential because it creates order and predictability in a society (Parsons, 1951). Social structure is also important for our human development. As we saw in Chapter 4, we develop a self-concept as we learn the attitudes, values, and behaviors of the people around us. When these attitudes and values are part of a predictable structure, it is easier to develop that self-concept.

Social structure gives us the ability to interpret the social situations we encounter. For example, we expect our families to care for us, our schools to educate us, and our police to protect us. When our circumstances change dramatically, most of us feel an acute sense of anxiety because we do not know what to expect or what is expected of us. For example, newly homeless individuals may feel disoriented because they do not know how to function in their new setting. The person

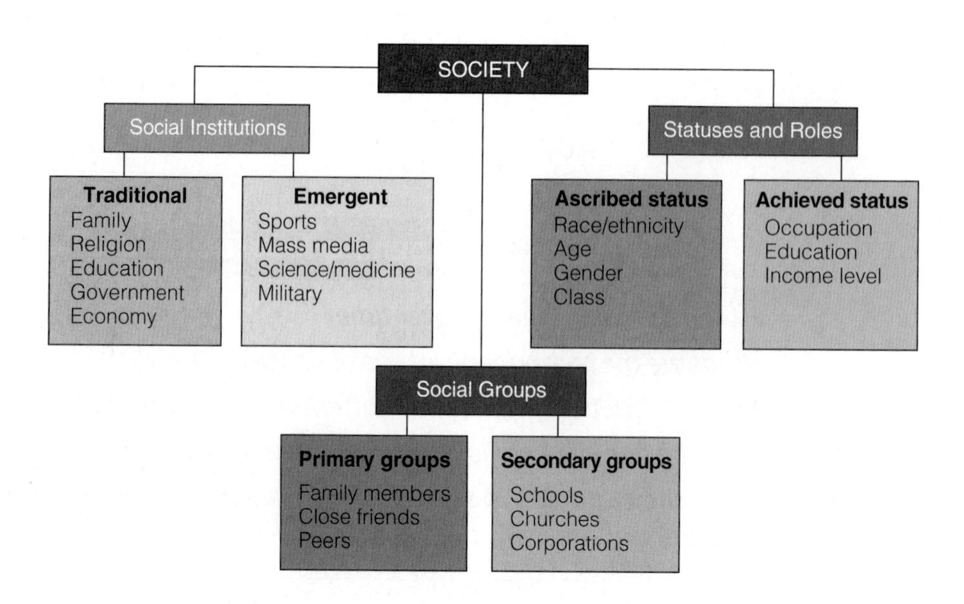

▶ **Figure 5.1 Social Structure Framework**

Box 5.1 Sociology and Everyday Life

How Much Do You Know About Homeless Persons?

True	False	
T	F	1. Most homeless people choose to be homeless.
T	F	2. Homelessness is largely a self-inflicted condition.
T	F	3. Homeless people do not work.
T	F	4. Most homeless people are mentally ill.
T	F	5. Homeless people typically panhandle (beg for money) so that they can buy alcohol or drugs.
T	F	6. Most homeless people are heavy drug users.
T	F	7. A large number of homeless persons are dangerous.
T	F	8. Homeless persons have existed throughout the history of the United States.
T	F	9. One out of every three homeless persons is a child.
T	F	10. Some homeless people have attended college and graduate school.

Answers on page 140

is likely to ask questions: "How will I survive on the streets?" "Where do I go to get help?" "Should I stay at a shelter?" "Where can I get a job?" Social structure helps people make sense out of their environment, even when they find themselves on the streets.

In addition to providing a map for our encounters with others, social structure may limit our options and place us in arbitrary categories not of our own choosing. Conflict theorists maintain that there is more to social structure than is readily visible and that we must explore the deeper, underlying structures that determine social relations in a society. Karl Marx suggested that the way economic production is organized is the most important structural aspect of any society. In capitalistic societies, where a few people control the labor of many, the social structure reflects a system of relationships of domination among categories of people (for example, owner–worker and employer–employee).

Social structure creates boundaries that define which persons or groups will be the "insiders" and which will be the "outsiders." *Social marginality* is the state of being part insider and part outsider in the social structure. Sociologist Robert Park (1928) coined this term to refer to persons (such as immigrants) who simultaneously share the life and traditions of two distinct groups. Social marginality results in stigmatization. A *stigma* is any physical or social attribute or sign that so devalues a person's social identity that it disqualifies that person from full social acceptance (Goffman, 1963b). A convicted criminal, wearing a prison uniform, is an example of a person who has been stigmatized; the uniform says that the person has done something wrong and should not be allowed unsupervised outside the prison walls.

Components of Social Structure

The social structure of a society includes its social positions, the relationships among those positions, and the kinds of resources attached to each of the positions. Social structure also includes all the groups that make up society and the relationships among those groups (Smelser, 1988). We begin by examining the social positions that are closest to the individual.

Status

A ***status* is a socially defined position in a group or society characterized by certain expectations, rights, and duties.** Statuses exist independently of the specific people occupying them (Linton, 1936); the statuses of professional athlete, rock musician, professor, college

social interaction the process by which people act toward or respond to other people; the foundation for all relationships and groups in society.

social structure the complex framework of societal institutions (such as the economy, politics, and religion) and the social practices (such as rules and social roles) that make up a society and that organize and establish limits on people's behavior.

status a socially defined position in a group or society characterized by certain expectations, rights, and duties.

Box 5.1 Sociology and Everyday Life

Answers to the Sociology Quiz on Homeless Persons

1. **False.** Less than 6 percent of all homeless people are that way by choice.
2. **False.** Most homeless persons did not inflict upon themselves the conditions that produced their homelessness. Some are the victims of child abuse or violence.
3. **False.** Many homeless people are among the working poor. Minimum-wage jobs do not pay enough for an individual to support a family or pay inner-city rent.
4. **False.** Most homeless people are not mentally ill; estimates suggest that about one-fourth of the homeless are emotionally disturbed.
5. **False.** Many homeless persons panhandle to pay for food, a bed at a shelter, or other survival needs.
6. **False.** Most homeless people are not heavy drug users. Estimates suggest that about one-third of the homeless are substance abusers. Many of these are part of the one-fourth of the homeless who are mentally ill.
7. **False.** Although an encounter with a homeless person occasionally ends in tragedy, most homeless persons are among the least threatening members of society. They are often the victims of crime, not the perpetrators.
8. **True.** Scholars have found that homelessness has always existed in the United States. However, the number of homeless persons has increased or decreased with fluctuations in the national economy.
9. **True.** Families with children are the fastest growing category of homeless persons in the United States. Many homeless children are alone. They may be runaways or "throwaways" whose parents do not want them to return home.
10. **True.** Some homeless persons have attended college and graduate school, and many have completed high school.

Sources: Based on Kroloff, 1993; Liebow, 1993; Snow and Anderson, 1993; U.S. Conference of Mayors, 2008; Vissing, 1996; and Waxman and Hinderliter, 1996.

student, and homeless person all exist exclusive of the specific individuals who occupy these social positions. For example, although thousands of new students arrive on college campuses each year to occupy the status of first-year student, the status of college student and the expectations attached to that position have remained relatively unchanged for the past one hundred years.

Does the term *status* refer only to high-level positions in society? No, not in a sociological sense. Although many people equate the term *status* with high levels of prestige, sociologists use it to refer to all socially defined positions—high rank and low rank. For example, both the position of director of the Department of Health and Human Services in Washington, D.C., and that of a homeless person who is paid about five dollars a week (plus bed and board) to clean up the dining room at a homeless shelter are social statuses (see Snow and Anderson, 1993).

Take a moment to answer the question "Who am I?" To determine who you are, you must think about your social identity, which is derived from the statuses you occupy and is based on your status set. A status set comprises all the statuses that a person occupies at a given time. For example, Maria may be a psychologist, a professor, a wife, a mother, a Catholic, a school volunteer, a Texas resident, and a Mexican American. All of these socially defined positions constitute her status set.

Ascribed and Achieved Status Statuses are distinguished by the manner in which we acquire them. An ***ascribed status*** **is a social position conferred at birth or received involuntarily later in life, based on attributes over which the individual has little or no control, such as race/ethnicity, age, and gender.** For example, Maria is a female born to Mexican American parents; she was assigned these statuses at birth. She is an adult and—if she lives long enough—will someday become an "older adult," which is an ascribed status received involuntarily later in life. An ***achieved status*** **is a social position that a person assumes voluntarily as a result of personal choice, merit, or direct effort.** Achieved statuses (such as occupation, education, and

© Jeff Brass/Getty Images

In the past, a person's status was primarily linked to his or her family background, education, occupation, and other sociological attributes. Today, some sociologists believe that celebrity status has overtaken the more traditional social indicators of status. The rock star Bono, shown here performing at a concert, is an example of celebrity status.

income) are thought to be gained as a result of personal ability or successful competition. Most occupational positions in modern societies are achieved statuses. For instance, Maria voluntarily assumed the statuses of psychologist, professor, wife, mother, and school volunteer. However, not all achieved statuses are positions that most people would want to attain; for example, being a criminal, a drug addict, or a homeless person is a negative achieved status.

Ascribed statuses have a significant influence on the achieved statuses we occupy. Race/ethnicity, gender, and age affect each person's opportunity to acquire certain achieved statuses. Those who are privileged by their positive ascribed statuses are more likely to achieve the more prestigious positions in a society. Those who are disadvantaged by their ascribed statuses may more easily acquire negative achieved statuses.

Master Status If we occupy many different statuses, how can we determine which is the most important? Sociologist Everett Hughes has stated that societies resolve this ambiguity by determining master statuses. A ***master status* is the most important status a person occupies;** it dominates all of the individual's other statuses and is the overriding ingredient in determining a person's general social position (Hughes, 1945). Being poor or rich is a master status that influences many other areas of life, including health, education, and life opportunities. Historically, the most common master statuses for women have related to positions in the family, such as daughter, wife, and mother. For men, occupation has usually been the most important status, although occupation is increasingly a master status for many women as well. "What do you do?" is one of the first questions many people ask when meeting another. Occupation provides important clues to a person's educational level, income, and family background. An individual's race/ethnicity may also constitute a master status in a society in which dominant-group members single out members of other groups as "inferior" on the basis of real or alleged physical, cultural, or nationality characteristics (see Feagin and Feagin, 2003).

Master statuses are vital to how we view ourselves, how we are seen by others, and how we interact with others. Justice Ruth Bader Ginsburg is both a U.S. Supreme Court justice and a mother. Which is her master status? Can you imagine how she would react if attorneys arguing a case before the Supreme Court treated her as if she were a mother rather than a justice? Lawyers wisely use "justice" as her master status and act accordingly.

Master statuses confer high or low levels of personal worth and dignity on people. Those are not characteristics that we inherently possess; they are derived from the statuses we occupy. For individuals who have no residence, being a homeless person readily becomes a master status regardless of the person's other attributes. Homelessness is a stigmatized master status that confers disrepute on its occupant because domiciled people often believe that a homeless person has a "character flaw." Sometimes this assumption is supported by how the media frame stories about homeless people (see Box 5.2). The circumstances under which

ascribed status a social position conferred at birth or received involuntarily later in life based on attributes over which the individual has little or no control, such as race/ethnicity, age, and gender.

achieved status a social position that a person assumes voluntarily as a result of personal choice, merit, or direct effort.

master status the most important status that a person occupies.

someone becomes homeless determine the extent to which that person is stigmatized. For example, individuals who become homeless as a result of natural disasters (such as a hurricane or a brush fire) are not seen as causing their homelessness or as being a threat to the community. Thus, they are less likely to be stigmatized. However, in cases in which homeless persons are viewed as the cause of their own problems, they are more likely to be stigmatized and marginalized by others. Snow and Anderson (1993: 199) observed the effects of homelessness as a master status:

> It was late afternoon, and the homeless were congregated in front of [the Salvation Army shelter] for dinner. A school bus approached that was packed with Anglo junior high school students being bused from an eastside barrio school to their upper-middle and upper-class homes in the city's northwest neighborhoods. As the bus rolled by, a fusillade of coins came flying out the windows, as the students made obscene gestures and shouted, "Get a job." Some of the homeless gestured back, some scrambled for the scattered coins—mostly pennies—others angrily threw the coins at the bus, and a few seemed oblivious to the encounter. For the passing junior high schoolers, the exchange was harmless fun, a way to work off the restless energy built up in school; but for the homeless it was a stark reminder of their stigmatized status and of the extent to which they are the objects of negative attention.

© Xinhua/Landov

© Rachel Epstein/PhotoEdit

Sociologists believe that being rich or poor may be a master status in the United States. How do the lifestyles of these two men differ based on their master statuses?

Status Symbols When people are proud of a particular social status that they occupy, they often choose to use visible means to let others know about their position. ***Status symbols*** **are material signs that inform others of a person's specific status.** For example, just as wearing a wedding ring proclaims that a person is married, owning a Rolls-Royce announces that one has "made it." As we saw in Chapter 3, achievement and success are core U.S. values. For this reason, people who have "made it" frequently want to display symbols to inform others of their accomplishments.

Status symbols for the domiciled and for the homeless may have different meanings. Among affluent persons, a full shopping cart in the grocery store and bags of merchandise from expensive department stores indicate a lofty financial position. By contrast, among the homeless, bulging shopping bags and overloaded grocery carts suggest a completely different status. Carts and bags are essential to street life; there is no other place to keep things, as shown by this description of Darian, a homeless woman in New York City:

> The possessions in her postal cart consist of a whole house full of things, from pots and pans to books, shoes, magazines, toilet articles, personal papers and clothing, most of which she made herself. . . .
>
> Because of its weight and size, Darian cannot get the cart up over the curb. She keeps it in the street near the cars. This means that as she pushes it slowly up and down the street all day long, she is living almost her entire life directly in traffic. She stops off along her route to sit or sleep for awhile and to be both stared at as a spectacle and to stare back. Every aspect of her life including sleeping, eating, and going to the bathroom is constantly in public view. . . . [S]he has no space to call her own and she never has a moment's privacy. Her privacy, her home, is her cart with all its possessions. (Rousseau, 1981: 141)

For homeless women and men, possessions are not status symbols as much as they are a link with the

Box 5.2 Framing Homelessness in the Media

Thematic and Episodic Framing

They live—and die—on a traffic island in the middle of a busy downtown street, surviving by panhandling drivers or turning tricks. Everyone in their colony is hooked on drugs or alcohol. They are the harsh face of the homeless in San Francisco.

The traffic island where these homeless people live is a 40-by-75 foot triangle chunk of concrete just west of San Francisco's downtown. . . . The little concrete divider wouldn't get a second glance, or have a name—if not for the colony that lives there in a jumble of shopping carts loaded with everything they own. It's called Homeless Island by the shopkeepers who work near it and the street sweepers who clean it; to the homeless, it is just the Island. The inhabitants live hand-to-mouth, sleep on the cement and abuse booze and drugs, mostly heroin. There are at least 3,000 others like them in San Francisco, social workers say. They are known as the "hard core," the people most visible on the streets, the most difficult to help. . . . (Fagan, 2003)

This news article is an example of typical media framing of stories about homeless people. The full article includes statements about how the homeless of San Francisco use drugs, lack ambition, and present a generally disreputable appearance on the streets. This type of framing of stories about the homeless is not unique. According to the media scholar Eungjun Min (1999: ix), media images typically portray the homeless as "drunk, stoned, crazy, sick, and drug abusers." Such representations of homeless people limit our understanding of the larger issues surrounding the problem of homelessness in the United States.

Most media framing of newspaper articles and television reports about the problem of homelessness can be classified into one of two major categories: thematic framing and episodic framing. *Thematic framing* refers to news stories that focus primarily on statistics about the homeless population and recent trends in homelessness. Examples include stories about changes in the U.S. poverty rate and articles about states and cities that have had the largest increases in poverty. Most articles of this type are abstract and impersonal, primarily presenting data and some expert's interpretation of what those data mean. Media representations of this type convey a message to readers that "the poor and homeless are faceless." According to some analysts, thematic framing of poverty is often dehumanizing because it "ignores the human tragedy of poverty—the suffering, indignities, and misery endured by millions of children and adults" (Mantsios, 2003: 101).

By contrast, *episodic framing* presents public issues such as poverty and homelessness as concrete events, showing them to be specific instances that occur more or less in isolation. For example, a news article may focus on the problems of one homeless family, describing how the parents and kids live in a car and eat meals from a soup kitchen. Often, what is not included is the *big picture of homelessness:* How many people throughout the city or nation are living in their cars or in shelters? What larger structural factors (such as reductions in public and private assistance to the poor, or high rates of unemployment in some regions) contribute to or intensify the problem of homelessness in this country?

For many years, the poor have been a topic of interest to journalists and social commentators. Between 1851 and 1995, the *New York Times* alone printed 4,126 articles that had the word *poverty* in the headline. How stories about the poor and homeless are framed in the media has been and remains an important concern for each of us because these reports influence how we view the less fortunate in our society. If we come to see the problem of homelessness as nothing more than isolated statistical data or as marginal situations that affect only a few people, then we will be unable to make a balanced assessment of the larger social problems involved.

Reflect & Analyze

How are the poor and homeless represented in the news reports and the television entertainment shows that you watch? Are the larger social issues surrounding homelessness discussed within the context of these shows? Should they be?

past, a hope for the future, and a potential source of immediate cash. As Snow and Anderson (1993: 147) note, selling personal possessions is not uncommon among most social classes; members of the working and middle classes hold garage sales, and those in the upper classes have estate sales. However, when homeless persons sell their personal possessions, they do so to meet their immediate needs, not because they want to "clean house."

status symbol a material sign that informs others of a person's specific status.

Roles

Role is the dynamic aspect of a status. Whereas we occupy a status, we play a role. A ***role*** **is a set of behavioral expectations associated with a given status.** For example, a carpenter (employee) hired to remodel a kitchen is not expected to sit down uninvited and join the family (employer) for dinner.

Role expectation **is a group's or society's definition of the way that a specific role ought to be played.** By contrast, ***role performance*** **is how a person actually plays the role.** Role performance does not always match role expectation. Some statuses have role expectations that are highly specific, such as that of surgeon or college professor. Other statuses, such as friend or significant other, have less-structured expectations. The role expectations tied to the status of student are more specific than those of being a friend. Role expectations are typically based on a range of acceptable behavior rather than on strictly defined standards.

Our roles are relational (or complementary); that is, they are defined in the context of roles performed by others. We can play the role of student because someone else fulfills the role of professor. Conversely, to perform the role of professor, the teacher must have one or more students.

Role ambiguity occurs when the expectations associated with a role are unclear. For example, it is not always clear when the provider–dependent aspect of the parent–child relationship ends. Should it end at age eighteen or twenty-one? When a person is no longer in school? Different people will answer these questions differently depending on their experiences and socialization, as well as on the parents' financial capability and psychological willingness to continue contributing to the welfare of their adult children.

Role Conflict and Role Strain Most people occupy a number of statuses, each of which has numerous role expectations attached. For example, Charles is a student who attends morning classes at the university, and he is an employee at a fast-food restaurant, where he works from 3:00 to 10:00 P.M. He is also Stephanie's boyfriend, and she would like to see him more often. On December 7, Charles has a final exam at 7:00 P.M., when he is supposed to be working. Meanwhile, Stephanie is pressuring him to take her to a movie. To top it off, his mother calls, asking him to fly home because his father is going to have emergency surgery. How can Charles be in all these places at once? Such experiences of role conflict can be overwhelming.

Role conflict **occurs when incompatible role demands are placed on a person by two or more statuses held at the same time.** When role conflict occurs, we may feel pulled in different directions. To deal with this problem, we may prioritize our roles and first complete the one we consider to be most important. Or we may compartmentalize our lives and "insulate" our various roles (Merton, 1968). That is, we may perform the activities linked to one role for part of the day and then engage in the activities associated with another role in some other time period or elsewhere. For example, under routine circumstances, Charles would fulfill his student role for part of the day and his employee role for another part of the day. In his current situation, however, he is unable to compartmentalize his roles.

Role conflict may occur as a result of changing statuses and roles in society. Research has found that women who engage in behavior that is gender-typed as "masculine" tend to have higher rates of role conflict than those who engage in traditional "feminine" behavior (Basow, 1992). According to the sociologist Tracey Watson (1987), role conflict can sometimes be attributed not to the roles themselves but to the pressures people feel when they do not fit into culturally prescribed roles. In her study of women athletes in college sports programs, Watson found role conflict in the traditionally incongruent identities of being a woman and being an athlete. Even though the women athletes in her study wore makeup and presented a conventional image when they were not on the bas-

Parents sometimes experience role conflict when they are faced with societal expectations that they will earn a living for their family and that they will also be good parents to their children. Obviously, this father needs to leave for work; however, his son has other needs.

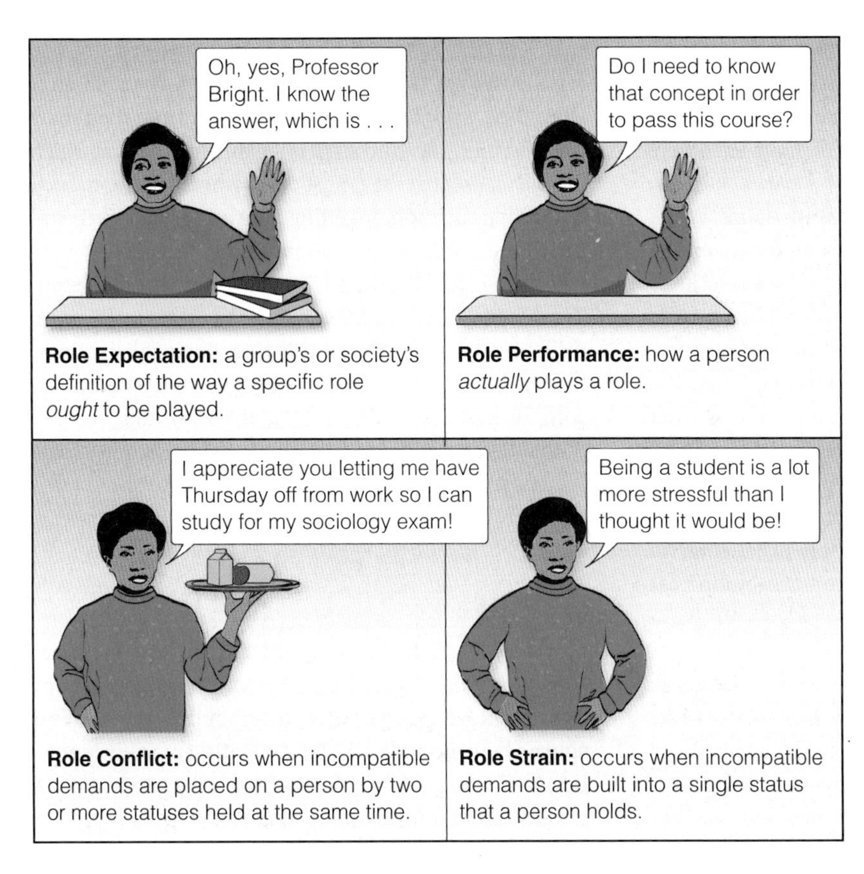

▶ **Figure 5.2 Role Expectation, Performance, Conflict, and Strain**
When playing the role of "student," do you sometimes personally encounter these concepts?

ketball court, their peers in school still saw them as "female jocks," thus leading to role conflict.

Whereas role conflict occurs between two or more statuses (such as being homeless and being a temporary employee of a social services agency), role strain takes place within one status. ***Role strain*** **occurs when incompatible demands are built into a single status that a person occupies** (Goode, 1960). For example, many women experience role strain in the labor force because they hold jobs that are "less satisfying and more stressful than men's jobs since they involve less money, less prestige, fewer job openings, more career roadblocks, and so forth" (Basow, 1992: 192). Similarly, married women may experience more role strain than married men because of work overload, marital inequality with their spouse, exclusive parenting responsibilities, unclear expectations, and lack of emotional support.

Recent social changes may have increased role strain in men. In the family, men's traditional position of dominance has eroded as more women have entered the paid labor force and demanded more assistance in child-rearing and homemaking responsibilities. Role strain may occur among African American men who have internalized North American cultural norms regarding masculinity yet find it very difficult (if not impossible) to attain cultural norms of achievement, success, and power because of racism and economic exploitation (Basow, 1992).

Sexual orientation, age, and occupation are frequently associated with role strain. Lesbians and gay men often experience role strain because of the pressures associated with having an identity heavily stigmatized by the dominant cultural group (Basow, 1992). Women in their thirties may experience the highest levels of role strain; they face a large amount of stress in terms of role demands and conflicting work and family expectations (Basow, 1992). Dentists, psychiatrists, and police officers have been found to experience high levels of occupation-related role strain, which may result in suicide. (The concepts of role expectation, role performance, role conflict, and role strain are illustrated in ▶ Figure 5.2.)

Individuals frequently distance themselves from a role they find extremely stressful or otherwise problematic. *Role distancing* occurs when people consciously

role a set of behavioral expectations associated with a given status.

role expectation a group's or society's definition of the way that a specific role *ought* to be played.

role performance how a person *actually* plays a role.

role conflict a situation in which incompatible role demands are placed on a person by two or more statuses held at the same time.

role strain a condition that occurs when incompatible demands are built into a single status that a person occupies.

foster the impression of a lack of commitment or attachment to a particular role and merely go through the motions of role performance (Goffman, 1961b). People use distancing techniques when they do not want others to take them as the "self" implied in a particular role, especially if they think the role is "beneath them." While Charles is working in the fast-food restaurant, for example, he does not want people to think of him as a "loser in a dead-end job." He wants them to view him as a college student who is working there just to "pick up a few bucks" until he graduates. When customers from the university come in, Charles talks to them about what courses they are taking, what they are majoring in, and what professors they have. He does not discuss whether the bacon cheeseburger is better than the chili burger. When Charles is really involved in role distancing, he tells his friends that he "works there but wouldn't eat there."

Role Exit ***Role exit*** **occurs when people disengage from social roles that have been central to their self-identity** (Ebaugh, 1988). Sociologist Helen Rose Fuchs Ebaugh studied this process by interviewing ex-convicts, ex-nuns, retirees, divorced men and women, and others who had exited voluntarily from significant social roles. According to Ebaugh, role exit occurs in four stages. The first stage is doubt, in which people experience frustration or burnout when they reflect on their existing roles. The second stage involves a search for alternatives; here, people may take a leave of absence from their work or temporarily separate from their marriage partner. The third stage is the turning point, at which people realize that they must take some final action, such as quitting their job or getting a divorce. The fourth and final stage involves the creation of a new identity.

Exiting the "homeless" role is often very difficult. The longer a person remains on the streets, the more difficult it becomes to exit this role. Personal resources diminish over time. Possessions are often stolen, lost, sold, or pawned. Work experience and skills become outdated, and physical disabilities that prevent individuals from working are likely to develop. However, a number of homeless people are able to exit this role.

Groups

Groups are another important component of social structure. To sociologists, a ***social group*** **consists of two or more people who interact frequently and share a common identity and a feeling of interdependence.** Throughout our lives, most of us participate in groups: our families and childhood friends, our college classes, our work and community organizations, and even society.

Primary and secondary groups are the two basic types of social groups. A ***primary group*** **is a small,**

© Gilles Mingasson

© MARIO ANZUONI/Reuters/Landov

Los Angeles Times columnist Steve Lopez (left) met a homeless man, Nathaniel Ayers (above) and learned that he had been a promising musician studying at the Juilliard School who had dropped out because of his struggle with mental illness. In his 2008 book titled *The Soloist*, Lopez chronicles the relationship he developed with Ayers and how he eventually helped get him off the street and be treated for his schizophrenia. This story is an example of role exit, and you can see it in the movie version of *The Soloist*, released in 2009.

less specialized group in which members engage in face-to-face, emotion-based interactions over an extended period of time. Primary groups include our family, close friends, and school- or work-related peer groups. By contrast, a ***secondary group*** **is a larger, more specialized group in which members engage in more impersonal, goal-oriented relationships for a limited period of time.** Schools, churches, and corporations are examples of secondary groups. In secondary groups, people have few, if any, emotional ties to one another. Instead, they come together for some specific, practical purpose, such as getting a degree or a paycheck. Secondary groups are more specialized than primary ones; individuals relate to one another in terms of specific roles (such as professor and student) and more-limited activities (such as course-related endeavors). Primary and secondary groups are further discussed in Chapter 6 ("Groups and Organizations").

Social solidarity, or cohesion, relates to a group's ability to maintain itself in the face of obstacles. Social solidarity exists when social bonds, attractions, or other forces hold members of a group in interaction over a period of time (Jary and Jary, 1991). For example, if a local church is destroyed by fire and congregation members still worship together in a makeshift setting, then they have a high degree of social solidarity.

Many of us build social networks that involve our personal friends in primary groups and our acquaintances in secondary groups. A *social network* is a series of social relationships that links an individual to others. Social networks work differently for men and women, for different races/ethnicities, and for members of different social classes. Traditionally, people of color and white women have been excluded from powerful "old-boy" social networks. At the middle- and upper-class levels, individuals tap social networks to find employment, make business deals, and win political elections. However, social networks typically do not work effectively for poor and homeless individuals. Snow and Anderson (1993) found that homeless men have fragile social networks that are plagued with instability. Homeless men often do not even know one another's "real" names.

Sociological research on the homeless has noted the social isolation experienced by people on the streets. Sociologist Peter H. Rossi (1989) found that a high degree of social isolation exists because the homeless are separated from their extended family and former friends. Rossi noted that among the homeless who did have families, most either did not wish to return or believed that they would not be welcome. Most of the avenues for exiting the homeless role and acquiring housing are intertwined with the large-scale, secondary groups that sociologists refer to as formal organizations.

A ***formal organization*** **is a highly structured group formed for the purpose of completing certain tasks or achieving specific goals.** Many of us spend most of our time in formal organizations, such as colleges, corporations, or the government. Chapter 6 ("Groups and Organizations") analyzes the characteristics of bureaucratic organizations; however, at this point we should note that these organizations are a very important component of social structure in all industrialized societies. We expect such organizations to educate us, solve our social problems (such as crime and homelessness), and provide work opportunities.

Today, formal organizations such as the National Law Center on Homelessness and Poverty work with groups around the country to make people aware that homelessness must be viewed within the larger context of poverty and to educate the public on the nature and extent of homelessness among various categories of people in the United States (see ▶ Figure 5.3).

Social Institutions

At the macrolevel of all societies, certain basic activities routinely occur—children are born and socialized, goods and services are produced and distributed, order is preserved, and a sense of purpose is maintained (Aberle et al., 1950; Mack and Bradford, 1979). Social institutions are the means by which these basic needs are met. A ***social institution*** **is a set of organized beliefs and rules that establishes how a society will attempt to meet its basic social needs.** In the past, these needs have centered around five basic social institutions: the

role exit a situation in which people disengage from social roles that have been central to their self-identity.

social group a group that consists of two or more people who interact frequently and share a common identity and a feeling of interdependence.

primary group Charles Horton Cooley's term for a small, less specialized group in which members engage in face-to-face, emotion-based interactions over an extended period of time.

secondary group a larger, more specialized group in which members engage in more-impersonal, goal-oriented relationships for a limited period of time.

formal organization a highly structured group formed for the purpose of completing certain tasks or achieving specific goals.

social institution a set of organized beliefs and rules that establishes how a society will attempt to meet its basic social needs.

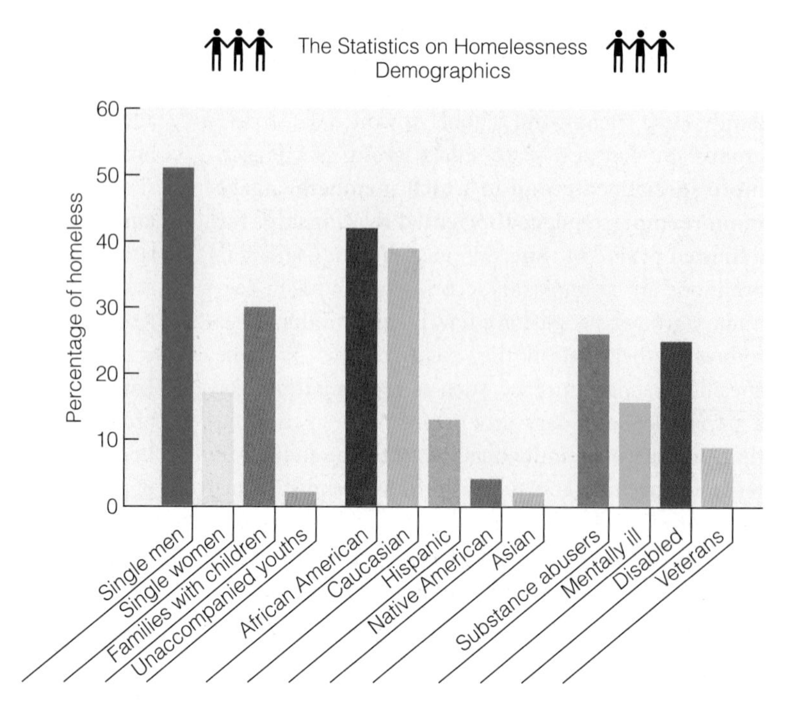

▶ **Figure 5.3 Who Are the Homeless?**
Source: National Law Center on Homelessness and Poverty, 2004. Reprinted courtesy of HowStuffWorks.com

family, religion, education, the economy, and the government or politics. Today, mass media, sports, science and medicine, and the military are also considered to be social institutions.

What is the difference between a group and a social institution? A group is composed of specific, identifiable people; an institution is a standardized way of doing something. The concept of "family" helps to distinguish between the two. When we talk about "your family" or "my family," we are referring to a specific family. When we refer to the family as a social institution, we are talking about ideologies and standardized patterns of behavior that organize family life. For example, the family as a social institution contains certain statuses organized into well-defined relationships, such as husband–wife, parent–child, and brother–sister. Specific families do not always conform to these ideologies and behavior patterns.

Functional theorists emphasize that social institutions exist because they perform five essential tasks:

1. *Replacing members.* Societies and groups must have socially approved ways of replacing members who move away or die. The family provides the structure for legitimated sexual activity—and thus procreation—between adults.
2. *Teaching new members.* People who are born into a society or move into it must learn the group's values and customs. The family is essential in teaching new members, but other social institutions educate new members as well.
3. *Producing, distributing, and consuming goods and services.* All societies must provide and distribute goods and services for their members. The economy is the primary social institution fulfilling this need; the government is often involved in the regulation of economic activity.
4. *Preserving order.* Every group or society must preserve order within its boundaries and protect itself from attack by outsiders. The government legitimates the creation of law enforcement agencies to preserve internal order and some form of military for external defense.
5. *Providing and maintaining a sense of purpose.* In order to motivate people to cooperate with one another, a sense of purpose is needed.

Although this list of functional prerequisites is shared by all societies, the institutions in each society perform these tasks in somewhat different ways depending on their specific cultural values and norms.

Conflict theorists agree with functionalists that social institutions are originally organized to meet basic social needs. However, they do not believe that social institutions work for the common good of everyone in society. For example, the homeless lack the power and resources to promote their own interests when they are opposed by dominant social groups. From the conflict perspective, social institutions such as the government maintain the privileges of the wealthy and powerful while contributing to the powerlessness of others (see Domhoff, 2002). For example, U.S. government policies in urban areas have benefited some people but exacerbated the problems of others. Urban renewal and transportation projects have caused the destruction of low-cost housing and put large numbers of people "on the street" (Katz, 1989). Similarly, the shift in governmental policies toward the mentally ill and welfare recipients has resulted in more people struggling—and often failing—to find affordable housing. Meanwhile, many wealthy and privileged bankers, investors, developers, and builders have benefited at the expense of the low-income casualties of those policies.

Societies, Technology, and Sociocultural Change

As we think about homeless people today, it is difficult to realize that for people in some societies, being without a place of residence is a way of life. Where people live and the mode(s) of production they use to generate a food supply are related to *subsistence technology*—the methods and tools that are available for acquiring the basic needs of daily life. Social scientists have identified five types of societies based on various levels of subsistence technology: hunting and gathering, horticultural and pastoral, agrarian, industrial, and postindustrial societies. The first three of these—hunting and gathering, horticultural and pastoral, and agrarian—are also referred to as preindustrial societies. According to the social scientists Gerhard Lenski and Jean Lenski, societies change over time through the process of *sociocultural evolution,* the changes that occur as a society gains new technology (see Nolan and Lenski, 1999). However, not all anthropologists and sociologists agree on the effects of new technology.

© Jason Laure/The Image Works

In contemporary hunting and gathering societies, women contribute to the food supply by gathering plants and sometimes hunting for small animals. These women of the Kalahari in Botswana gather and share edible roots.

Hunting and Gathering Societies

At present, fewer than 250,000 people support themselves solely through hunting, fishing, and gathering wild plant foods (Haviland, 1999). However, from the origins of human existence (several million years ago) until about 10,000 years ago, hunting and gathering societies were the only type of human society that existed. ***Hunting and gathering societies*** **use simple technology for hunting animals and gathering vegetation.** The technology in these societies is limited to tools and weapons that are used for basic subsistence, including spears, bows and arrows, nets, traps for hunting, and digging sticks for plant collecting. All tools and weapons are made of natural materials such as stone, bone, and wood.

In hunting and gathering societies, the basic social unit is the kinship group or family. People do not have private households or residences as we think of them. Instead, they live in small groups of about twenty-five to forty people. Kinship ties constitute the basic economic unit through which food is acquired and distributed. With no stable food supply, hunters and gatherers continually search for wild animals and edible plants. As a result, they remain on the move and seldom establish a permanent settlement (Nolan and Lenski, 1999).

Hunting and gathering societies are relatively egalitarian. Because it is impossible to accumulate a surplus of food, there are few resources upon which individuals or groups can build a power base. Some specialization (division of labor) occurs, primarily based on age and sex. Young children and older people are expected to contribute what they can to securing the food supply, but healthy adults of both sexes are expected to obtain most of the food. In some societies, men hunt for animals and women gather plants; in others, both women and men gather plants and hunt for wild game, with women more actively participating when smaller animals are nearby (Lorber, 1994; Volti, 1995).

In these societies, education, religion, and politics are not formal social institutions. Instead, their functions take place on an informal basis in the kinship group, which is responsible for teaching children basic survival skills such as how to hunt and gather food. Religion is based on *animism,* the belief that spirits inhabit virtually everything in the world. There is no organized religious body; the *shaman,* or religious leader, exercises some degree of leadership but receives no

hunting and gathering societies societies that use simple technology for hunting animals and gathering vegetation.

material rewards for his duties and is expected to work like everyone else to obtain food (Nolan and Lenski, 1999). Contemporary hunting and gathering societies are located in relatively isolated geographical areas. However, some analysts predict that these groups will soon cease to exist, as food producers with more dominating technologies usurp the geographic areas from which these groups have derived their food supply (Nolan and Lenski, 1999).

Horticultural and Pastoral Societies

The period between 13,000 and 7,000 B.C.E. marks the beginning of horticultural and pastoral societies. During this period, there was a gradual shift from *collecting* food to *producing* food, a change that has been attributed to three factors: (1) the depletion of the supply of large game animals as a source of food, (2) an increase in the size of the human population to feed, and (3) dramatic weather and environmental changes that probably occurred by the end of the Ice Age (Ferraro, 1992).

Why did some societies become horticultural while others became pastoral? Whether horticultural activities or pastoral activities became a society's primary mode of food production was related to water supply, terrain, and soils. ***Pastoral societies* are based on technology that supports the domestication of large animals to provide food** and emerged in mountainous regions and areas with low amounts of annual rainfall. Pastoralists—people in pastoral societies—typically remain nomadic as they seek new grazing lands and water sources for their animals. ***Horticultural societies* are based on technology that supports the cultivation of plants to provide food.** These societies emerged in more fertile areas that were better suited for growing plants through the use of hand tools.

The family is the basic unit in horticultural and pastoral societies. Because they typically do not move as often as hunter-gatherers or pastoralists, horticulturalists establish more permanent family ties and create complex systems for tracing family lineage. Some social analysts believe that the invention of a hoe with a metal blade was a contributing factor to the less nomadic lifestyle of the horticulturalists. Unlike the digging stick, use of the metal-blade hoe made planting more efficient and productive. Horticulturists using a hoe are able to cultivate the soil more deeply, and crops can be grown in the same area for longer periods. As a result, people become more *sedentary,* remaining settled for longer periods in the same location.

Unless there are fires, floods, droughts, or environmental problems, herding animals and farming are more reliable sources of food than hunting and gathering. When food is no longer in short supply, more infants are born, and children have a greater likelihood of surviving. When people are no longer nomadic, children are viewed as an economic asset: They can cultivate crops, tend flocks, or care for younger siblings.

Division of labor increases in horticultural and pastoral societies. As the food supply grows, not everyone needs to be engaged in food production. Some people can pursue activities such as weaving cloth or carpets, crafting jewelry, serving as priests, or creating the tools needed for building the society's structure. Horticultural and pastoral societies are less egalitarian than hunter-gatherers. Even though land is initially communally controlled (often through an extended kinship group), the idea of property rights emerges as people establish more-permanent settlements. At this stage, families with the largest surpluses not only have an economic advantage but also gain prestige and power, including the ability to control others. Slavery is a fairly common practice, and being a slave is a hereditary status in some pastoral societies.

In simple horticultural societies, a fairly high degree of gender equality exists because neither sex controls the food supply. Women contribute to food production because hoe cultivation is compatible with child care (Basow, 1992). In contemporary horticultural societies, women still do most of the farming while men hunt game, clear land, work with arts and crafts, make tools, participate in religious and ceremonial activities, and engage in war (Nielsen, 1990). Gender inequality is greater in pastoral societies because men herd the large animals and women contribute relatively little to subsistence production. In some herding societies, women's primary value is seen as their ability to produce male offspring so that the family lineage can be preserved and a sufficient number of males are available to protect the group against enemy attack (Nielsen, 1990).

Education, religion, and politics remain relatively informal in horticultural and pastoral societies. Boys learn how to plant and harvest crops, domesticate large animals, and fight. Girls learn how to do domestic chores, care for younger children, and, sometimes, cultivate the land. In horticultural societies, religion is based on ancestor worship; in pastoral societies, religion is based on belief in a god or gods, who are believed to take an active role in human affairs. Politics is based on a simple form of government that is backed up by military force.

Agrarian Societies

About five to six thousand years ago, agrarian (or agricultural) societies emerged, first in Mesopotamia and Egypt and slightly later in China. ***Agrarian societies* use the technology of large-scale farming, including**

© Steve Satushek/Getty Images

In the twenty-first century, most people around the globe still reside in agrarian societies that are in various stages of industrialization. Open-air markets such as this one in Bali, where people barter or buy their food from one another, are a common sight in agrarian societies.

animal-drawn or energy-powered plows and equipment, to produce their food supply. Farming made it possible for people to spend their entire lives in the same location, and food surpluses made it possible for people to live in cities, where they were not directly involved in food production. Unlike the digging sticks and hoes that had previously been used in farming, the use of animals to pull plows made it possible for people to generate a large surplus of food. In agrarian societies, land is cleared of all vegetation and cultivated with the use of the plow, a process that not only controls the weeds that might kill crops but also helps maintain the fertility of the soil. The land can be used more or less continuously because the plow turns the topsoil, thus returning more nutrients to the soil. In some cases, farmers reap several harvests each year from the same plot of land.

In agrarian societies, social inequality is the highest of all preindustrial societies in terms of both class and gender. The two major classes are the landlords and the peasants. The landlords own the fields and the harvests produced by the peasants. Inheritance becomes important as families of wealthy landlords own the same land for generations. By contrast, the landless peasants enter into an agreement with the landowners to live on and cultivate a parcel of land in exchange for part of the harvest or other economic incentives. Over time, the landlords grow increasingly wealthy and powerful as they extract labor, rent, and taxation from the landless workers. Politics is based on a feudal system controlled by a political–economic elite made up of the ruler, his royal family, and members of the landowning class. Peasants have no political power and may be suppressed through the use of force or military power.

Gender-based inequality grows dramatically in agrarian societies. Men gain control over both the disposition of the food surplus and the kinship system (Lorber, 1994). Because agrarian tasks require more labor and greater physical strength than horticultural ones, men become more involved in food production. Women may be excluded from these tasks because they are seen as too weak for the work or because it is believed that their child-care responsibilities are incompatible with the full-time labor that the tasks require (Nielsen, 1990). As more people own land or businesses, the rules pertaining to marriage become stronger, and women's lives become more restricted. Men demand that women practice premarital virginity and marital fidelity so that "legitimate" heirs can be produced to inherit the land and other possessions (Nielsen, 1990). This belief is supported by religion, which is a powerful force in agrarian societies. In simple agrarian societies, the gods are seen as being concerned about the individual's moral conduct. In advanced agrarian societies, monotheism (belief in one god) replaces a belief in multiple gods. Today, gender inequality continues in agrarian societies; the division of labor between women and men is very distinct in areas such as parts of the Middle East. Here, women's work takes place in the private sphere (inside the home), and men's work occurs in the public sphere, providing men with more recognition and greater formal status.

Industrial Societies

***Industrial societies* are based on technology that mechanizes production.** Originating in England during the Industrial Revolution, this mode of production dramatically transformed predominantly rural and

pastoral societies societies based on technology that supports the domestication of large animals to provide food, typically emerging in mountainous regions and areas with low amounts of annual rainfall.

horticultural societies societies based on technology that supports the cultivation of plants to provide food.

agrarian societies societies that use the technology of large-scale farming, including animal-drawn or energy-powered plows and equipment, to produce their food supply.

industrial societies societies based on technology that mechanizes production.

agrarian societies into urban and industrial societies. Chapter 1 describes how the revolution first began in Britain and then spread to other countries, including the United States.

Industrialism involves the application of scientific knowledge to the technology of production, thus making it possible for machines to do the work previously done by people or animals. New technologies, such as the invention of the steam engine and fuel-powered machinery, stimulated many changes. Before the invention of the steam engine, machines were run by natural power sources (such as wind or water mills) or harnessed power (either human or animal power). The steam engine made it possible to produce goods by machines powered by fuels rather than undependable natural sources or physical labor.

As inventions and discoveries build upon one another, the rate of social and technological change increases. For example, the invention of the steam engine brought about new types of transportation, including trains and steamships. Inventions such as electric lights made it possible for people to work around the clock without regard to whether it was daylight or dark outside. Take a look around you: Most of what you see would not exist if it were not for industrialization. Cars, computers, electric lights, televisions, telephones, and virtually every other possession we own are the products of an industrial society.

Industrialism changes the nature of subsistence production. In countries such as the United States, large-scale agribusinesses have practically replaced small, family-owned farms and ranches. However, large-scale agriculture has produced many environmental problems while providing solutions to the problem of food supply.

In industrial societies, a large proportion of the population lives in or near cities. Large corporations and government bureaucracies grow in size and complexity. The nature of social life changes as people come to know one another more as statuses than as individuals. In fact, a person's occupation becomes a key defining characteristic in industrial societies, whereas his or her kinship ties are most important in preindustrial societies. Although time is freed up for leisure activities, many people still work long hours or multiple jobs.

Social institutions are transformed by industrialism. The family diminishes in significance as the economy, education, and political institutions grow in size and complexity. Although the family is still a major social institution for the care and socialization of children, it loses many of its other production functions to businesses and corporations. The family is now a consumption unit, not a production unit. In advanced industrial societies such as the United States, families take on many diverse forms, including single-parent families, single-person families, and stepfamilies (see Chapter 15, "Families and Intimate Relationships"). Although the influence of traditional religion is diminished in industrial societies, religion remains a powerful institution. Religious organizations are important in determining what moral issues will be brought to the forefront (e.g., unapproved drugs, abortion, and violence and sex in the media) and in trying to influence lawmakers to pass laws regulating people's conduct. Politics in industrial societies is usually based on a democratic form of government. As nations such as South Korea, the People's Republic of China, and Mexico have become more industrialized, many people in these nations have intensified their demands for political participation.

Although the standard of living rises in industrial societies, social inequality remains a pressing problem. As societies industrialize, the status of women tends to decline further. For example, industrialization in the United States created a gap between the nonpaid work performed by women at home and the paid work that was increasingly performed by men and unmarried girls. The division of labor between men and women in the middle and upper classes also became much more distinct: Men were responsible for being "breadwinners"; women were seen as "homemakers" (Amott and Matthaei, 1996). This gendered division of labor increased the economic and political subordination of women. Likewise, although industrialization was a source of upward mobility for many whites, most people of color were left behind (Lorber, 1994).

In short, industrial societies have brought about some of the greatest innovations in all of human history, but they have also maintained and perpetuated some of the greatest problems, including violence; race-, class-, and gender-based inequalities; and environmental degradation.

Postindustrial Societies

A *postindustrial society* is one in which technology supports a service- and information-based economy. As discussed in Chapter 1, postmodern (or "postindustrial") societies are characterized by an *information explosion* and an economy in which large numbers of people either provide or apply information or are employed in service jobs (such as fast-food server or health care worker). For example, banking, law, and the travel industry are characteristic forms of employment in postindustrial societies, whereas producing steel or automobiles is representative of employment in industrial societies. There is a corresponding *rise of a consumer society* and the emergence of a *global village* in which people around the world communicate with one another by electronic technologies such as television, telephone, fax, e-mail, and the Internet.

Postindustrial societies produce knowledge that becomes a commodity. This knowledge can be leased or sold to others, or it can be used to generate goods, services, or more knowledge. In the previous types of societies we have examined, machinery or raw materials are crucial to how the economy operates. In postindustrial societies, the economy is based on involvement with people and communications technologies such as the mass media, computers, and the World Wide Web. For example, recent information from the U.S. Census Bureau indicates that more than three-quarters of all U.S. households have at least one computer (see "Census Profiles: Computer and Internet Access in U.S. Households"). Some analysts refer to postindustrial societies as "service economies," based on the assumption that many workers provide services for others. Examples include home health care workers and airline flight attendants. However, most of the new service occupations pay relatively low wages and offer limited opportunities for advancement.

Previous forms of production, including agriculture and manufacturing, do not disappear in postindustrial societies. Instead, they become more efficient through computerization and other technological innovations. Work that relies on manual labor is often shifted to less technologically advanced societies, where workers are paid low wages to produce profits for corporations based in industrial and postindustrial societies.

Knowledge is viewed as the basic source of innovation and policy formulation in postindustrial societies. As a result, education becomes one of the most impor-

© Nevada Wier/Corbis

In postindustrial economics, many service- and information-based jobs are located in countries far removed from where a corporation's consumers actually live. These call center employees in India are helping customers around the world.

CENSUS PROFILES

Computer and Internet Access in U.S. Households, 1984 to 2007

The U.S. Census Bureau collects extensive data on U.S. households in addition to the questions it used for Census 2000. For example, Current Population Survey data, collected from about 50,000 U.S. households, show an increase in the percentage of homes with computers and access to the Internet, as the following figure illustrates:

Computers and Internet Access in the Home: 1984 to 2007
(civilian noninstitutional population)

Percentage of households with a computer
Percentage of households with Internet access

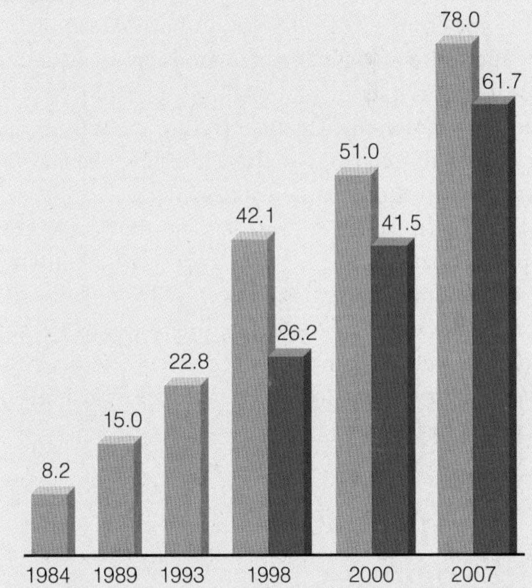

Note: Data on Internet access were not collected before 1997.

Since 1984, the first year in which the Census Bureau collected data on computer ownership and use, there has been more than a 700 percent increase in the percentage of households with computers. However, in Chapter 8 ("Class and Stratification in the United States") we will see that computer ownership varies widely by income and educational level.

Sources: Newburger, 2001; U.S. Census Bureau, 2007.

postindustrial societies societies in which technology supports a service- and information-based economy.

◆ Table 5.1 Technoeconomic Bases of Society

	Hunting and Gathering	Horticultural and Pastoral	Agrarian
Change from Prior Society	—	Use of hand tools, such as digging stick and hoe	Use of animal-drawn plows and equipment
Economic Characteristics	Hunting game, gathering roots and berries	Planting crops, domestication of animals for food	Labor-intensive farming
Control of Surplus	None	Men begin to control societies	Men own land or herds
Inheritance	None	Shared—patrilineal and matrilineal	Patrilineal
Control Over Procreation	None	Increasingly by men	Men—to ensure legitmacy of heirs
Women's Status	Relative equality	Decreasing in move to pastoralism	Low

Source: Adapted from Lorber, 1994: 140.

tant social institutions (Bell, 1973). Formal education and other sources of information become crucial to the success of individuals and organizations. Scientific research becomes institutionalized, and new industries—such as computer manufacturing and software development—come into existence that would not have been possible without the new knowledge and technological strategies. (The features of the different types of societies, distinguished by technoeconomic base, are summarized in ◆ Table 5.1.) Throughout this text, we will examine key features of postindustrial societies as well as the postmodern theoretical perspectives that have come to be associated with the process of postindustrialism.

Stability and Change in Societies

How do societies maintain some degree of social solidarity in the face of the changes we have described? As you may recall from Chapter 1, theorists using a functionalist perspective focus on the stability of societies and the importance of equilibrium even in times of rapid social change. By contrast, conflict perspectives highlight how societies go through continuous struggles for scarce resources and how innovation, rebellion, and conquest may bring about social change. Sociologists Emile Durkheim and Ferdinand Tönnies developed typologies to explain the processes of stability and change in the social structure of societies. A *typology* is a classification scheme containing two or more mutually exclusive categories that are used to compare different kinds of behavior or types of societies.

Durkheim: Mechanical and Organic Solidarity

Emile Durkheim (1933/1893) was concerned with the question "How do societies manage to hold together?" He asserted that preindustrial societies are held together by strong traditions and by the members' shared moral beliefs and values. As societies industrialized and developed more specialized economic activities, social solidarity came to be rooted in the members' shared dependence on one another. From Durkheim's perspective, social solidarity derives from a society's social structure, which, in turn, is based on the society's division of labor. *Division of labor* refers to how the various tasks of a society are divided up and performed. People in diverse societies (or in the same society at different points in time) divide their tasks somewhat differently, based on their own history, physical environment, and level of technological development.

To explain social change, Durkheim categorized societies as having either mechanical or organic solidarity. ***Mechanical solidarity* refers to the social cohesion of preindustrial societies, in which there is minimal division of labor and people feel united by shared values and common social bonds.** Durkheim used the term *mechanical solidarity* because he believed that people in such preindustrial societies feel a more or less automatic sense of belonging. Social interaction is characterized by face-to-face, intimate, primary-group relationships. Everyone is engaged in similar work, and little specialization is found in the division of labor.

***Organic solidarity* refers to the social cohesion found in industrial (and perhaps postindustrial) societies, in which people perform very specialized**

◆ Table 5.1 (continued)

	Industrial	Postindustrial
Change from Prior Society	Invention of steam engine	Invention of computer and development of "high-tech" society
Economic Characteristics	Mechanized production of goods	Information and service economy
Control of Surplus	Men own means of production	Corporate shareholders and high-tech entrepreneurs
Inheritance	Bilateral	Bilateral
Control Over Procreation	Men—but less so in later stages	Mixed
Women's Status	Low	Varies by class, race, and age

tasks and feel united by their mutual dependence. Durkheim chose the term *organic solidarity* because he believed that individuals in industrial societies come to rely on one another in much the same way that the organs of the human body function interdependently. Social interaction is less personal, more status oriented, and more focused on specific goals and objectives. People no longer rely on morality or shared values for social solidarity; instead, they are bound together by practical considerations. Which of Durkheim's categories most closely describes the United States today?

Tönnies: *Gemeinschaft* and *Gesellschaft*

Sociologist Ferdinand Tönnies (1855–1936) used the terms *Gemeinschaft* and *Gesellschaft* to characterize the degree of social solidarity and social control found in societies. He was especially concerned about what happens to social solidarity in a society when a "loss of community" occurs.

The ***Gemeinschaft* (guh-MINE-shoft) is a traditional society in which social relationships are based on personal bonds of friendship and kinship and on intergenerational stability.** These relationships are based on ascribed rather than achieved status. In such societies, people have a commitment to the entire group and feel a sense of togetherness. Tönnies (1963/1887) used the German term *Gemeinschaft* because it means "commune" or "community"; social solidarity and social control are maintained by the community. Members have a strong sense of belonging, but they also have very limited privacy.

By contrast, the ***Gesellschaft* (guh-ZELL-shoft) is a large, urban society in which social bonds are based on impersonal and specialized relationships, with little long-term commitment to the group or consensus on values.** In such societies, most people are "strangers" who perceive that they have very little in common with most other people. Consequently, self-interest dominates, and little consensus exists regarding values. Tönnies (1963/1887) selected the German term *Gesellschaft* because it means "association"; relationships are based on achieved statuses, and interactions among people are both rational and calculated.

Social Structure and Homelessness

In *Gesellschaft* societies such as the United States, a prevailing core value is that people should be able to take care of themselves. Thus, many people view the

mechanical solidarity Emile Durkheim's term for the social cohesion of preindustrial societies, in which there is minimal division of labor and people feel united by shared values and common social bonds.

organic solidarity Emile Durkheim's term for the social cohesion found in industrial societies, in which people perform very specialized tasks and feel united by their mutual dependence.

Gemeinschaft (guh-MINE-shoft) a traditional society in which social relationships are based on personal bonds of friendship and kinship and on intergenerational stability.

Gesellschaft (guh-ZELL-shoft) a large, urban society in which social bonds are based on impersonal and specialized relationships, with little long-term commitment to the group or consensus on values.

Box 5.3 Sociology and Social Policy

Homeless Rights Versus Public Space

> I had a bit of a disturbing experience yesterday as I was running errands downtown. First, I was glad to see the south Queen sidewalk east of University [in Toronto, Canada,] open. (Months of construction on the new opera house had blocked it off.) As I continued walking eastward past the acclaimed new structure (where I have enjoyed a performance or two) I wondered why the sidewalk was so narrow. It seems this stretch of Queen should feel a bit grander. When I reached the corner of Queen and Bay, I saw some police officers and city workers "taking action on sidewalk clearance." They were clearing a homeless person's worldly belongings off the sidewalk. Using shovels. And a pickup truck. . . .
>
> I think what I saw yesterday is unacceptable. Sure, the situation is complicated. Yes, there are a lot of stakeholders and stories to appreciate. But it's unfairness I want to see shoveled out of public space. Not people. Not blankets. Not kindness. And I hope I'm not alone. (Sandals, 2007)

"Public space protection" has become an issue in many cities, both in the United States and elsewhere. Record numbers of homeless individuals and families seek refuge on the streets and in public parks because they have nowhere else to go. However, this seemingly individualistic problem is actually linked to larger social concerns, including unemployment, lack of job training and education, lack of affordable housing, and cutbacks in social service agency budgets. The problem of homelessness also raises significant social policy issues, including the extent to which cities can make it illegal for people to remain for extended periods of time in public spaces.

Should homeless persons be allowed to sleep on sidewalks, in parks, and in other public areas? This issue has been the source

© Mark Ludak/The Image Works

Sidewalk clearance and public space protection are controversial topics in cities where law enforcement officials have been instructed to remove homeless individuals and their possessions from public spaces. What are the central issues in this social policy debate? Why should this problem be of concern to each of us?

homeless as "throwaways"—as beyond help or as having already had enough done for them by society. Some argue that the homeless made their own bad decisions, which led them into alcoholism or drug addiction, and should be held responsible for the consequences of their actions. In this sense, homeless people serve as a visible example to others to "follow the rules" lest they experience a similar fate.

Alternative explanations for homelessness in *Gesellschaft* societies have been suggested. Elliot Liebow (1993) notes that homelessness is rooted in poverty; overwhelmingly, homeless people are poor people who come from poor families. Homelessness is a "social class phenomenon, the direct result of a steady, across-the-board lowering of the standard of living of the working class and lower class" (Liebow, 1993: 224). As the standard of living falls, those at the bottom rungs of society are plunged into homelessness. The problem is exacerbated by a lack of jobs. Of those who find work, a growing number work full time, year-round, but remain poor because of substandard wages. Half of the households living below the poverty line pay more than 70 percent of their income for rent—if they are able to find accommodations that they can afford at all (Roob and McCambridge, 1992). Clearly, there is no simple answer to the question about what should be done to help the homeless. Nor, as discussed in Box 5.3, is there any consensus on what rights the homeless have in public spaces, such as parks and sidewalks. The answers we derive as a society and as individuals are often based on our social construction of this reality of life.

of controversy in a number of cities. As these cities have sought to improve their downtown areas and public spaces, they have taken measures to enforce city ordinances controlling loitering (standing around or sleeping in public spaces), "aggressive panhandling," and disorderly conduct. For example, Santa Monica, California, passed a law that makes it illegal for a person to occupy the doorway of a business between the hours of 11 P.M. and 7 A.M. if the owner has posted a sign to that effect (Wood, 2002).

Advocates for the homeless and civil liberties groups have filed lawsuits in several cities claiming that the rights of the homeless are being violated by the enforcement of these laws. The lawsuits assert that the homeless have a right to sleep in parks because no affordable housing is available for them. Advocates also argue that panhandling is a legitimate means of livelihood for some of the homeless and is protected speech under the First Amendment. In addition, they accuse public and law enforcement officials of seeking to punish the homeless on the basis of their "status," a cruel and unusual punishment prohibited by the Eighth Amendment.

The "homeless problem" is not a new one for city governments. Of the limited public funding that is designated for the homeless, most has been spent on shelters that are frequently overcrowded and otherwise inadequate. Officials in some cities have given homeless people a one-way ticket to another city. Still others have routinely run them out of public spaces.

What responsibility does society have to the homeless? Are laws restricting the hours that public areas or parks are open to the public unfair to homeless persons? Should city workers remove cardboard boxes, blankets, and other "makeshift" homes created by the homeless in parks? Some critics have argued that if the homeless and their advocates win these lawsuits, what they have won (at best) is the right for the homeless to live on the street, to slowly freeze to death, and to drink themselves into oblivion with the option of continuing to forgo seeking the help they need. Others have disputed this assertion and note that if society does not make available affordable housing and job opportunities, the least it can do is stop harassing homeless people who are getting by as best they can.

Contrary to a popular myth that most homeless people are single drifters, an increasing number of families are now homeless.

Reflect & Analyze

What do you think? What rights are involved? Whose rights should prevail?

Sources: Based on Kaufman, 1996; Sandals, 2007; and Wood, 2002.

Social Interaction: The Microlevel Perspective

So far in this chapter, we have focused on society and social structure from a macrolevel perspective, seeing how the structure of society affects the statuses we occupy, the roles we play, and the groups and organizations to which we belong. Functionalist and conflict perspectives provide a macrosociological overview because they concentrate on large-scale events and broad social features. For example, sociologists using the macrosociological approach to study the homeless might analyze how social institutions have operated to produce current conditions. By contrast, the symbolic interactionist perspective takes a microsociological approach, asking how social institutions affect our daily lives. We will now look at society from the microlevel perspective, which focuses on social interactions among individuals, especially face-to-face encounters.

Social Interaction and Meaning

When you are with other people, do you often wonder what they think of you? If so, you are not alone! Because most of us are concerned about the meanings that others ascribe to our behavior, we try to interpret their words and actions so that we can plan how we will react toward them (Blumer, 1969). We know that others have expectations of us. We also have certain expectations about them. For example, if we enter an elevator that has only one other person in it, we do not expect that individual to confront us and stare into our eyes. As a matter of fact, we would be quite upset if the person did so.

Social interaction within a given society has certain shared meanings across situations. For instance, our reaction would be the same regardless of *which* elevator we rode in *which* building. Sociologist Erving Goffman (1963b) described these shared meanings in his observation about two pedestrians approaching each other on a public sidewalk. He noted that each will tend to look at the other just long enough to acknowledge the other's presence. By the time they are about eight feet away from each other, both individuals will tend to look downward. Goffman referred to this behavior as *civil inattention*—the ways in which an individual shows an awareness that another is present without making this person the object of particular attention. The fact that people engage in civil inattention demonstrates that interaction does have a pattern, or *interaction order,* which regulates the form and processes (but not the content) of social interaction.

Does everyone interpret social interaction rituals in the same way? No. Race/ethnicity, gender, and social class play a part in the meanings we give to our interactions with others, including chance encounters on elevators or the street. Our perceptions about the meaning of a situation vary widely based on the statuses we occupy and our unique personal experiences. For example, sociologist Carol Brooks Gardner (1989) found that women frequently do not perceive street encounters to be "routine" rituals. They fear for their personal safety and try to avoid comments and propositions that are sexual in nature when they walk down the street. African Americans may also feel uncomfortable in street encounters. A middle-class African American college student described his experiences walking home at night from a campus job:

> So, even if you wanted to, it's difficult just to live a life where you don't come into conflict with others. . . . Every day that you live as a black person you're reminded how you're perceived in society. You walk the streets at night; white people cross the streets. I've seen white couples and individuals dart in front of cars to not be on the same side of the street. Just the other day, I was walking down the street, and this white female with a child, I saw her pass a young white male about 20 yards ahead. When she saw me, she quickly dragged the child and herself across the busy street. . . . [When I pass,] white men tighten their grip on their women. I've seen people turn around and seem like they're going to take blows from me. . . . So, every day you realize [you're black]. Even though you're not doing anything wrong; you're just existing. You're just a person. But you're a black person perceived in an unblack world. (qtd. in Feagin, 1991: 111–112)

As this passage indicates, social encounters have different meanings for men and women, whites and people of color, and individuals from different social classes. Members of the dominant classes regard the poor, unemployed, and working class as less worthy of attention, frequently subjecting them to subtle yet systematic "attention deprivation" (Derber, 1983). The same can certainly be said about how members of the dominant classes "interact" with the homeless.

The Social Construction of Reality

If we interpret other people's actions so subjectively, can we have a shared social reality? Some symbolic interaction theorists believe that there is very little shared reality beyond that which is socially created. Symbolic interactionists refer to this as the ***social construction of reality*—the process by which our perception of reality is largely shaped by the subjective meaning that we give to an experience** (Berger and Luckmann, 1967). This meaning strongly influences what we "see" and how we respond to situations.

When you watch a football game, do you "see" the same game as everyone else? The answer is no, according to researchers who asked Princeton and Dartmouth students to watch a film of a recent game between their two schools. The students were instructed to watch for infractions of the rules by each team. Although both groups saw the same film, the Princeton students saw twice as many rule infractions involving the Dartmouth team as the Dartmouth students saw. The researchers noted that one version of what transpired at the game was just as "real" to one person as another (entirely different) version was to another person (Hastorf and Cantril, 1954). When we see what we want or expect to see, we are engaged in the social construction of reality.

As discussed previously, our perceptions and behavior are influenced by how we initially define situations: We act on reality as we see it. Sociologists describe this process as the *definition of the situation,* meaning that we analyze a social context in which we find ourselves, determine what is in our best interest, and adjust our attitudes and actions accordingly. This can result in a ***self-fulfilling prophecy*—a false belief or prediction that produces behavior that makes the originally false belief come true** (Merton, 1968). An example would be a person who has been told repeatedly that she or he is not a good student; eventually, this person might come to believe it to be true, stop studying, and receive failing grades.

People may define a given situation in very different ways, a tendency demonstrated by the sociologist Jac-

People can have sharply contrasting perceptions of the same reality.

queline Wiseman (1970) in her study of "Pacific City's" skid row. She wanted to know how people who live or work on skid row (a run-down area found in all cities) felt about it. Wiseman found that homeless persons living on skid row evaluated it very differently from the social workers who dealt with them there. On the one hand, many of the social workers "saw" skid row as a smelly, depressing area filled with men who were "down-and-out," alcoholic, and often physically and mentally ill. On the other hand, the men who lived on skid row did not see it in such a negative light. They experienced some degree of satisfaction with their "bottle clubs [and a] remarkably indomitable and creative spirit"—at least initially (Wiseman, 1970: 18). As this study shows, we define situations from our own frame of reference, based on the statuses that we occupy and the roles that we play.

Dominant-group members with prestigious statuses may have the ability to establish how other people define "reality" (Berger and Luckmann, 1967: 109). Some sociologists have suggested that dominant groups, particularly higher-income white males in powerful economic and political statuses, perpetuate their own world view through ideologies that are frequently seen as "social reality." For example, the sociologist Dorothy E. Smith (1999) points out that the term "Standard North American Family" (meaning a heterosexual two-parent family) is an ideological code promulgated by the dominant group to identify how people's family life *should* be arranged. According to Smith (1999), this code plays a powerful role in determining how people in organizations such as the government and schools believe that a family should be. Likewise, the sociologist Patricia Hill Collins (1998) argues that "reality" may be viewed differently by African American women and other historically oppressed groups when compared to the perspectives of dominant-group members. However, according to Collins (1998), mainstream, dominant-group members sometimes fail to realize how much they could learn about "reality" from "outsiders." As these theorists state, social reality and social structure are often hotly debated issues in contemporary societies.

Ethnomethodology

How do we know how to interact in a given situation? What rules do we follow? Ethnomethodologists are interested in the answers to these questions. ***Ethnomethodology* is the study of the commonsense knowledge that people use to understand the situations in which they find themselves** (Heritage, 1984: 4). Sociologist Harold Garfinkel (1967) initiated this approach and coined the term: *ethno* for "people" or "folk" and *methodology* for "a system of methods." Garfinkel was critical of mainstream sociology for not recognizing the ongoing ways in which people create reality and produce their own world. Consequently, ethnomethodologists examine existing patterns of conventional

social construction of reality the process by which our perception of reality is shaped largely by the subjective meaning that we give to an experience.

self-fulfilling prophecy the situation in which a false belief or prediction produces behavior that makes the originally false belief come true.

ethnomethodology the study of the commonsense knowledge that people use to understand the situations in which they find themselves.

behavior in order to uncover people's background expectancies—that is, their shared interpretation of objects and events—as well as their resulting actions. According to ethnomethodologists, interaction is based on assumptions of shared expectancies. For example, when you are talking with someone, what expectations do you have that you will take turns? Based on your background expectancies, would you be surprised if the other person talked for an hour and never gave you a chance to speak?

To uncover people's background expectancies, ethnomethodologists frequently break "rules" or act as though they do not understand some basic rule of social life so that they can observe other people's responses. In a series of *breaching experiments,* Garfinkel assigned different activities to his students to see how breaking the unspoken rules of behavior created confusion.

The ethnomethodological approach contributes to our knowledge of social interaction by making us aware of subconscious social realities in our daily lives. However, a number of sociologists regard ethnomethodology as a frivolous approach to studying human behavior because it does not examine the impact of macrolevel social institutions—such as the economy and education—on people's expectancies. Women's studies scholars suggest that ethnomethodologists fail to do what they claim to do: look at how social realities are created. Rather, they take ascribed statuses (such as race, class, gender, and age) as "givens," not as *socially created* realities. For example, in the experiments that Garfinkel assigned to his students, he did not account for how gender affected their experiences. When Garfinkel asked students to reduce the distance between themselves and a nonrelative to the point that "their noses were almost touching," he ignored the fact that gender was as important to the encounter as was the proximity of the two persons. Scholars have recently emphasized that our expectations about reality are strongly influenced by our assumptions relating to gender, race, and social class (see Bologh, 1992).

Dramaturgical Analysis

Erving Goffman suggested that day-to-day interactions have much in common with being on stage or in a dramatic production. ***Dramaturgical analysis* is the study of social interaction that compares everyday life to a theatrical presentation.** Members of our "audience" judge our performance and are aware that we may slip and reveal our true character (Goffman, 1959, 1963a). Consequently, most of us attempt to play our role as well as possible and to control the impressions we give to others. ***Impression management (presentation of self)* refers to people's efforts to present themselves to others in ways that are most favorable to their own interests or image.**

For example, suppose that a professor has returned graded exams to your class. Will you discuss the exam and your grade with others in the class? If you are like most people, you probably play your student role differently depending on whom you are talking to and what grade you received on the exam. Your "presentation" may vary depending on the grade earned by the other person (your "audience"). In one study, students who all received high grades ("Ace–Ace encounters") willingly talked with one another about their grades and sometimes engaged in a little bragging about how they had "aced" the test. However, encounters between students who had received high grades and those who had received low or failing grades ("Ace–Bomber encounters") were uncomfortable. The Aces felt as if they had to minimize their own grade. Consequently, they tended to attribute their success to "luck" and were quick to offer the Bombers words of encouragement. On the other hand, the Bombers believed that they had to praise the Aces and hide their own feelings of frustration and disappointment. Students who received low or failing grades ("Bomber–Bomber encounters") were more comfortable when they talked with one another because they could share their negative emotions. They often indulged in self-pity and relied on face-saving excuses (such as an illness or an unfair exam) for their poor performances (Albas and Albas, 1988).

In Goffman's terminology, *face-saving behavior* refers to the strategies we use to rescue our performance when we experience a potential or actual loss of face. When the Bombers made excuses for their low scores, they were engaged in face-saving; the Aces attempted to help them save face by asserting that the test was unfair or that it was only a small part of the final grade. Why would the Aces and Bombers both participate in face-saving behavior? In most social interactions, all role players have an interest in keeping the "play" going so that they can maintain their overall definition of the situation in which they perform their roles.

Goffman noted that people consciously participate in *studied nonobservance,* a face-saving technique in which one role player ignores the flaws in another's performance to avoid embarrassment for everyone involved. Most of us remember times when we have failed in our role and know that it is likely to happen again; thus, we may be more forgiving of the role failures of others.

Social interaction, like a theater, has a front stage and a back stage. The *front stage* is the area where a player performs a specific role before an audience. The *back stage* is the area where a player is not required

Erving Goffman believed that people spend a great amount of time and effort managing the impression that they present. How do political candidates use impression management as they seek to accomplish their goal of being elected to public office?

to perform a specific role because it is out of view of a given audience. For example, when the Aces and Bombers were talking with each other at school, they were on the "front stage." When they were in the privacy of their own residences, they were in "back stage" settings—they no longer had to perform the Ace and Bomber roles and could be themselves.

The need for impression management is most intense when role players have widely divergent or devalued statuses. As we have seen with the Aces and Bombers, the participants often play different roles under different circumstances and keep their various audiences separated from one another. If one audience becomes aware of other roles that a person plays, the impression being given at that time may be ruined. For example, homeless people may lose jobs or the opportunity to get them when their homelessness becomes known. One woman had worked as a receptionist in a doctor's office for several weeks but was fired when the doctor learned that she was living in a shelter (Liebow, 1993). However, the homeless do not passively accept the roles into which they are cast. For the most part, they attempt—as we all do—to engage in impression management in their everyday life.

The dramaturgical approach helps us think about the roles we play and the audiences who judge our presentation of self. Today, many people are concerned not only about the impressions they make in face-to-face encounters but also in cyberspace (see "Sociology Works!"). However, the dramaturgical approach has been criticized for focusing on appearances and not the underlying substance. This approach may not place enough emphasis on the ways in which our everyday interactions with other people are influenced by occurrences within the larger society. For example, if some members of Congress belittle the homeless as being lazy and unwilling to work, it may become easier for people walking down a street to do likewise. Even so, Goffman's work has been influential in the development of the sociology of emotions, an important area of theory and research.

The Sociology of Emotions

Why do we laugh, cry, or become angry? Are these emotional expressions biological or social in nature? To some extent, emotions are a biologically given sense

dramaturgical analysis the study of social interaction that compares everyday life to a theatrical presentation.

impression management (presentation of self) Erving Goffman's term for people's efforts to present themselves to others in ways that are most favorable to their own interests or image.

Sociology *Works!*

Erving Goffman's Impression Management and Facebook

> Ethan [pseudonym] is in his early 20's and joined [a large software development company] as an entry level consultant six months ago. He joined Facebook in college to keep up with his current friends and used it primarily for getting to know new friends better. He now uses the site to keep in touch with these friends, but his usage has gone from an hour a week to 10 minutes a week. He has over 200 Facebook friends and most of the new employees he met at company orientation are listed as friends. Before starting his job, he purposefully "cleansed" all information about himself on the Internet: from Facebook, his blog, and his personal website. In particular, he removed all photos of himself involving "drinking alcohol." Because of that he is not concerned about strangers, managers, or mentors seeing his information online. (DiMicco and Millen, 2007)

In their recent study regarding how people engage in identity management on Facebook, the social networking website, research scientists Joan Morris DiMicco and David R. Millen (2007) found that Ethan was a good example of how individuals engage in impression management on such websites. When Erving Goffman defined impression management (presentation of self) as people's efforts to present themselves to others in ways that are most favorable to their own interests or image, he was thinking about the face-to-face encounters that each of us has in daily life. Today, many sociologists and social analysts believe that Goffman's ideas may also be applicable to social interactions that take place on the Internet, particularly social networking websites such as Facebook, MySpace, and LinkedIn. According to the journalist Stephanie Rosenbloom (2008: E1), "Now that first impressions are often made in cyberspace, not face-to-face, people are not only strategizing about how to virtually convey who they are, but also grappling with how to craft an e-version of themselves that appeals to multiple audiences—co-workers, fraternity brothers, Mom and Dad."

Although Facebook originated with—and remains popular among—college students, many people who are beyond their college years, including Ethan, still update their information on this networking website. A number of individuals employed by the software development company where DiMicco and Millen (2007) conducted their study stated that they attempt to balance the presentation of themselves as professionals versus nonprofessionals on the Web. They engage in a conscious process of determining what to include (or to exclude) from the personal profiles, photos, and blogs they post online.

Like Ethan, many of us believe that managing our personal identity on the Internet is important. We are concerned about what *people we know* might think if they view our personal information, yet we might be even more anxious about the impression we may make on *people we do not know* but who might become acquaintances or even prospective employers in the future. Looking at the dramaturgical perspective and Goffman's ideas of impression management in online communication offers rich new opportunities for application of classical sociological insights to our interactions with others.

Reflect & Analyze

How might you apply *impression management* and *face-saving behavior* to an analysis of your communications with others on the Internet?

(like hearing, smell, and touch), but they are also social in origin. We are socialized to feel certain emotions, and we learn how and when to express (or not express) those emotions (Hochschild, 1983).

How do we know which emotions are appropriate for a given role? Sociologist Arlie Hochschild (1983) suggests that we acquire a set of *feeling rules* that shapes the appropriate emotions for a given role or specific situation. These rules include how, where, when, and with whom an emotion should be expressed. For example, for the role of a mourner at a funeral, feeling rules tell us which emotions are required (sadness and grief, for example), which are acceptable (a sense of relief that the deceased no longer has to suffer), and which are unacceptable (enjoyment of the occasion expressed by laughing out loud) (see Hochschild, 1983: 63–68).

Feeling rules also apply to our occupational roles. For example, the truck driver who handles explosive cargos must be able to suppress fear. Although all jobs place some burden on our feelings, *emotional labor* occurs only in jobs that require personal contact with the public or the production of a state of mind (such as hope, desire, or fear) in others (Hochschild, 1983). With emotional labor, employees must display only certain carefully selected emotions. For example, flight attendants are required to act friendly toward passengers, to be helpful and open to requests, and to maintain an "omnipresent smile" in order to enhance the customers' status. By contrast, bill collectors are encouraged to show anger and make threats to customers, thereby supposedly deflating the customers' status and wearing down their presumed resistance to paying past-due bills. In both jobs, the employees are expected to show feelings that are often not their true ones (Hochschild, 1983).

Emotional labor may produce feelings of estrangement from one's "true" self. C. Wright Mills (1956) suggested that when we "sell our personality" in the course of selling goods or services, we engage in a seriously self-alienating process. In other words, the "commercialization" of our feelings may dehumanize our work role performance and create alienation and contempt that spill over into other aspects of our life (Hochschild, 1983; Smith and Kleinman, 1989).

Do all people experience and express emotions the same way? It is widely believed that women express emotions more readily than men; as a result, very little research has been conducted to determine the accuracy of this belief. In fact, women and men may differ more in the way they express their emotions than in their actual feelings. Differences in emotional expression may also be attributed to socialization, for the extent to which men and women have been taught that a given emotion is appropriate (or inappropriate) to their gender no doubt plays an important part in their perceptions.

Social class is also a determinant in managed expression and emotion management. Emotional labor is emphasized in middle- and upper-class families. Because middle- and upper-class parents often work with people, they are more likely to teach their children the importance of emotional labor in their own careers than are working-class parents, who tend to work with things, not people (Hochschild, 1983). Race is also an important factor in emotional labor. People of color spend much of their life engaged in emotional labor because racist attitudes and discrimination make it continually necessary to manage one's feelings.

Clearly, Hochschild's contribution to the sociology of emotions helps us understand the social context of our feelings and the relationship between the roles we play and the emotions we experience. However, her thesis has been criticized for overemphasizing the cost of emotional labor and the emotional controls that exist outside the individual (Wouters, 1989). The context in which emotions are studied and the specific emotions examined are important factors in determining the costs and benefits of emotional labor.

© Tom Prettyman/PhotoEdit

© David Silverman/Getty Images

Are there different gender-based expectations in the United States about the kinds of emotions that men, as compared with women, are supposed to show? What feeling rules shape the emotions of the men in these two roles?

Nonverbal Communication

In a typical stage drama, the players not only speak their lines but also convey information by nonverbal communication. In Chapter 3, we discussed the importance of language; now we will look at the messages we communicate without speaking. ***Nonverbal communication* is the transfer of information between persons without the use of words.** It includes not only visual cues (gestures, appearances) but also vocal features (inflection, volume, pitch) and environmental factors (use of space, position) that affect meanings (Wood, 1999). Facial expressions, head movements, body positions, and other gestures carry as much of the total meaning of our communication with others as our spoken words do (Wood, 1999).

Nonverbal communication may be intentional or unintentional. Actors, politicians, and salespersons may make deliberate use of nonverbal communication to convey an idea or "make a sale." We may also send nonverbal messages through gestures or facial expressions or even our appearance without intending to let other people know what we are thinking.

Functions of Nonverbal Communication Nonverbal communication often supplements verbal communication (Wood, 1999). Head and facial movements may provide us with information about other people's emotional states, and others receive similar information from us (Samovar and Porter, 1991a). We obtain first impressions of others from various kinds of nonverbal communication, such as the clothing they wear and their body positions.

Our social interaction is regulated by nonverbal communication. Through our body posture and eye contact, we signal that we do or do not wish to speak to someone. For example, we may look down at the sidewalk or off into the distance when we pass homeless persons who look as if they are going to ask for money.

© Dana White/PhotoEdit

© Richard Ross/Getty Images

© Krzysztof Mystkowski/AFP/Getty Images

Nonverbal communication may be thought of as an international language. What message do you receive from the facial expression, body position, and gestures of each of these people? Is it possible to misinterpret these messages?

Nonverbal communication establishes the relationship among people in terms of their responsiveness to and power over one another (Wood, 1999). For example, we show that we are responsive toward or like another person by maintaining eye contact and attentive body posture and perhaps by touching and standing close. By contrast, we signal to others that we do not wish to be near them or that we dislike them by refusing to look them in the eye or stand near them. We can even express power or control over others through nonverbal communication. Goffman (1956) suggested that *demeanor* (how we behave or conduct ourselves) is relative to social power. People in positions of dominance are allowed a wider range of permissible actions than are their subordinates, who are expected to show deference. *Deference* is the symbolic means by which subordinates give a required permissive response to those in power; it confirms the existence of inequality and reaffirms each person's relationship to the other (Rollins, 1985).

© David Young-Wolff/PhotoEdit

Have you watched other people's reactions to one another in an elevator? How might we explain the lack of eye contact and the general demeanor of the individuals pictured here?

Facial Expression, Eye Contact, and Touching Deference behavior is important in regard to facial expression, eye contact, and touching. This type of nonverbal communication is symbolic of our relationships with others. Who smiles? Who stares? Who makes and sustains eye contact? Who touches whom? All these questions relate to demeanor and deference; the key issue is the status of the person who is doing the smiling, staring, or touching relative to the status of the recipient (Goffman, 1967).

Facial expressions, especially smiles, also reflect gender-based patterns of dominance and subordination in society. Typically, white women have been socialized to smile and frequently do so even when they are not actually happy (Halberstadt and Saitta, 1987). Jobs held predominantly by women (including flight attendant, secretary, elementary schoolteacher, and nurse) are more closely associated with being pleasant and smiling than are "men's jobs." In addition to smiling more frequently, many women tend to tilt their heads in deferential positions when they are talking or listening to others. By contrast, men tend to display less emotion through smiles or other facial expressions and instead seek to show that they are reserved and in control (Wood, 1999).

Women are more likely to sustain eye contact during conversations (but not otherwise) as a means of showing their interest in and involvement with others. By contrast, men are less likely to maintain prolonged eye contact during conversations but are more likely to stare at other people (especially men) in order to challenge them and assert their own status (Pearson, 1985).

Eye contact can be a sign of domination or deference. For example, in a participant observation study of domestic (household) workers and their employers, the sociologist Judith Rollins (1985) found that the domestics were supposed to show deference by averting their eyes when they talked to their employers. Deference also required that they present an "exaggeratedly subservient demeanor" by standing less erect and walking tentatively.

Touching is another form of nonverbal behavior that has many different shades of meaning. Gender and power differences are evident in tactile communication from birth. Studies have shown that touching has variable meanings to parents: Boys are touched more roughly and playfully, whereas girls are handled more gently and protectively (Condry, Condry, and Pogatshnik, 1983). This pattern continues into adulthood, with women touched more frequently than men. Sociologist Nancy Henley (1977) attributed this pattern to power differentials between men and women

nonverbal communication the transfer of information between persons without the use of words.

CONCEPT QUICK REVIEW

Social Interaction: The Microlevel Perspective	
Social Interaction and Meaning	In a given society, forms of social interaction have shared meanings, although these may vary to some extent based on race/ethnicity, gender, and social class.
Social Construction of Reality	The process by which our perception of reality is largely shaped by the subjective meaning that we give to an experience.
Ethnomethodology	Studying the commonsense knowledge that people use to understand the situations in which they find themselves makes us aware of subconscious social realities in daily life.
Dramaturgical Analysis	The study of social interaction that compares everyday life to a theatrical presentation—it includes impression management (people's efforts to present themselves favorably to others).
Sociology of Emotions	We are socialized to feel certain emotions, and we learn how and when to express (or not express) them.
Nonverbal Conmunication	The transfer of information between persons without the use of speech, such as by facial expressions, head movements, and gestures.

wary of making value judgments—the differences are simply differences. Learning to understand and respect alternative styles of social interaction enhances our personal effectiveness by increasing the range of options we have for communicating with different people in diverse contexts and for varied reasons (Wood, 1999). (The Concept Quick Review summarizes the microlevel approach to social interaction.)

Future Changes in Society, Social Structure, and Interaction

The social structure in the United States has been changing rapidly in recent decades. Currently, there are more possible statuses for persons to occupy and roles to play than at any other time in history. Although achieved statuses are considered very important, ascribed statuses still have a significant effect on the options and opportunities that people have.

Ironically, at a time when we have more technological capability, more leisure activities and types of entertainment, and more quantities of material goods available for consumption than ever before, many people experience high levels of stress, fear for their lives because of crime, and face problems such as homelessness. In a society that can send astronauts into space to perform complex scientific experiments, is it impossible to solve some of the problems that plague us here on Earth?

Individuals and groups often show initiative in trying to solve some of our pressing problems (see Box 5.4). For example, Ellen Baxter has singlehandedly tried to create housing for hundreds of New York City's homeless by reinventing well-maintained, single-room-occupancy residential hotels to provide cheap lodging and social services (Anderson, 1993). However, individual initiative alone will not solve all our social problems in the future. Large-scale, formal organizations must become more responsive to society's needs.

At the microlevel, we need to regard social problems as everyone's problem; if we do not, they have a way of becoming everyone's problem anyway.

Chapter Review

- **How does social structure shape our social interactions?**

The stable patterns of social relationships within a particular society make up its social structure. Social structure is a macrolevel influence because it shapes and determines the overall patterns in which social interaction occurs. Social structure provides an ordered framework for society and for our interactions with others.

- **What are the main components of social structure?**

Social structure comprises statuses, roles, groups, and social institutions. A status is a specific position in a group or society and is characterized by certain expectations, rights, and duties. Ascribed statuses, such as gender, class, and race/ethnicity, are acquired at birth or involuntarily later in life. Achieved statuses, such as education and occupation, are assumed voluntarily as a result of personal choice, merit, or direct effort. We occupy a status, but a role is the set of behavioral expectations associated with a given status. A social group consists of two or more people who interact frequently and share a common identity and sense of interdependence. A formal organization is a highly structured group formed to complete certain tasks or achieve specific goals. A social institution is a set of organized beliefs and rules that establishes how a society attempts to meet its basic needs.

- **What are the functionalist and conflict perspectives on social institutions?**

According to functionalist theorists, social institutions perform several prerequisites of all societies: replace members; teach new members; produce, distribute, and consume goods and services; preserve order; and provide and maintain a sense of purpose. Conflict theorists suggest that social institutions do not work for the common good of all individuals. Institutions may enhance and uphold the power of some groups but exclude others, such as the homeless.

- **What are the major types of societies?**

Social scientists have identified five types of societies. Three of these are referred to as preindustrial societies—hunting and gathering, horticultural and pastoral, and agrarian societies. The other two are industrial and postindustrial societies. Industrial societies are characterized by mechanized production of goods. Postindustrial societies are based on technology that supports an information-based economy in which providing services is based on knowledge more than on the production of goods.

- **How do societies maintain stability in times of social change?**

According to Emile Durkheim, although changes in social structure may dramatically affect individuals and groups, societies manage to maintain some degree of stability. People in preindustrial societies are united by mechanical solidarity because they have shared values and common social bonds. Industrial societies are characterized by organic solidarity, which refers to the cohesion that results when people perform specialized tasks and are united by mutual dependence.

- **How do *Gemeinschaft* and *Gesellschaft* societies differ in social solidarity?**

According to Ferdinand Tönnies, the *Gemeinschaft* is a traditional society in which relationships are based on personal bonds of friendship and kinship and on intergenerational stability. The *Gesellschaft* is an urban society in which social bonds are based on impersonal and specialized relationships, with little group commitment or consensus on values.

- **What is the dramaturgical perspective?**

According to Erving Goffman's dramaturgical analysis, our daily interactions are similar to dramatic productions. Presentation of self refers to efforts to present our own self to others in ways that are most favorable to our own interests or self-image.

www.cengage.com/login

Want to maximize your online study time? Take this easy-to-use study system's diagnostic pre-test, and it will create a personalized study plan for you. By helping you to identify the topics that you need to understand better and then directing you to valuable online resources, it can speed up your chapter review. CengageNOW even provides a post-test so you can confirm that you are ready for an exam.

Key Terms

achieved status 140
agrarian societies 150
ascribed status 140
dramaturgical analysis 160
ethnomethodology 159
formal organization 147
Gemeinschaft 155
Gesellschaft 155
horticultural societies 150
hunting and gathering societies 149
impression management (presentation of self) 160
industrial societies 151
master status 141
mechanical solidarity 154
nonverbal communication 164
organic solidarity 154
pastoral societies 150
personal space 166
postindustrial societies 152
primary group 146
role 144
role conflict 144
role exit 146
role expectation 144
role performance 144
role strain 145
secondary group 147
self-fulfilling prophecy 158
social construction of reality 158
social group 146
social institution 147
social interaction 137
social structure 138
status 139
status symbol 142

Questions for Critical Thinking

1. Think of a person you know well who often irritates you or whose behavior grates on your nerves (it could be a parent, friend, relative, or teacher). First, list that person's statuses and roles. Then analyze the person's possible role expectations, role performance, role conflicts, and role strains. Does anything you find in your analysis help to explain the irritating behavior? How helpful are the concepts of social structure in analyzing individual behavior?
2. Are structural problems responsible for homelessness, or are homeless individuals responsible for their own situation? Use functionalist, conflict, symbolic interactionist, and postmodernist theoretical perspectives as tools for analyzing this issue.
3. You are conducting field research on gender differences in nonverbal communication styles. How are you going to account for variations among age, race, and social class?
4. When communicating with other genders, races, and ages, is it better to express and acknowledge different styles or to develop a common, uniform style?

The Kendall Companion Website

www.cengage.com/sociology/kendall

Visit this book's companion website, where you'll find more resources to help you study and successfully complete course projects. Resources include quizzes and flash cards, as well as special features such as an interactive sociology timeline, maps, General Social Survey (GSS) data, and Census 2000 data. The site also provides links to useful websites that have been selected for their relevance to the topics in this chapter and include those listed below. (*Note:* Visit the book's website for updated URLs.)

The Emile Durkheim Archive

http://durkheim.itgo.com/main.html

Visit this website to learn more about Emile Durkheim and his contributions to sociology. Click on "solidarity" for additional information about Durkheim's concepts of mechanical and organic solidarity.

The National Coalition for the Homeless

http://www.nationalhomeless.org

The mission of this organization is to end homelessness. Its website provides facts about homelessness, information on legislation and policy, personal stories, and additional resources pertinent to the study of homelessness.

The Psychology of Cyberspace

http://www.rider.edu/~suler/psycyber/psycyber.html

This site features an online hypertext book that explores social interaction and cyberspace, with links to additional sites that investigate the implications of cyberspace for interpersonal interaction.

chapter

6 Groups and Organizations

Chapter Focus Question

Why is it important for groups and organizations to enhance communication among participants and improve the flow of information while protecting the privacy of individuals?

Okay, I think I'll share why Facebook works for me and keeps me coming back. I was hesitant to sign up in the first place, I was afraid it would be a lame fad. . . . Since college is fairly dynamic (new classes every quarter), a directory of friends and students remains very dynamic and gives me a reason to come back (to see how friends are doing and what classes they are taking). Also it is cool to look up people you have in class and see what they are interested in. Who knows, it might help you start a conversation sometime (although, it might freak them out if you already know their interests). . . .

—a male college student in Washington state (in a comment posted at Linden, 2005) explaining why he likes Facebook

Do you think that even 5 percent of people [on

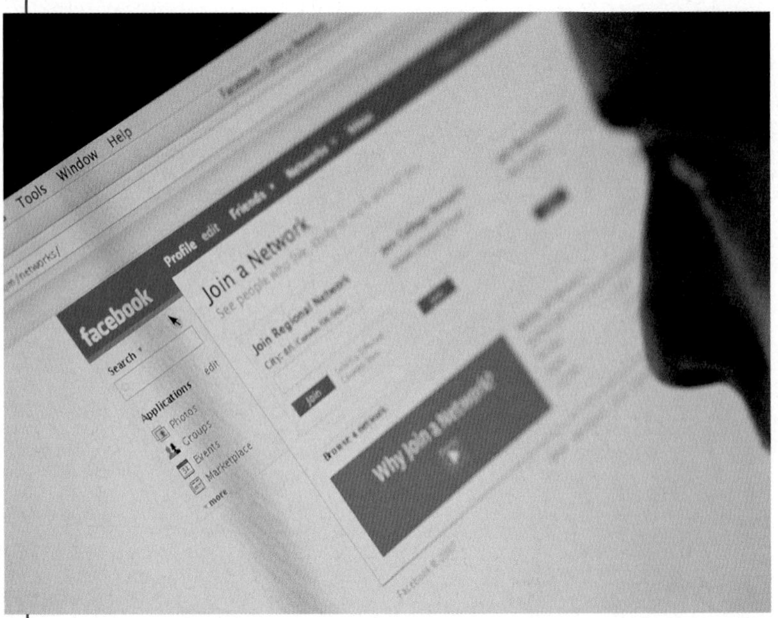

Have Facebook and other networking websites influenced our social interactions and group participation? Why are face-to-face encounters in groups and organizations still important in everyday life?

IN THIS CHAPTER

- Social Groups
- Group Characteristics and Dynamics
- Formal Organizations in Global Perspective
- Alternative Forms of Organization
- Organizations in the Future

Facebook and other networking sites] really connect with people in their network who want them to do the same kewl stuff as them? I doubt it, unless they already knew that person and just didn't know about their hobbies. The truth is none of these sites really connects people. That requires ongoing new information (like web bulletin boards, attending meetings). Or heaven forbid, actual human contact.

—another male college student (in a comment posted at Linden, 2005) claiming that individuals do not really connect with each other online despite the amount of time that they may spend on websites claiming to connect students through their common interests, lifestyles, and attitudes

The problem with facebook.com is people are putting their pictures, cell phone numbers and addresses on the Internet. The Internet is open to anyone. That's just asking for someone to knock on your door. . . . It may seem cool that you know a bunch of people, but it won't be cool if a strange person knows too much about you.

—a junior in a Texas college describing her concerns about facebook.com (Sheppard, 2005)

Sharpening Your Focus

- *What constitutes a social group?*
- *How are groups and their members shaped by group size, leadership style, and pressures to conform?*
- *What is the relationship between information and social organizations in societies such as ours?*
- *What purposes does bureaucracy serve?*
- *What alternative forms of organization exist as compared with the most widespread forms today?*

According to sociologists, we need groups and organizations—just as we need culture and socialization—to live and participate in a society. Historically, the basic premise of groups and organizations was that individuals engage in face-to-face interactions in order to be part of such a group; however, millions of people today communicate with others through the Internet, cell phones, and other forms of information technology that make it possible for them to "talk" with individuals they have never met and who may live thousands of miles away. A variety of networking websites, including Facebook, MySpace, Friendster, and xuqa, now compete with, or in some cases replace, live, person-to-person communications. For many college students, Facebook has become a fun way to get to know other people, to join online groups with similar interests or activities, and to plan "real-life" encounters. Despite the wealth of information and opportunities for new social connections that such websites offer, many of our daily activities require that we participate in social groups and formal organizations where *face-time*—time spent interacting with others on a face-to-face basis, rather than via Internet or cell phone—is necessary.

What do social groups and formal organizations mean to us in an age of rapid telecommunications? What is the relationship between information and social organizations in societies such as ours? How can we balance the information that we provide to other people about us with our own right to privacy and need for security? These questions are of interest to sociologists who seek to apply the sociological imagination to their studies of social groups, bureaucratic organizations, social networking, and virtual communities. Before we take a closer look at groups and organizations, take the quiz in Box 6.1 on personal privacy in groups, in formal organizations, and on the Internet.

Social Groups

Three strangers are standing at a street corner waiting for a traffic light to change. Do they constitute a group? Five hundred women and men are first-year graduate students at a university. Do they constitute a group? In everyday usage, we use the word *group* to mean any collection of people. According to sociologists, however, the answer to these questions is no; individuals who happen to share a common feature or to be in the same place at the same time do not constitute social groups.

Groups, Aggregates, and Categories

As we saw in Chapter 5, a *social group* is a collection of two or more people who interact frequently with one another, share a sense of belonging, and have a feeling of interdependence. Several people waiting for a traffic light to change constitute an ***aggregate*—a collection of people who happen to be in the same place at the same time but share little else in common.** Shoppers in a department store and passengers on an airplane flight are also examples of aggregates. People in aggregates share a common purpose (such as purchasing items or arriving at their destination) but generally do not interact with one another, except perhaps briefly. The first-year graduate students, at least initially, constitute a ***category*—a number of people who may never have met one another but share a similar characteristic** (such as education level, age, race, or gender). Men and women make up categories, as do Native Americans and Latinos/as, and victims of sexual or racial harassment. Categories are not social groups because the people in them usually do not create a social structure or have anything in common other than a particular trait.

Occasionally, people in aggregates and categories form social groups. For instance, people within the category known as "graduate students" may become an aggregate when they get together for an orientation to graduate school. Some of them may form social groups as they interact with one another in classes and seminars, find that they have mutual interests and concerns, and develop a sense of belonging to the group. Information technology raises new and interesting questions about what constitutes a group. For example, some people question whether we can form a social group on the Internet (see Box 6.2).

Types of Groups

As you will recall from Chapter 5, groups have varying degrees of social solidarity and structure. This structure is flexible in some groups and more rigid in others. Some groups are small and personal; others are large and impersonal. We more closely identify with the members of some groups than we do with others.

Cooley's Primary and Secondary Groups Sociologist Charles H. Cooley (1963/1909) used the term *primary group* to describe a small, less specialized group in which members engage in face-to-face, emotion-based interactions over an extended period of time. We have primary relationships with other individuals in our primary groups—that is, with our *significant others*, who frequently serve as role models.

Box 6.1 Sociology and Everyday Life

How Much Do You Know About Privacy in Groups, in Formal Organizations, and on the Internet?

True	False	
T	F	1. A fast-food restaurant can legally require all employees under the age of 21 to submit to periodic, unannounced drug testing.
T	F	2. Members of a high school football team can be required to submit to periodic, unannounced drug testing.
T	F	3. Parents of students at all U.S. colleges and universities are entitled to obtain a transcript of their children's college grades, regardless of the student's age.
T	F	4. A company has the right to keep its employees under video surveillance at all times while they are on company property—even in the company's restrooms.
T	F	5. A private club has the right to require an applicant for membership to provide his or her Social Security number as a condition of membership.
T	F	6. If a person applies for a job in a workplace that has more than 25 employees, the employer can require that person to provide medical information or take a physical examination prior to offering him or her a job.
T	F	7. Students at a church youth group meeting who hear one member of the group confess to an illegal act can be required to divulge what that member said.
T	F	8. A student's privacy is protected when using a computer, even if it is owned by the college or university, because deleting an e-mail or other document from a computer prevents anyone else from examining that document.

Answers on page 176.

In contrast, you will recall, a *secondary group* is a larger, more specialized group in which the members engage in more impersonal, goal-oriented relationships for a limited period of time. The size of a secondary group may vary. Twelve students in a graduate seminar may start out as a secondary group but eventually become a primary group as they get to know one another and communicate on a more personal basis. Formal organizations are secondary groups, but they also contain many primary groups within them. For example, how many primary groups do you think there are within the secondary-group setting of your college?

Sumner's Ingroups and Outgroups All groups set boundaries by distinguishing between insiders, who are members, and outsiders, who are not. Sociologist William Graham Sumner (1959/1906) coined the terms *ingroup* and *outgroup* to describe people's feelings toward members of their own and other groups. An ***ingroup*** **is a group to which a person belongs and with which the person feels a sense of identity.** An ***outgroup*** **is a group to which a person does not belong and toward which the person may feel a sense of competitiveness or hostility.** Distinguishing between our ingroups and our outgroups helps us establish our individual identity and self-worth. Likewise, groups are solidified by ingroup and outgroup distinctions; the presence of an enemy or hostile group binds members more closely together (Coser, 1956).

Group boundaries may be formal, with clearly defined criteria for membership. For example, a country club that requires an applicant for membership to be recommended by four current members and to pay a

aggregate a collection of people who happen to be in the same place at the same time but share little else in common.

category a number of people who may never have met one another but share a similar characteristic, such as education level, age, race, or gender.

ingroup a group to which a person belongs and with which the person feels a sense of identity.

outgroup a group to which a person does not belong and toward which the person may feel a sense of competitiveness or hostility.

Box 6.1 Sociology and Everyday Life

Answers to the Sociology Quiz on Privacy

1. True. In all but a few states, an employee in the private sector of the economy can be required to submit to a drug test even when nothing about the employee's job performance or history suggests illegal drug use. An employee who refuses can be terminated without legal recourse.

2. True. The U.S. Supreme Court has ruled that schools may require students to submit to random drug testing as a condition to participating in extracurricular activities such as sports teams, the school band, the future homemakers' club, the cheerleading squad, and the choir.

3. False. The Family Educational Right to Privacy Act, which allows parents of a student under age 18 to obtain their child's grades, requires the student's consent once he or she has attained age 18; however, that law applies only to institutions that receive federal educational funds.

4. False. An employer may not engage in video surveillance of its employees in situations where they have a reasonable right of privacy. At least in the absence of a sign warning of such surveillance, employees have such a right in company restrooms.

5. True. Although the Privacy Act of 1974 makes it illegal for federal, state, and local governmental agencies to deny rights, privileges, or benefits to individuals who refuse to provide their Social Security number unless disclosure is required by law, no federal law extends this prohibition to private groups and organizations.

6. False. The Americans with Disabilities Act prohibits employers in workplaces with more than 25 employees from asking job applicants about medical information or requiring a physical examination prior to employment.

7. True. Although confidential communications made privately to a minister, priest, rabbi, or other religious leader (or to an individual the person reasonably believes to hold such a position) generally cannot be divulged without the consent of the person making the communication, this does not apply when other people are present who are likely to hear the statement.

8. False. Deleting an e-mail or other document from a computer does not actually remove it from the computer's memory. Until other files are entered that write over the space where the document was located, experts can retrieve the document that was deleted.

$25,000 initiation fee has clearly set requirements for its members (see "Sociology Works!" on pages 180–181). However, group boundaries are not always that formal. For example, friendship groups usually do not have clear guidelines for membership; rather, the boundaries tend to be very informal and vaguely defined.

Ingroup and outgroup distinctions may encourage social cohesion among members, but they may also promote classism, racism, sexism, and ageism. Ingroup members typically view themselves positively and members of outgroups negatively. These feelings of group superiority, or *ethnocentrism,* are somewhat inevitable. However, members of some groups feel more free than others to act on their beliefs. If groups are embedded in larger groups and organizations, the large organization may discourage such beliefs and their consequences (Merton, 1968). Conversely, organizations may covertly foster these ingroup/outgroup distinctions by denying their existence or by failing to take action when misconduct occurs.

Reference Groups Ingroups provide us not only with a source of identity but also with a point of reference. A ***reference group* is a group that strongly influences a person's behavior and social attitudes, regardless of whether that individual is an actual member.** When we attempt to evaluate our appearance, ideas, or goals, we automatically refer to the standards of some group. Sometimes, we will refer to our membership groups, such as family or friends. Other times, we will rely on groups to which we do not currently belong but that we might wish to join in the future, such as a social club or a profession. We may also have negative reference groups. For many people, the Ku Klux Klan and neo-Nazi skinheads are examples of negative reference groups because most people's racial attitudes compare favorably with such groups' blatantly racist behavior.

Reference groups help explain why our behavior and attitudes sometimes differ from those of our membership groups. We may accept the values and norms of a group with which we identify rather than one to which we belong. We may also act more like members of a group we want to join than members of groups to which we already belong. In this case, reference groups are a source of anticipatory socialization. Many people have more than one reference group and often receive conflicting messages from these groups about how they should view themselves. For most of us, our reference-group attachments change many times during our life course, especially when we acquire a new status in a formal organization.

Networks A ***network* is a web of social relationships that links one person with other people and, through them, with other people they know.** Frequently, networks connect people who share common interests but who otherwise might not identify and interact with one another. For example, if A is tied to B, and B is tied to C, then a network is more likely to be formed among individuals A, B, and C. If this seems a little confusing at first, let's assume that Alice knows of Dolores and Eduardo only through her good friends Bill and Carolyn. For almost a year, Alice has been trying (without success) to purchase a house she can afford. Because large numbers of people are moving into her community, the real estate market is "tight," and houses frequently sell before a "for sale" sign goes up in the yard. However, through her friends Bill and Carolyn, Alice learns that their friends—Dolores and Eduardo—are about to put their house up for sale. Bill and Carolyn call Dolores and Eduardo to set up an appointment for Alice to see the house before it goes on the real estate market. Thanks to Alice's network, she is able to purchase the house before other people learn that it is for sale. Although Alice had not previously met Dolores and Eduardo, they are part of her network through her friendship with Bill and Carolyn. Scarce resources (in this case, the number of affordable houses available) are unequally distributed, and people often must engage in collaboration and competition in their efforts to deal with this scarcity. Another example of the use of networks to help overcome scarce resources is recent college graduates who seek help from friends and acquaintances in order to find a good job.

What are your networks? For a start, your networks consist of all the people linked to you by primary ties, including your relatives and close friends. Your networks also include your secondary ties, such as acquaintances, classmates, professors, and—if you are employed—your supervisor and co-workers. However, your networks actually extend far beyond these ties to include not only the people that you *know*, but also the people that you *know of*—and who know of you—through your primary and secondary ties. In fact, your networks potentially include a pool of between 500 and 2,500 acquaintances, if you count the connections of everyone in your networks (Milgram, 1967). Today, the term *networking* is widely used to describe the contacts that people make to find jobs or other opportunities; however, sociologists have studied social networks for many years in an effort to learn more about the linkages between individuals and their group memberships.

Group Characteristics and Dynamics

What purpose do groups serve? Why are individuals willing to relinquish some of their freedom to participate in groups? According to functionalists, people form groups to meet instrumental and expressive needs. *Instrumental,* or task-oriented, needs cannot always be met by one person, so the group works cooperatively to fulfill a specific goal. For example, think of how hard it would be to function as a one-person football team or to single-handedly build a skyscraper. Groups help members do jobs that are impossible to do alone or that would be very difficult and time-consuming at best. In addition to instrumental needs, groups also help people meet their *expressive,* or emotional, needs, especially those involving self-expression and support from family, friends, and peers.

Although not disputing that groups ideally perform such functions, conflict theorists suggest that groups also involve a series of power relationships whereby the needs of individual members may not be equally served. Symbolic interactionists focus on how the size of a group influences the kind of interaction that takes place among members. To many postmodernists, groups and organizations—like other aspects of postmodern societies—are generally characterized by superficiality and depthlessness in social relationships (Jameson, 1984). One postmodern thinker who focuses on this issue is the literary theorist Fredric Jameson, who believes that people experience a waning of emotion in organizations where fragmentation and superficiality are a way of life (Ritzer, 1997). For example, fast-food restaurant employees and

reference group a group that strongly influences a person's behavior and social attitudes, regardless of whether that individual is an actual member.

network a web of social relationships that links one person with other people and, through them, with other people they know.

Box 6.2 Framing Community in the Media

"Virtual Communities" on the Internet

Meeting new friends,
Imagining smiles . . .
Across the networks
Spanning the miles. . . .

From all walks of life
We come to the net.
A community of friends
Who have never met.
—from "Thoughts of Internet Friendships" by Jamie Wilkerson (1996)

As this excerpt from a poem posted on the Internet suggests, many people believe that they can make new friends and establish a community online. We are encouraged to establish such friendships by joining chat groups maintained by various Internet service providers. Chat groups are framed as a public service offered as part of the fee a subscriber pays for an Internet connection. To participate, people fill out a profile listing their hobbies and interests so that they can be matched with other participants. Many people hope to become part of the larger Internet community, which has been described as "a body of people looking for similar information, dealing with similar conditions, and abiding by the same general rules" (thewritemarket.com, 2003).

Although chat groups are framed as a new way to make friends, get dates, and establish a cyber community, as you study sociology you might ask whether this form of "community" is actually a true community. Because sociologists define a *social group* as a collection of two or more people who interact frequently with one another, share a sense of belonging, and have a feeling of interdependence, this definition suggests that people must have a sense of place (be in the same place at the same time at least part of the time) in order to establish a true social group or community. However, this definition was developed before the Internet provided people with the rapid communications that connect them with others around the world today. Are we able to form groups and establish communities with people whom we have never actually met?

Some social scientists believe that virtual communities established on the Internet constitute true communities (see Wellman, 2001). However, the sociologists Robyn Bateman Driskell and Larry Lyon examined existing theories and research on this topic and concluded that true communities cannot be established in the digital environment of cyberspace. According to Driskell and Lyon, although the Internet provides us with the opportunity to share interests with others whom we have not met (such as through chat groups) and to communicate with people we already know (such as by e-mail and instant messaging), the original concept of community, which "emphasized local place, common ties, and social interaction that is intimate, holistic, and all-encompassing," is lacking (Driskell and Lyon, 2002: 6). Virtual communities do not have geographic and social boundaries, are limited in their scope to specific areas of interest, are psychologically detached from close interpersonal ties, and have only limited concern for their "members" (Driskell and Lyon, 2002). In fact, if we spend many hours in social isolation doing impersonal searches for information, the Internet may *reduce* community, rather than enhance it.

The Internet provides a rapid means of communication among people who have computers and Internet access, but this may not add up to the establishment of true social groups and true communities in the traditional sociological sense of these terms. Even so, it is possible that the Internet will create a "weak community replacement" for people based on a virtual community of specialized ties developed by e-mail correspondence and chatroom discussions (Driskell and Lyon, 2002).

Chatrooms and other forms of communication on the Internet are extremely popular with millions of people; however, some sociologists question whether we can actually form social groups and true communities on the Internet. Is cyber chat different from our face-to-face interactions with others?

Reflect & Analyze

Do you think that chat groups are accurately framed in descriptions by Internet service providers, or are they overrated or misrepresented to potential participants?

customers interact in extremely superficial ways that are largely scripted: The employees follow scripts in taking and filling customers' orders ("Would you like fries and a drink with that?"), and the customers respond with their own "recipied" action. According to the sociologist George Ritzer (1997: 226), "[C]ustomers are mindlessly following what they consider tried-and-true social recipes, either learned or created by them previously, on how to deal with restaurant employees and, more generally, how to work their way through the system associated with the fast-food restaurant."

We will now look at certain characteristics of groups, such as how size affects group dynamics.

Group Size

The size of a group is one of its most important features. Interactions are more personal and intense in a ***small group,* a collectivity small enough for all members to be acquainted with one another and to interact simultaneously.**

Sociologist Georg Simmel (1950/1902–1917) suggested that small groups have distinctive interaction patterns that do not exist in larger groups. According to Simmel, in a ***dyad*—a group composed of two members**—the active participation of both members is crucial for the group's survival. If one member withdraws from interaction or "quits," the group ceases to exist. Examples of dyads include two people who are best friends, married couples, and domestic partnerships. Dyads provide members with a more intense bond and a sense of unity not found in most larger groups.

When a third person is added to a dyad, a ***triad,* a group composed of three members,** is formed. The nature of the relationship and interaction patterns changes with the addition of the third person. In a triad, even if one member ignores another or declines to participate, the group can still function. In addition, two members may unite to create a coalition that can subject the third member to group pressure to conform. A *coalition* is an alliance created in an attempt to reach a shared objective or goal. If two members form a coalition, the other member may be seen as an outsider or intruder. Like dyads, triads can exist as separate entities or be contained within formal organizations.

As the size of a group increases beyond three people, members tend to specialize in different tasks, and everyday communication patterns change. For instance, in groups of more than six or seven people, it becomes increasingly difficult for everyone to take part in the same conversation; therefore, several conversations will probably take place simultaneously. Members are also likely to take sides on issues and form a number of coalitions. In groups of more than ten or twelve people, it becomes virtually impossible for all members to participate in a single conversation unless one person serves as moderator and guides the discussion. As shown in ▶ Figure 6.1, when the size of the group increases, the number of possible social interactions also increases.

Although large groups typically have less social solidarity than small ones, they may have more power. However, the relationship between size and power is more complicated than it might initially seem. The power relationship depends on both a group's *absolute* size and its *relative* size (Simmel, 1950/1902–1917; Merton, 1968). The absolute size is the number of members the group actually has; the relative size is the

According to the sociologist Georg Simmel, interaction patterns change when a third person joins a dyad—a group composed of two members. How might the conversation between these two women change when another person arrives to talk with them?

small group a collectivity small enough for all members to be acquainted with one another and to interact simultaneously.

dyad a group composed of two members.

triad a group composed of three members.

complete a task or reach a particular goal. ***Expressive leadership*** **provides emotional support for members;** this type of leadership is most appropriate when the group is dealing with emotional issues, and harmony, solidarity, and high morale are needed. Both kinds of leadership are needed for groups to work effectively. Traditionally, instrumental and expressive leadership roles have been limited by gender socialization. Instrumental leadership has been linked with men, whereas expressive leadership has been linked with women. Social change in recent years has somewhat blurred the distinction between gender-specific leadership characteristics, but these outdated stereotypes have not completely disappeared (Basow, 1992).

Leadership Styles Three major styles of leadership exist in groups: authoritarian, democratic, and laissez-faire. ***Authoritarian leaders*** **make all major group decisions and assign tasks to members.** These leaders focus on the instrumental tasks of the group and demand compliance from others. In times of crisis, such as a war or natural disaster, authoritarian leaders may be commended for their decisive actions. In other situations, however, they may be criticized for being dictatorial and for fostering intergroup hostility. By contrast, ***democratic leaders*** **encourage group discussion and decision making through consensus building.** These leaders may be praised for their expressive, supportive behavior toward group members, but they may also be blamed for being indecisive in times of crisis.

Laissez-faire literally means "to leave alone." ***Laissez-faire leaders*** **are only minimally involved in decision making and encourage group members to make their own decisions.** On the one hand, laissez-faire leaders may be viewed positively by group members because they do not flaunt their power or position. On the other hand, a group that needs active leadership is not likely to find it with this style of leadership, which does not work vigorously to promote group goals.

Studies of kinds of leadership and decision-making styles have certain inherent limitations. They tend to focus on leadership that is imposed externally on a group (such as bosses or political leaders) rather than leadership that arises within a group. Different decision-making styles may be more effective in one setting than another. For example, imagine attending a college class in which the professor asked the students to determine what should be covered in the course, what the course requirements should be, and how students should be graded. It would be a difficult and cumbersome way to start the semester; students might spend the entire term negotiating these matters and never actually learn anything.

Organizations have different leadership styles based on the purpose of the group. How do leadership styles in the military differ from those on college and university campuses?

Group Conformity

To what extent do groups exert a powerful influence in our lives? Groups have a significant amount of influence on our values, attitudes, and behavior. In order to gain and then retain our membership in groups, most of us are willing to exhibit a high level of conformity to the wishes of other group members. ***Conformity*** **is the process of maintaining or changing behavior to comply with the norms established by a society, subculture, or other group.** We often experience powerful pressure from other group members to conform. In some situations, this pressure may be almost overwhelming.

In several studies (which would be impossible to conduct today for ethical reasons), researchers found that the pressure to conform may cause group members to say they see something that is contradictory to what they are actually seeing or to do something that they would otherwise be unwilling to do. As we look at two of these studies, ask yourself what you might have done if you had been involved in this research.

Asch's Research Pressure to conform is especially strong in small groups in which members want to fit in with the group. In a series of experiments conducted by Solomon Asch (1955, 1956), the pressure toward group conformity was so great that participants were willing to contradict their own best judgment if the rest of the group disagreed with them.

One of Asch's experiments involved groups of undergraduate men (seven in each group) who were allegedly recruited for a study of visual perception. All the men were seated in chairs. However, the person in the sixth chair did not know that he was the only actual subject; all the others were assisting the researcher. The participants were first shown a large card with a vertical line on it and then a second card with three vertical lines (see ▶ Figure 6.2). Each of the seven participants was asked to indicate which of the three lines on the second card was identical in length to the "standard line" on the first card.

In the first test with each group, all seven men selected the correct matching line. In the second trial, all seven still answered correctly. In the third trial, however, the actual subject became very uncomfortable when all the others selected the incorrect line. The subject could not understand what was happening and became even more confused as the others continued to give incorrect responses on eleven out of the next fifteen trials.

If you had been in the position of the subject, how would you have responded? Would you have continued to give the correct answer, or would you have been swayed by the others? When Asch (1955) averaged the responses of all fifty actual subjects who participated in the study, he found that about 33 percent routinely chose to conform to the group by giving the same (incorrect) responses as Asch's assistants. Another 40 percent gave incorrect responses in about half of the trials. Although 25 percent always gave correct responses, even they felt very uneasy and "knew that something was wrong." In discussing the experiment afterward, most of the subjects who gave incorrect responses indicated that they had known the answers were wrong but decided to go along with the group in order to avoid ridicule or ostracism.

After conducting additional research, Asch concluded that the size of the group and the degree of social cohesion felt by participants were important influences on the extent to which individuals respond to group pressure. In dyads, for example, the subject was much less likely to conform to an incorrect response from one assistant than in four-member groups. This effect peaked in groups of approximately seven members and then leveled off (see ▶ Figure 6.3). Not surprisingly, when groups were not cohesive (when more than one member dissented), group size had less effect. If even a single assistant did not agree with the others, the subject was reassured by hearing someone else question the accuracy of incorrect responses and was much less likely to give a wrong answer himself.

One contribution of Asch's research is the dramatic way in which it calls our attention to the power that groups have to produce a certain type of conformity. *Compliance* is the extent to which people say (or do) things so that they may gain the approval of other people. Certainly, Asch demonstrated that people will bow to social pressure in small-group settings. From a sociological perspective, however, the study was flawed because it involved deception about the purpose of the study and about the role of individual group members. Moreover, the study included only male college students, thus making it impossible for us to generalize its findings to other populations, including women and

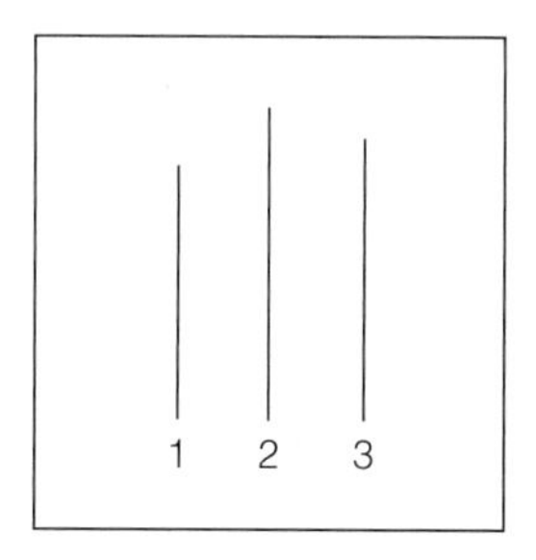

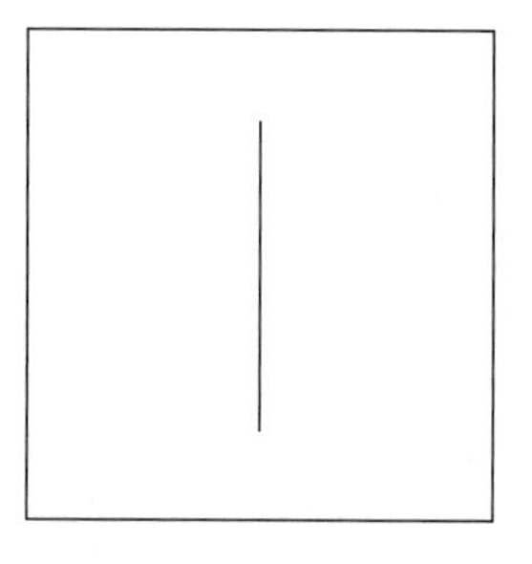

▶ **Figure 6.2 Asch's Cards**
Although Line 2 is clearly the same length as the line in the lower card, Solomon Asch's research assistants tried to influence "actual" participants by deliberately picking Line 1 or Line 3 as the correct match. Many of the participants went along rather than risk the opposition of the "group."

Source: Asch, 1955.

expressive leadership an approach to leadership that provides emotional support for members.

authoritarian leaders people who make all major group decisions and assign tasks to members.

democratic leaders leaders who encourage group discussion and decision making through consensus building.

laissez-faire leaders leaders who are only minimally involved in decision making and who encourage group members to make their own decisions.

conformity the process of maintaining or changing behavior to comply with the norms established by a society, subculture, or other group.

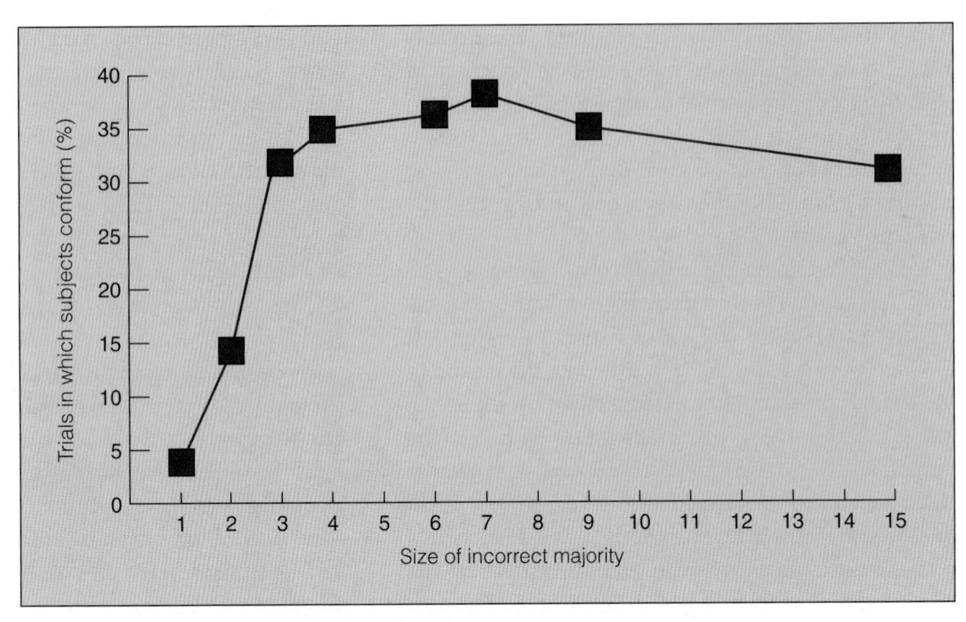

▶ **Figure 6.3 Effect of Group Size in the Asch Conformity Studies**
As more people are added to the "incorrect" majority, subjects' tendency to conform by giving wrong answers increases—but only up to a point. Adding more than seven people to the incorrect majority does not further increase subjects' tendency to conform—perhaps because subjects are suspicious about why so many people agree with one another.

Source: Asch, 1955.

people who were not undergraduates. Would Asch's conclusions have been the same if women had participated in the study? Would the same conclusions be reached if the study were conducted today? We cannot answer these questions with certainty, but the work of Solomon Asch and his student, Stanley Milgram, has had a lasting impact on social science perceptions about group conformity and obedience to authority.

Milgram's Research How willing are we to do something because someone in a position of authority has told us to do it? How far are we willing to go in following the demands of that individual? Stanley Milgram (1963, 1974) conducted a series of controversial experiments to find answers to these questions about people's obedience to authority. *Obedience* is a form of compliance in which people follow direct orders from someone in a position of authority.

Milgram's subjects were men who had responded to an advertisement for participants in an experiment. When the first (actual) subject arrived, he was told that the study concerned the effects of punishment on learning. After the second subject (an assistant of Milgram's) arrived, the two men were instructed to draw slips of paper from a hat to get their assignments as either the "teacher" or the "learner." Because the drawing was rigged, the actual subject always became the teacher, and the assistant the learner. Next, the learner was strapped into a chair with protruding electrodes that looked something like an electric chair. The teacher was placed in an adjoining room and given a realistic-looking but nonoperative shock generator. The "generator's" control panel showed levels that went from "Slight Shock" (15 volts) on the left, to "Intense Shock" (255 volts) in the middle, to "DANGER: SEVERE SHOCK" (375 volts), and finally to "XXX" (450 volts) on the right.

The teacher was instructed to read aloud a pair of words and then repeat the first of the two words. At that time, the learner was supposed to respond with the second of the two words. If the learner could not provide the second word, the teacher was instructed to press the lever on the shock generator so that the learner would be punished for forgetting the word. Each time the learner gave an incorrect response, the teacher was supposed to increase the shock level by 15 volts. The alleged purpose of the shock was to determine if punishment improves a person's memory.

What was the maximum level of shock that a "teacher" was willing to inflict on a "learner"? The learner had been instructed (in advance) to beat on the wall between him and the teacher as the experiment continued, pretending that he was in intense pain. The teacher was told that the shocks might be "extremely painful" but that they would cause no permanent damage. At about 300 volts, when the learner quit responding at all to questions, the teacher often turned to the experimenter to see what he should do next. When the experimenter indicated that the teacher should give

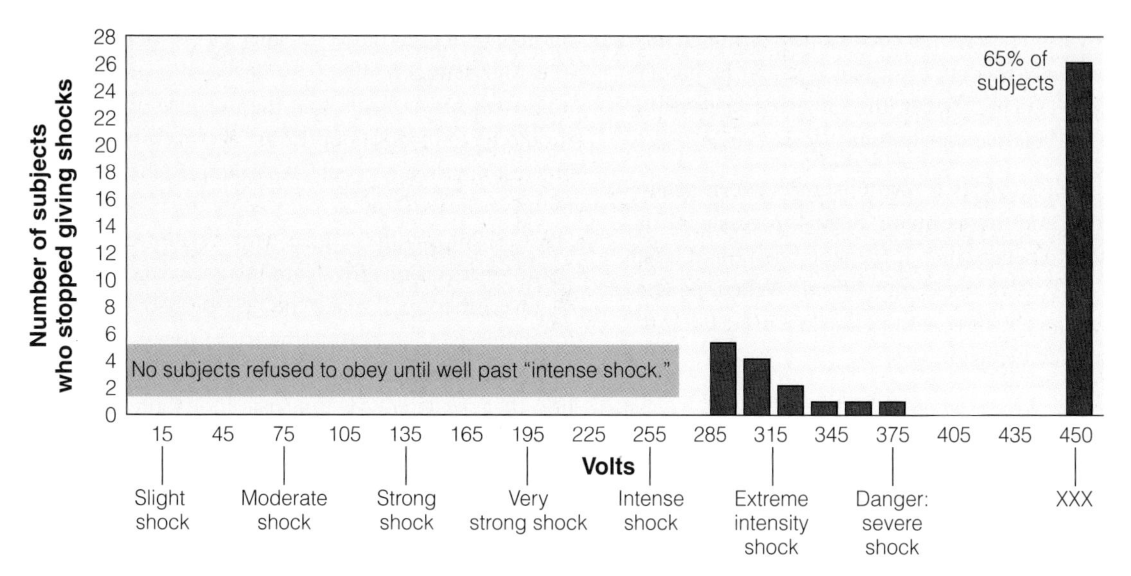

▶ **Figure 6.4 Results of Milgram's Obedience Experiment**
Even Milgram was surprised by subjects' willingness to administer what they thought were severely painful and even dangerous shocks to a helpless "learner."

Source: Milgram, 1963.

increasingly painful shocks, 65 percent of the teachers administered shocks all the way up to the "XXX" (450-volt) level (see ▶ Figure 6.4). By this point in the process, the teachers were frequently sweating, stuttering, or biting on their lip. According to Milgram, the teachers (who were free to leave whenever they wanted to) continued in the experiment because they were being given directions by a person in a position of authority (a university scientist wearing a white coat).

What can we learn from Milgram's study? The study provides evidence that obedience to authority may be more common than most of us would like to believe. None of the "teachers" challenged the process before they had applied 300 volts. Almost two-thirds went all the way to what could have been a deadly jolt of electricity if the shock generator had been real. For many years, Milgram's findings were found to be consistent in a number of different settings and with variations in the research design (Miller, 1986).

This research once again raises some questions originally posed in Chapter 2 concerning research ethics. As was true of Asch's research, Milgram's subjects were deceived about the nature of the study in which they were asked to participate. Many of them found the experiment extremely stressful. These conditions cannot be ignored by social scientists because subjects may receive lasting emotional scars from such research. It would be virtually impossible today to obtain permission to replicate this experiment in a university setting.

Group Conformity and Sexual Harassment Let's look at a more contemporary example of how social science research can help us learn about the ways in which group conformity may contribute to a complex social problem such as sexual harassment, which consists of unwanted sexual advances, requests for sexual favors, or other verbal or physical conduct of a sexual nature.

Psychologist John Pryor (Pryor and McKinney, 1991) has conducted behavioral experiments on college campuses to examine the social dynamics of harassment. In one of his studies, a graduate student (who was actually a member of the research team) led research subjects to believe that they would be training undergraduate women to use a computer. The actual purpose of the experiment was to observe whether the trainers (subjects) would harass the women if given the opportunity and encouraged to do so. By design, the graduate student purposely harassed the women (who were also part of the research team), setting an example for the subjects to follow.

Pryor found that when the "trainers" were led to believe that sexual harassment was condoned and then were left alone with the women, they took full advantage of the situation in 90 percent of the experiments. Shannon Hoffman, one of the women who participated in the research, felt vulnerable because of the permissive environment created by the men in charge:

> It was very uncomfortable for me. I realized that had it been out of the experimental setting that, as a woman, I would have been very nervous with someone that close to me and reaching around me. So it kind of made me feel a little bit powerless as far as that goes because there was nothing I could do about

it. But I also realized that in a business setting, if this person really was my boss, that it would be harder for me to send out the negative signals or whatever to try to fend off that type of thing. (PBS, 1992)

This research suggests a relationship between group conformity and harassment. Sexual harassment is more likely to occur when it is encouraged (or at least not actively discouraged) by others. When people think they can get away with it, they are more likely to engage in such behavior.

Groupthink

As we have seen, individuals often respond differently in a group context than they might if they were alone. Social psychologist Irving Janis (1972, 1989) examined group decision making among political experts and found that major blunders in U.S. history may be attributed to pressure toward group conformity. To describe this phenomenon, he coined the term ***groupthink*—the process by which members of a cohesive group arrive at a decision that many individual members privately believe is unwise.** Why not speak up at the time? Members usually want to be "team players." They may not want to be the ones who undermine the group's consensus or who challenge the group's leaders. Consequently, members often limit or withhold their opinions and focus on consensus rather than on exploring all of the options and determining the best course of action. ▶ Figure 6.5 summarizes the dynamics and results of groupthink.

The tragic 2003 explosion of the space shuttle *Columbia* while preparing to land has been cited as an example of this process. During takeoff, a chunk of insulated foam fell off the bipod ramp of the external fuel tank, striking and damaging the shuttle's left wing. Although some NASA engineers had previously raised concerns that hardened foam popping off the fuel tank could cause damage to the ceramic tiles protecting the shuttle, and although their concerns were again raised following *Columbia*'s liftoff, these concerns were overruled by NASA officials prior to and during the flight (Glanz and Wong, 2003; Schwartz, 2003). One analyst subsequently described the way that NASA dealt with these concerns as an example of "the ways that smart people working collectively can be dumber than the sum of their brains" (Schwartz and Wald, 2003: WK3).

Social Exchange/Rational Choice Theories

Social exchange/rational choice theories focus on the process by which actors—individuals, groups, corporations, or societies, for example—settle on one optimal outcome out of a range of possible choices. The foundation of this approach is the doctrine of *utilitarianism*—a belief that the purpose of all action should be to bring about the greatest happiness to the greatest number of people. An example of this belief is found in the assertion of early economist Adam Smith (1976/1776) that individuals who are allowed to make economic decisions free from the external constraints of government will make the best decisions not only for themselves but also for the entire society.

Social exchange theories are based on the assumption that *self-interest* is the basic motivating factor in people's interactions. According to this approach, people learn to adjust their behavior so that they receive rewards from others rather than negative responses or punishment (Homans, 1974). When people do not *give* and *take* in a manner that is deemed appropriate by other group members, conflict often ensues, and relationships among people may be destabilized (Gouldner, 1960). Consider this example of give and take: A good friend offers you a gift, but you decide to refuse it. What factors contribute to your decision? Self-interest—such as keeping the friendship or acquiring a possession of some worth—might dictate that you accept the gift. However, other factors may also be involved. What if you do not like the gift or do not want to feel obligated to reciprocate in some manner? The offer of the gift forces you to assess your self-interest in the situation: What will you gain or lose by accepting or rejecting the gift? Ultimately, your decision may cause solidarity or conflict; it may stabilize or destabilize your relationship with the other person.

Now, if we think of a similar exchange involving *words* rather than tangible objects, a similar process occurs. In work settings, for example, an exchange might involve conferring a reward (prestige) on someone in return for a valuable contribution (such as expert advice) (see Homans, 1958, 1974; and Blau, 1964, 1975). Based on self-interest, a person may accept or reject the statements or implicit assumptions of another person. In fact, people often compete with one another as they seek to maximize their rewards and minimize their punishments (Blau, 1964, 1975). People do not always gain reciprocal benefits from exchanges with others, particularly in situations where one person in the exchange occupies a dominant power position over another person (Emerson, 1962).

Rational choice theorists have analyzed situations in which the actors have differing amounts of power. *Rational choice theories* are based on the assumption that social life can be explained by using models of rational individual action (Outhwaite and Bottomore, 1994). Accordingly, rational choice theorists are more concerned with explaining *social outcomes* than in predicting what an individual will do in a particular

Process of Groupthink	Example: *Columbia* Explosion
PRIOR CONDITIONS Isolated, cohesive, homogeneous decision-making group Lack of impartial leadership High stress	NASA had previously orchestrated many successful shuttle missions and was under pressure to complete additional space missions that would fulfill agency goals and keep its budget intact.
SYMPTOMS OF GROUPTHINK Closed-mindedness Rationalization Squelching of dissent "Mindguards" Feelings of righteousness and invulnerability	Although *Columbia*'s left wing had been damaged on takeoff when a chunk of insulated foam from the external fuel tank struck it, NASA did not regard this as a serious problem because it had occurred on previous launches. Some NASA engineers stated that they did not feel free to raise questions about problems.
DEFECTIVE DECISION MAKING Incomplete examination of alternatives Failure to examine risks and contingencies Incomplete search for information	The debate among engineers regarding whether the shuttle had been damaged to the extent that the wing might burn off on reentry was not passed on to the shuttle crew or to NASA's top officials in a timely manner because the engineers either harbored doubts about their concerns or were unwilling to believe that the mission was truly imperiled.
CONSEQUENCES Poor decisions	The shuttle *Columbia* was destroyed during reentry into the Earth's atmosphere, killing all seven crew members and strewing debris across large portions of the United States.

© AP Images/Chris O'Meara

NASA Kennedy Space Center (NASA-KSC)

© AP Images/Dr. Scott Lieberman

▶ Figure 6.5 Janis's Description of Groupthink

In Janis's model, prior conditions such as a highly homogeneous group with committed leadership can lead to potentially disastrous "groupthink," which short-circuits careful and impartial deliberation. Events leading up to the tragic 2003 explosion of the space shuttle *Columbia* have been cited as an example of this process.

Sources: Broder, 2003; Glanz and Wong, 2003; Schwartz (with Wald), 2003; Schwartz and Broder, 2003; Schwartz and Wald, 2003.

situation (Hechter and Kanazawa, 1997). Rational choice theory assumes that *actors* are purposeful or intentional in their decisions; however, many theorists acknowledge that not all actions are necessarily rational and that people do not always act rationally (Hechter and Kanazawa, 1997).

What are the key elements of rational choice theories? According to the sociologist James S. Coleman (1990), *actors* and *resources* are important factors in rational choice. Actors may include individuals, groups, corporations, and societies. Resources comprise the things over which actors have control and in which they have some interest. Major constraints on actors' choices are the scarcity of resources and structural

groupthink the process by which members of a cohesive group arrive at a decision that many individual members privately believe is unwise.

restrictions (Marsden, 1983). Actors with fewer resources are less likely to pursue the most highly valued goal or end. They may decide to go for the next-most-attractive goal or end out of fear that they will lose the chance to acquire even the next-most-attractive end if they initially pursue an unrealistic goal or expectation. For example, suppose that Zoe, who has very few economic resources, wants to attend an Ivy League university (her most-highly-valued goal). Although her first-choice school sends her a letter of acceptance, it does not offer her a scholarship. Meanwhile, a large state university (her next-most-attractive goal) admits her and offers her a full four-year scholarship. When Zoe weighs her options—attending "Ivy U" by taking out large student loans and getting a job or attending "State U" on a full scholarship that gives her time to study—she may select her next-most-attractive goal, whereas individuals with greater economic resources would probably attend their first-choice institution.

In addition to availability of resources, a second major constraint on actors' choices is social institutions (Friedman and Hechter, 1988). Institutional constraints such as rules, laws, ordinances, corporate policies, and religious doctrines limit actors' available choices. Using a rational choice approach, the sociologist Michael Hechter (1987) studied what happens when people "do their own thing" in an organization. Among other findings, Hechter concluded that organizations with large numbers of employees hire others (including security guards, managers, and inspectors) to control people's behavior so that they will not unduly pursue their propensity to maximize their own gain or pleasure, particularly at the expense of the organization.

Formal Organizations in Global Perspective

Over the past century, the number of formal organizations has increased dramatically in the United States and other industrialized nations. Previously, everyday life was centered in small, informal, primary groups, such as the family and the village. With the advent of industrialization and urbanization (as discussed in Chapter 1), people's lives became increasingly dominated by large, formal secondary organizations. A *formal organization,* you will recall, is a highly structured secondary group formed for the purpose of achieving specific goals in the most efficient manner. Formal organizations (such as corporations, schools, and government agencies) usually keep their basic structure for many years in order to meet their specific goals.

Types of Formal Organizations

We join some organizations voluntarily and others out of necessity. Sociologist Amitai Etzioni (1975) classified formal organizations into three categories—normative, coercive, and utilitarian—based on the nature of membership in each.

Normative Organizations We voluntarily join *normative organizations* when we want to pursue some common interest or gain personal satisfaction or prestige from being a member. Political parties, ecological activist groups, religious organizations, parent–teacher associations, and college sororities and fraternities are examples of normative, or voluntary, associations.

Class, gender, and race are important determinants of a person's participation in a normative association. Class (socioeconomic status based on a person's education, occupation, and income) is the most significant predictor of whether a person will participate in mainstream normative organizations; membership costs may exclude some from joining. Those with higher socioeconomic status are more likely to be not only members but also active participants in these groups. Gender is also an important determinant. Half of the voluntary associations in the United States have all-female memberships; one-fifth are all male. However, all-male organizations usually have higher levels of prestige than all-female ones (Odendahl, 1990).

Throughout history, people of all racial–ethnic categories have participated in voluntary organizations, but the involvement of women in these groups has largely gone unrecognized. For example, African American women were actively involved in antislavery societies in the nineteenth century and in the civil rights movement in the twentieth century (see Scott, 1990). Other normative organizations focusing on civil rights, self-help, and philanthropic activities in which African American women and men have been involved include the National Association for the Advancement of Colored People (NAACP) and the Urban League. Similarly, Native American women have participated in the American Indian Movement, a group organized to fight problems ranging from police brutality to housing and employment discrimination (Feagin and Feagin, 2003). Mexican American women (as well as men) have held a wide range of leadership positions in La Raza Unida Party and the League of United Latin American Citizens, organizations oriented toward civic activities and protest against injustices (Amott and Matthaei, 1996).

Coercive Organizations People do not voluntarily become members of *coercive organizations*—associations that people are forced to join. Total institutions, such as boot camps, prisons, and some mental hospitals, are examples of coercive organizations. As dis-

© AP Images/Vincent Thian

© A. Ramey/PhotoEdit

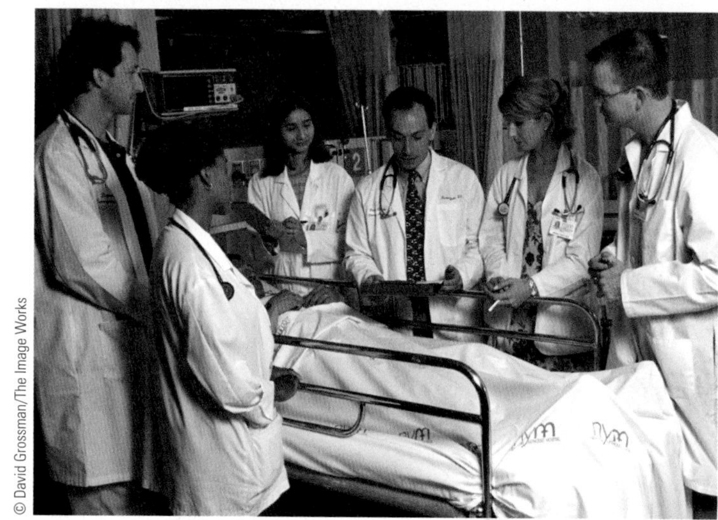
© David Grossman/The Image Works

Normative organizations rely on volunteers to fulfill their goals; for example, Red Cross workers in Sri Lanka aided the relief efforts in that country following a deadly tsunami. Coercive organizations rely on involuntary recruitment; these prison inmates in Alabama are being resocialized in a total institution. Utilitarian organizations provide material rewards to participants; in teaching hospitals such as this one, medical students and patients hope that they may benefit from involvement within the organization.

cussed in Chapter 4, the assumed goal of total institutions is to resocialize people through incarceration. These environments are characterized by restrictive barriers (such as locks, bars, and security guards) that make it impossible for people to leave freely. When people leave without being officially dismissed, their exit is referred to as an "escape."

Utilitarian Organizations We voluntarily join *utilitarian organizations* when they can provide us with a material reward we seek. To make a living or earn a college degree, we must participate in organizations that can provide us these opportunities. Although we have some choice regarding where we work or attend school, utilitarian organizations are not always completely voluntary. For example, most people must continue to work even if the conditions of their employment are less than ideal. (The Concept Quick Review summarizes types of groups, sizes of groups, and types of formal organizations.)

Bureaucracies

The bureaucratic model of organization remains the most universal organizational form in government, business, education, and religion. A ***bureaucracy* is an organizational model characterized by a hierarchy of authority, a clear division of labor, explicit rules and procedures, and impersonality in personnel matters.**

Sociologist Max Weber (1968/1922) was interested in the historical trend toward bureaucratization that accelerated during the Industrial Revolution. To Weber, the bureaucracy was the most "rational" and efficient means of attaining organizational goals because it contributed to coordination and control. According to Weber, ***rationality* is the process by which traditional methods of social organization, characterized by informality and spontaneity, are gradually replaced by efficiently administered formal rules and**

bureaucracy an organizational model characterized by a hierarchy of authority, a clear division of labor, explicit rules and procedures, and impersonality in personnel matters.

rationality the process by which traditional methods of social organization, characterized by informality and spontaneity, are gradually replaced by efficiently administered formal rules and procedures.

CONCEPT QUICK REVIEW

Characteristics of Groups and Organizations

Types of Social Groups	Primary group	Small, less specialized group in which members engage in face-to-face, emotion-based interaction over an extended period of time
	Secondary group	Larger, more specialized group in which members engage in more impersonal, goal-oriented relationships for a limited period of time
	Ingroup	A group to which a person belongs and with which the person feels a sense of identity
	Outgroup	A group to which a person does not belong and toward which the person may feel a sense of competitiveness or hostility
	Reference group	A group that strongly influences a person's behavior and social attitudes, regardless of whether the person is actually a member
Group Size	Dyad	A group composed of two members
	Triad	A group composed of three members
	Formal organization	A highly structured secondary group formed for the purpose of achieving specific goals
Types of Formal Organizations	Normative	Organizations we join voluntarily to pursue some common interest or gain personal satisfaction or prestige by joining
	Coercive	Associations that people are forced to join (total institutions such as boot camps and prisons are examples)
	Utilitarian	Organizations we join voluntarily when they can provide us with a material reward that we seek

procedures. Bureaucracy can be seen in all aspects of our lives, from small colleges with perhaps a thousand students to multinational corporations employing many thousands of workers worldwide.

In his study of bureaucracies, Weber relied on an ideal-type analysis, which he adapted from the field of economics. An ***ideal type*** **is an abstract model that describes the recurring characteristics of some phenomenon** (such as bureaucracy). To develop this ideal type, Weber abstracted the most characteristic bureaucratic aspects of religious, educational, political, and business organizations. For example, to develop an ideal type for bureaucracy in higher education, you would need to include the relationships among governing bodies (such as boards of regents or trustees), administrators, faculty, staff, and students. You would also have to include the rules and policies that govern the school's activities (such as admissions criteria, grading policies, and graduation requirements). Although no two schools would have exactly the same criteria, the ideal-type constructs would be quite similar. Weber acknowledged that no existing organization would exactly fit his ideal type of bureaucracy (Blau and Meyer, 1987).

Ideal Characteristics of Bureaucracy Weber set forth several ideal-type characteristics of bureaucratic organizations. Weber's model (see ▶ Figure 6.6) highlights the organizational efficiency and productivity that bureaucracies strive for in these five central elements of the ideal organization:

Division of Labor Bureaucratic organizations are characterized by specialization, and each member has highly specialized tasks to fulfill.

Hierarchy of Authority In a bureaucracy, each lower office is under the control and supervision of a higher one. Those few individuals at the top of the hierarchy have more power and exercise more control than do the many at the lower levels. Those who are lower in the hierarchy report to (and often take orders from) those above them in the organizational pyramid. Persons at the upper levels are responsible not only for

Characteristics	Effects
• Division of labor • Hierarchy of authority • Rules and regulations • Qualification-based employment • Impersonality	• Inefficiency and rigidity • Resistance to change • Perpetuation of race, class, and gender inequalities

▶ **Figure 6.6 Characteristics and Effects of Bureaucracy**
The very characteristics that define Weber's idealized bureaucracy can create or exacerbate the problems that many people associate with this type of organization. Can you apply this model to an organization with which you are familiar?

their own actions but also for those of the individuals they supervise.

Rules and Regulations Rules and regulations establish authority within an organization. These rules are typically standardized and provided to members in a written format. In theory, written rules and regulations offer clear-cut standards for determining satisfactory performance so that each new member does not have to reinvent the rules.

Qualification-Based Employment Bureaucracies require competence and hire staff members and professional employees based on specific qualifications. Individual performance is evaluated against specific standards, and promotions are based on merit as spelled out in personnel policies.

Impersonality Bureaucracies require that everyone must play by the same rules and be treated the same. Personal feelings should not interfere with organizational decisions.

Contemporary Applications of Weber's Theory How well do Weber's theory of rationality and his ideal-type characteristics of bureaucracy withstand the test of time? More than a century later, many organizational theorists still apply Weber's perspective. For example, the sociologist George Ritzer used Weber's theories to examine fast-food restaurants such as McDonald's. According to Ritzer, the process of "McDonaldization" has become a global phenomenon as four elements of rationality can be found in fast-food restaurants and other "speedy" or "jiffy" businesses (such as Sir Speedy Printing and Jiffy Lube). Ritzer (2000a: 433) identifies four dimensions of formal rationality—efficiency, predictability, emphasis on quantity rather than quality, and control through nonhuman technologies—that are found in today's fast-food restaurants:

> Efficiency means the search for the best means to the end; in the fast-food restaurant, the drive-through window is a good example of heightening the efficiency of obtaining a meal. Predictability means a world of no surprises; the Big Mac in Los Angeles is indistinguishable from the one in New York; similarly, the one we consume tomorrow or next year will be just like the one we eat today. Rational systems tend to emphasize quantity, usually large quantities, rather than quality. The Big Mac is a good example of this emphasis on quantity rather than quality. Instead of the human qualities of a chef, fast-food restaurants rely on nonhuman technologies like unskilled cooks following detailed directions and assembly-line methods applied to the cooking and serving of food. Finally, such a formally rational system brings with it various irrationalities, most notably the demystification and dehumanization of the dining experience.

Although still useful today, Weber's ideal type largely failed to take into account the informal side of bureaucracy.

The Informal Side of Bureaucracy When we look at an organizational chart, the official, formal structure of a bureaucracy is readily apparent. In practice, however, a bureaucracy has patterns of activities and interactions that cannot be accounted for by its organizational chart. These have been referred to as *bureaucracy's other face* (Page, 1946).

The ***informal side of a bureaucracy* is composed of those aspects of participants' day-to-day activities and interactions that ignore, bypass, or do not correspond with the official rules and procedures of the bureaucracy.** An example is an informal "grapevine" that spreads information (with varying degrees of accuracy) much faster than do official channels of communication, which tend to be slow and unresponsive. The informal structure has also been referred to as *work*

ideal type an abstract model that describes the recurring characteristics of some phenomenon (such as bureaucracy).

informal side of a bureaucracy those aspects of participants' day-to-day activities and interactions that ignore, bypass, or do not correspond with the official rules and procedures of the bureaucracy.

culture because it includes the ideology and practices of workers on the job. Workers create this work culture in order to confront, resist, or adapt to the constraints of their jobs, as well as to guide and interpret social relations on the job (Zavella, 1987). Today, computer networks and e-mail offer additional opportunities for workers to enhance or degrade their work culture. Some organizations have sought to control offensive communications so that workers will not be exposed to a hostile work environment brought about by colleagues, but such control has raised significant privacy issues (see Box 6.3).

Positive and Negative Aspects of Informal Structure Is informal structure good or bad? Should it be controlled or encouraged? Two schools of thought have emerged with regard to these questions. One approach emphasizes control (or eradication) of informal groups; the other suggests that they should be nurtured. Traditional management theories are based on the assumption that people are basically lazy and motivated by greed. Consequently, informal groups must be controlled (or eliminated) in order to ensure greater worker productivity.

By contrast, the other school of thought asserts that people are capable of cooperation. Thus, organizations should foster informal groups that permit people to work more efficiently toward organizational goals. Chester Barnard (1938), an early organizational theorist, focused on the functional aspects of informal groups. He suggested that organizations are cooperative systems in which informal groups "oil the wheels" by providing understanding and motivation for participants. In other words, informal networks serve as a means of communication and cohesion among individuals, as well as protect the integrity of the individual (Barnard, 1938; Perrow, 1986).

© Sean Justice/Getty Images

How do people use the informal "grapevine" to spread information? Is this faster than the organization's official communications channels? Is it more or less accurate than official channels?

The *human relations approach,* which is strongly influenced by Barnard's model, views informal networks as a type of adaptive behavior that workers engage in because they experience a lack of congruence between their own needs and the demands of the organization (Argyris, 1960). Organizations typically demand dependent, childlike behavior from their members and strive to thwart the members' ability to grow and achieve "maturity" (Argyris, 1962). At the same time, members have their own needs to grow and mature. Informal networks help workers fill this void. Large organizations would be unable to function without strong informal norms and relations among participants (Blau and Meyer, 1987).

More-recent studies have confirmed the importance of informal networks in bureaucracies. Whereas some scholars have argued that women and people of color receive fairer treatment in larger bureaucracies than they do in smaller organizations, others have stressed that they may be categorically excluded from networks that are important for survival and advancement in the organization (Kanter, 1993/1977; South et al., 1982; Benokraitis and Feagin, 1995; Feagin, 1991).

Informal networks thrive in contemporary organizations because e-mail and websites have made it possible for people to communicate throughout the day without ever having to engage in face-to-face interaction. The need to meet at the water fountain or the copy machine in order to exchange information is long gone: Workers now have an opportunity to tell one another—and higher-ups, as well—what they think.

Problems of Bureaucracies

The characteristics that make up Weber's "rational" model of bureaucracy have a dark side that has frequently given this type of organization a bad name. Three of the major problems of bureaucracies are (1) inefficiency and rigidity, (2) resistance to change, and (3) perpetuation of race, class, and gender inequalities.

Inefficiency and Rigidity Bureaucracies experience inefficiency and rigidity at both the upper and lower levels of the organization. The self-protective behavior of officials at the top may render the organization inefficient. One type of self-protective behavior is the

Box 6.3 Sociology and Social Policy

Computer Privacy in the Workplace

I think I'm getting paranoid. When I'm at work, I think somebody is watching me. When I send an e-mail or search the Web, I wonder who knows about it besides me. Don't get me wrong. I get all my work done first, but when I have some spare time, I may play a computer game or check out some web site I'm interested in. I know that when I'm at work, my time belongs to the company, but somehow I still feel like it's an invasion of my privacy for some computer to monitor every single thing I do. I mean, I work for a company that makes cardboard boxes, not the CIA!

—a student in one of the author's classes, expressing her irritation over computer surveillance at work

Do employers really have the right to monitor everything that their employees do on company-owned computers? Generally speaking, the answer is yes, and the practice is widespread. A recent survey found that about one-half of all U.S. companies monitor their employees' e-mail and more than 60 percent of employers monitor Internet connections (American Management Association, 2001).

Employers assert not only that they have the right to engage in such surveillance but also that it may be necessary for them to do so for their own protection. As for their right to do so, they note that they own the computer, pay for the Internet service, and pay the employee to spend his or her time on company business. As for it possibly being necessary for them to monitor their employees' computer use, employers argue that they may be held legally responsible for harassing or discriminatory e-mail sent on company computers and that surveillance is the only way to protect against such liability. As a result, many employers take the position, according to the Privacy Foundation's Stephen Keating (qtd. in Agonafir, 2002), that "You leave your First Amendment [privacy] rights at the door when you work for a private employer. That's the way it has always been."

In most instances, courts have upheld monitoring even when the employees were not aware of the surveillance, and according to the American Management Association (2001), about one-third of all U.S. employers who engage in computer surveillance do not advise their employees that they are doing so. (At the time of this writing, Connecticut is the only state which requires that employees be notified of monitoring before it can legally occur.) In several high-profile cases, the firing of an employee based on his or her "inappropriate" e-mail has been upheld even when the employer's stated policy was *not* to monitor employee e-mail.

Yet there are valid arguments against computer surveillance as well, and invasion of a worker's privacy is certainly one of them. When an employee makes a personal phone call while at work, and it is a local or toll-free call, the employee usually has a reasonable expectation of privacy—a reasonable belief that neither fellow workers nor his or her employer is eavesdropping on that call. How about "snail mail"? An employee has a reasonable expectation of privacy that the employer will not steam open a personal letter addressed to the employee, read it, and reseal the envelope. Why should an e-mail exchange with friends or relatives not be equally private and protected? If the employer is going to read an employee's e-mail or track the person's Internet activities, shouldn't the employer at least have to make sure that its workers are aware of that policy?

Regarding the argument that personal e-mail wastes the employer's time, privacy advocates state that most employees who exchange personal e-mails or surf the Internet while at work are either doing so on their own time during breaks or at least at times when they have a "free" moment or two. If an employee's productivity falls below expected levels, shouldn't that be obvious to his or her boss without snooping on the employee's e-mail?

Finally, with regard to the employer possibly being held responsible for its employees' actions, Chief Judge Edith H. Jones (qtd. in Gordon, 2001) of the U.S. Fifth Circuit Court has observed that "It seems highly disproportionate to inflict a monitoring program that may invade thousands of people's privacy for the sake of exposing a handful of miscreants." The need to prevent a crime or to protect a company against potential liability must be balanced against each individual's privacy rights.

Ultimately, this balancing must be done by the legislature or the courts. There are no easy answers to this pressing social policy issue, but it should remain a concern for all who live in a democratic society.

Reflect & Analyze

Are you concerned about computer and cell phone privacy in your own life? Should businesses and colleges have the right to monitor our conversations? Why or why not?

to all people regardless of race, gender, or class. Workers and students alike can benefit from organizational environments that make it possible for people to explore their joint interests without fear of losing their privacy or being pitted against one another in a competitive struggle for advantage. (For an example of students working together on a meaningful activity that benefits others, see Box 6.4.)

Chapter Review

• How do sociologists distinguish among social groups, aggregates, and categories?

Sociologists define a social group as a collection of two or more people who interact frequently, share a sense of belonging, and depend on one another. People who happen to be in the same place at the same time are considered an aggregate. Those who share a similar characteristic are considered a category. Neither aggregates nor categories are considered social groups.

• How do sociologists classify groups?

Sociologists distinguish between primary and secondary groups. Primary groups are small and personal, and members engage in emotion-based interactions over an extended period. Secondary groups are larger and more specialized, and members have less personal and more formal, goal-oriented relationships. Sociologists also divide groups into ingroups, outgroups, and reference groups. Ingroups are groups to which we belong and with which we identify. Outgroups are groups we do not belong to or perhaps feel hostile toward. Reference groups are groups that strongly influence the behavior of people whether or not they are actually members.

• What is the significance of group size?

In small groups, all members know one another and interact simultaneously. In groups with more than three members, communication dynamics change, and members tend to assume specialized tasks.

• What are the major styles of leadership?

Leadership may be authoritarian, democratic, or laissez-faire. Authoritarian leaders make major decisions and assign tasks to individual members. Democratic leaders encourage discussion and collaborative decision making. Laissez-faire leaders are minimally involved and encourage members to make their own decisions.

• What do experiments on conformity show us about the importance of groups?

Groups may have significant influence on members' values, attitudes, and behaviors. In order to maintain ties with a group, many members are willing to conform to norms established and reinforced by group members.

• What are the strengths and weaknesses of bureaucracies?

A bureaucracy is a formal organization characterized by hierarchical authority, division of labor, explicit procedures, and impersonality. According to Max Weber, bureaucracy supplies a rational means of attaining organizational goals because it contributes to coordination and control. A bureaucracy also has an informal structure, which includes the daily activities and interactions that bypass the official rules and procedures. The informal structure may enhance productivity or may be counterproductive to the organization. A bureaucracy may be inefficient, resistant to change, and a vehicle for perpetuating class, race, and gender inequalities.

Key Terms

aggregate 174
authoritarian leaders 182
bureaucracy 189
bureaucratic personality 194
category 174
conformity 182
democratic leaders 182
dyad 179
expressive leadership 182
goal displacement 194
groupthink 186
ideal type 190
informal side of a bureaucracy 191
ingroup 175
instrumental leadership 181
iron law of oligarchy 195
laissez-faire leaders 182
network 177
outgroup 175
rationality 189
reference group 176
small group 179
triad 179

Questions for Critical Thinking

1. Who might be more likely to conform in a bureaucracy, those with power or those wanting more power?
2. Although there has been much discussion recently concerning what is and what is not sexual harassment, it has been difficult to reach a clear consensus on what behaviors and actions are acceptable. What are some specific ways that both women and men can avoid contributing to an atmosphere of sexual harassment in organizations? Consider team relationships, management and mentor relationships, promotion policies, attitudes, behavior, dress and presentation, and after-work socializing.
3. Do the insights gained from Milgram's research on obedience outweigh the elements of deception and stress that were forced on its subjects?
4. If you were forming a company based on humane organizational principles, would you base the promotional policies on merit and performance or on affirmative action goals?

The Kendall Companion Website

www.cengage.com/sociology/kendall

Visit this book's companion website, where you'll find more resources to help you study and successfully complete course projects. Resources include quizzes and flash cards, as well as special features such as an interactive sociology timeline, maps, General Social Survey (GSS) data, and Census 2000 data. The site also provides links to useful websites that have been selected for their relevance to the topics in this chapter and include the one listed below. (*Note:* Visit the book's website for updated URLs.)

Sociologyindex: Organization
http://sociologyindex.com/organization.htm

This website provides information on how to study formal organizations and features links to other sites that provide access to primary writings by scholars in sociology and the management sciences.

McDonaldization
http://www.mcdonaldization.com

The objective of this site is to educate the public about the McDonaldization of America. In addition to presenting information on George Ritzer's text *The McDonaldization of Society,* the site features an introduction to the concept of McDonaldization, related topics, articles, news, and interesting links

7 Deviance and Crime

Chapter Focus Question

What do studies of peer cliques and youth gangs tell us about deviance?

"What worries me," said Marlene [a Southside Chicago block-club president], "is that there's about seventy children on my block who use that park—and that's not counting the ones who live on the other side. Can't have them around your boys [gang members]."

"You all are something else," Big Cat [a gang leader for the "Black Kings"] said, shaking his head. "I been cooperating with you all for years now, never complaining that I'm losing money. . . . I don't get no respect for that?"

"If you're in our park, we can't be. It's as simple as that," Marlene replied. "I'll give you the nighttime. Maybe I can convince folks that you all need to work at night, but that's going to be tough. But, bottom line, baby, is we can't have you all there during the day. . . ."

© A. Ramey/PhotoEdit

Members of the California group known as the Culver City Boyz typify how gang members use items of clothing and gang signs made with their hands to assert their membership in the group and solidarity with one another. Some people might view this conduct as deviant behavior, whereas many gang members view it as an act of conformity.

IN THIS CHAPTER

"Okay," interjected [local pastor] Wilkins. "Now you have to stop for the summer, Big Cat. We're not asking for a two-year thing, or nothing like that. Just when the kids are outside."

"I guess I could work it on 59th, but that [business owner] keeps telling us he doesn't want us around, keeps calling the cops. . . ."

"If I get him to leave you alone during the day, and you can hang out in the parking lot on the other side of the store, you'll leave the park for the summer."

"Yeah," Big Cat replied, dejected at the compromise. "Okay, we'll be gone."

—sociologist Sudhir Alladi Venkatesh (2006: 294–295) describing a conversation among three people he interviewed during his research into the underground economy in Chicago and the gangs that constitute part of that economy

Sociologists and criminologists typically define a *gang* as a group of people, usually young, who band together for purposes generally considered to be deviant or criminal by the larger society. Throughout the past century, gang behavior has been of special interest to sociologists (see Puffer, 1912), who generally agree that youth gangs can be found in many settings and among all racial and ethnic categories. U.S. government sources estimate that about 26,500 gangs, containing approximately 785,000 gang members, have been active in the United States in recent years.

As unusual as it initially may sound, some important similarities exist between youth gangs and peer cliques, which are typically viewed as conforming to

Sharpening Your Focus

- *What is deviant behavior?*
- *When is deviance considered a crime?*
- *What are the major theoretical perspectives on deviance?*
- *How are crimes classified?*
- *How does the criminal justice system deal with crime?*

most social norms. At the most basic level, *cliques* are friendship circles, whose members identify one another as mutually connected (Adler and Adler, 1998). However, cliques are much more complex than this definition suggests. According to the sociologists Patricia A. Adler and Peter Adler (1998: 56), cliques "have a hierarchical structure, being dominated by leaders, and are exclusive in nature, so that not all individuals who desire membership are accepted." Moreover, sociologists have found that cliques function as "bodies of power" in schools by "incorporating the most popular individuals, offering the most exciting social lives, and commanding the most interest and attention from classmates" (Adler and Adler, 1998: 56).

Although cliques may have some similarities with gangs, there are also significant differences: Gangs play a large role in the economy of many low-income urban neighborhoods, where residents often believe that they must do whatever is necessary to survive. Some activities in the underground economy include the performance of unregulated, unreported, and untaxed work, whereas others involve more widely recognized criminal activities such as the sale of drugs by gang members. According to Venkatesh (2006), one remarkable thing about studying deviance and crime in settings such as "Marquis Park" (a pseudonym for a real Southside Chicago neighborhood) was learning that residents and gang members sometimes forge temporary alliances and engage in self-initiated policing so that neighborhood children may play safely at the park or enjoy other everyday activities without fear of harm. Venkatesh's study reveals people's efforts to survive with the resources that they amass in the underground economy, as well as residents' willingness to negotiate with gang members if it will help restore a sense of order to their neighborhood. This unique form of community policing often takes place without the assistance of law enforcement officials.

In this chapter, we look at the relationship among conformity, deviance, and crime; even in times of national crisis and war, "everyday" deviance and crime occur as usual. People do not stop activities that might be viewed by others—or by law enforcement officials—as violating social norms. An example is gang behavior, which is used in this chapter as an example of deviant behavior. For individuals who find a source of identity, self-worth, and a feeling of protection by virtue of gang membership, no radical change occurs in daily life even as events around them may change. Youth gangs have been present in the United States for many years because they meet perceived needs of members. Some gangs may be thought of as being very similar to youth cliques, whereas other gangs engage in activities that constitute crime. Before reading on, take the quiz on peer cliques, youth gangs, and deviance in Box 7.1.

What Is Deviance?

***Deviance* is any behavior, belief, or condition that violates significant social norms in the society or group in which it occurs.** We are most familiar with *behavioral* deviance, based on a person's intentional or inadvertent actions. For example, a person may engage in intentional deviance by drinking too much or robbing a bank, or in inadvertent deviance by losing money in a Las Vegas casino or laughing at a funeral.

Although we usually think of deviance as a type of behavior, people may be regarded as deviant if they express a radical or unusual *belief system.* Members of cults (such as Moonies and satanists) and of far-right-wing or far-left-wing political groups may be considered deviant when their religious or political beliefs become known to people with more-conventional cultural beliefs. However, individuals who are considered to be "deviant" by one category of people may be seen as conformists by another group. For example, adolescents in some peer cliques and youth gangs may shun mainstream cultural beliefs and values but routinely conform to subcultural codes of dress, attitude (such as defiant individualism), and behavior (Jankowski, 1991). Those who think of themselves as "Goths" may wear black trench coats, paint their fingernails black, and listen to countercultural musicians.

In addition to their behavior and beliefs, individuals may also be regarded as deviant because they possess a specific *condition* or characteristic. A wide range of conditions have been identified as "deviant," including being obese (Degher and Hughes, 1991; Goode, 1996) and having AIDS (Weitz, 2004). For example, research by the sociologist Rose Weitz (2004) has shown that persons with AIDS live with a stigma that affects their relationships with other people, including family members, friends, lovers, colleagues, and health care workers. Chapter 5 defines a *stigma* as any physical or social attribute or sign that so devalues a person's social identity that it disqualifies the person from full social acceptance (Goffman, 1963b). Based on this definition, the stigmatized person has a "spoiled identity" as a result of being negatively evaluated by others (Goffman, 1963b). To avoid or reduce stigma, many people seek to conceal the characteristic or condition that might lead to stigmatization.

Box 7.1 Sociology and Everyday Life

How Much Do You Know About Peer Cliques, Youth Gangs, and Deviance?

True	False	
T	F	1. According to some sociologists, deviance may serve a useful purpose in society.
T	F	2. Peer cliques on high school campuses have few similarities to youth gangs.
T	F	3. Most people join gangs to escape from broken homes caused by divorce or the death of a parent.
T	F	4. Juvenile gangs are an urban problem; few rural areas have problems with gangs.
T	F	5. Street crime has a much higher economic cost to society than crimes committed in executive suites or by government officials.
T	F	6. Persons age 15 to 24 account for more than half of all arrests for property crimes such as burglary, larceny, arson, and vandalism.
T	F	7. Studies have shown that peer cliques have become increasingly important to adolescents over the past two decades.
T	F	8. Gangs are an international problem.

Answers on page 204.

Who Defines Deviance?

Are some behaviors, beliefs, and conditions inherently deviant? In commonsense thinking, deviance is often viewed as inherent in certain kinds of behavior or people. For sociologists, however, deviance is a formal property of social situations and social structure. As the sociologist Kai T. Erikson (1964: 11) explains,

> Deviance is not a property inherent in certain forms of behavior; it is a property conferred upon these forms by the audiences which directly or indirectly witness them. The critical variable in the study of deviance, then, is the social audience rather than the individual actor, since it is the audience which eventually determines whether or not any episode of behavior or any class of episodes is labeled deviant.

Based on this statement, we can conclude that deviance is *relative*—that is, an act becomes deviant when it is socially defined as such. Definitions of deviance vary widely from place to place, from time to time, and from group to group (see "Sociology Works!"). Today, for example, some women wear blue jeans and very short hair to college classes; some men wear an earring and long hair. In the past, such looks violated established dress codes in many schools, and administrators probably would have asked these students to change their appearance or leave school.

Deviant behavior also varies in its *degree of seriousness,* ranging from mild transgressions of folkways, to more serious infringements of mores, to quite serious violations of the law. Have you kept a library book past its due date or cut classes? If so, you have violated folkways. Others probably view your infraction as relatively minor; at most, you might have to pay a fine or receive a lower grade. Violations of mores—such as falsifying a college application or cheating on an examination—are viewed as more serious infractions and are punishable by stronger sanctions, such as academic probation or expulsion. Some forms of deviant behavior violate the criminal law, which defines the behaviors that society labels as criminal. A ***crime*** **is a behavior that violates criminal law and is punishable with fines, jail terms, and/or other negative sanctions.** Crimes range from minor offenses (such as traffic violations) to major offenses (such as murder). A subcategory, ***juvenile delinquency,*** **refers to a violation of law or the commission of a status offense by young people.** Note that the legal concept of juvenile delinquency includes not only crimes but also status offenses (such as

deviance any behavior, belief, or condition that violates significant cultural norms in the society or group in which it occurs.

crime behavior that violates criminal law and is punishable with fines, jail terms, and other sanctions.

juvenile delinquency a violation of law or the commission of a status offense by young people.

Box 7.1 Sociology and Everyday Life

Answers to the Sociology Quiz on Peer Cliques, Youth Gangs, and Deviance

1. True. From Durkheim to contemporary functionalists, theorists have regarded some degree of deviance as functional for societies.

2. False. Many social scientists believe that there are striking similarities between adolescent cliques and youth gangs, including the demands that are placed on members in each category to conform to group norms pertaining to behavior, appearance, and other people with whom one is allowed to associate.

3. False. Recent studies have found that people join gangs for a variety of reasons, including the desire to gain access to money, recreation, and protection.

4. False. Gangs are frequently thought of as an urban problem because central-city gangs organized around drug dealing have become prominent in recent years; however, gangs are found in rural areas throughout the country as well.

5. False. Although street crime—such as assault and robbery—often has a greater psychological cost, crimes committed by persons in top positions in business (such as accounting and tax fraud) or government (including the Pentagon) have a far greater economic cost, especially for U.S. taxpayers.

6. True. This age group accounts for about 54 percent of all arrests for property crimes, the most common crimes committed in the United States.

7. True. As more youths grow up in single-parent households or in households where both parents are employed, many adolescents have turned to members of their peer cliques to satisfy their emotional needs and to gain information.

8. True. Gangs are found in nations around the world. In countries such as Japan, youth gangs are often points of entry for adult crime organizations.

Sources: Based on Adler and Adler, 1998, 2003; Inciardi, Horowitz, and Pottieger, 1993; and Jankowski, 1991.

cutting school or running away from home), which are illegal only when committed by younger people.

What Is Social Control?

Societies not only have norms and laws that govern acceptable behavior; they also have various mechanisms to control people's behavior. ***Social control*** **refers to the systematic practices that social groups develop in order to encourage conformity to norms, rules, and laws and to discourage deviance.** Social control mechanisms may be either internal or external. Internal social control takes place through the socialization process: Individuals *internalize* societal norms and values that prescribe how people should behave and then follow those norms and values in their everyday lives. By contrast, external social control involves the use of negative sanctions that proscribe certain behaviors and set forth the punishments for rule breakers and nonconformists. In contemporary societies, the criminal justice system, which includes the police, the courts, and the prisons, is the primary mechanism of external social control.

For some social analysts, maintaining social control is critical for the stability of society. Political scientist James Q. Wilson (1996: xv) uses the image of broken windows to explain how neighborhoods may decay into disorder and crime if no one maintains social control:

> If a factory or office window is broken, passersby observing it will conclude that no one cares and no one is in charge. In time, a few will begin throwing rocks to break more windows. Soon all the windows will be broken, and now passersby will think that, not only is no one in charge of the building, no one is in charge of the street on which it faces. Only the young, the criminal, or the foolhardy have any business on an unprotected avenue, and so more and more citizens will abandon the street to those they assume prowl it. Small disorders lead to larger and larger ones, and perhaps even to crime.

But if most actions deemed deviant do little or no direct harm to society or its members, why is social control so important to groups and societies? Why are some actions punished whereas others are not? Why

Sociology *Works!*

Social Definitions of Deviance: Have You Seen Bigfoot or a UFO Lately?

> My mind's open to anything. After all, they just found another planet. So, who knows? Anything's possible.
>
> —Jim Maier, a resident of Seneca, Illinois, explaining why he thinks it might be possible that a group of observers actually saw Bigfoot (a so-called "wild man" who is allegedly covered in hair, stands about eight feet tall, has a strong odor, and walks on much larger feet than those of a typical human being) near his community, about seventy miles southwest of Chicago (qtd. in Wischnowsky, 2005)

> Bigfoot is one of those things that people like to believe in.... Regardless of whether there are such things as Bigfoot, people like that thrill of uncertainty, that sense of danger. It's exciting to try and discover the unknown. And it's a lot more fun to have that little bit of doubt when you're sitting out in the woods.
>
> —sociologist Christopher Bader describing why tales of the improbable, such as sightings of Bigfoot (also known as Sasquatch), are exciting and believable to some individuals, whereas others think that people who spend countless hours waiting in a densely wooded area to catch sight of Bigfoot are engaged in deviant behavior (qtd. in Wischnowsky, 2005)

Sociology contributes to our thinking about conformity and deviance by making us aware that the people we are around help us define what we think of as "normal" beliefs and actions. If we are surrounded by individuals who believe that a Bigfoot or UFO (unidentified flying object) sighting is just around the corner, we may think of such beliefs as normal and gain a personal sense of belonging when we go out and wait with these individuals for Bigfoot or a flying saucer to show up. For this reason, some people join groups such as the Bigfoot Field Researchers Organization (http://www.bfro.net) so that they can share their outings, compare field notes on recent sightings, and feel that they are part of an important group or a clique. Among other Bigfoot believers, followers are treated with respect when they record sightings rather than receiving blank stares or comments like "You've got to be kidding?" all the while they are being labeled as "weird" or "deviant" by outsiders who are nonbelievers (Wischnowsky, 2005).

Looking at the seemingly deviant behavior of going out on Bigfoot or UFO sightings from a sociological perspective, researchers such as Christopher Bader place these actions within a larger social context. One context that Bader uses for studying people's fascination with Bigfoot sightings and other paranormal occurrences is the sociology of religion. According to Bader, many people who believe in Bigfoot or UFOs "believe without the kinds of evidence that would convince outsiders—it's a matter of faith" (qtd. in Weiss, 2004). This faith may be intensified by use of the Internet, where true believers may easily report their sightings without fear of ridicule or being identified as deviant by outsiders.

Reflect & Analyze

At your college or university, what beliefs and actions of individuals and groups might be classified as conformity by some people but identified as deviance by others? For example, do some students and/or professors believe that certain buildings are haunted and stay away from those areas?

is the same belief or action punished in one group or society and not in another? These questions pose interesting theoretical concerns and research topics for sociologists and criminologists who examine issues pertaining to law, social control, and the criminal justice system. ***Criminology* is the systematic study of crime and the criminal justice system, including the police, courts, and prisons.**

The primary interest of sociologists and criminologists is not questions of how crime and criminals can best be controlled but rather on social control as a social product. Sociologists do not judge certain kinds of behavior or people as being "good" or "bad." Instead, they attempt to determine what types of behavior are defined as deviant, who does the defining, how and why people become deviants, and how society deals with deviants. Although sociologists have developed a number of theories to explain deviance and crime, no one perspective is a comprehensive explanation of all deviance. Each theory provides a different lens through which we can examine aspects of deviant behavior.

social control systematic practices developed by social groups to encourage conformity to norms, rules, and laws and to discourage deviance.

criminology the systematic study of crime and the criminal justice system, including the police, courts, and prisons.

Functionalist Perspectives on Deviance

As we have seen in previous chapters, functionalists focus on societal stability and the ways in which various parts of society contribute to the whole. According to functionalists, a certain amount of deviance contributes to the smooth functioning of society.

What Causes Deviance, and Why Is It Functional for Society?

Sociologist Emile Durkheim believed that deviance is rooted in societal factors such as rapid social change and lack of social integration among people. As you will recall, Durkheim attributed the social upheaval he saw at the end of the nineteenth century to the shift from mechanical to organic solidarity, which was brought about by rapid industrialization and urbanization. Although many people continued to follow the dominant morals (norms, values, and laws) as best they could, rapid social change contributed to *anomie*—a social condition in which people experience a sense of futility because social norms are weak, absent, or conflicting. According to Durkheim, as social integration (bonding and community involvement) decreased, deviance and crime increased. However, from his perspective, this was not altogether bad because he believed that deviance has positive social functions in terms of its consequences. For Durkheim (1964a/1895), deviance is a natural and inevitable part of all societies. Likewise, contemporary functionalist theorists suggest that deviance is universal because it serves three important functions:

1. *Deviance clarifies rules.* By punishing deviant behavior, society reaffirms its commitment to the rules and clarifies their meaning.
2. *Deviance unites a group.* When deviant behavior is seen as a threat to group solidarity and people unite in opposition to that behavior, their loyalties to society are reinforced.
3. *Deviance promotes social change.* Deviants may violate norms in order to get them changed. For example, acts of civil disobedience—including lunch counter sit-ins and bus boycotts—were used to protest and eventually correct injustices such as segregated buses and lunch counters in the South. Students periodically stage campus demonstrations to call attention to perceived injustices, such as a tuition increase or the firing of a popular professor.

Functionalists acknowledge that deviance may also be dysfunctional for society. If too many people violate the norms, everyday existence may become unpredictable, chaotic, and even violent. If even a few people commit acts that are so violent that they threaten the survival of a society, then deviant acts move into the realm of the criminal and even the unthinkable. Of course, the example that stands out in everyone's mind is terrorist attacks around the world and the fear that remains constantly present as a result.

Although there are a wide array of contemporary functionalist theories regarding deviance and crime, many of these theories focus on social structure. For this reason, the first theory we will discuss is referred to as a structural functionalist approach. It describes the relationship between the society's economic structure and the reasons that people might engage in various forms of deviant behavior.

Strain Theory: Goals and Means to Achieve Them

Modifying Durkheim's (1964a/1895) concept of *anomie*, the sociologist Robert Merton (1938, 1968) developed strain theory. According to ***strain theory*, people feel strain when they are exposed to cultural goals that they are unable to obtain because they do not have access to culturally approved means of achieving those goals.** The goals may be material possessions and money; the approved means may include an education and jobs. When denied legitimate access to these goals, some people seek access through deviant means. Strain theory is often used to explain deviance by people from lower-income neighborhoods, who are typically depicted as being left out of the economic mainstream, feeling hopeless, and sometimes turning their anger and rage toward other people or things. In this way, the structure of the society and the economic status of the people involved are major factors in why some people commit deviant and/or criminal acts.

Merton identified five ways in which people adapt to cultural goals and approved ways of achieving them: conformity, innovation, ritualism, retreatism, and rebellion (see ◆ Table 7.1). According to Merton, *conformity* occurs when people accept culturally approved goals and pursue them through approved means. Persons who want to achieve success through conformity work hard, save their money, and so on. Even people who find that they are blocked from achieving a high level of education or a lucrative career may take a lower-paying job and attend school part time, join the military, or seek alternative (but legal) avenues, such as playing the lottery, to "strike it rich."

Conformity is also crucial for members of middle- and upper-class teen cliques, who often gather in small groups to share activities and confidences. Some youths are members of a variety of cliques, and peer

◆ **Table 7.1 Merton's Strain Theory of Deviance**

Mode of Adaptation	Method of Adaptation	Seeks Culture's Goals	Follows Culture's Approved Ways
Conformity	Accepts culturally approved goals; pursues them through culturally approved means	Yes	Yes
Innovation	Accepts culturally approved goals; adopts disapproved means of achieving them	Yes	No
Ritualism	Abandons society's goals but continues to conform to approved means	No	Yes
Retreatism	Abandons both approved goals and the approved means to achieve them	No	No
Rebellion	Challenges both the approved goals and the approved means to achieve them	No—seeks to replace	No—seeks to replace

approval is of crucial significance to them—being one of the "in" crowd, not an "outcast" or a "loner," is a significant goal for many teenagers. In the aftermath of the recent school shootings, for example, numerous journalists trekked to school campuses to report that athletes ("jocks"), cheerleaders, and other "popular" students enforce the social code at high schools (Adler, 1999; Cohen, 1999). One report suggested that "from who's in which clique to where you sit in the cafeteria, every day [high school] can be a struggle to fit in" (Adler, 1999: 56). A comparison of appearance norms held by some teen peer groups and some juvenile gang members shows that what constitutes conformity within one group may be viewed as deviance within another (see ◆ Table 7.2).

Merton classified the remaining four types of adaptation as deviance. *Innovation* occurs when people accept society's goals but adopt disapproved means of achieving them. Innovations for acquiring material possessions or money cover a wide variety of illegal activities, including theft and drug dealing. For example, the journalist Nathan McCall (1994: 6) describes how his innovative behavior took the form of hustling when he was a gang member:

> Hustling seemed like the thing to do. With Shell Shock [a fellow gang member] as my main partner, I tried every nickel-and-dime hustle I came across, focusing mainly on stealing. We stole everything that wasn't nailed down, from schoolbooks, which

strain theory the proposition that people feel strain when they are exposed to cultural goals that they are unable to obtain because they do not have access to culturally approved means of achieving those goals.

◆ **Table 7.2 Deviants or Conformists? High School "Uniforms"**

Group	Appearance
Jocks, Cheerleaders, and the "In" Crowd	Clothes from Tommy Hilfiger, Abercrombie & Fitch, the Gap, Old Navy, "letter" jackets, white baseball caps. Short hair for boys; long, "frosted," or "streaked" hair for girls.
"Hicks" or "Kickers"	Cowboy boots, big hats, and oversize belt buckles (regional).
"Surfers"	Sun-bleached hair, tropical or other light clothing (regional).
"Skaters" (as in Skateboards)	Grunge look (regional).
"Freaks," "Punks," and "Ravers"	Spiky and/or brightly colored hair (Kool Aid used in some cases), black clothing, extensive body piercing all over face and body, numerous tattoos.
"Outcasts"—Boys	Makeup, including face powder and black eyeliner; dress in feminine ways or in black leather and chains; black T-shirts, trench coats, Doc Martens.
"Outcasts"—Girls	Black nail polish and lipstick; black leather, chains; visible tattoos; Doc Martens or high-platform shoes.

Sources: Based on Cohen, 1999; and *Newsweek*, 1999.

we sold at half price, to wallets, which we lifted from guys' rear pockets. We even stole gifts from under the Christmas tree of a girl we visited. . . .

Although Merton primarily focused on deviance committed by persons from lower-income backgrounds, innovation is also used by middle- and upper-income people. For example, affluent adults may cheat on their income taxes or embezzle money from their employer to maintain an expensive lifestyle. Students from middle- and high-income families may cheat on exams in hopes of receiving higher grades and ensuring their admission to a top college.

Merton's third mode of adaptation is *ritualism,* which occurs when people give up on societal goals but still adhere to the socially approved means of achieving them. Ritualism is the opposite of innovation; persons who cannot obtain expensive material possessions or wealth may nevertheless seek to maintain the respect of others by being a "hard worker" or "good citizen." *Retreatism* occurs when people abandon both the approved goals and the approved means of achieving them. Merton included persons such as skid-row alcoholics and drug addicts in this category; however, not all retreatists are destitute. Some may be middle- or upper-income individuals who see themselves as rejecting the conventional trappings of success or the means necessary to acquire them.

The fifth type of adaptation, *rebellion,* occurs when people challenge both the approved goals and the approved means for achieving them, and advocate an alternative set of goals or means. To achieve their alternative goals, rebels may engage in acts of violence such as rioting or may register their displeasure with society through acts of vandalism or graffiti (as further discussed in Box 7.2, "You Can Make a Difference").

Sociologist Robert Merton identified five ways in which people adapt to cultural goals and approved ways of achieving them. Consider the young woman shown here. Which of Merton's modes of adaptation might best explain her views on social life?

Opportunity Theory: Access to Illegitimate Opportunities

Expanding on Merton's strain theory, sociologists Richard Cloward and Lloyd Ohlin (1960) suggested that for deviance to occur, people must have access to ***illegitimate opportunity structures*—circumstances that provide an opportunity for people to acquire through illegitimate activities what they cannot achieve through legitimate channels.** For example, gang members may have insufficient legitimate means to achieve conventional goals of status and wealth but have illegitimate opportunity structures—such as theft, drug dealing, or robbery—through which they can achieve these goals. In his study of the "Diamonds," a Chicago street gang whose members are second-generation Puerto Rican youths, sociologist Felix M. Padilla (1993) found that gang membership was linked to the members' belief that they might reach their aspirations by transforming the gang into a business enterprise. Coco, one of the Diamonds, explains the importance of sticking together in the gang's income-generating business organization:

> We are a group, a community, a family—we have to learn to live together. If we separate, we will never have a chance. We need each other even to make sure that we have a spot for selling our supply [of drugs]. You know, there is people around here, like some opposition, that want to take over your *negocio* [business]. And they think that they can do this very easy. So we stick together, and that makes other people think twice about trying to take over what is yours. In our case, the opposition has never tried messing with our hood, and that's because

Box 7.2 You Can Make a Difference

Seeing the Writing on the Wall—and Doing Something About It!

"Pirus Rule the Streets of Bompton Fools"

What does this graffiti—written on a Compton, California, wall—mean? To figure out the message, it helps to know that Pirus is a collective made up of several Blood gangs (street gangs identified, among other things, by the red color worn by members) in Los Angeles County. Gang graffiti such as "Pirus Rule" is one way the gang claims its supremacy not only over rival gangs but also over the city as a whole, as indicated by the way the letter "C" in Compton was replaced with a "B" to emphasize the gang's Blood/Pirus identity (Alonso, 1998: 17).

Although not all graffiti is done by street gangs, some gang members use graffiti as an illegal form of communication. Although "taggers" primarily use graffiti as (at least in their opinion) an art form, street gangs use graffiti to increase their visibility, mark their territory, threaten rival gangs, and intimidate local residents (Salt Lake City Sheriff's Department, 2007). According to law enforcement officials, taggers are usually less violent than are members of traditional street gangs, and their "art" is usually more "artistic" and less threatening than street-gang graffiti. However, the work of both taggers and street-gang members defaces walls, buses, subways, and other public areas.

Is there anything we can do when we see graffiti to get it removed and to improve the appearance of our community? How might we lessen the opportunities for gang members to use graffiti to communicate with one another and to threaten outsiders? Here are some suggestions from law enforcement officials:

- Do not confront or challenge a person who is tagging a wall or writing graffiti on a public space. Whether they are gang members or not, some taggers are armed, and even if unarmed, they may assault a challenger with spray paint or a physical attack.
- Make sure that owners of private property or public officials are notified about the graffiti because it is important that the graffiti be painted over immediately. Studies show that if graffiti is left up, it becomes a status symbol, and the area is likely to be hit again and again.
- Look for adopt-a-wall programs or other groups in which volunteers assist in cleaning off or painting over graffiti.
- Find out if your city has a graffiti hotline where you can report graffiti. Many cities have instituted these hotlines so that graffiti can be quickly removed from both public and private property.
- Be aware of graffiti done by children and young people that might indicate that they are thinking about becoming, or have become, involved with gangs. Look for graffiti on or around a residence, such as drawings or "doodles" of gang-related figures, themes of violence, or gang symbols. Also look for the use of substitute letters, such as replacing the "C" in Compton with a "B" (as discussed above) or intentionally misspelling a word (such as when "cigarette" becomes "bigarette") because the removed letter is in a rival gang's name. (Center for Problem-Oriented Policing, 2002; NAGIA, 2007)

Although graffiti may appear to be a small issue when it is examined within the larger context of gang-related activity and urban crimes, this kind of behavior is one telling sign that law enforcement officials use when identifying possible criminal trends. A dramatic increase in the amount of graffiti in a community is often a sign that gang membership is growing and that gang activities are becoming more confrontational toward rival gangs and toward society as a whole.

For additional information on graffiti and gang indicators, conduct searches on these websites:

- Center for Problem-Oriented Policing: **http://www.popcenter.org**
- National Alliance of Gang Investigators Association: **http://www.nagia.org**

they know it's protected real good by us fellas. (qtd. in Padilla, 1993: 104)

Based on their research, Cloward and Ohlin (1960) identified three basic gang types—criminal, conflict, and retreatist—which emerge on the basis of what type of illegitimate opportunity structure is available in a specific area. The *criminal gang* is devoted to theft, extortion, and other illegal means of securing an income. For young men who grow up in a criminal gang, running drug houses and selling drugs on street corners make it possible for them to support themselves and their families as well as purchase material possessions to impress others. By contrast, *conflict gangs* emerge

illegitimate opportunity structures circumstances that provide an opportunity for people to acquire through illegitimate activities what they cannot achieve through legitimate channels.

in communities that do not provide either legitimate or illegitimate opportunities. Members of conflict gangs seek to acquire a "rep" (reputation) by fighting over turf (territory) and adopting a value system of toughness, courage, and similar qualities. Unlike criminal and conflict gangs, members of *retreatist gangs* are unable to gain success through legitimate means and are unwilling to do so through illegal ones. As a result, the consumption of drugs is stressed, and addiction is prevalent.

Sociologist Lewis Yablonsky (1997) has updated Cloward and Ohlin's findings on delinquent gangs. According to Yablonsky, today's gangs are more likely to use and sell drugs, and carry more lethal weapons than gang members did in the past. Today's gangs have become more varied in their activities and are more likely to engage in intraracial conflicts, with "black on black and Chicano on Chicano violence," whereas minority gangs in the past tended to band together to defend their turf from gangs of different racial and ethnic backgrounds (Yablonsky, 1997: 3).

How useful are social structural approaches such as opportunity theory and strain theory in explaining deviant behavior? Although there are weaknesses to these approaches, they focus our attention on one crucial issue: the close association between certain forms of deviance and social class position. According to criminologist Anne Campbell (1984: 267), gangs are a "microcosm of American society, a mirror image in which power, possession, rank, and role . . . are found within a subcultural life of poverty and crime."

However, the social scientists Charles Tittle and Robert Meier (1990) dispute the proposition that class position is the most important factor in explaining why some people commit crimes. According to Tittle and Meier, most people from low-income backgrounds *do not* commit crimes, whereas some people from middle- and upper-income backgrounds *do* commit crimes. Likewise, some activities of gang members from low-income neighborhoods have commonalities with actions taken by nondelinquent youths in suburban cliques. Consider the practice of guarding one's "turf." Both adolescent gang members and high school clique participants often "guard" their favorite location, and a group may be known to others by the place that its members have chosen. Journalists recently described how clique members at Glenbrook, a suburban Chicago high school, jealously guard their turf. Moreover, the cliques are named for their favorite perches: The fashionable "wall people" favor a bench along the wall outside the cafeteria, whereas the punkish "trophy-case" kids sit on the floor under a display of memorabilia (Adler, 1999: 58). The relationship between conventional behavior and deviance is obviously much more complex than either opportunity theory or strain theory might suggest.

Conflict theorists suggest that criminal law is unequally enforced along class lines. Consider this setting, in which low-income defendants are arraigned by a judge who sees them only on a television monitor. Do you think, as a rule, that these defendants will be as well represented by attorneys as a wealthier defendant might be?

Conflict Perspectives on Deviance

Who determines what kinds of behavior are deviant or criminal? Different branches of conflict theory offer somewhat divergent answers to this question. One branch emphasizes power as the central factor in defining deviance and crime: People in positions of power maintain their advantage by using the law to protect their own interests. Another branch emphasizes the relationship between deviance and capitalism, whereas a third focuses on feminist perspectives and the confluence of race, class, and gender issues in regard to deviance and crime.

Deviance and Power Relations

Conflict theorists who focus on power relations in society suggest that the lifestyles considered deviant by political and economic elites are often defined as illegal. From this perspective, the law defines and controls two distinct categories of people: (1) *social dynamite*—persons who have been marginalized (including rioters, labor organizers, gang members, and criminals)—and (2) *social junk*—members of stigmatized groups (such as welfare recipients, the homeless, and persons with disabilities) who are costly to society but relatively harmless (Spitzer, 1975). According to this approach, norms and laws are established for the benefit of those in power and do not reflect any absolute standard of right and wrong (Turk, 1969, 1977). As a result, the activities of poor and lower-income individuals are more likely to be defined as criminal than those of persons from middle- and upper-income backgrounds. Moreover, the criminal justice system is more focused on, and is less forgiving of, deviant and criminal behavior engaged in by people in specific categories. For example, research shows that young, single, urban males are more likely to be perceived as members of the *dangerous classes* and receive stricter sentences in criminal courts (Miethe and Moore, 1987). Power differentials are also evident in how victims of crime are treated. When the victims are wealthy, white, and male, law enforcement officials are more likely to put forth more extensive efforts to apprehend the perpetrator as contrasted with cases in which the victims are poor, black, and female (Smith, Visher, and Davidson, 1984). Recent research generally supports this assertion (Wonders, 1996). This branch of conflict theory shows how power relations in society influence the law and the criminal justice system, often to the detriment of people who are at the bottom of the social structure hierarchy, and it questions functionalist views on conformity and deviance that are based on the assumption that laws reflect a consensus among the majority of people.

Deviance and Capitalism

A second branch of conflict theory—Marxist/critical theory—views deviance and crime as a function of the capitalist economic system. Although the early economist and social thinker Karl Marx wrote very little about deviance and crime, many of his ideas are found in a critical approach that has emerged from earlier Marxist and radical perspectives on criminology. The critical approach is based on the assumption that the laws and the criminal justice system protect the power and privilege of the capitalist class. According to the social scientist Barry Krisberg (1975), *privilege* is the possession of what is most valued by a particular social group in a given historical period. As such, privilege includes not only rights such as life, liberty, and happiness, but also material possessions such as money, luxury items, land, and houses.

As you may recall from Chapter 1, Marx based his critique of capitalism on the inherent conflict that he believed existed between the capitalists (*bourgeoisie*) and the working class (*proletariat*). In a capitalist society, social institutions (such as law, politics, and education, which make up the superstructure) legitimize existing class inequalities and maintain the capitalists' superior position in the class structure. According to Marx, capitalism produces haves and have-nots, who engage in different forms of deviance and crime.

Why do people commit crimes? Some critical theorists believe that members of the capitalist class commit crimes because they are greedy and want more than they have. Corporate or white-collar crimes such as stock market manipulation, land speculation, fraudulent bankruptcies, and crimes committed on behalf of organizations often involve huge sums of money and harm many people. By contrast, street crimes such as robbery and aggravated assault generally involve small sums of money and cause harm to limited numbers of victims. According to these theorists, the poor commit street crimes in order to survive; they find that they cannot afford the essentials, such as food, clothing, shelter, and health care. Thus, some crime represents a rational response by the poor to the unequal distribution of resources in society (Gordon, 1973). Further, living in poverty may lead to violent crime and victimization *of the poor by the poor.* For example, violent gang activity may be a collective response of young people to seemingly hopeless poverty (Quinney, 1979).

According to the sociologist Richard Quinney (2001/1974), people with economic and political power define as criminal any behavior that threatens their own interests. The powerful use law to control those who are without power. For example, drug laws

According to Karl Marx, capitalism produces haves and have-nots, and each group engages in different types of crime. Statistically, the man being arrested here is much more likely to be suspected of a financial crime than a violent crime.

enacted early in the twentieth century were actively enforced in an effort to control immigrant workers, especially the Chinese, who were being exploited by the railroads and other industries (Tracy, 1980). By contrast, antitrust legislation passed at about the same time was seldom enforced against large corporations owned by prominent families such as the Rockefellers, Carnegies, and Mellons. Having antitrust laws on the books merely shored up the government's legitimacy by making it appear responsive to public concerns about big business (Barnett, 1979).

In sum, the Marxist/critical approach argues that criminal law protects the interests of the affluent and powerful. The way that laws are written and enforced benefits the capitalist class by ensuring that individuals at the bottom of the social class structure do not infringe on the property or threaten the safety of those at the top (Reiman, 1998). However, critics assert that critical theorists have not shown that powerful economic and political elites actually manipulate lawmaking and law enforcement for their own benefit. Rather, people of all classes share a consensus about the criminality of certain acts. For example, laws that prohibit murder, rape, and armed robbery protect not only middle- and upper-income people but also low-income people, who are frequently the victims of such violent crimes. To shift the focus from seeing law as always working for the rich and against the poor, structural Marxist theorists such as John Hagan (1989) argue that law can be used to control the members of *any* class who pose a threat to the existence of capitalism. For example, the excessive greed of a few affluent capitalists might "rock the boat" and bring into question the business practices of many others if their behavior is not targeted and sanctioned.

Feminist Approaches

Can theories developed to explain male behavior be used to understand female deviance and crime? According to feminist scholars, the answer is no. A new interest in women and deviance developed in 1975 when two books—Freda Adler's *Sisters in Crime* and Rita James Simons's *Women and Crime*—declared that women's crime rates were going to increase significantly as a result of the women's liberation movement. Although this so-called *emancipation theory* of female crime has been refuted by subsequent analysts, Adler's and Simons's works encouraged feminist scholars (both women and men) to examine more closely the relationship among gender, deviance, and crime. More recently, feminist scholars such as Kathleen Daly and Meda Chesney-Lind (1988) have developed theories and conducted research to fill the void in our knowledge about gender and crime. For example, in a study of the female offender, Chesney-Lind (1997) examined the cultural factors in women's lives that may contribute to their involvement in criminal behavior. Although there is no single feminist perspective on deviance and crime, three schools of thought have emerged.

Why do women engage in deviant behavior and commit crimes? According to the *liberal feminist approach,* women's deviance and crime are a rational response to the gender discrimination that women experience in families and the workplace. From this view, lower-income and minority women typically have fewer opportunities not only for education and good jobs but also for "high-end" criminal endeavors. As some feminist theorists have noted, a woman is no more likely to be a big-time drug dealer or an organized crime boss than she is to be a corporate director (Daly and Chesney-Lind, 1988; Simpson, 1989).

By contrast, the *radical feminist approach* views the cause of women's crime as originating in patriarchy (male domination over females). This approach focuses on social forces that shape women's lives and experiences and shows how exploitation may trigger deviant behavior and criminal activities. From this view, arrests and prosecution for crimes such as prostitution reflect our society's sexual double standard whereby it is acceptable for a man to pay for sex but unacceptable for a woman to accept money for such services. Although state laws usually view both the female prostitute and the male customer as violating the law, in most states the woman is far more likely than the man to be arrested, brought to trial, convicted, and sentenced.

The third school of feminist thought, the *Marxist (socialist) feminist approach,* is based on the assumption that women are exploited by both capitalism and patriarchy. From this approach, women's criminal behavior is linked to gender conflict created by the economic and social struggles that often take place in postindustrial societies such as ours. According to the social scientist James Messerschmidt (1986), men control women biologically and economically just as members of the capitalist class control the labor of workers. As a result, women experience "double marginality," which provides them with fewer opportunities to commit certain types of deviance and crime. Because most females have relatively low-wage jobs (if any) and few economic resources, crimes such as prostitution and shoplifting become a means to earn money or acquire consumer goods. However, instead of freeing women from their problems, prostitution institutionalizes women's dependence on men and results in a form of female sexual slavery (Vito and Holmes, 1994). Lower-income women are further victimized by the fact that they are often the targets of

After this young prostitute advertised her services on Craigslist, the Sacramento vice squad and the FBI arranged a meeting that led to her arrest. Which of the feminist theories of women's crime best explains this young woman's offense?

violent acts by lower-class males, who perceive themselves as being powerless in the capitalist economic system. Because Western societies value aggressive male behavior, whether in sports or business pursuits, men who feel powerless may "prove" their manliness by *doing gender*—attempting to improve their male self-image through acts of violence or abuse against women or children (Siegel, 2007).

Some feminist scholars have noted that these approaches to explaining deviance and crime neglect the centrality of race and ethnicity and focus on the problems and perspectives of women who are white, middle- and upper-income, and heterosexual without taking into account the views of women of color, lesbians, and women with disabilities (Martin and Jurik, 1996).

Approaches Focusing on Race, Class, and Gender

Some recent studies have focused on the simultaneous effects of race, class, and gender on deviant behavior. In one study, the sociologist Regina Arnold (1990) examined the relationship between women's earlier victimization in their family and their subsequent involvement in the criminal justice system. Arnold interviewed African American women serving criminal sentences and found that adolescent females are often "labeled and processed as deviants—and subsequently as criminals—for refusing to accept or participate in their own victimization." Arnold attributes many of the women's offenses to living in families in which sexual abuse, incest, and other violence left them few choices except to engage in deviance. Economic marginality and racism also contributed to their victimization: "To be young, Black, poor, and female is to be in a high-risk category for victimization and stigmatization on many levels" (Arnold, 1990: 156). According to Arnold, the criminal behavior of the women in her study was linked to class, gender, and racial oppression, which they experienced daily in their families and at school and work.

Feminist sociologists and criminologists believe that research on women as both victims and perpetrators of crime is long overdue. For example, few studies of violent crime—such as robbery and aggravated assault—have included women as subjects or respondents. Some scholars have argued that women are less motivated to commit such crimes, are not as readily exposed to attractive targets, and are more protected from being the victims of such crimes than men are (see Sommers and Baskin, 1993). Other scholars have stressed that research should integrate women into the larger picture of criminology (Simpson, 1989). For example, sociologists Susan Ehrlich Martin and Nancy C. Jurik (1996) examined the role of women in the criminal justice system and found that women continue to experience significant barriers in justice occupations ranging from law enforcement to the legal profession.

Symbolic Interactionist Perspectives on Deviance

As we discussed in Chapter 4, symbolic interactionists focus on *social processes,* such as how people develop a self-concept and learn conforming behavior through socialization. According to this approach, deviance is learned in the same way as conformity—through interaction with others. Although there are a number of symbolic interactionist perspectives on deviance, we will examine four major approaches—differential association and differential reinforcement theories, rational choice theory, control theory, and labeling theory.

Differential Association Theory and Differential Reinforcement Theory

How do people learn deviant behavior through their interactions with others? According to the sociologist Edwin Sutherland (1939), people learn the necessary techniques and the motives, drives, rationalizations, and attitudes of deviant behavior from people with whom they associate. ***Differential association theory* states that people have a greater tendency to deviate from societal norms when they frequently associate with individuals who are more favorable toward deviance than conformity.** From this approach, criminal behavior is learned within intimate personal groups such as one's family and peer groups. Learning criminal behavior also includes learning the techniques of committing crimes, as former gang member Nathan McCall explains:

> Sometimes I picked up hustling ideas at the 7-Eleven, which was like a criminal union hall: Crapshooters, shoplifters, stickup men, burglars, everybody stopped off at the store from time to time. While hanging up there one day, I ran into Holt. . . . He had a pocketful of cash, even though he had quit school and was unemployed. I asked him, "Yo, man, what you been into?"
>
> "Me and my partner kick in cribs and make a killin'. You oughta come go with us sometimes." . . . I hooked school one day, went with them, and pulled my first B&E [breaking and entering]. Before we went to the house, [Holt] . . . explained his system: "Look, man, we gonna split up and go to each house on the street. Knock on the door. If somebody answers, make up a name and act like you at the wrong crib. If nobody answers, we mark it for a hit." . . .
>
> After I learned the ropes, Shell Shock [another gang member] and I branched out, doing B&Es on our own. We learned to get in and out of houses in no time flat. (McCall, 1994: 93–94)

As McCall's orientation to breaking and entering shows, learning deviance may involve the acquisition of certain attitudes and the mastery of specialized techniques.

Differential association theory contributes to our knowledge of how deviant behavior reflects the individual's learned techniques, values, attitudes, motives, and rationalizations. It calls attention to the fact that criminal activity is more likely to occur when a person has frequent, intense, and long-lasting interactions with others who violate the law. However, it does not explain why many individuals who have been heavily exposed to people who violate the law still engage in conventional behavior most of the time.

Criminologist Ronald Akers (1998) has combined differential association theory with elements of psychological learning theory to create *differential reinforcement theory,* which suggests that both deviant behavior and conventional behavior are learned through the same social processes. Akers starts with the fact that people learn to evaluate their own behavior through interactions with significant others. If the

Is this example of graffiti likely to be the work of an isolated artist or of a gang member? In what ways do gangs reinforce such behavior?

persons and groups that a particular individual considers most significant in his or her life define deviant behavior as being "right," that individual is more likely to engage in deviant behavior; likewise, if the person's most significant friends and groups define deviant behavior as "wrong," the person is less likely to engage in that behavior. This approach helps explain not only juvenile gang behavior but also how peer cliques on high school campuses have such a powerful influence on people's behavior. Returning to our example of how clique members at Glenbrook High School turfed out territory at the school, notice how one student responded to powerful pressures to conform:

> As an experiment . . . Lauren Barry, a pink-haired trophy-case kid at Glenbrook, switched identities with a well-dressed girl from "the wall." Barry walked around all day in the girl's expensive jeans and Doc Martens, carrying a shopping bag from Abercrombie & Fitch. "People kept saying, 'Oh, you look so pretty,'" she recalls. "I felt really uncomfortable." It was interesting, but the next day, and ever since, she's been back in her regular clothes. (qtd. in Adler, 1999: 58)

Another approach to studying deviance is rational choice theory, which suggests that people weigh the rewards and risks involved in certain types of behavior and then decide which course of action to follow.

Rational Choice Theory

As you may recall from Chapter 6, rational choice theory is based on the assumption that when people are faced with several courses of action, they will usually do what they believe is likely to have the best overall outcome (Elster, 1989). The ***rational choice theory of***

***deviance* states that deviant behavior occurs when a person weighs the costs and benefits of nonconventional or criminal behavior and determines that the benefits will outweigh the risks involved in such actions.** Rational choice approaches suggest that most people who commit crimes do not engage in random acts of antisocial behavior. Instead, they make careful decisions based on weighing the available information regarding *situational factors,* such as the place of the crime, suitable targets, and the availability of people to deter the behavior, and *personal factors,* such as what rewards they may gain from their criminal behavior (Siegel, 2007).

How useful is rational choice theory in explaining deviance and crime? A major strength of this theory is that it explains why high-risk youths do not constantly engage in delinquent acts: They have learned to balance risk against the potential for criminal gain in each situation. Moreover, rational choice theory is not limited by the underlying assumption of most social structural theories, which is that the primary participants in deviant and criminal behaviors are people in the lower classes. Rational choice theory also has important policy implications regarding crime reduction or prevention, suggesting that people must be taught that the risks of engaging in criminal behavior far outweigh any benefits they may gain from their actions. Thus, people should be taught *not* to engage in crime.

Control Theory: Social Bonding

Control theories focus on another aspect of why some people do not engage in deviant behavior. According to the sociologist Walter Reckless (1967), society produces pushes and pulls that move people toward criminal behavior; however, some people "insulate" themselves from such pressures by having positive self-esteem and good group cohesion. Reckless suggests that many people do not resort to deviance because of *inner containments*—such as self-control, a sense of responsibility, and resistance to diversions—and *outer containments*—such as supportive family and friends, reasonable social expectations, and supervision by others. Those with the strongest containment mechanisms are able to withstand external pressures that might cause them to participate in deviant behavior.

Extending Reckless's containment theory, sociologist Travis Hirschi's (1969) social control theory is based on the assumption that deviant behavior is minimized when people have strong bonds that bind them to families, schools, peers, churches, and other social institutions. ***Social bond theory* holds that the probability of deviant behavior increases when a person's ties to society are weakened or broken.** According to Hirschi, social bonding consists of (1) *attachment* to other people, (2) *commitment* to conformity, (3) *involvement* in conventional activities, and (4) *belief* in the legitimacy of conventional values and norms. Although Hirschi did not include females in his study, others who have replicated that study with both females and males have found that the theory appears to apply to each (see Naffine, 1987).

What does control theory have to say about delinquency and crime? Control theories suggest that the probability of delinquency increases when a person's social bonds are weak and when peers promote antisocial values and violent behavior. However, some critics assert that Hirschi was mistaken in his assumption that a weakened social bond leads to deviant behavior. The chain of events may be just the opposite: People who routinely engage in deviant behavior may find that their bonds to people who would be positive influences are weakened over time (Agnew, 1985; Siegel, 2007). Or, as labeling theory suggests, people may engage in deviant and criminal behavior because of destructive social interactions and encounters (Siegel, 2007).

Labeling Theory

***Labeling theory* states that deviance is a socially constructed process in which social control agencies designate certain people as deviants, and they, in turn, come to accept the label placed upon them and begin to act accordingly.** Based on the symbolic interaction theory of Charles H. Cooley and George H. Mead (see Chapter 4), labeling theory focuses on the variety of symbolic labels that people are given

differential association theory the proposition that individuals have a greater tendency to deviate from societal norms when they frequently associate with persons who are more favorable toward deviance than conformity.

rational choice theory of deviance the belief that deviant behavior occurs when a person weighs the costs and benefits of nonconventional or criminal behavior and determines that the benefits will outweigh the risks involved in such actions.

social bond theory the proposition that the probability of deviant behavior increases when a person's ties to society are weakened or broken.

labeling theory the proposition that deviants are those people who have been successfully labeled as such by others.

in their interactions with others. Sociologist Larry J. Siegel (1998: 212) explains the link between labeling and deviance as follows:

> Labels imply a variety of behaviors and attitudes; labels thus help define not just one trait but the whole person. For example, people labeled "insane" are also assumed to be dangerous, dishonest, unstable, violent, strange, and otherwise unsound. Valued labels, including "smart," "honest," and "hard worker," which suggest overall competence, can improve self-image and social standing. Research shows that people who are labeled with one positive trait, such as being physically attractive, are assumed to maintain others, such as intelligence and competence. In contrast, negative labels, including "troublemaker," "mentally ill," and "stupid," help stigmatize their targets and reduce their self-image.

How does the process of labeling occur? The act of fixing a person with a negative identity, such as "criminal" or "mentally ill," is directly related to the power and status of those persons who *do* the labeling and those who are *being labeled.* Behavior, then, is not deviant in and of itself; it is defined as such by a social audience (Erikson, 1962). According to the sociologist Howard Becker (1963), *moral entrepreneurs* are often the ones who create the rules about what constitutes deviant or conventional behavior. Becker believes that moral entrepreneurs use their own perspectives on "right" and "wrong" to establish the rules by which they expect other people to live. They also label others as deviant. Often, these rules are enforced on persons with less power than the moral entrepreneurs. Becker (1963: 9) concludes that the deviant is "one to whom the label has successfully been applied; deviant behavior is behavior that people so label."

As the definition of labeling theory suggests, several stages may occur in the labeling process. ***Primary deviance* refers to the initial act of rule breaking** (Lemert, 1951). However, if individuals accept the negative label that has been applied to them as a result of the primary deviance, they are more likely to continue to participate in the type of behavior that the label was initially meant to control. ***Secondary deviance* occurs when a person who has been labeled a deviant accepts that new identity and continues the deviant behavior.** For example, a person may shoplift an item of clothing from a department store but not be apprehended or labeled as a deviant. The person may subsequently decide to forgo such behavior in the future. However, if the person shoplifts the item, is apprehended, is labeled as a "thief," and subsequently accepts that label, then the person may shoplift items from stores on numerous occasions. A few people engage in ***tertiary deviance,* which occurs when a person who has been labeled a deviant seeks to normalize the behavior by relabeling it as nondeviant** (Kitsuse, 1980). An example would be drug users who believe that using marijuana or other illegal drugs is no more deviant than drinking alcoholic beverages and therefore should not be stigmatized. (See ▶ Figure 7.1, which takes a closer look at labeling theory.)

According to control theory, strong bonds—including close family ties—are a factor in explaining why many people do not engage in deviant behavior. Why do some sociologists believe that quality family time is more important in discouraging delinquent behavior than is time spent with other young people?

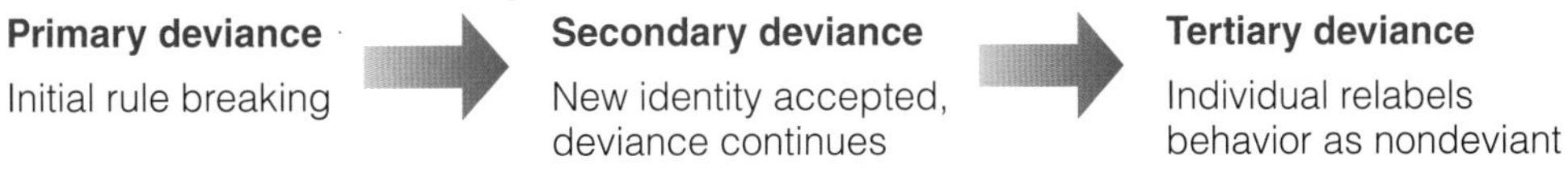

Primary deviance
Initial rule breaking

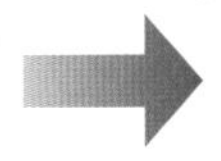

Secondary deviance
New identity accepted, deviance continues

Tertiary deviance
Individual relabels behavior as nondeviant

▶ **Figure 7.1 A Closer Look at Labeling Theory**

Can labeling theory be applied to high school peer groups and gangs? In a classic study, the sociologist William Chambliss (1973) documented how the labeling process works in some high schools when he studied two groups of adolescent boys: the "Saints" and the "Roughnecks." Members of both groups were constantly involved in acts of truancy, drinking, wild parties, petty theft, and vandalism. Although the Saints committed more offenses than the Roughnecks, the Roughnecks were the ones who were labeled as "troublemakers" and arrested by law enforcement officials. By contrast, the Saints were described as being the "most likely to succeed," and none of them were ever arrested. According to Chambliss (1973), the Roughnecks were more likely to be labeled as deviants because they came from lower-income families, did poorly in school, and were generally viewed negatively, whereas the Saints came from "good families," did well in school, and were generally viewed positively. Although both groups engaged in similar behavior, only the Roughnecks were stigmatized by a deviant label.

Another study of juvenile offenders also found that those from lower-income families were more likely to be arrested and indicted than were middle-class juveniles who participated in the same kinds of activities (Sampson, 1986). In determining how to deal with youthful offenders, the criminal justice system frequently takes into account such factors as the offender's family life, educational achievement (or lack thereof), and social class. The individuals most likely to be apprehended, labeled as delinquent, and prosecuted are people of color who are young, male, unemployed, and undereducated, and who reside in urban high-crime areas (Vito and Holmes, 1994). Why might this be true? According to the criminologist Robert J. Sampson (1997), family and neighborhood, more than the individual characteristics of people involved in deviance and crime, are important factors in determining variations in crime rates. For example, parents with the lowest incomes may have the most difficulty with parenting, which may result in young people receiving harsh or erratic discipline and poor supervision (Sampson and Laub, 1993). However, even young people who have been chronically involved in delinquent behavior may reach certain *turning points* in their life, such as marriage or a career, which may cause them to decide against crime (Sampson and Laub, 1993).

How successful is labeling theory in explaining deviance and social control? One contribution of labeling theory is that it calls attention to the way in which social control and personal identity are intertwined: Labeling may contribute to the acceptance of deviant roles and self-images. Critics argue that this does not explain what caused the original acts that constituted primary deviance, nor does it provide insight into why some people accept deviant labels and others do not (Cavender, 1995).

Postmodernist Perspectives on Deviance

Departing from other theoretical perspectives on deviance, some postmodern theorists emphasize that the study of deviance reveals how the powerful exert control over the powerless by taking away their free will to think and act as they might choose. From this approach, institutions such as schools, prisons, and mental hospitals use knowledge, norms, and values to categorize people into "deviant" subgroups such as slow learners, convicted felons, or criminally insane individuals, and then to control them through specific patterns of discipline.

An example of this idea is found in social theorist Michel Foucault's *Discipline and Punish* (1979), in which Foucault examines the intertwining nature of power, knowledge, and social control. In this study of prisons from the mid-1800s to the early 1900s, Foucault found that many penal institutions ceased torturing prisoners who disobeyed the rules and began using new surveillance techniques to maintain social control. Although the prisons appeared to be more humane in the post-torture era, Foucault contends that the new means of surveillance impinged more on prisoners and

primary deviance the initial act of rule-breaking.

secondary deviance the process that occurs when a person who has been labeled a deviant accepts that new identity and continues the deviant behavior.

tertiary deviance deviance that occurs when a person who has been labeled a deviant seeks to normalize the behavior by relabeling it as nondeviant.

brought greater power to prison officials. To explain, he described the *Panoptican*—a structure that gives prison officials the possibility of complete observation of criminals at all times. Typically, the Panoptican was a tower located in the center of a circular prison from which guards could see all the cells. Although the prisoners knew they could be observed at any time, they did not actually know when their behavior was being scrutinized. As a result, prison officials were able to use their knowledge as a form of power over the inmates. Eventually, the guards did not even have to be present all the time because prisoners believed that they were under constant scrutiny by officials in the observation post. If we think of this in contemporary times, we can see how cameras, computers, and other devices have made continual surveillance quite easy in virtually all institutions. In such cases, social control and discipline are based on the use of knowledge, power, and technology.

Foucault's view on deviance and social control has influenced other social analysts, including Shoshana Zuboff (1988), who views the computer as a modern Panoptican that gives workplace supervisors virtually unlimited capabilities for surveillance over subordinates. Today, cell phones and the Internet provide new opportunities for surveillance by government officials

CONCEPT QUICK REVIEW

Theoretical Perspectives on Deviance

	Theory	Key Elements
Functionalist Perspectives		
Robert Merton	Strain theory	Deviance occurs when access to the approved means of reaching culturally approved goals is blocked. Innovation, ritualism, retreatism, or rebellion may result.
Richard Cloward/Lloyd Ohlin	Opportunity theory	Lower-class delinquents subscribe to middle-class values but cannot attain them. As a result, they form gangs to gain social status and may achieve their goals through illegitimate means.
Conflict Perspectives		
Karl Marx Richard Quinney	Critical approach	The powerful use law and the criminal justice system to protect their own class interests.
Kathleen Daly Meda Chesney-Lind	Feminist approach	Historically, women have been ignored in research on crime. Liberal feminism views women's deviance as arising from gender discrimination, radical feminism focuses on patriarchy, and socialist feminism emphasizes the effects of capitalism and patriarchy on women's deviance.
Symbolic Interactionist Perspectives		
Edwin Sutherland	Differential association	Deviant behavior is learned in interaction with others. A person becomes delinquent when exposure to law-breaking attitudes is more extensive than exposure to law-abiding attitudes.
Travis Hirschi	Social control/social bonding	Social bonds keep people from becoming criminals. When ties to family, friends, and others become weak, an individual is most likely to engage in criminal behavior.
Howard Becker	Labeling theory	Acts are deviant or criminal because they have been labeled as such. Powerful groups often label less-powerful individuals.
Edwin Lemert	Primary/secondary deviance	Primary deviance is the initial act. Secondary deviance occurs when a person accepts the label of "deviant" and continues to engage in the behavior that initially produced the label.
Postmodernist Perspective		
Michel Foucault	Knowledge as power	Power, knowledge, and social control are intertwined. In prisons, for example, new means of surveillance that make prisoners think they are being watched all the time give officials knowledge that inmates do not have. Thus, the officials have a form of power over the inmates.

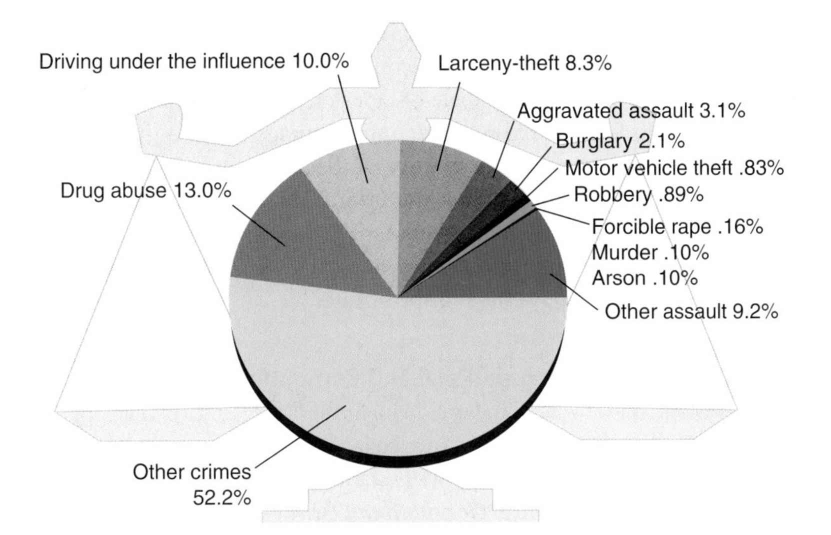

▶ **Figure 7.2 Distribution of Arrests by Type of Offense, 2007**

Source: FBI, 2008.

and others who are not visible to the individuals who are being watched.

We have examined functionalist, conflict, interactionist, and postmodernist perspectives on social control, deviance, and crime (see the Concept Quick Review). All of these explanations contribute to our understanding of the causes and consequences of deviant behavior; however, we now turn to the subject of crime itself.

Crime Classifications and Statistics

Crime in the United States can be divided into different categories. We will look first at the legal classifications of crime and then at categories typically used by sociologists and criminologists.

How the Law Classifies Crime

Crimes are divided into felonies and misdemeanors. The distinction between the two is based on the seriousness of the crime. A *felony* is a major crime such as rape, homicide, or aggravated assault, for which punishment typically ranges from more than a year's imprisonment to death. A *misdemeanor* is a minor crime that is typically punished by less than one year in jail. In either event, a fine may be part of the sanction as well. Actions that constitute felonies and misdemeanors are determined by the legislatures in the various states; thus, their definitions vary from jurisdiction to jurisdiction.

Other Crime Categories

The *Uniform Crime Report* (UCR) is the major source of information on crimes reported in the United States. The UCR has been compiled since 1930 by the Federal Bureau of Investigation based on information filed by law enforcement agencies throughout the country. When we read that the rate of certain types of crimes has increased or decreased when compared with prior years, for example, this information is usually based on UCR data. The UCR focuses on violent crime and property crime (which, prior to 2004, were jointly referred to in that report as "index crimes"), but also contains data on other types of crime (see ▶ Figure 7.2). The FBI estimates that in 2007, about 14.2 million arrests were made in the United States for all criminal infractions (excluding traffic violations). Although the UCR gives some indication of crime in the United States, the figures do not reflect the actual number and kinds of crimes, as will be discussed later.

Violent Crime ***Violent crime*** **consists of actions—murder, forcible rape, robbery, and aggravated assault—involving force or the threat of force against others.** Although only 4 percent (597,447 out of 14,209,365) of all arrests in the United States in 2007 were for violent crimes, this category is probably the most anxiety-provoking of all criminal behavior: Most of us know someone who has been a victim of violent crime, or we have been so ourselves. Victims are often physically injured or even lose their lives; the psychological trauma may last for years after the event (Parker and Anderson-Facile, 2000). Violent crime receives the most sustained attention from law enforcement officials and the media (see Warr, 2000).

Nationwide, there is growing concern over juvenile violence. Beginning in the late 1980s and continuing to the present, juvenile violent-crime arrest rates have

violent crime actions—murder, forcible rape, robbery, and aggravated assault—involving force or the threat of force against others.

continued to rise, a trend that some scholars link to a corresponding increase in gang membership (see Inciardi, Horowitz, and Pottieger, 1993; Thornberry et al., 1993). Fear of violence is felt not only by the general public but by gang members themselves, as Charles Campbell commented:

> The generation I'm in is going to be lost. Of the circle of friends I grew up in, three are dead, four are in jail, and another is out of school and just does nothing. When he runs out of money he'll sell a couple bags of weed. . . .
>
> I would carry a gun because I am worried about that brother on the fringe. There are some people, there is nothing out there for them. They will blow you away because they have nothing to lose. There are no jobs out there. It's hard to get money to go to school. (qtd. in Lee, 1993: 21)

Property Crime ***Property crimes*** **include burglary (breaking into private property to commit a serious crime), motor vehicle theft, larceny-theft (theft of property worth $50 or more), and arson.** Some offenses, such as robbery, are both violent crimes and property crimes. In the United States, a property crime occurs, on average, once every three seconds; a violent crime occurs, on average, once every twenty-two seconds (see ▶ Figure 7.3). In most property crimes, the primary motive is to obtain money or some other desired valuable.

Public Order Crime *Public order crimes* involve an illegal action voluntarily engaged in by the participants, such as prostitution, illegal gambling, the private use of illegal drugs, and possession of illegal pornography. Many people assert that such conduct should not be labeled as a crime; these offenses are often referred to as ***victimless crimes*** because they **involve a willing exchange of illegal goods or services among adults.** However, morals crimes can include children and adolescents as well as adults. Young children and adolescents may unwillingly become child pornography "stars" or prostitutes.

Occupational and Corporate Crime Although the sociologist Edwin Sutherland (1949) developed the theory of white-collar crime sixty years ago, it was not until the 1980s that the public became fully aware of its nature. ***Occupational (white-collar) crime*** **comprises illegal activities committed by people in the course of their employment or financial affairs.**

In addition to acting for their own financial benefit, some white-collar offenders become involved in criminal conspiracies designed to improve the market share or profitability of their companies. This is known as ***corporate crime*****—illegal acts committed by corporate employees on behalf of the corporation and with its support.** Examples include antitrust violations; tax evasion; misrepresentations in advertising; infringements on patents, copyrights, and trademarks; price fixing; and financial fraud. These crimes are a result of deliberate decisions made by corporate personnel to enhance resources or profits at the expense of competitors, consumers, and the general public.

Although people who commit occupational and corporate crimes can be arrested, fined, and sent to prison, many people often have not regarded such

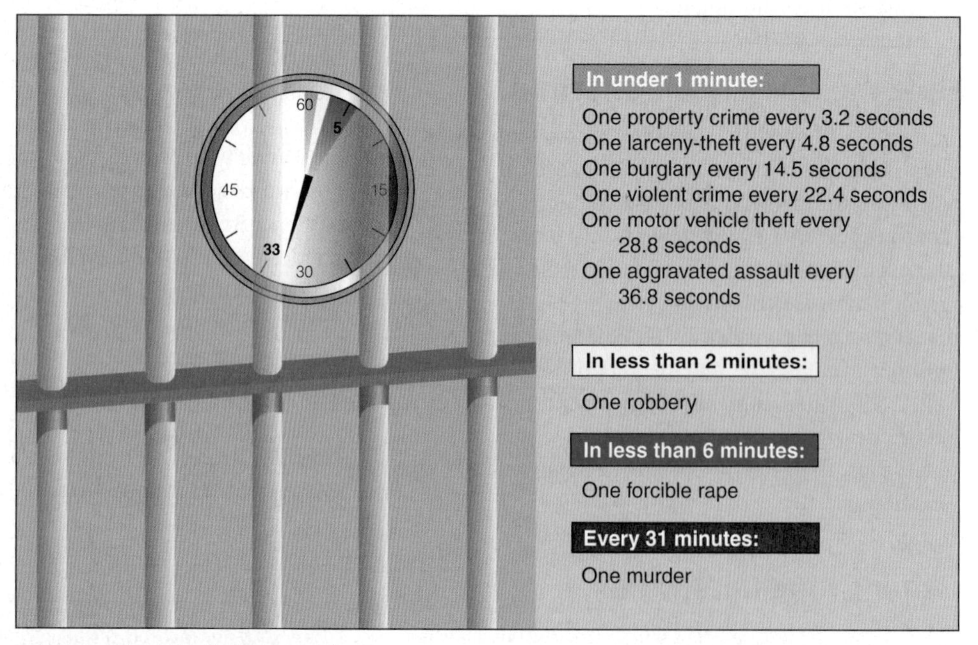

▶ **Figure 7.3 The FBI Crime Clock**

Source: FBI, 2008.

behavior as "criminal." People who tend to condemn street crime are less sure of how their own (or their friends') financial and corporate behavior should be judged. At most, punishment for such offenses has usually been a fine or a relatively brief prison sentence.

Until recently, public concern and media attention focused primarily on the street crimes disproportionately committed by persons who were poor, powerless, and nonwhite. Today, however, part of our focus has shifted to crimes committed in corporate suites, such as fraud, tax evasion, and insider trading by executives at large and well-known corporations. Bernard Madoff, the former chairperson of NASDAQ, admitted to defrauding his clients of up to $50 billion in a massive scheme that took place over a number of years. Madoff used his social connections to raise large sums of money for a fund that he used for his own gain. Clients invested in the fund in hopes that Madoff would manage their money wisely and that they would earn large returns on their investments. Instead, he lived lavishly and used new money that came in from investors to pay off existing clients who wanted to cash out of his fund rather than using the new money for the purpose intended. However, Madoff is not an isolated example of such criminal endeavors. Over the past decade, numerous occupational and corporate criminals, including Dennis Kozlowski, the former chief executive of Tyco International, who was convicted of looting more than $600 million from his company, and former Enron executives Ken Lay and Jeff Skilling, who were convicted of corporate conspiracy and fraud in connection with the collapse of one-time energy giant Enron, have all engaged in activities that have cost other people billions of dollars and, in some cases, their life savings.

Corporate crimes are often more costly in terms of money and lives lost than street crimes. Thousands of jobs and billions of dollars were lost as a result of corporate crime in the year 2005 alone. Deaths resulting from corporate crimes such as polluting the air and water, manufacturing defective products, and selling unsafe foods and drugs far exceed the number of deaths due to homicides each year. Other costs include the effect on the moral climate of society (Clinard and Yeager, 1980; Simon, 1996). Throughout the United States, the confidence of everyday people in the nation's economy has been shaken badly by the greedy and illegal behavior of corporate insiders.

Organized Crime ***Organized crime* is a business operation that supplies illegal goods and services for profit.** Premeditated, continuous illegal activities of organized crime include drug trafficking, prostitution, loan-sharking, money laundering, and large-scale theft such as truck hijackings (Simon, 1996). No single organization controls all organized crime; rather, many groups operate at all levels of society. In recent decades, organized crime in the United States has become increasingly transnational in nature. Globalization of the economy and the introduction of better communications technology have made it possible for groups around the world to operate in the United States and other nations. The Federal Bureau of Investigation has identified three major categories of organized crime groups: (1) Italian organized crime and racketeering, (2) Eurasian/Middle Eastern organized crime, and (3) Asian and African criminal enterprises, as shown in ▶ Figure 7.4.

Organized crime thrives because there is great demand for illegal goods and services. Criminal organizations initially gain control of illegal activities by combining threats and promises. For example, small-time operators running drug or prostitution rings may be threatened with violence if they compete with organized crime or fail to make required payoffs (Cressey, 1969).

Apart from their illegal enterprises, organized crime groups have infiltrated the world of legitimate business. Known linkages between legitimate businesses and organized crime exist in banking, hotels and motels, real estate, garbage collection, vending machines, construction, delivery and long-distance hauling, garment manufacture, insurance, stocks and bonds, vacation resorts, and funeral parlors (National Council on Crime and Delinquency, 1969). In addition, some law enforcement and government officials are corrupted through bribery, campaign contributions, and favors intended to buy them off.

Based on current economic problems in the United States, some criminologists believe that organized crime will have an even greater effect on our nation in the future. According to these analysts, organized crime may further weaken the U.S. economy because

property crime crimes including burglary (breaking into private property to commit a serious crime), motor vehicle theft, larceny-theft (theft of property worth $50 or more), and arson.

victimless crimes crimes that involve a willing exchange of illegal goods or services among adults.

occupational (white-collar) crime illegal activities committed by people in the course of their employment or financial affairs.

corporate crime illegal acts committed by corporate employees on behalf of the corporation and with its support.

organized crime a business operation that supplies illegal goods and services for profit.

▶ **Figure 7.4 Organized Crime Threats in the United States**

Source: http://www.fbi.gov/hq/cid/orgcrime/ocshome.htm.

illegal activities such as tax evasion scams and cigarette trafficking bring about greater losses in tax revenue for state and federal governments. Organized crime groups that are involved in areas such as commodities, credit, insurance, stocks, securities, and investments will also have the ability to further weaken already-troubled financial and housing markets (see Finklea, 2009).

Political Crime The term ***political crime*** **refers to illegal or unethical acts involving the usurpation of power by government officials, or illegal/unethical acts perpetrated against the government by outsiders seeking to make a political statement, undermine the government, or overthrow it.** Government officials may use their authority unethically or illegally for the purpose of material gain or political power (Simon, 1996). They may engage in graft (taking advantage of political position to gain money or property) through bribery, kickbacks, or "insider" deals that financially benefit them. For example, in the late 1980s, several top Pentagon officials were found guilty of receiving bribes for passing classified information on to major defense contractors that had garnered many lucrative contracts from the government (Simon, 1996).

Other types of corruption have been costly for taxpayers, including dubious use of public funds and public property, corruption in the regulation of commercial activities (such as food inspection), graft in zoning and land-use decisions, and campaign contributions and other favors to legislators that corrupt the legislative process. Whereas some political crimes are for personal material gain, others (such as illegal wiretapping and political "dirty tricks") are aimed at gaining or maintaining political office or influence.

Some acts committed by agents of the government against persons and groups believed to be threats to national security are also classified as political crimes. Four types of political deviance have been attributed to some officials: (1) secrecy and deception designed to manipulate public opinion, (2) abuse of power, (3) prosecution of individuals due to their political activities, and (4) official violence, such as police brutality against people of color or the use of citizens as unwilling guinea pigs in scientific research (Simon, 1996).

Political crime frequently involves the use of an office for personal material gain. When he was governor of Illinois, Rod Blagojevich, shown here, was accused of influence peddling and was eventually impeached and then arraigned under federal racketeering and fraud charges.

Political crimes also include illegal or unethical acts perpetrated against the government by outsiders seeking to make a political statement or to undermine or overthrow the government. Examples include treason, acts of political sabotage, and terrorist attacks on public buildings.

Crime Statistics

How useful are crime statistics as a source of information about crime? As mentioned previously, official crime statistics provide important information on crime; however, the data reflect only those crimes that have been reported to the police. The number of violent-crime arrests decreased slightly (1.1 percent) in 2007 while the number of arrests for property crime increased 5.4 percent. These statistics may show that property crimes are up, or they may reflect (at least partially) an increase in the number of crimes *reported.* Why are some crimes not reported? People are more likely to report crime when they believe that something can be done about it (apprehension of the perpetrator or retrieval of their property, for example). About half of all assault and robbery victims do not report the crime because they may be embarrassed or fear reprisal by the perpetrator. Thus, the number of crimes reported to police represents only the proverbial "tip of the iceberg" when compared with all offenses actually committed. Official statistics are problematic in social science research because of these limitations.

The National Crime Victimization Survey was developed by the Bureau of Justice Statistics as an alternative means of collecting crime statistics. In this annual survey, the members of 100,000 randomly selected households are interviewed to determine whether they have been the victims of crime, even if the crime was not reported to the police. The most recent victimization survey indicates that 50 percent of all violent crimes and 61 percent of all property crimes are not reported to the police and are thus not reflected in the UCR (U.S. Bureau of Justice Statistics, 2007).

Studies based on anonymous self-reports of criminal behavior also reveal much higher rates of crime than those found in official statistics. For example, self-reports tend to indicate that adolescents of all social classes violate criminal laws. However, official statistics show that those who are arrested and placed in juvenile facilities typically have limited financial resources, have repeatedly committed serious offenses, or both (Steffensmeier and Allan, 2000). Data collected for the Juvenile Court Statistics Program also reflect class and racial bias in criminal justice enforcement. Not all children who commit juvenile offenses are apprehended and referred to court. Children from white, affluent families are more likely to have their cases handled outside the juvenile justice system (for example, a youth may be sent to a private school or hospital rather than to a juvenile correctional facility).

Many crimes committed by persons of higher socioeconomic status in the course of business are handled by administrative or quasi-judicial bodies, such as the Securities and Exchange Commission or the Federal Trade Commission, or by civil courts. As a result, many elite crimes are never classified as "crimes," nor are the businesspeople who commit them labeled as "criminals."

Terrorism and Crime

In the twenty-first century, the United States and other nations are confronted with a difficult prospect: how to deal with terrorism. ***Terrorism* is the calculated, unlawful use of physical force or threats of violence against persons or property in order to intimidate or coerce a government, organization, or individual for the purpose of gaining some political, religious, economic, or social objective.** A frequently asked question today is this: What is the difference between terrorism and organized crime? According to authorities, the principal distinction between organized crime groups and terrorist groups is motivation: "Money motivates organized crime, and ideology motivates terrorism" (Finklea, 2009: 23). However, money is still the linking element between organized crime and terrorism because terrorist organizations typically obtain money for their activities from criminal acts such as money laundering and drug trafficking.

How are sociologists and criminologists to explain world terrorism, which may have its origins in more than one nation and include diverse "cells" of terrorists who operate in a somewhat ganglike manner but are believed to be following directives from leaders elsewhere? In order to deal with the aftermath of terrorist attacks, government officials typically focus on

political crime illegal or unethical acts involving the usurpation of power by government officials, or illegal/unethical acts perpetrated against the government by outsiders seeking to make a political statement, undermine the government, or overthrow it.

terrorism the calculated unlawful use of physical force or threats of violence against persons or property in order to intimidate or coerce a government, organization, or individual for the purpose of gaining some political, religious, economic, or social objective.

"known enemies" such as Osama bin Laden. The nebulous nature of the "enemy" and the problems faced by any one government trying to identify and apprehend the perpetrators of acts of terrorism have resulted in a global "war on terror." Social scientists who use a rational choice approach suggest that terrorists are rational actors who constantly calculate the gains and losses of participation in violent—and sometimes suicidal—acts against others. Chapter 14 ("Politics and Government in Global Perspective") further discusses the issue of terrorism.

Street Crimes and Criminals

Given the limitations of official statistics, is it possible to determine who commits crimes? We have much more information available about conventional (street) crime than elite crime; therefore, statistics concerning street crime do not show who commits all types of crime. Gender, age, class, and race are important factors in official statistics pertaining to street crime.

Gender and Crime In 2007, almost 76 percent of all persons arrested were male. Males made up about 82 percent of persons arrested for violent crime and 66.6 percent of persons arrested for property crime (FBI, 2008).

Before considering differences in crime rates by males and females, three similarities should be noted. First, the three most common arrest categories for both men and women are driving under the influence of alcohol or drugs (DUI), larceny, and minor or criminal mischief types of offenses. These three categories account for about 47 percent of all male arrests and about 49 percent of all female arrests. Second, liquor law violations (such as underage drinking), simple assault, and disorderly conduct are middle-range offenses for both men and women. Third, the rate of arrests for murder, arson, and embezzlement is relatively low for both men and women.

The most important gender differences in arrest rates are reflected in the proportionately greater involvement of men in major property crimes (such as robbery) and violent crime (particularly murder and non-negligent manslaughter), as shown in ▶ Figure 7.5. In 2007, men accounted for about 88 percent of robberies and murders and 60 percent of all larceny-theft arrests in the United States. Of those types of offenses, males under age 18 accounted for approximately 20 percent of the 2007 arrests. The property crimes for which women are most frequently arrested are nonviolent in nature, including shoplifting, theft of services, passing bad checks, credit card fraud, and employee pilferage. Often when women are arrested for serious violent and property crimes, they are typically seen as accomplices to men who planned the crime and instigated its commission; however, this assumption frequently does not prove true today. Studies have found that some women play an active role in planning and carrying out robberies and other major crimes.

Age and Crime Of all factors associated with crime, the age of the offender is one of the most significant. Arrest rates for violent crime and property crime are highest for people between the ages of 13 and 25, with the peak being between ages 16 and 17. In 2007, persons under age 25 accounted for more than 44 percent of all arrests for violent crime and almost 54 percent of all arrests for property crime (FBI, 2008). Individuals under age 18 accounted for over 25 percent of all arrests for robbery, burglary, and larceny-theft.

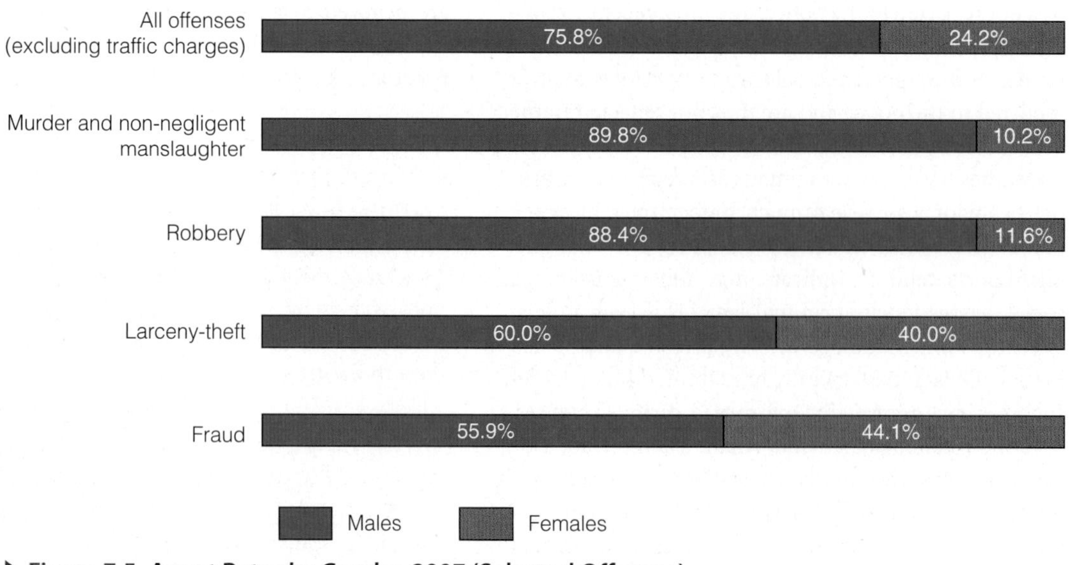

▶ **Figure 7.5 Arrest Rates by Gender, 2007 (Selected Offenses)**

Source: FBI, 2008.

© Sean Cayton/The Image Works

Most of the crimes that women commit are nonviolent ones. Nevertheless, many women are incarcerated. What effects might a mother's imprisonment have on the future of her infant?

Scholars do not agree on the reasons for this age distribution. In one earlier study, the sociologist Mark Warr (1993) found that peer influences (defined as exposure to delinquent peers, time spent with peers, and loyalty to peers) tend to be more significant in explaining delinquent behavior than age itself. More recent studies have tended to confirm this finding.

The median age of those arrested for aggravated assault and homicide is somewhat older, generally in the late twenties. Typically, white-collar criminals are even older because it takes time to acquire both a high-ranking position and the skills needed to commit this type of crime.

Rates of arrest remain higher for males than females at every age and for nearly all offenses. This female-to-male ratio remains fairly constant across all age categories. The most significant gender difference in the age curve is for prostitution. In 2007, almost 60 percent of all women arrested for prostitution were under age 35. For individuals over age 45, many more men than women are arrested for sex-related offenses (including procuring the services of a prostitute). This difference has been attributed to a more stringent enforcement of prostitution statutes when young females are involved (Chesney-Lind, 1997). It has also been suggested that opportunities for prostitution are greater for younger women. This age difference may not have the same impact on males, who continue to purchase sexual services from young females or males (see Steffensmeier and Allan, 2000).

Social Class and Crime Individuals from all social classes commit crimes; they simply commit different kinds of crimes. Persons from lower socioeconomic backgrounds are more likely to be arrested for violent and property crimes. By contrast, persons from the upper part of the class structure generally commit white-collar or elite crimes, although only a very small proportion of these individuals will ever be arrested or convicted of a crime.

What about social class and violence by youths? Between 1992 and 2006, there were 617 violent deaths in U.S. schools (U.S. Department of Education, 2007). Most of these deaths were not attributed to lower-income, inner-city youths, as popular stereotypes might suggest. Instead, a number of these acts of violence were perpetrated by young people from affluent families.

Similarly, membership in today's youth gangs cannot be identified with just one social class. Across class lines, the percentage of students reporting the presence of gangs at their school increased from 21 to 24 percent between 2003 and 2005. Twenty-five percent of students at public schools reported gang activity at their school, but only four percent of private school students reported that they had knowledge of gang members at their school. This is not surprising, given the fact that the U.S. Department of Education estimates that 50 percent of gang members are part of the nation's underclass—the class comprising families whose members are poor, seldom employed, and caught in patterns of long-term deprivation. According to studies from the Department of Education, however, about 35 percent of gang members are working class, whereas 15 percent are middle or upper-middle class. Today, females are more visible in both female gangs and in groups that previously were known as all-male gangs.

In any case, official statistics are not always an accurate reflection of the relationship between social class and crime. Self-report data from offenders themselves may be used to gain information on family income, years of education, and occupational status; however, such reports rely on respondents to report information accurately and truthfully.

Race and Crime In 2007, whites (including Latinos/as) accounted for almost 70 percent of all arrests, as shown in ▶ Figure 7.6. Compared with African Americans, arrest rates for whites were higher for nonviolent property crimes such as fraud and larceny-theft but were lower for violent crimes such as robbery. In 2007, whites accounted for almost 60 percent of all arrests for property crimes and about 59 percent of arrests for violent crimes. African Americans accounted for almost 39 percent of arrests for violent crimes and 28 percent of arrests for property crimes (FBI, 2008).

Although official arrest records reveal certain trends, these data tell us very little about the actual dynamics of crime by racial–ethnic category. According to official statistics, African Americans are overrepresented in arrest data. In 2007, African Americans made up about 12 percent of the U.S. population but accounted for 28 percent of all arrests. Latinos/as made up about 13 percent of the U.S. population and accounted for about 13 percent of all arrests. Native Americans (designated in the UCR as "American Indian" or "Alaskan Native") made up 1.3 percent of all arrests; however, most of their offenses were for alcohol-related crimes or disorderly conduct. In 2007, less than 1 percent (0.8%) of all arrests were of Asian Americans or Pacific Islanders (FBI, 2008).

Criminologist Coramae Richey Mann (1993) has argued that arrest statistics are not an accurate reflection of the crimes actually committed in our society. Reporting practices differ in accordance with race and social class. Arrest statistics reflect the UCR's focus on violent and property crimes, especially property crimes, which are committed primarily by low-income people. This emphasis draws attention away from the white-collar and elite crimes committed by middle- and upper-income people (Harris and Shaw, 2000). Police may also demonstrate bias and racism in their decisions regarding whom to question, detain, or arrest under certain circumstances (Mann, 1993). Some law enforcement officials believe that problems such as these primarily occurred in the past; however, issues still arise in the twenty-first century about police brutality against persons of color and unequal treatment of individuals who reside in racially segregated, low-income areas of urban centers and rural communities.

Another reason that statistics may show a disproportionate number of people of color being arrested is because of the focus of law enforcement on certain types of crime and certain neighborhoods in which crime is considered more prevalent. As discussed previously, many poor, young, central-city males turn to forms of criminal activity due to their belief that no opportunities exist for them to earn a living wage through legitimate employment. Because of the trend of law enforcement efforts to focus on drug-related offenses, arrest rates for young people of color have risen rapidly. These young people are also more likely to live in central-city areas, where there are more police patrols to make arrests.

Finally, arrest should not be equated with guilt: Being arrested does not mean that a person is guilty of the crime with which he or she has been charged. In the United States, individuals accused of crimes are, at least theoretically, "innocent until proven guilty" (Mann, 1993).

Crime Victims

Based on the National Crime Victimization Survey (NCVS), men are more likely to be victimized by crime, although women tend to be more fearful of crime, par-

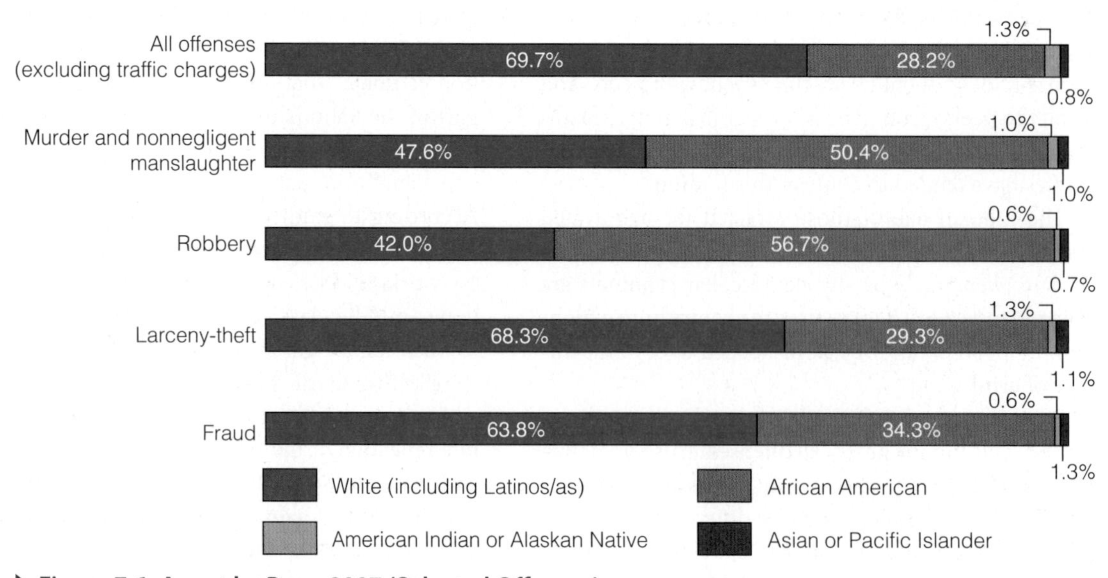

▶ **Figure 7.6 Arrest by Race, 2007 (Selected Offenses)**

Note: Classifications as used in Uniform Crime Report.
Source: FBI, 2008.

ticularly crimes directed toward them, such as forcible rape (Warr, 2000). Victimization surveys indicate that men are the most frequent victims of most crimes of violence and theft. Among males who are now 12 years old, an estimated 89 percent will be the victims of a violent crime at least once during their lifetime, as compared with 73 percent of females. The elderly also tend to be more fearful of crime but are the least likely to be victimized. Young men of color between the ages of 12 and 24 have the highest criminal victimization rates. In 2007, African American males were more likely to be victimized than African American females, younger African Americans were more likely to be the victims of violent crime than older African Americans, African Americans with lower annual incomes were at greater risk of violence than those in households with higher annual incomes, and African Americans living in urban areas were more likely than those in suburban or rural areas to be victims of crime (U.S. Department of Justice Statistics, 2008).

A study by the Justice Department found that Native Americans are more likely to be victims of violent crimes than are members of any other racial category and that the rate of violent crimes against Native American women was about 50 percent higher than that for African American men (Perry, 2004). During the period covered in the study (from 1992 to 2002), Native Americans were the victims of violent crimes at a rate more than twice the national average. They were also more likely to be the victims of violent crimes committed by members of a race other than their own (Perry, 2004). There has been a shift over the past twenty years in which more Native Americans have moved from reservations to urban areas. In the cities they do not tend to live in segregated areas, so they come into contact more often with people of other racial and ethnic groups, whereas African Americans and whites are more likely to live in segregated areas of the city and commit violent crimes against other people in their same racial or ethnic category. According to the survey, the average annual rate at which Native Americans were victims of violent crime—101 crimes per 1,000 people, ages 12 or older—is about two-and-a-half times the national average of 41 crimes per 1,000 people who are above the age of 12. By comparison, the average annual rate for whites was 41 crimes per 1,000 people; for African Americans, 50 per 1,000; and for Asian Americans, 22 per 1,000 (Perry, 2004).

The burden of robbery victimization falls more heavily on some categories of people than others. NCVS data indicate that males are robbed at almost twice the rate of females. African Americans are more than twice as likely to be robbed as whites. Young people have a much greater likelihood of being robbed than middle-aged and older persons. Persons from lower-income families are more likely to be robbed than people from higher-income families (U.S. Bureau of Justice Statistics, 2008).

The Criminal Justice System

Of all of the agencies of social control (including families, schools, and churches) in contemporary societies, only the criminal justice system has the power to control crime and punish those who are convicted of criminal conduct. The *criminal justice system* refers to the more than 55,000 local, state, and federal agencies that enforce laws, adjudicate crimes, and treat and rehabilitate criminals. The system includes the police, the courts, and corrections facilities, and it employs more than 2 million people in 17,000 police agencies, nearly 17,000 courts, more than 8,000 prosecutorial agencies, about 6,000 correctional institutions, and more than 3,500 probation and parole departments. More than $150 billion is spent annually for civil and criminal justice, which amounts to more than $500 for every person living in the United States (Siegel, 2006).

The term *criminal justice system* is somewhat misleading because it implies that law enforcement agencies, courts, and correctional facilities constitute one large, integrated system, when, in reality, the criminal justice system is made up of many bureaucracies that have considerable discretion in how decisions are made. *Discretion* refers to the use of personal judgment by police officers, prosecutors, judges, and other criminal justice system officials regarding whether and how to proceed in a given situation (see ▶ Figure 7.7). The police are a prime example of discretionary processes because they have the power to selectively enforce the law and have on many occasions been accused of being too harsh or too lenient on alleged offenders.

The Police

The role of the police in the criminal justice system continues to expand. The police are responsible for crime control and maintenance of order, but local police departments now serve numerous other human service functions, including improving community relations, resolving family disputes, and helping people during emergencies. It should be remembered that not all "police officers" are employed by local police departments; they are employed in more than 25,000 governmental agencies ranging from local jurisdictions to federal levels. However, we will focus primarily on metropolitan police departments because they constitute the vast majority of the law enforcement community.

▶ Figure 7.7 Discretionary Powers in Law Enforcement

Metropolitan police departments are made up of a chain of command (similar to the military) with ranks such as officer, sergeant, lieutenant, and captain, and each rank must follow specific rules and procedures. However, individual officers maintain a degree of discretion in the decisions they make as they respond to calls and try to apprehend fleeing or violent offenders. The problem of police discretion is most acute when decisions are made to use force (such as grabbing, pushing, or hitting a suspect) or deadly force (shooting and killing a suspect). Generally, deadly force is allowed only in situations in which a suspect is engaged in a felony, is fleeing the scene of a felony, or is resisting arrest and has endangered someone's life.

Although many police departments have worked to improve their public image in recent years, the practice of *racial profiling*—the use of ethnic or racial background as a means of identifying criminal suspects—remains a highly charged issue. Officers in some police departments have singled out for discriminatory treatment African Americans, Latinos/Latinas, and other people of color, treating them more harshly than white (Euro-American) individuals. However, police department officials typically contend that race is only one factor in determining why individuals are questioned or detained as they go about everyday activities such as driving a car or walking down the street. By contrast, equal-justice advocacy groups argue that differential treatment of minority-group members amounts to a race-based double standard, which they believe exists not only in police work but throughout the criminal justice system (see Cole, 2000).

The belief that differential treatment takes place on the basis of race contributes to a negative image of police among many people of color who believe that they have been hassled by police officers, and this assumption is intensified by the fact that police departments have typically been made up of white male personnel at all levels. In recent years, this situation has slowly begun to change. Currently, about 22 percent of all *sworn officers*—those who have taken an oath and been given the powers to make arrests and use necessary force in accordance with their duties—are women and minorities (Cole and Smith, 2004). The largest percentages of minority and women police officers are located in cities with a population of 250,000 or more. African Americans make up a larger percentage of the police department in cities with a larger proportion of African American residents (such as Detroit), and Latinos/Latinas constitute a larger percentage in cities such as San Antonio and El Paso, Texas, where Latinos/Latinas make up a larger proportion of the population. Women officers of all races are more likely to be employed in larger departments in cities of more than 250,000 (where they make up 16 percent of all officers) as compared with smaller communities (cities of less than 50,000), where women officers make up only 2 to 5 percent of the force (Cole and Smith, 2004). In the past, women were excluded from police departments and other law enforcement careers largely because of stereotypical beliefs that they were not physically and psychologically strong enough to enforce the law. However, studies have indicated that as more females have entered police work, they receive similar evaluations to male officers from their administrators and that fewer complaints are filed against women officers, which some researchers believe is a function of how female officers more effectively control potentially violent encounters (Brandl, Stroshine, and Frank, 2001).

In the future, the image of police departments may change as greater emphasis is placed on *community-oriented policing*—an approach to law enforcement in

© AP Images/Bebeto Matthews

The problems that some police departments have with racial issues can generate tragic consequences. Omar Edwards, an off-duty African American New York City police officer, was running after a man who had tried to break into Edwards's car. Three Caucasian plainclothes officers noticed the altercation and ordered the men to stop. When Edwards turned to them with his gun in his hand, one of the officers shot him fatally.

which officers maintain a presence in the community, walking up and down the streets or riding bicycles, getting to know people, and holding public service meetings at schools, churches, and other neighborhood settings. Community-oriented policing is often limited by budget constraints and lack of available personnel to conduct this type of "hands-on" community involvement. In many jurisdictions, police officers believe that they have only enough time to keep up with reports of serious crime and life-threatening occurrences and that the level of available personnel and resources does not allow officers to take on a greatly expanded role in the community.

The Courts

Criminal courts determine the guilt or innocence of those persons accused of committing a crime. In theory, justice is determined in an adversarial process in which the prosecutor (an attorney who represents the state) argues that the accused is guilty, and the defense attorney asserts that the accused is innocent. In reality, judges wield a great deal of discretion. Working with prosecutors, they decide whom to release and whom to hold for further hearings, and what sentences to impose on those persons who are convicted.

Prosecuting attorneys also have considerable leeway in deciding which cases to prosecute and when to negotiate a plea bargain with a defense attorney. As cases are sorted through the legal machinery, a steady attrition occurs. At each stage, various officials determine what alternatives will be available for those cases still remaining in the system. These discretionary decisions often have a disproportionate impact on youthful offenders who are poor (see Box 7.3).

About 90 percent of criminal cases are never tried in court; instead, they are resolved by *plea bargaining*, a process in which the prosecution negotiates a reduced

Box 7.3 Sociology and Social Policy

Juvenile Offenders and "Equal Justice Under Law"

When you walk into the U.S. Supreme Court building in Washington, D.C., it is impossible to miss the engraved statement overhead: "Equal Justice Under Law." Do young people, regardless of race, class, or gender, receive the same treatment under the law?

In courtrooms throughout the nation, judges have a wide range of discretion in their decisions regarding juveniles alleged to have committed some criminal or status offense. Whereas judges in television courtroom dramas are often African Americans, women, or members of other subordinate groups, "real-life" judges typically come from capitalist or managerial and professional backgrounds. Because more than 90 percent are white and most are male, their decisions may reflect a built-in class, racial, and gender bias.

Juvenile courts were established under a different premise than courts for adults. Under the doctrine of *parens patriae* (the state as parent), the official purpose of juvenile courts has been to care for, rather than punish, youthful offenders. In theory, less weight is given to offenses and more weight to the youth's physical, mental, or social condition. The juvenile court seeks to change or resocialize offenders through treatment or therapy, not to punish them. Consequently, judges in juvenile courts are given relatively wide latitude, or discretion, in the decisions they mete out regarding young offenders.

Unlike adult offenders, juveniles are not always represented by legal counsel. A juvenile hearing is not a trial but rather an informal private hearing before a judge or probation officer, with only the young person and a parent or guardian present. No jury is convened, and the juvenile offender does not cross-examine her or his accusers. In addition, the offender is not "sentenced"; rather, the case is "adjudicated" or "disposed of." Finally, the offender is not "punished" but instead may be "remanded to the custody" of a youth authority in order to receive training, treatment, or care.

Because of judicial discretion, courts may treat juveniles differently based on gender. Considerable disparity exists in the disposition of juvenile cases, with much of the variation thought to result from judges' beliefs rather than objective facts in the case. Female offenders are more likely than males to be institutionalized for committing status offenses such as truancy, running away from home, and other offenses that serve as "buffer charges" for suspected sexual misconduct (Chesney-Lind, 1989).

Disparity also exists on the basis of race and class. Judges tend to see youths from white, middle- or upper-class families as being very much like their own children and to believe that the families will take care of the problem on their own. They may view juveniles from lower-income families or other racial–ethnic groups as delinquents in need of attention from authorities. Furthermore, some judges view gang members from impoverished central cities as "guilty by association" because of their companions.

The political climate may have an effect on how judges dispose of juvenile cases. In the process of dealing with the public perception that the juvenile justice system is too lenient, some judges may have inadvertently contributed to other problems. Many more youths have been remanded to overcrowded juvenile detention facilities that are unable to provide necessary educational, health, and social services. Based on a judge's discretion, many juvenile offenders are incarcerated under indeterminate sentences and placed in a detention facility that may serve merely as a school for adult criminality.

Reflect & Analyze

Do you believe that the approach to juvenile justice that is used in many U.S. states is fair? Why or why not? Have you or someone you know had a direct experience with the juvenile courts? If so, what was the effect of that experience?

Sources: Based on Barlow and Kauzlarich, 2002; Chesney-Lind, 1989; and Inciardi, Horowitz, and Pottieger, 1993.

sentence for the accused in exchange for a guilty plea (Senna and Siegel, 2002). Defendants (especially those who are poor and cannot afford to pay an attorney) may be urged to plead guilty to a lesser crime in return for not being tried for the more serious crime for which they were arrested. Prison sentences given in plea bargains vary widely from one region to another and even from judge to judge within one state.

Those who advocate the practice of plea bargaining believe that it allows for individualized justice for alleged offenders because judges, prosecutors, and defense attorneys can agree to a plea and punishment that best fits the offense and the offender. They also believe that this process helps reduce the backlog of criminal cases in the court system as well as the lengthy process often involved in a criminal trial. However, those who seek to abolish plea bargaining believe that this practice leads to innocent people pleading guilty to crimes they have not committed or pleading guilty to a crime other than the one they actually committed because they are offered a lesser sentence (Cole and Smith, 2004).

More serious crimes, such as murder, felonious assault, and rape, are more likely to proceed to trial than other forms of criminal conduct; however, many of these cases do not reach the trial stage. For example, one study of 75 of the largest counties in the United States found that only 26 percent of murder cases actually went to trial. By contrast, only about 6 percent of all other cases proceeded to trial (Reaves, 2001).

One of the most important activities of the court system is establishing the sentence of the accused after he or she has been found guilty or has pleaded guilty. Typically, sentencing involves the following kinds of sentences or dispositions: fines, probation, alternative or intermediate sanctions (such as house arrest or electronic monitoring), incarceration, and capital punishment (Siegel, 2006).

Punishment and Corrections

***Punishment* is any action designed to deprive a person of things of value (including liberty) because of some offense the person is thought to have committed** (Barlow and Kauzlarich, 2002). Historically, punishment has had four major goals:

1. *Retribution* is punishment that a person receives for infringing on the rights of others (Cole and Smith, 2004). Retribution imposes a penalty on the offender and is based on the premise that the punishment should fit the crime: The greater the degree of social harm, the more the offender should be punished. For example, an individual who murders should be punished more severely than one who shoplifts. This function has received renewed interest over the past three decades as some critics have argued that the concept of rehabilitation is not working to reduce criminal behavior.
2. *General deterrence* seeks to reduce criminal activity by instilling a fear of punishment in the general public. However, we most often focus on *specific deterrence,* which inflicts punishment on specific criminals to discourage them from committing future crimes. Recently, criminologists have debated whether imprisonment has a deterrent effect, given the fact that many of those who are released from prison become recidivists (previous offenders who commit new crimes).
3. *Incapacitation* is based on the assumption that offenders who are detained in prison or are executed will be unable to commit additional crimes. This approach is often expressed as "lock 'em up and throw away the key!" In recent years, more emphasis has been placed on *selective incapacitation,* which means that offenders who repeat certain kinds of crimes are sentenced to long prison terms (Cole and Smith, 2004).
4. *Rehabilitation* seeks to return offenders to the community as law-abiding citizens by providing therapy or vocational or educational training. Based on this approach, offenders are treated, not punished, so that they will not continue their criminal activity. However, many correctional facilities are seriously understaffed and underfunded in the rehabilitation programs that exist. The job skills (such as agricultural work) that many offenders learn in prison do not transfer to the outside world, nor are offenders given any assistance in finding work that fits their skills once they are released.

Recently, newer approaches have been advocated for dealing with criminal behavior. Key among these is the idea of *restoration,* which is designed to repair the damage done to the victim and the community by an offender's criminal act (Cole and Smith, 2004). This approach is based on the *restorative justice perspective,* which states that the criminal justice system should promote a peaceful and just society; therefore, the system should focus on peacemaking rather than on punishing offenders. Advocates of this approach believe that punishment of offenders actually encourages crime rather than deterring it and are in favor of approaches such as probation with treatment. Opponents of this approach suggest that increased punishment of offenders leads to lower crime rates and that the restorative justice approach amounts to "coddling criminals." However, numerous restorative justice programs are now in operation, and many are associated with community policing programs as they seek to help offenders realize the damage that they have done to their victims and the community and to be reintegrated into society (Senna and Siegel, 2002).

Instead of the term *punishment,* the term *corrections* is often used. Criminologists George F. Cole and Christopher E. Smith (2004: 409) explain corrections as follows:

> Corrections refers to the great number of programs, services, facilities, and organizations responsible for the management of people accused or convicted of criminal offenses. In addition to prisons and jails, corrections includes probation, halfway houses,

punishment any action designed to deprive a person of things of value (including liberty) because of some offense the person is thought to have committed.

© Robin Nelson/PhotoEdit

In recent years, military-style boot camps such as this one have been used as an alternative to prison and long jail terms for nonviolent offenders under age 30. Critics argue that structural solutions—not stopgap measures such as these camps—are needed to reduce crime.

> education and work release programs, parole supervision, counseling, and community service. Correctional programs operate in Salvation Army hostels, forest camps, medical clinics, and urban storefronts.

As Cole and Smith (2004) explain, corrections is a major activity in the United States today. Consider the fact that about 6.5 million adults (more than one out of every twenty men and one out of every hundred women) are under some form of correctional control. The rate of African American males under some form of correctional supervision is even greater (one out of every six African American adult men and one out of three African American men in their twenties). Some analysts believe that these figures are a reflection of centuries of underlying racial, ethnic, and class-based inequalities in the United States as well as sentencing disparities that reflect race-based differences in the criminal justice system. However, others argue that newer practices such as determinate or mandatory sentences may help to reduce such disparities over time. A *determinate sentence* sets the term of imprisonment at a fixed period of time (such as three years) for a specific offense. *Mandatory sentencing guidelines* are established by law and require that a person convicted of a specific offense or series of offenses be given a penalty within a fixed range. Although these practices limit judicial discretion in sentencing, many critics are concerned about the effects of these sentencing approaches. Another area of great discord within and outside the criminal justice system is the issue of the death penalty.

The Death Penalty

Historically, removal from the group has been considered one of the ultimate forms of punishment. For many years, capital punishment, or the death penalty, has been used in the United States as an appropriate and justifiable response to very serious crimes. In 2008, 37 inmates were executed (as contrasted with 98 in 1999), and about 3,300 people awaited execution, having received the death penalty under federal law or the laws of one of the states that have the death penalty (Death Penalty Information Center, 2009). By far, the largest number of people on death row are in states such as California, Florida, Texas, Pennsylvania, and Alabama.

Because of the finality of the death penalty, it has been a subject of much controversy and numerous Supreme Court debates about the decision-making process involved in capital cases. In 1972 the U.S. Supreme Court ruled (in *Furman v. Georgia*) that *arbitrary* application of the death penalty violates the Eighth Amendment to the Constitution but that the death penalty itself is not unconstitutional. In other words, capital punishment is legal if it is fairly imposed. Although there have been a number of cases involving death penalty issues before the Supreme Court since that time, the court typically has upheld the constitutionality of this practice. Yet the fact remains that racial disparities are highly evident in the death row census. African Americans make up about 42 percent of the death row population but less than 13 percent of the U.S. population. The ex-slave states are more likely to execute criminals than are other states (see ▶ Figure 7.8). African Americans are eight to ten times more likely to be sentenced to death for homicidal rape than are whites (non-Latinos/as) who have committed the same crime (Marquart, Ekland-Olson, and Sorensen, 1994).

People who have lost relatives and friends as a result of criminal activity often see the death penalty as justified. However, capital punishment raises many doubts for those who fear that innocent individuals may be executed for crimes they did not commit. For still others, the problem of racial discrimination in the sentencing process poses troubling questions. Other questions that remain today involve the execution of those who are believed to be insane and of those defendants who did not have effective legal counsel during their trial. In 2002, for example, the Supreme Court ruled (in *Atkins v. Virginia*) that executing the mentally retarded is unconstitutional. In another landmark case (*Ring v. Arizona*), the Court

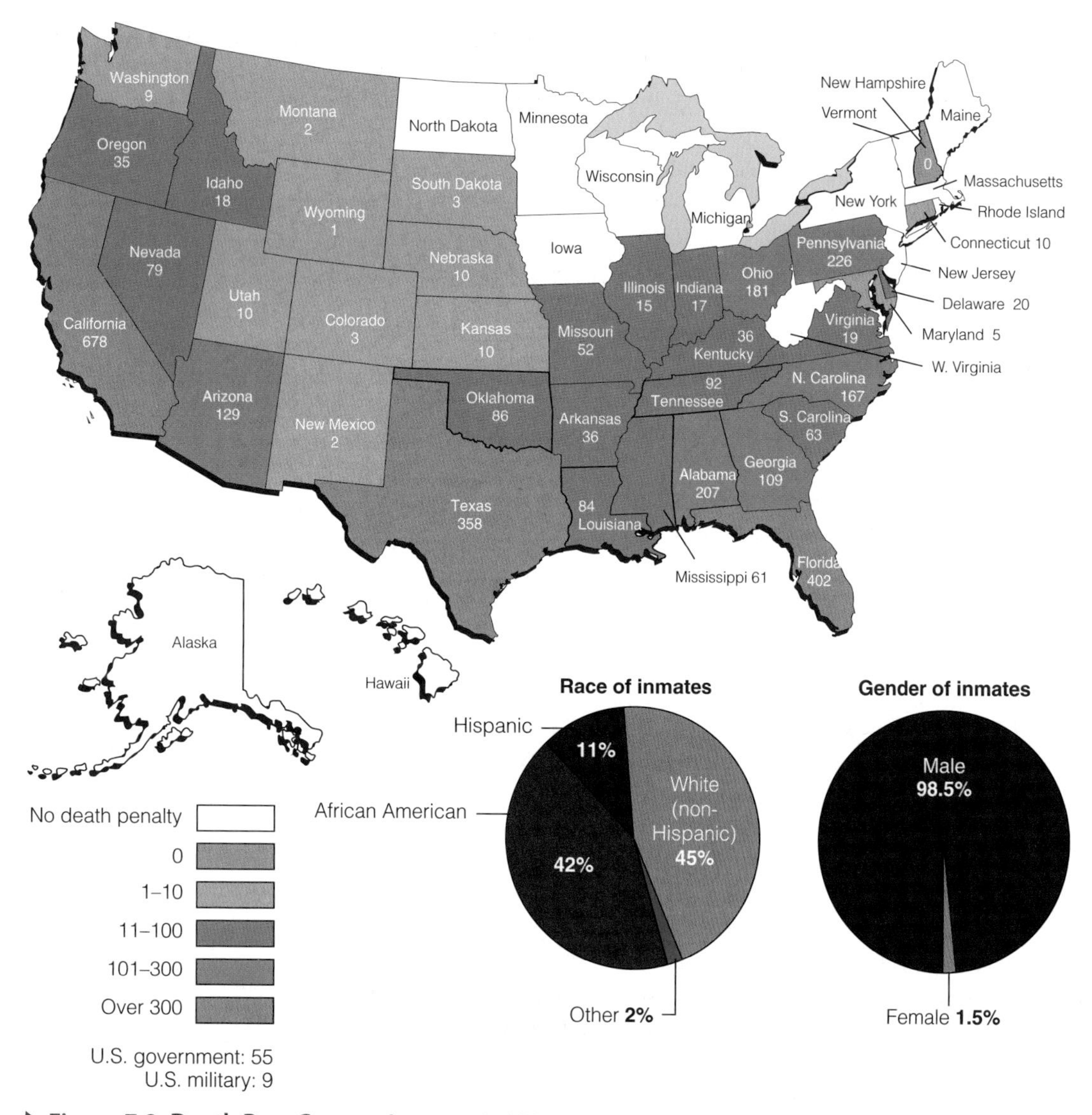

▶ **Figure 7.8 Death Row Census, January 1, 2009**
Death row inmates are heavily concentrated in certain states. African Americans, who make up about 13 percent of the U.S. population, account for approximately 42 percent of inmates on death row. In addition to those persons held by state governments, 64 inmates were held on death row by the federal government or U.S. military.

Sources: Death Penalty Information Center, 2009.

ruled that juries, not judges, must decide whether a convicted murderer should receive the death penalty (Cole and Smith, 2004).

Executions resumed in 2008 after a *de facto* moratorium was lifted by the Supreme Court when the Justices decided to uphold lethal injection. However, only southern states returned to regular executions, and executions in the South accounted for 95 percent of all executions in 2008. With 18 executions, Texas accounted for 49 percent of all executions for that year (Death Penalty Information Center, 2009). Although only 111 new death sentences were handed down in 2008 (down from 284 in 1999), the issue of the death penalty is far from resolved; the debate, which has taken place for more than two hundred years, no doubt will continue through much of the twenty-first century.

Deviance and Crime in the United States in the Future

Two pressing questions pertaining to deviance and crime will face us in the future: Is the solution to our "crime problem" more law and order? Is equal justice under the law possible?

Although many people in the United States agree that crime is one of the most important problems in this country, they are divided over what to do about it. Some of the frustration about crime might be based on unfounded fears; studies show that the overall crime rate has been decreasing slightly in recent years.

One thing is clear: The existing criminal justice system cannot solve the "crime problem." If roughly 20 percent of all crimes result in arrest, only half of those lead to a conviction in serious cases, and less than 5 percent of those result in a jail term, the "lock 'em up and throw the key away" approach has little chance of succeeding. Nor does the high rate of recidivism among those who have been incarcerated speak well for the rehabilitative efforts of our existing correctional facilities. Reducing street crime may hinge on finding ways to short-circuit criminal behavior.

One of the greatest challenges is juvenile offenders, who may become the adult criminals of tomorrow. However, instead of military-style boot camps or other stopgap measures, *structural solutions*—such as more and better education and jobs, affordable housing, more equality and less discrimination, and socially productive activities—are needed to reduce street crime. In the past, structural solutions such as these have made it possible for immigrants who initially committed street crimes to leave the streets, get jobs, and lead productive lives. Ultimately, the best approach for reducing delinquency and crime would be prevention: to work with young people *before* they become juvenile offenders to help them establish family relationships, build self-esteem, choose a career, and get an education that will help them pursue that career. Sociologist Elliott Currie (1998) has proposed that an initial goal in working to prevent delinquency and crime is to pinpoint specifically what kinds of preventive programs work and to establish priorities that make prevention possible. Among these priorities are preventing child abuse and neglect, enhancing children's intellectual and social development, providing support and guidance to vulnerable adolescents, and working intensively with juvenile offenders (Currie, 1998).

Is equal justice under the law possible? As long as racism, sexism, classism, and ageism exist in our society, people will see deviant and criminal behavior through a selective lens. To solve the problems addressed in this chapter, we must ask ourselves what we can do to ensure the rights of everyone, including the poor, people of color, and women and men alike. Many of us can counter classism, racism, sexism, and ageism where they occur. Perhaps the only way that the United States can have equal justice under the law in the future (and, perhaps, less crime as a result) is to promote social justice for individuals regardless of their race, class, gender, or age.

The Global Criminal Economy

Consider this scenario:

> Con men operating out of Amsterdam sell bogus U.S. securities by telephone to Germans; the operation is controlled by an Englishman residing in Monaco, with his profits in Panama. Which police force should investigate? In which jurisdiction should a prosecution be mounted? There may even be a question about whether a crime has been committed, although if all the actions had taken place in a single country there would be little doubt. (United Nations Development Programme, 1999: 104)

As this example shows, international criminal activity poses new and interesting questions not only for those who are the victims of such actions but also for governmental agencies mandated to control crime.

Global crime—the networking of powerful criminal organizations and their associates in shared activities around the world—is a relatively new phenomenon (Castells, 1998). However, it is an extremely lucrative endeavor as criminal organizations have increasingly set up their operations on a transnational basis, using the latest communication and transportation technologies.

How much money and other resources change hands in the global criminal economy? Although the exact amount of profits and financial flows originating in the global criminal economy is impossible to determine, the 1994 United Nations Conference on Global Organized Crime estimated that about $500 billion (in U.S. currency) per year is accrued in the global trade in drugs alone. Today, profits from all kinds of global criminal activities are estimated to range from $750 billion to over $1.5 trillion per year (United Nations Development Programme, 1999). Some analysts believe that even these figures underestimate the true nature and extent of the global criminal economy (Castells, 1998). The highest income-producing activities of global criminal organizations include trafficking in drugs, weapons, and nuclear material; smuggling of things and people (including many migrants); trafficking in women and children for the sex industry; and trafficking in body parts such as corneas and major organs for the medical industry. Undergirding the entire criminal system is money laundering and various complex financial schemes and international trade

Box 7.4 Sociology in Global Perspective

The Global Reach of Russian Organized Crime

> On a spring day when warm sunshine flooded the narrow, potholed streets, I took a taxi to Metropolitan Correctional Center (MCC), an imposing collection of tomblike cinder block towers in lower Manhattan, to interview Monya Elson—one of the most dangerous Russian mobsters the feds ever netted. I passed through several layers of security before I was shepherded by an armed guard up an elevator and deposited in a small, antiseptic cubicle with booming acoustics where lawyers meet their clients. . . . At least half a dozen armed guards stood outside the door, which was closed but had an observation window. Elson, an edgy man with a dark mien, was brought into the room, his hands and feet chained. He is considered a maximum-security risk, and for good reason: a natural-born extortionist and killing machine, Elson is perhaps the most prolific hit man in Russian mob history. (Friedman, 2000: 1)

Does this passage sound like the beginning of a best-selling novel? Although it has all the mystery and intrigue of a good novel, this paragraph is the beginning of a nonfiction book, *Red Mafiya: How the Russian Mob Has Invaded America* by the journalist Robert I. Friedman (2000). Friedman and other analysts have documented how more than thirty Russian crime syndicates operate in the United States, particularly in cities such as New York, Miami, San Francisco, Los Angeles, and Denver. And, as one New York state tax agent stated, "The Russian [mob] didn't come here to enjoy the American dream. They came here to steal it" (Friedman, 2000: Intro).

When and how did the Russian mob first become an international problem? Although Russian organized crime first gained a stronghold in the United States in the 1970s—a period of détente between this country and the Soviet Union, when Moscow allowed many Soviet Jews to emigrate to the United States and is believed to have sent many hard-core Russian criminals with them (*CBS News*, 2000)—many experts believe that the breakup of the former Soviet Union in the late 1980s created a weakened and impoverished Russia that produced an active and vigorous Russian mob that rapidly extended its grip to more than fifty countries worldwide.

Today, in cities ranging from New York and San Francisco to Toronto and Hong Kong, the powerful Russian mafia traffics in prostitution and drugs such as heroin; commits acts of extortion, arson, murder, burglary, and money laundering; and engages in the illegal sale of guns and missiles (Lindberg and Markovic, 2001). It maintains its position by a combination of three attributes that are characteristic of other major international crime groups, as well: (1) *non-ideology*—an absence of political motivations or goals, (2) *hierarchy*—a well-organized structure of specialized criminal cells controlled by a boss, and (3) *limited membership*—membership restrictions based on ethnicity, kinship, race, criminal record, and other factors deemed relevant by the group (Lindberg and Markovic, 2001).

What is the future of Russian organized crime on a global basis? Many analysts believe that organized crime not only remains out of control in Russia but that it is also growing in power and strength in the United States and other nations, where it has infiltrated major financial institutions such as banks and brokerage firms. Is it something that we should be concerned about? Probably so. As Monya Elson, the mobster described at the beginning of this feature, admits, "I am a criminal. And for you this is bad. . . . I am proud of what I am" (qtd. in Friedman, 2000: 1). Elson purportedly killed more than one hundred people before he was caught and sent to prison.

Reflect & Analyze

Since 2001, most of U.S. attention to international crime has focused on terrorism. Should we turn some of our attention to the threat posed by international organized crime? Which type of crime concerns you more?

networks that make it possible for people to use the resources they obtain through illegal activity for the purposes of consumption and investment in the ("legitimate") formal economy.

Who engages in global criminal activities? According to the sociologist Manuel Castells (1998), the following groups are the major players in the global criminal economy:

> The Sicilian *Costa Nostra* (and its associates, *La Camorra, 'Ndrangheta,* and *Sacra Corona Unita*), the American Mafia, the Colombian cartels, the

> Mexican cartels, the Nigerian criminal networks, the Japanese *Yakuza,* the Chinese Triads, the constellation of Russian *Mafiyas,* the Turkish heroin traffickers, the Jamaican Posses, and a myriad of regional and local criminal groupings in all countries that come together in a global, diversified network that permeates boundaries and links up ventures of all sorts.

Although these groups have existed for many years in their countries of origin and perhaps surrounding territories, they have expanded rapidly in the era of global communications and rapid transportation networks. In some countries, criminal organizations maintain a quasi-legal existence and are visible in "legitimate" business and political activities. For example, in Japan the *Yakuza* have operated for many years without much scrutiny by the Japanese government. Today, the *Yakuza* have exported their practice of blackmail and extortion of corporations to the United States and other nations, where they send in violent *provocateurs,* known as the *Sokaiya,* to intimidate Japanese executives living abroad (Castells, 1998). Among other global criminal organizations, these Japanese gang members have been able to operate in the United States by investing heavily in real estate and engaging in agreements with other organized crime groups in this country. Recently, law enforcement officials have become increasingly concerned about the operations of criminal networks in the United States that trace their origins to Russia (see Box 7.4).

According to Castells (1998), networking and strategic alliances between criminal networks have been key factors in the success of the criminal organizations that have sought to expand their criminal activities over the past two decades. It is through such networks that the criminal economy thrives on a global basis, allowing participants to escape police control and live beyond the laws of any one nation.

Can anything be done about global crime? Recent studies have concluded that reducing global crime will require a global response, including the cooperation of law enforcement agencies, prosecutors, and intelligence services across geopolitical boundaries. However, this approach is problematic because countries such as the United States often have difficulty getting the various law enforcement agencies to cooperate within their own nation. Similarly, law enforcement agencies in high-income nations such as the United States and Canada are often suspicious of law enforcement agencies in low-income countries, believing that their officers are corrupt. Regulation by the international community (for example, through the United Nations) would also be necessary to control global criminal activities such as international money laundering and trafficking in people and controlled substances such as drugs and weapons. However, development and enforcement of international agreements on activities such as the smuggling of migrants or the trafficking of women and children for the sex industry have been extremely limited thus far. Many analysts acknowledge that economic globalization has provided greater opportunities for wealth through global organized crime (Castells, 1998; United Nations Development Programme, 1999).

Chapter Review

• How do sociologists view deviance?

Sociologists are interested in what types of behavior are defined by societies as "deviant," who does that defining, how individuals become deviant, and how those individuals are dealt with by society.

• What are the main functionalist theories for explaining deviance?

Functionalist perspectives on deviance include strain theory and opportunity theory. Strain theory focuses on the idea that when people are denied legitimate access to cultural goals, such as a good job or a nice home, they may engage in illegal behavior to obtain them. Opportunity theory suggests that for deviance to occur, people must have access to illegitimate means to acquire what they want but cannot obtain through legitimate means.

• How do conflict and feminist perspectives explain deviance?

Conflict perspectives on deviance focus on inequalities in society. Marxist conflict theorists link deviance and crime to the capitalist society, which divides people into haves and have-nots, leaving crime as the only source of support for those at the bottom of the economic ladder. Feminist approaches to deviance focus on the relationship between gender and deviance.

- **How do symbolic interactionists view deviance?**

According to symbolic interactionists, deviance is learned through interaction with others. Differential association theory states that individuals have a greater tendency to deviate from societal norms when they frequently associate with persons who tend toward deviance instead of conformity. According to social control theories, everyone is capable of committing crimes, but social bonding (attachments to family and to other social institutions) keeps many from doing so. According to labeling theory, deviant behavior is that which is labeled deviant by those in powerful positions.

- **What is the postmodernist view on deviance?**

Postmodernist views on deviance focus on how the powerful control others through discipline and surveillance. This control may be maintained through largely invisible forces such as the Panoptican, as described by Michel Foucault, or by newer technologies that place everyone—not just "deviants"—under constant surveillance by authorities who use their knowledge as power over others.

- **How do sociologists classify crime?**

Sociologists identify six main categories of crime: violent crime (murder, forcible rape, robbery, and aggravated assault), property crime (burglary, motor vehicle theft, larceny-theft, and arson), public order crimes (sometimes referred to as "morals" crimes), occupational (white-collar) crime, organized crime, and political crime.

- **What are the main sources of crimes statistics?**

Official crime statistics are taken from the Uniform Crime Report, which lists crimes reported to the police, and the National Crime Victimization Survey, which interviews households to determine the incidence of crimes, including those not reported to police. Studies show that many more crimes are committed than are officially reported.

- **How are age and class related to crime statistics?**

Age is the key factor in crime statistics. In 2007, persons under age 25 accounted for more than 44 percent of all arrests for violent crime and almost 54 percent of all arrests for property crime. Persons from lower socioeconomic backgrounds are more likely to be arrested for violent and property crimes; white-collar crime is more likely to occur among the upper socioeconomic classes.

- **Who are the most frequent victims of crime?**

Young males of color between ages 12 and 24 have the highest criminal victimization rates. The elderly tend to be fearful of crime but are the least likely to be victimized.

- **How is discretion used in the criminal justice system?**

The criminal justice system, including the police, the courts, and prisons, often has considerable discretion in dealing with offenders. The police often use discretion in deciding whether to act on a situation. Prosecutors and judges use discretion in deciding which cases to pursue and how to handle them.

CENGAGENOW™

www.cengage.com/login

Want to maximize your online study time? Take this easy-to-use study system's diagnostic pre-test, and it will create a personalized study plan for you. By helping you to identify the topics that you need to understand better and then directing you to valuable online resources, it can speed up your chapter review. CengageNOW even provides a post-test so you can confirm that you are ready for an exam.

Key Terms

labeling theory 215
occupational (white-collar) crime 220
organized crime 221
political crime 222
primary deviance 216
property crime 220
punishment 231
rational choice theory of deviance 214
secondary deviance 216
social bond theory 215
social control 204
strain theory 206
terrorism 223
tertiary deviance 216
victimless crime 220
violent crime 219

Questions for Critical Thinking

1. Does public toleration of deviance lead to increased crime rates? If people were forced to conform to stricter standards of behavior, would there be less crime in the United States?
2. Should so-called victimless crimes, such as prostitution and recreational drug use, be decriminalized? Do these crimes harm society?
3. As a sociologist armed with a sociological imagination, how would you propose to deal with the problem of crime in the United States? What programs would you suggest enhancing? What programs would you reduce?

The Kendall Companion Website

www.cengage.com/sociology/kendall

Visit this book's companion website, where you'll find more resources to help you study and successfully complete course projects. Resources include quizzes and flash cards, as well as special features such as an interactive sociology timeline, maps, General Social Survey (GSS) data, and Census 2000 data. The site also provides links to useful websites that have been selected for their relevance to the topics in this chapter and include those listed below. (*Note:* Visit the book's website for updated URLs.)

Federal Bureau of Investigation (FBI)

http://www.fbi.gov

Visit the FBI's home page to learn more about the agency and to access an array of information and resources on crime. The site features an easy-to-use keyword search engine that allows you to quickly locate information on topics of interest.

Bureau of Justice Statistics (BJS)

http://www.ojp.usdoj.gov/bjs

This site provides interesting statistics, reports, and analyses on a number of crime-related topics, including crime victimization, courts and sentencing, firearms and crime, homicide trends, international crime statistics, and crime expenditures.

truTV

http://www.trutv.com

Formerly known as Court TV, truTV has a website that combines comprehensive legal coverage and analysis with current events coverage, information on recent criminal and civil cases, legal news about public figures, unusual crime stories, streaming video feeds of trials, legal help links, a searchable crime library, chat rooms, and message boards. Specific topics in the criminal justice system are typically well researched, objectively presented, and very interesting.

chapter

8 Class and Stratification in the United States

Chapter Focus Question

How is the American Dream influenced by social stratification?

President Barack Obama: Of the many responsibilities granted to a President by our Constitution, few are more serious or more consequential than selecting a Supreme Court Justice. The members of our highest court are granted life tenure, often serving long after the Presidents who appointed them. . . . So I don't take this decision lightly. . . .

After completing this exhaustive process, I have decided to nominate an inspiring woman who I believe will make a great Justice: Judge Sonia Sotomayor of the great state of New York. Over a distinguished career that spans three decades, Judge Sotomayor has worked at almost every level of our judicial system, providing her with a depth of experience and a breadth of perspective that will be invaluable as a Supreme Court Justice. . . .

On August 8, 2009, Sonia Sotomayor, shown here with President Barack Obama, was sworn in as the first Latina/o member of the U.S. Supreme Court. To many observers, Sotomayor's career is emblematic of the American Dream.

IN THIS CHAPTER

- What Is Social Stratification?
- Systems of Stratification
- Classical Perspectives on Social Class
- Contemporary Sociological Models of the U.S. Class Structure
- Inequality in the United States
- Poverty in the United States
- Sociological Explanations of Social Inequality in the United States
- U.S. Stratification in the Future

But as impressive and meaningful as Judge Sotomayor's sterling credentials in the law is her own extraordinary journey. Born in the South Bronx, she was raised in a housing project not far from Yankee Stadium, making her a lifelong Yankees fan. I hope this will not disqualify her—in the eyes of the New Englanders in the Senate.

Sonia's parents came to New York from Puerto Rico during the Second World War, her mother as part of the Women's Army Corps. . . . Sonia's father was a factory worker with a third-grade education who didn't speak English. But like Sonia's mother, he had a willingness to work hard, a strong sense of family, and a belief in the American Dream. . . .

Judge Sonia Sotomayor: Thank you, Mr. President, for the most humbling honor of my life. You have nominated me to serve on the country's highest court, and I am deeply moved. . . . Although I grew up in very modest and challenging circumstances, I consider my life to be immeasurably rich. I was raised in a Bronx public housing project, but studied at two of the nation's *(Continued)*

Sharpening Your Focus

- *How do prestige, power, and wealth determine social class?*
- *What role does occupational structure play in a functionalist perspective on class structure?*
- *What role does ownership of resources play in a conflict perspective on class structure?*
- *How are social stratification and poverty linked?*

finest universities. I did work as an assistant district attorney, prosecuting violent crimes that devastate our communities. But then I joined a private law firm and worked with international corporations doing business in the United States. I have had the privilege of serving as a Federal District trial judge, and am now serving as a Federal Appellate Circuit Court judge. . . .

It is a daunting feeling to be here. Eleven years ago, during my confirmation process for appointment to the Second Circuit, I was given a private tour of the White House. It was an overwhelming experience for a kid from the South Bronx. Yet never in my wildest childhood imaginings did I ever envision that moment, let alone did I ever dream that I would live this moment.

—remarks by President Barack Obama and Judge Sonia Sotomayor upon her nomination to the U.S. Supreme Court (White House.gov, 2009b)

The remarkable success of Judge Sonia Sotomayor has been described by media analysts and by the judge herself as a contemporary version of the American Dream, which culminated in her nomination by the Obama administration to serve as a Justice of the U.S. Supreme Court. Like Judge Sotomayor, many other highly successful people in this nation started in humble origins, living in housing projects or inner-city slums, and eventually rising to top positions in their respective fields. Among the many people who grew up in one of the 400,000 public housing residences owned by the New York City Housing Authority, for example, are past and current members of Congress, top executives who head up corporations such as Starbucks and Xerox, and highly paid athletes and entertainers, including Jay-Z, Wesley Snipes, Marc Anthony, and Mike Tyson (Alvarez and Wilson, 2009).

Judge Sotomayor and many other upwardly mobile individuals describe their life experiences as "living the American Dream." This brings us to an important question in studying class and stratification in the United States: What is the American Dream? Simply stated, the American Dream is the belief that if people work hard and play by the rules, they will have a chance to get ahead. Moreover, each generation will be able to have a higher standard of living than that of its parents (Danziger and Gottschalk, 1995). The American Dream is based on the assumption that people in the United States have equality of opportunity regardless of their race, creed, color, national origin, gender, or religion.

For middle- and upper-income people, the American Dream typically means that each subsequent generation will be able to acquire more material possessions and wealth than people in the preceding generation. To some people, achieving the American Dream means having a secure job, owning a home, and getting a good education for their children. To others, it is the promise that anyone may rise from poverty to wealth (from "rags to riches") if he or she works hard enough. In the case of Judge Sotomayor and other high-level judges or members of Congress, achieving the American Dream means that they have acquired a unique opportunity to influence the course of legal and/or governmental history in the United States.

When we talk about the American Dream, it is important to realize that not all people, even those who work hard for many years—or even for their entire lifetime—will achieve the sort of success that Judge Sotomayor and others who have risen to the tops of their respective professions have accomplished. The way a society is stratified has a major influence on a person's position in the class structure. In this chapter, we examine systems of social stratification and how the U.S. class system may make it easier for some individuals to attain (or maintain) top positions in society while others have great difficulty moving up from poverty or low-income origins. Before we explore class and strati-

Box 8.1 Sociology and Everyday Life

How Much Do You Know About Wealth, Poverty, and the American Dream?

True	False	
T	F	1. People no longer believe in the American Dream.
T	F	2. Individuals over age 65 have the highest rate of poverty.
T	F	3. Men account for two out of three impoverished adults in the United States.
T	F	4. About 5 percent of U.S. residents live in households whose members sometimes do not get enough to eat.
T	F	5. Income is more unevenly distributed than wealth.
T	F	6. People who are poor usually have personal attributes that contribute to their impoverishment.
T	F	7. A number of people living below the official poverty line have full-time jobs.
T	F	8. One in three U.S. children will be poor at some point in their childhood.

Answers on page 244.

fication, test your knowledge of wealth, poverty, and the American Dream by taking the quiz in Box 8.1.

What Is Social Stratification?

***Social stratification* is the hierarchical arrangement of large social groups based on their control over basic resources** (Feagin and Feagin, 2008). Stratification involves patterns of structural inequality that are associated with membership in each of these groups, as well as the ideologies that support inequality. Sociologists examine the social groups that make up the hierarchy in a society and seek to determine how inequalities are structured and persist over time.

Max Weber's term ***life chances* refers to the extent to which individuals have access to important societal resources such as food, clothing, shelter, education, and health care.** According to sociologists, more-affluent people typically have better life chances than the less-affluent because they have greater access to quality education, safe neighborhoods, high-quality nutrition and health care, police and private security protection, and an extensive array of other goods and services. In contrast, persons with low- and poverty-level incomes tend to have limited access to these resources. *Resources* are anything valued in a society, ranging from money and property to medical care and education; resources are considered to be scarce because of their unequal distribution among social categories. If we think about the valued resources available in the United States, for example, the differences in life chances are readily apparent. As one analyst suggested, "Poverty narrows and closes life chances. The victims of poverty experience a kind of arteriosclerosis of opportunity. Being poor not only means economic insecurity, it also wreaks havoc on one's mental and physical health" (Ropers, 1991: 25). Our life chances are intertwined with our class, race, gender, and age.

All societies distinguish among people by age. Young children typically have less authority and responsibility than older persons. Older persons, especially those without wealth or power, may find themselves at the bottom of the social hierarchy. Similarly, all societies differentiate between females and males: Women are often treated as subordinate to men. From society to society, people are treated differently as a result of their religion, race/ethnicity, appearance, physical strength, disabilities, or other distinguishing characteristics. All of these differentiations result in inequality. However, systems of stratification are also linked to the specific economic and social structure of a society and to a nation's position in the system of global stratification, which is so significant for understanding social inequality that we will devote the next chapter to this topic (Chapter 9).

social stratification the hierarchical arrangement of large social groups based on their control over basic resources.

life chances Max Weber's term for the extent to which individuals have access to important societal resources such as food, clothing, shelter, education, and health care.

Box 8.1 Sociology and Everyday Life

Answers to the Sociology Quiz on Wealth, Poverty, and the American Dream

1. False. The American Dream appears to be alive and well in the minds of many people despite recent economic problems in this country. U.S. culture places a strong emphasis on the goal of monetary success, and many people use legal or illegal means to attempt to achieve that goal.

2. False. As a group, children have a higher rate of poverty than the elderly. Government programs such as Social Security have been indexed for inflation in some years, whereas many of the programs for the young have been scaled back or eliminated. However, many elderly individuals still live in poverty.

3. False. Women, not men, account for two out of three impoverished adults in the United States. Reasons include the lack of job opportunities for women, lower pay than men for comparable jobs, lack of affordable day care for children, sexism in the workplace, and a number of other factors.

4. True. It is estimated that about 5 percent of the U.S. population (1 in 20 people) resides in household units where members do not get enough to eat.

5. False. Wealth is more unevenly distributed among the U.S. population than is income. However, both wealth and income are concentrated in very few hands compared with the size of the overall population.

6. False. According to one widely held stereotype, the poor are lazy and do not want to work. Rather than looking at the structural characteristics of society, people cite the alleged personal attributes of the poor as the reason for their plight.

7. True. Many of those who fall below the official poverty line are referred to as the "working poor" because they work full time but earn such low wages that they are still considered to be impoverished.

8. True. According to data from the Children's Defense Fund, one in three U.S. children will live in a family that is below the official poverty line at some point in their childhood. For some of these children, poverty will be a persistent problem throughout their childhood and youth.

Sources: Based on Children's Defense Fund, 2001; Gilbert, 2008; and U.S. Census Bureau, 2009.

Systems of Stratification

Around the globe, one of the most important characteristics of systems of stratification is their degree of flexibility. Sociologists distinguish among such systems based on the extent to which they are open or closed. In an *open system,* the boundaries between levels in the hierarchies are more flexible and may be influenced (positively or negatively) by people's achieved statuses. Open systems are assumed to have some degree of social mobility. ***Social mobility* is the movement of individuals or groups from one level in a stratification system to another** (Rothman, 2005). This movement can be either upward or downward. ***Intergenerational mobility* is the social movement experienced by family members from one generation to the next.** For example, Sarah's father is a carpenter who makes good wages in good economic times but is often unemployed when the construction industry slows to a standstill. Sarah becomes a neurologist, earning $350,000 a year, and moves from the working class to the upper-middle class. Between her father's generation and her own, Sarah has experienced upward social mobility.

By contrast, ***intragenerational mobility* is the social movement of individuals within their own lifetime.** Consider, for example, RaShandra, who began her career as a high-tech factory worker and through increased experience and taking specialized courses in her field became an entrepreneur, starting her own highly successful Internet-based business. RaShandra's advancement is an example of upward intragenerational social mobility. However, both intragenerational mobility and intergenerational mobility may be downward as well as upward.

In a *closed system,* the boundaries between levels in the hierarchies of social stratification are rigid, and people's positions are set by ascribed status. Open and closed systems are ideal-type constructs; no actual stratification system is completely open or closed. The systems of stratification that we will examine—slavery, caste, and class—are characterized by different hierarchical structures and varying degrees of mobility. Let's examine these three systems of stratification to determine how people acquire their positions in each and what potential for social movement they have.

Slavery

***Slavery* is an extreme form of stratification in which some people are owned by others.** It is a closed system in which people designated as "slaves" are treated as property and have little or no control over their lives. According to some social analysts, throughout recorded history only five societies have been slave societies—those in which the social and economic impact of slavery was extensive: ancient Greece, the Roman Empire, the United States, the Caribbean, and Brazil (Finley, 1980). Others suggest that slavery also existed in the Americas prior to European settlement, and throughout Africa and Asia (Engerman, 1995).

Those of us living in the United States are most aware of the legacy of slavery in our own country. Beginning in the 1600s, slaves were forcibly imported to the United States as a source of cheap labor. Slavery was defined in law and custom by the 1750s, making it possible for one person to own another person (Healey, 2002). In fact, early U.S. presidents including George Washington, James Madison, and Thomas Jefferson owned slaves. As practiced in the United States, slavery had four primary characteristics: (1) it was for life and was inherited (children of slaves were considered to be slaves); (2) slaves were considered property, not human beings; (3) slaves were denied rights; and (4) coercion was used to keep slaves "in their place" (Noel, 1972). Although most slaves were powerless to bring about change, some were able to challenge slavery—or at least their position in the system—by engaging in activities such as sabotage, intentional carelessness, work slowdowns, or running away from owners and working for the abolition of slavery (Healey, 2002). Despite the fact that slavery in this country officially ended many years ago, sociologists such as Patricia Hill Collins (1990) believe that its legacy is deeply embedded in current patterns of prejudice and discrimination against African Americans.

Slavery is not simply an unfortunate historical legacy. Although legal slavery and serfdom, involving private ownership of people, no longer exist at the beginning of the twentieth-first century, economist Stanley L. Engerman (1995: 175) believes that the world will not be completely free of slavery as long as there are "debt bondage, child labor, contract labor, and other varieties of coerced work for limited periods of time, with limited opportunities for mobility, and with limited political and economic power."

The Caste System

Like slavery, caste is a closed system of social stratification. **A *caste system* is a system of social inequality in which people's status is permanently determined at birth based on their parents' ascribed characteristics.** Vestiges of caste systems exist in contemporary India and South Africa.

In India, caste is based in part on occupation; thus, families typically perform the same type of work from generation to generation. By contrast, the caste system of South Africa was based on racial classifications and the belief of white South Africans (Afrikaners) that they were morally superior to the black majority. Until the 1990s, the Afrikaners controlled the government, the police, and the military by enforcing *apartheid*—the separation of the races. Blacks were denied full citizenship and restricted to segregated hospitals, schools, residential neighborhoods, and other facilities. Whites held almost all of the desirable jobs; blacks worked as manual laborers and servants.

In a caste system, marriage is endogamous, meaning that people are allowed to marry only within their own group. In India, parents traditionally have selected marriage partners for their children. In South Africa, interracial marriage was illegal until 1985.

Cultural beliefs and values sustain caste systems. Hinduism, the primary religion of India, reinforced the caste system by teaching that people should accept their fate in life and work hard as a moral duty. Caste systems grow weaker as societies industrialize: The values reinforcing the system break down, and people start to focus on the types of skills needed for industrialization.

As we have seen, in closed systems of stratification, group membership is hereditary, and it is almost impossible to move up within the structure. Custom and

social mobility the movement of individuals or groups from one level in a stratification system to another.

intergenerational mobility the social movement (upward or downward) experienced by family members from one generation to the next.

intragenerational mobility the social movement (upward or downward) of individuals within their own lifetime.

slavery an extreme form of stratification in which some people are owned by others.

caste system a system of social inequality in which people's status is permanently determined at birth based on their parents' ascribed characteristics.

Although some of Karl Marx's ideas have been discredited, his concept of class conflict between the capitalist and working classes continues to be visible in events such as strikes. Mexican, Mexican American, and many other workers in the United States took a day off from work to express their concern about stricter immigration laws in this country.

Karl Marx: Relationship to the Means of Production

According to Karl Marx, class position and the extent of our income and wealth are determined by our work situation, or our relationship to the means of production. As we have previously seen, Marx stated that capitalistic societies consist of two classes—the capitalists and the workers. The ***capitalist class (bourgeoisie)* consists of those who own the means of production**—the land and capital necessary for factories and mines, for example. The ***working class (proletariat)* consists of those who must sell their labor to the owners in order to earn enough money to survive** (see ▶ Figure 8.1).

According to Marx, class relationships involve inequality and exploitation. The workers are exploited as capitalists maximize their profits by paying workers less than the resale value of what they produce but do not own. Marx believed that a deep level of antagonism exists between capitalists and workers because of extreme differences in the *material interests* of the people in these two classes. According to the sociologist Erik O. Wright (1997: 5), material interests are "the interests people have in their material standard of living, understood as the package of toil, consumption and leisure. Material interests are thus not interests of maximizing consumption *per se,* but rather interests in the trade-off between toil, leisure and consumption." Wright suggests that *exploitation* is the key concept for understanding Marx's assertion that *interests* are generated by class relations: "In an exploitative relation, the exploiter *needs* the exploited since the exploiter depends upon the effort of the exploited." In other words, the capitalists *need* the workers to derive profits; therefore, capitalists benefit when workers do not have adequate resources to provide for themselves and hence must sell their labor power to the capitalist class. As Marx suggests, exploitation involves ongoing interactions between the two antagonistic classes, which are structured by a set of social relations that binds the exploiter and the exploited together (E. Wright, 1997).

Continual exploitation results in workers' ***alienation*—a feeling of powerlessness and estrangement from other people and from oneself.** In Marx's view, alienation develops as workers manufacture goods that embody their creative talents but the goods do not belong to them. Workers are also alienated from the work itself because they are forced to perform it in order to live. Because the workers' activities are not their own, they feel self-estrangement. Moreover, the workers are separated from others in the factory because they individually sell their labor power to the capitalists as a commodity.

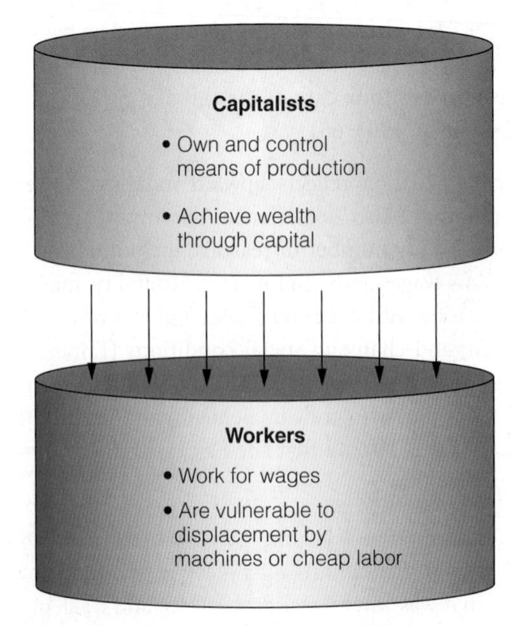

▶ **Figure 8.1 Marx's View of Stratification**

© LYNN ISCHAY/The Plain Dealer/Landov

Although unions have lost some of their importance over the last 50 years, this 2009 United Auto Workers rally in support of the Employee Free Choice Act—legislation that would make it easier for employees to join a union—shows that many workers still believe in a collective response to workplace issues.

In Marx's view, the capitalist class maintains its position at the top of the class structure by control of the society's *superstructure,* which is composed of the government, schools, churches, and other social institutions that produce and disseminate ideas perpetuating the existing system of exploitation. Marx predicted that the exploitation of workers by the capitalist class would ultimately lead to ***class conflict*—the struggle between the capitalist class and the working class.** According to Marx, when the workers realized that capitalists were the source of their oppression, they would overthrow the capitalists and their agents of social control, leading to the end of capitalism. The workers would then take over the government and create a more egalitarian society.

Why has no workers' revolution occurred? According to the sociologist Ralf Dahrendorf (1959), capitalism may have persisted because it has changed significantly since Marx's time. Individual capitalists no longer own and control factories and other means of production; today, ownership and control have largely been separated. For example, contemporary transnational corporations are owned by a multitude of stockholders but run by paid officers and managers. Similarly, many (but by no means all) workers have experienced a rising standard of living, which may have contributed to a feeling of complacency. Moreover, as discussed in Chapter 1, many people have become so engrossed in the process of consumption—including acquiring more material possessions and going on outings to shopping malls, movie theaters, and amusement parks such as Disney World—that they are less likely to engage in workers' rebellions against the system that has brought them a relatively high standard of living (Gottdiener, 1997). During the twentieth century, workers pressed for salary increases and improvements in the workplace through their activism and labor union membership. They also gained more legal protection in the form of workers' rights and benefits such as workers' compensation insurance for job-related injuries and disabilities (Dahrendorf, 1959). For these reasons, and because of a myriad of other complex factors, the workers' revolution predicted by Marx never came to pass. However, the failure of his prediction does not mean that his analysis of capitalism and his theoretical contributions to sociology are without validity.

Marx had a number of important insights into capitalist societies. First, he recognized the economic basis of class systems (Gilbert, 2008). Second, he noted the relationship between people's social location in the class structure and their values, beliefs, and behavior. Finally, he acknowledged that classes may have opposing (rather than complementary) interests. For example, capitalists' best interests are served by a decrease in labor costs and other expenses and a corresponding increase in profits; workers' best interests are served by well-paid jobs, safe working conditions, and job security.

capitalist class (or **bourgeoisie**) Karl Marx's term for the class that consists of those who own and control the means of production.

working class (or **proletariat**) those who must sell their labor to the owners in order to earn enough money to survive.

alienation a feeling of powerlessness and estrangement from other people and from oneself.

class conflict Karl Marx's term for the struggle between the capitalist class and the working class.

Box 8.2 Framing Class in the Media

Taking the TV Express to Wealth and Upward Social Mobility

Hello fans,

I often get asked what it is I have on my computer monitors up in my booth.

Well, one of the things I have is a DVR deck that allows me to record and replay moments in the [*Deal or No Deal*] game almost like an instant replay in sports.

Sometimes I like to replay the moment the player opens the million dollar case in slow motion. It is actually possible to find the exact frame of video where you see the player's heart snap in two like a stale tortilla.

The Banker

—The Banker's Blog from *Deal or No Deal,* February 4, 2007 (nbc.com, 2007)

NBC's popular game show *Deal or No Deal* is one of many television entertainment programs that offers contestants instant wealth if they are lucky and make the right choices. Shows that give away large sums of money or other substantial prizes are extremely popular with audiences because these programs suggest that riches, fame, and happiness are just around the corner and that the American Dream can be achieved quickly and without much effort. This type of media framing would have us believe that it is fairly easy to find a shortcut to wealth and a higher position in the class structure through playing a simple game or winning a talent competition (such as *American Idol*). This sort of framing depicts upward mobility as being similar to riding an express elevator to the top of a high-rise building: You can bypass many floors (or economic levels) in one swift move to the top. Game shows provide the illusion that a few lucky contestants can win instant riches without having to make the usual stops (such as acquiring a good education or accumulating years of work-related experience) that typically are necessary for an individual to experience upward social mobility. As defined in this chapter, *social mobility* is the movement of individuals or groups from one level in a stratification system to another. According to conflict theorists, upward social mobility has become increasingly rare over the past century; however, this reality does not stop people from wishing and hoping that they will find the next express elevator to the top. The media (like state lotteries and other "giveaways") often foster the notion that upward mobility is more easily attained than is actually the case.

If you are wondering about the blog (that you read above) by the "banker" on *Deal or No Deal,* he is describing how he thinks the contestants on the show feel when they realize that they have just lost the opportunity to win an easy one million dollars for less than an hour's "work" (on the other hand, some contestants do leave the show with much more money than they previously possessed). If you are not familiar with the show, here's a brief overview: A contestant is confronted with 26 sealed briefcases, each of which contains a card representing an amount of money ranging from one penny to one million dollars. The contestant must choose one case to keep (as the final prize) and then, one at a time, select additional cases to eliminate. As various sums of money are eliminated through the case-selection process, the "banker" calls down from a hidden booth and makes monetary settlement offers to the contestant to "buy" the chosen case. If the contestant takes the banker's offer, and that sum of money is less than the amount in the chosen case, the contestant has lost money. If the contestant refuses the banker's offer, future offers may either go up or down, depending on which sums

Max Weber: Wealth, Prestige, and Power

Max Weber's analysis of class builds upon earlier theories of capitalism (particularly those by Marx) and of money (particularly those by Simmel, as discussed in Chapter 1). Living in the late nineteenth and early twentieth centuries, Weber was in a unique position to see the transformation that occurred as individual, competitive, entrepreneurial capitalism went through the process of shifting to bureaucratic, industrial, corporate capitalism. As a result, Weber had more opportunity than Marx to see how capitalism changed over time.

Weber agreed with Marx's assertion that economic factors are important in understanding individual and group behavior. However, Weber emphasized that no single factor (such as economic divisions between capitalists and workers) was sufficient for defining the location of categories of people within the class structure. According to Weber, the access that people have to important societal resources (such as economic, social, and political power) is crucial in determining people's life chances. To highlight the importance of life chances for categories of people, Weber developed a multidimensional approach to social stratification that reflects the interplay among wealth, prestige, and power. In his analysis of these dimensions of class

Trae Patton/© NBC/Courtesy: Everett Collection

According to popular media framing, instant wealth and fame can be yours on television shows such as *Deal or No Deal* and *American Idol.* By selecting the right cases or being voted the best at singing a variety of songs, contestants gain large sums of money and public acclaim. Some say that media representations of easy upward mobility present a deceiving picture about the realities of social class and the relative permanence of vast economic and social inequality. What do you think?

have been eliminated through the case-selection process. Each time the banker makes an offer, the contestant has to decide whether it is a "deal" or "no deal." If no "deal" is made, the process of eliminating cases continues until the contestant either accepts a subsequent "deal" from the banker or the contestant eliminates all of the cases except the one that the contestant originally chose to keep—which may contain a much lower amount than the banker's earlier offers. Some contestants leave the game ecstatic over their large winnings; others leave with a good sum of money but with some displeasure that they were outsmarted by the banker because they turned down more money from him than they "earned" through the case-selection process. Some contestants leave with little or nothing.

Deal or No Deal became a highly successful game show because its premise is framed in such a manner as to highlight the notion that the American Dream is just around the corner. Even if you have not seen *Deal or No Deal,* similar framing has been used on television entertainment shows for many years, and you can no doubt provide other examples of this approach. In each case, the key message is that a quick and easy way exists to gain vast sums of money and, in some situations, to instantly achieve the American Dream if you are lucky, talented, or simply able to convince other people to vote for you.

Reflect & Analyze

Do you think that television shows *mirror* people's perceptions of wealth and class in our society? Do the media help to *shape* our society and *create* cultural perceptions that we hold to be true about ourselves or individuals in our social world? Or are media representations merely a reflection of the nation and world in which we live?

structure, Weber viewed the concept of "class" as an *ideal type* (that can be used to compare and contrast various societies) rather than as a specific social category of "real" people (Bourdieu, 1984).

***Wealth* is the value of all of a person's or family's economic assets, including income, personal property, and income-producing property.** Weber placed categories of people who have a similar level of wealth and income in the same class. For example, he identified a privileged commercial class of *entrepreneurs*—wealthy bankers, ship owners, professionals, and merchants who possess similar financial resources. He also described a class of *rentiers*—wealthy individuals who live off their investments and do not have to work. According to Weber, entrepreneurs and rentiers have much in common. Both are able to purchase expensive consumer goods, control other people's opportunities to acquire wealth and property, and monopolize costly status privileges (such as education) that provide contacts and skills for their children.

Weber divided those who work for wages into two classes: the middle class and the working class. The

wealth the value of all of a person's or family's economic assets, including income, personal property, and income-producing property.

middle class consists of white-collar workers, public officials, managers, and professionals. The working class consists of skilled, semiskilled, and unskilled workers.

The second dimension of Weber's system of stratification is ***prestige*—the respect or regard with which a person or status position is regarded by others.** Fame, respect, honor, and esteem are the most common forms of prestige. A person who has a high level of prestige is assumed to receive deferential and respectful treatment from others. Weber suggested that individuals who share a common level of social prestige belong to the same status group regardless of their level of wealth. They tend to socialize with one another, marry within their own group of social equals, spend their leisure time together, and safeguard their status by restricting outsiders' opportunities to join their ranks (Beeghley, 2008).

The other dimension of Weber's system is ***power*—the ability of people or groups to achieve their goals despite opposition from others.** The powerful can shape society in accordance with their own interests and direct the actions of others (Tumin, 1953). According to Weber, social power in modern societies is held by bureaucracies; individual power depends on a person's position within the bureaucracy. Weber suggested that the power of modern bureaucracies was so strong that even a workers' revolution (as predicted by Marx) would not lessen social inequality (Hurst, 2007).

Weber stated that wealth, power, and prestige are separate continuums on which people can be ranked from high to low, as shown in ▶ Figure 8.2. Individuals may be high on one dimension while being low on another. For example, people may be very wealthy but have little political power (for example, a recluse who has inherited a large sum of money). They may also have prestige but not wealth (for instance, a college professor who receives teaching excellence awards but lives on a relatively low income). In Weber's multidimensional approach, people are ranked on all three dimensions. Sociologists often use the term ***socioeconomic status (SES)* to refer to a combined measure that attempts to determine class location by classifying individuals, families, or households in terms of factors such as income, occupation, and education.**

What important insights does Weber provide in regard to social stratification and class? Weber's analysis of social stratification contributes to our understanding by emphasizing that people behave according to both their economic interests and their values. He also added to Marx's insights by developing a multidimensional explanation of the class structure and by identifying additional classes.

A substantial advantage of Weber's theory is that it has made empirical investigation of the U.S. class structure possible. Through his distinctions among wealth, power, and prestige, Weber makes it possible for researchers to examine the different dimensions of social stratification. Weber's enlarged conceptual formulation of stratification is the theoretical foundation for mobility research by sociologists such as Peter Blau and Otis Duncan. Blau and Duncan (1967) measured the three dimensions from Weber's theory through a study of the occupational positions that individuals

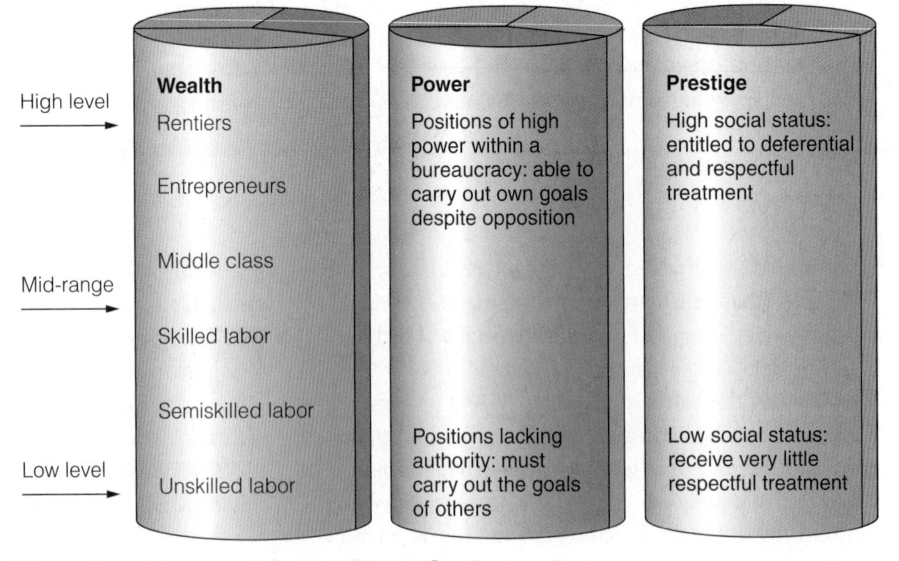

▶ Figure 8.2 Weber's Multidimensional Approach to Social Stratification
According to Max Weber, wealth, power, and prestige are separate continuums. Individuals may rank high in one dimension and low in another, or they may rank high or low in more than one dimension. Also, individuals may use their high rank in one dimension to achieve a comparable rank in another. How does Weber's model compare with Marx's approach as shown in Figure 8.1?

◆ **Table 8.1 Prestige Ratings for Selected Occupations in the United States** Respondents were asked to evaluate a list of occupations according to their prestige; the individual rankings were averaged and then converted into scores, with 1 the lowest possible score and 99 the highest possible score. How would you rate the prestige level of these various jobs?

Occupation	Score	Occupation	Score
Physician	86	Airplane pilot	61
Attorney	75	Police officer	60
College professor	74	Electrician	51
Architect	73	Funeral director	49
Aerospace engineer	72	Mail carrier	47
Dentist	72	Secretary	46
Clergy	69	Butcher	35
Pharmacist	68	Baker	35
Petroleum engineer	66	Garbage collector	28
Registered nurse	66	Bill collector	24
Accountant	65	Janitor	22
Grade school teacher	64	Maid	20

Source: Hauser and Warren, 1996.

hold. According to Blau and Duncan, a person's occupational position is not identical to either economic class or prestige, but is closely related to both. As you might expect, different occupations have significantly different levels of status or prestige (see ◆ Table 8.1). For the past fifty years, occupational ratings by prestige have been remarkably consistent in the United States (Gilbert, 2008). These rankings have become the foundation for status attainment research, which uses sophisticated statistical measurements to assess the influence of family background and education on people's occupational mobility and success (see Blau and Duncan, 1967; Duncan, 1968). *Status attainment research* focuses on the process by which people ultimately reach their position in the class structure. Based largely on studies of men, this research uses the father's occupation and the son's education and first job as primary determinants of the eventual class position of the son. Obviously, family background is the central factor in this process because the son's education and first job are linked to the family's economic status. In addition, the family's location in the class system is related to the availability of social ties that may open occupational doors for the son.

Although they have been widely employed in some prestigious sociological research, status attainment models have several serious limitations. One is the focus of this research on the occupational prestige of traditionally male jobs and the exclusion of women's work, which has often been unpaid. In regard to African Americans, Patricia Hill Collins (1990: 45) noted that "the higher rates of Black male unemployment, the racial discrimination that has crowded all African-Americans into a narrow set of occupations, and the existence of household arrangements other than two-parent nuclear families . . . have all combined to make status attainment models less suitable for explaining Black social class dynamics." Moreover, the status attainment model is unable to take into account power differentials rooted in inequalities based on race, ethnicity, or gender.

A significant limitation of occupational prestige rankings is that the level of prestige accorded to a position may not actually be based on the importance of the position to society. The highest ratings may be given

prestige the respect or regard with which a person or status position is regarded by others.

power according to Max Weber, the ability of people or groups to achieve their goals despite opposition from others.

socioeconomic status (SES) a combined measure that, in order to determine class location, attempts to classify individuals, families, or households in terms of factors such as income, occupation, and education.

Photo Essay

What Keeps the American Dream Alive?

Although the American Dream of rags to riches may be an illusive goal for many people, most of us still believe that a person in the United States can get ahead by gaining a good education, through hard work, by marketing a creative idea, by winning the lottery, or by some other means. Whether or not they can ultimately rise to the top economic and social tiers of society, many people still strive to attain their personal—although perhaps scaled down—version of the American Dream.

Some sociological perspectives suggest that vast inequalities between the rich and the poor create such a large divide that upward mobility is virtually impossible for those in the lower economic tiers of society. However, other perspectives are based on the assumption that human capital—in the form of education, hard work, and outstanding achievement—can help a person move up the socioeconomic ladder.

Regardless of which perspective you or I might subscribe to, millions of people in the United States and around the world see this country as the land in which dreams can come true and in which a person can create a better life for his or her family.

As you view the pictures on these two pages, think about the ways in which various people seek out their own American Dream. Doing so helps us gain a better understanding of some of the issues relating to social stratification.

Quick and easy ways to attain the American Dream—such as winning a very large lottery drawing—have great appeal to many people in the United States and throughout the world. However, despite the widespread publicity that winners receive, only a very small fraction of those persons who attain the American Dream of wealth do so through lotteries or gambling.

© Frances M. Roberts/Alamy

Small Businesses and the American Dream

We often think of "big-ticket" entrepreneurs such as Bill Gates or Michael Dell as having achieved the American Dream. However, sidewalk vendors who own their own business may believe that they, too, have achieved their dream, especially when they come from nations where no similar dream would have even been possible.

Long Hours and Hard Work

This single mother with four children seeks the American Dream by working three different jobs as a practical nurse while attending college. For some people, getting ahead requires 24/7 commitment, but she believes that it will be worth the effort to become a registered nurse and earn better pay with fewer hours than she now works.

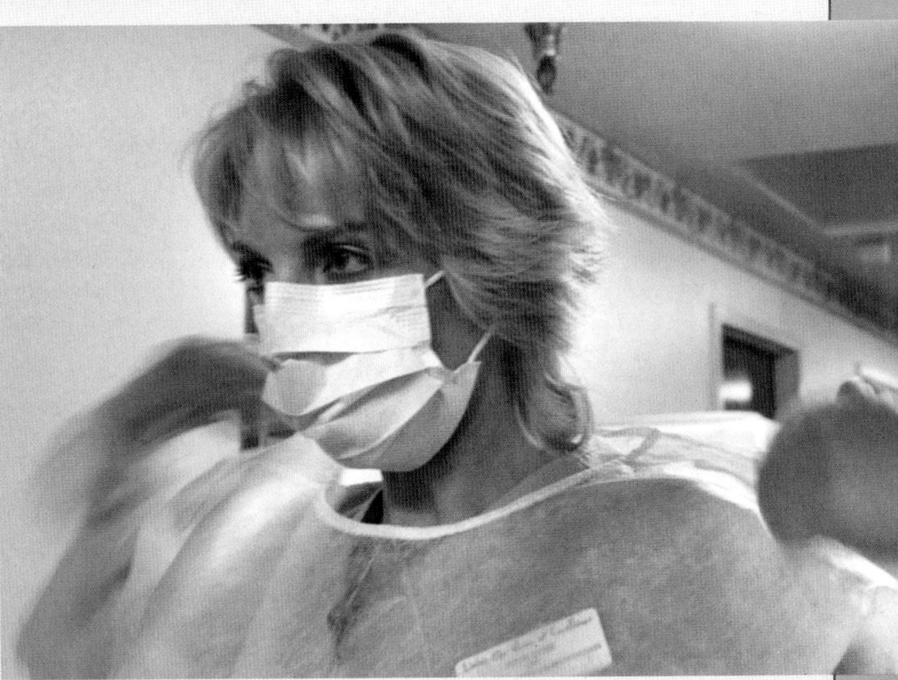

© AP Images/James Branaman/The Kitsap Sun

© Liz Hafalia/San Francisco Chronicle/CORBIS

Rags to Riches

From homeless person to millionaire stockbroker sounds like the plot line of a movie, which it is: *The Pursuit of Happyness* (2006), starring Will Smith. But it is also the real-life story of Chris Gardner, chief executive of Gardner Rich LLC, a multimillion-dollar Chicago brokerage firm, on whose autobiography *The Pursuit of Happyness* is based. Although Gardner never attended college, he has been highly successful in his financial endeavors and is now hoping to get investors to help him create a billion-dollar investment fund to promote economic opportunities for South Africans. Here you see him on a visit to the soup kitchen at Glide Memorial Church in San Francisco, California, where he used to eat.

© AP Images/Tim Boyd

The Cost of an Education

For individuals who were not born into affluent families, education is important for attaining the American Dream of upward mobility. Here you see a Latina high school student looking over a federal application for student aid during a conference held at the University of New Hampshire. The university hosted the conference to encourage minority students to consider attending there. Students often turn to student aid programs as a source of funding in the hope that they can obtain a college education. What will happen to the American Dream for students such as the woman shown here if this type of funding is reduced or eliminated in the future?

Reflect & Analyze

1. **Since the U.S. economic crisis in 2008–2009, the number of undocumented workers in this country has shrunk dramatically. Do you think that the workers have given up their hope of achieving the American Dream, or have they deferred it?**
2. **Does the U.S. government have the obligation to keep the American Dream alive for people of lower incomes? Are there alternatives to government-funded student aid programs, for example?**
3. **Do you know anyone whose life was changed as a result of playing a state lottery? Was the change positive or detrimental to the person's life? What conclusions can you draw from this person's experience?**

Turning to Video

Watch the ABC video *India Inc: Economic Explosion* (run time 2:38), available on the Kendall Companion Website and through Cengage Learning eResources accounts. Because of the country's growing economy (it's the second-fastest-growing economy in the world), there are more middle-class Indians with buying power than the entire U.S. population. Many American jobs are being outsourced to India, and India has even begun hiring out-of-work Americans. As you watch this news report, think about the photographs, commentary, and questions you encountered in this photo essay. After you've watched the video, consider two more questions: How does another country's growth affect the American Dream, and to what degree is it possible that India will replace the United States as the land of opportunity?

Today, great disparities exist in the distribution of educational resources. Because funding for education comes primarily from local property taxes, school districts in wealthy suburban areas generally pay higher teachers' salaries, have newer buildings, and provide state-of-the-art equipment. By contrast, schools in poorer areas have a limited funding base. Students in central-city schools and poverty-stricken rural areas often attend dilapidated schools that lack essential equipment and teaching materials. Author Jonathan Kozol (1991, qtd. in Feagin and Feagin, 1994: 191) documented the effect of a two-tiered system on students:

> Kindergartners are so full of hope, cheerfulness, high expectations. By the time they get into fourth grade, many begin to lose heart. They see the score, understanding they're not getting what others are getting. . . . They see suburban schools on television. . . . They begin to get the point that they are not valued much in our society. By the time they are in junior high, they understand it. "We have eyes and we can see; we have hearts and we can feel. . . . We know the difference."

Poverty extracts such a toll that many young people will not have the opportunity to finish high school, much less enter college.

Crime and Lack of Safety Along with diminished access to quality health care, nutrition, and housing, and unequal educational opportunities, crime and lack of safety are other consequences of inequality. As discussed in Chapter 7 ("Deviance and Crime"), although people from all classes commit crimes, they commit different kinds of crime. Capitalism and the rise of the consumer society may be factors in the criminal behavior of some upper-middle-class and upper-class people, who can be motivated by greed or the competitive desire to stay ahead of others in their reference group. By contrast, crimes committed by people in the lower classes may be motivated by feelings of anger, frustration, and hopelessness.

According to Marxist criminologists, capitalism produces social inequalities that contribute to criminality among people, particularly those who are outside the economic mainstream. Poverty and violence are also linked. In his recent ethnographic study of inner-city life in Philadelphia, the sociologist Elijah Anderson (1999: 33) suggests that what some people refer to as "random, senseless street violence" is often not random at all, but instead a response to profound social inequalities in the inner city:

> The inclination to violence springs from the circumstances of life among the ghetto poor—the lack of jobs that pay a living wage, limited basic public services (police response in emergencies, building maintenance, trash pickup, lighting, and other services that middle-class neighborhoods take for granted), the stigma of race, the fallout from rampant drug use and drug trafficking, and the resulting alienation and absence of hope for the future. Simply living in such an environment places young people at special risk of falling victim to aggressive behavior. . . . [T]his environment means that even youngsters whose home lives reflect mainstream values—and most of the homes in the community do—must be able to handle themselves in a street-oriented environment. (Anderson, 1999: 32–33)

As Anderson states, consequences of inequality include both crime and lack of safety on the streets, particularly for people who feel a profound sense of alienation from mainstream society and its institutions. Those who are able to take care of themselves and protect their loved ones against aggression are accorded deference and regard by others. However, Anderson believes that it

Conflict theorists see schools as agents of the capitalist class system that perpetuate social inequality: Upper-class students are educated in well-appointed environments such as the one shown here, whereas children of the poor tend to go to antiquated schools with limited facilities.

is wrong to place blame solely on individuals for the problems that exist in urban ghettos; instead, he asserts that the focus should be on the socioeconomic structure and public policy that have threatened the well-being of people who live in poverty.

Poverty in the United States

When many people think about poverty, they think of people who are unemployed or on welfare. However, many hardworking people with full-time jobs live in poverty. The U.S. Social Security Administration has established an *official poverty line,* which is based on what is considered to be the minimum amount of money required for living at a subsistence level. The poverty level is computed by determining the cost of a minimally nutritious diet (a low-cost food budget on which a family could survive nutritionally on a short-term, emergency basis) and multiplying this figure by three to allow for nonfood costs. In 2007, 37.3 million people lived below the official government poverty level of $21,200 for a family of four (DeNavas-Walt, Proctor, and Smith, 2008). Many of the people below the poverty line hold full-time jobs at low wages (see Box 8.3).

When sociologists define poverty, they distinguish between absolute and relative poverty. ***Absolute poverty* exists when people do not have the means to secure the most basic necessities of life.** This definition comes closest to that used by the federal government. Absolute poverty often has life-threatening consequences, such as when a homeless person freezes to death on a park bench. By comparison, ***relative poverty* exists when people may be able to afford basic necessities but are still unable to maintain an average standard of living** (Ropers, 1991). A family must have income substantially above the official poverty line in order to afford the basic necessities, even when these are purchased at the lowest possible cost. At about 155 percent of the official poverty line, families could live on an economy budget. What is it like to live on the economy budget? John Schwarz and Thomas Volgy (1992: 43) offer the following distressing description:

> Members of families existing on the economy budget never go out to eat, for it is not included in the food budget; they never go out to a movie, concert, or ball game or indeed to any public or private establishment that charges admission, for there is no entertainment budget; they have no cable television, for the same reason; they never purchase alcohol or cigarettes; never take a vacation or holiday that involves any motel or hotel or, again, any meals out; never hire a baby-sitter or have any other paid child care; never give an allowance or other spending money to the children; never purchase any lessons or home-learning tools for the children; never buy books or records for the adults or children, or any toys, except in the small amounts available for birthday or Christmas presents ($50 per person over the year); never pay for a haircut; never buy a magazine; have no money for the feeding or veterinary care of any pets; and, never spend any money for preschool for the children, or educational trips for them away from home, or any summer camp or other activity with a fee.

Who Are the Poor?

Poverty in the United States is not randomly distributed, but rather is highly concentrated according to age and race/ethnicity, as indicated in ◆ Table 8.2 (page 270), as well as gender.

Age Today, children are at a much greater risk of living in poverty than are older persons. A generation ago, persons over age 65 were at the greatest risk of being poor; however, government programs such as Social Security and pension plans have been indexed for inflation and thus provide for something closer to an adequate standard of living than do other social welfare programs. Even so, older women are twice as likely to be poor as older men; older African Americans and Latinos/as are much more likely to live below the poverty line than are non-Latino/a whites.

The age category most vulnerable to poverty today is the very young. In 2007, both the poverty rate and the number in poverty increased for children under 18 years old, rising to 18.0 percent from 17.2 percent in 2004. The number of children in poverty increased to 13.3 million in 2007, as compared with 12.8 million in 2006. One out of every three persons below the poverty line is under 18 years of age, and a large number of children hover just above the official poverty line. The precarious position of African American and Latino/a children is even more striking. In 2007, almost 35 percent of all African Americans under age 18 lived in poverty; 29 percent of Latino/a children were also poor, as compared with almost 15 percent of non-Latino/a white children (U.S. Census Bureau, 2009).

What do such statistics indicate about the future of our society? Children as a group are poorer now than they were at the beginning of the 1980s, whether they live in one- or two-parent families. The majority live in two-parent families in which one or both parents

Box 8.3 Sociology and Social Policy

Should Our Laws Guarantee People a Living Wage?

> One [of the most surprising things I learned] is just how difficult it is, how stressful it is to live check to check. It was an incredible strain on my relationship with [my fiancée,] Alex. Suddenly we were exhausted when we were around each other. We had no energy to really give to one another. We were so tired at the end of the day. We ate dinner together, and then we were just done. You know, you see how the quality of your life devoted to relationships can really deteriorate quickly. One thing we talk about on [FX Network's *30 Days*] is that it's no surprise that families making less than $25,000 a year are twice as likely to get divorced as a family that makes $50,000 a year.
>
> —Morgan Spurlock, the producer and director of *Super Size Me,* discussing what it was like when he and his fiancée tried working for and living on the minimum wage for thirty days. (qtd. in Campus Progress, 2006)

At some point in our lives, most of us have held a job paying the minimum wage, and we know the limitations of trying to survive on such low earnings. The federal *minimum wage* is the hourly rate that (with certain exceptions) is the lowest amount an employer can legally pay its employees (each state may adopt a higher minimum wage, but not a lower one). In 2007 the minimum wage was set at $5.85 per hour, and the rate increased to $6.55 per hour in July 2008 and to $7.25 per hour in July 2009. Although that represents a substantial increase, a person earning minimum wage and working forty hours every week, fifty-two weeks per year (in other words, no time off, no vacation) would still earn only $15,080 per year—an amount just slightly above the *official poverty line* (and slightly below that line for a person with two children). The low hourly rate paid by many employers to their employees and the high compensation "packages" received by many companies' chief executive officers—who often earn nearly 400 times as much money as their employees (Mintz, 2007)—are one of the major causes of social inequality in this country.

Morgan Spurlock and other social analysts have called our attention to the fact that living at or below the poverty line (the minimum amount of money required for living at a subsistence level) can be difficult and stressful. According to some analysts, we should do away with the idea of a minimum wage (a "poverty wage," as some describe it) and instead focus on a minimum *living wage*—a wage sufficient, based on a forty-hour week, to provide the necessities and comforts essential to an acceptable standard of living in the community in which the individual lives.

Proponents of a living wage assert that it is not only more humane (eliminating the necessity of many low-wage workers to hold more than one job in order to "make ends meet" and reducing the number of families living in poverty) but also that it makes good sense financially: A living wage would reduce taxes by reducing the amount of money the government must pay for services provided to poverty-level families such as food stamps, emergency medical treatment, and low-income housing. Critics of the living wage argue that a living-wage requirement—or, for that matter, any increase in the federal minimum wage—causes inflation, increases the cost of living for everyone, and increases unemployment because employers are not able to afford this added cost of doing business.

Reflect & Analyze

What do you think: Should your city, your state, or the federal government require that employers pay their employees, at the very least, a living wage? Why or why not?

are employed. However, children in single-parent households headed by women have a much greater likelihood of living in poverty: Approximately 30 percent of white (non-Latino/a) children under age 18 in female-headed households live below the poverty line, as sharply contrasted with more than 50 percent of Latina/o and 49 percent of African American children in the same category (DeNavas-Walt, Proctor, and Mills, 2004). Nor does the future look bright: Many governmental programs established to alleviate childhood poverty and malnutrition have been seriously cut back or eliminated altogether.

Gender About two-thirds of all adults living in poverty are women. In 2007, single-parent families headed by women had a 28 percent poverty rate as compared

absolute poverty a level of economic deprivation that exists when people do not have the means to secure the most basic necessities of life.

relative poverty a condition that exists when people may be able to afford basic necessities but are still unable to maintain an average standard of living.

◆ **Table 8.2 Percentage Distribution of Poverty in the United States**

	All Races[a]	White[b]	African American	Asian American	Hispanic[c]
By Age					
Under 18 years	18.0	14.9	34.5	12.5	28.6
18–24 years	17.3	15.3	27.6	13.6	22.0
25–34	12.3	10.7	21.8	10.1	18.6
35–44	9.4	8.2	17.0	7.4	16.7
45–64	8.5	7.3	15.6	7.5	15.6
65 and above	9.7	8.1	23.2	11.3	17.1
By Education					
No high school diploma	22.4	19.8	35.9	18.8	27.0
4 years of high school	12.0	10.0	22.6	12.8	16.2
Some college (no degree)	8.2	7.0	14.7	10.0	10.1
College degree or more	3.9	3.6	5.6	5.5	6.0

[a]Includes other races/ethnicities not shown separately.
[b]Non-Hispanic white.
[c]Includes Hispanic persons of any race.
Source: U.S. Census Bureau, 2009.

with a 5 percent rate for two-parent families. Sociologist Diana Pearce (1978) coined a term to describe this problem: The ***feminization of poverty*** **refers to the trend in which women are disproportionately represented among individuals living in poverty.** According to Pearce (1978), women have a higher risk of being poor because they bear the major economic and emotional burdens of raising children when they are single heads of households but earn between 70 and 80 cents for every dollar a male worker earns. More women than men are unable to obtain regular, full-time, year-round employment, and lack of adequate, affordable day care exacerbates this problem.

Does the feminization of poverty explain poverty in the United States today? Is poverty primarily a women's issue? On the one hand, this thesis highlights a genuine problem—the link between gender and poverty. On the other hand, several major problems exist with this argument. First, women's poverty is not a new phenomenon. Women have always been more vulnerable to poverty (see Katz, 1989). Second, all women are not equally vulnerable to poverty. Many in the upper and upper-middle classes have the financial resources, education, and skills to support themselves regardless of the presence of a man in the household. Third, event-driven poverty does not explain the realities of poverty for many women of color, who instead may experience "reshuffled poverty"—a condition of deprivation that follows them regardless of their marital status or the type of family in which they live. Research by

Many women are among the "working poor," who, although employed full time, have jobs in service occupations that are typically lower paying and less secure than jobs in other sectors of the labor market. Does the nature of women's work contribute to the feminization of poverty in the United States?

Mary Jo Bane (1986; Bane and Ellwood, 1994) demonstrates that two out of three African American families headed by a woman were poor before the family event that made the woman a single mother. In addition, the poverty risk for a two-parent African American family is more than twice that for a white two-parent family.

Finally, poverty is everyone's problem, not just women's. When women are impoverished, so are their children. Moreover, many of the poor in our society are men, especially the chronically unemployed, older persons, the homeless, persons with disabilities, and men of color who have spent their adult lives without hope of finding work.

Race/Ethnicity According to some stereotypes, most of the poor and virtually all welfare recipients are people of color. However, this stereotype is false; white Americans (non-Latinos/as) account for approximately two-thirds of those below the official poverty line. However, such stereotypes are perpetuated because a disproportionate percentage of the impoverished in the United States are made up of African Americans, Latinos/as, and Native Americans. About 24 percent of African Americans and 22 percent of Latinas/os were among the officially poor in 2007, as compared with about 8 percent of non-Latino/a whites (DeNavas-Walt, Proctor, and Smith, 2008).

Native Americans are among the most severely disadvantaged persons in the United States. About one-third live below the poverty line, and some of these individuals live in conditions of extreme poverty. For example, a congressional study found that 70 percent of the Navajo households in Arizona, New Mexico, and Utah lived in houses without running water, sewer facilities, or electricity (Benokraitis, 1999). Some Native Americans receive governmental assistance; many others do not fall within the officially defined criteria for social welfare assistance.

Economic and Structural Sources of Poverty

Social inequality and poverty have both economic and structural sources. Unemployment is a major cause of contemporary poverty. Tough economic times provide fewer opportunities for individuals to get an entry-level position that may help them to gain a toehold in American society. Massive plant closings, as in the aftermath of the General Motors bankruptcy filing in 2009, contribute to a trickle-down effect that causes workers both within the auto industry and in other employment sectors throughout the nation to lose their jobs or take severe pay cuts.

Low wages paid for many jobs is another major cause: Half of all families living in poverty are headed by someone who is employed, and one-third of those family heads work full time. A person with full-time employment in a minimum-wage job cannot keep a family of four from sinking below the official poverty line.

Structural problems contribute to both unemployment and underemployment. In addition to massive layoffs due to economic difficulties in recent years, corporations for a number of years have been disinvesting in the United States, displacing millions of people from their jobs. Economists refer to this displacement as the *deindustrialization* of America (Bluestone and Harrison, 1982). Even as they have closed their U.S. factories and plants, many corporations have opened new facilities in other countries where "cheap labor" exists because people will, of necessity, work for lower wages. ***Job deskilling*—a reduction in the proficiency needed to perform a specific job that leads to a corresponding reduction in the wages for that job**—has resulted from the introduction of computers and other technology (Hodson and Parker, 1988). The shift from manufacturing to service occupations has resulted in the loss of higher-paying positions and their replacement with lower-paying and less secure positions that do not offer the wages, job stability, or advancement potential of the disappearing manufacturing jobs. Many of the new jobs are located in the suburbs, thus making them inaccessible to central-city residents.

The problems of unemployment, underemployment, and poverty-level wages are even greater for people of color and young people in declining central cities. The unemployment rate for African Americans is almost double that of whites (U.S. Bureau of Labor Statistics, 2007b). African Americans have also experienced gender differences in employment that may produce different types of economic vulnerability. African American men who find employment typically earn more than African American women; however, the men's employment is often less secure. In recent decades, African American men have been more likely to lose their jobs because of declining employment in the manufacturing sector (Bane, 1986; Collins, 1990; Bane and Ellwood, 1994).

feminization of poverty the trend in which women are disproportionately represented among individuals living in poverty.

job deskilling a reduction in the proficiency needed to perform a specific job that leads to a corresponding reduction in the wages for that job.

Solving the Poverty Problem

The United States has attempted to solve the poverty problem in several ways. One of the most enduring is referred to as social welfare. When most people think of "welfare," they think of food stamps and programs such as Temporary Assistance for Needy Families (TANF) or the earlier program it replaced, Aid to Families with Dependent Children (AFDC). However, the primary beneficiaries of social welfare programs are not poor. Some analysts estimate that approximately 80 percent of all social welfare benefits are paid to people who do not qualify as "poor." For example, many recipients of Social Security are older people in middle- and upper-income categories.

When older persons, including members of Congress, accept Social Security payments, they are not stigmatized. Similarly, veterans who receive benefits from the Veterans Benefits Administration are not viewed as "slackers," and farmers who profit because of price supports are not considered to be lazy and unwilling to work. Unemployed workers who receive unemployment compensation are viewed with sympathy because of the financial plight of their families. By contrast, poor women and children who receive minimal benefits from welfare programs tend to be stigmatized and sometimes humiliated, even when our nation describes itself as having compassion for the less fortunate (see "Sociology Works!").

© AP Images/Paul Beaty

This Illinois Link card represents a modern approach to helping people of limited income purchase groceries. Data-encoded cards such as this one were developed to prevent the trading or selling of traditional food stamps.

Sociological Explanations of Social Inequality in the United States

Obviously, some people are disadvantaged as a result of social inequality. Therefore, is inequality always harmful to society?

Functionalist Perspectives

According to the sociologists Kingsley Davis and Wilbert Moore (1945), inequality is not only inevitable but also necessary for the smooth functioning of society. The Davis–Moore thesis, which has become the definitive functionalist explanation for social inequality, can be summarized as follows:

1. All societies have important tasks that must be accomplished and certain positions that must be filled.
2. Some positions are more important for the survival of society than others.
3. The most important positions must be filled by the most qualified people.
4. The positions that are the most important for society and that require scarce talent, extensive training, or both must be the most highly rewarded.
5. The most highly rewarded positions should be those that are functionally unique (no other position can perform the same function) and on which other positions rely for expertise, direction, or financing.

Davis and Moore use the physician as an example of a functionally unique position. Doctors are very important to society and require extensive training, but individuals would not be motivated to go through years of costly and stressful medical training without incentives to do so. The Davis–Moore thesis assumes that social stratification results in ***meritocracy*—a hierarchy in which all positions are rewarded based on people's ability and credentials.**

Critics have suggested that the Davis–Moore thesis ignores inequalities based on inherited wealth and intergenerational family status (Rossides, 1986). The thesis assumes that economic rewards and prestige are the only effective motivators for people and fails to take into account other intrinsic aspects of work, such as self-fulfillment (Tumin, 1953). It also does not ad-

Sociology *Works!*

Reducing Structural Barriers to Achieving the American Dream

> Our society recognizes a moral obligation to provide a helping hand to those in need, but those in poverty have been getting only the back of the hand. They receive little or no public assistance. Instead, they are scolded and told that they have caused their own misfortunes. This is our "compassion gap"—a deep divide between our moral commitments and how we actually treat those in poverty.

In this statement, the sociologists Fred Block, Anna C. Korteweg, and Kerry Woodward (2008: 166) describe the contradiction between our nation's alleged moral commitment to alleviating poverty and how we actually treat people who live in poverty. Children, single mothers with children, and people of color (particularly African Americans and Latinos/as) make up a disproportionate segment of the nation's poorest groups, and individuals in these categories are the persons most disadvantaged by arguments asserting that the poor have no one but themselves to blame for their poverty.

Numerous sociological studies regarding wealth and poverty demonstrate how structural factors contribute to the ability of some individuals to achieve the American Dream whereas other people are hampered in achieving that goal by factors that are beyond their control. Yet, according to the Economic Mobility Project, policy makers do little to alleviate poverty in the United States because of the widely held belief that the American Dream should provide everyone with equality of *opportunity* but not necessarily equality of *outcome:* "The belief in America as a land of opportunity may also explain why rising inequality in the United States has yielded so little in terms of responsiveness from policy makers: if the American Dream is alive and well, then there is no need for government intervention to smooth the rough edges of capitalism. Diligence and skill, the argument goes, will yield a fair distribution of rewards" (Pew Charitable Trusts, 2007). However, research continues to reveal that structural factors beyond the control of individuals are important in determining where a person's place will be in the U.S. class system.

Can we keep the American Dream alive for all? According to Block, Korteweg, and Woodward (2008), we must take a number of specific steps to revitalize the American Dream for more people in this country and to reverse the compassion gap. We must make people aware of how far social reality has departed from the ideals of the American Dream. As a nation, we must also take action to deal with the costs of four critical services that have risen much more rapidly than people's wages and the rate of inflation. These four critical services are *health care, higher education, high-quality child care,* and *housing.* As sociologists and other social analysts have suggested, if we as a nation are to claim that we have a commitment to compassion, we must make it our collective responsibility to help remove the structural barriers that currently reduce opportunities, mobility, and a chance for a better way of life for millions of Americans: "True compassion requires that we build a society in which every person has a first chance, a second chance, and, if needed, a third and fourth chance, to achieve the American Dream. We . . . need to use every instrument we have—faith groups, unions, community groups, and most of all government programs—to address the structural problems that reproduce poverty in our affluent society" (Block, Korteweg, and Woodward, 2008: 175).

Reflect & Analyze

Consider the area where you live. Do you know of people there whose situations could be improved if they were given a greater opportunity to take charge of their lives? How could such a change happen?

equately explain how such a reward system guarantees that the most qualified people will gain access to the most highly rewarded positions.

Conflict Perspectives

From a conflict perspective, people with economic and political power are able to shape and distribute the rewards, resources, privileges, and opportunities in society for their own benefit. Conflict theorists do not believe that inequality serves as a motivating force for people; they argue that powerful individuals and groups use ideology to maintain their favored positions at the expense of others. Core values in the United States emphasize the importance of material possessions, hard work, individual initiative to get ahead, and behavior that supports the existing social structure. These same values support the prevailing resource distribution system and contribute to social inequality.

meritocracy a hierarchy in which all positions are rewarded based on people's ability and credentials.

Are wealthy people smarter than others? According to conflict theorists, certain stereotypes suggest that this is the case; however, the wealthy may actually be "smarter" than others only in the sense of having "chosen" to be born to wealthy parents from whom they could inherit assets. Conflict theorists also note that laws and informal social norms support inequality in the United States. For the first half of the twentieth century, both legalized and institutionalized segregation and discrimination reinforced employment discrimination and produced higher levels of economic inequality. Although laws have been passed to make these overt acts of discrimination illegal, many forms of discrimination still exist in educational and employment opportunities.

Symbolic Interactionist Perspectives

Symbolic interactionists focus on microlevel concerns and usually do not analyze larger structural factors that contribute to inequality and poverty. However, many significant insights on the effects of wealth and poverty on people's lives and social interactions can be derived from applying a symbolic interactionist approach. Using qualitative research methods and influenced by a symbolic interactionist approach, researchers have collected the personal narratives of people across all social classes, ranging from the wealthiest to the poorest people in the United States.

© Steven Rubin/The Image Works

According to a functional perspective, people such as these Harvard Law School graduates attain high positions in society because they are the most qualified and they work the hardest. Is our society a meritocracy? How would conflict theorists answer this question?

A few studies provide rare insights into the social interactions between people from vastly divergent class locations. Sociologist Judith Rollins's (1985) study of the relationship between household workers and their employers is one example. Based on in-depth interviews and participant observation, Rollins examined rituals of deference that were often demanded by elite white women of their domestic workers, who were frequently women of color. According to the sociologist Erving Goffman (1967), *deference* is a type of ceremonial activity that functions as a symbolic means whereby appreciation is regularly conveyed to a recipient. In fact, deferential behavior between nonequals (such as employers and employees) confirms the inequality of the relationship and each party's position in the relationship relative to the other. Rollins identified three types of linguistic deference between domestic workers and their employers: use of the first names of the workers, contrasted with titles and last names (Mrs. Adams, for example) of the employers; use of the term *girls* to refer to female household workers regardless of their age; and deferential references to employers, such as "Yes, ma'am." Spatial demeanor, including touching and how close one person stands to another, is an additional factor in deference rituals across class lines. Rollins (1985: 232) concludes that

> The employer, in her more powerful position, sets the essential tone of the relationship; and that tone . . . is one that functions to reinforce the inequality of the relationship, to strengthen the employer's belief in the rightness of her advantaged class and racial position, and to provide her with justification for the inegalitarian social system.

Many concepts introduced by the sociologist Erving Goffman (1959, 1967) could be used as springboards for examining microlevel relationships between inequality and people's everyday interactions. What could you learn about class-based inequality in the United States by using a symbolic interactionist approach to examine a setting with which you are familiar?

The Concept Quick Review summarizes the three major perspectives on social inequality in the United States.

U.S. Stratification in the Future

The United States is facing one of the greatest economic challenges it has experienced since the Great

CONCEPT QUICK REVIEW

Sociological Explanations of Social Inequality in the United States	
Functionalist Perspectives	Some degree of social inequality is necessary for the smooth functioning of society (in order to fill the most important functions) and thus is inevitable.
Conflict Perspectives	Powerful individuals and groups use ideology to maintain their favored positions in society at the expense of others, and wealth is not necessary in order to motivate people.
Symbolic Interactionist Perspectives	The beliefs and actions of people reflect their class location in society.

Depression, in the 1930s. Although we have strong hopes that the American Dream will remain alive and well, many people are concerned that the current economic slump will create a lack of upward mobility for Americans. The nationwide slump in housing and jobs has distressed people across all income levels, and rising rates of unemployment and a shifting stock market bring about weekly predictions that things are either getting better or becoming worse.

So this brings us to an important final question in this chapter: Will social inequality in the United States increase, decrease, or remain the same in the future? Many social scientists believe that existing trends point to an increase in social inequality. First, the purchasing power of the dollar has stagnated or declined since the early 1970s. As families started to lose ground financially, more family members (especially women) entered the labor force in an attempt to support themselves and their families (Gilbert, 2008). Economist and former Secretary of Labor Robert Reich (1993) has noted that in recent years the employed have been traveling on two escalators—one going up and the other going down. The gap between the earnings of workers and the income of managers and top executives has widened even in an era when large salaries and even larger bonuses for CEOs have been frowned on by political leaders and everyday people.

Second, wealth continues to become more concentrated at the top of the U.S. class structure. As the rich have grown richer, more people have found themselves among the ranks of the poor. Third, federal tax laws in recent years have benefited corporations and wealthy families at the expense of middle- and lower-income families, and even if changes are made in the tax code, wealthier individuals and corporations typically find new ways in which to shelter their incomes. Finally, as previously mentioned, structural sources of upward mobility are shrinking, whereas the rate of downward mobility has increased.

Are we sabotaging our future if we do not work constructively to eliminate poverty? It has been said that a chain is no stronger than its weakest link. If we apply this idea to the problem of poverty, then it is to our advantage to see that those who cannot find work or do not have a job that provides a living wage receive adequate training and employment. Innovative programs can combine job training with producing something useful to meet the immediate needs of people living in poverty. Children of today—the adults of tomorrow—need nutrition, education, health care, and safety as they grow up (see Box 8.4).

Some social analysts believe that the United States will become a better nation if it attempts to regain the American Dream by attacking poverty. According to the sociologist Michael Harrington (1985: 13), if we join in solidarity with the poor, we will "rediscover our own best selves . . . we will regain the vision of America."

Box 8.4 You Can Make a Difference

Feeding the Hungry

The great fear among us all is that we are going to have to feed even more people. . . . It's not enough to just hand food out anymore.

—Robert Egger, director of the nonprofit Central Kitchen in Washington, D.C. (qtd. in Clines, 1996)

Egger is one of the people responsible for an innovative chef's training program that feeds hope as well as hunger.

© Joe Sohm/The Image Works

Many community volunteers try to make a difference during holiday seasons by providing food for people who otherwise might have none. Serving dinner for the homeless at a Los Angeles mission is an example. However, some programs empower homeless persons by teaching them kitchen arts so that they can prepare food for themselves and for others.

At the Central Kitchen, located in the nation's capital, staff and guest chefs annually train around 48 homeless persons in three-month-long kitchen-arts courses. While the trainees are learning about food preparation, which will help them get starting jobs in the restaurant industry, they are also helping feed about 3,000 homeless persons each day. Much of the food is prepared using donated goods such as turkeys that people have received as gifts at office parties and given to the kitchen, and leftover food from grocery stores including 7-Eleven stores, restaurants such as Pizza Hut, hotel food services, and college cafeterias. Central Kitchen got its start using leftovers from then-President George H. W. Bush's inaugural banquet in the late 1980s (Clines, 1996). Recently, donated food has gotten a boost from the Good Samaritan law passed in 1996, which exempts nonprofit organizations and *gleaners*—volunteers who collect what is left in the field after harvesting—from liability for problems with food that they contribute in good faith (see Burros, 1996).

Can you think of ways that leftover food could be recovered from places where you eat so the food could be redistributed to persons in need? Have you thought about suggesting that members of an organization to which you belong might donate their time to help the Salvation Army, Red Cross, or other voluntary organization to collect, prepare, and serve food to others? If you would like to know more,

- "A Citizens Guide to Food Recovery" is available to help individuals participate in food recovery. Call 800-GLEAN-IT, or call the Salvation Army in your community. On the Internet: **http://www.salvationarmy.org**
- World Hunger Year has projects such as Reinvesting in America that try to end hunger: **http://www.whyhunger.org**
- Contact the American Red Cross: **http://www.redcross.org**

Chapter Review

- **What is social stratification, and how does it affect our daily life?**

Social stratification is the hierarchical arrangement of large social groups based on their control over basic resources. People are treated differently based on where they are positioned within the social hierarchies of class, race, gender, and age.

- **What are the major systems of stratification?**

Stratification systems include slavery, caste, and class. Slavery, an extreme form of stratification in which people are owned by others, is a closed system. The caste system is also a closed one in which people's status is determined at birth based on their parents' position in society. The class system, which exists in the United States, is a type of stratification based on ownership of resources and on the type of work people do.

- **How did classical sociologists such as Karl Marx and Max Weber view social class?**

Karl Marx and Max Weber acknowledged social class as a key determinant of social inequality and social change. For Marx, people's relationship to the means of production determines their class position. Weber developed a multidimensional concept of stratification that focuses on the interplay of wealth, power, and prestige.

- **What are some of the consequences of inequality in the United States?**

The stratification of society into different social groups results in wide discrepancies in income and wealth and in variable access to available goods and services. People with high income or wealth have greater opportunity to control their own lives. People with less income have fewer life chances and must spend their limited resources to acquire basic necessities.

- **How do sociologists view poverty?**

Sociologists distinguish between absolute poverty and relative poverty. Absolute poverty exists when people do not have the means to secure the basic necessities of life. Relative poverty exists when people may be able to afford basic necessities but still are unable to maintain an average standard of living.

- **Who are the poor?**

Age, gender, and race tend to be factors in poverty. Children have a greater risk of being poor than do the elderly, while women have a higher rate of poverty than do men. Although whites account for approximately two-thirds of those below the poverty line, people of color account for a disproportionate share of the impoverished in the United States.

- **What is the functionalist view on class?**

Functionalist perspectives view classes as broad groupings of people who share similar levels of privilege on the basis of their roles in the occupational structure. According to the Davis–Moore thesis, stratification exists in all societies, and some inequality is not only inevitable but also necessary for the ongoing functioning of society. The positions that are most important within society and that require the most talent and training must be highly rewarded.

- **What is the conflict view on class?**

Conflict perspectives on class are based on the assumption that social stratification is created and maintained by one group (typically the capitalist class) in order to enhance and protect its own economic interests. Conflict theorists measure class according to people's relationships with others in the production process.

- **What is the symbolic interactionist view on class?**

Unlike functionalist and conflict perspectives that focus on macrolevel inequalities in societies, symbolic interactionist views focus on microlevel inequalities such as how class location may positively or negatively influence one's identity and everyday social interactions. Symbolic interactionists use terms such as *social cohesion* and *deference* to explain how class binds some individuals together while categorically separating out others.

www.cengage.com/login

Want to maximize your online study time? Take this easy-to-use study system's diagnostic pre-test, and it will create a personalized study plan for you. By helping you to identify the topics that you need to understand better and then directing you to valuable online resources, it can speed up your chapter review. CengageNOW even provides a post-test so you can confirm that you are ready for an exam.

Key Terms

absolute poverty 268
alienation 248
capitalist class (bourgeoisie) 248
caste system 245
class conflict 249
class system 247
feminization of poverty 270
income 260
intergenerational mobility 244
intragenerational mobility 244
job deskilling 271
life chances 243
meritocracy 272
pink-collar occupations 255
power 252
prestige 252
relative poverty 268
slavery 245
social mobility 244
social stratification 243
socioeconomic status (SES) 252
wealth 251
working class (proletariat) 248

Questions for Critical Thinking

1. Based on the Weberian and Marxian models of class structure, what is the class location of each of your ten closest friends or acquaintances? What is their location in relationship to yours? To one another's? What does their location tell you about friendship and social class?
2. Should employment be based on meritocracy, need, or affirmative action policies?
3. What might happen in the United States if the gap between rich and poor continues to widen?

The Kendall Companion Website

www.cengage.com/sociology/kendall

Visit this book's companion website, where you'll find more resources to help you study and successfully complete course projects. Resources include quizzes and flash cards, as well as special features such as an interactive sociology timeline, maps, General Social Survey (GSS) data, and Census 2000 data. The site also provides links to useful websites that have been selected for their relevance to the topics in this chapter and include those listed below. (*Note:* Visit the book's website for updated URLs.)

U.S. Census Bureau

http://www.census.gov

The U.S. Census Bureau publishes numerous reports that provide comprehensive national data on income, poverty, and other measures that depict the class structure of the United States. In addition, the much-used reference *Statistical Abstract of the United States* is available online at this site.

Inequality.org

http://www.demos.org/inequality

Inequality.org is a nonprofit organization whose goal is to disseminate information and ideas that are not widely publicized in the media. The site provides a wealth of information on inequality in the United States with regard to health care, technology, economics, education, and other contemporary issues.

Institute for Research on Poverty (IRP)

http://www.irp.wisc.edu

Established in 1966 at the University of Wisconsin–Madison by the U.S. Office of Economic Opportunity, IRP is a national nonprofit and nonpartisan organization dedicated to researching the causes and consequences of poverty and social inequality in the United States. The site is an excellent source for statistics, links, news, and research on national issues such as health, education, welfare reform, low-wage workers, and child support.

chapter

9 Global Stratification

Chapter Focus Question

How do global stratification and economic inequality affect the life chances of people around the world?

Marathon running has taken me a long way from my roots in the small town of Baringo in Kenya's Rift Valley. I grew up knowing what it was like to be poor and hungry. Whenever I come to London or other cities in the developed world to compete in marathons, I enter a different universe where choice, opulence and opportunity characterize people's lives.

It has been fascinating to follow the debate in Britain about school meals. I have listened to the arguments about whether children should be allowed to eat Turkey Twizzlers, or beefburgers and chips. I wish it could be the same the world over. While nutrition is a serious matter for any child, for me and my classmates [in Kenya] it was never really a case of what we might choose to eat, but rather whether we would eat at all.

The success story of marathon world-record-holder Paul Tergat, who grew up poor and hungry, calls our attention to issues of global stratification and inequality.

IN THIS CHAPTER

- Wealth and Poverty in Global Perspective
- Problems in Studying Global Inequality
- Classification of Economies by Income
- Measuring Global Wealth and Poverty
- Global Poverty and Human Development Issues
- Theories of Global Inequality
- Global Inequality in the Future

Most kids in Baringo had to help their families earn a living. Education was out of the question or, at best, something only one child in the family could pursue. For the lucky ones like me, who could go to school, the three-mile trek each morning on an empty stomach made it difficult, and sometimes impossible, to concentrate on lessons.

When I was eight, that changed. The United Nations began distributing food at the schools in the area and a heavy burden was lifted from our shoulders. My friends and I no longer worried about being hungry in class. We ate a simple meal each day and could stay focused during lessons. . . . I often ask myself: without the benefit of school meals, would I have become a literate, healthy, successful long-distance runner?

—Paul Tergat (2005), a world record holder in the marathon and winner of two silver Olympic medals, describing his early childhood, marked by poverty and hunger in Kenya

Marathoner Paul Tergat speaks for millions of people around the world who have experienced poverty and hunger. In his role as Ambassador Against Hunger for the World Food Program, Tergat encourages others to get involved in campaigns against hunger, illiteracy, pollution, homelessness, and other problems that limit people's life chances and opportunities. He also highlights the fact that although students in

Sharpening Your Focus

- *What is global stratification, and how does it contribute to economic inequality?*
- *How are global poverty and human development related?*
- *What is modernization theory, and what are its stages?*
- *How do conflict theorists explain patterns of global stratification?*

high-income nations have many food choices, some of which may be bad for them, students in low-income nations have very little food and extremely limited choices in life without intervention from the outside (Hattori, 2006).

Regardless of where people live in the world, social and economic inequalities are pressing daily concerns. Poverty and inequality know no political boundaries or national borders. In this chapter, we examine global stratification and inequality, and discuss perspectives that have been developed to explain the nature and extent of this problem. Before reading on, test your knowledge of global wealth and poverty by taking the quiz in Box 9.1.

Wealth and Poverty in Global Perspective

What do we mean by global stratification? *Global stratification* refers to the unequal distribution of wealth, power, and prestige on a global basis, resulting in people having vastly different lifestyles and life chances both within and among the nations of the world. Just as the United States is divided into classes, the world is divided into unequal segments characterized by extreme differences in wealth and poverty. For example, the income gap between the richest and the poorest 20 percent of the world population continues to widen (see ▶ Figure 9.1). However, when we compare social and economic inequality within other nations, we find gaps that are more pronounced than they are in the United States.

As previously defined, *high-income countries* are nations characterized by highly industrialized economies; technologically advanced industrial, administrative, and service occupations; and relatively high levels of national and per capita (per person) income. In contrast, *middle-income countries* are nations with industrializing economies, particularly in urban areas, and moderate levels of national and personal income. *Low-income countries* are primarily agrarian nations with little industrialization and low levels of national and personal income. Within some nations, the poorest one-fifth of the population has an income that is only a slight fraction of the overall average per capita income for that country. For example, in Brazil, Bolivia, and Honduras, less than 3 percent of total national income accrues to the poorest one-fifth of the population (World Bank, 2005).

Just as the differences between the richest and poorest people in the world have increased, the gap

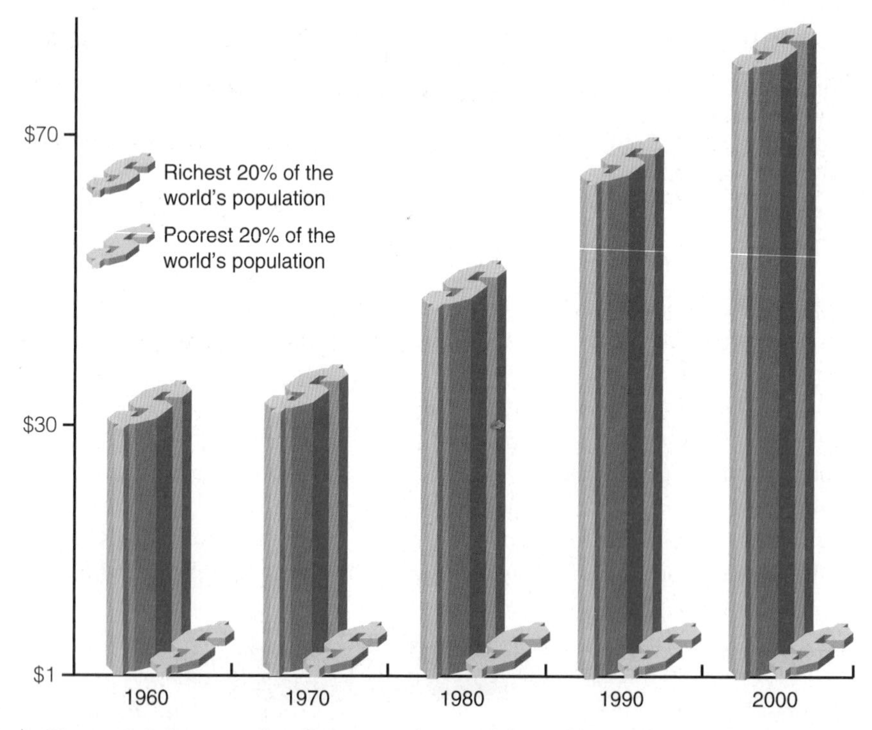

▶ **Figure 9.1 Income Gap Between the World's Richest and Poorest People**
The income gap between the richest and poorest people in the world continued to grow between 1960 and 2000. As this figure shows, in 1960 the highest-income 20 percent of the world's population received $30 for each dollar received by the lowest-income 20 percent. By 2000, the disparity had increased: $74 to $1.

Sources: International Monetary Fund, 1992; United Nations Development Programme, 2003.

Box 9.1 Sociology and Everyday Life

How Much Do You Know About Global Wealth and Poverty?

True	False	
T	F	1. The world's ten richest people are U.S. citizens.
T	F	2. The assets of the 200 richest people are more than the combined income of over 40 percent of the world's population.
T	F	3. More than one billion people worldwide live below the international poverty line, earning less than $1.25 each day.
T	F	4. Although poverty is a problem in most areas of the world, relatively few people die of causes arising from poverty.
T	F	5. In low-income countries, the problem of poverty is unequally shared between men and women.
T	F	6. The majority of people with incomes below the poverty line live in urban areas of the world.
T	F	7. Most analysts agree that the World Bank was created to serve the poor of the world and their borrowing governments.
T	F	8. Poor people in low-income countries meet most of their energy needs by burning wood, dung, and agricultural wastes, which increases health hazards and environmental degradation.

Answers on page 284.

in global income differences between rich and poor countries has continued to widen over the past fifty years. In 1960, the wealthiest 20 percent of the world population had more than thirty times the income of the poorest 20 percent. By 2000, the wealthiest 20 percent of the world population had almost eighty times the income of the poorest 20 percent (United Nations Development Programme, 2008). Income disparities *within* countries are even more pronounced.

Dramatic changes in the global economy in recent years have brought about an economic crisis that affects all of the nations of the world; however, this financial crisis had its origins in the United States, where a real estate asset bubble, fed by a boom in subprime mortgage lending, burst. At the same time, a number of major banks experienced liquidity and solvency problems. Prior to the 2008 global crisis, economic growth had been strong across many low- and middle-income countries even though the growth rate for high-income nations had continued to move downward. As a result of these economic problems, many analysts believe that the world may experience a global recession that will greatly affect rich and poor nations alike. As is generally true, however, they believe that the poorer nations will bear the larger impact of the financial crisis. For example, low-income economies are the most vulnerable to potential losses of official aid, pay for workers, and foreign direct investment. The slowdown in trade harms low-income nations such as Nigeria, Mongolia, Papua New Guinea, and Zimbabwe, which export commodities to middle- and high-income nations. According to the World Bank (2009), the global economic crisis may trap 46 million more people below the poverty line, which this organization now defines as living on less than $1.25 per day (the previous povery line was $1.00 per day). An additional 53 million people will be living on less than $2.00 a day. Overall, nearly 40 percent of low- and middle-income nations are highly exposed to the poverty effects of this crisis, and most of them do not have the ability to raise funds within their own economies or from international financial sources to reduce the severe effects that this downturn will have on their economies and on the lives of their people.

Many people have sought to address the issue of world poverty and to determine ways in which resources can be used to meet the urgent challenge of poverty. However, not much progress has been made on this front despite a great deal of talk and billions of dollars in "foreign aid" flowing from high-income to low-income nations. The idea of "development" has become the primary means used in attempts to reduce social and economic inequalities and alleviate the worst effects of poverty in the less industrialized nations of the world. Often, the nations that have not been able to reduce or eliminate poverty are chastised

Box 9.3 Sociology and Social Policy

Should We in the United States Do Something About Child Labor in Other Nations?

One of the half-dozen men and women sitting on a bench eating was a sinewy, bare-chested laborer in his late 30's named Mongkol Latlakorn. It was a hot, lazy day, and so we started chatting idly about the food and, eventually, our families. Mongkol mentioned that his daughter, Darin, was 15, and his voice softened as he spoke of her. She was beautiful and smart, and her father's hopes rested on her.

"Is she in school?" we asked.

"Oh, no," Mongkol said, his eyes sparkling with amusement. "She's working in a factory in Bangkok. She's making clothing for export to America." He explained that she was paid $2 a day for a nine-hour shift, six days a week.

"It's dangerous work," Mongkol added. "Twice the needles went right through her hands. But the managers bandaged up her hands, and both times she got better again and went back to work."

"How terrible," we murmured sympathetically.

Mongkol looked up, puzzled. "It's good pay," he said. "I hope she can keep that job. There's all this talk about factories closing now, and she said there are rumors that her factory might close. I hope that doesn't happen. I don't know what she would do then."

He was not, of course, indifferent to his daughter's suffering; he simply had a different perspective from ours—not only when it came to food but also when it came to what constituted desirable work.

—*New York Times* journalists Nicholas Kristof and Sheryl WuDunn (2000) explaining why some analysts believe that the campaign against sweatshops risks harming the very people it is intended to help

In the United States, as in most other high-income countries, most people believe that children as young as Darin should not be working in factories—whether in this country or in some low-income nation—and that they certainly should not be working long hours, six or seven days each week. Rather, we believe that (although they may hold a part-time job, such as delivering newspapers, mowing lawns, or babysitting) children should be in school, obtaining an education that will allow them to better themselves, and that they should have plenty of spare time to be, well, just to be *kids*.

Our nation's laws—at both federal and state levels—restrict the number of hours that people under age sixteen can work and also list certain occupations that are deemed to be too hazardous for young workers to perform. Those laws are designed to ensure that young people in this country have time to attend school and are not employed in jobs that are dangerous. Many other high- and middle-income nations have similar laws, although the ages and types of proscribed work vary from one nation to another.

But are we engaging in ethnocentrism if we try to impose such laws on people in other nations? Are we viewing their labor practices from our own high-income-nation

© Jenny Matthews/Alamy

In an effort to reduce poverty, some nations have developed adult literacy programs so that people can gain an education that will help lift them out of poverty. In regions such as Solomuna, Eritrea, women's literacy is a particularly crucial issue.

What is literacy, and why is it important for human development? The United Nations Educational, Scientific and Cultural Organization (UNESCO) defines a literate person as "someone who can, with understanding, both read and write a short, simple statement on their everyday life" (United Nations, 1997: 89). Based on this definition, people who can write only with figures, their name, or a memorized phrase are not considered literate. The adult literacy rate in the low-income countries (55.2 percent) is far less than that of the high-income countries (99 percent), and for women the rate is even lower. Women in the poorest nations have a literacy rate of 45.9 percent, as compared to 64.5 percent for men. By sharp contrast, women in high-income nations have a 98.7 percent literacy rate while men have a 99.3 percent rate (United Nations Development Programme, 2008). Women constitute about two-thirds of those who are illiterate: There are approximately 75 literate women for every 100 literate men (United Nations, 1997). Literacy is crucial for

sense of right and wrong rather than in the context of the economies in which these children labor? In some nations, child labor—even in factories and under dangerous conditions—is viewed as a necessity for the family's and the country's economic survival. Do we really know what is best for children in other nations? And even if we *would* impose our beliefs about child labor on people elsewhere, what can we do about child labor in those countries, anyway?

The United Nations has attempted to do something about exploitative or dangerous child labor. Article 32 of the Convention on the Rights of the Child recognizes the right of all children to be protected from any work that threatens their health, education, or development and requires nations to set minimum ages for employment and to regulate working conditions—and to enforce those requirements with appropriate penalties. The problem is that each nation sets its own minimum age and its own definition of exploitative working conditions, and each nation acts on its own in deciding how to enforce those laws. In many nations, the minimum age for full-time work is lower than the required age for compulsory education, producing an inherent conflict with the goal of protecting children from work that interferes with their education, and if a child's parents don't see as much value in an education as in the child producing income for the family, they may prefer for the child to earn even a small income that the family needs in order to survive (Siddiqi and Patrinos, 1995). As a result, in many nations children as young as eight or nine years old still work twelve- to fourteen-hour days in garment and other factories, on coffee plantations and in rice fields, and at looms making rugs for the homes and offices of more-affluent people in other nations (Sloan, 2005).

One way that people in many nations, including the United States, have attempted to reduce or eliminate exploitative or dangerous child labor is by boycotting—refusing to buy—products from companies and nations that permit such practices. Boycotts and public pressure in general can produce results, but the results can be both good and bad. For example, when the U.S. Congress considered a bill that would have taken punitive action against companies benefiting from child labor, the threat of such a measure panicked the garment industry in Bangladesh, where child workers were fired by many garment manufacturers in order to avoid possible sanctions.

Some analysts believe that the best way to end abusive child labor is to increase the educational opportunities for children in low-income countries so that parents realize that having their children attend school instead of working full time will result in those children having the chance to get a better-paying job than the parents hold. To accomplish such a goal, however, we would have to be willing to pay more for the goods we buy that are manufactured or grown in low-income nations so that the parents could earn enough not to need the children's wages for survival. It would also mean sending money to other countries—money that would be earmarked solely for use in educating the children in those nations.

Reflect & Analyze

Do you think we should be concerned about child labor practices in the United States and other nations? Is it important for us to reflect on what effects our purchasing habits might have on other people worldwide?

women because it has been closely linked to decreases in fertility, improved child health, and increased earnings potential.

Persistent Gaps in Human Development

Some middle- and lower-income countries have made progress in certain indicators of human development. The gap between some richer and middle- or lower-income nations has narrowed significantly for life expectancy, adult literacy, and daily calorie supply; however, the overall picture for the world's poorest people remains dismal. The gap between the poorest nations and the middle-income nations has continued to widen. Poverty, food shortages, hunger, and rapidly growing populations are pressing problems for at least 1.8 billion people, most of them women and children living in a state of absolute poverty. Although more women around the globe have paid employment than in the past, more and more women are still finding themselves in poverty because of increases in single-person and single-parent households headed by women and the fact that low-wage work is often the only source of livelihood available to them. According to an analyst for the Inter-American Development Bank, women experience sexual discrimination not only in terms of employment but also in wages:

> In Honduras, for example, coffee and tobacco farmers prefer to hire girls and women as laborers because they are willing to accept low wages and are more reliable workers. Especially in poor countries, female labor is primarily sought for low-paid positions in services, agriculture, small-scale commerce, and in the growing, unregulated manufacturing and agribusiness industries, which pay their workers individual rather than family wages, offer seasonal or part-time employment, and carry few or no benefits. Hence, this explains the seemingly

contradictory trends of women's increased economic participation alongside their growing impoverishment. (Buvinić, 1997: 47)

Theories of Global Inequality

Why is the majority of the world's population growing richer while the poorest 20 percent—more than one billion people—are so poor that they are effectively excluded from even a moderate standard of living? Social scientists have developed a variety of theories that view the causes and consequences of global inequality somewhat differently. We will examine the development approach and modernization theory, dependency theory, world systems theory, and the new international division of labor theory.

Development and Modernization Theory

According to some social scientists, global wealth and poverty are linked to the level of industrialization and economic development in a given society. Although the process by which a nation industrializes may vary somewhat, industrialization almost inevitably brings with it a higher standard of living in a nation and some degree of social mobility for individual participants in the society. Specifically, the traditional caste system becomes obsolete as industrialization progresses. Family status, race/ethnicity, and gender are said to become less significant in industrialized nations than in agrarian-based societies. As societies industrialize, they also urbanize as workers locate their residences near factories, offices, and other places of work. Consequently, urban values and folkways overshadow the beliefs and practices of the rural areas. Analysts using a development framework typically view industrialization and economic development as essential steps that nations must go through in order to reduce poverty and increase life chances for their citizens.

Earlier in the chapter, we discussed the post–World War II Marshall Plan, under which massive financial aid was provided to the European nations to help rebuild infrastructure lost in the war. Based on the success of this infusion of cash in bringing about modernization, President Truman and many other politicians and leaders in the business community believed that it should be possible to help so-called underdeveloped nations modernize in the same manner.

The most widely known development theory is ***modernization theory*—a perspective that links global inequality to different levels of economic development and suggests that low-income economies can move to middle- and high-income economies by achieving self-sustained economic growth.** According to modernization theory, the low-income, less-developed nations can improve their standard of living only with a period of intensive economic growth and accompanying changes in people's beliefs, values, and attitudes toward work. As a result of modernization, the values of people in developing countries supposedly become more similar to those of people in high-income nations. The number of hours that people work at their jobs each week is one measure of the extent to which individuals subscribe to the *work ethic,* a core value widely believed to be of great significance in the modernization process.

Perhaps the best-known modernization theory is that of Walt W. Rostow (1971, 1978), who, as an economic advisor to U.S. President John F. Kennedy, was highly instrumental in shaping U.S. foreign policy toward Latin America in the 1960s. To Rostow, one of the largest barriers to development in low-income nations was the traditional cultural values held by people, particularly beliefs that are fatalistic, such as viewing extreme hardship and economic deprivation as inevitable and unavoidable facts of life. In cases of fatalism, people do not see any need to work in order to improve their lot in life: If it is predetermined for them, why bother? Based on modernization theory, poverty can be attributed to people's cultural failings, which are further reinforced by governmental policies interfering with the smooth operation of the economy.

Rostow suggested that all countries go through four stages of economic development, with identical content, regardless of when these nations started the process of industrialization. He compared the stages of economic development to an airplane ride. The first stage is the *traditional stage,* in which very little social change takes place, and people do not think much about changing their current circumstances. According to Rostow, societies in this stage are slow to change because the people hold a fatalistic value system, do not subscribe to the work ethic, and save very little money. The second stage is the *take-off stage*—a period of economic growth accompanied by a growing belief in individualism, competition, and achievement. During this stage, people start to look toward the future, to save and invest money, and to discard traditional values. According to Rostow's modernization theory, the development of capitalism is essential for the transformation from a traditional, simple society to a modern, complex one. With the financial help and advice of the high-income countries, low-income countries will eventually be able to "fly" and enter the third stage of economic development. In the third stage, the country moves toward *technological maturity.* At this point, the

© Chris Hondros/Getty Images

Poverty and war continue to devastate low-income countries such as Afghanistan. Many low-income countries receive aid from industrialized nations through initiatives such as the World Food Program. Modernization theory links global inequality to levels of economic development, but factors such as war and internal conflict also greatly contribute to patterns of global inequality.

country will improve its technology, reinvest in new industries, and embrace the beliefs, values, and social institutions of the high-income, developed nations. In the fourth and final stage, the country reaches the phase of *high mass consumption* and a correspondingly high standard of living.

Modernization theory has had both its advocates and its critics. According to proponents of this approach, studies have supported the assertion that economic development occurs more rapidly in a capitalist economy. In fact, the countries that have been most successful in moving from low- to middle-income status typically have been those that are most centrally involved in the global capitalist economy. For example, the nations of East Asia have successfully made the transition from low-income to higher-income economies through factors such as a high rate of savings, an aggressive work ethic among employers and employees, and the fostering of a market economy.

Critics of modernization theory point out that it tends to be Eurocentric in its analysis of low-income countries, which it implicitly labels as backward (see Evans and Stephens, 1988). In particular, modernization theory does not take into account the possibility that all nations do not industrialize in the same manner. In contrast, some analysts have suggested that modernization of low-income nations today will require novel policies, sequences, and ideologies that are not accounted for by Rostow's approach.

Which sociological perspective is most closely associated with the development approach? Modernization theory is based on a market-oriented perspective which assumes that "pure" capitalism is good and that the best economic outcomes occur when governments follow the policy of laissez-faire (or hands-off) business, giving capitalists the opportunity to make the "best" economic decisions, unfettered by government restraints or cumbersome rules and regulations (see Chapter 13, "The Economy and Work in Global Perspective"). In today's global economy, however, many analysts believe that national governments are no longer central corporate decision makers and that transnational corporations determine global economic expansion and contraction. Therefore, corporate decisions to relocate manufacturing processes around the world make the rules and regulations of any one nation irrelevant and national boundaries obsolete (Gereffi, 1994). Just as modernization theory most closely approximates a functionalist approach to explaining inequality, dependency theory, world systems theory, and the new international division of labor theory are perspectives rooted in the conflict approach. All four of these approaches are depicted in ▶ Figure 9.3.

Dependency Theory

***Dependency theory* states that global poverty can at least partially be attributed to the fact that the low-income countries have been exploited by the high-income countries.** Analyzing events as part of a particular historical process—the expansion of global capitalism—dependency theorists see the greed of the rich countries as a source of increasing impoverishment of the poorer nations and their people. Dependency theory disputes the notion of the development approach, and modernization theory specifically, that economic growth is the key to meeting important human needs in societies. In contrast, the poorer nations are trapped in a cycle of structural dependency on the

modernization theory a perspective that links global inequality to different levels of economic development and suggests that low-income economies can move to middle- and high-income economies by achieving self-sustained economic growth.

dependency theory the belief that global poverty can at least partially be attributed to the fact that the low-income countries have been exploited by the high-income countries.

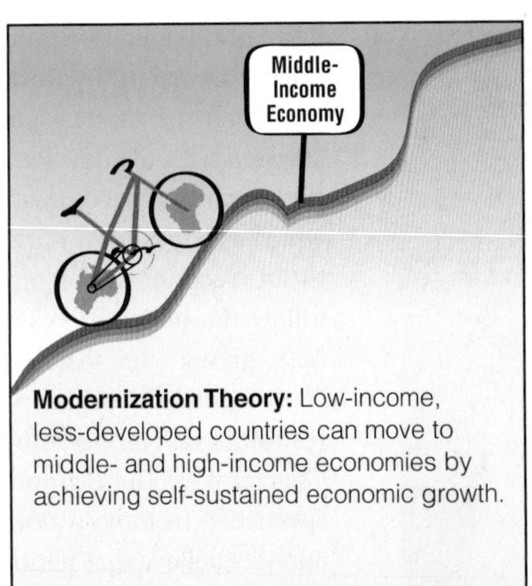

Modernization Theory: Low-income, less-developed countries can move to middle- and high-income economies by achieving self-sustained economic growth.

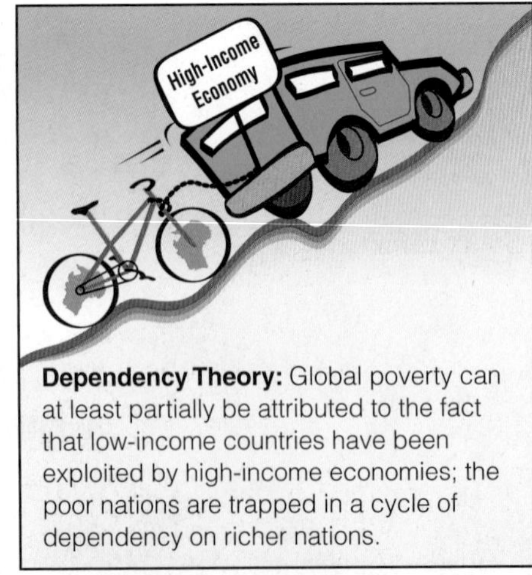

Dependency Theory: Global poverty can at least partially be attributed to the fact that low-income countries have been exploited by high-income economies; the poor nations are trapped in a cycle of dependency on richer nations.

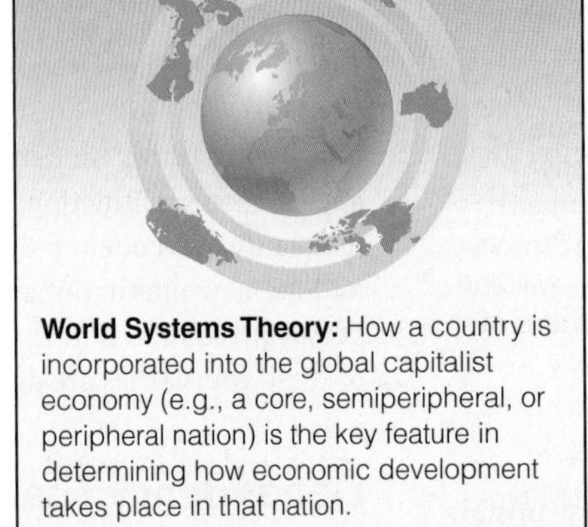

World Systems Theory: How a country is incorporated into the global capitalist economy (e.g., a core, semiperipheral, or peripheral nation) is the key feature in determining how economic development takes place in that nation.

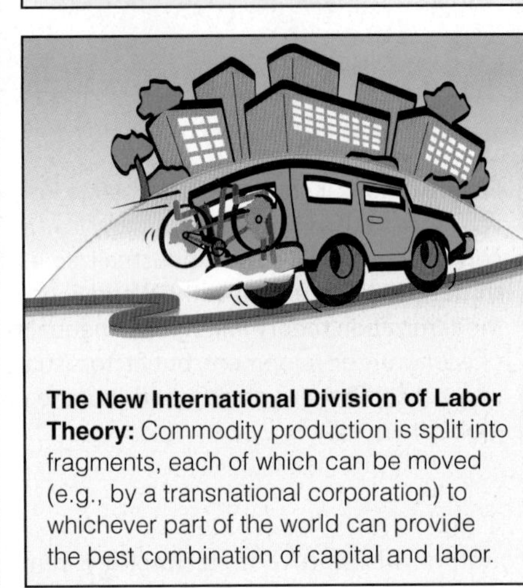

The New International Division of Labor Theory: Commodity production is split into fragments, each of which can be moved (e.g., by a transnational corporation) to whichever part of the world can provide the best combination of capital and labor.

▶ **Figure 9.3 Approaches to Studying Global Inequality**
What causes global inequality? Social scientists have developed a variety of explanations, including the four theories shown here.

richer nations due to their need for infusions of foreign capital and external markets for their raw materials, making it impossible for the poorer nations to pursue their own economic and human development agendas. For this reason, dependency theorists believe that countries such as Brazil, Nigeria, India, and Kenya cannot reach the sustained economic growth patterns of the more-advanced capitalist economies.

Dependency theory has been most often applied to the newly industrializing countries (NICs) of Latin America, whereas scholars examining the NICs of East Asia found that dependency theory had little or no relevance to economic growth and development in that part of the world. Therefore, dependency theory had to be expanded to encompass transnational economic linkages that affect developing countries, including foreign aid, foreign trade, foreign direct investment, and foreign loans. On the one hand, in Latin America and sub-Saharan Africa, transnational linkages such as foreign aid, investments by transnational corporations, foreign debt, and export trade have been significant impediments to development within a country. On the other hand, East Asian countries such as Taiwan, South Korea, and Singapore have historically also had high rates of dependency on foreign aid, foreign trade, and interdependence with transnational corporations but have still experienced high rates of economic growth despite dependency. According to the sociologist Gary Gereffi (1994), differences in outcome are probably associated with differences in the timing and sequencing of a nation's relationship with external entities such as foreign governments and transnational corporations. However, in her study of the satellite factory system in Taiwan, the sociologist Ping-Chun Hsiung found that managers in the plants use the norms and behavior patterns of Chinese society to derive high profits for the capitalists. As one manager explained to Hsiung,

> We managers are the mediators between the boss and the workers. We have to communicate [the workers'] opinion to the boss, while at the same time watching out for the boss's pocket. . . . It isn't really a bad

thing if the workers are not happy with their wages. It implies that they have the potential to be worth more. How to manipulate them all depends on us, the managers. . . . Chinese are humble. We seldom talk about how good we are, not to mention boast. When things come down to wage conflicts, I always turn the issue around by asking the workers to give me a figure. That is, I ask them to tell me exactly how much more they believe their labor is worth. If they can't come out with a concrete price, then, they have to listen to me. I may decide to give them a raise. I may not. It's all up to me. Even if they do give me a price, the chances are it will always be lower than the real value of their labor. For example, if what they really want is a one-hundred-dollar raise, as Chinese they will only say that their work is worth eighty dollars more, at the most. When this happens, I can really cut it to fifty dollars. . . . By handling them this way, their productivity will go up because I do show them that I did recognize their unrest [and give them a raise]. . . . It is a win–win battle for the company when things come down to wage conflict, you know. (qtd. in Hsiung, 1996: 62)

Dependency theory makes a positive contribution to our understanding of global poverty by noting that "underdevelopment" is not necessarily the cause of inequality. Rather, it points out that exploitation not only of one country by another but of countries by transnational corporations may limit or retard economic growth and human development in some nations. However, what remains unexplained is how East Asia and India had successful "dependency management" whereas many Latin American countries did not (Gereffi, 1994). Although some analysts have found dependency theory to be useful, other conflict theorists have sought to explain global inequality by using the framework of world systems theory or the new international division of labor theory.

World Systems Theory

World systems theory suggests that what exists under capitalism is a truly global system that is held together by economic ties. From this approach, global inequality does not emerge solely as a result of the exploitation of one country by another. Instead, economic domination involves a complex world system in which the industrialized, high-income nations benefit from other nations and exploit their citizens. This theory is most closely associated with the sociologist Immanuel Wallerstein (1979, 1984), who believed that a country's mode of incorporation into the capitalist work economy is the key feature in determining how economic development takes place in that nation. According to world systems theory, the capitalist world economy is a global system divided into a hierarchy of three major types of nations—core, semiperipheral, and peripheral—in which upward or downward mobility is conditioned by the resources and obstacles that characterize the international system. ***Core nations* are dominant capitalist centers characterized by high levels of industrialization and urbanization.** Core nations such as the United States, Japan, and Germany possess most of the world's capital and technology. Even more importantly for their position of domination, they exert massive control over world trade and economic agreements across national boundaries. Some cities in core nations are referred to as *global cities* because they serve as international centers for political, economic, and cultural concerns. New York, Tokyo, and London are the largest global cities, and they are often referred to as the "command posts" of the world economy.

Even within core nations such as Germany, which typically is viewed as a large, wealthy nation that represents Europe's largest economy, many people suffer from poverty. Rising rates of unemployment during the global economic crisis have brought about increases in the number of individuals living in poverty, with children especially hard hit by the economic downturn. It is important to note that even though millions of people live very well in core nations, others within those countries lead lives that are more comparable to their counterparts in semiperipheral nations.

***Semiperipheral nations* are more developed than peripheral nations but less developed than core nations.** Nations in this category typically provide labor and raw materials to core nations within the world system. These nations constitute a midpoint between the core and peripheral nations that promotes the stability and legitimacy of the three-tiered world economy. These nations include South Korea and Taiwan in East Asia, Mexico and Brazil in Latin America, India in South Asia, and Nigeria and South Africa in Africa. Only two global cities are located in semiperipheral nations: São Paulo, Brazil, which is the center of the Brazilian economy, and Singapore, which is the economic center of a multicountry region in Southeast Asia. According to Wallerstein, semiperipheral nations exploit

core nations according to world systems theory, dominant capitalist centers characterized by high levels of industrialization and urbanization.

semiperipheral nations according to world systems theory, nations that are more developed than peripheral nations but less developed than core nations.

© James Marshall/The Image Works

A variety of factors—such as foreign investment and the presence of transnational corporations—have contributed to the economic growth of nations such as Singapore.

peripheral nations, just as the core nations exploit both the semiperipheral and the peripheral.

Most low-income countries in Africa, South America, and the Caribbean are ***peripheral nations*—nations that are dependent on core nations for capital, have little or no industrialization (other than what may be brought in by core nations), and have uneven patterns of urbanization.** According to Wallerstein (1979, 1984), the wealthy in peripheral nations benefit from the labor of poor workers and from their own economic relations with core nation capitalists, whom they uphold in order to maintain their own wealth and position. At a global level, uneven economic growth results from capital investment by core nations; disparity between the rich and the poor within the major cities in these nations is increased in the process. The U.S./Mexican border is an example of disparity and urban growth: Transnational corporations have built *maquiladora* plants so that goods can be assembled by low-wage workers to keep production costs down. ▶ Figure 9.4 describes this process. In 2001 there were almost 3,800 maquiladora plants in Mexico, employing 1.3 million people (*Austin American-Statesman,* 2001). Because of a demand for a large supply of low-wage workers, thousands of people moved from the rural regions of Mexico to urban areas along the border in hope of earning a higher wage. This influx pushed already overcrowded cities far beyond their capacity. Many people live on the edge of the city in *shantytowns* made from discarded materials or in low-cost rental housing in central-city slums because their wages are low and affordable housing is nonexistent (Flanagan, 2002). In fact, housing shortages are among the most pressing problems in many peripheral nations. Recently, the government of Mexico discontinued reporting on maquiladora plants as separate from other companies doing business in that country. As a result, it is more difficult for social scientists to study the effect of these plants on workers and on the national economies in which these businesses are located.

Not all social analysts agree with Wallerstein's perspective on the hierarchical position of nations in the global economy. However, most scholars acknowledge that nations throughout the world are influenced by a relatively small number of cities and transnational corporations that have prompted a shift from an international to a more global economy (see Knox and Taylor, 1995; Wilson, 1997). Even Wallerstein (1991) acknowledges that world systems theory is an "incomplete, unfinished critique" of long-term, large-scale social change that influences global inequality.

The New International Division of Labor Theory

Although the term *world trade* has long implied that there is a division of labor between societies, the nature and extent of this division have recently been reassessed based on the changing nature of the world economy. According to the *new international division of labor theory,* commodity production is being split into fragments that can be assigned to whichever part of the world can provide the most profitable combination of capital and labor. Consequently, the new international division of labor has changed the pattern of geographic specialization between countries, whereby high-income countries have now become dependent on low-income countries for labor. The low-income countries provide transnational corporations with a situation in which they can pay lower wages and taxes and face fewer regulations regarding workplace conditions and environmental protection (Waters, 1995). Overall, a global manufacturing system has emerged in which transnational corporations establish labor-intensive, assembly-oriented export production, ranging from textiles and clothing to technologically sophisticated exports such as computers, in middle- and lower-income nations (Gereffi, 1994). At the same time, manufacturing technologies are shifting from the large-scale, mass-production assembly lines of the past toward a more flexible production process involving microelectronic technologies. Even service industries—such as processing insurance claims forms—that were formerly thought to be less mobile have become exportable through electronic transmis-

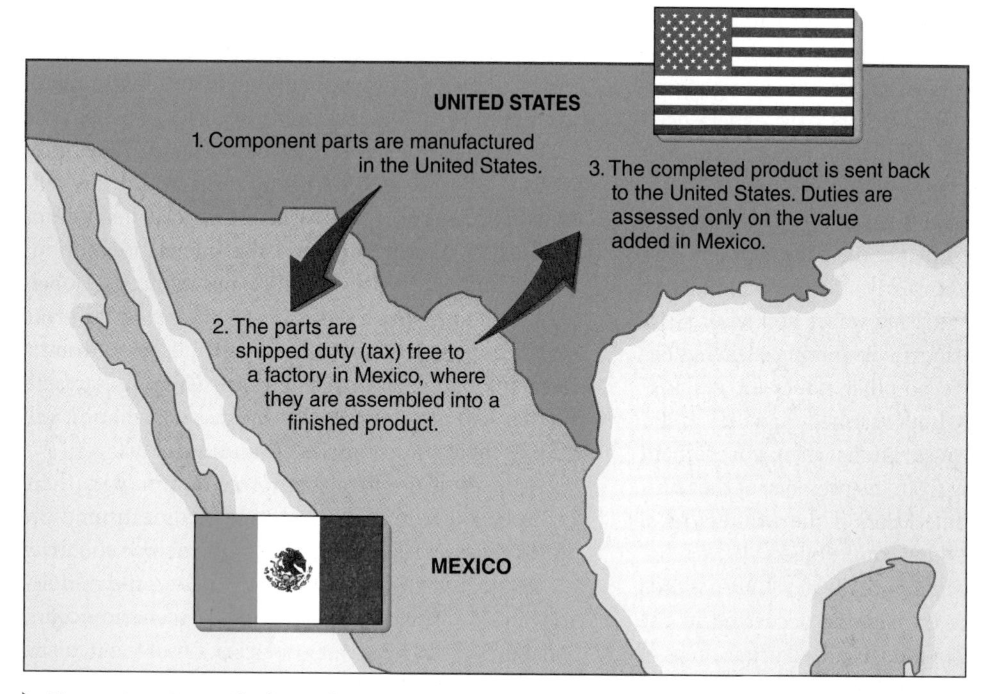

▶ **Figure 9.4 Maquiladora Plants**
Here is the process by which transnational corporations establish plants in Mexico so that profits can be increased by using low-wage workers there to assemble products that are then brought into the United States for sale.

sion and the Internet. The global nature of these activities has been referred to as *global commodity chains,* a complex pattern of international labor and production processes that results in a finished commodity ready for sale in the marketplace.

Some commodity chains are producer-driven, whereas others are buyer-driven. *Producer-driven commodity chains* is the term used to describe industries in which transnational corporations play a central part in controlling the production process. Industries that produce automobiles, computers, and other capital- and technology-intensive products are typically producer-driven. In contrast, *buyer-driven commodity chains* is the term used to refer to industries in which large retailers, brand-name merchandisers, and trading companies set up decentralized production networks in various middle- and low-income countries. This type of chain is most common in labor-intensive, consumer-goods industries such as toys, garments, and footwear (Gereffi, 1994). Athletic footwear companies such as Nike and Reebok are examples of the buyer-driven model. Because these products tend to be labor intensive at the manufacturing stage, the typical factory system is very competitive and globally decentralized. Workers in buyer-driven commodity chains are often exploited by low wages, long hours, and poor working conditions. In fact, most workers cannot afford the products they make. Tini Heyun Alwi, who works on the assembly line of the shoe factory in Indonesia that makes Reebok sneakers, is an example: "I think maybe I could work for a month and still not be able to buy one pair" (qtd. in Goodman, 1996: F1). Because Tini earns only 2,600 Indonesian rupiah ($1.28) per day working a ten-hour shift six days a week, her monthly income would fall short of the retail price of the athletic shoes (Goodman, 1996). Sociologist Gary Gereffi (1994: 225) explains the problem with studying the new global patterns as follows:

> The difficulty may lie in the fact that today we face a situation where (1) the political unit is *national,* (2) industrial production is *regional,* and (3) capital movements are *international.* The rise of Japan and the East Asian [newly industrializing countries] in the 1960s and 1970s is the flip side of the "deindustrialization" that occurred in the United States and much of Europe. Declining industries in North America have been the growth industries in East Asia.

peripheral nations according to world systems theory, nations that are dependent on core nations for capital, have little or no industrialization (other than what may be brought in by core nations), and have uneven patterns of urbanization.

As other analysts suggest, these changes have been a mixed bag for people residing in these countries. For example, Indonesia has been able to woo foreign business into the country, but workers have experienced poverty despite working full time in factories making such consumer goods as Nike tennis shoes (Gargan, 1996). As employers feel pressure from workers to raise wages, clashes erupt between the workers and managers or owners. Similarly, the governments in these countries fear that rising wages and labor strife will drive away the businesses, sometimes leaving behind workers who have no other hopes for employment and become more impoverished than they previously were. Moreover, in situations where government officials were benefiting from the presence of the companies, they also become losers if the workers rebel against their pay or working conditions.

Although most discussions of the new international division of labor focus on changes occurring in the lives of people residing in industrialized urban areas of developing nations, millions of people continue to live in grinding poverty in rural regions of these countries (see "Sociology Works!").

Global Inequality in the Future

As we have seen, social inequality is vast both within and among the countries of the world. Even in high-income nations where wealth is highly concentrated, many poor people coexist with the affluent. In middle- and low-income countries, there are small pockets of wealth in the midst of poverty and despair.

What are the future prospects for greater equality across and within nations? Not all social scientists agree on the answer to this question. Depending on the theoretical framework they apply in studying global inequality, social analysts may describe either an optimistic or a pessimistic scenario for the future. Moreover, some analysts highlight the human rights issues embedded in global inequality, whereas others focus primarily on an economic framework.

In some regions, persistent and growing poverty continues to undermine human development and future possibilities for socioeconomic change. Gross inequality has high financial and quality-of-life costs to people, even among those who are not the poorest of the poor. In the future, continued population growth, urbanization, environmental degradation, and violent conflict threaten even the meager living conditions of those residing in low-income nations. From this approach, the future looks dim not only for people in low-income and middle-income countries but also for those in high-income countries, who will see their quality of life diminish as natural resources are depleted, the environment is polluted, and high rates of immigration and global political unrest threaten the high standard of living that many people have previously enjoyed. According to some social analysts, transnational corporations and financial institutions such as the World Bank and the International Monetary Fund will further solidify and control a globalized economy, which will transfer the power to make significant choices to these organizations and away from the people and their governments. As a result, further loss of resources and means of livelihood will affect people and countries around the globe.

As a result of global corporate domination, there could be a leveling out of average income around the world, with wages falling in the high-income countries and wages increasing significantly in low- and middle-income countries. If this pessimistic scenario occurs, there is likely to be greater polarization of the rich and the poor and more potential for ethnic and national conflicts over such issues as worsening environmental degradation and who has the right to natural resources. For example, pulp-and-paper companies in Indonesia, along with palm oil plantation owners, have continued clearing land for crops by burning off vast tracts of jungle, producing high levels of smog and pollution across seven Southeast Asian nations and creating havoc for millions of people (Mydans, 1997b).

On the other hand, a more optimistic scenario is also possible. With modern technology and worldwide economic growth, it might be possible to reduce absolute

The current trend of using grain-based ethanol to power cars and trucks has proved to be a financial boon to some U.S. farmers but has had another effect as well: The price of grain has increased worldwide, making it even harder for people in developing countries to procure enough food to eat.

Sociology *Works!*

Why *Place* Matters in Global Poverty

We're deadly poor. We grow just enough food for ourselves to eat, with no surplus grain. We don't have to pay the grain tax anymore, but our lives aren't much better.

—Zhou Zhiwen, a woman who lives in Yangmiao, China, describing what rural poverty is like in her village (qtd. in French, 2008: YT4)

Zhou Zhiwen lives in an area of China that has largely been untouched by the economic boom in her country. Even with the recent abolition of agricultural taxes for people who are impoverished, local villagers such as Zhou continue to live in abject poverty. As more people have risen out of poverty in China's urban centers in recent decades, poverty in the rural areas, mountainous regions, and deserts remains severe. According to some villagers, the central government is "out of touch with rural realities in places like this," and officials have made little effort to take care of the rural poor (French, 2008). China's poverty is widespread, and the income gap between rural and urban residents has widened over the past three decades (IFAD, 2002). Many people live close to, or below, the minimum standard for poverty: Approximately 350 million people in China live below the international poverty line of $1.25 per day (World Bank, 2007, 2009).

For many years, sociologists studying poverty have focused on differences in rural and urban poverty. Throughout the world, they have found that *place* does matter when it comes to finding the deepest pockets of poverty (Rural Policy Research Institute, 2004). Where people live strongly influences how much money they will make, and income inequalities are important indicators of the life chances of entire families. In some developing countries, the rural poor rely primarily on agriculture, fishing, forestry, and sometimes small-scale industries and services for their livelihood (Khan, 2001). When they are unable to derive sufficient economic resources from these endeavors, little else is available to them. Some migrate to urban centers in hopes of finding new opportunities, but many remain behind, living in grinding poverty. When sociologists speak of "place," they are referring to such things as an area's natural environment, which includes its climate, natural resources, and degree of isolation (Rural Policy Research Institute, 2004). Place also involves the economic structure in the area, such as the extent to which adequate amounts of food can be raised to meet people's needs, or whether an individual can earn sufficient money to purchase food.

Can rural poverty be reduced? According to some social policy analysts, broad economic stability, competitive markets, and public investment in *physical* and *social* infrastructure are important prerequisites for a reduction in rural poverty in developing nations (Khan, 2001). From this perspective, a major reduction in rural poverty in China will occur only if people have access to land and credit, education, health care, support services, and entitlements to food through well-designed public works programs and other transfer mechanisms.

Reflect & Analyze

Whether changes that reduce poverty will occur in China's future remains to be seen, but the sociological premise that *place* matters in regard to poverty remains a valid assumption in helping us explain global poverty and inequality. Can you apply this idea to rural and urban areas with which you are familiar in the United States? Why are issues such as this important to each of us even if we do not live in a rural area and have no personal experience with poverty?

poverty and to increase people's opportunities. Among the trends cited by the Human Development Report (United Nations Development Programme, 2008) that have the potential to bring about more sustainable patterns of development are the socioeconomic progress made in many low- and middle-income countries over the past thirty years as technological, social, and environmental improvements have occurred. For example, technological innovation continues to improve living standards for some people. Fertility rates are declining in some regions (but remain high in others, where there remains grave cause for concern about the availability of adequate natural resources for the future). Finally, health and education may continue to improve in lower-income countries. According to the Human Development Report (United Nations Development Programme, 2008), healthy, educated populations are crucial for the future in order to reduce global poverty. The education of women is of primary importance in the future if global inequality is to be reduced. As one analyst stated, "If you educate a boy, you educate a human being. If you educate a girl, you educate generations" (Buvinić, 1997: 49). All aspects of schooling and training are crucial for the future, including agricultural extension services in rural areas to help women farmers in regions such as western Kenya produce more crops to feed their families. As we saw earlier in the chapter, easier access to water can make a crucial

Box 9.4 You Can Make a Difference

Global Networking to Reduce World Hunger and Poverty

We, the people of the world, will mobilize the forces of transnational civil society behind a widely shared agenda that binds our many social movements in pursuit of just, sustainable, and participatory human societies. In so doing we are forging our own instruments and processes for redefining the nature and meaning of human progress and for transforming those institutions that no longer respond to our needs. We welcome to our cause all people who share our commitment to peaceful and democratic change in the interest of our living planet and the human societies it sustains.

—International NGO Forum, United Nations Conference on Environment and Development, Rio de Janeiro, Brazil, June 12, 1992 (qtd. in Korten, 1996: 333)

If everyone lit just one little candle, what a bright world this would be.

—line from the 1950s theme song for Bishop Fulton J. Sheen's television series *Life Is Worth Living* (Sheen, 1995: 245)

Willie Colon, a Puerto Rican salsa star and spokesperson for CARE, visited this Bolivian classroom as part of that international relief organization's project to reduce women's poverty.

When many of us think about problems such as world poverty, we tend to see ourselves as powerless to bring about change in so vast an issue. However, a recurring message from social activists and religious leaders is that each person can contribute something to the betterment of other people and sometimes the entire world.

An initial way for each of us to become involved is to become more informed about global issues and to learn how we can contribute time and resources to organizations seeking to address social issues such as illiteracy and hunger. We can also find out about meetings and activities of organizations and participate in online discussion forums where we can express our opinions, ask questions, share information, and interact with other people interested in topics such as international relief and development. At first, it may not feel like you are doing much to address global problems; however, information and education are the first steps to promoting greater understanding of social problems and of the world's people, whether they reside in high-, middle-, or low-income countries and regardless of their individual socioeconomic position. Likewise, it is important to help our own nation's children understand that they can make a difference in ending hunger in the United States and other nations.

Would you like to function as a catalyst for change? You can learn how to proceed by gathering information from organizations that seek to reduce problems such as poverty and to provide forums for interacting with other people. Here are a few starting points for your search:

- CARE International is a confederation of 10 national members in North America, Europe, Japan, and Australia. CARE assists the world's poor in their efforts to achieve social and economic well-being. Its work reaches 25 million people in 53 nations in Africa, Asia, Latin America, and Eastern Europe. Programs include emergency relief, education, health and population, children's health, reproductive health, water and sanitation, small economic activity development, agriculture, community development, and environment. Contact CARE at 151 Ellis Street, NE, Atlanta, GA 30303-2439. On the Internet: **http://www.care.org**

Other organizations fighting world hunger and health problems include the following:

- World Hunger Year (WHY): **http://www.whyhunger.org**
- "Kids Can Make a Difference," an innovative guide developed by WHY: **http://www.kidscanmakeadifference.org**
- World Health Organization: **http://www.who.int**

difference in people's lives. Mayra Buvinić, of the Inter-American Development Bank, puts global poverty in perspective for people living in high-income countries by pointing out that their problems are our problems. She provides the following example:

> Reina is a former guerilla fighter in El Salvador who is being taught how to bake bread under a post-civil war reconstruction program. But as she says, "the only thing I have is this training and I don't want to be a baker. I have other dreams for my life."
>
> Once upon a time, women like Reina . . . only migrated [to the United States] to follow or find a husband. This is no longer the case. It is likely that Reina, with few opportunities in her own country, will sooner or later join the rising number of female migrants who leave families and children behind to seek better paying work in the United States and other industrial countries. Wisely spent foreign aid can give Reina the chance to realize her dreams in her *own* country. (Buvinić, 1997: 38, 52)

From this viewpoint, we can enjoy prosperity only by ensuring that other people have the opportunity to survive and thrive in their own surroundings (see Box 9.4). The problems associated with global poverty are therefore of interest to a wide-ranging set of countries and people.

Chapter Review

- **What is global stratification, and how does it contribute to economic inequality?**

Global stratification refers to the unequal distribution of wealth, power, and prestige on a global basis, which results in people having vastly different lifestyles and life chances both within and among the nations of the world. Today, the income gap between the richest and the poorest 20 percent of the world population continues to widen, and within some nations the poorest one-fifth of the population has an income that is only a slight fraction of the overall average per capita income for that country.

- **How are global poverty and human development related?**

Income disparities are not the only factor that defines poverty and its effect on people. The United Nations' Human Development Index measures the level of development in a country through indicators such as life expectancy, infant mortality rate, proportion of underweight children under age five (a measure of nourishment and health), and adult literacy rate for low-income, middle-income, and high-income countries.

- **What is modernization theory?**

Modernization theory is a perspective that links global inequality to different levels of economic development and suggests that low-income economies can move to middle- and high-income economies by achieving self-sustained economic growth.

- **How does dependency theory differ from modernization theory?**

Dependency theory states that global poverty can at least partially be attributed to the fact that the low-income countries have been exploited by the high-income countries. Whereas modernization theory focuses on how societies can reduce inequality through industrialization and economic development, dependency theorists see the greed of the rich countries as a source of increasing impoverishment of the poorer nations and their people.

- **What is world systems theory, and how does it view the global economy?**

According to world systems theory, the capitalist world economy is a global system divided into a hierarchy of three major types of nations: Core nations are dominant capitalist centers characterized by high levels of industrialization and urbanization, semiperipheral nations are more developed than peripheral nations but less developed than core nations, and peripheral nations are those countries that are dependent on core nations for capital, have little or no industrialization (other than what may be brought in by core nations), and have uneven patterns of urbanization.

- **What is the new international division of labor theory?**

The new international division of labor theory is based on the assumption that commodity production is split into fragments that can be assigned to whichever part of the world can provide the most profitable combination of capital and labor. This division of labor has changed the pattern of geographic specialization between countries, whereby high-income countries have become dependent on low-income countries for labor. The low-income countries provide transnational corporations with a situation in which they can pay lower wages and taxes, and face fewer regulations regarding workplace conditions and environmental protection.

www.cengage.com/login

Want to maximize your online study time? Take this easy-to-use study system's diagnostic pre-test, and it will create a personalized study plan for you. By helping you to identify the topics that you need to understand better and then directing you to valuable online resources, it can speed up your chapter review. CengageNOW even provides a post-test so you can confirm that you are ready for an exam.

Key Terms

core nations 299
dependency theory 297
modernization theory 296
peripheral nations 300
semiperipheral nations 299
social exclusion 285

Questions for Critical Thinking

1. You have decided to study global wealth and poverty. How would you approach your study? What research methods would provide the best data for analysis? What might you find if you compared your research data with popular presentations—such as films and advertising—of everyday life in low- and middle-income countries?
2. How would you compare the lives of poor people living in the low-income nations of the world with those in central cities and rural areas of the United States? In what ways are their lives similar? In what ways are they different?
3. Should U.S. foreign policy include provisions for reducing poverty in other nations of the world? Should U.S. domestic policy include provisions for reducing poverty in the United States? How are these issues similar? How are they different?
4. Using the theories discussed in this chapter, devise a plan to alleviate global poverty. Assume that you have the necessary wealth, political power, and other resources necessary to reduce the problem. Share your plan with others in your class, and create a consolidated plan that represents the best ideas and suggestions presented.

The Kendall Companion Website

www.cengage.com/sociology/kendall

Visit this book's companion website, where you'll find more resources to help you study and successfully complete course projects. Resources include quizzes and flash cards, as well as special features such as an interactive sociology timeline, maps, General Social Survey (GSS) data, and Census 2000 data. The site also provides links to useful websites that have been selected for their relevance to the topics in this chapter and include those listed below. (*Note:* Visit the book's website for updated URLs.)

National Labor Committee
http://www.nlcnet.org

The mission of the National Labor Committee is to educate the public on human and labor rights abuses by corporations. Its website features numerous reports on contemporary slavery around the world, a photo gallery, information on how to become involved, and numerous links to sites displaying violations of human rights and labor rights.

The United Nations
http://www.un.org

The United Nations' website contains a wealth of information regarding social, economic, and political conditions around the world. The United Nations' publications on economic and social development are a good source of data about global stratification.

The World Bank Group
http://www.worldbank.org

The World Bank Group helps the poorest people and countries around the world by providing development assistance in areas such as education, health, debt relief, and the environment. Click on "Development Topics" and scroll down to "Poverty" to access publications and other information related to global poverty.

chapter 10 Race and Ethnicity

Chapter Focus Question

How have racial and ethnic relations improved in the United States over the past fifty years?

People could not even register to vote when we came back here in 1963 for the March on Washington. And I was here when Dr. King stood and said, "I have a dream today," a dream deeply rooted in the American dream. And to come back [to Washington for the inauguration of President Barack Obama] 45 years later, it is almost too much. It is almost too much.

—Representative John Lewis, Democrat of Georgia, recalling the struggles that African Americans have had in gaining a greater semblance of equality in the United States (qtd. in *New York Times,* 2009: P6)

The reality is just as it would not have happened without the sacrifices and struggles of the past, it also would not have happened without the idealism and the audacity of the present generation. The reality is

On January 20, 2009, Barack Obama was sworn in as the President of the United States, becoming the first person of color ever to hold that office. However, does his victory indicate that racism and discrimination have disappeared from U.S. society?

IN THIS CHAPTER

- Race and Ethnicity
- Prejudice
- Discrimination
- Sociological Perspectives on Race and Ethnic Relations
- Racial and Ethnic Groups in the United States
- Global Racial and Ethnic Inequality in the Future

that Obama was embraced by young people of all colors, before he was embraced by their parents and their elders.

—Ben Jealous, president of the NAACP, describing the election of Mr. Obama as the nation's first African American President (qtd. in *New York Times,* 2009: P6)

Stephen Johns, known as "Big John," was opening the door for a man he thought was just an elderly visitor to the U.S. Holocaust Museum in Washington when he was shot dead. . . . Mr. Johns was a security guard. The bullet that killed him was a reminder of the continuing menace of bigotry and violence that pervades this country—and that we insist on underestimating. The authorities have identified an 88-year-old, hard-core white supremacist . . . as the killer. Our knee-jerk tendency is to comfort ourselves by declaring that this guy is so freakish, so far out of the American mainstream, that he is not representative of much of anything. Sane people are not violently obsessed with blacks and Jews. The murder was a tragic aberration. After all, this is a country that only recently elected an African-American president. So, let's mop up the blood from the museum floor, and try to keep matters in perspective.

(Continued)

Sharpening Your Focus

- *How do race and ethnicity differ?*
- *How do discrimination and prejudice differ?*
- *How are racial and ethnic relations analyzed according to sociological perspectives?*
- *What are the unique experiences of racial and ethnic groups in the United States?*

. . . But we should not pretend that things are better than they are. Racism is still a powerful force in the U.S. . . . [A]nd murderous violence is as much of a problem as ever. . . . We need to be vigilant. When I first heard about the murder of Mr. Johns and the violent desecration of the Holocaust Museum, I thought of the Rev. Dr. Martin Luther King, Jr., and how much we miss his moral leadership. He fought not just for civil rights, but against violence and injustice of all kinds, and he warned us of the debilitating effects of unnecessary warfare. "He who passively accepts evil is as much involved in it as he who helps to perpetrate it," said Dr. King. The bullet that silenced him seemed to come out of nowhere, suddenly, aberrationally, like the bullet that destroyed Stephen Johns.

—Bob Herbert, a *New York Times* columnist, discussing a recent murder at the Holocaust Museum and explaining why he believes that the problem of racism is far from over in the United States despite the recent election of this country's first African American president (Herbert, 2009: A17)

What is the best reflection of contemporary racial and ethnic relations in the United States? Have things really changed since the 1950s and 1960s, when racism and overt discrimination were a way of life in many communities in this nation? The 2009 inauguration of Barack Obama as the first African American U.S. president is often referred to as a milestone in history, and newspaper articles frequently highlight the "rising sense of racial optimism" (Saulny, 2009: A1) that exists among people across racial/ethnic and class lines because of Mr. Obama's election. However, violent acts such as the murder at the Holocaust Museum continue to be perpetrated by people who apparently have racially biased motives, and many individuals around the nation indicate that they harbor no illusions that problems associated with race are over in this nation. Some people define recent changes as progress; as expressed by one individual, "I'm not saying that the playing field is even, but having elected a black president has done a lot [to improve race relations]" (qtd. in Saulny, 2009: A26). In this chapter, we examine prejudice, discrimination, sociological perspectives on race and ethnic relations, and commonalities and differences in the experiences of racial and ethnic groups in the United States. In the process, sports is used as an example of the effects of race and ethnicity on people's lives because this area of social life shows us how, for more than a century, people who have been singled out for negative treatment on the basis of their perceived race or ethnicity have sought to overcome prejudice and discrimination through their determination in endeavors that can be judged based on objective standards ("What's the final score?") rather than subjective standards ("Do I like this person based on characteristics such as race or ethnicity?"). Before reading on, test your knowledge about race, ethnicity, and sports by taking the quiz in Box 10.1.

Race and Ethnicity

What is race? Some people think it refers to skin color (the Caucasian "race"); others use it to refer to a religion (the Jewish "race"), nationality (the British "race"), or the entire human species (the human "race") (Marger, 2009). Popular usages of *race* have been based on the assumption that a race is a grouping or classification based on *genetic* variations in physical appearance, particularly skin color. However, social scientists and biologists dispute the idea that biological race is a

Box 10.1 Sociology and Everyday Life

How Much Do You Know About Race, Ethnicity, and Sports?

True	False	
T	F	1. Because sports are competitive and fans, coaches, and players want to win, the color of the players has not been a factor, only their performance.
T	F	2. In the late 1800s and early 1900s, boxing provided social mobility for some Irish, Jewish, and Italian immigrants.
T	F	3. African Americans who competed in boxing matches in the late 1800s often had to agree to lose before they could obtain a match.
T	F	4. Racially linked genetic traits explain many of the differences among athletes.
T	F	5. All racial and ethnic groups have viewed sports as a means to become a part of the mainstream.
T	F	6. Until recently, the positions of quarterback and kicker in the National Football League have been held almost exclusively by white players.
T	F	7. In recent years, players of color have moved into coaching, management, and ownership positions in professional sports.
T	F	8. The odds are good that many outstanding high school and college athletes will make the pros if they do not get injured.
T	F	9. Racism and sexism appear to be on the decline in sports in the United States.

Answers on page 312.

meaningful concept. In fact, the idea of race has little meaning in a biological sense because of the enormous amount of interbreeding that has taken place within the human population. For these reasons, sociologists sometimes place "race" in quotation marks to show that categorizing individuals and population groups on biological characteristics is neither accurate nor based on valid distinctions between the genetic makeup of differently identified "races." Today, sociologists emphasize that race is a *socially constructed reality,* not a biological one. From this approach, the social significance that people accord to race is more significant than any biological differences that might exist among people who are placed in arbitrary categories.

A ***race*** **is a category of people who have been singled out as inferior or superior, often on the basis of real or alleged physical characteristics such as skin color, hair texture, eye shape, or other subjectively selected attributes** (Feagin and Feagin, 2008). Categories of people frequently thought of as racial groups include Native Americans, Mexican Americans, African Americans, and Asian Americans. This classification is rooted in nineteenth-century distinctions made by some biologists, who divided the world's population into three racial categories: *Caucasian*—people characterized as having relatively light skin and fine hair; *Negroid*—people with darker skin and coarser, curlier hair; and *Mongoloid*—people with yellow or brown skin and distinctively shaped eyelids. However, racial categorization based on phenotypical differences (such as facial characteristics or skin color) does not correlate with genotypical differences (differences in genetic makeup). As anthropologists and biologists have acknowledged for some time, most of us have more in common genetically with individuals from another "race" than we have with the genetic average of people from our own "race." Moreover, throughout human history, extensive interbreeding has made such classifications unduly simplistic.

As compared with race, *ethnicity* defines individuals who are believed to share common characteristics that differentiate them from the other collectivities in a society. An ***ethnic group*** **is a collection of people distinguished, by others or by themselves, primarily on the basis of cultural or nationality characteristics** (Feagin and Feagin, 2008). Examples of ethnic groups include Jewish Americans, Irish Americans, and Italian Americans. Ethnic groups share five main

race a category of people who have been singled out as inferior or superior, often on the basis of physical characteristics such as skin color, hair texture, and eye shape.

ethnic group a collection of people distinguished, by others or by themselves, primarily on the basis of cultural or nationality characteristics.

Box 10.1 Sociology and Everyday Life

Answers to the Sociology Quiz on Race, Ethnicity, and Sports

1. False. Discrimination has been pervasive throughout the history of sports in the United States. For example, African American athletes, regardless of their abilities, were excluded from white teams for many years.

2. True. Irish Americans were the first to dominate boxing, followed by Jewish Americans and then Italian Americans. Boxing, like other sports, was a source of social mobility for some immigrants.

3. True. Promoters, who often set up boxing matches that pitted fighters by race, assumed that white fans were more likely to buy tickets if the white fighters frequently won.

4. False. Although some scholars and journalists have used biological or genetic factors to explain the achievements of athletes, sociologists view these explanations as being based on the inherently racist assumption that people have "natural" abilities (or disabilities) because of their race or ethnicity.

5. False. Some racial and ethnic groups—including Chinese Americans and Japanese Americans—have not viewed sports as a means of social mobility.

6. True. As late as the 1990s, whites accounted for about 90 percent of the quarterbacks and kickers on NFL teams. However, this changed early in the twenty-first century, and today there are some African Americans playing virtually every position on all professional football teams.

7. False. Although more African American players are employed by these teams (especially in basketball), their numbers have not increased significantly in coaching and management positions. Only one professional sports team is currently owned by an African American.

8. False. The odds of becoming a professional athlete are very low. The percentage of high school football players who make it to the pros is estimated at about .09 percent; for basketball, about .03 percent; and for baseball, about .5 percent. The percentages of college athletes who make it to the pros are a little higher: 5.8 percent for football, 2.9 percent for basketball, and 5.6 percent for baseball.

9. False. Even as people of color and white women have made gains on collegiate and professional teams, scholars have documented the continuing significance of racial and gender discrimination in sports.

Sources: Based on Coakley, 2004; Hight, 1994; Messner, Duncan, and Jensen, 1993; Nelson, 1994; and www.coasports.org, 2009.

characteristics: (1) *unique cultural traits,* such as language, clothing, holidays, or religious practices; (2) *a sense of community;* (3) *a feeling of ethnocentrism;* (4) *ascribed membership from birth;* and (5) *territoriality,* or the tendency to occupy a distinct geographic area (such as Little Italy or Little Havana) by choice and/or for self-protection. Although some people do not identify with any ethnic group, others participate in social interaction with individuals in their ethnic group and feel a sense of common identity based on cultural characteristics such as language, religion, or politics. However, ethnic groups are not only influenced by their own past history but also by patterns of ethnic domination and subordination in societies.

The Social Significance of Race and Ethnicity

Race and ethnicity take on great social significance because how people act in regard to these terms drastically affects other people's lives, including what opportunities they have, how they are treated, and even how long they live. According to the sociologists Michael Omi and Howard Winant, race "permeates every institution, every relationship, and every individual" in the United States:

> As we . . . compare real estate prices in different neighborhoods, select a radio channel to enjoy while we drive to work, size up a potential client, customer, neighbor, or teacher, stand in line at the unemployment office, or carry out a thousand other normal tasks, we are compelled to think racially, to use the racial categories and meaning systems into which we have been socialized. (Omi and Winant, 1994: 158)

Historically, stratification based on race and ethnicity has pervaded all aspects of political, economic, and social life. Consider sports as an example. Throughout the early history of the game of baseball, many African Americans had outstanding skills as players but were categorically excluded from Major League teams

© Jeff Greenberg/PhotoEdit

Miami's Little Havana is an ethnic enclave where people participate in social interaction with other individuals in their ethnic group and feel a sense of shared identity. Ethnic enclaves provide economic and psychological support for recent immigrants as well as for those who were born in the United States.

because of their skin color. Even after Jackie Robinson broke the "color line" to become the first African American in the Major Leagues in 1947, his experience was marred by racial slurs, hate letters, death threats against his infant son, and assaults on his wife (Ashe, 1988; Peterson, 1992). With some professional athletes from diverse racial–ethnic categories having multimillion-dollar contracts and lucrative endorsement deals, it is easy to assume that racism in sports is a thing of the past. However, this *commercialization* of sports does not mean that racial prejudice and discrimination no longer exist (Coakley, 2004).

Racial Classifications and the Meaning of Race

If we examine racial classifications throughout history, we find that in ancient Greece and Rome a person's race was the group to which she or he belonged, associated with an ancestral place and culture. From the Middle Ages until about the eighteenth century, a person's race was based on family and ancestral ties, in the sense of a *line,* to a national group. During the eighteenth century, physical differences such as the darker skin hues of Africans became associated with race, but racial divisions were typically based on differences in religion and cultural tradition rather than on human biology. With the intense (though misguided) efforts that surrounded the attempt to justify black slavery and white dominance in all areas of life during the second half of the nineteenth century, *races* came to be defined as distinct biological categories of people who were not all members of the same family but who shared inherited physical and cultural traits that were alleged to be different from those traits shared by people in other races. Hierarchies of races were established, placing the "white race" at the top, the "black race" at the bottom, and others in between.

However, racial classifications in the United States have changed over the past century. If we look at U.S. Census Bureau classifications, for example, we can see how the meaning of race continues to change. First, race is defined by perceived skin color: white or nonwhite. Whereas one category exists for "whites" (who vary considerably in actual skin color and physical appearance), all of the remaining categories are considered "nonwhite."

Second, racial purity is assumed to exist. Prior to the 2000 census, for example, the true diversity of the U.S. population was not revealed in census data because multiracial individuals were forced to either select a single race as being their "race" or to select the vague category of "other." Professional golfer Tiger Woods is an example of how people often have a mixed racial heritage. Woods describes himself as one-half Asian American (one-fourth Thai and one-fourth Chinese), one-eighth white, one-eighth Native American, and one-fourth African American (White, 1997). Census 2000 made it possible—for the first time—for individuals to classify themselves as being of more than one race (see "Census Profiles: Percentage Distribution of Persons Reporting Two or More Races").

Third, categories of official racial classifications may (over time) create a sense of group membership or "consciousness of kind" for people within a somewhat arbitrary classification. When people of European descent were classified as "white," some began to see themselves as different from "nonwhite." Consequently, Jewish, Italian, and Irish immigrants may have felt more a part of the Northern European white mainstream in the late eighteenth and early nineteenth centuries. Whether Chinese Americans, Japanese Americans, Korean Americans, and Filipino Americans come to think of themselves collectively as "Asian Americans" because of official classifications remains to be seen.

In the future, increasing numbers of children in the United States will be likely to have a mixed racial or ethnic heritage such as this student describes:

> I am part French, part Cherokee Indian, part Filipino, and part black. Our family taught us to be aware of all these groups, and just to be ourselves. But I have never known what I am. People have asked if I am a Gypsy, or a Portuguese, or a Mexican, or lots of other things. It seems to make people curious, uneasy, and sometimes belligerent. Students I don't even know stop me on campus and ask, "What are you anyway?" (qtd. in Davis, 1991: 133)

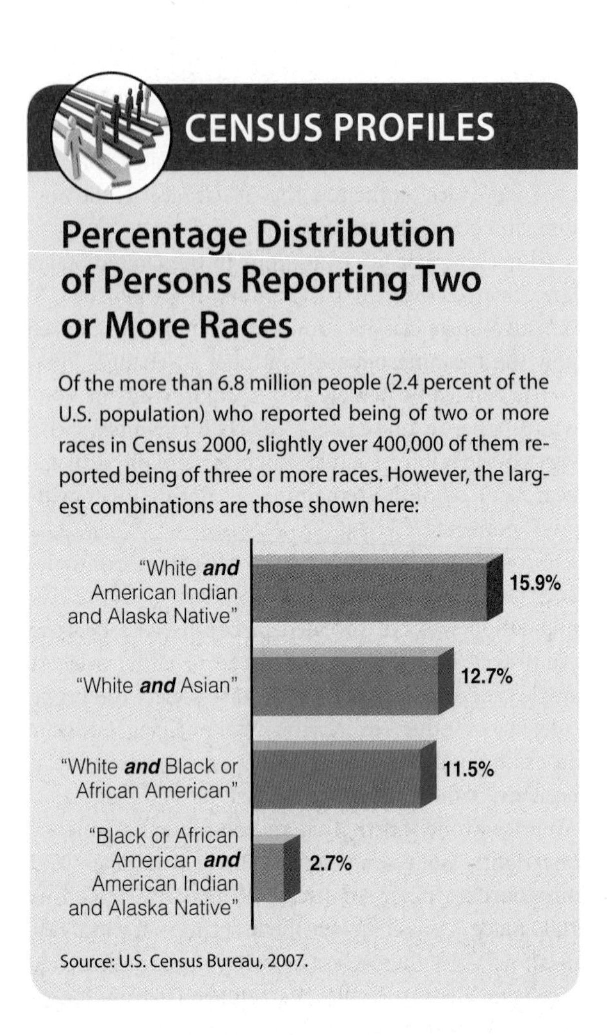

The way people are classified remains important because such classifications affect their access to employment, housing, social services, federal aid, and many other "publicly or privately valued goods" (Omi and Winant, 1994: 3). However, recent publicity about celebrities of mixed-race heritage, such as the golfer Tiger Woods, together with the change in the way that the Census Bureau asks about race, may lead more people to identify themselves as having more than one racial heritage.

Dominant and Subordinate Groups

The terms *majority group* and *minority group* are widely used, but their meanings are less clear as the composition of the U.S. population continues to change. Accordingly, many sociologists prefer the terms *dominant* and *subordinate* to identify power relationships that are based on perceived racial, ethnic, or other attributes and identities. To sociologists, a ***dominant group* is one that is advantaged and has superior resources and rights in a society** (Feagin and Feagin, 2008). In the United States, whites with Northern European ancestry (often referred to as Euro-Americans, white Anglo-Saxon Protestants, or WASPs) are considered to be a dominant group. A ***subordinate group* is one whose members, because of physical or cultural characteristics, are disadvantaged and subjected to unequal treatment by the dominant group and who regard themselves as objects of collective discrimination.** Persons of color and white women are considered to be subordinate-group members in the United States, particularly when these individuals are from lower-income categories.

It is important to note that, in the sociological sense, *group* as used in these two terms is misleading because people who merely share ascribed racial or ethnic characteristics do not constitute a group. However, the terms *dominant group* and *subordinate group* do give us a way to describe relationships of advantage/disadvantage and power/exploitation that exist in contemporary nations.

Prejudice

Although there are various meanings of the word *prejudice,* sociologists typically use it in a specialized sense. From a sociological perspective, ***prejudice* is a negative attitude based on faulty generalizations about members of selected racial and ethnic groups.** The term *prejudice* is from the Latin words *prae* ("before") and *judicium* ("judgment"), which means that people may be biased either for or against members of other groups even before they have had any contact with them. Although prejudice can be either *positive* (bias in favor of a group—often our own) or *negative* (bias against a group—one we deem less worthy than our own), it most often refers to the negative attitudes that people may have about members of other racial or ethnic groups.

Stereotypes

Prejudice is rooted in ethnocentrism and stereotypes. When used in the context of racial and ethnic relations, *ethnocentrism* refers to the tendency to regard one's own culture and group as the standard—and thus superior—whereas all other groups are seen as inferior. Ethnocentrism is maintained and perpetuated by ***stereotypes*—overgeneralizations about the appearance, behavior, or other characteristics of members of particular categories.** The term *stereotype* comes from the Greek word *stereos* ("solid") and refers to a fixed mental impression. Although stereotypes can be either positive or negative, examples of negative stereotyping abound in sports. Think about the Native American names, images, and mascots used by sports teams such as the Atlanta Braves, Cleveland Indians,

Contemporary prejudice and discrimination cannot be understood without taking into account the historical background. School integration in the 1950s was accomplished despite white resistance. Today, integration in education, housing, and many other areas of social life remains a pressing social issue.

Golden State Warriors, Kansas City Chiefs, and Washington Redskins. Members of Native American groups have been actively working to eliminate the use of stereotypic mascots (with feathers, buckskins, beads, spears, and "warpaint"), "Indian chants," and gestures (such as the "tomahawk chop"), which they claim trivialize and exploit Native American culture. College and university sports teams with Native American names and logos also remain the subject of controversy in the twenty-first century. The National Collegiate Athletic Association (NCAA) recently adopted a policy prohibiting colleges with hostile or abusive mascots, nicknames, or imagery on their team uniforms from hosting any NCAA championship competitions. According to sociologist Jay Coakley (2004), the use of stereotypes and words such as *redskin* symbolize a lack of understanding of the culture and heritage of native peoples and are offensive to many Native Americans. Although some people see the use of these names and activities as "innocent fun," others view them as a form of racism.

Racism

What is racism? ***Racism* is a set of attitudes, beliefs, and practices that is used to justify the superior treatment of one racial or ethnic group and the inferior treatment of another racial or ethnic group.** As discussed later, some forms of racism are based on biological arguments about the innate inferiority of members of certain racial and ethnic groups; however, more-recent forms of racism have attempted to justify the unequal treatment of the same groups of people on other grounds. The world has seen a long history of racism: It can be traced from the earliest civilizations. At various times throughout U.S. history, various categories of people, including Irish Americans, Italian Americans, Jewish Americans, African Americans, and Latinos/as, have been the objects of racist ideology. However, not everyone is equally racist. Studies have shown that the underlying reasoning behind racism differs according to factors such as gender, age, class, and geography (see Cashmore, 1996; Feagin and Vera, 1995).

Racism may be overt or subtle. Overt racism is more blatant and may take the form of public statements about the "inferiority" of members of a racial or ethnic group. In sports, for example, calling a player of color a derogatory name, participating in racist chanting during a sporting event, and writing racist graffiti in a team's locker room are all forms of overt racism. These racist actions are blatant, but subtle forms of racism are often hidden from sight and more difficult to prove. Examples of subtle racism in sports include those descriptions of African American athletes which suggest that they have "natural" abilities and are better suited for team positions requiring speed and agility. By contrast, whites are described as having the intelligence, dependability, and leadership and decision-making skills needed in positions requiring higher levels of responsibility and control.

dominant group a group that is advantaged and has superior resources and rights in a society.

subordinate group a group whose members, because of physical or cultural characteristics, are disadvantaged and subjected to unequal treatment by the dominant group and who regard themselves as objects of collective discrimination.

prejudice a negative attitude based on faulty generalizations about members of selected racial and ethnic groups.

stereotypes overgeneralizations about the appearance, behavior, or other characteristics of members of particular categories.

racism a set of attitudes, beliefs, and practices that is used to justify the superior treatment of one racial or ethnic group and the inferior treatment of another racial or ethnic group.

Box 10.2 Sociology in Global Perspective

An Update on Racism and Antiracism in European Football

I just wanted to make it public. People have got to be aware that certain things in sport are not acceptable. . . . I just think there are certain boundaries that shouldn't be crossed. Race is not a topic to make fun of. From what I gathered, [the offending player on the other team] was just trying to provoke me. There's other means of provoking a player without crossing that threshold.

—Oguchi Onyewu, a professional soccer player in Europe who traces his origins to South Africa, has been subjected to monkey noises and racial remarks by fans in the past but recently has also experienced such harsh actions by another player (qtd. in Longman, 2009: Y1)

You have players of African origin who are regularly pushed by opponents with words to provoke or unsettle them. Most don't have the professional status to come in the open and say, "I've had enough of that." Oguchi is a big guy, mentally and physically. He's well established. With [a lawsuit demanding "moral compensation"—in other words, an apology], he is saying, "If I don't do it, who is going to do it?"

—Jean-Louis Dupont, an attorney, is representing Onyewu in court to call attention to racism in sports and to demand an apology from the offending player (qtd. in Longman, 2009: Y8)

For a number of years, the problem of racism has been a real issue for the European football family and for the image of football in general. Although some have declared that the problem has become much less intense, situations such as the one discussed above continue to occur at a surprising rate. According to a spokesperson for the UEFA (Union of European Football Association), "Racism is a sad reflection of society in general, but, because of the high-profile nature of football on our continent, we have a

Sporting events, such as soccer in Europe and football in the United States, are sometimes tainted by the actions of some fans who demonstrate racist behavior by their comments about or actions toward players, officials, or other fans. What can be done to reduce such harmful behavior?

Racism tends to intensify in times of economic uncertainty and high rates of immigration. Recently, relatively high rates of immigration in nations such as the United States, Canada, England, France, and Germany have been accompanied by an upsurge in racism and racial conflict (see Box 10.2). Sometimes, intergroup racism and conflicts further exacerbate strained relationships between dominant-group members and subordinate racial and ethnic group members. For example, when animosities have run very high among African American and Salvadoran groups in Long Island, New York, some white Americans have pointed to those hostilities as evidence that both groups are inferior and not deserving of assistance from the U.S. government or from charitable organizations such as the Catholic church (Mahler, 1995).

Theories of Prejudice

Are some people more prejudiced than others? To answer this question, some theories focus on how individuals may transfer their internal psychological problem onto an external object or person. Others look at factors such as social learning and personality types.

The *frustration–aggression hypothesis* states that people who are frustrated in their efforts to achieve a highly desired goal will respond with a pattern of aggression toward others (Dollard et al., 1939). The object of their aggression becomes the ***scapegoat*—a person or group that is incapable of offering resistance to the hostility or aggression of others** (Marger, 2009). Scapegoats are often used as substitutes for the actual source of the frustration. For example, members of subordinate racial and ethnic groups are often

particular responsibility to take steps to stamp it out and prevent it occurring in the future" (uefa.com, 2002).

Soccer is Europe's equivalent of football in the United States, and European fans are at least equally as passionate about sports as are fans in the United States. However, racism is an issue in European soccer just as it is an issue in U.S. sports. The UEFA, which is European soccer's ruling body, receives numerous reports of racist behavior at soccer matches throughout Europe, and officials have decided that the UEFA must work to reduce racism. Accusations of racist behavior against players of other nationalities or ethnic groups have been lodged against sports fans and some team members at various athletic events throughout Europe. At some soccer matches, fans shout racist comments during the game; throw bottles, cans, and other missiles at players; and pass out derogatory literature about other ethnic groups or nationalities outside the stadium.

Acknowledging that racism is a problem that "is created and stimulated outside of [soccer], but is often given expression and public focus through the game," UEFA officials recently issued a call for unity and developed a plan to reduce racist behavior at athletic competitions (sportsnetwork .com, 2003). According to the call for unity issued by the UEFA, "Racism is a problem for all of us, which must be faced. We hope that all parts of European football can come together to unite against racism and do all we can to eradicate it from our game both on and off the field" (uefa.com, 2002). Among the things that the plan calls for are these actions on the part of soccer clubs:

- Soccer clubs should issue a statement in printed programs and on the grounds where the competition takes place that the club will not tolerate racism and should set forth specific actions that the club will take against people who engage in racist chanting.
- Regular public address announcements should be made condemning racist chanting at matches.
- To become a season-ticket holder, people should agree not to take part in any racist abuse.
- All racist graffiti should be removed quickly from the grounds.
- Soccer clubs should work with other organizations, including schools, youth clubs, and local businesses, to raise public awareness of the importance of eliminating racial abuse and discrimination. (adapted from uefa.com, 2002)

Will these steps help to reduce racism in European sporting events? Analysts believe that these actions are at least a step in the right direction. If the UEFA enforces the rules that it has established, fans and players will be more aware of the negative sanctions that might occur if they do not "clean up their act." UEFA officials and others have acknowledged that a problem does exist and that the problem is not limited to a small handful of fans and players. But, in the final analysis, as the UEFA officials stated, "Of course no one organization can solve this problem. Everyone involved, including the clubs, fans, players, police and those responsible for stewarding, has a responsibility here" (uefa .com, 2002).

Reflect & Analyze

Have you ever been at a sporting event and witnessed truly obnoxious verbal behavior by other spectators? How did you react? Also, remember that some people believe that such behavior is constitutionally protected—a matter of "free speech." Do you agree? Why or why not?

blamed for societal problems (such as unemployment or an economic recession) over which they have no control.

According to some symbolic interactionists, prejudice results from social learning; in other words, it is learned from observing and imitating significant others, such as parents and peers. Initially, children do not have a frame of reference from which to question the prejudices of their relatives and friends. When they are rewarded with smiles or laughs for telling derogatory jokes or making negative comments about out-group members, children's prejudiced attitudes may be reinforced.

Psychologist Theodor W. Adorno and his colleagues (1950) concluded that highly prejudiced individuals tend to have an ***authoritarian personality,*** **which is characterized by excessive conformity, submissiveness to authority, intolerance, insecurity, a high level of superstition, and rigid, stereotypic thinking** (Adorno et al., 1950). It is most likely to develop in a family environment in which dominating parents who are anxious about status use physical discipline but show very little love in raising their children (Adorno et al., 1950). Other scholars have linked prejudiced

scapegoat a person or group that is incapable of offering resistance to the hostility or aggression of others.

authoritarian personality a personality type characterized by excessive conformity, submissiveness to authority, intolerance, insecurity, a high level of superstition, and rigid, stereotypic thinking.

© Michael Greenlar/The Image Works

According to the frustration–aggression hypothesis, members of white supremacy groups such as the Ku Klux Klan often use members of subordinate racial and ethnic groups as scapegoats for societal problems over which they have no control.

attitudes to traits such as submissiveness to authority, extreme anger toward outgroups, and conservative religious and political beliefs (Altemeyer, 1981, 1988; Weigel and Howes, 1985).

Measuring Prejudice

To measure levels of prejudice, some sociologists use the concept of ***social distance*—the extent to which people are willing to interact and establish relationships with members of racial and ethnic groups other than their own** (Park and Burgess, 1921). Sociologist Emory Bogardus (1925, 1968) developed a scale to measure social distance in specific situations, ranging from minimal contact to marriage. He concluded that some groups were consistently ranked as more desirable than others for close interpersonal contacts. More recently, analysts have found that whites who accept racial stereotypes desire greater social distance from people of color than do whites who reject negative stereotypes (Krysan and Farley, 1993). But can prejudice really be measured? Existing research does not provide us with a conclusive answer to this question. Most social-distance research has examined the perceptions of whites; few studies have measured the perceptions of people of color about their interactions with members of the dominant group.

Discrimination

Whereas prejudice is an attitude, ***discrimination* involves actions or practices of dominant-group members (or their representatives) that have a harmful impact on members of a subordinate group** (Feagin and Feagin, 2008). Prejudiced attitudes do not always lead to discriminatory behavior. As shown in ▶ Figure 10.1, the sociologist Robert Merton (1949) identified four combinations of attitudes and responses. Unprejudiced nondiscriminators are not personally prejudiced and do not discriminate against others. For example, two players on a professional sports team may be best friends although they are of different races. Unprejudiced discriminators may have no personal prejudice but still engage in discriminatory behavior because of peer-group pressure or economic, political, or social interests. For example, in some sports a coach might feel no prejudice toward African American players but believe that white fans will accept only a certain percentage of people of color on the team. Prejudiced nondiscriminators hold personal prejudices but do not discriminate due to peer pressure, legal demands, or a desire for profits. For example, a coach with prejudiced beliefs may hire an African American player to enhance the team's ability to win (Coakley, 2004). Finally, prejudiced discriminators hold personal prejudices and actively discriminate against others. For example, an umpire who is personally preju-

	Prejudiced attitude?	Discriminatory behavior?
Unprejudiced nondiscriminator	No	No
Unprejudiced discriminator	No	Yes
Prejudiced nondiscriminator	Yes	No
Prejudiced discriminator	Yes	Yes

▶ **Figure 10.1 Merton's Typology of Prejudice and Discrimination**

Merton's typology shows that some people may be prejudiced but not discriminate against others. Do you think that it is possible for a person to discriminate against some people without holding a prejudiced attitude toward them? Why or why not?

diced against African Americans may intentionally call a play incorrectly based on that prejudice.

Discriminatory actions vary in severity from the use of derogatory labels to violence against individuals and groups. The ultimate form of discrimination occurs when people are considered to be unworthy to live because of their race or ethnicity. ***Genocide* is the deliberate, systematic killing of an entire people or nation.** Examples of genocide include the killing of thousands of Native Americans by white settlers in North America and the extermination of six million European Jews by Nazi Germany. More recently, the term *ethnic cleansing* has been used to define a policy of "cleansing" geographic areas by forcing persons of other races or religions to flee—or die.

Discrimination also varies in how it is carried out. Individuals may act on their own, or they may operate within the context of large-scale organizations and institutions, such as schools, churches, corporations, and governmental agencies. How does individual discrimination differ from institutional discrimination? ***Individual discrimination* consists of one-on-one acts by members of the dominant group that harm members of the subordinate group or their property.** For example, a person may decide not to rent an apartment to someone of a different race. By contrast, ***institutional discrimination* consists of the day-to-day practices of organizations and institutions that have a harmful impact on members of subordinate groups.** For example, a bank might consistently deny loans to people of a certain race. Institutional discrimination is carried out by the individuals who implement the policies and procedures of organizations.

Sociologist Joe R. Feagin has identified four major types of discrimination:

1. *Isolate discrimination* is harmful action intentionally taken by a dominant-group member against a member of a subordinate group. This type of discrimination occurs without the support of other members of the dominant group in the immediate social or community context. For example, a prejudiced judge may give harsher sentences to all African American defendants but may not be supported by the judicial system in that action.
2. *Small-group discrimination* is harmful action intentionally taken by a limited number of dominant-group members against members of subordinate groups. This type of discrimination is not supported by existing norms or other dominant-group members in the immediate social or community context. For example, a small group of white students may deface a professor's office with racist epithets without the support of other students or faculty members.
3. *Direct institutionalized discrimination* is organizationally prescribed or community-prescribed action that intentionally has a differential and negative impact on members of subordinate groups. These actions are routinely carried out by a number of dominant-group members based on the norms of the immediate organization or community (Feagin and Feagin, 2008). Intentional exclusion of people of color from public accommodations is an example of this type of discrimination.
4. *Indirect institutionalized discrimination* refers to practices that have a harmful impact on subordinate-group members even though the organizationally or community-prescribed norms or regulations guiding these actions were initially established with no intent to harm. For example, special education classes were originally intended to provide extra educational opportunities for children with various types of disabilities. However, critics claim that these programs have amounted to racial segregation in many school districts.

Various types of racial and ethnic discrimination call for divergent remedies if we are to reduce discriminatory actions and practices in contemporary social life. Since the 1950s and 1960s, many U.S. sociologists have analyzed the complex relationship between prejudice and discrimination. Some have reached the conclusion that prejudice is difficult, if not seemingly impossible, to eradicate because of the deeply held racist beliefs and attitudes that are often passed on from person to person and from one generation to the next. However, the persistence of prejudicial attitudes and beliefs does

social distance the extent to which people are willing to interact and establish relationships with members of racial and ethnic groups other than their own.

discrimination actions or practices of dominant-group members (or their representatives) that have a harmful effect on members of a subordinate group.

genocide the deliberate, systematic killing of an entire people or nation.

individual discrimination behavior consisting of one-on-one acts by members of the dominant group that harm members of the subordinate group or their property.

institutional discrimination the day-to-day practices of organizations and institutions that have a harmful impact on members of subordinate groups.

not mean that racial and ethnic discrimination should be allowed to flourish until such a time as prejudice is effectively eliminated. From this approach, discrimination must be tackled aggressively through demands for change and through policies that specifically target patterns of discrimination (see "Sociology Works!").

Sociological Perspectives on Race and Ethnic Relations

Symbolic interactionist, functionalist, and conflict analysts examine race and ethnic relations in different ways. Functionalists focus on the macrolevel intergroup processes that occur between members of dominant groups and subordinate groups in society. Conflict theorists analyze power and economic differentials between the dominant group and subordinate groups. Symbolic interactionists examine how microlevel contacts between people may produce either greater racial tolerance or increased levels of hostility.

Symbolic Interactionist Perspectives

What happens when people from different racial and ethnic groups come into contact with one another? In the *contact hypothesis,* symbolic interactionists point out that contact between people from divergent groups should lead to favorable attitudes and behavior when certain factors are present. Members of each group must (1) have equal status, (2) pursue the same goals, (3) cooperate with one another to achieve their goals, and (4) receive positive feedback when they interact with one another in positive, nondiscriminatory ways (Allport, 1958; Coakley, 2004).

What happens when individuals meet someone who does not conform to their existing stereotype? Frequently, they ignore anything that contradicts the stereotype, or they interpret the situation to support their prejudices (Coakley, 2004). For example, a person who does not fit the stereotype may be seen as an exception: "You're not like other [persons of a particular race]."

When a person is seen as conforming to a stereotype, he or she may be treated simply as one of "you people." Former Los Angeles Lakers basketball star Earvin "Magic" Johnson (1992: 31–32) described how he was categorized along with all other African Americans when he was bused to a predominantly white school:

> On the first day of [basketball] practice, my teammates froze me out. Time after time I was wide open, but nobody threw me the ball. At first I thought they just didn't see me. But I woke up after a kid named Danny Parks looked right at me and then took a long jumper. Which he missed.
>
> I was furious, but I didn't say a word. Shortly after that, I grabbed a defensive rebound and took the ball all the way down for a basket. I did it again and a third time, too.
>
> Finally Parks got angry and said, "Hey, pass the [bleeping] ball."
>
> That did it. I slammed down the ball and glared at him. Then I exploded. "I *knew* this would happen!" I said. "That's why I didn't want to come to this [bleeping] school in the first place!"
>
> "Oh, yeah? Well, you people are all the same," he said. "You think you're gonna come in here and do whatever you want? Look, hotshot, your job is to get the rebound. Let us do the shooting."

The interaction between Johnson and Parks demonstrates that when people from different racial and ethnic groups come into contact with one another, they may treat one another as stereotypes, not as individuals. Eventually, Johnson and Parks were able to work out most of their differences. "There's nothing like winning to help people get along," Johnson explained (1992: 32). Although we might hope that nothing like this happens today, there is much evidence that covert discrimination occurs in many sports and social settings.

Symbolic interactionist perspectives make us aware of the importance of intergroup contact and the fact that it may either intensify or reduce racial and ethnic stereotyping and prejudice.

Functionalist Perspectives

How do members of subordinate racial and ethnic groups become a part of the dominant group? To answer this question, early functionalists studied immigration and patterns of majority- and minority-group interactions.

Assimilation ***Assimilation*** **is a process by which members of subordinate racial and ethnic groups become absorbed into the dominant culture.** To some analysts, assimilation is functional because it contributes to the stability of society by minimizing group differences that might otherwise result in hostility and violence.

Assimilation occurs at several distinct levels, including the cultural, structural, biological, and psychological stages. *Cultural assimilation,* or *acculturation,* occurs when members of an ethnic group adopt dominant-group traits, such as language, dress, values,

Sociology *Works!*

Attacking Discrimination to Reduce Prejudice?

Question: Do you think it is possible to reduce racial and ethnic prejudice in the United States by attacking discrimination at the societal level?

Answer: Well, I think since it's individuals who hate each other and do mean stuff, they should work it out for themselves. We don't need the government telling us what to do or how to behave. Like, my dad runs a company, and he says that the government should "butt out" of our business.

—"Brian," an introductory sociology student, stating why he believes that the government has "no business" trying to reduce discrimination (author's notes)

Brian's comment is typical of how many people feel about court rulings and government initiatives over the past sixty years that have sought to reduce the corrosive effects of racial and ethnic discrimination in American life. Based on the widely held axiom that "prejudice causes discrimination," many people argue that discrimination can be alleviated only through changing the attitudes of individuals. From this perspective, discrimination will go away over time if people are encouraged to shed their negative attitudes about members of other racial or ethnic groups. Discarding negative stereotypes about other groups and bringing to light the truth about popular myths regarding the superiority of one's own race, ethnic group, or nationality are widely seen as the best ways of bringing about positive social change. Diversity training sessions held in schools and at the workplace are a classic example of this approach.

Since the civil rights era in the 1960s, however, sociologists have demonstrated that discrimination can be tackled up front and now—rather than waiting for prejudice to diminish—so that people in subordinate racial/ethnic categories can gain a greater measure of human dignity and a variety of opportunities that they otherwise would not have. Establishing social policies and laws to eliminate specific practices of segregation and discrimination in education, employment, housing, health care, law enforcement, and other areas of public life, for example, has served as a significant starting point for social change that has positively affected generations of people of color, women of all racial and ethnic categories, religious minorities, and many others who have lived outside the mainstream of social life. By enacting legislation that prohibits discrimination based on race, ethnic origin, and color, our nation has sought to provide people with greater equality before the law and access to crucial opportunities and social resources. These changes have, over time, reduced the prejudiced attitudes of many people on a wide variety of issues.

Although progress has been made in reducing some aspects of overt prejudice and institutional discrimination, it is clear that racism is not a thing of the past. However, if previous sociological research tells us anything about the future, it is that we must continue to tackle not only individual prejudices and discriminatory conduct but also the larger, societal patterns of discrimination—embedded in the organizations and institutions of which we are a part—that restrict freedom, opportunities, and quality of life for all people.

Reflect & Analyze

Using sports as an example, let's think about these questions: How might college sporting events serve to reduce prejudice and discrimination on campus and beyond? How might these same events serve to perpetuate negative stereotypes and popular myths about racial and ethnic "differences"? What do you think? (See Box 10.4, "You Can Make a Difference," for additional discussion on this topic.)

religion, and food preferences. Cultural assimilation in this country initially followed an "Anglo conformity" model; members of subordinate ethnic groups were expected to conform to the culture of the dominant white Anglo-Saxon population (Gordon, 1964). However, members of some groups refused to be assimilated and sought to maintain their unique cultural identity.

Structural assimilation, or *integration,* occurs when members of subordinate racial or ethnic groups gain acceptance in everyday social interaction with members of the dominant group. This type of assimilation typically starts in large, impersonal settings such as schools and workplaces, and only later (if at all) results in close friendships and intermarriage. (Box 10.3 on pages 324–325 examines the extent to which television advertisements correctly reflect the extent of structural

assimilation a process by which members of subordinate racial and ethnic groups become absorbed into the dominant culture.

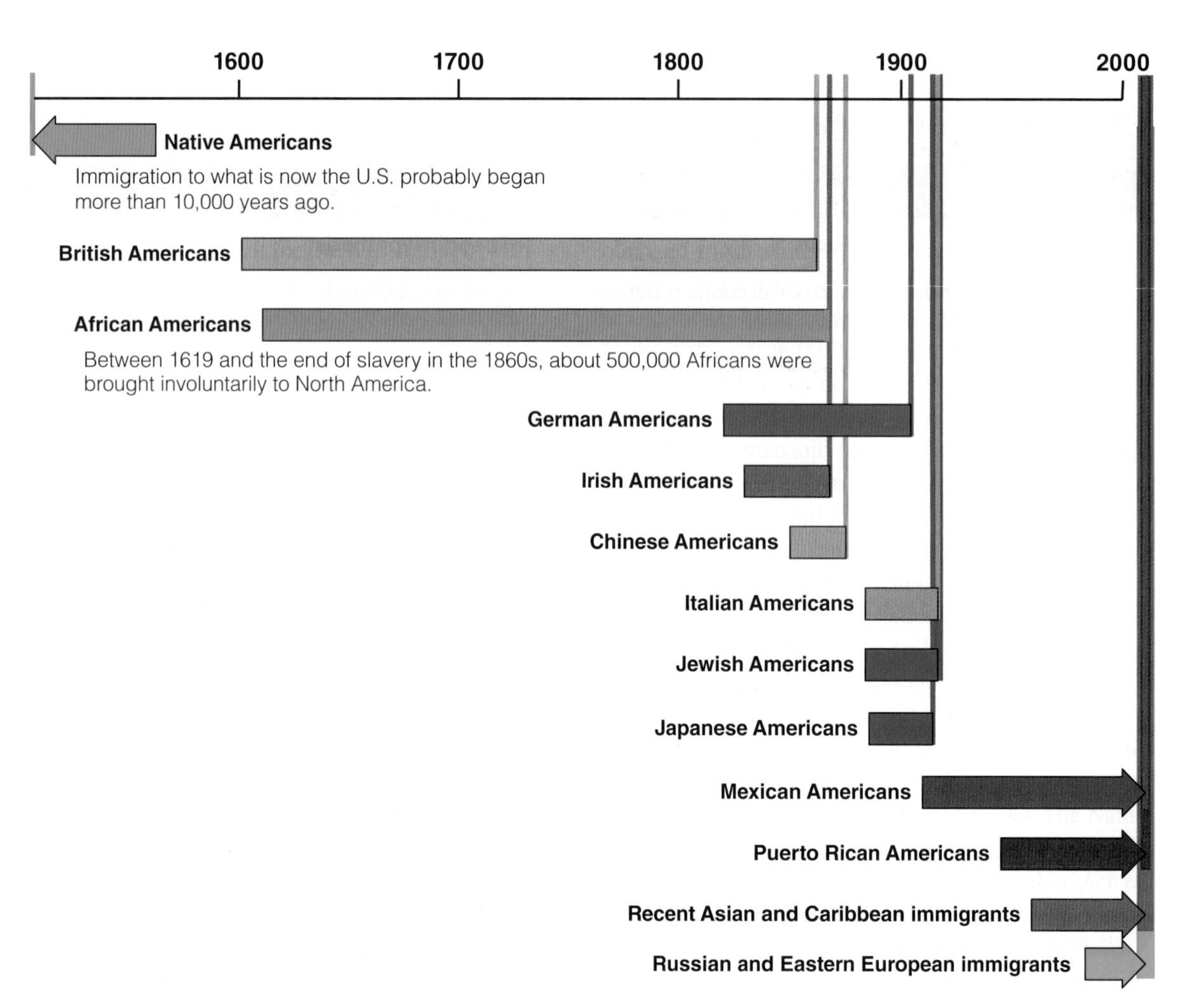

Figure 10.2 Time Line of Racial and Ethnic Groups in the United States

way as far south as the tip of South America (Cashmore, 1996).

As schoolchildren are taught, Spanish explorer Christopher Columbus first encountered the native inhabitants in 1492 and referred to them as "Indians." When European settlers (or invaders) arrived on this continent, the native inhabitants' way of life was changed forever. Experts estimate that approximately two million native inhabitants lived in North America at that time; however, their numbers had been reduced to less than 240,000 by 1900.

Genocide, Forced Migration, and Forced Assimilation Native Americans have been the victims of genocide and forced migration. Although the United States never had an official policy that set in motion a pattern of deliberate extermination, many Native Americans were either massacred or died from European diseases (such as typhoid, smallpox, and measles) and starvation (Wagner and Stearn, 1945; Cook, 1973). In battle, Native Americans were often no match for the Europeans, who had "modern" weaponry (Amott and Matthaei, 1996). Europeans justified their aggression by stereotyping the Native Americans as "savages" and "heathens" (Takaki, 1993). Early atrocities perpetrated against the Native Americans were described by one social historian:

> The degree of violence that was woven into the texture of early frontier life fairly boggles the mind of our, in some ways, far more delicate age. In the 1650s, Dutch colonists brought back eighty decapitated Indian heads from a massacre and used them as kickballs in the streets of New Amsterdam. . . . An English traveler in the northern colonies casually recorded in his diary, in 1760, that "some people have an Indian's Skin for a Tobacco Pouch," while a Revolutionary War soldier campaigning against the Iroquois could note with equal dispassion that he had been given a pair of boot-tops made from the freshly skinned legs of two enemy braves. (Bordewich, 1996: 35–36)

After the Revolutionary War, the federal government offered treaties to the Native Americans so that more of their land could be acquired for the growing white population. Scholars note that the government broke treaty after treaty as it engaged in a policy of wholesale removal of indigenous nations in order to clear the land for settlement by Anglo-Saxon "pioneers" (Green, 1977). Entire nations were forced to move in order to accommodate the white settlers. The "Trail of Tears" was one of the most disastrous of the forced migrations. In the coldest part of the winter of 1832, over half of the Cherokee Nation died during or as a result of their forced relocation from the southeastern United States to the Indian Territory in Oklahoma (Thornton, 1984).

Native Americans were made wards of the government (meaning they had a legal status similar to that of minors and incompetents) and were subjected to forced assimilation on the reservations after 1871 (Takaki, 1993). Native American children were placed in boarding schools operated by the Bureau of Indian Affairs to hasten their assimilation into the dominant culture. About 98 percent of native lands had been expropriated by 1920 (see McDonnell, 1991). Even after Native Americans received full citizenship and the right to vote in 1924, the Supreme Court continued to hold that they were wards of the government.

Native Americans Today Currently, about 2.5 million Native Americans (1.5 percent of the U.S. population) live in the United States, including Aleuts, Inuit (Eskimos), Cherokee, Navajo, Choctaw, Chippewa, Sioux, and over 500 other nations of varying sizes and different locales. There is a wide diversity among the people in this category: Each nation has its own culture, history, and unique identity, and more than 250 Native American languages are spoken today. Most Native Americans live in the Southwest, and about one-third live on reservations. Native Americans are the most disadvantaged racial or ethnic group in the United States in terms of income, employment, housing, nutrition, and health. The life chances of Native Americans who live on reservations are especially limited. They have the highest rates of infant mortality and death by exposure and malnutrition. They also have high rates of suicide, substance abuse, and school violence (Kershaw, 2005).

Historically, Native Americans have had very limited educational opportunities and a very high rate of unemployment. In recent years, however, a network of tribal colleges has been successful in providing some Native Americans with the education they need to move into the ranks of the skilled working class and beyond (Bordewich, 1996). Across the nation, Native Americans own and operate many types of enterprises, such as construction companies, computer graphic design firms, grocery stores, and management consulting businesses. Casino gambling operations and cigarette shops on Native American reservations—resulting from a reinterpretation of federal law in the 1990s—have brought more income to some of the nations, but this change has not been without its critics, many of whom believe that these businesses result in new problems for Native Americans.

Native Americans are currently in a transition from a history marked by prejudice and discrimination to a contemporary life in which they may find new opportunities. Many see the challenge for Native Americans today as erasing negative stereotypes while maintaining their heritage and obtaining recognition for their contributions to this nation's development and growth.

Native Americans and Sports Early in the twentieth century, Native Americans such as Jim Thorpe gained national visibility as athletes in football, baseball, and track and field. Teams at boarding schools such as the Carlisle Indian Industrial School in Pennsylvania and the Haskell Institute in Kansas were well-known. However, after the first three decades of the twentieth century, Native Americans became much less prominent in sports. Although some Navajo athletes have been very successful in basketball and some Choctaws have excelled in baseball, Native Americans have seldom been able to compete at the college, professional, or Olympic level (Blanchard, 1980; Oxendine, 2003). Native American scholar Joseph B. Oxendine (2003) attributes the lack of athletic participation to these factors: (1) a reduction in opportunities for developing sports skills, (2) restricted opportunities for participation, and (3) a lessening of Native Americans' interest in competing with and against non-Native Americans. For example, Native Americans have fewer opportunities to participate in college athletics, which serves

Native Americans have historically had a low rate of college attendance, but the development of a network of tribal colleges has provided them a local source of upward mobility. However, although a few Native Americans early in the twentieth century were well-known athletes, Native Americans today have little opportunity to compete in sports at the college, professional, or Olympic level.

and know only the English language. Although they constitute only 1 percent of the U.S. population, their numbers have increased significantly in some communities (such as Pikesville, Maryland, and Roslyn Estates, New York), where they make up slightly less than 20 percent of the population. Since the beginning of the post-Soviet era, which is marked by the fall of the former Soviet Union in 1991, many people who immigrated to the United States from Russia have been political refugees, particularly a large number of scientists and engineers who have been employed in U.S. high-tech industries.

Discrimination against White Ethnics Many white ethnic immigrants entered the United States between 1830 and 1924. Irish Catholics were among the first to arrive, with over four million Irish fleeing the potato famine and economic crisis in Ireland and seeking jobs in the United States (Feagin and Feagin, 2008). When they arrived, they found that British Americans controlled the major institutions of society. The next arrivals were Italians who had been recruited for low-wage industrial and construction jobs. British Americans viewed Irish and Italian immigrants as "foreigners": The Irish were stereotyped as ape-like, filthy, bad-tempered, and heavy drinkers, and the Italians were depicted as lawless, knife-wielding thugs looking for a fight—"dagos" and "wops" (short for "without papers") (Feagin and Feagin, 2008).

Both Irish Americans and Italian Americans were subjected to institutionalized discrimination in employment. Employment ads read "Help Wanted—No Irish Need Apply" and listed daily wages at $1.30–$1.50 for "whites" and $1.15–$1.25 for "Italians" (Gambino, 1975: 77). In spite of discrimination, white ethnics worked hard to establish themselves in the United States, often establishing mutual self-help organizations and becoming politically active (Mangione and Morreale, 1992).

Between 1880 and 1920, over two million Jewish immigrants arrived in the United States and settled in the Northeast. Jewish Americans differ from other white ethnic groups in that some focus their identity primarily on their religion whereas others define their Jewishness in terms of ethnic group membership (Feagin and Feagin, 2008). In any case, Jews continued to be the victims of *anti-Semitism*—prejudice, hostile attitudes, and discriminatory behavior targeted at Jews. For example, signs in hotels read "No Jews Allowed," and some help-wanted ads stated "Christians Only" (Levine, 1992: 55). In spite of persistent discrimination, Jewish Americans achieved substantial success in many areas, including business, education, the arts and sciences, law, and medicine.

As previously discussed, more-recent white ethnic immigration has involved Russia and Eastern Europe, with large numbers of political refugees coming to the United States to seek freedom from persecution. Because the Soviet Union was well-known for sports, a number of prominent athletes living in the United States and other nations trace their roots to that part of the world.

White Ethnics and Sports Sports provided a pathway to assimilation for many white ethnics. The earliest collegiate football players who were not white Anglo-Saxon Protestants were of Irish, Italian, and Jewish ancestry. Sports participation provided edu-

After coming to the United States in the nineteenth century, many white ethnic immigrants faced severe poverty, an issue explored by director Martin Scorsese in his 2002 film *Gangs of New York*.

cational opportunities that some white ethnics would not have had otherwise.

Boxing became a way to make a living for white ethnics who did not participate in collegiate sports. Boxing promoters encouraged ethnic rivalries to increase their profits, pitting Italians against Irish or Jews, and whites against African Americans (Levine, 1992; Mangione and Morreale, 1992). Eventually, Italian Americans graduated from boxing into baseball and football. Jewish Americans found that sports lessened the shock of assimilation and gave them an opportunity to refute stereotypes about their physical weaknesses and counter anti-Semitic charges that they were "unfit to become Americans" (Levine, 1992: 272).

Today, assimilation is so complete that little attention is paid to the origins of white ethnic athletes. As former Pittsburgh Steeler running back Franco Harris stated, "I didn't know I was part Italian until I became famous" (qtd. in Mangione and Morreale, 1992: 384). In another example, some white ethnics in sports today are not usually thought of as Russian Americans, but when they win tennis matches or medals at the Olympics, their ethnic origins are often mentioned admiringly because of the troubled history of the nation from which their ancestors came. Among these Russian American athletes are the professional tennis player Maria Sharapova, professional ice hockey players Alexander Ovechkin and Alexandre Volchkov, professional basketball player Andrei Kirilenko, and Olympic gymnast Nastia Liukin.

Asian Americans

The U.S. Census Bureau uses the term *Asian Americans* to designate the many diverse groups with roots in Asia. Chinese and Japanese immigrants were among the earliest Asian Americans. Many Filipinos, Asian Indians, Koreans, Vietnamese, Cambodians, Pakistani, and Indonesians have arrived more recently. Today, Asian Americans belong to the fastest-growing ethnic minority group in the United States and constitute about 5 percent of the nation's population. About 13.1 million people in the United States identified themselves as Asian American in 2007. At the beginning of the twenty-first century, the Asian American population grew by 3.7 percent per year; however, recent figures show less growth because of the economic recession.

Chinese Americans The initial wave of Chinese immigration occurred between 1850 and 1880, when more than 200,000 Chinese men were "pushed" from China by harsh economic conditions and "pulled" to the United States by the promise of gold in California and employment opportunities in the construction of transcontinental railroads. Far fewer Chinese women immigrated; however, many of them were brought to the United States against their will and forced into prostitution, where they were treated like slaves (Takaki, 1993).

Chinese Americans were subjected to extreme prejudice and stereotyped as "coolies," "heathens," and "Chinks." The late historian Ronald Takaki (1989: 13) described the economic context in which this discrimination occurred:

> Unlike the Irish and other groups from Europe, Asian immigrants could not become "mere individuals, indistinguishable in the cosmopolitan mass of the population." Regardless of their personal merits, they sadly discovered, they could not gain acceptance in the larger society. They were judged not by the content of their character but by their complexion. . . . "Color" in America operated within an economic context. Asian immigrants came here to meet demands for labor—plantation workers, railroad crews, miners, factory operatives, cannery workers, and farm laborers. Employers developed a dual-wage system to pay Asian laborers less than white workers and pitted the groups against each other in order to depress wages for both. Ethnic antagonism . . . led white laborers to demand the restriction of Asian workers already here in a segregated labor market of low-wage jobs and the exclusion of future Asian immigrants. Thus the class interests of white capital as well as white labor needed Asians as "strangers."

Historically, Chinatowns in major U.S. cities have provided a safe haven and an economic enclave for many Asian immigrants. Some contemporary Chinese Americans reside in these neighborhoods, whereas others visit to celebrate cultural diversity and ethnic pride.

As Takaki noted, prejudice and discrimination ran high against Asian immigrants, leading to Congress's passage of the Chinese Exclusion Act of 1882, which brought Chinese immigration to a halt. The Exclusion Act was not repealed until World War II, when Chinese Americans who were contributing to the war effort by working in defense plants pushed for its repeal (Takaki, 1993). After immigration laws were further relaxed in the 1960s, the second and largest wave of Chinese immigration occurred, with immigrants coming primarily from Hong Kong and Taiwan. These recent immigrants have had more education and workplace skills than earlier arrivals, and brought families and capital with them to pursue the American Dream (Chen, 1992).

Today, many Chinese Americans live in large urban enclaves in California, New York, Hawaii, Illinois, and Texas. As a group, they have enjoyed considerable upward mobility. Some own laundries, restaurants, and other businesses; others have professional careers (Chen, 1992). However, many Chinese Americans, particularly recent immigrants, remain in the lower tier of the working class—providing low-wage labor in garment and knitting factories and Chinese restaurants.

Japanese Americans Most of the early Japanese immigrants were men who worked on sugar plantations in the Hawaiian Islands in the 1860s. Like Chinese immigrants, the Japanese American workers were viewed as a threat by white workers, and immigration of Japanese men was curbed in 1908. However, Japanese women were permitted to enter the United States for several years thereafter because of the shortage of women on the West Coast. Although some Japanese women married white men, this practice was stopped by laws prohibiting interracial marriage.

With the exception of the enslavement of African Americans, Japanese Americans experienced one of the most vicious forms of discrimination ever sanctioned by U.S. laws. During World War II, when the United States was at war with Japan, nearly 120,000 Japanese Americans were placed in internment camps, where they remained for more than two years despite the total lack of evidence that they posed a security threat to this country (Takaki, 1993). This action was a direct violation of the citizenship rights of many *Nisei* (second-generation Japanese Americans), who were born in the United States (see Daniels, 1993). Ironically, only Japanese Americans were singled out for such harsh treatment; German Americans avoided this fate even though the United States was also at war with Germany. Four decades later, the U.S. government issued an apology for its actions and eventually paid $20,000 each to some of those who had been placed in internment camps (Daniels, 1993; Takaki, 1993).

Since World War II, many Japanese Americans have been very successful. The median income of Japanese Americans is more than 30 percent above the national average. However, most Japanese Americans (and many other Asian Americans) live in states that not only have higher incomes but also higher costs of living than the national average. In addition, many Asian American families have more persons in the paid labor force than do other families (Takaki, 1993).

Korean Americans The first wave of Korean immigrants were male workers who arrived in Hawaii between 1903 and 1910. The second wave came to the U.S. mainland following the Korean War in 1954 and was made up primarily of the wives of servicemen and Korean children who had lost their parents in the war. The third wave arrived after the Immigration Act of 1965 permitted well-educated professionals to migrate to the United States. Korean Americans have helped one another open small businesses by pooling money through the *kye*—an association that grants members money on a rotating basis to gain access to more capital. According to Takaki (1989), Korean Americans were a hidden minority before 1965 because so few lived in the United States. After that time, however, Korean Americans have become a very visible group in this country.

Today, many Korean Americans live in California and New York, where there is a concentration of Korean-owned grocery stores, businesses, and churches. Unlike earlier Korean immigrants, more-recent arrivals have come as settlers and have brought their families with them. However, their experiences with other subordinate racial and ethnic groups have not always been harmonious. Ongoing discord has existed between African Americans and Korean Americans in New York and among African Americans, Latinos, and Korean Americans in California.

Filipino Americans Today, Filipino Americans constitute the second-largest category of Asian Americans, with a population of almost three million in the United States. To understand the status of Filipino Americans, it is important to look at the complex relationship between the Philippine Islands and the United States government. After Spain lost the Spanish-American War, the United States established colonial rule over the islands, a rule that lasted from 1898 to 1946 (Feagin and Feagin, 2008). Despite control by the United States, Filipinos were not granted U.S. citizenship. But, like the Chinese and the Japanese, male Filipinos were allowed to migrate to Hawaii and the U.S. mainland to work in agriculture and in fish canneries in Seattle and Alaska. Like other Asian Americans, Filipino Americans were accused of taking jobs away from white workers and suppressing wages, and Congress restricted Filipino immigration to fifty people per year between the Great Depression and the aftermath of World War II.

The second wave of Filipino immigrants came following the Immigration Act of 1965, when large numbers of physicians, nurses, technical workers, and other professionals moved to the U.S. mainland. Most Filipinos have not had the start-up capital necessary to open their own businesses, and many have been employed in the low-wage sector of the service economy. However, the average household income of Filipino American families is relatively high because about 75 percent of Filipino American women are employed, and nearly half have a four-year college degree (Espiritu, 1995).

Indochinese Americans Indochinese Americans include people from Vietnam, Cambodia, Thailand, and Laos, most of whom have come to the United States in the past three decades. Vietnamese refugees who had the resources to flee at the beginning of the Vietnam War were the first to arrive. Next came Cambodians and lowland Laotians, referred to as "boat people" by the media. Many who tried to immigrate did not survive at sea; others were turned back when they reached this country or were kept in refugee camps for long periods of time. When they arrived in the United States, inflation was high, the country was in a recession, and many native-born citizens feared that they would lose their jobs to these new refugees, who were willing to work very hard for low wages. The frustrations of many Indochinese American immigrants were expressed by a Hmong refugee from Laos:

> In our old country, whatever we had was made or brought in by our own hands; we never had any doubts that we would not have enough for our mouths. But from now on to the future, that time is over. We are so afraid and worried that there will be one day when we will not have anything for eating or paying the rent, and these days these things are always in our minds. . . . Don't know how to read or write, don't know how to speak the language. My life is only to live day by day until the last day I live, and maybe that is the time when my problems will be solved. (qtd. in Portes and Rumbaut, 1996: 155)

Today, many Indochinese Americans are foreign born; about half live in the western states, especially California. Even though most Indochinese immigrants spoke no English when they arrived in this country, some of their children have done very well in school and have been stereotyped as "brains."

Asian Americans and Sports Until recently, Asian Americans received little recognition in sports. However, as women's athletic events, including ice skating and gymnastics, have garnered more media coverage, names of persons such as Kristi Yamaguchi, Michelle Kwan, and Amy Chow have become widely known and often idolized by fans. As one sports analyst stated, "[These athletes] are of Asian descent, but more importantly they are Asian Americans whose actions reflect upon the United States. . . . As role models, particularly for the Asian-American community, they exemplify success, integrity, discipline and a dedicated work ethic" (Shum, 1997).

Recent studies show that Asian Americans remain rare in certain sports such as men's college basketball (Chu, 2008). Although basketball is popular among Asian American youths, of 4,814 Division I men's college basketball players in 2006–2007, only 0.4 percent (19) were Asian Americans (including Pacific Islanders and ethnically mixed Asian Americans). Some believe that fewer Asian Americans are involved in certain types of sports because of family pressure for

Increasing numbers of Asian Americans are distinguishing themselves in college and professional athletics. Champion figure skater Michelle Kwan is one of the best-known recent Asian American sports heroes.

high academic achievement and good earning potential, whereas others believe that it is difficult for Asian American athletes to move into established sports networks such as college and professional programs because of assumptions that they do not have the best body types for athletics and that their interests are focused elsewhere (Chu, 2008). In the future, however, there may be many more well-known Asian American men and women participating in college and professional sporting events.

Latinos/as (Hispanic Americans)

The terms *Latino* (for males), *Latina* (for females), and *Hispanic* are used interchangeably to refer to people who trace their origins to Spanish-speaking Latin America and the Iberian peninsula (Cashmore, 1996). However, as racial–ethnic scholars have pointed out, the label *Hispanic* was first used by the U.S. government to designate people of Latin American and Spanish descent living in the United States (Oboler, 1995), and it has not been fully accepted as a source of identity by the more than 44.3 million Latinos/as who live in the United States today (Oboler, 1995; Romero, 1997; U.S. Census Bureau, 2009). Instead, many of the people who trace their roots to Spanish-speaking countries think of themselves as Mexican Americans, Chicanos/as, Puerto Ricans, Cuban Americans, Salvadorans, Guatemalans, Nicaraguans, Costa Ricans, Argentines, Hondurans, Dominicans, or members of other categories. Many also think of themselves as having a combination of Spanish, African, and Native American ancestry.

Mexican Americans or Chicanos/as Mexican Americans—including both native- and foreign-born people of Mexican origin—are the largest segment (approximately two-thirds) of the Latino population in the United States. Most Mexican Americans live in the southwestern region of the United States, although more have moved throughout the United States in recent years.

Immigration from Mexico is the primary vehicle by which the Mexican American population grew in this country. Initially, Mexican-origin workers came to work in agriculture, where they were viewed as a readily available cheap and seasonal labor force. Many initially entered the United States as undocumented workers ("illegal aliens"); however, they were more vulnerable to deportation than other illegal immigrants because of their visibility and the proximity of their country of origin. For more than a century, there has been a "revolving door" between the United States and Mexico that has been open when workers were needed and closed during periods of economic recession and high rates of U.S. unemployment.

The early experiences of Mexican Americans in this country were not always positive. In fact, many have experienced disproportionate poverty as a result of internal colonialism. Following the Mexican-American War, the United States seized land that had previously belonged to Mexico, and many formerly wealthy Mexican ranchers became impoverished farmhands. As sociologist Mary Romero (1997: 12) explains,

> The structure of opportunity for Chicanos . . . is rooted in the history of westward expansion, the geographical proximity and poverty of Mexico that facilitate continued immigration, and the historical labor functions of Mexican workers in the U.S. economy. Capitalist penetration of the Southwest dispossessed Chicanos of their land, created a cheap labor force and brought about the eventual destruction or transformation of the indigenous social system governing the lives of the Mexican residents.

Mexican Americans have long been seen as a source of cheap labor, while—ironically—at the same time, they have been stereotyped as lazy and unwilling to work. As has been true of other groups, when white workers viewed Mexican Americans as a threat to their jobs, they demanded that the "illegal aliens" be sent back to Mexico. Consequently, U.S. citizens who happen to be Mexican American have been asked for proof of their citizenship, especially when anti-immigration sentiments are running high. Many Mexican American families have lived in the United States for four or five generations—they have fought in wars, made educational and political gains, and consider themselves to be solid U.S. citizens. Thus, it is a great source of frustration for them to be viewed as illegal immigrants or to be asked "How long have you been in this country?"

Puerto Ricans When Puerto Rico became a territory of the United States in 1917, Puerto Ricans acquired U.S. citizenship and the right to move freely to and from the mainland. In the 1950s, many migrated to the mainland when the Puerto Rican sugar industry collapsed, settling primarily in New York and New Jersey. Although living conditions have improved substantially for some Puerto Ricans, life has been difficult for the many living in poverty in Spanish Harlem and other barrios. Nevertheless, in recent years Puerto Ricans have made dramatic advances in education, the arts, and politics. Increasing numbers have become lawyers, physicians, and college professors.

Cuban Americans Cuban Americans live primarily in the Southeast, especially Florida. As a group, they have fared somewhat better than other Latinos/as because many Cuban immigrants were affluent professionals and businesspeople who fled Cuba after Fidel Castro's 1959 Marxist revolution. This early wave of Cuban immigrants has median incomes well above those of other Latinos/as; however, this group is still below the national average. The second wave of Cuban Americans, arriving in the 1970s, has fared worse. Many had been released from prisons and mental hospitals in Cuba, and their arrival fueled an upsurge in prejudice against all Cuban Americans. The more-recent arrivals have developed their own ethnic and economic enclaves in Miami's Little Havana, and many of the earlier immigrants have become mainstream professionals and entrepreneurs.

Latinos/as and Sports Since the early 1900s, Latinos have played Major League Baseball. Originally, Cubans, Puerto Ricans, and Venezuelans were selected for their light skin as well as for their skill as players (Hoose, 1989). Today, Latinos represent more than 20 percent of all major leaguers. If not for a 1974 U.S. Labor Department quota limiting how many foreign-born players can play professional baseball, this number might be even larger (Hoose, 1989).

Recently, Latinos in sports have gained more recognition as books and websites have been created to describe their accomplishments. For example, the website Latino Legends in Sports was created in 1999 to inform people about the contributions of Latino and Latina athletes (see **http://www.latinosportslegends.com**). Recently, the website has featured interviews with boxer John Ruiz, baseball Hall of Famer Tony Perez, and Olympic speed skater Derek Parra (Latino Legends in Sports, 2007).

Education is a crucial issue for Latinos/as. Because of past discrimination and unequal educational opportunities, many Latinos/as currently have low levels of educational attainment. Many are unable to attend college or participate in collegiate sports, which is essential for being drafted in professional sports other than baseball. Consequently, the overall number of Latinas/os in college and professional sports is low compared to the rest of the U.S. population who are in this age bracket.

Middle Eastern Americans

Since 1970, many immigrants have arrived in the United States from countries located in the "Middle East," which is the geographic region from Afghanistan to Libya and includes Arabia, Cyprus, and Asiatic Turkey. Placing people in the "Middle Eastern" American category is somewhat like placing wide diversities of people in the categories of Asian American or Latino/a; some U.S. residents trace their origins to countries such as Bahrain, Egypt, Iran, Iraq, Kuwait, Lebanon, Oman, Qatar, Saudi Arabia, Syria, UAE (United Arab Emirates), and Yemen. Middle Eastern Americans speak a variety of languages and have diverse religious backgrounds: Some are Muslim, some are Coptic Christian, and others are Melkite Catholic. Although some are from working-class families, Lebanese Americans, Syrian Americans, Iranian Americans, and Kuwaiti Americans primarily come from middle- and upper-income family backgrounds. For example, Iranian Americans are scientists, professionals, and entrepreneurs.

In cities across the United States, Muslims have established social, economic, and ethnic enclaves. On the Internet, they have created websites that provide information about Islamic centers, schools, and lists of businesses and services available from those who adhere to Islam, one of the fastest-growing religions in this country (see Chapter 17, "Religion"). In cities such as Seattle, incorporation into the economic mainstream has been relatively easy for Palestinian immigrants who left their homeland in the 1980s. Some have found well-paid employment with corporations such as Microsoft because they bring educational skills and talents to the information-based economy, including the ability to translate software into Arabic for Middle Eastern markets (M. Ramirez, 1999). In the United States, Islamic schools and centers often bring together people from a diversity of countries such as Egypt and Pakistan. Many Muslim leaders and parents focus on how to raise children to be good Muslims and good U.S. citizens. However, recent immigrants continue to be torn between establishing roots in the United States and the continuing divisions and strife that exist in their homelands. Some Middle Eastern Americans experience prejudice and discrimination based on their speech patterns, appearance (such as the *hijabs,* or "head-to-toe covering" that leaves only the face exposed, which many girls and women wear), or the assumption that "all Middle Easterners" are somehow associated with terrorism.

Following the September 11, 2001, attacks on the United States by terrorists whose origins were traced to the Middle East, hate crimes and other forms of discrimination escalated in this country against people who were assumed to be Arabs, Arab Americans, or Muslims. With the passage of the U.S. Patriot Act—a law giving the federal government greater authority to engage in searches and surveillance with less judicial review than previously—in the aftermath of the terrorist attacks, many Arab Americans have expressed concern that this law is being used to target people who appear to be of Middle Eastern origins. To counter this potential oppression and loss of rights, some Arab Americans have engaged in social activism to highlight their concerns (Feagin and Feagin, 2008).

© Jim West/The Image Works

Muslims in the United States who wear traditional attire may face prejudice and/or discrimination as they go about their daily lives.

Middle Eastern Americans and Sports Although more Islamic schools are beginning to focus on sports, particularly for teenage boys, there has been less emphasis on competitive athletics among many Middle Eastern Americans. Because of popular sporting events in their countries of origin, some Middle Eastern Americans play golf or soccer. As well, some Iranian Americans follow the soccer careers of professional players from Iran, who now play for German, Austrian, Belgian, and Greek clubs. Keeping up with global sporting events is easy with all-sports television cable channels and websites that provide up-to-the-minute information about players and competitions. Over time, there will probably be greater participation by Middle Eastern American males in competitions such as soccer and golf; however, girls and women in Muslim families are typically not allowed to engage in athletic activities. Although little research has been done on this issue in the United States, one study of Islamic countries in the Middle East found that female athletes face strong cultural opposition to their sports participation (Dupre and Gains, 1997).

Global Racial and Ethnic Inequality in the Future

Throughout the world, many racial and ethnic groups seek *self-determination*—the right to choose their own way of life. As many nations are currently structured, however, self-determination is impossible.

Worldwide Racial and Ethnic Struggles

The cost of self-determination is the loss of life and property in ethnic warfare. In recent years, the Cold War has given way to dozens of smaller wars over ethnic dominance. In Europe, for example, ethnic violence has persisted in Yugoslavia, Spain, Britain (between the Protestant majority and the Catholic minority in Northern Ireland), Romania, Russia, Moldova, and Georgia. Ethnic violence continues in the Middle East, Africa, Asia, and Latin America. Hundreds of thousands have died from warfare, disease (such as the cholera epidemic in war-torn Rwanda), and refugee migration.

Ethnic wars have a high price even for survivors, whose life chances can become bleaker even after the violence subsides. In ethnic conflict between Abkhazians and Georgians in the former Soviet Union, for example, as many as two thousand people have been killed and over eighty thousand displaced. More recently, ethnic hatred has devastated the province of Kosovo, which is located in Serbia—Yugoslavia's dominant republic—and brought about the deaths of thousands of ethnic Albanians (Bennahum, 1999).

In the twenty-first century, the struggle between the Israeli government and various Palestinian factions over the future and borders of Palestine continues to make headlines. Discord in this region has heightened tensions among people not only in Israel and Palestine but also in the United States and around the world as deadly clashes continue and political leaders are apparently unable to reach a lasting solution to the decades-long strife.

Growing Racial and Ethnic Diversity in the United States

Racial and ethnic diversity is increasing in the United States. African Americans, Latinos/as, Asian Americans, and Native Americans constitute one-fourth of the U.S. population, whereas whites are a shrinking percentage of the population. In 2007, white Americans made up about 70 percent of the population, in contrast to 80 percent in 1980. It is predicted that by

Box 10.4 You Can Make a Difference

Working for Racial Harmony

Suppose that you are talking with several friends about a series of racist incidents at your college. Having studied the sociological imagination, you decide to start an organization similar to No Time to Hate, which was started at Emory University several years ago to reduce racism on campus. In analyzing racism, your group identifies factors contributing to the problem: (1) divisiveness between different cultural and ethnic communities, (2) persistent lack of trust, (3) the fact that many people never really communicate with one another, (4) the need to bring different voices into the curriculum and college life generally, and (5) the need to learn respect for people from different backgrounds (Loeb, 1994). Your group also develops a set of questions to be answered regarding racism on campus:

- *Encouraging inclusion and acceptance.* Do members of our group reflect the college's racial and ethnic diversity? How much do I know about other people's history and culture? How can I become more tolerant—or accepting—of people who are different from me?
- *Raising consciousness.* What is racism? What causes it? Can people participate in racist language and behavior without realizing what they are doing? What is our college or university doing to reduce racism?
- *Becoming more self-aware.* How much do I know about my own family roots and ethnic background? How do the families and communities in which we grow up affect our perceptions of racial and ethnic relations?
- *Using available resources.* What resources are available for learning more about working to reduce racism? Here are some agencies to contact:
  - ACLU (American Civil Liberties Union), 132 West 43rd Street, New York, NY 10036. (212) 944-9800. Online: **http://www.aclu.org**
  - ADL (Anti-Defamation League), 823 United Nations Plaza, New York, NY 10017. (212) 490-2525. Online: **http://www.adl.org**
  - NAACP (National Association for the Advancement of Colored People), 4805 Mt. Hope Drive, Baltimore, MD 21215. (410) 358-8900. Online: **http://www.naacp.org**
  - National Council of La Raza, 1119 19th, NW, Suite 1000, Washington, DC 20036. Online: **http://www.nclr.org**

What additional items would you add to the list of problem areas on your campus? How might your group's objective be reached? Over time, many colleges and universities have been changed as a result of involvement by students like you!

2056, the roots of the average U.S. resident will be in Africa, Asia, Hispanic countries, the Pacific islands, and the Middle East—not white Europe.

What effect will these changes have on racial and ethnic relations? Several possibilities exist. On the one hand, conflicts may become more overt and confrontational as people continue to use *sincere fictions*—personal beliefs that reflect larger societal mythologies, such as "I am not a racist" or "I have never discriminated against anyone"—even when these are inaccurate perceptions (Feagin and Vera, 1995). Interethnic tensions may increase as competition for education, jobs, and other resources continues to grow.

On the other hand, there is reason for cautious optimism. Throughout U.S. history, members of diverse racial and ethnic groups have struggled to gain the freedom and rights that were previously withheld from them. Today, minority grassroots organizations are pressing for affordable housing, job training, and educational opportunities (Feagin and Feagin, 2008). As discussed in Box 10.4, movements composed of both whites and people of color continue to oppose racism in everyday life, to seek to heal divisions among racial groups, and to teach children about racial tolerance (Rutstein, 1993). Many groups hope not only to affect their own microcosm but also to contribute to worldwide efforts to end racism (Ford, 1994).

To eliminate racial discrimination, it will be necessary to equalize opportunities in schools and workplaces. As Michael Omi and Howard Winant (1994: 158) have emphasized,

> Today more than ever, opposing racism requires that we notice race, not ignore it, that we afford it the recognition it deserves and the subtlety it embodies. By noticing race we can begin to challenge racism, with its ever-more-absurd reduction of human experience to an essence attributed to all without regard for historical or social context. . . . By noticing race we can develop the political insight and mobilization necessary to make the U.S. a more racially just and egalitarian society.

Chapter Review

• How do race and ethnicity differ?

A race is a category of people who have been singled out as inferior or superior, often on the basis of physical characteristics such as skin color, hair texture, or eye shape. An ethnic group is a collection of people distinguished primarily by cultural or national characteristics, including unique cultural traits, a sense of community, a feeling of ethnocentrism, ascribed membership, and territoriality.

• What are dominant and subordinate groups?

A dominant group is an advantaged group that has superior resources and rights in society. A subordinate group is a disadvantaged group whose members are subjected to unequal treatment by the majority group. Use of the terms *dominant* and *subordinate* reflects the importance of power in relationships.

• How is prejudice related to discrimination?

Prejudice is a negative attitude often based on stereotypes, which are overgeneralizations about the appearance, behavior, or other characteristics of all members of a group. Discrimination involves actions or practices of dominant-group members that have a harmful impact on members of a subordinate group.

• What are the major psychological explanations of prejudice?

According to the frustration–aggression hypothesis of prejudice, people frustrated in their efforts to achieve a highly desired goal may respond with aggression toward others, who then become scapegoats. Another theory of prejudice focuses on the authoritarian personality, marked by excessive conformity, submissiveness to authority, intolerance, insecurity, superstition, and rigid thinking.

• How do individual discrimination and institutional discrimination differ?

Individual discrimination involves actions by individual members of the dominant group that harm members of subordinate groups or their property. Institutional discrimination involves day-to-day practices of organizations and institutions that have a harmful impact on members of subordinate groups.

• How do sociologists view racial and ethnic group relations?

Symbolic interactionists suggest that increased contact between people from divergent groups should lead to favorable attitudes and behavior when members of each group (1) have equal status, (2) pursue the same goals, (3) cooperate with one another to achieve goals, and (4) receive positive feedback when they interact with one another. Functionalists stress that members of subordinate groups become a part of the mainstream through assimilation, the process by which members of subordinate groups become absorbed into the dominant culture. Conflict theorists focus on economic stratification and access to power in race and ethnic relations. The caste perspective views inequality as a permanent feature of society, whereas class perspectives focus on the link between capitalism and racial exploitation. According to racial formation theory, the actions of the U.S. government substantially define racial and ethnic relations.

• How have the experiences of various racial–ethnic groups differed in the United States?

Native Americans suffered greatly from the actions of European settlers, who seized their lands and made them victims of forced migration and genocide. Today, they lead lives characterized by poverty and lack of opportunity. White Anglo-Saxon Protestants are the most privileged group in the United States, although social class and gender affect their life chances. White ethnic Americans, whose ancestors migrated from Southern and Eastern European countries, have gradually made their way into the mainstream of U.S. society. Following the abolishment of slavery through the Emancipation Proclamation in 1863, African Americans were still subjected to segregation, discrimination, and lynchings. More recently, despite civil rights legislation and economic and political gains by many African Americans, racial prejudice and discrimination still exist. Asian American immigrants as a group have enjoyed considerable upward mobility in U.S. society in recent decades, but many Asian Americans still struggle to survive by working at low-paying jobs and living in urban ethnic enclaves. Although some Latinos/as have made substantial political, economic, and professional gains in U.S. society, as a group they still are subjected to anti-immigration sentiments. Middle Eastern immigrants to the United States speak a variety of languages and have diverse religious backgrounds. Because they generally come from middle-class backgrounds, they have made inroads into mainstream U.S. society.

www.cengage.com/login

Want to maximize your online study time? Take this easy-to-use study system's diagnostic pre-test, and it will create a personalized study plan for you. By helping you to identify the topics that you need to understand better and then directing you to valuable online resources, it can speed up your chapter review. CengageNOW even provides a post-test so you can confirm that you are ready for an exam.

Key Terms

assimilation 320
authoritarian personality 317
discrimination 318
dominant group 314
ethnic group 311
ethnic pluralism 322
genocide 319
individual discrimination 319
institutional discrimination 319
internal colonialism 325
prejudice 314
race 311
racism 315
scapegoat 316
segregation 322
social distance 318
split labor market 326
stereotypes 314
subordinate group 314

Questions for Critical Thinking

1. Do you consider yourself defined more strongly by your race or by your ethnicity? How so?
2. Given that subordinate groups have some common experiences, why is there such deep conflict between some of these groups?
3. What would need to happen in the United States, both individually and institutionally, for a positive form of ethnic pluralism to flourish in the twenty-first century?

The Kendall Companion Website

www.cengage.com/sociology/kendall

Visit this book's companion website, where you'll find more resources to help you study and successfully complete course projects. Resources include quizzes and flash cards, as well as special features such as an interactive sociology timeline, maps, General Social Survey (GSS) data, and Census 2000 data. The site also provides links to useful websites that have been selected for their relevance to the topics in this chapter and include those listed below. (*Note:* Visit the book's website for updated URLs.)

Race & Ethnicity: Resources
http://eserver.org/race/resources.html

This site consists of reference materials, articles, literary works, and additional resources that address issues of race and ethnicity in the United States. The site also includes a useful search engine to help you locate specific material of interest.

American Studies Crossroads Project
http://crossroads.georgetown.edu

Maintained by Crossroads, a project of the American Studies Association (sponsored by Georgetown University), this site houses the largest bibliography of Web-based resources in the field of American studies. On the opening page, click on "American Studies Web," scroll down the page, and select "Ethnicity, Race, and Religious Cultures." This link will connect you to a well-organized selection of resources related to the study of race and ethnicity in the United States.

chapter 11 Sex and Gender

Chapter Focus Question

How do expectations about female and male appearance, and especially weight, reflect gender inequality?

As I sat in the theater at the Aladdin Hotel on the Strip [in Las Vegas, Nevada], I was enclosed by a sea of crowns. Little girls and teenagers attended the [Miss America] Pageant in droves, many wearing the crowns and sashes that represented their biggest pageant victories.

Sitting in the middle of the cheering section for Miss Kansas, surrounded by cardboard daisies on wooden sticks with Miss Kansas's face in the center and shouts of "You go girl!" I felt as if I were at a political convention or a religious revival. Also marooned in the Miss Kansas section, without any daisies, were a little girl, her mother (who had twice competed in the Miss Nevada state pageant), and her grandmother, who sat in front of me. The twelve-year-old watched

Jennifer Berry, Miss Oklahoma, accepts her crown as the 2006 Miss America at the Aladdin Casino in Las Vegas. What is your opinion of pageants such as this one?

IN THIS CHAPTER

- Sex: The Biological Dimension
- Gender: The Cultural Dimension
- Gender Stratification in Historical and Contemporary Perspective
- Gender and Socialization
- Contemporary Gender Inequality
- Perspectives on Gender Stratification
- Gender Issues in the Future

the Pageant with wide eyes the whole night. She wore no crown or sash, but she looked smart in a black velvet dress.

After the talent segment, the girl turned to me and asked, "When I'm in the Miss America Pageant, I want to play the piano and the saxophone for my talent. I can switch back and forth. Do you think that would work?"

"Well, it would certainly be different," I replied.

"Good, then that would help me win."

"So, you really want to be Miss America someday?"

The girl nodded her head, her face solemn. Before replying, I paused. "Well, you can do that. But, you know, there are so many other things to do besides being a beauty queen."

The little girl did not hear me. She was rapturously watching as Miss Oklahoma was crowned Miss America 2006. All the other girls in the audience, those with crowns and those without, stood together, mouthing the words to the famous theme song as the new Miss America was serenaded by the voice of the great Pageant

Sharpening Your Focus

- *How do a society's resources and economic structure influence gender stratification?*
- *What are the primary agents of gender socialization?*
- *How does the contemporary workplace reflect gender stratification?*
- *How do functionalist, conflict, and feminist perspectives on gender stratification differ?*

emcee, Bert Parks, who died in 1992: *"There she is, Miss America, there she is, your ideal. . . ."*

—Hilary Levey (2007: 72), a graduate student in sociology at Princeton University, describes her thoughts upon attending a Miss America Pageant. Although Ms. Levey never competed in a beauty pageant, her mother was named Miss America (from Michigan) in 1970.

Many little girls are similar to the one that Hilary Levey encountered at the Miss America Pageant: They have their hearts set on being chosen as the winner of a beauty and/or talent competition such as Miss America or Miss USA. Tens of thousands of beauty pageants—ranging from local beach bikini pageants to international scholarship competitions—are held annually. Two competitions—the Miss America Scholarship program and the Miss USA Pageant—involve more than 7,500 local and regional pageants across the country each year (Banet-Weiser, 1999).

Of course, all pageants are not identical. For example, organizers of the Miss America Pageant claim that their competition focuses on both talent and beauty and that it exists "to provide personal and professional opportunities for young women and to promote their voices in culture, politics and the community" (Miss America, 2008). By contrast, Miss USA, which is part of the Miss Universe system and partly owned by Donald Trump, originated as a "bathing beauty" competition that was sponsored by a swimwear company. Today, Miss USA and its younger counterpart, Miss Teen USA, continue to look for female models who look outstanding in swimsuits and evening gowns, and who can promote a variety array of products ranging from suntan lotion to flashy diamonds (Angelotti, 2006). Regardless of somewhat different stated goals, these talent and beauty competitions are really about physical beauty and appearance.

For this reason, competitions such as Miss America, Miss USA, Miss Universe, and Miss Teen USA have been the subject of both praise and criticism for the ways in which they portray girls and young women. Some individuals believe that beauty pageants are good for women because they encourage individual achievement and promote self-confidence. Pageant winners are often praised for being positive role models for young women, particularly if the title holder remains scandal free during the year of her reign. However, some critics of beauty pageants claim that these events promote an unrealistic beauty ideal that is not attainable for most people and that is not necessarily desirable in the real world (see Banet-Weiser, 1999). Other critics believe that pageants are degrading to women because the contestants are ranked "like prize horses" and given a sash to put around their neck (Corsbie-Massay, 2005: 1). Some feminist analysts argue that beauty pageants objectify women (Watson and Martin, 2004).

What is objectification? *Objectification* is the process whereby some people treat other individu-

◆ **Table 11.1 The Objectification of Women**

General Aspects of Objectification	Objectification Based on Cultural Preoccupation with "Looks"
Women are responded to primarily as "females," whereas their personal qualities and accomplishments are of secondary importance.	Women are often seen as the objects of sexual attraction, not full human beings—for example, when they are stared at.
Women are seen as "all alike."	Women are seen by some as depersonalized body parts—for example, "a piece of ass."
Women are seen as being subordinate and passive, so things can easily be "done to a woman"—for example, discrimination, harassment, and violence.	Depersonalized female sexuality is used for cultural and economic purposes—such as in the media, advertising, the fashion and cosmetics industries, and pornography.
Women are seen as easily ignored or trivialized.	Women are seen as being "decorative" and status-conferring objects to be bought (sometimes collected) and displayed by men and sometimes by other women.
	Women are evaluated according to prevailing, narrow "beauty" standards and often feel pressure to conform to appearance norms.

Source: Schur, 1983.

Box 11.1 Sociology and Everyday Life

How Much Do You Know About Body Image and Gender?

True	False	
T	F	1. Most people have an accurate perception of their physical appearance.
T	F	2. Recent studies show that up to 95 percent of men express dissatisfaction with some aspect of their bodies.
T	F	3. Many young girls and women believe that being even slightly "overweight" makes them less "feminine."
T	F	4. Physical attractiveness is a more central part of self-concept for women than for men.
T	F	5. Contestants in beauty pageants such as Miss America have remained about the same in body size throughout the history of these competitions.
T	F	6. Thinness has always been the "ideal" body image for women.
T	F	7. Women bodybuilders have gained full acceptance in society.
T	F	8. The media play a significant role in shaping societal perceptions about the ideal female body.

Answers on page 348.

als as if they were objects or things, not human beings. For example, we objectify women—or men—when we judge them strictly on the basis of their physical appearance rather than on their individual qualities, attributes, or actions (Schur, 1983). Although men may be objectified in some societies, the objectification of girls and women is widespread and particularly common in the United States and many other nations (see ◆ Table 11.1). In regard to beauty pageants, organizers seek to deflect this criticism by providing contestants with an opportunity to talk about themselves and their interests or to answer questions that supposedly will show that they are intelligent and knowledgeable about current events. At the end of each pageant, however, the winner's physical attractiveness is most often highlighted rather than the true substance of her life (Angelotti, 2006). Although some people think of the Miss America Pageant and similar competitions as a vestige of the past, many women and men in the twenty-first century are strongly influenced by the images that these competitions project regarding beauty, body image, race/ethnicity, identity, and consumerism (Watson and Martin, 2004).

Some differences between men and women are biological in nature; however, many differences between the sexes are socially constructed. Studying sociology makes us aware of differences that relate to gender (a social concept) as well as differences that are based on a person's biological makeup, or sex. In this chapter, we examine the issue of gender: what it is and how it affects us. Before reading on, test your knowledge about body image and gender by taking the quiz in Box 11.1.

Sex: The Biological Dimension

Whereas the word *gender* is often used to refer to the distinctive qualities of men and women (masculinity and femininity) that are culturally created, ***sex* refers to the biological and anatomical differences between females and males.** At the core of these differences is the chromosomal information transmitted at the moment a child is conceived. The mother contributes an X chromosome and the father either an X (which produces a female embryo) or a Y chromosome (which produces a male embryo). At birth, male and female infants are distinguished by ***primary sex characteristics:* the genitalia used in the reproductive process.** At puberty, an increased production of hormones results in the development of ***secondary sex characteristics:* the physical traits (other than reproductive organs) that identify an individual's sex.** For women, these include larger breasts, wider hips, narrower shoulders, a layer of fatty tissue throughout the body, and menstruation. For

sex the biological and anatomical differences between females and males.

primary sex characteristics the genitalia used in the reproductive process.

secondary sex characteristics the physical traits (other than reproductive organs) that identify an individual's sex.

Box 11.1 Sociology and Everyday Life

Answers to the Sociology Quiz on Body Image and Gender

1. False. Many people do not have a very accurate perception of their own bodies. For example, many young girls and women think of themselves as "fat" when they are not. Some young boys and men tend to believe that they need a well-developed chest and arm muscles, broad shoulders, and a narrow waist.

2. True. In recent studies, up to 95 percent of men believed they needed to improve some aspect of their bodies.

3. True. More than half of all adult women in the United States are currently dieting, and over three-fourths of normal-weight women think they are "too fat." Recently, very young girls have developed similar concerns. For example, 80 percent of fourth-grade girls in one study were watching their weight.

4. True. Women have been socialized to believe that being physically attractive is very important. Studies have found that weight and body shape are the central determinants of women's perception of their physical attractiveness.

5. False. Contestants in Miss America and other national beauty pageants decreased in body size, becoming much thinner and less curvaceous, during the 1980s and 1990s. Today, more emphasis is place on being physically fit, but winning contestants typically have much lower body weight than the average woman of their height and age.

6. False. The "ideal" body image for women has changed a number of times. A positive view of body fat has prevailed for most of human history; however, in the twentieth century in the United States, this view gave way to "fat aversion."

7. False. Although bodybuilding among women has gained some degree of acceptance, women bodybuilders are still expected to be very "feminine" and not to overdevelop themselves.

8. True. Women in the United States are bombarded by advertising, television programs, and films containing images of women that typically represent an ideal that most real women cannot attain.

Sources: Based on Fallon, Katzman, and Wooley, 1994; Kilbourne, 1999; Seid, 1994; and Turner, 1997.

men, they include development of enlarged genitals, a deeper voice, greater height, a more muscular build, and more body and facial hair.

Hermaphrodites/Transsexuals

Sex is not always clear-cut. Occasionally, a hormone imbalance before birth produces a ***hermaphrodite*—a person in whom sexual differentiation is ambiguous or incomplete.** Hermaphrodites tend to have some combination of male and female genitalia. In one case, for example, a chromosomally normal (XY) male was born with a penis just one centimeter long and a urinary opening similar to that of a female. Some people may be genetically of one sex but have a gender identity of the other. That is true for a ***transsexual*, a person in whom the sex-related structures of the brain that define gender identity are opposite from the physical sex organs of the person's body.** Consequently, transsexuals often feel that they are the opposite sex from that of their sex organs. Transsexuals may become aware of this conflict between gender identity and physical sex as early as the preschool years. Some transsexuals take hormone treatments or have a sex-change operation to alter their genitalia in order to achieve a body congruent with their sense of sexual identity. Many transsexuals who receive hormone treatments or undergo surgical procedures go on to lead lives that they view as being compatible with their true sexual identity.

Western societies acknowledge the existence of only two sexes; some other societies recognize three—men, women, and *berdaches* (or *hijras* or *xaniths*), biological males who behave, dress, work, and are treated in most respects as women. The closest approximation of a third sex in Western societies is a ***transvestite*, a male who lives as a woman or a female who lives as a man but does not alter the genitalia.** Although transvestites are not treated as a third sex, they often "pass" for members of that sex because their appearance and mannerisms fall within the range of what is expected from members of the other sex.

Transsexuality may occur in conjunction with homosexuality, but this is frequently not the case. Some researchers believe that both transsexuality and homosexuality have a common prenatal cause such as a critically timed hormonal release due to stress in the mother or the presence of certain hormone-mimicking chemicals during critical steps of fetal development. Researchers continue to examine this issue and debate the origins of transsexuality and homosexuality.

Sexual Orientation

Sexual orientation **refers to an individual's preference for emotional–sexual relationships with members of the opposite sex (heterosexuality), the same sex (homosexuality), or both (bisexuality).** Some scholars believe that sexual orientation is rooted in biological factors that are present at birth; others believe that sexuality has both biological and social components and is not preordained at birth.

The terms *homosexual* and *gay* are most often used in association with males who prefer same-sex relationships; the term *lesbian* is used in association with females who prefer same-sex relationships. Heterosexual individuals, who prefer opposite-sex relationships, are sometimes referred to as *straight*. However, it is important to note that heterosexual people are much less likely to be labeled by their sexual orientation than are people who are gay, lesbian, or bisexual.

What criteria do social scientists use to classify individuals as gay, lesbian, or homosexual? In a definitive study of sexuality in the mid-1990s, researchers at the University of Chicago established three criteria for identifying people as homosexual or bisexual: (1) *sexual attraction* to persons of one's own gender, (2) *sexual involvement* with one or more persons of one's own gender, and (3) *self-identification* as a gay, lesbian, or bisexual (Michael et al., 1994). According to these criteria, then, having engaged in a homosexual act does not necessarily classify a person as homosexual. In fact, many respondents in the University of Chicago study indicated that although they had at least one homosexual encounter when they were younger, they were no longer involved in homosexual conduct and never identified themselves as gay, lesbian, or bisexual.

Studies have examined how sexual orientation is linked to identity. Sociologist Kristin G. Esterberg (1997) interviewed lesbian and bisexual women to determine how they "perform" lesbian or bisexual identity through daily activities such as choice of clothing and hairstyles, as well as how they use body language and talk. According to Esterberg (1997), some of the women viewed themselves as being "lesbian from birth" whereas others had experienced shifts in their identities, depending on social surroundings, age, and political conditions at specific periods in their lives. Another study looked at gay and bisexual men. Human development scholar Ritch C. Savin-Williams (2004) found that gay/bisexual youths often believe from an early age that they are different from other boys:

> The pattern that most characterized the youths' awareness, interpretation, and affective responses to childhood attractions consisted of an overwhelming desire to be in the company of men. They wanted to touch, smell, see, and hear masculinity. This awareness originated from earliest childhood memories; in this sense, they "always felt gay."

However, most of the boys and young men realized that these feelings were not typical of other males and were uncomfortable when others attempted to make them conform to the established cultural definitions of masculinity, such as showing a great interest in team sports, competition, and aggressive pursuits.

The term *transgender* was created to describe individuals whose appearance, behavior, or self-identification does not conform to common social rules of gender expression. Transgenderism is sometimes used to refer to those who cross-dress, to transsexuals, and to others outside mainstream categories. Although some gay and lesbian advocacy groups oppose the concept of transgender as being somewhat meaningless, others applaud the term as one that might help unify diverse categories of people based on sexual identity. Various organizations of gays, lesbians, and transgendered persons have been unified in their desire to reduce hate crimes and other forms of ***homophobia*—extreme prejudice directed at gays, lesbians, bisexuals, and others who are perceived as not being heterosexual.**

hermaphrodite a person in whom sexual differentiation is ambiguous or incomplete.

transsexual a person in whom the sex-related structures of the brain that define gender identity are opposite from the physical sex organs of the person's body.

transvestite a male who lives as a woman or a female who lives as a man but does not alter the genitalia.

sexual orientation a person's preference for emotional–sexual relationships with members of the opposite sex (heterosexuality), the same sex (homosexuality), or both (bisexuality).

homophobia extreme prejudice directed at gays, lesbians, bisexuals, and others who are perceived as not being heterosexual.

Gender: The Cultural Dimension

***Gender* refers to the culturally and socially constructed differences between females and males found in the meanings, beliefs, and practices associated with "femininity" and "masculinity."** Although biological differences between women and men are very important, in reality most "sex differences" are socially constructed "gender differences." According to sociologists, social and cultural processes, not biological "givens," are the most important factors in defining what females and males are, what they should do, and what sorts of relations do or should exist between them. Sociologist Judith Lorber (1994: 6) summarizes the importance of gender:

> Gender is a human invention, like language, kinship, religion, and technology; like them, gender organizes human social life in culturally patterned ways. Gender organizes social relations in everyday life as well as in the major social structures, such as social class and the hierarchies of bureaucratic organizations.

Virtually everything social in our lives is *gendered:* People continually distinguish between males and females and evaluate them differentially. Gender is an integral part of the daily experiences of both women and men (Kimmel and Messner, 2004).

A microlevel analysis of gender focuses on how individuals learn gender roles and acquire a gender identity. ***Gender role* refers to the attitudes, behavior, and activities that are socially defined as appropriate for each sex and are learned through the socialization process.** For example, in U.S. society, males are traditionally expected to demonstrate aggressiveness and toughness, whereas females are expected to be passive and nurturing. ***Gender identity* is a person's perception of the self as female or male.** Typically established between eighteen months and three years of age, gender identity is a powerful aspect of our self-concept. Although this identity is an individual perception, it is developed through interaction with others. As a result, most people form a gender identity that matches their biological sex: Most biological females think of themselves as female, and most biological males think of themselves as male. Body consciousness is a part of gender identity. ***Body consciousness* is how a person perceives and feels about his or her body;** it also includes an awareness of social conditions in society that contribute to this self-knowledge (Thompson, 1994). Consider, for example, these comments by Steve Michalik, a former Mr. Universe:

> I was small and weak, and my brother Anthony was big and graceful, and my old man made no bones about loving him and hating me. . . . The minute I walked in from school, it was, "You worthless little s--t, what are you doing home so early?" His favorite way to torture me was to tell me he was going to put me in a home. We'd be driving along in Brooklyn somewhere, and we'd pass a building with iron bars on the windows, and he'd stop the car and say to me,

© Alex Farnsworth/The Image Works

© MOHAMMED AMEEN/Reuters/Landov

In what ways do these two figures contradict traditional gender expectations? Do you see this trend as a healthy one?

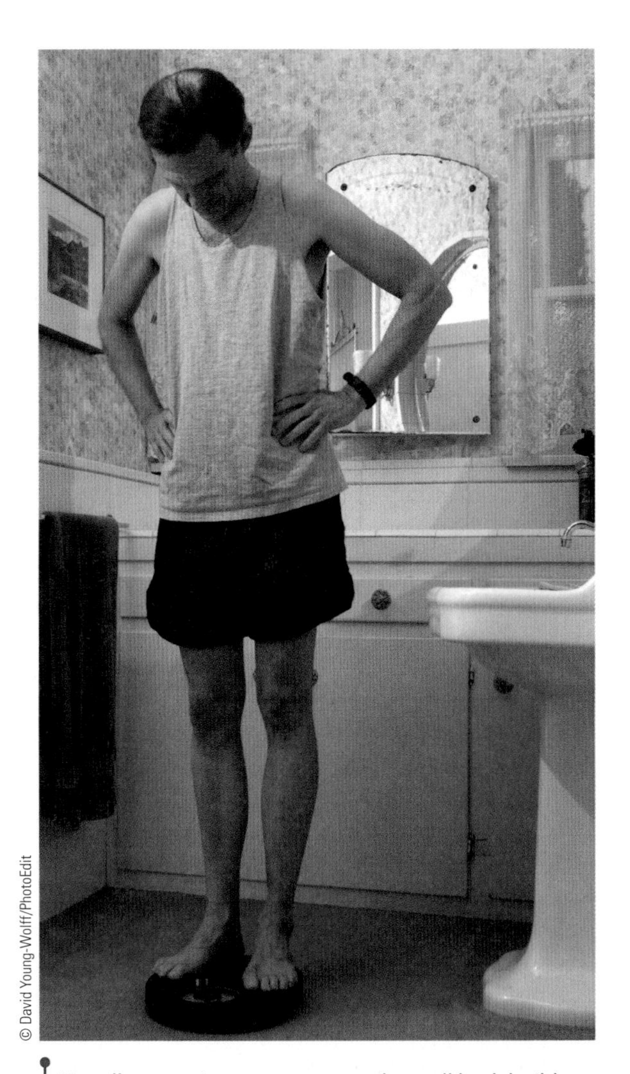
© David Young-Wolff/PhotoEdit

© A. Ramey/PhotoEdit

Not all anorectics are women, and not all bodybuilders are men. However, Susan Bordo argues that these two issues are manifestations of the same desire: to avoid having soft, flabby flesh.

> "Get out. This is the home we're putting you in." I'd be standing there sobbing on the curb—I was maybe eight or nine at the time. (qtd. in Klein, 1993: 273)

As we grow up, we become aware, as Michalik did, that the physical shape of our bodies subjects us to the approval or disapproval of others. Being small and weak may be considered positive attributes for women, but they are considered negative characteristics for "true men."

A macrolevel analysis of gender examines structural features, external to the individual, that perpetuate gender inequality. These structures have been referred to as *gendered institutions,* meaning that gender is one of the major ways by which social life is organized in all sectors of society. Gender is embedded in the images, ideas, and language of a society and is used as a means to divide up work, allocate resources, and distribute power. For example, every society uses gender to assign certain tasks—ranging from child rearing to warfare—to females and to males, and differentially rewards those who perform these duties.

These institutions are reinforced by a *gender belief system,* which includes all the ideas regarding masculine and feminine attributes that are held to be valid in a society. This belief system is legitimated by religion, science, law, and other societal values (Lorber,

gender the culturally and socially constructed differences between females and males found in the meanings, beliefs, and practices associated with "femininity" and "masculinity."

gender role the attitudes, behavior, and activities that are socially defined as appropriate for each sex and are learned through the socialization process.

gender identity a person's perception of the self as female or male.

body consciousness a term that describes how a person perceives and feels about his or her body.

1994, 2005). For example, gendered belief systems may change over time as gender roles change. Many fathers take care of young children today, and there is a much greater acceptance of this change in roles. However, popular stereotypes about men and women, as well as cultural norms about gender-appropriate appearance and behavior, serve to reinforce gendered institutions in society.

The Social Significance of Gender

Gender is a social construction with important consequences in everyday life. Just as stereotypes regarding race/ethnicity have built-in notions of superiority and inferiority, gender stereotypes hold that men and women are inherently different in attributes, behavior, and aspirations. Stereotypes define men as strong, rational, dominant, independent, and less concerned with their appearance. Women are stereotyped as weak, emotional, nurturing, dependent, and anxious about their appearance.

The social significance of gender stereotypes is illustrated by eating problems. The three most common eating problems are anorexia, bulimia, and obesity. With *anorexia,* a person has lost at least 25 percent of body weight due to a compulsive fear of becoming fat (Lott, 1994). With *bulimia,* a person binges by consuming large quantities of food and then purges the food by induced vomiting, excessive exercise, laxatives, or fasting. With *obesity,* individuals are 20 percent or more above their desirable weight, as established by the medical profession. For a 5-foot-4-inch woman, that is about twenty-five pounds; for a 5-foot-10-inch man, it is about thirty pounds (Burros, 1994: 1).

Sociologist Becky W. Thompson argues that, based on stereotypes, the primary victims of eating problems are presumed to be white, middle-class, heterosexual women. However, such problems also exist among women of color, working-class women, lesbians, and some men. According to Thompson, explanations regarding the relationship between gender and eating problems must take into account a complex array of social factors, including gender socialization and women's responses to problems such as racism and emotional, physical, and sexual abuse (Thompson, 1994; see also Wooley, 1994).

Bodybuilding is another gendered experience. *Bodybuilding* is the process of deliberately cultivating an increase in the mass and strength of the skeletal muscles by means of lifting and pushing weights (Mansfield and McGinn, 1993). In the past, bodybuilding was predominantly a male activity; musculature connoted power, domination, and virility (Klein, 1993). Today, an increasing number of women engage in this activity. As gendered experiences, eating problems and bodybuilding have more in common than we might think. Women's studies scholar Susan Bordo (2004) has noted that the anorexic body and the muscled body are not opposites but instead are both united against the common enemy of soft, flabby flesh. In other words, the *body* may be objectified through both compulsive dieting and compulsive bodybuilding.

Sexism

***Sexism* is the subordination of one sex, usually female, based on the assumed superiority of the other sex.** Sexism directed at women has three components: (1) negative attitudes toward women; (2) stereotypical beliefs that reinforce, complement, or justify the prejudice; and (3) discrimination—acts that exclude, distance, or keep women separate (Lott, 1994).

Can men be victims of sexism? Although women are more often the target of sexist remarks and practices, men can be victims of sexist assumptions. According to the social psychologist Hilary M. Lips (2001), an example of sexism directed against men is the mistaken idea that it is more harmful for female soldiers to be killed in battle than male soldiers.

Like racism, sexism is used to justify discriminatory treatment. When women participate in what is considered gender-inappropriate endeavors in the workplace, at home, or in leisure activities, they often find that they are the targets of prejudice and discrimination. Obvious manifestations of sexism are found in the undervaluing of women's work and in hiring and promotion practices that effectively exclude women from an organization or confine them to the bottom of the organizational hierarchy. Even today, some women who enter nontraditional occupations (such as firefighting and welding) or professions (such as dentistry and architecture) encounter hurdles that men do not face (see "Sociology Works!").

Sexism is interwoven with ***patriarchy*—a hierarchical system of social organization in which cultural, political, and economic structures are controlled by men.** By contrast, ***matriarchy* is a hierarchical system of social organization in which cultural, political, and economic structures are controlled by women;** however, few (if any) societies have been organized in this manner. Patriarchy is reflected in the way that men may think of their position as men as a given whereas women may deliberate on what their position in society should be (see Box 11.2 for an example). As the sociologist Virginia Cyrus (1993: 6) explains, "Under patriarchy, men are seen as 'natural' heads of households, Presidential candidates, corporate executives, college presidents, etc. Women, on the other hand, are men's subordinates, playing such supportive roles as housewife, mother, nurse, and secretary." Gender inequality

Sociology *Works!*

Institutional Discrimination: Women in a Locker-Room Culture

News Item: Morgan Stanley, a bank holding company, to pay $46 million in a case charging systemic sex discrimination

Responses: "I'm so happy that there's a settlement that's good for everybody," said lead plaintiff Allison Schieffelin, a former bond trader who was awarded a settlement of $12 million. (qtd. in Shell, 2007)

"It sends a message to employers everywhere that allegations of discrimination need to be taken seriously," said Elizabeth Grossman, an Equal Employment Opportunity Commission lawyer. (qtd. in Shell, 2007)

For decades, sociologists have called attention to the fact that both individual discrimination and institutional discrimination—based on sex, race/ethnicity, and other devalued characteristics and attributes of subordinate-group members—are widespread (Benokraitis and Feagin, 1995). Previous sex-discrimination lawsuits typically involved details about the crude behavior of male employees toward women, or pornography in the workplace, or male bosses' demands that women employees accompany them to strip clubs and other objectionable locations as part of their work-related duties (Anderson, 2007). However, the Morgan Stanley case involved none of these issues and instead focused on women's opportunities for training, gaining new clients, promotion, and pay equity—all factors related to institutional discrimination, previously defined (in Chapter 9) as the day-to-day-practices of organizations and institutions that have a harmful impact on members of subordinate groups.

In the case of Morgan Stanley, the Wall Street investment bank, female employees were passed over for new clients, paid less than male employees, and often passed over for promotion. Although Morgan Stanley denied all charges, the firm agreed to pay at least $46 million to settle this class-action suit filed by eight current and former female brokers. A primary issue in the case involved the way in which accounts were distributed in the firm's retail branches. According to the lawsuit, a "locker-room culture" prevailed in which accounts were often given to golf buddies based on a "power ranking" system that dictated how the accounts of retiring and departing brokers were to be distributed and how new accounts were to be assigned. A new system put in place after the lawsuit will automate the distribution of accounts, brokers will receive written information about the distribution of accounts, and data will be more readily available to individuals who believe they are being discriminated against (Anderson, 2007).

Sociological theorizing and research have increased public awareness that the day-to-day practices of organizations and institutions may have a negative and differential effect on individuals who have historically been excluded from workplace settings such as prestigious Wall Street firms, which traditionally have been dominated by white males. Although many gains have been made through legislation and litigation to reduce institutional discrimination, recent cases such as the Morgan Stanley sex-bias lawsuit demonstrate that much remains to be done before women truly have equal opportunities in the workplace.

Reflect & Analyze

Why is it important for all employees to feel that they are being treated fairly at work? Are some employment settings more resistant to change than others? What do you think?

and a division of labor based on male dominance are nearly universal, as we will see in the following discussion on the origins of gender-based stratification.

Gender Stratification in Historical and Contemporary Perspective

How do tasks in a society come to be defined as "men's work" or "women's work"? Three factors are important in determining the gendered division of labor in a society: (1) the type of subsistence base, (2) the supply of and demand for labor, and (3) the extent to which women's child-rearing activities are compatible with certain types of work. As defined in Chapter 5, *subsistence* refers

sexism the subordination of one sex, usually female, based on the assumed superiority of the other sex.

patriarchy a hierarchical system of social organization in which cultural, political, and economic structures are controlled by men.

matriarchy a hierarchical system of social organization in which cultural, political, and economic structures are controlled by women.

Box 11.2 Sociology in Global Perspective

The Rise of Islamic Feminism in the Middle East?

I would like for all of the young Muslim girls to be able to relate to Iman, whether they wear the hijab [head scarf] or not. Boys will also enjoy Iman's adventures because she is one tough, smart girl! Iman gets her super powers from having very strong faith in Allah, or God. She solves many of the problems by explaining certain parts of the Koran that relate to the story.

—Rima Khoreibi, an author from Dubai (United Arab Emirates), explaining that she has written a book about an Islamic superhero who is female because she would like to dispel a widely held belief that sexism in her culture is deeply rooted in Islam (see theadventuresofiman.com, 2007; Kristof, 2006)

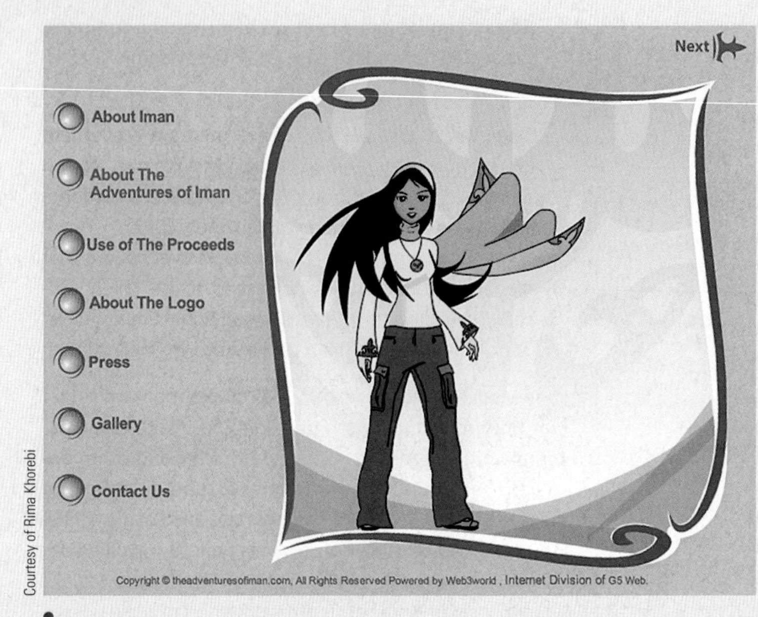

Courtesy of Rima Khorebi

The home page of *The Adventures of Iman*.

Although Rima Khoreibi and many others who have written fictional and nonfictional accounts of girls and women living in the Middle East typically do not deny that sexism exists in their region or that sexism is deeply interwoven with patriarchy around the world, they dispute the perception that Islam is inherently misogynistic (possessing hatred or strong prejudice toward women). As defined in this chapter, *patriarchy* is a hierarchical system of social organization in

to the means by which a society gains the basic necessities of life, including food, shelter, and clothing. You may recall that societies are classified, based on subsistence, as hunting and gathering societies, horticultural and pastoral societies, agrarian societies, industrial societies, and postindustrial societies. The first three of these categories are all *preindustrial* societies.

Preindustrial Societies

The earliest known division of labor between women and men is in hunting and gathering societies. While the men hunt for wild game, women gather roots and berries. A relatively equitable relationship exists because neither sex has the ability to provide all the food necessary for survival. When wild game is nearby, both men and women may hunt. When it is far away, hunting becomes incompatible with child rearing (which women tend to do because they breast-feed their young), and women are placed at a disadvantage in terms of contributing to the food supply (Lorber, 1994). In most hunting and gathering societies, women are full economic partners with men; relations between them tend to be cooperative and relatively egalitarian (Chafetz, 1984; Bonvillain, 2001). Little social stratification of any kind is found because people do not acquire a food surplus.

In horticultural societies, which first developed ten to twelve thousand years ago, a steady source of food becomes available. People are able to grow their own food because of hand tools, such as the digging stick and the hoe. Women make an important contribution to food production because hoe cultivation is compatible with child care. A fairly high degree of gender equality exists because neither sex controls the food supply.

When inadequate moisture in an area makes planting crops impossible, *pastoralism*—the domestication of large animals to provide food—develops. Herding is primarily done by men, and women contribute relatively little to subsistence production in such societies. In some herding societies, women have relatively low status; their primary value is their ability to produce male offspring so that the family lineage can be preserved and enough males will exist to protect the group against attack (Nielsen, 1990). Even so, the relationship between men and women is more equitable than it is in agrarian societies, which first developed about eight thousand to ten thousand years ago.

which cultural, political, and economic structures are controlled by men. The influence of religion on patriarchy is a topic of great interest to contemporary scholars, particularly those applying a feminist approach to their explanations of why persistent social inequalities exist between women and men and how these inequalities are greater in some regions of the world than in others.

According to some gender studies specialists, a newer form of feminist thinking is emerging among Muslim women. Often referred to as "feminist Islam" or "Islamic feminism," this approach is based on the belief that greater gender equality may be possible in the Muslim world if the teachings of Islam, as set forth in the Koran (or *Qur'an*)—the Islamic holy book—are followed more closely. Islamic feminism is based on the principle that Muslim women should retain their allegiance to Islam as an essential part of their self-determination and identity, but that they should also work to change patriarchal control over the basic Islamic world view (Wadud, 2002). According to the journalist Nicholas D. Kristof (2006), both Islam and evangelical Christianity have been on the rise in recent years because both religions provide "a firm moral code, spiritual reassurance and orderliness to people vexed by chaos and immorality around them, and they offer dignity to the poor."

Islamic feminists believe that the rise of Islam might contribute to greater, rather than less, equality for women. From this perspective, stories about characters such as Iman may help girls and young women realize that they can maintain their deep religious convictions and their head scarf (*hijab*) while, at the same time, working for greater equality for women and more opportunities for themselves. In *The Adventures of Iman,* the female hero always wears a pink scarf around her neck, and she uses the scarf to cover her hair when she is praying to Allah. Iman quotes the Qur'an when she is explaining to others that Muslims are expected to be tolerant, kind, and righteous. For Iman, religion is a form of empowerment, not an extension of patriarchy.

Islamic feminism is quite different from what most people think of as Western feminism (particularly in regard to issues such as the wearing of the *hijab* or the fact that in Saudi Arabia, a woman may own a motor vehicle but may not legally drive it). However, change is clearly under way in many regions of the Middle East and other areas of the world as rapid economic development and urbanization quickly change the lives of many people.

Reflect & Analyze

Why is women's inequality a complex issue to study across nations? What part does culture play in defining the roles of women and men in various societies? How do religious beliefs influence what we think of as "appropriate" or "inappropriate" behaviors for men, women, and children? What do you think?

In agrarian societies, gender inequality and male dominance become institutionalized. Agrarian societies rely on agriculture—farming done by animal-drawn or mechanically powered plows and equipment. Because agrarian tasks require more labor and greater physical strength than horticultural ones, men become more involved in food production. It has been suggested that women are excluded from these tasks because they are viewed as too weak for the work and because child-care responsibilities are considered incompatible with the full-time labor that the tasks require (Nielsen, 1990). Most of the world's population currently lives in agrarian societies in various stages of industrialization.

Why does gender inequality increase in agrarian societies? Scholars cannot agree on an answer; however, some suggest that it results from private ownership of property. When people no longer have to move continually in search of food, they can acquire a surplus. Men gain control over the disposition of the surplus and the kinship system, and this control serves men's interests (Lorber, 1994). The importance of producing "legitimate" heirs to inherit the surplus increases significantly, and women's lives become more secluded and restricted as men attempt to ensure the legitimacy of their children. Premarital virginity and marital fidelity are required; indiscretions are punished (Nielsen, 1990). However, some scholars argue that male dominance existed before the private ownership of property (Firestone, 1970; Lerner, 1986).

Industrial Societies

An *industrial society* is one in which factory or mechanized production has replaced agriculture as the major form of economic activity. As societies industrialize, the status of women tends to decline further. Industrialization in the United States created a gap between the nonpaid work performed by women at home and the paid work that increasingly was performed by men and unmarried girls (Amott and Matthaei, 1996). When families needed extra money, their daughters worked in the textile mills until they married. In 1900, for example, 22 percent of single white women who had been born in the United States were in the paid labor force. Because factory work was not compatible with child-care responsibilities, only 3 percent of married women were so employed, and only when they were extremely poor (Amott and Matthaei, 1996). As it became more difficult to make a living by farming, many men found work in the factories, where their primary responsibility was often supervising the work of women and children. Men began to press for a clear division between "men's work" and "women's work," as well as corresponding pay differentials (higher for men, lower for women).

In the United States, the division of labor between men and women in the middle and upper classes became much more distinct with industrialization. The men were responsible for being "breadwinners"; the women were seen as "homemakers." In this new "cult of domesticity" (also referred to as the "cult of true womanhood"), the home became a private, personal sphere in which women created a haven for the family. Those who supported the cult of domesticity argued that women were the natural keepers of the domestic sphere and that children were the mother's responsibility. Meanwhile, the "breadwinner" role placed enormous pressures on men to support their families—being a good provider was considered to be a sign of manhood (Amott and Matthaei, 1996). However, this gendered division of labor increased the economic and political subordination of women. As a result, many white women focused their efforts on acquiring a husband who was capable of bringing home a good wage. Single women and widows and their children tended to live a bleak existence, crowded into run-down areas of cities, where they were often unable to support themselves on their meager wages.

The cult of true womanhood not only increased white women's dependence on men but also became a source of discrimination against women of color, based on both their race and the fact that many of them had to work in order to survive. Employed, working-class white women were similarly stereotyped at the same time that they became more economically dependent on their husbands because their wages were so much lower.

As people moved from a rural, agricultural lifestyle to an urban existence, body consciousness increased. People who worked in offices often became sedentary and exhibited physical deterioration from their lack of activity. As gymnasiums were built to fight this lack of physical fitness, a new image of masculinity developed. Whereas the "burly farmer" or "robust workman" had previously been the idealized image of masculinity, now the middle-class man who exercised and lifted weights came to embody this ideal (Klein, 1993).

In the late nineteenth century, middle-class women started to become preoccupied with body fitness. As industrialization progressed and food became more plentiful, the social symbolism of body weight and size changed. Previously, it had been considered a sign of high status to be somewhat overweight, but now a slender body reflected an enhanced social status. To the status-seeking middle-class man, a slender wife became a symbol of the husband's success. At the same time, excess body weight was seen as a reflection of moral or personal inadequacy, or a lack of willpower (Bordo, 2004).

Postindustrial Societies

Chapter 5 defines *postindustrial societies* as ones in which technology supports a service- and information-based economy. In such societies, the division of labor in paid employment is increasingly based on whether people provide or apply information or are employed in service jobs such as fast-food restaurant counter help or health care workers. For both women and men in the labor force, formal education is increasingly crucial for economic and social success. However, as some women have moved into entrepreneurial, managerial, and professional occupations, many others have remained in the low-paying service sector, which affords few opportunities for upward advancement.

Will technology change the gendered division of labor in postindustrial societies? Scholars do not agree on the effects of computers, the Internet, the World Wide Web, cellular phones, and many newer forms of communications technology on the role of women in society. For example, some feminist writers had a pessimistic view of the impact of computers and monitors on women's health and safety, predicting that women in secretarial and administrative roles would experience an increase in eyestrain, headaches, and problems such as carpal tunnel syndrome. However, some medical experts now believe that such problems extend to both men and women, as computers have become omnipresent in more people's lives. The term "24/7" has come to mean that a person is available "twenty-four hours a day, seven days a week" via cell phones, pocket pagers, fax machines, e-mail, and other means of communication, whether the individual is at the office or four thousand miles away on "vacation."

How do new technologies influence gender relations in the workplace? Although some analysts presumed that technological developments would reduce the boundaries between women's and men's work, researchers have found that the gender stereotyping associated with specific jobs has remained remarkably stable even when the nature of work and the skills required to perform it have been radically transformed. Today, men and women continue to be segregated into different occupations, and this segregation is particularly visible within individual workplaces (as discussed later in the chapter).

How does the division of labor change in families in postindustrial societies? For a variety of reasons, more households are headed by women with no adult male present. As shown in the Census Profiles feature, the percentage of U.S. households headed by a single mother with children under eighteen has increased. Chapter 15 ("Families and Intimate Relationships") discusses a number of reasons why the current division of labor in household chores in some families is between a woman and her children rather than between women and men. Consider, for example, that almost one-fourth (23 percent) of all U.S. children live with their mother only (as contrasted with just 5 percent who reside with their father only); among African American children, 48 percent live with their mother only (U.S. Census Bureau, 2007). This means that women in these households truly have a double burden, both from family responsibilities and from the necessity of holding gainful employment in the labor force.

Even in single-person or two-parent households, programming "labor-saving" devices (if they can be afforded) often means that a person must have some leisure time to learn how to do the programming. According to analysts, leisure is deeply divided along gender lines, and women have less time to "play in the house" than do men and boys. Some websites seek to appeal to women who have economic resources but are short on time, making it possible for them to shop, gather information, "telebank," and communicate with others at all hours of the day and night.

In postindustrial societies such as the United States, more than 60 percent of adult women are in the labor force, meaning that finding time to care for children, help aging parents, and meet the demands of the workplace will continue to place a heavy burden on women, despite living in an information- and service-oriented economy.

How people accept new technologies and the effect these technologies have on gender stratification are related to how people are socialized into gender roles. However, gender-based stratification remains rooted in the larger social structures of society, which individuals have little ability to control.

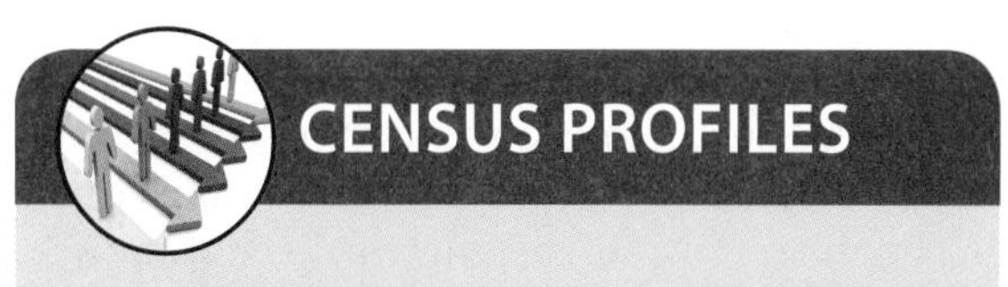

Single Mothers with Children Under 18

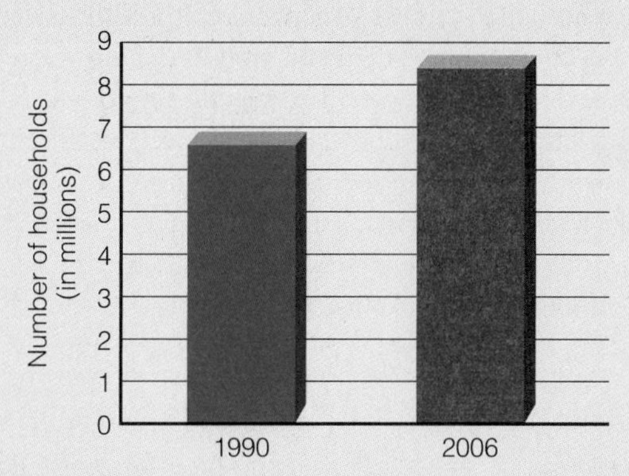

As shown above, the number of U.S. families headed by single mothers increased from almost 6,600,000 in 1990 to about 8,400,000 in 2006. For census purposes, a single mother is identified as a woman who is widowed, divorced, separated, or never married and who has children under 18 living at home. In contrast, only 2,095,000 single male householders have children under age 18 living at home. Of the total number of 36,468,000 family households with children under 18 reported in 2006, 25,982,000 of these households were composed of married couples and their children. In your opinion, what are the implications for these figures for families in the future?

Source: U.S. Census Bureau, 2009.

Gender and Socialization

We learn gender-appropriate behavior through the socialization process. Our parents, teachers, friends, and the media all serve as gendered institutions that communicate to us our earliest, and often most lasting, beliefs about the social meanings of being male or female and about thinking and behaving in masculine or feminine ways. Some gender roles have changed dramatically in recent years; others remain largely unchanged over time.

Many parents prefer boys to girls because of stereotypical ideas about the relative importance of males and females to the future of the family and society. Although some parents prefer boys to girls because these parents believe old myths about the biological inferiority of females, research suggests that social expectations also play a major role in this preference. We are socialized to believe that it is important to have a son, especially for a first or only child. For many years, it was assumed that a male child could support his parents in their later years and carry on the family name.

Across cultures, boys are preferred to girls, especially when the number of children that parents can have is limited by law or economic conditions. For example, in China, which strictly regulates the allowable number of children to one per family, a disproportionate number of female fetuses are aborted. In the United States, as well, one recently study found that Chinese Americans also strongly favored male children over female children: Some families had additional children in an effort to produce male offspring while other parents engaged in selective fertilization techniques in hopes of producing a male heir. In India, the practice of aborting female fetuses is widespread, and female infanticide occurs frequently. As a result, both India and China have a growing surplus of young men who will face a shortage of women their own age, and Chinese Americans in the United States may experience a similar situation in the future.

In this country, some sex selection no doubt takes place through abortion. However, most women seek abortions because of socioeconomic factors, problematic relationships with partners, health-related concerns, and lack of readiness or ability to care for a child (or another child) (Lott, 1994).

Parents and Gender Socialization

From birth, parents act toward children on the basis of the child's sex. Baby boys are perceived to be less fragile than girls and tend to be treated more roughly by their parents. Girl babies are thought to be "cute, sweet, and cuddly" and receive more gentle treatment. Parents strongly influence the gender-role development of children by passing on—both overtly and covertly—their own beliefs about gender. When girl babies cry, parents respond to them more quickly, and parents are more prone to talk and sing to girl babies (Wharton, 2004).

Children's toys reflect their parents' gender expectations. Gender-appropriate toys for boys include computer games, trucks and other vehicles, sports equipment, and war toys such as guns and soldiers. Girls' toys include "Barbie" dolls, play makeup, and homemaking items. Parents' choices of toys for their children are not likely to change in the near future. A group of college students in one study were shown slides of toys and asked to decide which ones they would buy for girls and boys. Most said they would buy guns, soldiers, jeeps, carpenter tools, and red bicycles for boys; girls would get baby dolls, dishes, sewing kits, jewelry boxes, and pink bicycles (Fisher-Thompson, 1990).

When children are old enough to help with household chores, they are often assigned different tasks. Maintenance chores (such as mowing the lawn) are assigned to boys, whereas domestic chores (such as shopping, cooking, and clearing the table) are assigned to girls. Chores may also become linked with future occupational choices and personal characteristics. Girls who are responsible for domestic chores such as caring for younger brothers and sisters may learn nurturing behaviors that later translate into employment as a nurse or schoolteacher. Boys may learn about computers and other types of technology that lead to different career options.

In the past, most studies of gender socialization focused on white, middle-class families and paid little attention to ethnic differences (Raffaelli and Ontai, 2004). According to earlier studies, children from middle- and upper-income families are less likely to be assigned gender-linked chores than children from lower-income backgrounds. In addition, gender-linked chore assignments occur less frequently in African American families, where both sons and daughters tend to be socialized toward independence, employment, and child care (Bardwell, Cochran, and Walker, 1986; Hale-Benson, 1986). Sociologist Patricia Hill Collins (1991) suggests that African American mothers are less likely to socialize their daughters into roles as subordinates; instead, they are likely to teach them a critical posture that allows them to cope with contradictions.

In contrast, more-recent studies of gender socialization in U.S. Latino/a families suggest that adolescent females of Mexican, Puerto Rican, Cuban, or other Central or South American descent receive different gender socialization from their parents than do their male siblings (Raffaelli and Ontai, 2004). Latinas are given more-stringent curfews and are allowed less interaction with members of the opposite sex than are the adolescent males in their families. Rules for dating, school activities, and part-time jobs are more stringent for the girls because many parents want to protect their daughters and keep them closer to home.

Across classes and racial/ethnic categories, mothers typically play a stronger role in gender socialization of daughters, whereas fathers do more to socialize sons than daughters (McHale, Crouter, and Tucker, 1999). However, many parents are aware of the effect that gender socialization has on their children and make a conscientious effort to provide nonsexist experiences for them. For example, one study found that mothers with nontraditional views encourage their daughters to be independent (Brooks-Gunn, 1986). Many fathers also take an active role in socializing their sons to be thoughtful and caring individuals who do not live by traditional gender stereotypes. However, peers often make nontraditional gender socialization much more difficult for parents and children (see Rabinowitz and Cochran, 1994).

Peers and Gender Socialization

Peers help children learn prevailing gender-role stereotypes, as well as gender-appropriate and gender-inappropriate behavior (Hibbard and Buhrmester, 1998). During the preschool years, same-sex peers have a powerful effect on how children see their gender roles (Maccoby and Jacklin, 1987); children are more socially acceptable to their peers when they conform to implicit societal norms governing the "appropriate" ways that girls and boys should act in social situations and what prohibitions exist in such cases (Martin, 1989).

Male peer groups place more pressure on boys to do "masculine" things than female peer groups place on girls to do "feminine" things. For example, girls wear jeans and other "boy" clothes, play soccer and softball, and engage in other activities traditionally associated with males. By contrast, if a boy wears a dress, plays hopscotch with girls, and engages in other activities associated with being female, he will be ridiculed by his peers. This distinction between the relative value of boys' and girls' behaviors strengthens the cultural message that masculine activities and behavior are more important and more acceptable (Wood, 1999).

During adolescence, peers are often stronger and more effective agents of gender socialization than adults (Hibbard and Buhrmester, 1998). Peers are thought to be especially important in boys' development of gender identity (Maccoby and Jacklin, 1987). Male bonding that occurs during adolescence is believed to reinforce masculine identity (Gaylin, 1992) and to encourage gender-stereotypical attitudes and behavior (Huston, 1985; Martin, 1989). For example, male peers have a tendency to ridicule and bully others about their appearance, size, and weight. Aleta Walker painfully recalls walking down the halls at school when boys would flatten themselves against the lockers and cry, "Wide load!" At lunchtime, the boys made a production of watching her eat lunch and frequently made sounds like grunts or moos (Kolata, 1993). Because peer acceptance is extremely important for both males and females during their first two decades, such actions can have very harmful consequences for the victims.

© Barbara Campbell

Parents, peers, and the larger society all influence our perceptions about gender-appropriate behavior.

As young adults, men and women still receive many gender-related messages from peers. Among college students, for example, peer groups are organized largely around gender relations and play an important role in career choices and the establishment of long-term, intimate relationships (Holland and Eisenhart, 1990). In a study of women college students at two universities (one primarily white, the other predominantly African American), anthropologists Dorothy C. Holland and Margaret A. Eisenhart (1990) found that the peer system propelled women into a world of romance in which their attractiveness to men counted most. Although peers did not initially influence the women's choices of majors and careers, they did influence whether the women continued to pursue their original goals, changed their course of action, or were "derailed" (Holland and Eisenhart, 1981, 1990).

Teachers, Schools, and Gender Socialization

From kindergarten through college, schools operate as a gendered institution. Teachers provide important messages about gender through both the formal content of classroom assignments and informal interactions with students. Sometimes, gender-related messages from teachers and other students reinforce gender roles that have been taught at home; however, teachers may also contradict parental socialization. During the early years of a child's schooling, teachers' influence is very powerful; many children spend more hours per day with their teachers than they do with their own parents.

According to some researchers, the quantity and quality of teacher–student interactions often vary between the education of girls and that of boys (Wellhousen and Yin, 1997). One of the messages that teachers may communicate to students is that boys are more important than girls. Research spanning the past thirty years shows that unintentional gender bias occurs in virtually all educational settings. ***Gender bias* consists of showing favoritism toward one gender over the other.** Researchers consistently find that teachers devote more

gender bias behavior that shows favoritism toward one gender over the other.

time, effort, and attention to boys than to girls (Sadker and Sadker, 1994). Males receive more praise for their contributions and are called on more frequently in class, even when they do not volunteer.

Teacher–student interactions influence not only students' learning but also their self-esteem (Sadker and Sadker, 1985, 1986, 1994). A comprehensive study of gender bias in schools suggested that girls' self-esteem is undermined in school through such experiences as (1) a relative lack of attention from teachers; (2) sexual harassment by male peers; (3) the stereotyping and invisibility of females in textbooks, especially in science and math texts; and (4) test bias based on assumptions about the relative importance of quantitative and visual–spatial ability, as compared with verbal ability, where girls typically excel. White males may have better self-esteem because they receive more teacher attention than all other students (Sadker and Sadker, 1994).

Teachers also influence how students treat one another during school hours. Many teachers use sex segregation as a way to organize students, resulting in unnecessary competition between females and males. In one study, for example, a teacher divided her class into the "Beastly Boys" and the "Gossipy Girls" for a math game and allowed her students to do the "give me five" hand-slapping ritual when one group outscored the other (Thorne, 1995). Competition based on gender often reinforces existing misconceptions about the skills and attributes of boys and girls, and may contribute to overt and subtle discrimination in the classroom and beyond.

The effect of gender bias is particularly problematic if teachers take a "boys will be boys" attitude when boys and young men make derogatory remarks or demonstrate aggressive behavior against girls and young women. When girls complain of ***sexual harassment*—unwanted sexual advances, requests for sexual favors, or other verbal or physical conduct of a sexual nature**—their concerns are sometimes overlooked or downplayed by teachers and school administrators. Sexual harassment is prohibited by law, and teachers and administrators are obligated to investigate such incidents, as well as issues pertaining to women's equal opportunities to play sports.

Sports and Gender Socialization

Children spend more than half of their nonschool time in play and games, but the type of games played sometimes differs with the child's sex. Studies indicate that boys are socialized to participate in highly competitive, rule-oriented games with a larger number of participants than games played by girls. Girls typically are socialized to play with other girls, in groups of two or three, in activities such as hopscotch and jump rope that involve a minimum of competitiveness. Other research shows that boys express more favorable attitudes toward games and sports that involve physical exertion and competition than girls do. Some analysts believe this difference in attitude is linked to ideas about what is gender-appropriate behavior for boys and girls (Brustad, 1996).

For males, competitive sport becomes a means of "constructing a masculine identity, a legitimated outlet for violence and aggression, and an avenue for upward mobility" (Lorber, 1994: 43). For females, being an athlete and a woman may constitute contradictory statuses. One study during the 1980s found that women college basketball players dealt with this contradiction by dividing their lives into segments. On the basketball court, the women "did athlete"; they pushed, shoved,

Teachers often use competition between boys and girls because they hope to make a learning activity more interesting. Here, a middle-school girl leads other girls against boys in a Spanish translation contest. What are the advantages and disadvantages of gender-based competition in classroom settings?

fouled, ran hard, sweated, and cursed. Off the court, they "did woman"; after the game, they showered, dressed, applied makeup, and styled their hair, even if they were only getting on a van for a long ride home (Watson, 1987). However, researchers in a more recent study concluded that perceptions about women athletes may be changing. Specifically, ideas about what constitutes the ideal body image for girls and women are changing as more females become involved in physical fitness activities and athletic competitions. Young women and men in one poll rated the athletic female body higher than that of the anorexic model (Heywood and Dworkin, 2003).

Since passage in 1972 of Title IX, which mandates equal opportunities in academic and athletic programs for females, girls' and young women's participation in athletics has increased substantially. More girls play soccer and softball and participate in other sports formerly regarded as "male" activities. However, even with these changes over the past three decades, only about 42 percent of high school and college athletes are female. According to the sociologist Michael A. Messner (2002), girls and women have been empowered by their entry into sports; however, sex segregation of female and male athletes, as well as coaches, persists.

Most sports are rigidly divided into female and male events, and funding of athletic programs is often unevenly divided between men's and women's programs. Assumptions about male and female physiology and athletic capabilities influence the types of sports in which members of each sex are encouraged to participate. For example, women who engage in activities that are assumed to be "masculine" (such as bodybuilding) may either ignore their critics or attempt to redefine the activity or its result as "feminine" or womanly (Klein, 1993; Lowe, 1998). Some women bodybuilders do not want their bodies to get "overbuilt." They have learned that they are more likely to win women's bodybuilding competitions if they look and pose "more or less along the lines of fashion models" (Klein, 1993: 179). In her study of more than 100 people connected with women's bodybuilding, the sociologist Maria R. Lowe (1998) found that "women of steel" (the female bodybuilders) live in a world where size and strength must regularly be balanced with a nod toward grace and femininity.

Cautious optimism is possible regarding the changing nature of sports and gender socialization, based on several studies of women in sports (Heywood and Dworkin, 2003; Messner, 2002). Clearly, changes have occurred that might positively influence the gender socialization of both girls and boys; however, it appears that much remains to be done to bring about greater gender equity in the area of sport. One such area is how the media report on women's and men's sporting events, and the attributes (such as physical attractiveness) that they highlight regarding female competitors while they emphasize the athletic skills of male competitors.

Recently, courts in states such as California have told some colleges and universities that they must increase their proportion of female athletes, which in certain cases means bringing female participation in varsity sports to within 1.5 percent of their proportion of the overall student population. Universities are also being advised to add women's sports teams, such as women's field hockey, in order to offer a wider diversity of sports in which female athletes and coaches might participate (Thomas, 2009).

In recent years, women have expanded their involvement in professional sports through organizations such as the WNBA; however, most sports remain rigidly divided into female events and male events. Do you think that media coverage of women's and men's college and professional sporting events differs?

Mass Media and Gender Socialization

The media, including newspapers, magazines, television, and movies, are powerful sources of gender stereotyping. Although some critics argue that the media simply reflect existing gender roles in society, others point out that the media have a unique ability to shape ideas (see Box 11.3). Think of the impact that television

sexual harassment unwanted sexual advances, requests for sexual favors, or other verbal or physical conduct of a sexual nature.

How Do We "Do Gender" in the Twenty-First Century?

What distinctive ways of acting and feeling are characteristic of women? Of men? For centuries, people have used a male/female dichotomy to answer these questions and, in the process, have identified women's and men's behaviors as opposites in many respects: Men are supposed to be "real men" and meet the normative conception of *masculinity* by being aggressive, independent, and powerful, whereas women are supposed to demonstrate *femininity* by being passive, dependent, and weak.

However, many theorists using a symbolic interactionist perspective suggest that gender is something that we *do* rather than being a set of masculine or feminine traits that resides within the individual. Sociologists Candace West and Don H. Zimmerman (1987) coined the term "doing gender" to refer to the process by which we *socially* create differences between males and females that are *not* based on natural, essential, or biological factors but instead are based on the things we do in our social interactions. According to West and Zimmerman (1987), *accountability* is involved in the process of doing gender: We know that our actions will be evaluated by others based on how well they think we meet the normative conceptions of appropriate attitudes and activities that are expected of people in our sex category (the socially required displays that identify a person as being either male or female).

What is the primary difference in these two approaches? These viewpoints are based on different assumptions. The male/female dichotomy is based on the assumption that women and men have inherently different traits, whereas the concept of *doing gender* is based on the assumption that, through our interactions with others, we produce, reproduce, and sustain the social meanings that are accorded to gender in any specific society at any specific point in time (in other words, those meanings may change from time to time and from place to place). By focusing on gender as an *accomplishment* (rather than something that is previously established), symbolic interactionists emphasize that some people's resistance to existing gendered norms is probable and that social change is possible. Symbolic interactionist theories also make us aware that change is less likely to take place when people feel constrained to be accountable to others for their behavior as women or men (Fenstermaker and West, 2002).

When you look at the pictures on these three pages, think about how the people in each setting are "doing gender" in everyday life. Are they doing gender based on what they perceive to be the normative expectations of others? Or are they doing gender as they see fit? What do you think?

© Image Source Pink/Getty Images

Gender and Appearance

How do we do gender on a shopping trip? Consider, for example, the differences in clothing shown in this photo. For the man, a casual outing means wearing jeans, sandals, and a partially unbuttoned shirt. For the woman, a casual outing means a "pulled together" outfit with heels, a sexy dress, her hair in place, and carefully positioned sunglasses.

© Ariel Skelley/Getty Images

Gender and Home Life

Although in recent years more men have assumed greater responsibility at home with regard to household tasks and child rearing, women are more likely to be seen as the primary caregivers for children. Does our society still hold women, more than men, accountable for the well-being of children?

Gender and Careers

In recent decades, more women have become doctors and lawyers than in the past. How has this affected the way that people do gender in settings that reflect their profession? Do professional women look and act more like their male colleagues, or have men changed their appearance and activities at work as a result of having female colleagues?

© Tim Pannell/Corbis

© Ed Bock/CORBIS

Gender and Social Interaction

How are the men shown here doing gender after their game of pick-up basketball? Consider, for example, what they are wearing, how they are sitting, and how they are communicating with one another. If three women were in this photo instead of three men, in what ways would their attire, actions, and expressions be different?

© moodboard/Alamy

Gender and Success

When you hear someone referred to as a "wealthy entrepreneur," do you tend to think of a man or a woman? Are the typical trappings of success—such as luxury cars and expensive private airplanes—more associated with how men or how women do gender? Do you believe that this situation will change in the future?

Reflect & Analyze

1. **If you went to a medical clinic for the first time and was told that your physician was "Dr. Smith," would you automatically assume that this person was a man? If so, why?**
2. **Is your social behavior dependent on gender context? In other words, do you behave differently when around members of your own sex than you do in mixed situations? Come up with a specific example of a situation where you felt the need to adapt your behavior because of social and gender expectations.**
3. **Do you think that wide differences in how two people look when they go out on a date can affect the outcome of their encounter? Why or why not?**

Turning to Video

Watch the ABC video *Are Men Smarter? Sexism in the Classroom* (run time 4:33), available on the Kendall Companion Website and through Cengage Learning eResources accounts. This special report focuses on the experience of leading neurobiologist Ben Barres, who became a female-to-male transsexual at the age of forty. Having been active in his field as both Barbara and Ben, Dr. Barres has gained unique perspective on the biased perceptions of men and women in science. As you watch the video, think about the photographs, commentary, and questions that you encountered in this photo essay. After you've watched the video, consider another question: How do Ben Barres's life experiences and professional accomplishments reflect West and Zimmerman's concept of "doing gender"?

the most power (Richardson, 1993). Most of these are men, however.

Conflict theorists in the Marxist tradition assert that gender stratification results from private ownership of the means of production; some men not only gain control over property and the distribution of goods but also gain power over women. According to Friedrich Engels and Karl Marx, marriage serves to enforce male dominance. Men of the capitalist class instituted monogamous marriage (a gendered institution) so that they could be certain of the paternity of their offspring, especially sons, whom they wanted to inherit their wealth. Feminist analysts have examined this theory, among others, as they have sought to explain male domination and gender stratification.

Feminist Perspectives

***Feminism*—the belief that women and men are equal and that they should be valued equally and have equal rights**—is embraced by many men as well as women. It holds in common with men's studies the view that gender is a socially constructed concept that has important consequences in the lives of all people (Craig, 1992). According to the sociologist Ben Agger (1993), men can be feminists and propose feminist theories; both women and men have much in common as they seek to gain a better understanding of the causes and consequences of gender inequality. Over the past three decades, many different organizations have been formed to advocate causes uniquely affecting women or men and to help people gain a better understanding of gender inequality (see Box 11.4).

Feminist theory seeks to identify ways in which norms, roles, institutions, and internalized expectations limit women's behavior. It also seeks to demonstrate how women's personal control operates even within the constraints of relative lack of power (Stewart, 1994).

Liberal Feminism In liberal feminism, gender equality is equated with equality of opportunity. The roots of women's oppression lie in women's lack of equal civil rights and educational opportunities. Only when these constraints on women's participation are removed will women have the same chance for success as men. This approach notes the importance of gender-role socialization and suggests that changes need to be made in what children learn from their families, teachers, and the media about appropriate masculine and feminine attitudes and behavior. Liberal feminists fight for better child-care options, a woman's right to choose an abortion, and elimination of sex discrimination in the workplace.

Radical Feminism According to radical feminists, male domination causes all forms of human oppression, including racism and classism (Tong, 1989). Radical feminists often trace the roots of patriarchy to women's childbearing and child-rearing responsibilities, which make them dependent on men (Firestone, 1970; Chafetz, 1984). In the radical feminist view, men's oppression of women is deliberate, and ideological justification for this subordination is provided by other institutions such as the media and religion. For women's condition to improve, radical feminists claim, patriarchy must be abolished. If institutions are currently gendered, alternative institutions—such as women's organizations seeking better health care, day care, and shelters for victims of domestic violence and rape—should be developed to meet women's needs.

Socialist Feminism Socialist feminists suggest that women's oppression results from their dual roles as paid *and* unpaid workers in a capitalist economy. In the workplace, women are exploited by capitalism; at home, they are exploited by patriarchy (Kemp, 1994). Women are easily exploited in both sectors; they are paid low wages and have few economic resources. Gendered job segregation is "the primary mechanism in capitalist society that maintains the superiority of men over women, because it enforces lower wages for women in the labor market" (Hartmann, 1976: 139). As a result, women must do domestic labor either to gain a better-paid man's economic support or to stretch their own wages (Lorber, 1994). According to socialist feminists, the only way to achieve gender equality is to eliminate capitalism and develop a socialist economy that would bring equal pay and rights to women.

Multicultural Feminism Recently, academics and activists have been rethinking the experiences of women of color from a feminist perspective. The experiences of African American women and Latinas/Chicanas have been of particular interest to some social analysts. Building on the civil rights and feminist movements of the late 1960s and early 1970s, some contemporary black feminists have focused on the cultural experiences of African American women. A central assumption of this analysis is that race, class, and gender are forces that simultaneously oppress African American women (Hull, Bell-Scott, and Smith, 1982). The effects of these three statuses cannot be adequately explained as "double" or "triple" jeopardy (race + class + gender = a poor African American woman) because these ascribed characteristics are not simply added to one another. Instead, they are multiplicative in nature (race × class × gender); different characteristics may be more significant in one

feminism the belief that all people—both women and men—are equal and that they should be valued equally and have equal rights.

Box 11.4 You Can Make a Difference

Joining Organizations to Overcome Sexism and Gender Inequality

Over the past four decades, many college students—like you—have been actively involved in organizations seeking to dismantle sexism and reduce gender-based inequalities in the United States and other nations. Two people speak about the importance of feminist advocacy for women and men:

> I was raised on a pure, unadulterated feminist ethic. . . . [But] the feminism I was raised on was very cerebral. It forced a world full of people to change the way they think about women. I want more than their minds. I want to see them do it. . . . I know that sitting on the sidelines will not get me what I want from my movement. And it is mine. . . . Don't be fooled into thinking that feminism is old-fashioned. The movement is ours and we need it. . . . The next generation is coming. (Neuborne, 1995: 29–35)
>
> —Ellen Neuborne, a reporter for *USA Today*

> For men who are messed up (that is, facing problems related to their emotional lives, sexuality, their place in society, and gender politics—in other words, me and virtually every other man I have ever met) feminism offers the best route to understanding the politics of such personal problems and coming to terms with those problems. (Jensen, 1995: 111)
>
> —Robert Jensen, a journalism professor at the University of Texas at Austin

If you are interested in joining an organization that deals with the problem of sexism or organizes activities such as Take Back the Night, an annual event that promotes awareness of violence against women, contact your college's student activities office.

Opportunities for involvement in local, state, and national organizations that promote women's and/or men's rights are available. Here are a few examples:

- The National Organization for Women (NOW), 1100 H St. NW, 3rd floor, Washington, DC 20005. (202) 628-8669. NOW works to end gender bias and seeks greater representation of women in all areas of public life. On the Internet, NOW provides links to other feminist resources: **http://www.now.org**
- National Organization for Men Against Sexism (NOMAS), P.O. Box 455, Louisville, CO 80027. (303) 666-7043. NOMAS has a *profeminist stance* that seeks to end sexism and an *affirmative stance* on the rights of gay men and lesbians.

On the Internet:

- *MenWeb (Men's Voices Magazine)* is an e-zine that offers support and advocacy for men. It features articles on the men's movement and a national calendar of men's events, as well as links to other sites: **http://www.menweb.org**

situation than another. For example, a well-to-do white woman (class) may be in a position of privilege when compared to people of color (race) and men from lower socioeconomic positions (class), yet be in a subordinate position as compared with a white man (gender) from the capitalist class (Andersen and Collins, 1998). In order to analyze the complex relationship among these characteristics, the lived experiences of African American women and other previously "silenced people" must be heard and examined within the context of particular historical and social conditions.

Another example of multicultural feminist studies is the work of the psychologist Aida Hurtado (1996), who explored the cultural identification of Latinas/Chicanas. According to Hurtado, distinct differences exist between the world views of the white (non-Latina) women who participate in the women's movement and many Chicanas, who have a strong sense of identity with their own communities. Hurtado (1996) suggests that women of color do not possess the "relational privilege" that white women have because of their proximity to white patriarchy through husbands, fathers, sons, and others. Like other multicultural feminists, Hurtado calls for a "politics of inclusion," creating social structures that lead to positive behavior and bring more people into a dialogue about how to improve social life and reduce inequalities. Feminists who analyze race, class, and gender suggest that equality will occur only when all women, regardless of race/ethnicity, class, age, religion, sexual orientation, or ability (or disability), are treated more equitably (Andersen and Collins, 1998).

Evaluation of Conflict and Feminist Perspectives

Conflict and feminist perspectives provide insights into the structural aspects of gender inequality in society. These approaches emphasize factors external to individuals that contribute to the oppression of white women and people of color; however, they have been criticized for emphasizing the differences between men and women without taking into account the commonalities they share. Feminist approaches have also been criticized for their emphasis on male dominance without a corre-

© Larry Kolvoord/The Image Works

Latinas have become increasingly involved in social activism for causes that they believe are important. This woman is showing her support for amnesty for undocumented workers in the United States.

sponding analysis of the ways in which some men may also be oppressed by patriarchy and capitalism.

Recently, the debate has continued about whether the feminist movement has diminished the well-being of boys and men. Sociologist William J. Goode (1982) suggests that some men have felt "under attack" by women's demands for equality because the men do not see themselves as responsible for societal conditions such as patriarchy but attribute their own achievements to hard work and intelligence, not to built-in societal patterns of male domination and female subordination.

In *Stiffed: The Betrayal of the American Man,* the author and journalist Susan Faludi (1999) argues that men are not as concerned about the possibility of the feminist movement diminishing their own importance as they are experiencing an identity crisis brought about by the current societal emphasis on wealth, power, fame, and looks (often to the exclusion of significant social values). According to Faludi, men are increasingly aware of their body image and are spending ever-increasing sums of money on men's cosmetics, health and fitness gear and classes, and cigar bars and "gentlemen's clubs." The popularity of her book, among others, suggests that issues of gender inequality and of men's and women's roles in society are far from resolved.

Gender Issues in the Future

Over the past century, women made significant progress in the labor force (Reskin and Padavic, 2002). Laws were passed to prohibit sexual discrimination in the workplace and school. Affirmative action programs helped make women more visible in education, government, and the professional world. More women entered the political arena as candidates instead of as volunteers in the campaign offices of male candidates (Lott, 1994).

Many men joined movements to raise their consciousness, realizing that what is harmful to women may also be harmful to men. For example, women's lower wages in the labor force suppress men's wages as well; in a two-paycheck family, women who are paid less contribute less to the family's finances, thus placing a greater burden on men to earn more money.

In the midst of these changes, however, many gender issues remain unresolved. In the labor force, gender segregation and the wage gap are still problems. Although women continue to narrow the wage gap, women earn about 80 cents for every dollar compared with men, and analysts believe that the narrowing can partly be attributed to the economic boom (for some) of the 1990s. Employers have had to look for employees based on merit (rather than race, class, and gender) in order to have the number and types of employees they need to meet the demands of global competition (Barakat, 2000).

Although some employers have implemented family-leave policies, these do not relieve women's domestic burden in the family. Analysts believe that the burden of the "double day" or "second shift" will probably preserve women's inequality at home and in the workplace for another generation (Reskin and Padavic, 2002).

© Mauritius/Photolibrary

Some analysts believe that contemporary approaches to education do not fit the needs of boys and are perhaps responsible for their decline in academic achievement. What do you think about this issue?

How will the economic problems of the early twenty-first century affect gender inequality in the United States and around the world? It is too soon to tell, but the fallout from the economic crisis will probably be more strongly felt by women than men, particularly for those individuals who were already in the greatest social and financial peril. What do you think might be done to provide more equal opportunities for girls and women even in difficult economic times?

Chapter Review

- **How do sex and gender differ?**

Sex refers to the biological categories and manifestations of femaleness and maleness; *gender* refers to the socially constructed differences between females and males. In short, sex is what we (generally) are born with; gender is what we acquire through socialization.

- **How do gender roles and gender identity differ from gendered institutions?**

Gender role encompasses the attitudes, behaviors, and activities that are socially assigned to each sex and that are learned through socialization. Gender identity is an individual's perception of self as either female or male. By contrast, gendered institutions are those structural features that perpetuate gender inequality.

- **How does the nature of work affect gender equity in societies?**

In most hunting and gathering societies, fairly equitable relationships exist between women and men because neither sex has the ability to provide all of the food necessary for survival. In horticultural societies, a fair degree of gender equality exists because neither sex controls the food supply. In agrarian societies, male dominance is very apparent; agrarian tasks require more labor and physical strength, and females are often excluded from these tasks because they are viewed as too weak or too tied to child-rearing activities. In industrialized societies, a gap exists between nonpaid work performed by women at home and paid work performed by men and women. A wage gap also exists between women and men in the marketplace.

- **What are the key agents of gender socialization?**

Parents, peers, teachers and schools, sports, and the media are agents of socialization that tend to reinforce stereotypes of appropriate gender behavior.

- **What causes gender inequality in the United States?**

Gender inequality results from economic, political, and educational discrimination against women. In most workplaces, jobs are either gender segregated or the majority of employees are of the same gender. Although the degree of gender segregation in the professional workplace has declined since the 1970s, racial and ethnic segregation remains deeply embedded.

- **How is occupational segregation related to the wage gap?**

Many women work in lower-paying, less-prestigious jobs than men. This occupational segregation leads to a disparity, or wage gap, between women's and men's earnings. Even when women are employed in the same job as men, on average they do not receive the same, or comparable, pay.

- **How do functionalists and conflict theorists differ in their view of division of labor by gender?**

According to functionalist analysts, women's roles as caregivers in contemporary industrialized societies are crucial in ensuring that key societal tasks are fulfilled. The husband performs the instrumental tasks of economic support and decision making; the wife assumes the expressive tasks of providing affection and emotional support for the family. According to conflict analysts, the gendered division of labor within families and the workplace—particularly in agrarian and industrial societies—results because of male control and dominance over women and resources.

Key Terms

body consciousness 350
comparable worth 367
feminism 373
gender 350
gender bias 359
gender identity 350
gender role 350
hermaphrodite 348
homophobia 349
matriarchy 352
patriarchy 352
primary sex characteristics 347
secondary sex characteristics 347
sex 346
sexism 352
sexual harassment 360
sexual orientation 349
transsexual 348
transvestite 348
wage gap 365

Questions for Critical Thinking

1. Do the media reflect societal attitudes on gender, or do the media determine and teach gender behavior? (As a related activity, watch television for several hours and list the roles for women and men depicted in programs and those represented in advertising.)
2. Review the concept of cultural relativism in Chapter 3. Should the U.S. State Department and human rights groups such as Amnesty International protest genital mutilation in other nations and withhold any funding or aid for those nations until they cease the practice?
3. Examine the various academic departments at your college. What is the gender breakdown of the faculty in selected departments? What is the gender breakdown of undergraduates and graduate students in those departments? Are there major differences among social science, science, and humanities departments? What hypothesis can you come up with to explain your observations?

The Kendall Companion Website

www.cengage.com/sociology/kendall

Visit this book's companion website, where you'll find more resources to help you study and successfully complete course projects. Resources include quizzes and flash cards, as well as special features such as an interactive sociology timeline, maps, General Social Survey (GSS) data, and Census 2000 data. The site also provides links to useful websites that have been selected for their relevance to the topics in this chapter and include those listed below. (*Note:* Visit the book's website for updated URLs.)

Women's Studies Database: Gender Issues
http://www.mith2.umd.edu/WomensStudies/GenderIssues

This University of Maryland website is a gateway to numerous online resources concerning gender issues. Information is readily available on many topics, including women in the work force, sex discrimination, women in sports, and the glass ceiling.

American Men's Studies Association (AMSA)
http://mensstudies.org

This site is maintained by the American Men's Studies Association, a nonprofit professional organization of scholars, therapists, and other individuals interested in the examination of masculinity in modern society. The site houses information about the AMSA, access to the *Journal of Men's Studies*, and interesting Web links.

WomenWatch
http://www.un.org/womenwatch

Visit the United Nations' "Internet Gateway on the Advancement and Empowerment of Women" to access information on global women's issues. The site features links, reports, analyses, databases, and online forums that address the most current information on women's rights and issues around the world.

chapter

12 Aging and Inequality Based on Age

Chapter Focus Question

Given the fact that aging is an inevitable consequence of living—unless a person dies young—why do many people in the United States devalue older persons?

The women in my family, at least on my mother's side, seem to live long and well. My grandmother Pearl's "third act" was one of worldwide travel and voracious learning. . . . I am a good 20-something years from retirement, yet I find myself thinking often lately of my own third act. In part that is because I am watching my mother . . . tinker with her personal life's script. In keeping with her family legacy, Mom is not one to stand still, and over the years has accumulated a couple of master's degrees, a Ph.D. and a law degree, along with a kaleidoscope of work experience. After my father died two years ago, it seemed only logical that my mother would mourn, then take a few exotic trips and find a more challenging job.

But just because Mom was ready for her third act didn't mean the working world was ready for Mom.

© AP Images/Army Golden Knights, SSG Joe Abeln

Many older individuals actively challenge the stereotypes and assumptions of others about what it means to be a "senior citizen." Here, on the occasion of his 85th birthday in 2009, former President George H. W. Bush, parachutes in tandem over Kennebunkport, Maine.

IN THIS CHAPTER

- The Social Significance of Age
- Age in Global Perspective
- Age and the Life Course in Contemporary Society
- Inequalities Related to Aging
- Living Arrangements for Older Adults
- Sociological Perspectives on Aging
- Death and Dying
- Aging in the Future

Unlike so many career shifts she had made over the years, this one did not go smoothly. Her calls to prospective employers often went unanswered. Her résumé did not always open doors. She looks as if she's in her 50's, but her résumé makes it clear that she's in her 60's, and suddenly the years of experience that have been her greatest strength somehow disqualified her.

Mom did not take this quietly. "I know more than I did 20 years ago, and my brain works as well as it did 30 years ago," she said. The problem, she points out, is not with her generation, but with mine. "Employers—who are your age—have dismissed people who are my age," she said.

—Journalist Lisa Belkin (2006) describes the problem that many older workers, including her mother, encounter in finding new jobs. However, the story had a positive outcome for Belkin's mother: She eventually found the fulfilling job that she was seeking.

If we apply our sociological imagination to the issue of aging and work as described by Lisa Belkin, we see that views on age and the problems that people experience in growing older are not just personal problems but also public issues of concern to everyone. Eventually, all of us will be affected by aging. ***Aging* is the physical, psychological, and social processes associated with growing older** (Atchley and Barusch, 2004). In the United States and some other high-income countries, older people are the targets of prejudice and discrimination based on myths about aging. For example, older persons may be viewed as incompetent solely on the basis of their age. Although some older people may need assistance from others and support

Sharpening Your Focus

- *How does functional age differ from chronological age?*
- *How does age determine a person's roles and statuses in society?*
- *What actions can be taken to bring about a more equitable society for older people?*

from society, many others are physically, socially, and financially independent.

Today, people age 65 and older make up 13 percent of the total population in the United States, and it is estimated that people in this age group will constitute 20 percent of the total population by 2050. Almost 90 percent of the people living in the high-income countries today are expected to survive to age 65. However, by sharp contrast, in low-income countries such as Zambia, Uganda, and Rwanda, nearly half of the total population in each nation is not expected to survive to *age 40* (United Nations Development Programme, 2003). In this chapter, we examine the sociological aspects of aging. We will also examine how older people seek dignity, autonomy, and empowerment in societies such as the United States that may devalue those who do not fit the ideal norms of youth, beauty, physical fitness, and self-sufficiency. Before reading on, test your knowledge about aging and age-based discrimination in the United States by taking the quiz in Box 12.1.

Childhood is one of the most significant periods in the life course. For some children, these are carefree years; for others, these are times of powerlessness and vulnerability.

The Social Significance of Age

"How old are you?" This is one of the most frequently asked questions in the United States. Beyond indicating how old or young a person is, age is socially significant because it defines what is appropriate for or expected of people at various stages. For example, child development specialists have identified stages of cognitive development based on children's ages:

> [W]e do not expect our preschool children, much less our infants, to have adultlike memories or to be completely logical. We are seldom surprised when a 4-year-old is misled by appearances; we express little dismay when our 2½ year old calls a duck a chicken. . . . But we would be surprised if our 7-year-olds continued to think segmented routes were shorter than other identical routes or if they continued to insist on calling all reasonably shaggy-looking pigs "doggy." We expect some intellectual (or cognitive) differences between preschoolers and older children. (Lefrançois, 1996: 196)

At the other end of the age continuum, a 75-year-old grandmother who travels through her neighborhood on in-line skates will probably raise eyebrows and perhaps garner media coverage about her actions because she is defying norms regarding age-appropriate behavior.

When people say "Act your age," they are referring to ***chronological age*—a person's age based on date of birth** (Atchley and Barusch, 2004). However, most of us actually estimate a person's age on the basis of ***functional age*—observable individual attributes such as physical appearance, mobility, strength, coordination, and mental capacity that are used to assign people to age categories** (Atchley and Barusch, 2004). Because we typically do not have access to other people's birth certificates to learn their chronological age, visible characteristics—such as youthful appearance or gray hair and wrinkled skin—may become our criteria for determining whether someone is "young" or "old." According to the historian Lois W. Banner (1993: 15), "Appearance, more than any other factor, has occasioned the objectification of aging. We define someone as old because he or she looks old." In fact, feminist scholars believe that functional age is so subjective that it is evaluated differently for women and men—as men age, they are believed to become more distinguished or powerful, whereas when women grow older, they are thought to be "over-the-hill" or grandmotherly (Banner, 1993).

Trends in Aging

Over the past 25 years, the U.S. population has been aging. The median age (the age at which half the people are younger and half are older) has increased by slightly more than 6 years—from 30 in 1980 to 36.4 in 2006 (U.S. Census Bureau, 2009). This change was partly a result of the Baby Boomers (people born between 1946 and 1964) moving into middle age and partly a result of more people living longer. As shown in ▶ Figure 12.1a, the number of older persons—age 65 and above—increased significantly between 1980 and 2000. The population over age 85 has been growing especially fast. When new census data for the 2010 census are available,

Box 12.1 Sociology and Everyday Life

How Much Do You Know About Aging and Age-Based Discrimination?

True	False	
T	F	1. The U.S. Supreme Court has recently issued rulings that make it easier for individuals to show that they have been discriminated against based on their age.
T	F	2. Women in the United States have a longer life expectancy than do men.
T	F	3. Scientific studies have documented the fact that women age faster than men do.
T	F	4. Most older persons are economically secure today as a result of Social Security, Medicare, and retirement plans.
T	F	5. Studies show that advertising no longer stereotypes older persons.
T	F	6. After women reach menopause, they may enjoy sexual activity more than when they were younger.
T	F	7. Organizations representing older individuals have demanded the same rights and privileges as those accorded to younger persons.
T	F	8. The rate of elder abuse in the United States has been greatly exaggerated by the media.

Answers on page 383.

we will no doubt learn of additional changes in the age composition of the U.S. population.

An increase in the number of older people living in the United States and other high-income nations resulted from an increase in life expectancy (greater longevity) combined with a stabilizing of birth rates (Atchley and Barusch, 2004). ***Life expectancy* is the average number of years that a group of people born in the same year could expect to live.** Based on the death rates in the year of birth, life expectancy shows the average length of life of a ***cohort*—a group of people born within a specified period of time.** Cohorts may be established on the basis of one-, five-, or ten-year intervals; they may also be defined by events taking place at the time of their birth, such as Depression-era babies or Baby Boomers (Moody, 2002). For the cohort born in 2004, as an example, life expectancy at birth was 77.9 years for the overall U.S. population—75.2 for males and 80.4 for females. However, as Figure 12.1b shows, there are significant racial–ethnic and sex differences in life expectancy. Although the life expectancy of people of color has improved over the past 50 years, higher rates of illness and disability—attributed to poverty, inadequate health care, and greater exposure to environmental risk factors—still persist.

Today, a much larger percentage of the U.S. population is over age 65 than in the past. One of the fastest-growing segments of the population is made up of people age 85 and above. This cohort is expected to almost double in size between 2000 and 2025 (U.S. Census Bureau, 2009); by the year 2050, the Census Bureau predicts that the number of persons age 85 and over will have increased to about 20 million (almost 5 percent of the population). Even more astonishing is the fact that the number of centenarians (persons 100 years of age and above) in this country will increase more than 12 times, from 66,000 in 1999 to about 834,000 in 2050 (L. Ramirez, 1999).

The current distribution of the U.S. population is depicted in the "age pyramid" in Figure 12.1d. If, every year, the same number of people are born as in the previous year and a certain number die in each age group, the rendering of the population distribution should be pyramid shaped. As you will note, however, Figure 12.1d is not a perfect pyramid, but instead reflects declining birth rates among post-Baby Boomers.

aging the physical, psychological, and social processes associated with growing older.

chronological age a person's age based on date of birth.

functional age a term used to describe observable individual attributes such as physical appearance, mobility, strength, coordination, and mental capacity that are used to assign people to age categories.

life expectancy an estimate of the average lifetime of people born in a specific year.

cohort a group of people born within a specified period in time.

a. U.S. Population Growth, 1980–2000
The percentage of persons 65 years of age and above increased dramatically between 1980 and 2000.

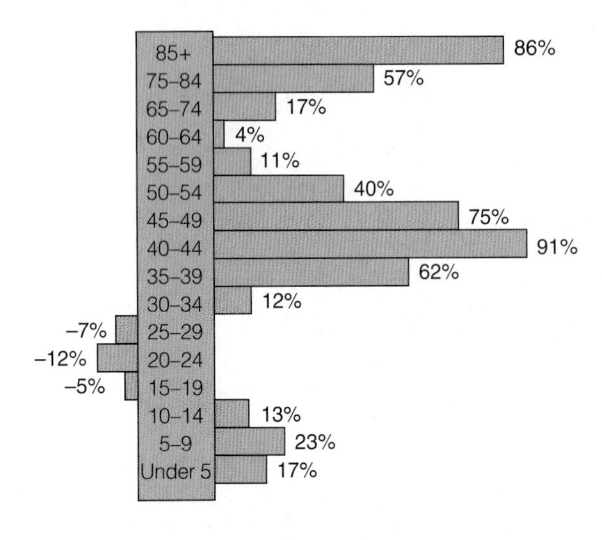

b. Selected Life Expectancies by Race, Ethnicity, and Sex, 2000
There are significant racial–ethnic and sex differences in life expectancy.

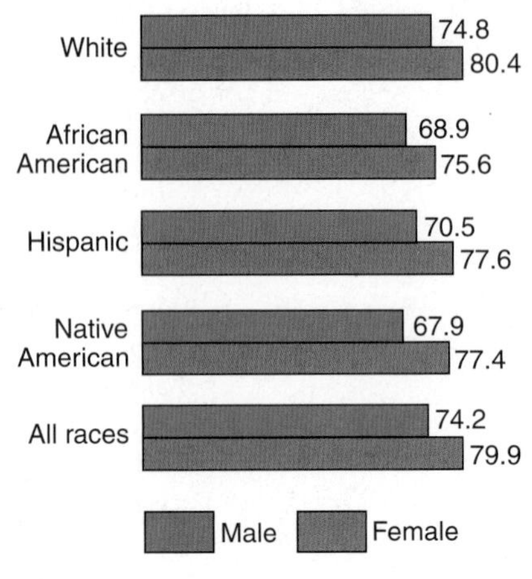

c. Percentage Distribution of U.S. Population by Age, 2000–2050 (projected)
Projections indicate that an increasing percentage of the U.S. population will be over age 65; one of the fastest-growing categories is persons age 85 and over.

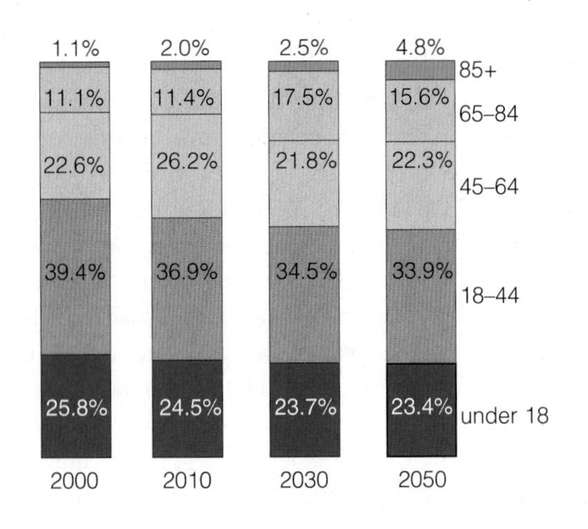

d. U.S. Age Pyramid by Age and Sex, 2000 (in millions)
Declining birth rates and increasing aging of the population are apparent in this age pyramid.

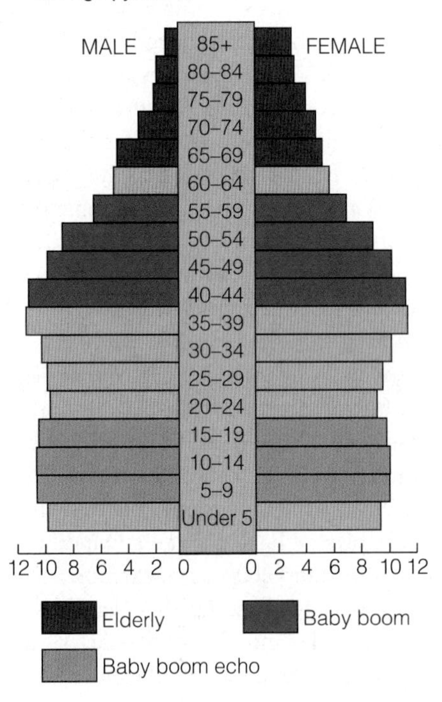

▶ **Figure 12.1 Trends in Aging and Life Expectancy**

Source: U.S. Census Bureau, 2007.

As a result of changing population trends, research on aging has grown dramatically in the past 60 years. ***Gerontology*** **is the study of aging and older people.** A subfield of gerontology, *social gerontology,* is the study of the social (nonphysical) aspects of aging, including such topics as the societal consequences of an aging population and the personal experience of aging (Atchley and Barusch, 2004). According to gerontologists, age is viewed differently from society to society, and its perception changes over time.

Age in Historical and Contemporary Perspectives

People are assigned to different roles and positions based on the age structure and role structure in a par-

Box 12.1 Sociology and Everyday Life

Answers to the Sociology Quiz on Aging and Age-Based Discrimination

1. False. In 2009 the U.S. Supreme Court ruled that persons claiming age discrimination must prove that age was the determining factor in adverse job decisions.

2. True. In 2009 female life expectancy (at birth) was 80.8 years, as compared with 75.7 for males. These figures vary by race and ethnicity.

3. False. No studies have documented that women actually age faster than men. However, some scholars have noted a "double standard" of aging that places older women at a disadvantage with respect to older men because women's worth in the United States is often defined in terms of physical appearance.

4. False. Although some older persons are economically secure, persons who rely solely on Social Security, Medicare, and/or pensions tend to live on low, fixed incomes that do not adequately meet their needs. Slightly less than 10 percent of people 65 and older live below the official poverty line.

5. False. Studies have shown that advertisements frequently depict older persons negatively—for example, as chronically ill or absentminded.

6. True. Women may enjoy sexual activity more after menopause because their sexual enjoyment is now separated from the possibility of pregnancy.

7. True. Organizations such as the Gray Panthers and AARP have been instrumental in the enactment of legislation beneficial to older persons.

8. False. Although cases of abuse and neglect of older persons are highly dramatized in the media, most coverage pertains to problems in hospitals, nursing homes, or other long-term care facilities. We know very little about the nature and extent of abuse that occurs in private homes.

Sources: Based on Atchley and Barusch, 2004; and U.S. Census Bureau, 2009.

ticular society. *Age structure* is the number of people of each age level within the society; *role structure* is the number and type of positions available to them (Riley and Riley, 1994). Over the years, the age continuum has been chopped up into finer and finer points. Two hundred years ago, people divided the age spectrum into "babyhood," a *very* short childhood, and then adulthood. What we would consider "childhood" today was quite different two hundred years ago, when agricultural societies needed a large number of strong arms and backs to work on the land to ensure survival. When 95 percent of the population had to be involved in food production, categories such as toddlers, preschoolers, preteens, teenagers, young adults, the middle-aged, and older persons did not exist.

If the physical labor of young people is necessary for society's survival, then young people are considered "little adults" and are expected to act like adults and do adult work. Older people are also expected to continue to be productive for the benefit of the society as long as they are physically able. In preindustrial societies, people of all ages help with the work, and little training is necessary for the roles that they fill. During the seventeenth and eighteenth centuries in the United States, for example, older individuals helped with the work and were respected because they were needed—and because few people lived that long.

Age in Global Perspective

Physical and sociocultural environments have different effects on how people experience aging and old age. In fact, concepts such as *young* and *old* may vary considerably from culture to culture. Unlike the sophisticated data-gathering techniques used to determine the number of older people in high-income and middle-income nations, we know less about the life expectancies and the aging populations in hunting and gathering, horticultural, pastoral, and agrarian societies. However, reaching the age of 30 or 40 is less likely for people in low-income (less-developed) nations than reaching

gerontology the study of aging and older people.

the age of 70 or 80 in many high-income (developed) countries.

Preindustrial Societies

People in hunting and gathering societies are not able to accumulate a food surplus and must spend much of their time seeking food. They do not have permanent housing that protects them from the environment. In such societies, younger people may be viewed as a valuable asset in hunting and gathering food, whereas older people may be viewed as a liability because they typically move more slowly, are less agile, and may be perceived as being less productive.

Although more people reach older ages in horticultural, pastoral, and agrarian societies, life is still very hard for most people. It is possible to accumulate a surplus, so older individuals, particularly men, are often the most privileged in a society because they have the most wealth, power, and prestige. In agrarian societies, farming makes it possible for more people to live to adulthood and to more-advanced years. In these societies, the proportion of older people living with other family members is extremely high, with few elderly living alone.

© Jacob Halaska/Photolibrary

In lower-income nations, older people may still perform useful economic functions such as gathering food or farming.

In recent years, a growing number of people are reaching age 60 and above in some middle- and lower-income nations. Consider that India, for example, has about a billion people in its population. If only 5 percent of the population reaches age 60 or above, there will still be a significant increase in the number of older people in that country. Because so much of the world's population resides in India and other low-income nations, the proportion of older people in these countries will increase dramatically during the twenty-first century.

Industrial and Postindustrial Societies

In industrial societies, living standards improve and advances in medicine contribute to greater longevity for more people. Although it is often believed that less-industrialized countries accord greater honor, prestige, and respect to older people, some studies have found that the stereotypical belief that people in such nations will be taken care of by their relatives, particularly daughters and sons, is not necessarily true today (Martin and Kinsella, 1994).

In postindustrial societies, information technologies are extremely important, and a large proportion of the working population is employed in service-sector occupations in the fields of education and health care, both of which may benefit older people. Some more-affluent older people may move away from family and friends upon retirement in pursuit of recreational facilities or a better climate (such as the popular move from the northeastern United States to the southern "sunbelt" states). Others may relocate to be closer to children or other relatives. The shift from a society that was primarily young to a society that is older will bring about major changes in societal patterns and in the needs of the population. Issues that must be addressed include the health care system, the Social Security system, transportation, housing, and recreation.

A Case Study: Aging, Gender, and Japanese Society

A significant increase in the older population in Japan has occurred over the past 30 years, whereas it took almost a century in North America and Europe. If the present trend continues in Japan, by 2025 people age 65 and over will make up about 25 percent of the total population, and more than half of the older population will be over 75 years of age. It is widely assumed that older people are respected and revered in Japan; however, several recent studies suggest that sociocultural changes and population shifts may be bringing about a gradual change in the social importance of the elderly in that nation. Until recently, most of the focus on the aging population in Japan has been on men and the workplace. Currently, feminist activists in that country are questioning why there has been so little attention given to the health and aging of women who are age 40 and above. When women have been discussed in regard to the aging population, the subject has primarily been caregiving for elderly relatives. For example, the "homebody" or "professional housewife" who cares not only for her husband and children but also for other relatives, particularly the ill and the aged, has been used as the "ideal" model of the Japanese woman, against whom all others are measured. However, with over 60 percent of Japanese women in the labor force, greater pressure is being placed on Japanese policy makers to consider how the government can play a larger role in the care of the aging population, rather than placing the burden completely on families, particularly women. As one analyst explained,

> Because financially secure middle-class women are assumed to represent Japanese women as a whole, the situation of the majority who must give up work to look after their relatives, often at great cost to the well-being of the entire family, is usually erased from national consciousness. Moreover, many live to be well over ninety years of age, and daughters-in-law in their seventies find themselves . . . caring for one or more incontinent, immobile, and sometimes senile relatives. Furthermore, because stroke is the usual cause of disability among the elderly population in Japan, intensive nursing is often required, but men assist very rarely with this onerous duty. It is, therefore, the debate about home nursing and living together as a three-generation family that takes up most of the energy of activist women in Japan today. . . . (Lock, 1999: 61)

Challenges such as these are presented by the aging of the population not only in Japan but also in the United States and other nations. Social policy issues about older people reflect the intertwining nature of class, gender, and age as individuals and nations adapt to the "graying" of their populations. Those persons who are growing older, whether in Japan or any other nation, seek to have both greater longevity and a high quality of life during their later years.

Age and the Life Course in Contemporary Society

During the twentieth century, life expectancy steadily increased as industrialized nations developed better water and sewage systems, improved nutrition, and made tremendous advances in medical science. However, children today are often viewed as an economic liability; they cannot contribute to the family's financial well-being and must be supported. In industrialized and postindustrial societies, the skills necessary for many roles are more complex and the number of unskilled positions is more limited. Consequently, children are expected to attend school and learn the necessary skills for future employment rather than perform unskilled labor. Further, older people are typically expected to retire so that younger people can assume their economic, political, and social roles. However, when economic crises occur and many jobs are lost, age-based inequality tends to increase. For example, the current trend to close down businesses and lay off workers, or to otherwise "downsize" the work force, has contributed to the unemployment of some older workers who would have liked to continue their employment for additional years and to the forced retirement of others. Many employers believe that they can save money by offering early retirement packages to older workers, thereby saving their company money and preserving jobs for younger, lower-paid workers. Such "early retirement" is not always voluntary and may pose significant economic risks for individuals who find that they cannot live on their Social Security and existing retirement plans or who find that their health benefits have been reduced. However, even for the financially secure, retirement often presents emotional issues, as Jack Culberg explains:

> When you suddenly leave [the corporate jungle], life is pretty empty. I was sixty-five, the age people are supposed to retire. I started to miss it quite a bit. The phone stops ringing. The king is dead. You start wanting to have lunch with old friends. At the beginning, they're nice to you, but then you realize that they're busy, they're working. They've got a job to do and just don't have the time to talk to anybody where it doesn't involve their business. I could be nasty and say, "Unless they can make a buck out of it"—but I won't. . . . You hesitate to call them. (qtd. in Terkel, 1996: 9–10)

As Culberg's statement indicates, people tend to think of age in narrowly defined categories, and people believe that reaching "retirement age" places many of them out of the economic and social mainstream of society.

In the United States, age differentiation is typically based on categories such as infancy, childhood, adolescence, young adulthood, middle adulthood, and later adulthood. In Chapter 4 ("Socialization"), we examined the socialization process that occurs when people are in various stages of the life course. However, these narrowly defined age categories have had a profound effect on our perceptions of people's capabilities, responsibilities, and entitlement. In this chapter, we will look at what is considered appropriate for or expected of people at various ages. These expectations are somewhat arbitrarily determined and produce ***age stratification*—inequalities, differences, segregation, or conflict between age groups** (Atchley and Barusch, 2004). We will now examine some of those age groups and the unique problems associated with each one.

Infancy and Childhood

Infancy (birth to age 2) and childhood (ages 3 to 12) are typically thought of as carefree years; however, children are among the most powerless and vulnerable people in society. Historically, children were seen as the property of their parents, who could do with them as they pleased (Tower, 1996). In fact, whether an infant survives the first year of life depends on a wide variety of parental factors, as a community health scholar explains:

> All infants are not created equal. Those born to teenage mothers or to mothers who smoke cigarettes, drink alcohol, or take drugs are at higher risk for death in their first year. Those born in very rural areas or in inner cities are more likely to die as infants. Those born to black women are at twice the risk as those born to white women. Older mothers carry a high risk for conceiving an infant with Down's syndrome, and Native American women carry a high risk for having a baby with a serious birth defect. Add to the mix a mother's education; her economic, marital, and nutritional status; and whether she had adequate prenatal care, which all play into whether her infant will make it through the first year of life.
>
> But surviving the first year is only one piece of the equation. Quality of life is another. Infants who survive the first year can have lives so compromised that their future is seriously limited. . . . We cannot always predict which infants will survive, and we certainly cannot predict who will be happy. (Schneider, 1995: 26)

Moreover, early socialization plays a significant part in children's experiences and their quality of life. Many children are confronted with an array of problems in their families because of marital instability, an increase in the number of single-parent households, and the percentage of families in which both parents are employed full time. These factors have heightened the need for high-quality, affordable child care for infants and young children. However, many parents have few options regarding who will take care of their children while they work. These statistics from the Children's Defense Fund (2002) point out the perils of infancy and childhood:

> Every day in the United States, 1 out of every 5 infants are born into poverty; over 3,000 children die from gunshot wounds every year, 7 million children are at home alone on a regular basis without adult supervision, and every 11 seconds a child is reported abused or neglected.

As these statistics show, the risk of death or permanent disability from accidents is a major concern in childhood. In fact, two-thirds of all childhood deaths are caused by injuries. (The other third are caused by cancers, birth defects, heart disease, pneumonia, and HIV/AIDS.) Although many previous childhood killers such as polio, measles, and diphtheria are now

© George Shelley/Masterfile

People in all cultures understand their responsibility to socialize the next generation. However, the physical arrangements associated with rearing children, including how they are moved from place to place, vary from culture to culture.

© Olivier Martel/Corbis

controlled through immunizations and antibiotics, motor vehicle accidents have become a major source of injury and death for infants and children. (As compared with all other racial–ethnic categories, Native American children have the highest rate of motor vehicle deaths.) The childhood motor vehicle fatality rate is higher in the South and Southwest, where more parents own pickup trucks and allow children to ride in the truck bed (Schneider, 1995). Despite laws and protective measures implemented to protect infants and children, far too many lose their lives at an early age due to the abuse, neglect, or negligence of adults.

Adolescence

In contemporary industrialized countries, adolescence roughly spans the teenage years, although some analysts place the lower and upper ages at 15 and 24. Before the twentieth century, adolescence did not exist as an age category. Today, it is a period in which young people are expected to continue their education and perhaps hold a part-time job.

What inequalities based on age are experienced by adolescents in our society? Adolescents are not granted full status as adults in most societies, but they are not allowed to act "childish" either. Early teens are considered too young to do "adult" things, such as stay out late at night, vote, drive, use tobacco, or consume alcoholic beverages. Many adolescents also face conflicting demands to attend school and to make money. Most states have compulsory school attendance laws requiring young people between the ages of 6 and 16 or 18 to attend school regularly; however, students who see no benefit from school or believe that the money they make working is more important may find themselves labeled as juvenile offenders for missing school. Moreover, juvenile laws define behavior such as truancy or running away from home as forms of delinquency—which would not be offenses if they were committed by an adult. Despite child labor laws implemented to control working conditions for young employees, many adolescents of today are employed in settings with hazardous working conditions, low wages, no benefits, and long work hours.

A variety of reports have labeled contemporary U.S. teenagers as a "generation at risk" because of the many problems that social analysts believe to be profound among today's adolescents (see Zill and Nord, 1994; Carnegie Council on Adolescent Development, 1995). Among the most pressing adolescent problems identified were crime and violence, teen pregnancy, suicide, drug abuse, and excessive peer pressure. However, other social analysts dispute these claims and suggest that teenagers have become the "scapegoat generation" and are widely viewed as being a problem for society. Although defining the "youth problem" in this manner may be harmful for all adolescents, it could be especially harmful for young people of color from low-income families. Without educational and economic opportunities, they are the most likely to constitute the majority of young people in jails, prisons, and detention facilities, where it is believed that they can do less harm to other people or to the nation as a whole. One analyst finds this trend especially troublesome:

> American adults have regarded adolescents with hope and foreboding [for many decades]. What is transpiring is new and ominous. A particular danger attends older generations indulging "they-deserve-it" myths to justify enriching ourselves at the expense of younger ones. The message . . . adults have spent two decades sending to youths is: *You are not our kids. We don't care about you.* (Males, 1996: 43)

In Males's opinion, the primary way to save the adolescent generation of today is to reduce poverty among children, teenagers, and young families, as well as to move away from the large number of age-based laws that restrict adolescents' opportunities for employment and freedom (Males, 1996).

Young Adulthood

Young adulthood, which follows adolescence and lasts to about age 39, is socially significant because during this time people are expected to get married, have children, and get a job. People who do not fulfill these activities during young adulthood tend to be viewed negatively. Individuals who do not get married by age 39 are often quizzed by relatives and friends about their intentions and sometimes their sexual orientation. Those who do not have a first child by the time they reach the end of their thirties are likely to experience questions from others about whether they plan to become parents or not. Even more significantly, those who are unable to find steady employment tend to become suspect because they have not "settled down" and "taken life seriously" or are viewed as being "lazy and unwilling to work." However, for some young adults, finding a job may be more difficult than for others. As previously discussed, race/ethnicity and gender strongly influence people's opportunities. For example, the sociologist William J. Wilson examined employment opportunities available to young adults living in Chicago's South Side and found that many who wanted

age stratification the inequalities, differences, segregation, or conflict between age groups.

to work could find no source of employment. According to one 32-year-old woman in his study,

> There's not enough jobs. . . . There's not enough factories, there's not enough work. Most all the good jobs are in the suburbs. Sometimes it's hard for the people in the city to get to the suburbs, because everybody don't own a car. Everybody don't drive. (qtd. in W. Wilson, 1996: 39)

Problems such as the one described here have become worse since the recent economic problems have left cities such as Chicago hard hit by the loss of factory jobs and a downturn in employment in the service sector.

People who are unable to earn income and pay into Social Security or other retirement plans in early adulthood typically find themselves disadvantaged as they enter middle and late adulthood.

Middle Adulthood

Prior to the twentieth century, life expectancy in the United States was only about 47 years, so the concept of middle adulthood—people between the ages of 40 and 65—did not exist until fairly recently. Normal changes in appearance occur during these years, and although these changes have little relationship to a person's health or physical functioning, they are socially significant to many people (Lefrançois, 1999).

As people progress through middle adulthood, they experience *senescence* (primary aging) in the form of molecular and cellular changes in the body. Wrinkles and gray hair are visible signs of senescence. Less-visible signs include arthritis and a gradual dulling of the senses of taste, smell, touch, and vision. Typically, reflexes begin to slow down, but the actual nature and extent of change vary widely from person to person. And stereotypes and self-stereotypes about aging may often be important in determining changes in a person's work ability (see "Sociology Works!").

People also experience a change of life in this stage. Women undergo *menopause*—the cessation of the menstrual cycle caused by a gradual decline in the body's production of the "female" hormones estrogen and progesterone. Menopause typically occurs between the mid-forties and the early fifties and signals the end of a woman's childbearing capabilities. Some women may experience irregular menstrual cycles for several years, followed by hot flashes, retention of body fluids, swollen breasts, and other aches and pains. Other women may have few or no noticeable physical symptoms. The psychological aspects of menopause are often as important as any physical effects. In one study, Anne Fausto-Sterling (1985) concluded that many women respond negatively to menopause because of negative stereotypes associated with menopausal and postmenopausal women. These stereotypes make the natural process of aging in women appear abnormal when compared with the aging process of men. Actually, many women experience a new interest in sexual activity because they no longer have to worry about the possibility of becoming pregnant. On the other hand, a few women have recently chosen to produce children using new medical technologies long after they have undergone menopause.

Men undergo a *climacteric,* in which the production of the "male" hormone testosterone decreases. Some have argued that this change in hormone levels produces nervousness and depression in men; however, it is not clear whether these emotional changes are due to biological changes or to a more general "midlife crisis," in which men assess what they have accomplished (Benokraitis, 2002). Ironically, even as such biological changes may have a liberating effect on some people, they may also reinforce societal stereotypes of older people, especially women, as "sexless." Recently, intensive marketing of products such as Viagra for erectile dysfunction has made people of all ages more aware not only of the potential sexual problems associated with aging but also with the possibility of reducing or solving these problems with the use of prescription drugs.

Along with primary aging, people in middle adulthood also experience *secondary aging,* which occurs as a result of environmental factors and lifestyle choices. For example, smoking, drinking heavily, and engaging in little or no physical activity are factors that affect the aging process. People who live in regions with high levels of environmental degradation and other forms of pollution are also at greater risk of aging more rapidly and having chronic illnesses and diseases associated with these external factors.

About 30 percent of all people in middle adulthood find that they are part of the "sandwich generation," having not only the responsibility of raising their children but also of simultaneously taking care of their aging parents—often while holding a full-time job, as well. The burden of these responsibilities disproportionately falls on women: "It's not like your husband is going to wash his own mother," a female forty-two-year-old former lawyer noted. "There's just no amount of feminism that is going to change that" (Vincent, 2004).

On the positive side, middle adulthood for some people represents the time during which (1) they have the highest levels of income and prestige, (2) they leave the problems of child rearing behind them and are content with their spouse of many years, and (3) they may have grandchildren, who give them another tie to the future. Even so, persons during middle adulthood know that, given society's current structure, their status may begin to change signifi-

Sociology *Works!*

"I Think I Can't, I Think I Can't!": The Self-Fulfilling Prophecy and Older People

Most of us are familiar with the children's story of *The Little Engine That Could,* in which the engine on the train said, "I think I can, I think I can!" when he was trying to pull the train up a steep hill. Used for years as an example of the power of positive thinking, the reverse side of the story might be that if people think that they cannot do something, that task becomes more difficult, and perhaps even impossible, for them to accomplish. Sociologists have a similar corollary to this story of the train, and the idea has widely referred to as the Thomas theorem or the self-fulfilling prophecy. As defined in Chapter 5, a *self-fulfilling prophecy* is a false belief or prediction that produces behavior that makes the originally false belief come true. The original statement of this prophecy, known as the "Thomas theorem" because of its possible originator, W. I. Thomas, is as follows: "If [people] define situations as real, they are real in their consequences."

How does this statement apply to the lives of older people? Researchers have found a number of ways in which this statement is applicable to the way that many people perceive the aging process and the way that many older people may view themselves in nations that more highly value youth over older people. The Norwegian sociologist Per Erik Solem (2008) explored the influence of the work environment on how age-related subjective changes occurred in people's work ability. He found that although age and physical health are obviously associated with decline in the work ability of older individuals, stereotypes and self-stereotypes (negative assumptions about themselves that are embraced by individuals who are being stereotyped) about aging are also important in producing a decline in a person's work ability. According to Solem (2008: 44), objective capacities such as physical strength and endurance may be important when it comes to performing certain tasks, such as lifting heavy equipment or nursing bedridden patients; however, "what workers believe they are able to do, influences to what extent they use their potential of objective abilities." In other words, if older individuals *subjectively* define themselves as less able than they actually are, given their *objective* potential, then they may be less likely to perform a task that might be quite within their capabilities to do.

Although changing associated with aging will always be a factor in some types of job performance, including those that require quick reactions or heavy physical work, many occupations can benefit from the experience and expertise of older workers because of their job-relevant knowledge or skills. For this reason, Solem (2008) suggests that older workers should be provided with new learning opportunities and a chance to maintain their subjective work ability throughout their careers.

Sociology not only helps us to see that cultural changes are needed if we are to rid ourselves and our society of age-based negative stereotypes about older individuals, but it also reminds us that the self-fulfilling prophecy is applicable at any stage of life. "I think I can" is a positive psychosocial force, whereas "I think I can't" is a negative force.

Reflect & Analyze

Can you think of periods of time in your life when a negative stereotype and/or self-stereotype may have contributed to your not doing something that you had the objective ability to do? In addition to issues pertaining to age, do you think the Thomas theorem might also be applicable to problems associated with race, class, and/or gender?

cantly when they reach the end of this period of their lives. As previously discussed, those who had few opportunities available earlier in life tend to become increasingly disadvantaged as they grow older.

Late Adulthood

Late adulthood is generally considered to begin at age 65—which formerly was referred to as the "normal" retirement age. However, with changes in Social Security regulations that provide for full retirement benefits to be paid only after a person reaches 66 or 67 years of age (based upon the individual's year of birth), many older persons have chosen to retire after the traditional age of 65. *Retirement* is the institutionalized separation of an individual from an occupational position, with continuation of income through a retirement pension based on prior years of service (Atchley and Barusch, 2004). Retirement means the end of a status that has long been a source of income and a means of personal identity. Perhaps the loss of a valued status explains why many retired persons introduce themselves by saying "I'm retired now, but I was a (banker, lawyer, plumber, supervisor, and so on) for forty years." As shown by ▶ Map 12.1, the percentage of the population age 65 and above varies from state to state, with Florida, Pennsylvania, and West Virginia having the highest proportion of people age 65 and over.

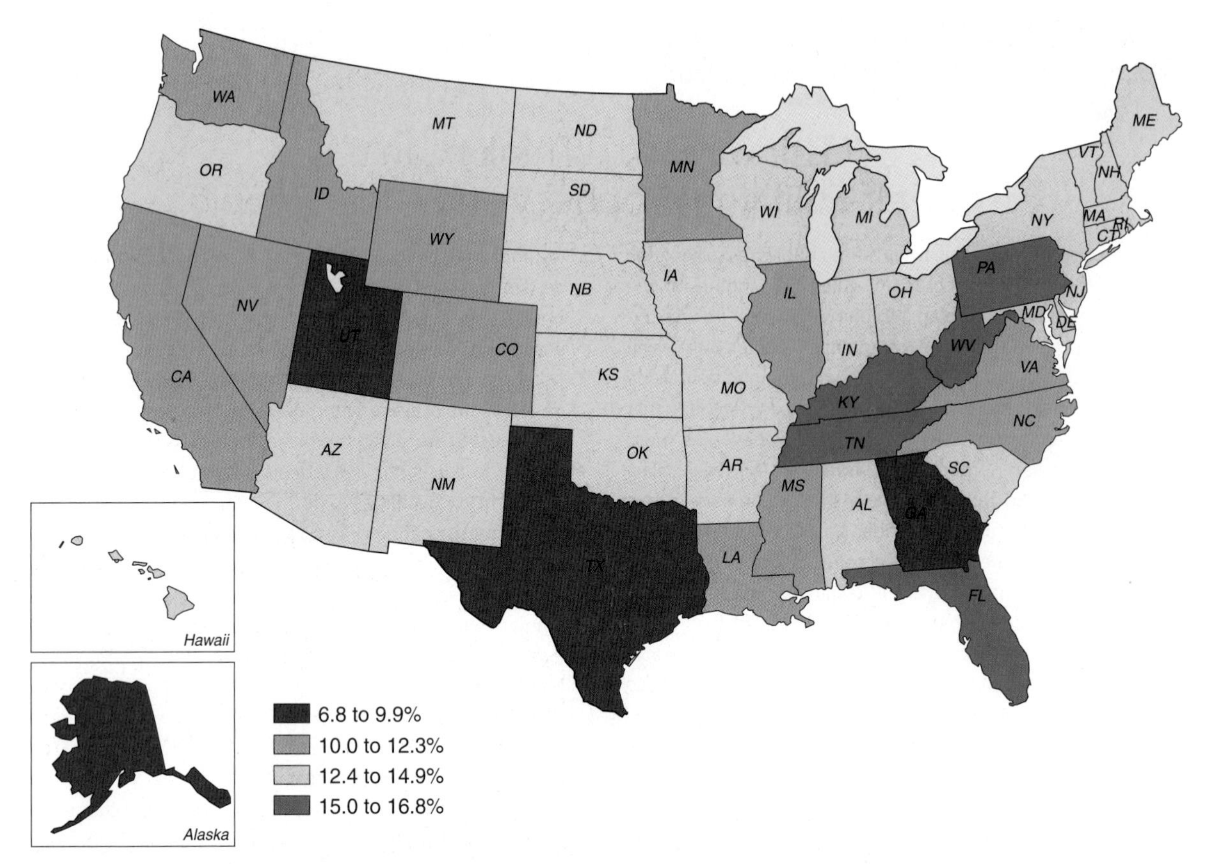

▶ **Map 12.1 Percentage of Resident Population Age 65 and Older by State**

Source: U.S. Census Bureau, 2007.

Some gerontologists subdivide late adulthood into three categories: (1) the "young-old" (ages 65 to 74), (2) the "old-old" (ages 75 to 85), and (3) the "oldest-old" (over age 85) (see Moody, 2002). Although these are somewhat arbitrary divisions, the "young-old" are less likely to suffer from disabling illnesses, whereas some of the "old-old" are more likely to suffer such illnesses (Belsky, 1999). However, one study found that the prevalence of disability among those 85 and over decreased during the 1980s due to better health care. In fact, it was reported that Jeanne Calment of Paris, France, who died in 1997 at age 122, rode a bicycle until she was 100 (Whitney, 1997).

The rate of biological and psychological changes in older persons may be as important as their chronological age in determining how they are perceived by themselves and others. As adults grow older, they actually become shorter, partly because bones that have become more porous with age develop curvature. A loss of three inches in height is not uncommon. As bones become more porous, they also become more brittle; simply falling may result in broken bones that take longer to heal. With age, arthritis increases, and connective tissue stiffens joints. Wrinkled skin, "age spots," gray (or white) hair, and midriff bulge appear; however, people sometimes use Oil of Olay, Clairol, or Buster's Magic Tummy Tightener in the hope of avoiding looking older (Atchley and Barusch, 2004).

Older persons also have increased chances of heart attacks, strokes, and cancer, and some diseases affect virtually only persons in late adulthood. Alzheimer's disease (a progressive and irreversible deterioration of brain tissue) is an example; about 55 percent of all organic mental disorders in the older population are caused by Alzheimer's (Atchley and Barusch, 2004). Persons with this disease have an impaired ability to function in everyday social roles; eventually, they cease to be able to recognize people they have always known and lose all sense of their own identity. Finally, they may revert to a speechless, infantile state such that others must feed them, dress them, sit them on the toilet, and lead them around. The disease can last up to 20 years; currently, there is no cure. Alzheimer's affects an estimated 3 percent of people over 65, and nearly half of those over age 85 may have the disease (Alzheimer's Disease Education and Referral Center, 2005).

Fortunately, most older people do not suffer from Alzheimer's and are not incapacitated by their physical condition. Only about 5 percent of older people live in nursing homes, about 10 percent have trouble walking, and about 30 percent have hearing problems. Although most older people experience some decline in strength, flexibility, stamina, and other physical capabilities, much of that decline does not result simply from the aging process and is avoidable; with proper exercise, some of it is even reversible.

With the physical changes come changes in the roles that older adults are expected (or even allowed) to perform. For example, people may lose some of the abilities necessary for driving a car safely, such as vision or reflexes. Although it is not true of all older persons, the average individual over age 65 does not react as rapidly as the average person who is younger than 65 (Lefrançois, 1999). In 2007 the issue of elderly drivers was widely debated in the media and political arenas after several incidents in which drivers over 80 years of age caused accidents in which other people were killed or injured (see Box 12.2).

The physical and psychological changes that come with increasing age can cause stress. According to Erik Erikson (1963), older people must resolve a tension of "integrity versus despair." They must accept that the life cycle is inevitable, that their lives are nearing an end, and that achieving inner harmony requires accepting both one's past accomplishments and past disappointments. As Hardy Howard, Sr., an 86-year-old African American man, explains,

> I feel good to be as old as I am, and when I think back I never had any problem walking or going up stairs, no false teeth or hearing aids, no nothing like that. . . . You know, what you get out of life is what you put in. Measure unto others as I would have them measure unto me. Anybody I can help, I don't care if they're white, black, green, or gray—anything you want, if I can do you a favor, I'll do it. And this has been my logic since I was a little boy, and I haven't changed from it even today. (qtd. in Mucciolo, 1992: 91–92)

Like many older people, Howard has worked to maintain his dignity and autonomy. In fact, a lot of older people are able to maintain their activities for many years beyond when younger people think it might be possible.

A survey conducted for the AARP (formerly known as the American Association of Retired Persons) found that more than half of U.S. people age 45 or over believed that they had satisfactory sex lives but that sexual activity and contentment diminished with age (Toner, 1999). The most frequent reasons for lack of sexual fulfillment were declining health and lack of a sexual partner. The partner gap was widest for women age 75 or older. Among women and men age 75 and up who did have partners, more than a fourth reported that they had sexual intercourse once a week (Toner, 1999). Other recent surveys of people age 100 and over found that most centenarians have a sense of having "enjoyed the journey to 100" (L. Ramirez, 1999: A4). For example, Ben Levinson, age 100, who regularly worked out at a gym and set a world record for the shot put in the Nike World Masters Games, appeared on *The Tonight Show* and other television programs talking about his longevity. *Masters games* and *masters athletes* are terms referring to the growing number of adults who are past the average peak performance age of 35 in sporting events but who continue to train with high intensity and compete at various levels (Morgan and Kunkel, 1998).

Will the life stages as we currently understand them accurately reflect aging in the future? Research continues to show that there are limited commonalities between those who are age 65 and those who are centenarians; however, many people tend to place everyone from 65 upward in categories such as "old," "elderly," or "senior citizen." In the future, we will probably see such categorizations revised or deleted as growing numbers of older people reject such labels as forms of "ageism." Some analysts believe that the existing life-course and life-stages models will be modified to reflect a sense of "old age" beginning at age 75 or 80 and that new stages will be added for those who reach age 90 and 100 (Morgan and Kunkel, 1998).

Inequalities Related to Aging

In previous chapters, we have seen how prejudice and discrimination may be directed toward individuals based on ascribed characteristics—such as race/ethnicity or gender—over which they have no control. The same holds true for age.

Ageism

Stereotypes regarding older persons reinforce *ageism,* defined in Chapter 4 as prejudice and discrimination against people on the basis of age, particularly against older persons. Ageism against older persons is rooted in the assumption that people become unattractive, unintelligent, asexual, unemployable, and mentally incompetent as they grow older.

Ageism is reinforced by stereotypes, whereby people have narrow, fixed images of certain groups. One-sided and exaggerated images of older people are used repeatedly in everyday life. Older persons are often stereotyped as thinking and moving slowly; as being bound to themselves and their past, unable to change and grow; as being unable to move forward and often moving backward. They are viewed as cranky, sickly, and lacking in social value; as egocentric and demanding; as shallow, enfeebled, aimless, and absentminded.

Box 12.2 Sociology and Social Policy

Driving While Elderly: Policies Pertaining to Age and Driving

> You are 84 years old and yesterday you killed my son. I don't know if you're even sound enough to understand what you've done but I have no sympathy for you or any of your family. And I pray EVERY day that you remain on this planet you see my son when you close your eyes. . . . We very much support a mandatory limit on the driving age for seniors.
>
> —Amanda Wesling, the mother of an eight-year-old boy killed by an elderly driver who accidentally drove a car into the boy's elementary school, expressing her anger at what the woman had done and at the fact that the woman was still allowed to drive (qtd. in Suhr, 2007)

Ironically, the elderly driver involved in this deadly crash was on her way to a senior citizens' center for a driving class because it was time to renew her driver's license. This accident followed in the aftermath of a number of other tragic accidents involving older drivers, including a 2003 accident in Santa Monica, California, where an 86-year-old man's car hurtled through a farmers' market, killing 10 people and injuring more than 70 others (CNN.com, 2003a). Other accidents involving drivers over the age of 85 have included a man driving a sport utility vehicle into a crowd of pedestrians and vendors at an open-air market in Rochester, New York, injuring 10 people, and an 87-year-old North Dakota woman, who was on her way to a medical appointment, smashing her car into a hospital lobby, injuring 5 other women (*CBS News*, 2007a).

Shortly after each of these incidents—and others involving elderly drivers—occurred, newspapers and cable TV channels began on-the-site reporting of the accident and lengthy discussions about the possible implications of these tragic incidents for other older drivers. Should there be a mandatory age at which people must quit driving? Should there be more frequent testing of older drivers? What could be done to prevent future occurrences of this nature? All of these questions buzzed around, but no answers were readily apparent. Advocates for older people emphasize that these tragedies seldom happen and that all older drivers should not be penalized for the actions of a very few people.

Most of us can recall the time when we anxiously awaited our next birthday so that we would be old enough to get our first driver's license. All states in the United States have a minimum age for getting a learner's permit or a first driver's license, but far fewer states have policies regarding drivers over the age of 65. However, this is likely to change in the near future with the aging of the U.S. population. By 2030, one in four drivers in this country will be over age 65. Some people will voluntarily give up their car keys when they believe that they can no longer drive safely. The National Institute on Aging estimates that about 600,000 people age 70 or over give up their driving privileges each year (Neergaard, 2003). But the question remains: What, if anything, should social policy makers, such as state legislators, do about those who either do not know that it is time to quit driving or who simply refuse to give up this privilege, which they often equate to giving up their freedom?

Clearly, there are competing rights and concerns involved in social policies pertaining to older drivers. On the one hand, older drivers point out that most cities lack adequate public transportation and that not being able to drive makes them a burden on other people, if other people are even available to drive them around to doctors' appointments, for groceries and other errands, and for recreational activities. On the other hand, pedestrians, other drivers, and

The media contribute to negative images of older persons, many of whom are portrayed as doddering, feebleminded, wrinkled, and laughable men and women, literally standing on their last legs (see Box 12.3). This is especially true with regard to advertising. In one survey, 40 percent of respondents over age 65 agreed that advertising portrays older people as unattractive and incompetent (Pomice, 1990). According to the advertising director of one magazine, "Advertising shows young people at their best and most beautiful, but it shows older people at their worst" (qtd. in Pomice, 1990: 42). Of older persons who do appear on television, most are male; only about one in ten characters appearing to be age 65 or older is a woman, conveying a subtle message that older women are especially unimportant (Pomice, 1990).

Fortunately, in recent years there has been a growing effort by the media to draw attention to the contributions, talents, and stamina of older persons rather than showing only stereotypical and negative portrayals. For example, the *New York Times*, CNN (Cable News Network), and other news sources highlighted more than a dozen runners over age 80 in the New York City Marathon. The media pointed out that some runners were in their nineties and that records are maintained of such accomplishments as being the fastest 85-year-old ever to complete the race (Barron, 1997).

Despite some changes in media coverage of older people, many younger individuals still hold negative stereotypes of "the elderly." In one study, William C. Levin (1988) showed photographs of the same man (disguised to appear as ages 25, 52, and 73 in various

© tom carter/Alamy

Although the vast majority of motor vehicle accidents are caused by younger drivers, when drivers in their seventies or eighties are involved in major accidents, widespread media coverage typically focuses on the driver's age. What effect, if any, will the aging baby boomer population have on social policies relating to driving? Keep in mind that many baby boomers are already in their sixties.

the general public have a vested interest in not having individuals of *any* age drive who are unsafe and who might, for whatever reason, constitute a threat to others.

What is happening at this time? The American Medical Association has developed a set of guidelines to help doctors tell when older patients' driving is questionable. The National Older Drivers Research Center offers driving-rehabilitation specialists who can instruct older drivers on additional safety measures they can take (such as avoiding unprotected left turns, unfamiliar roads, and night driving). However, at the bottom line, many decisions will be made at the state level. These may include whether to require more frequent license renewals for older drivers, whether drivers over a certain age should be retested before their license is renewed, which tests (such as eye exams) should be performed for license renewal, and what should be done if a person is unable to pass a vision test and show driving competency.

As we think about social policy pertaining to older people, particularly in regard to the right to drive, it is important to have empathy for people who will quickly be affected by new rules requiring them to prove their competence, but it is also important to remember that if we are fortunate to live long enough, each of us will also face such questions as we reach various stages in the course of our lives!

Reflect & Analyze

What safety issues are similar for young and older drivers? How can a nation balance the rights of individuals to engage in activities such as driving a motor vehicle while protecting the safety of others? What part do social rules play in dealing with issues such as these?

photos) to a group of college students and asked them to evaluate these (apparently different) men for employment purposes. Based purely on the photographs, the "73-year-old" was viewed by many of the students as being less competent, less intelligent, and less reliable than the "25-year-old" and the "52-year-old."

Many older people resist ageism by continuing to view themselves as being in middle adulthood long after their actual chronological age would suggest otherwise. In one study of people aged 60 and over, 75 percent of the respondents stated that they thought of themselves as middle-aged and only 10 percent viewed themselves as being old. When the same people were interviewed again 10 years later, one-third still considered themselves to be middle-aged. Even at age 80, one out of four men and one out of five women said that the word *old* did not apply to them; this lack of willingness to acknowledge having reached an older age is a consequence of ageism in society (Belsky, 1999).

Wealth, Poverty, and Aging

Many of the positive images of aging and suggestions on how to avoid the most negative aspects of ageism are based on an assumption of class privilege, meaning that people can afford plastic surgery, exercise classes, and social activities such as ballroom dancing or golf, and that they have available time and facilities to engage in pursuits that will "keep them young." However, many older people with meager incomes, little savings, and poor health, as well as those who are isolated in rural areas or high-crime sections of central cities, do

Box 12.3 Framing Aging in the Media

"Just When You Thought You Were Too Old for Romance"

People think when you get to a certain age, there's no more sex. That ain't true.

—"Harry" (played by Joseph Bologna) makes this statement in the film *Boynton Beach Club,* and moviegoers over the age of fifty often cheer loudly in their local theater because they are tired of seeing older people being portrayed as sexless beings.

Films that involve romance between older individuals, such as *Boynton Beach Club, Something's Gotta Give,* and *Calendar Girls,* have become more popular with film makers and movie audiences in recent years. These movies have come in the aftermath of a drought of many years in which older actors found it increasingly difficult to find mature roles in films. If they did land a part, the characters they portrayed typically were chronically ill or dying—although that genre of films has not disappeared, as evidenced by movies such as *Aurora Borealis,* in which the actor Donald Sutherland plays the role of a suicidal Parkinson's disease patient. By contrast, however, *Boynton Beach Club* is about five people living in an adult community in Florida who want to get back into the dating game and to have ongoing adventures in their lives. Although the characters meet at a local bereavement group, they soon begin to enjoy a new era of their lives as they interact with new people and rediscover an interest in their own sexuality. For the Baby Boomer generation, films such as this show that what it means to be older has changed dramatically, and this is also an important message for younger film audiences.

According to some analysts, films such as *Boynton Beach Club* reflect a shift in the attitudes of producers, directors, and Hollywood financial backers as they begin to realize that older moviegoers (who constitute a lucrative market for films and DVDs) are not going to pay money to watch plots that revolve entirely around adolescents and young people. As Susan Seidelman, the writer-director of *Boynton Beach Club,* stated, "There are a lot of people over 40 and 50 who would go to the movies, but they want to see a movie that reflects their experiences as an older person. That's a tough sell for your average Hollywood movie executive, a fetus in a suit" (qtd. in Newcott, 2007). Even if this comment is an overstatement, it highlights the point that the manner in which the media frame stories and articles about older people is important in our society.

For many years, older individuals have either been stereotyped or ignored by the mainstream media, which have been accused of presenting distorted images of the elderly (see Delloff, 1987) or of "geezer bashing" (see Hess, 1991). Even today, some media reports on Social Security, Medicare, and other issues concerning older Americans are framed in such a way that older people appear to be pitted against younger people in a battle over money and other scarce resources. When news reports suggest that the money spent for programs for older individuals is money that is taken away from affordable housing, better schools, fighting infant mortality, and investing in the future, this zero-sum-game framing ("if you win, I lose") unnecessarily pits older people against youth. However, this type of framing does not show the big picture of how other activities (such as military spending and war) might be an even greater drain on such domestic spending programs (Hess, 1991).

Perhaps it is important for books and movies to tell us stories about the feelings of older individuals and to show their daily struggle for love, hope, and dignity in a world that often does not respect the contributions of the elderly. Because the process of aging has been considered boring or downright depressing by many media analysts, older people have been neglected in much media coverage. However, we need to know more about the aging process, particularly its emotional components. We also need to see how older people conduct their lives, review their past experiences, and deal with the aging process because each of us may live to see the day when older people are *us,* not *them*—the other or the "outsider"—in what is likely to remain a youth-oriented nation.

Reflect & Analyze

Have you seen films or watched television programs that are a good example of how stories about older people might be framed? Have you also seen "geezer bashing"? For example, what about the portrayal of older people in some comedies and late-night television shows? What effect does media framing of aging have on your opinions of the elderly?

not have the same opportunities to follow popular recommendations about "successful aging" (Stoller and Gibson, 1997: 76). In fact, for many older people of color, aging is not so much a matter of seeking to defy one's age but rather of attempting to survive in a society that devalues both old age and minority status.

If we compare wealth (all economic resources of value, whether they produce cash or not) with income (available money or its equivalent in purchasing power), we find that older people tend to have more wealth but less income than younger people. Older people are more likely to own a home that has increased

For many years, advertisers have bombarded women with messages about the importance of a youthful appearance. TV Land's *She's Got the Look* is a reality show that features a competition for models over the age of 35. This youthful grandmother was one of the contestants.

substantially in market value; however, some may not have the available cash to pay property taxes, to buy insurance, and to maintain the property (Moody, 2002). Among older persons, a wider range of assets and income is seen than in other age categories. For example, many of the wealthiest people on the *Forbes* list of the richest people in this country (see Chapter 8) are over 65 years of age. On the other hand, more than 10 percent of all people over 65 live in poverty, as shown in ▶ Figure 12.2 on page 396.

Age, Gender, and Inequality

Age, gender, and poverty are intertwined. Although middle-aged and older women make up an increasing portion of the work force, they are paid substantially less than men their age, receive raises at a slower pace, and still work largely in gender-segregated jobs (see Chapter 11). As a result, women do not garner economic security for their retirement years at the same rate that many men do. These factors have contributed to the economic marginality of the current cohort of older women (Gonyea, 1994).

In one study, the gerontologists Melissa A. Hardy and Lawrence E. Hazelrigg (1993) found that gender was more directly related to poverty in older persons than was race/ethnicity, educational background, or occupational status. Hardy and Hazelrigg (1993) suggested that many women who are now age 65 and over spent their early adult lives as financial dependents of husbands or as working nonmarried women trying to support themselves in a culture that did not see women as the heads of households or as sole providers of family income. Because they were not viewed as being responsible for a family's financial security, women were paid less; therefore, older women may have to rely on inadequate income replacement programs originally designed to treat them as dependents. Women also have a greater risk of poverty in their later years; statistically, women tend to marry men who are older than themselves, and women live longer than men. Consequently, nearly half of all women over age 65 are widowed and living alone on fixed incomes (Smith, 2003).

As shown in Figure 12.2, the percentage of persons aged 65 and older living below the poverty line decreased significantly between 1980 and 2000. This largely resulted from increasing benefit levels of ***entitlements*—certain benefit payments paid by the government,** including Social Security, Supplemental Social Income (SSI), Medicare, Medicaid, and civil service pensions, which are the primary source of income for many persons over age 65 (Moody, 2002). Ninety percent of all people in the United States over age 65 draw Social Security benefits (AARP, 2002). Social Security and SSI provide virtually all of the financial support available to 25 percent of people over age 65.

Analysts note that Social Security keeps more white men out of poverty than it does white women and people of color (Browne and Broderick, 1994). The effectiveness of Social Security as a means of escaping poverty declines as age increases for all groups except white men (Axinn, 1989). Even for white men, Social Security does not provide the financial security in their later years that many of them thought it would when they were younger. For example, when Harold Milton was asked if he had ever worried about how he

entitlements certain benefit payments made by the government (Social Security, for example).

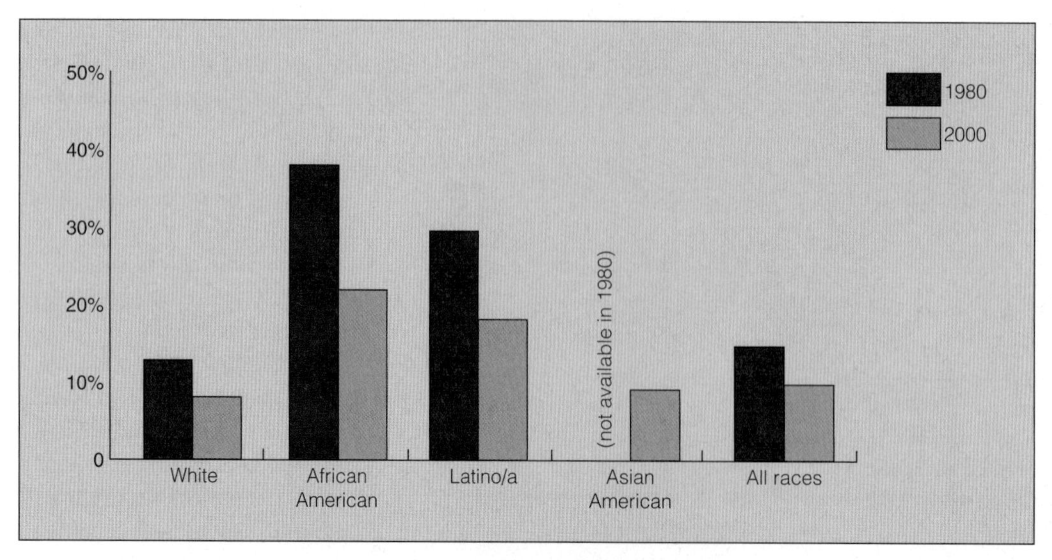

▶ **Figure 12.2 Percentage of Persons Age 65+ Below Poverty Level**
Although many of the wealthiest people are over 65 years of age, more than 10 percent of all people over 65 live in poverty. Note the disparities by racial–ethnic group.

Source: U.S. Census Bureau, 2007.

would manage after he retired, he responded, "No, it never entered my mind. You see, I was counting on my Social Security." When he retired, Milton found that he could not live on his Social Security benefits of $250 and SSI of $86 per month, so he took a part-time job as a handyman to earn another $100 per month (Margolis, 1990). According to many social analysts, Social Security is in need of change and might fail outright before many of today's young people reach retirement age.

Medicare, the other of the two largest entitlement programs, is a nationwide health care program for persons aged 65 and older who are covered by Social Security or who are eligible to "buy into" the program by paying a monthly premium (Atchley and Barusch, 2004). Although Medicare's primary purpose is to provide long-term support, it only partially does so. The reasons for this apparent contradiction are (1) skilled nursing care other than in a hospital is covered only if it immediately follows at least three days in a hospital (and many people are not sick enough to qualify for hospital care), and (2) Medicare reimburses only the first ninety days of hospital care or the first hundred days of skilled nursing care (Atchley and Barusch, 2004).

Age and Race/Ethnicity

Age, race/ethnicity, and economic inequality are closely intertwined. Inequalities that exist later in life originate in individuals' early participation in the labor force and are amplified in late adulthood. For example, older African Americans continue to feel the impact of segregated schools and overt patterns of job discrimination that were present during their early years. Although African Americans constitute only about 8.4 percent of the population age 65 and over, they account for 26 percent of the low-income older population. Among persons age 65 and over, 8.3 percent of whites reported poverty-level incomes in 2002, as compared with 23.8 percent of African Americans and 21.4 percent of Latinos/as (Federal Interagency Forum on Aging-Related Statistics, 2005). As you will recall from Chapter 8, the poverty line is determined by estimating how much a low-budget family must spend annually for groceries and then multiplying that amount by three. Based on the assumption that persons in late adulthood eat less than younger people, the poverty line is placed at a lower dollar amount ($9,367 for a single person in 2005) for persons 65 and older than for people below that age ($10,160 in 2005). In other words, at age 64 a person with an income of $10,000 is considered to be below the poverty line (and thus entitled to assistance), but one year later, on the same income, the person is no longer "poor." Although caloric demand does decrease for older people, their nutritional needs remain the same, and no data indicate that their food costs are less (Margolis, 1990).

As previously noted, the primary reason for the lower income status of many older African Americans can be traced to a pattern of limited employment opportunities and periods of unemployment throughout their lives, combined with their concentration in secondary-sector jobs, which pay lower wages, are sporadic, have few benefits, and were not covered by Social Security prior to the 1950s (Hooyman and

Kiyak, 2002). Moreover, health problems may force some African Americans out of the labor force earlier than other workers due to a higher rate of chronic diseases such as hypertension, diabetes, and kidney failure (Hooyman and Kiyak, 2002).

Similarly, older Latinas/os have higher rates of poverty than whites (Anglos) because of lack of educational and employment opportunities. Some older Latinos/as entered the country illegally and have had limited opportunities for education and employment, leaving them with little or no Social Security or other benefits in their old age. High rates of poverty among older Latinas/os are associated with poor health conditions, lack of regular care by a physician, and fewer trips to the hospital for medical treatment of illness, disease, or injury.

Older Native Americans are among the most disadvantaged of all categories. Older Native Americans are more likely to live in high-poverty, rural areas than are other minority older populations. Some studies have found that about 50 percent of all older Native Americans live in poverty, having incomes that are between 40 and 60 percent less than those of older white Americans. In addition to experiencing educational and employment discrimination similar to that of African Americans and Latinos/as, older Native Americans were also the objects of historical oppression and federal policies toward the native nations that exacerbated patterns of economic impoverishment. Consequently, some older Native Americans have the worst living conditions and poorest health of all older people in this country. Research findings regarding the mental health of older Native Americans also indicate a high rate of depression, alcoholism and other drug abuse, and suicide (see Chapter 18, "Health, Health Care, and Disability").

Among older Asian Americans, many who arrived in the United States prior to 1930 have fared less well in their old age than those who were native-born or were more recent arrivals. As discussed in Chapter 10, many older Asian Americans from Japan and China received less education and experienced more economic deprivation than did later cohorts of Japanese Americans and Chinese Americans. Today, many older Asian Americans remain in Chinatown, Japantown, or Koreatown, where others speak their language and provide goods and services that help them maintain their culture, and where mutual-aid and benevolent societies and recreational clubs provide them with social contacts and delivery of services. As a result, many older Asian Americans who qualify for various forms of entitlements, such as Supplemental Security Income, do not apply for it. Moreover, cultural values, including traditional healing practices, may help explain why many Asian Americans, particularly Chinese American elders, do not use physical and mental health services that are available to them.

For many years, studies in sociology and gerontology primarily focused on the attributes and needs of older white Americans from middle- and low-income backgrounds. Overall, more research is needed on the unique needs of older people from diverse racial and ethnic categories.

Older People in Rural Areas

The lives of many older people differ based on whether they reside in urban or rural areas. Despite the stereotypical image of the rural elderly living in a pleasant home located in an idyllic country setting, rural elders, as compared with older urban residents, typically have lower incomes, are more likely to be poor, and have fewer years of schooling. The rural elderly also tend to be in poorer health, and many of them are less likely to receive needed health care because many rural areas lack adequate medical and long-term care facilities (Coburn and Bolda, 1999). With the "graying of America," the population of older adults (age 65 and over) has continued to grow in rural areas; however, this growth varies from region to region, with the Midwest and the South having a greater concentration of older persons in rural areas than the West and the Northeast have.

Why are older people in rural areas more likely to have lower incomes and tend to be classified as "poor"? Some social analysts attribute lower income

© AP Images/Marcio Jose Sanchez

Older people have actively protested changes that they view as detrimental to Social Security, Medicare, and other programs for senior citizens. What part will older Americans play in political activism in the future?

two million jobs were lost, and the unemployment rate soared to 9.5 percent (Mahler, 2009).

Where have so many jobs gone? In some cases, the jobs were eliminated, and the work of former employees was heaped on other workers (Chitale, 2009). In other situations, tens of thousands of jobs left the country and are now performed by workers in nations such as China and India. Many major corporations thinned out their work force to save money; others filed for bankruptcy and reorganized their operations to do more with fewer employees. In the automobile, construction, and trade industries, a sharp decline in demand resulted in major layoffs and firings. Who is to blame? Although some unemployed individuals blame their personal problems on the overall downturn in the global economy, others blame greedy Wall Street executives or lackadaisical government watchdogs for the financial crisis and economic recession. Regardless of where the blame should be placed, many people have lost their homes through foreclosures, and millions of people are unemployed. Bureau of Labor Statistics data indicate that 14.5 million people were unemployed in the first half of 2009, which is twice as many as the 7.2 million who were unemployed in 2007.

What are the connections between the unemployment problem described here and the larger economic structure of our society and the global economy? In this chapter, we will discuss the economy as a social institution and explain how our current problems are linked to the larger world of work—what people are paid for their labor, who gains and who loses in tough economic times, how people feel about their work, and what impact recent changes in the economy may have on your future. Before reading on, test your knowledge about the economy and the world of work by taking the quiz in Box 13.1.

Comparing the Sociology of Economic Life with Economics

Perhaps you are wondering how a sociological perspective on the economy differs from the study of economics. Although aspects of the two disciplines overlap, each provides a unique perspective on economic institutions. Economists attempt to explain how the limited resources and efforts of a society are allocated among competing ends. To economists, an imbalance exists between people's wants and society's ability to meet those wants. To illustrate, think about college registration. Would you like to have the "perfect" schedule—with the classes you want, at the times you want, and with "preferred" professors? How many students at your school actually manage to arrange such a schedule? What organizational constraints make it impossible for all people to have what they need or want? Some economists believe that the most important fact of economics is the law of scarcity, which means that there will never be enough resources to meet everyone's wants (Ruffin and Gregory, 2000). Colleges do not have the financial or human resources to provide everything that students (or faculty) want.

Economists focus on the complex workings of economic systems (such as monetary policy, inflation, and the national debt), whereas sociologists focus on interconnections among the economy, other social institutions, and the social organization of work. At the macrolevel, sociologists may study the impact of transnational corporations on industrialized and low-income nations. At the microlevel, sociologists might study people's satisfaction with their jobs. To better understand these issues, we will examine how economic systems came into existence and how they have changed over time.

Economic Systems in Global Perspective

The ***economy* is the social institution that ensures the maintenance of society through the production, distribution, and consumption of goods and services.** *Goods* are tangible objects that are necessary (such as food, clothing, and shelter) or desired (such as DVDs and electric toothbrushes). *Services* are intangible activities for which people are willing to pay (such as dry cleaning, a movie, or medical care). In high-income nations today, many of the goods and services that we consume are information goods. Examples include databases and surveys ("intermediate products") and the mass media, computer software, and the Internet ("information goods").

Some goods and services are produced by human labor (the plumber who unstops your sink, for example); others are primarily produced by capital (such as Internet access available through a service provider). *Labor* refers to the group of people who contribute their physical and intellectual services to the production process in return for wages that they are paid by firms (Boyes and Melvin, 2002). Capital is wealth (money or property) owned or used in business by a person or corporation. Obviously, money, or financial capital, is needed to invest in the physical capital (such as machinery, equipment, buildings, warehouses, and factories) used in production. For example, a person

Box 13.1 Sociology and Everyday Life

How Much Do You Know About the Economy and the World of Work?

True	False	
T	F	1. In today's economy, professions are largely indistinguishable from other occupations.
T	F	2. Workers' skills are usually upgraded when new technology is introduced in the workplace.
T	F	3. Many of the new jobs being created in the service sector pay poorly and offer little job security.
T	F	4. In the United States and other nations, most assembly-line workers are white men.
T	F	5. Labor unions will probably cease to exist sometime during this century.
T	F	6. Few workers resist work conditions that they consider to be oppressive.
T	F	7. Around the world, positions with the most job security are located in large, transnational corporations.
T	F	8. Assembly lines are rapidly disappearing from all sectors of the U.S. economy.

Answers on page 412.

who owns a thousand shares of Dell Computer stock owns financial capital, but these shares also represent an ownership interest in Dell's physical capital.

To better understand the economy in the United States today, let's briefly look at three broad categories of economies: preindustrial, industrial, and postindustrial economies.

Preindustrial Economies

Hunting and gathering, horticultural and pastoral, and agrarian societies are all preindustrial economic structures. Most workers in these societies engage in ***primary sector production*—the extraction of raw materials and natural resources from the environment.** These materials and resources are typically consumed or used without much processing. For example, portions of contemporary sub-Saharan Africa have a relatively high rate of exports in primary commodities, and foreign direct investment is concentrated in mineral extraction. Consequently, most of sub-Saharan Africa is highly dependent on primary sector production, which, in turn, is highly vulnerable to the whims of the primary commodity markets (United Nations Development Programme, 2003). In recent years, many people have lost economic ground in sub-Saharan Africa, where per-capita incomes are lower than they were in 1970 (United Nations Development Programme, 2003).

Throughout history, the production units in hunting and gathering societies have been small; most goods are produced by family members. The division of labor is by age and gender (Hodson and Sullivan, 2008). The potential for producing surplus goods increases as people learn to domesticate animals and grow their own food. In horticultural and pastoral societies, the economy becomes distinct from family life. The distribution process becomes more complex, with the accumulation of a *surplus* such that some people can engage in activities other than food production. In agrarian societies, production is primarily related to producing food. However, workers have a greater variety of specialized tasks, such as warlord or priest; for example, warriors are necessary to protect the surplus goods from plunder by outsiders (Hodson and Sullivan, 2008). Once a surplus is accumulated, more people can also engage in trade. Initially, the surplus goods are distributed through a system of *barter*—the direct exchange of goods or services considered of equal value by the traders. However, bartering is limited as a method of distribution; equivalencies are difficult to determine (how many fish equal one rabbit?) because there is no way to assign a set value to the items being traded. As a result, *money,* a medium of exchange with a relatively fixed value, came into use in order to aid the distribution of goods and services in society.

economy the social institution that ensures the maintenance of society through the production, distribution, and consumption of goods and services.

primary sector production the sector of the economy that extracts raw materials and natural resources from the environment.

Box 13.1 Sociology and Everyday Life

Answers to the Sociology Quiz on the Economy and the World of Work

1. False. Even in tough economic times, professions have four characteristics that distinguish them from other occupations: (1) abstract, specialized knowledge; (2) autonomy; (3) authority over clients and subordinate occupational groups; and (4) a degree of altruism.

2. False. Jobs are often deskilled when new technology (such as bar code scanners or computerized cash registers) is installed in the workplace. Some of the workers' skills are no longer needed because a "smart machine" now provides the answers (such as how much something costs or how much change a customer should receive).

3. True. Many of the new jobs being created in the service sector, such as nurse's aide, child-care worker, hotel maid, and fast-food server, offer little job security and low pay.

4. False. Today, most assembly-line workers are young girls and women in developing nations, and women or men of color in the United States.

5. False. Sociologists who have examined organized labor generally predict that unions will continue to exist; however, their strength may wane in the global economy.

6. False. Many workers resist work conditions that they believe are unjust or oppressive. Some engage in sabotage; others join unions or participate in other types of pro-worker organizations.

7. False. Although it is difficult to determine which types of jobs are the most secure, many positions in transnational corporations have been lost through downsizing and plant relocations and closings.

8. False. According to some scholars, assembly lines will remain a fact of life for businesses ranging from fast-food restaurants to high-tech semiconductor plants. However, in certain sectors of the economy, such as the automobile industry, the number of jobs is rapidly constricting.

Sources: Based on Hodson and Sullivan, 2008; Isidore, 2009; and Mahler, 2009.

What was the U.S. economy like in the preindustrial era? As previous chapters have stated, the agricultural revolution brought about dramatic changes in the nature of work, including the increasing division between work and home. In the preindustrial economy of the colonial period (from the 1600s to the early 1700s), white men earned a livelihood through agricultural work or as small-business owners who ran establishments such as inns, taverns, and shops. During this period, white women worked primarily in their homes, doing such tasks as cooking, cleaning, and child care. Some also developed *cottage industries*—producing goods in their homes that could be sold to nonfamily members. However, a number of white women also worked outside their households as midwifes, physicians, nurses, teachers, innkeepers, and shopkeepers (Hesse-Biber and Carter, 2000). By contrast, the experiences of people of color were quite different in preindustrial America. As the sociologists Sharlene Hesse-Biber and Gregg Lee Carter (2000) explain, the institution of slavery, which came about largely as a result of the demand for cheap agricultural labor, was a major force in the exploitation of many people of color, but it was particularly the women of color who suffered a double burden of oppression in the form of both sexism and racism. For example, African American women were exploited as workers, as breeders of slaves, and sometimes as sex objects for white men (Hesse-Biber and Carter, 2000). By contrast, Native American women in some agricultural communities held greater power because they were able to maintain control over land, tools, and surplus food (Hesse-Biber and Carter, 2000).

Do preindustrial forms of work still exist in contemporary high-income nations? In short, yes. Even in high-income nations such as the United States, entire families work in the agricultural sector of the economy, performing tasks such as picking ripened cherries. Here is one journalist's description of "home life" among some seasonal cherry pickers in the state of Washington:

> At the height of cherry-picking season, 12 men were making a patch of woods . . . into a home. Plastic grocery bags hung from nails in the Ponderosa pine bark: the pantry. A shard of mirror was wedged into the trunk of another pine: the bathroom. The kitchen was a Sunbeam propane grill, the bushes served as

© Mark Richards/PhotoEdit

© AP Images/Alan Marler

Even as the United States and other high-income countries have increasingly relied on high-tech economies, other forms of work still exist in the agricultural and industrial sectors of the economy. Here, workers hand-pick a strawberry crop and work in a paper mill.

> toilets, and six cheap tents formed the bedroom. . . . Cherry picking requires special timing and skills; the fruit must be picked just as soon as it is ripe, and workers must be careful not to separate the stems from the fruit or damage the branches where next year's buds will form. . . . Working from 4 A.M. until noon, when heat can damage the fruit, pickers can make $4 for filling a 40-pound box, or up to $80 a day. Many travel for weeks at a time following the ripening from California to Oregon, then from Washington to Montana. (Kelley, 1999: A11)

As this example shows, some parts of the agricultural sector of the U.S. economy have not been changed by industrialization or postindustrialization. The cherry pickers described above are employed in the same region as many high-tech information employees who work for Microsoft or other computer manufacturers or software designers.

Industrial Economies

Industrialization brings sweeping changes to the system of production and distribution of goods and services. Drawing on new forms of energy (such as steam, gasoline, and electricity) and machine technology, factories proliferate as the primary means of producing goods. Most workers engage in ***secondary sector production*—the processing of raw materials (from the primary sector) into finished goods.** For example, steelworkers process metal ore; autoworkers then convert the ore into automobiles, trucks, and buses. In industrial economies, work becomes specialized and repetitive, activities become bureaucratically organized, and workers primarily work with machines instead of with one another.

Mass production results in larger surpluses that benefit some people and organizations but not others. Goods and services become more unequally distributed because some people can afford anything they want and others can afford very little. Nations engaging primarily in secondary sector production also have some primary sector production, but they rely on less-industrialized nations for the raw materials from which to make many products. In sum, the typical characteristics of industrial economies include the following:

1. *New forms of energy, mechanization, and the growth of the factory system.* With the introduction of the steam engine and steam-powered machines, work becomes centered in factories, which are viewed as separate and distinct spheres from the home and the earlier cottage industries that had been located in the home.
2. *Increased division of labor and specialization among workers.* With industrialization, people carry out a wider diversity of jobs which have different but integrated activities that contribute to the production of specific goods.

secondary sector production the sector of the economy that processes raw materials (from the primary sector) into finished goods.

to the tune of billions of dollars in taxpayers' monies and other funds. Some analysts believe that these bailouts literally saved the U.S. economy from collapsing, but others argue that this form of extreme government intervention helped many of the individuals and corporations that initially caused the nation's economic crisis.

Socialism

***Socialism* is an economic system characterized by public ownership of the means of production, the pursuit of collective goals, and centralized decision making.** Like "pure" capitalism, "pure" socialism does not exist. Karl Marx described socialism as a temporary stage en route to an ideal communist society. Although the terms *socialism* and *communism* are associated with Marx and are often used interchangeably, they are not identical. Marx defined communism as an economic system characterized by common ownership of all economic resources. In the *Communist Manifesto* and *Das Kapital,* he predicted that the working class would become increasingly impoverished and alienated under capitalism. As a result, the workers would become aware of their own class interests, revolt against the capitalists, and overthrow the entire system (see Turner, Beeghley, and Powers, 2002). After the revolution, private property would be abolished, and capital would be controlled by collectives of workers who would own the means of production. The government (previously used to further the interests of the capitalists) would no longer be necessary. People would contribute according to their abilities and receive according to their needs (Marx and Engels, 1967/1848; Marx, 1967/1867). Over the years, state control was added as an organizing principle for communist societies. This structure is referred to as a system of "state socialism," and the former Soviet Union (the "Union of Soviet Socialist Republics," what today is Russia and certain other nearby countries) became the best example.

Why did state socialism in the Soviet Union not evolve into a communist economic system? For one thing, Marx's ideas about a truly communistic society were based on his perceptions of *industrialized* nations. At the time that an attempt was made to implement communism in Russia, however, it was largely a feudal society that had not experienced the type of conditions Marx envisioned would be necessary for the development of communism (Rosengarten, 1995). The system of state socialism failed for other reasons as well, as discussed later in this chapter. However, many of the ideas of Marx have had a profound effect on how sociologists and other researchers view our contemporary economic and social problems (see "Sociology Works!" on page 424).

Public Ownership of the Means of Production In a truly socialist economy, the means of production are owned and controlled by a collectivity or the state, not by private individuals or corporations. Prior to the early 1990s, the state owned all the natural resources and almost all the capital in the Soviet Union. In the 1980s, for example, state-owned enterprises produced more than 88 percent of agricultural output and 98 percent of retail trade, and owned 75 percent of the urban housing space (Boyes and Melvin, 2002). At least in theory, goods were produced to meet the needs of the people. Access to housing and medical care was considered to be a right.

Leaders of the Soviet Union and some Eastern European nations decided to abandon government ownership and control of the means of production because the system was unresponsive to the needs of the marketplace and offered no incentive for increased efficiency (Boyes and Melvin, 2002). Shortages and widespread unrest led to a reform movement by then-President Mikhail Gorbachev in the late 1980s.

Since the 1990s, Russia and other states in the former Soviet Union have attempted to privatize ownership of production. China—previously the world's other major communist economy—announced in 1997 that it would privatize most state industries (Serrill, 1997). In *privatization,* resources are converted from state ownership to private ownership; the government takes an active role in developing, recognizing, and protecting private property rights (Boyes and Melvin, 2002).

Pursuit of Collective Goals Ideal socialism is based on the pursuit of collective goals rather than on personal profits. Equality in decision making replaces hierarchical relationships (such as between owners and workers or political leaders and citizens). Everyone shares in the goods and services of society, especially necessities such as food, clothing, shelter, and medical care, based on need, not on ability to pay. In reality, however, few societies can or do pursue purely collective goals.

Centralized Decision Making Another tenet of socialism is centralized decision making. In theory, economic decisions are based on the needs of society; the government is responsible for aiding the production and distribution of goods and services. Central planners set wages and prices to ensure that the production process works. When problems such as shortages and unemployment arise, they can be dealt with quickly and effectively by the central government (Boyes and Melvin, 2002).

Centralized decision making is hierarchical. In the former Soviet Union, for example, broad economic policy decisions were made by the highest authorities of the Communist Party, who also held political power. The production units (the enterprises and farms) at the bottom of the structure had little voice in the decision-making process. Wages and prices were based on political priorities and eventually came to be completely unrelated to actual supply and demand.

The collapse of state socialism in the former Soviet Union was due partly to the declining ability of the Communist Party to act as an effective agent of society and partly to the growing incompatibility of central planning with the requirements of a modern economy (see Misztal, 1993). More than a decade after centralized decision making was abolished in Russia, people are still faced with poverty, soaring unemployment, and high crime rates. Organized criminal groups have muscled their way into business and trade; many workers feel that their future is very dim.

Mixed Economies

As we have seen, no economy is truly capitalist or socialist; most economies are mixtures of both. A ***mixed economy*** **combines elements of a market economy (capitalism) with elements of a command economy (socialism).** Sweden, Great Britain, and France have mixed economies, sometimes referred to as ***democratic socialism*****—an economic and political system that combines private ownership of some of the means of production, governmental distribution of some essential goods and services, and free elections.** For example, government ownership in Sweden is limited primarily to railroads, mineral resources, a public bank, and liquor and tobacco operations (Feagin, Baker, and Feagin, 2006). Compared with capitalist economies, however, the government in a mixed economy plays a larger role in setting rules, policies, and objectives.

The government is also heavily involved in providing services such as medical care, child care, and transportation. In Sweden, for example, all residents have health insurance, housing subsidies, child allowances, paid parental leave, and day-care subsidies. National insurance pays medical bills associated with work-related injuries, and workplaces are specially adapted for persons with disabilities. College tuition is free, and public funds help subsidize cultural institutions such as theaters and orchestras ("General Facts on Sweden," 1988; Kelman, 1991). Recently, some analysts have suggested that the United States has assumed many of the characteristics of a *welfare state* (a state that uses extensive government action to provide support and services to its citizens) as it has attempted to meet the basic needs of older persons, young children, unemployed people, and persons with a disability.

Perspectives on Economy and Work in the United States

Functionalists, conflict theorists, and symbolic interactionists view the economy and the nature of work from a variety of perspectives. We first examine functionalist and conflict views of the economy; then we focus on the symbolic interactionist perspective on job satisfaction and alienation.

Functionalist Perspective

Functionalists view the economy as a vital social institution because it is the means by which needed goods and services are produced and distributed. When the economy runs smoothly, other parts of society function more effectively. However, if the system becomes unbalanced, such as when demand does not keep up with production, a maladjustment occurs (in this case, a surplus). Some problems can be easily remedied in the marketplace (through "free enterprise") or through government intervention (such as paying farmers *not* to plant when there is an oversupply of a crop). However, other problems, such as periodic *peaks* (high points) and *troughs* (low points) in the business cycle, are more difficult to resolve. The *business cycle* is the rise and fall of economic activity relative to long-term growth in the economy (McEachern, 2003).

From this perspective, peaks occur when "business" has confidence in the country's economic future. During a peak, or *expansion period,* the economy thrives: Plants are built, raw materials are ordered, workers are hired,

socialism an economic system characterized by public ownership of the means of production, the pursuit of collective goals, and centralized decision making.

mixed economy an economic system that combines elements of a market economy (capitalism) with elements of a command economy (socialism).

democratic socialism an economic and political system that combines private ownership of some of the means of production, governmental distribution of some essential goods and services, and free elections.

Sociology *Works!*

Marx: Not Completely Right, but Not Completely Wrong Either

The legacy of Karl Marx, the German economist and philosopher, has been a mixed one. Some people believe that his works should have no influence on contemporary thinking about the economy and political life because some of his predictions about capitalism proved to be incorrect; however, others emphasize that Marx was not completely wrong either, particularly when we look at his ideas in light of the current economic crisis. Slightly over 160 years ago, Karl Marx wrote *The Communist Manifesto,* and the journalist and social commentator Barbara Ehrenreich (2008) argues that, despite changing times, this document has retained relevance for today:

> The Manifesto makes for quaint reading today. All that talk about "production," for example: Did they actually make things in those days? Did the proletariat really slave away in factories instead of call centers? But on one point Marx and Engels proved right: Within capitalist societies, or at least the kind of wildly unregulated capitalism America has had, the rich got richer, the workers got poorer, and the erstwhile middle class has been sliding toward ruin. The last two outcomes are what Marx called "immiseration," which, in translation, is the process you're undergoing when you have cancer and no health insurance or a mortgage payment due and no paycheck coming in.

You may recall that Ehrenreich is a journalist and social activist who has written books such as *Nickel and Dimed* (previously mentioned in Chapter 8) and speaks on the behalf of the working class and the poor. After her time spent working as a manual laborer and hourly-wage earner at fast-food chains, hotels, and big-box stores such as Wal-Mart, Ehrenreich described how difficult it is for people earning the minimum wage or less to eke out a living as they become increasingly impoverished under capitalism. Some of her research and writing has been strongly influenced by conflict theorists such as Karl Marx.

Although the revolution of the workers and the fall of capitalism that Marx predicted (based on the workers becoming fed up with immiseration and "revolting, seizing the means of production, and insisting on running the show themselves") did not occur, Ehrenreich (2008) argues that this does not mean that Marx was incorrect in his assessment about the problems inherent in advanced capitalism:

> The revolution didn't happen, of course, at least not here. For the last several years, American workers have sweetly acquiesced to declining wages, rising prices, speed-ups at work, disappearing pensions, and increasingly threadbare health insurance. While CEO pay escalated to the 8-figure range and above, so-called ordinary Americans took on second jobs and crowded into multi-generational households with uncomfortable long waits for the bathroom. But all this immiseration—combined with fabulous enrichment at the top—did end up destabilizing the capitalist system, if only because, in the last few years, America's substitute

and production increases. In addition, upward social mobility for workers and their families becomes possible. As a result, workers may hope their children will not have to follow their footsteps into the factory. Ben Hamper (1992: 13) describes how GM workers felt:

> Being a factory worker in Flint, Michigan, wasn't something purposely passed on from generation to generation. To grow up believing that you were brought into this world to follow in your daddy's footsteps, just another chip-off-the-old-shoprat, was to engage in the lowest possible form of negativism. Working the line for GM was something fathers did so that their offspring wouldn't have to.

The American Dream of upward mobility is linked to peaks in the business cycle. Once the peak is reached, however, the economy turns down because too large a surplus of goods has been produced. In part, this is due to *inflation*—a sustained and continuous increase in prices (McEachern, 2003). Inflation erodes the value of people's money, and they are no longer able to purchase as high a percentage of the goods that have been produced. Because of this lack of demand, fewer goods are produced, workers are laid off, credit becomes difficult to obtain, and people cut back on their purchases even more, fearing unemployment. Eventually, this produces a distrust of the economy, resulting in a *recession*—a decline in an economy's total production that lasts six months or longer. To combat a recession, the government lowers interest rates (to make borrowing easier and to get more money back into circulation) in an attempt to spur the beginning of the next expansion period.

Conflict Perspective

Conflict theorists have a different view of business cycles and the economic system. From a conflict perspective, business cycles are the result of capitalist greed. In

Karl Marx's beliefs about capitalism, socialism, and communism are widely known today, although it has been more than 100 years since his famous proclamation—as emblazoned on his tombstone outside London—"Workers of all lands unite."

for decent wages has been easy credit. . . . Marx's argument was that the coexistence of great wealth for the few and growing poverty for the many is not only morally objectionable, it's also inherently unstable. [Marx] may have been wrong about the reasons for the instability, but no one can any longer deny it's there. When the greed of the rich collided with the needs of the poor—for a home, for example—the result was a global credit meltdown. Obviously, the way to address the crisis is to deal with the poverty and inequality that led to it. . . .

All sociologists do not agree with the ideas of Marx or with Ehrenreich's explanation of them, but the 160-year-old *Communist Manifesto* and the current writings of Ehrenreich do reflect the importance of taking into account the poor and acknowledging that everyone does not share equally in the hardships associated with tough economic times. While the super-rich may have to give up their personal jets and the upper-middle class their private exercise classes, it quite another thing for the poor to lose their low-wage jobs, a place to live, and what remains of their dignity.

Reflect & Analyze

What alternate explanations do sociologists give for the current global economic crisis? What might we learn about the economy as a social institution by conducting research at locations frequented by the working class and the poor, such as food banks and homeless shelters? Is there a place for social activism in the study of sociology?

order to maximize profits, capitalists suppress the wages of workers. As the prices of products increase, workers are not able to purchase them in the quantities that have been produced. The resulting surpluses cause capitalists to reduce production, close factories, and lay off workers, thus contributing to the growth of the reserve army of the unemployed, whose presence helps reduce the wages of the remaining workers. In some situations, workers are replaced with machines or nonunionized workers. In 1994, for example, some trucking companies claimed that they could not afford to pay union truck drivers over $17 an hour, so they hired nonunion freight carriers who made about $9 an hour. Drivers such as Tim Hart claimed that pay was the only difference between his work and that of a union driver: "I bust my tail, doing the exact same thing the union drivers do. And they make double what I do. It bothers me. I mean, if those companies can afford to pay that much, why can't mine?" (qtd. in D. Johnson, 1994: 10).

Karl Marx referred to the propensity of capitalists to maximize profits by reducing wages as the *falling rate of profit,* which he believed to be one of the inherent contradictions of capitalism that would produce its eventual downfall. According to the political sociologist Michael Parenti, business *is* the economic system. Parenti believes that political leaders treat the health of the capitalist economy as a necessary condition for the health of the nation and that the goals of big business (rapid growth, high profits, and secure markets at home and abroad) become the goals of government (Parenti, 1996). In sum, to some conflict theorists, capitalism is the problem; to some functionalist theorists, capitalism is the solution to society's problems.

Symbolic Interactionist Perspective

Sociologists who focus on microlevel analyses are interested in how the economic system and the social

organization of work affect people's attitudes and behavior. Symbolic interactionists, in particular, have examined the factors that contribute to job satisfaction.

According to symbolic interactionists, work is an important source of self-identity for many people; it can help people feel positive about themselves, or it can cause them to feel alienated. *Job satisfaction* refers to people's attitudes toward their work, based on (1) their job responsibilities, (2) the organizational structure in which they work, and (3) their individual needs and values (Hodson and Sullivan, 2008). Studies have found that worker satisfaction is highest when employees have some degree of control over their work, when they are part of the decision-making process, when they are not too closely supervised, and when they feel that they play an important part in the outcome (Kohn et al., 1990). Persons in management and supervisory positions may feel more of a sense of ownership of the products or services produced by their organization. For example, Deborah Kent, the first African American woman to head a vehicle assembly plant at the Ford Motor Company, feels such a sense of ownership that she sometimes asks strangers on the street how they like their vehicle if it was produced in her plant:

> I feel this sense of ownership, responsibility and pride about every Econoline, Villager, and Quest on the road. Particularly since I know we're the only manufacturing plant producing those vehicles. And if something's wrong, I want to know about it. (qtd. in Williams, 1995: F7)

Job satisfaction is often related to both intrinsic and extrinsic factors. Intrinsic factors pertain to the nature of the work itself, whereas extrinsic factors include such things as vacation and holiday policies, parking privileges, on-site day-care centers, and other amenities that contribute to workers' overall perception that their employer cares about them. In one study, the sociologist Ruth Milkman (1997) found that autoworkers were ambivalent about their jobs. Although they hated the chronic stress and humiliation of factory work, they liked the high pay and good benefits.

Alienation occurs when workers' needs for self-identity and meaning are not met and when work is done strictly for material gain, not a sense of personal satisfaction. According to Marx, workers are resistant to having very little power and no opportunities to make workplace decisions. This lack of control contributes to an ongoing struggle between workers and employers. Job segmentation, isolation of workers, and the discouragement of any type of pro-worker organizations (such as unions) further contribute to feelings of helplessness and frustration. Some occupations may be more closely associated with high levels of alienation than others.

The Concept Quick Review summarizes the three major sociological perspectives on the economy and work in the United States.

© AP Images/Victoria Arocho

Workplace anger and violence are of great concern in the twenty-first century. In one recent case, Michael McDermott was found guilty in the shooting deaths of seven of his co-workers at a Massachusetts Internet company. However, the rate of such violent attacks appears to be declining.

The Social Organization of Work

How do societies organize work? As societies grow larger in population and more complex in their division of labor, different categories of work are identified by government bureaucracies. In the United States, this is done by agencies such as the U.S. Department of Labor and the U.S. Census Bureau. These classifications are often somewhat arbitrary and may not reflect the actual job descriptions and tasks associated with the work that many people do. For example, sociologists often find that Census Bureau categories such as "executive, administrators, and managerial" or "technical and related support" encompass such a wide range of jobs and differences in compensation that it is difficult to make generalizations from statistical data about workers in these employment categories.

Occupations

***Occupations* are categories of jobs that involve similar activities at different work sites** (Reskin and Padavic, 2002). Over 500 different occupational categories and 31,000 occupation titles, ranging from motion picture cartoonist to drop-hammer operator, are currently listed by the U.S. Census Bureau. Historically, occupa-

CONCEPT QUICK REVIEW

Perspectives on Economy and Work in the United States

Key Concept	
Functionalist Perspective	The economy is a vital social institution because it is the means by which needed goods and services are produced and distributed.
Conflict Perspective	The capitalist economy is based on greed. In order to maximize profits, capitalists suppress the wages of workers, who, in turn, cannot purchase products, making it necessary for capitalists to reduce production, close factories, lay off workers, and adopt other remedies that are detrimental to workers and society.
Symbolic Interactionist Perspective	Many workers experience job satisfaction when they like their job responsibilities and the organizational structure in which they work and when their individual needs and values are met. Alienation occurs when workers do not gain a sense of self-identity from their jobs and when their work is done completely for material gain and not for personal satisfaction.

tions were classified as blue collar and white collar. Blue-collar workers were primarily factory and craft workers who did manual labor; white-collar workers were office workers and professionals. However, contemporary workers in the service sector do not easily fit into either of these categories; neither do the so-called pink-collar workers, primarily women, who are employed in occupations such as preschool teacher, dental assistant, secretary, and clerk (Hodson and Sullivan, 2008).

Sociologists establish broad occupational categories by distinguishing between employment in the primary labor market and in the secondary labor market. The ***primary labor market*** **consists of high-paying jobs with good benefits that have some degree of security and the possibility of future advancement.** By contrast, the ***secondary labor market*** **consists of low-paying jobs with few benefits and very little job security or possibility for future advancement** (Bonacich, 1972). Some jobs in the secondary labor market are also characterized by hazardous working conditions and unscrupulous employers. Jane H. Lii, a Chinese-speaking journalist on an undercover assignment, described such problems when she briefly worked in a garment factory in Brooklyn, New York:

> Seven days later, after 84 hours of work, I got my reward, in the form of a promise that in three weeks I would be paid $54.24, or 65 cents an hour [much less than the minimum wage]. I also walked away from the lint-filled factory with aching shoulders, a stiff back, a dry cough, and a burning sore throat. (Lii, 1995: 1)

Like many workers in the lower tier of the secondary labor market, garment factory workers have relatively little formal education, may be working in violation of child labor laws, and may be recent immigrants who feel powerless to call attention to their plight. By contrast, professionals in the upper tier of the primary labor market are characterized by extensive education or training and some degree of control over their work.

Professions

What occupations are professions? Athletes who are paid for playing sports are referred to as "professional athletes." Dog groomers, pest exterminators, automobile mechanics, and nail technicians (manicurists) also refer to themselves as professionals. Although sociologists do not always agree on exactly which occupations are professions, they do agree that the number of people categorized as "professionals" has grown dramatically since World War II. According to the sociologist Steven Brint (1994), the contemporary professional middle class includes most doctors, natural scientists,

occupations categories of jobs that involve similar activities at different work sites.

primary labor market the sector of the labor market that consists of high-paying jobs with good benefits that have some degree of security and the possibility of future advancement.

secondary labor market the sector of the labor market that consists of low-paying jobs with few benefits and very little job security or possibility for future advancement.

engineers, computer scientists, certified public accountants, economists, social scientists, psychotherapists, lawyers, policy experts of various sorts, professors, at least some journalists and editors, some clergy, and some artists and writers.

Characteristics of Professions ***Professions*** **are high-status, knowledge-based occupations that have five major characteristics** (Freidson, 1970, 1986; Larson, 1977):

1. *Abstract, specialized knowledge.* Professionals have abstract, specialized knowledge of their field based on formal education and interaction with colleagues. Education provides the credentials, skills, and training that allow professionals to have job opportunities and assume positions of authority within organizations (Brint, 1994).
2. *Autonomy.* Professionals are autonomous in that they can rely on their own judgment in selecting the relevant knowledge or the appropriate technique for dealing with a problem. Consequently, they expect patients, clients, or students to respect that autonomy.
3. *Self-regulation.* In exchange for autonomy, professionals are theoretically self-regulating. All professions have licensing, accreditation, and regulatory associations that set professional standards and that require members to adhere to a code of ethics as a form of public accountability. Realistically, however, professionals are often constrained by organizational power structures. Many work within large-scale bureaucracies that have rules, policies, and procedures to which professionals, like everyone else, must adhere. For example, physicians who are salaried employees in for-profit hospital corporations must follow bureaucratic rules, regulations, and controls (Ritzer, 2000a).
4. *Authority.* Because of their authority, professionals expect compliance with their directions and advice. Their authority is based on mastery of the body of specialized knowledge and on their profession's autonomy: Professionals do not expect the client to argue about the professional advice rendered. Professionals also have authority over persons in subordinate occupations; for example, attorneys may delegate routine legal work, such as the completion of standard forms or documents, to paralegals or legal secretaries (Hodson and Sullivan, 2008).
5. *Altruism.* Ideally, professionals have concern for others, not just their own self-interest. The term *altruism* implies some degree of self-sacrifice whereby professionals go beyond their self-interest or personal comfort so that they can help a patient or client (Hodson and Sullivan, 2008). Professionals also have a responsibility to protect and enhance their knowledge and to use it for the public interest.

In the past, job satisfaction among professionals has generally been very high because of relatively elevated levels of income, autonomy, and authority. In the future, professionals may either become the backbone of a postindustrial society or suffer from "intellectual obsolescence" if they cannot keep up with the knowledge explosion (Leventman, 1981).

Reproduction of Professionals Although higher education is one of the primary qualifications for a profession, the emphasis on education gives children whose parents are professionals a disproportionate advantage early in life (Brint, 1994). There is a direct linkage between parental education and income and children's scores on college admissions tests such as the SAT, as shown in ▶ Figure 13.3. In turn, test scores are directly related to students' ability to gain admission to colleges and universities, which serve as springboards to most professions.

Race and gender are also factors in access to the professions. Historically, people of color have been underrepresented. Today, as more persons from underrepresented groups have gained access to professions such as law and medicine, their children have also gained the educational and mentoring opportunities necessary for professional careers. Between 1980 and 2000, there was a 96 percent increase in the number of African Americans, age 25 and above, with at least 4 years of college education (U.S. Census Bureau, 2007). However, by 2004, only 18 percent of African Americans and 12 percent of Latinos/as age 25 and above had graduated from college, as compared with 28 percent of whites and 50 percent of Asian Americans (U.S. Census Bureau, 2007).

Deprofessionalization Certain professions are undergoing a process of *deprofessionalization,* in which some of the characteristics of a profession are eliminated (Hodson and Sullivan, 2008). Occupations such as pharmacist have already been *deskilled,* as Nino Guidici explains:

> In the old days [people] took druggists as doctors. . . . All we do [now] is count pills. Count out twelve on the counter, put 'em in here, count out twelve more. . . . Doctors used to write out their own formulas and we made most of these things. Most of the work is now done in the laboratory. The real druggist is found in the manufacturing firms. They're the factory workers and they're the pharmacists. We just get the name of the drugs and the number and the directions. It's a lot easier. (qtd. in Terkel, 1990/1972)

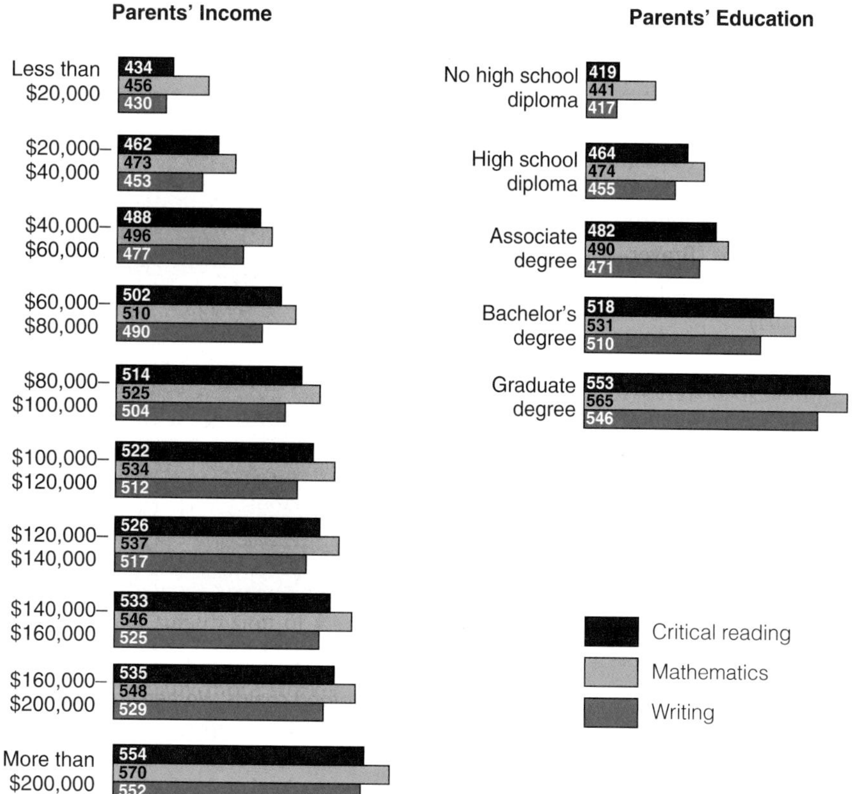

▶ **Figure 13.3 SAT Scores by Parents' Income and Education, 2008**

Source: SAT, 2009.

However, colleges of pharmacy in many universities have fought against deprofessionalization by upgrading the degrees awarded to pharmacy graduates from the traditional B.S. in pharmacy to a Ph.D. This upgrading of degrees has also occurred over the past three decades in law schools, where the Bachelor of Laws (LL.B.) has been changed to the Juris Doctor (J.D.) degree.

Managers and the Managed

A wide variety of occupations are classified as "management" positions. The generic term *manager* is often used to refer to executives, managers, and administrators (Hodson and Sullivan, 2008). At the upper level of a workplace bureaucracy are *executives* (Hodson and Sullivan, 2008). *Administrators* often work for governmental bureaucracies or organizations dealing with health, education, or welfare (such as hospitals, colleges and universities, and nursing homes) and are usually appointed. *Managers* typically have responsibility for workers, physical plants, equipment, and the financial aspects of a bureaucratic organization.

Management in Bureaucracies Managers are essential in contemporary bureaucracies, in which work is highly specialized and authority structures are hierarchical (see Chapter 6). Managers often control workers by applying organizational rules. Workers at each level of the hierarchy take orders from their immediate superiors and perhaps give orders to a few subordinates. Upper-level managers are typically responsible for coordination of activities and control of workers. The *span of control,* or the number of workers a manager supervises, is affected by the organizational structure and by technology. Some analysts believe that hierarchical organization is necessary to coordinate the activities of a large number of people; others suggest that it produces apathy and alienation among workers (Blauner, 1964). Lack of worker control over the labor process was built into the earliest factory systems through techniques known as scientific management (Taylorism) and mass production (Fordism).

Scientific Management (Taylorism) At the beginning of the twentieth century, industrial engineer Frederick Winslow Taylor revolutionized management with a system he called *scientific management.* In an effort to increase productivity in factories, Taylor did numerous *time-and-motion* studies of workers whom he considered to be reasonably efficient. From these studies, he broke down each task into its smallest components to determine the "one best way" of doing

professions high-status, knowledge-based occupations.

Seeking to improve economic and social opportunities for farmworkers, the late César Chávez held rallies and engaged in other protest activities in an effort to better the workers' lives.

have used a number of methods to resist workplace alienation. Many have also joined labor unions to gain strength through collective actions.

Labor Unions

U.S. labor unions came into being in the mid-nineteenth century, as previously discussed. Unions have been credited with gaining an eight-hour workday, a five-day workweek, health and retirement benefits, sick leave and unemployment insurance, and workplace health and safety standards for many employees. Most of these gains have occurred through *collective bargaining*—negotiations between employers and labor union leaders on behalf of workers. In some cases, union leaders have called strikes to force employers to accept the union's position on wages and benefits. While on strike, workers may picket in front of the workplace to gain media attention, to fend off "scabs" (nonunion workers) who might take over their jobs, and in some cases to discourage customers from purchasing products made or sold by their employer. In recent years, strike activity has diminished significantly as many workers have feared losing their jobs. As ▶ Figure 13.4 shows, in 2007 only 21 strikes, or work stoppages, involving more than 1,000 workers were reported, as compared with significantly higher numbers in the 1960s and 1970s—especially compared with 1970, when there were more than eighteen times as many strikes as in 2007. The number of workers involved in the actions declined from a peak of more than 2.5 million in 1971 to 339,000 in 1997 (U.S. Census Bureau, 2007).

Although the overall *number* of union members in the United States has increased since the 1960s, primarily as a result of the growth of public employee unions (such as the American Federation of Teachers), the *proportion* of all employees who are union members has declined. Today, about 15 percent of all U.S. employees belong to unions or employee associations. Part of the decline in union participation has been attributed to the structural shift away from manufacturing and manual work and toward the service sector, where employees have been less able to unionize.

Historically, labor unions contributed to the employment gains of people of color. In 2005, for example, the highest rates of union affiliation were among African American men (16 percent) and women (14 percent), as compared with white men (13 percent) and women (11 percent), and Latinos and Latinas (10 percent) (U.S. Census Bureau, 2007). African Americans have higher rates of union affiliation because of extensive public sector employment, where workers

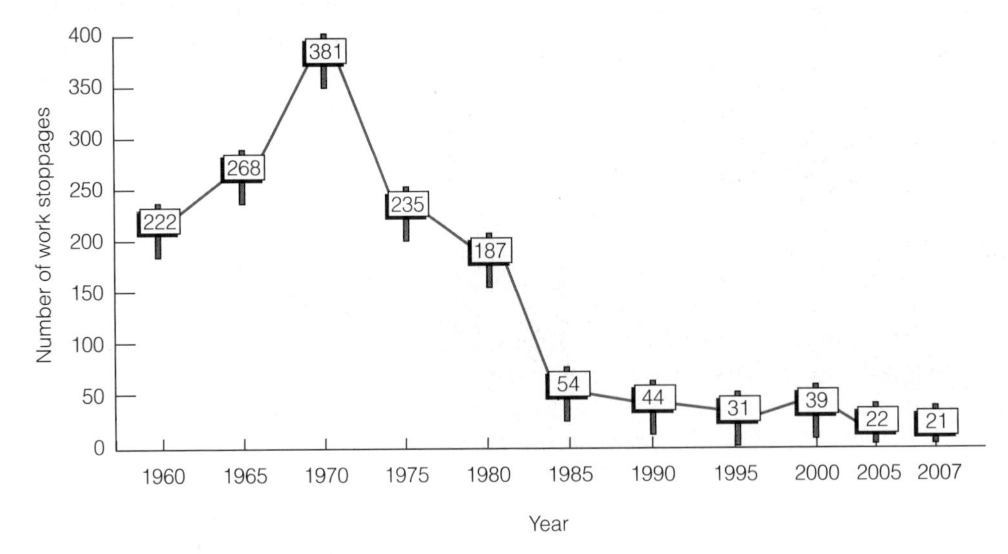

▶ **Figure 13.4 Major Work Stoppages in the United States, 1960–2007**

Source: U.S. Census Bureau, 2009.

are more likely to be unionized (Costello and Stone, 1994). In the past, many African Americans moved into the middle class through unionized jobs in heavy industry. More recently, the loss of many of these jobs has been especially damaging for people of color (Higginbotham, 1994).

A "spillover" effect exists between union and nonunion firms in the area of wages and benefits; nonunion firms must compete for labor with union firms. Some nonunion firms pay higher wages in order to dissuade union membership. However, the spillover effect may work in the opposite direction as well: If unionized workers lose wages and benefits, nonunion workers may also see a lowering of their wages and benefits (Amott, 1993). Usually, union members earn higher wages than nonunion workers in comparable jobs. In 2003 union workers averaged over $9,000 a year more in wages than did nonunion workers (U.S. Census Bureau, 2007). Women in unions earn an average weekly wage that is 1.3 times that of nonunionized women.

Although labor union membership has declined in most industrialized countries, including the United States, unions are still highly influential in labor negotiations involving wages and working conditions in Scandinavia, Germany, the Netherlands, Ireland, and much of Eastern Europe (Greenhouse, 1997). In most industrialized countries, collective bargaining by unions has been dominated by men. However, in countries such as Sweden, Germany, Austria, and Great Britain, women workers have gained some important concessions as a result of labor union participation.

Absenteeism, Sabotage, and Resistance

Absenteeism is one means by which workers resist working conditions that they consider to be oppressive. Ben Hamper (1992: 47) describes how employees were chastised in a meeting for their absenteeism at the Flint, Michigan, GM plant:

> The Plant Manager introduced the man in charge of overseeing worker attendance. . . . He didn't seem happy at all. The attendance man unveiled a large chart illustrating the trends in absenteeism. . . . He pointed to Monday. . . . Monday was an unpopular day attendance-wise. He moved the pointer over to Tuesday and Wednesday which showed a significant gain in attendance. The chart peaked way up high on Thursday. Thursday was pay night. Everybody showed up on Thursday.
>
> "Then we arrive at Friday," the attendance man announced. A guilty wave of laughter spread through the workers. None of the bossmen appeared at all amused. Friday was an unspoken Sabbath for many of the workers. Paychecks in their pockets, the leash was temporarily loosened. To get a jump on the weekend was often a temptation too difficult to resist.

Other workers use sabotage to bring about informal work stoppages. The phrase "throwing a monkey wrench in the gears" originated with the practice of workers "losing" a tool in assembly-line machinery. Sabotage of this sort effectively brought the assembly line to a halt until the now-defective piece of machinery could be repaired. Industrial sabotage has been described as an expression of workers' deep-seated frustrations about poor work environments (Feagin, Baker, and Feagin, 2006).

Although most workers do not sabotage machinery, a significant number do resist what they perceive to be oppression from supervisors and employers. In a study of Asian American women who work in low-status occupations such as hotel housekeepers, the sociologist Esther Ngan-Ling Chow (1994) found that many of the women felt they had little to lose if they chose to resist or rebel. Onyoung, one of the Korean American women interviewed by Chow, described how she felt:

> Many of us depend on this job to support our families. When I am blamed for the things that I do not do and have no part of it, I will yell my guts out to protest. At most I will lose this job. I want my dignity, to be an honorable person, which my parents taught me. (qtd. in Chow, 1994: 214)

Resistance helps people in lower-tier service jobs survive at work (Dill, 1988; Chow, 1994). In interactions with their employers, domestic workers attempt to maintain their personal dignity and to gain mastery of situations in which they are defined as objects, not as human beings. Scholars have documented resistance by African American domestic workers (Rollins, 1985; Dill, 1988), Chicana private housekeepers (Romero, 1992; Hondagneu-Sotelo, 2001), and Japanese domestic helpers (Glenn, 1986). According to Chow (1994: 218), many Asian American women become "active agents and goal-oriented actors, capable of taking charge of their own lives" in their interactions with supervisors and co-workers. Workers who feel that they can take charge of their own lives may be less likely to experience high levels of alienation than workers who feel powerless over their plight.

Studies of worker resistance are very important to our understanding of how people deal with the social organization of work. Previously, workers (especially women) have been portrayed as passive "victims" of their work environment. As we have seen, this is not always the case for either women or men; many workers resist problematic situations and demand better wages and work conditions.

Employment Opportunities for Persons with a Disability

An estimated 48 million persons in the United States have one or more physical or mental disabilities that differentially affect their opportunities for employment. This number continues to increase for several reasons. First, with advances in medical technology, many people who formerly would have died from an accident or illness now survive, although with an impairment. Second, as more people live longer, they are more likely to experience diseases that may have disabling consequences. Third, persons born with serious disabilities are more likely to survive infancy because of medical technology. However, less than 15 percent of persons with a disability today were born with it; accidents, disease, and war account for most disabilities in this country.

In 1990 the United States became the first nation to formally address the issue of equality for persons with a disability. When Congress passed the Americans with Disabilities Act (ADA), this law established "a clear and comprehensive prohibition of discrimination on the basis of disability." Combined with previous disability rights laws (such as those that provide for the elimination of architectural barriers from new federally funded buildings and for the maximum integration of schoolchildren with disabilities), the ADA is a legal mandate for the full equality of people with disabilities.

Despite this law, about two-thirds of working-age persons with a disability are unemployed today. Most persons with a disability believe they could work if they were offered the opportunity. However, even when persons with a disability are able to find jobs, they earn less than persons without a disability (Yelin, 1992). On the average, workers with a disability make 85 percent (for men) and 70 percent (for women) of what their co-workers without disabilities earn, and the gap is growing (Shapiro, 1993). Studies have shown a thirty-year overall decline in the economic condition of persons with disabilities (Yelin, 1992; Burkhauser, Haveman, and Wolfe, 1993). The problem has been particularly severe for African Americans and Latinos/as with disabilities (Kirkpatrick, 1994). Among Latinos/as with disabilities, only 10 percent are employed full time; those who work full time earn 73 percent of what white (non-Latino) persons with disabilities earn.

What does it cost to "mainstream" persons with disabilities? A survey of personnel directors and other executives responsible for making hiring decisions for their companies found that the average cost of workplace modifications to accommodate employees with a disability was less than $500 (Heldrich Center for Workforce Development, 2003). Other accommodations cost about $1,000, with the largest single cost being $14,500 each for special Braille computer displays for visually impaired employees (Noble, 1995). As employment opportunities change in the global economy, it is important that employers use common sense and a

© Michael Newman/PhotoEdit

Workers with a disability are able to engage in a wide variety of occupations when they are offered the opportunity to do so.

Box 13.3 You Can Make a Difference

Creating Access to Information Technologies for Workers with Disabilities

People are shocked I can do this. They have never imagined a blind person could find a use for a computer, much less make a business out of it.

—Cathy Murtha, a partner in Magical Mist Creations, a company that designs and evaluates sites on the World Wide Web (qtd. in Sreenivasan, 1996: C7)

If you are a person with a visual impairment or a person who plans to supervise workers—some of whom may be visually impaired—you can make a difference in the workplace by learning about new technologies that make jobs more accessible for workers with disabilities. For example, new computer hardware and software programs and the Internet are having a profound effect on workers who are visually impaired. According to Larry Scadden, technological consultant for persons with disabilities at the National Science Foundation, "The Internet has changed forever the lives of blind people, mainly because it provides independent access to information" (qtd. in Sreenivasan, 1996: C7). The goal of making the Internet accessible to visually impaired persons requires several things:

- Text must be in such a form that it can be read aloud by a speech synthesizer with text-to-speech software or be output in a Braille format.
- The user must be able to navigate the screen with a keyboard, not a mouse.
- Good descriptions must be given of photos and graphics that the user will not be able to see.

To find out more about creating greater access to information for persons with visual impairment and other disabilities, contact the following agencies:

- American Council of the Blind
  http://www.acb.org
- American Foundation for the Blind
  http://www.afb.org

These resources are also available for further information:

- The Americans with Disabilities Act page provides information on the act and links to disability-related sites, organized by subject:
  http://www.ada.gov
- The International Center for Disability Resources on the Internet provides resources for persons with disabilities and includes links to other sites:
  http://www.icdri.org
- The Laurent Clerc National Deaf Education Center website contains a list of national organizations for deaf and hard-of-hearing persons, with links to those organizations' sites:
  http://clerccenter.gallaudet.edu

To reach websites that are "user friendly" for persons with visual impairments, contact the following sites:

- Caldwell College:
  http://www.caldwell.edu
- On Island Communications:
  http://www.onisland.com
- Cathy's Newstand:
  http://orsaminore.dreamhosters.com/handy/documenti/uk_blind.html
- Ann Morris Enterprises:
  http://www.annmorris.com

creative approach to thinking about how persons with disabilities can fit into the marketplace (see Box 13.3).

The Global Economy in the Future

How will the U.S. economy look in the future? What about the global economy? Although sociologists do not have a crystal ball with which to predict the future, some general trends can be suggested.

The U.S. Economy

Many of the trends we examined in this chapter will produce dramatic changes in the organization of the economy and work in the twenty-first century. U.S. workers may find themselves fighting for a larger piece of an ever-shrinking economic pie that includes a trade deficit and a national debt of almost $9 trillion. In April 2007, our national debt was $8,851,983,831,350, making each citizen's share more than $29,000! Even as the gender and racial–ethnic composition of the labor force continues to diversify, many workers will remain in race- and/or gender-segregated occupations and industries. At the same time, workers may increasingly be fragmented into two major labor market divisions: (1) those who work in the innovative, primary sector and (2) those whose jobs are located in the growing secondary, marginal sector. In the innovative sector, increased productivity will be the watchword as corporations respond to heightened international competition. In the marginal sector, alienation will grow as temporary workers, sometimes professionals, look for avenues of upward mobility or at least a chance to make their work life more tolerable. Labor unions will be unable to help workers unless the unions embark on innovative programs to recruit new members, improve their image, and recover their former political clout (Hodson and Sullivan, 2008). However, in spite of these problems, most analysts predict that the United States will remain a major player in the world economy.

Global Economic Interdependence and Competition

Most social analysts predict that transnational corporations will become even more significant in the global economy of this century. As they continue to compete for world market share, these corporations will become even less aligned with the values of any one nation. However, those who advocate increased globalization typically focus on its potential impact on industrialized nations, not the effect it may have on the 80 percent of the world's population that resides in low-income and middle-income nations. Persons in low-income nations may become increasingly baffled and resentful when they are bombarded with media images of Western affluence and consumption; billions of "have-nots" may feel angry at the "haves"—including the engineers and managers of transnational companies living and working in their midst.

In recent years, the average worker in the United States and other high-income countries has benefited more from global economic growth than have workers in lower-income countries. For example, the average citizen of Switzerland has an income several hundred times that of a resident of Ethiopia. More than a billion of the world's people live in abject poverty; for many, this means attempting to survive on less than $400 a year.

A global workplace is emerging in which telecommunications networks link workers in distant locations. In the developed world, the skills of some professionals will transcend the borders of their countries. For example, there is a demand for the services of international law specialists, engineers, and software designers across countries. Even as nations become more dependent on one another, they will also become more competitive in the economic sphere. Changes in the global economy may require people in all nations—including the United States—to make changes in the way that things are done on the individual, regional, and national levels (see Box 13.4).

The fate of high-income nations is increasingly linked to the globalization of markets. Dramatic changes in stock markets in nations such as Japan and the United States create reverberations that are felt in financial markets around the world.

Box 13.4 Sociology and Social Policy

Does Globalization Change the Nature of Social Policy?

In 1492 Christopher Columbus set sail for India, going west. He had the Niña, the Pinta and the Santa Maria. He never did find India, but he called the people he met "Indians" and came home and reported to his king and queen: "The world is round." I set off for India 512 years later. I knew just which direction I was going. I went east. I had Lufthansa business class, and I came home and reported only to my wife and only in a whisper: "The world is flat."

—author and newspaper columnist Thomas L. Friedman (2005a) describing how the global economy is changing all areas of economic, political, and social life, including how we formulate social policy

What does Thomas Friedman mean when he states that "the world is flat"? As discussed in Friedman's (2005b) best-selling book, *The World Is Flat: A Brief History of the Twenty-First Century,* "flat" means "level" or "connected" because (in his opinion) there is a more level global playing field in business (and almost any other endeavor) in the twenty-first century. According to Friedman, global telecommunications and the lowering of many trade and political barriers have brought about a new global era driven by *individuals,* not just by major corporations or giant trade organizations such as the World Bank. These individuals include entrepreneurs who create startup ventures around the world and computer freelancers whose work knows no boundaries (based on the idea of the older nation-state borders) when it comes to the transfer of information. Other factors that Friedman also believes have contributed to the "flattening" of the world include the fall of the Berlin Wall, the emergence of Netscape, the streaming of the supply chain (Wal-Mart, for example), and the organization of information on the Internet by Google and Yahoo (Zakaria, 2005). Worldwide, many freelancers and business entrepreneurs are not in the United States: They reside in nations such as India and China, where it is now possible to do more than merely compete in low-wage manufacturing and routine information labor (such as workers in call centers), but also in the top levels of research and design work.

If Friedman's assertions are correct that the world has become flat and the United States is losing dominance in the global political and economic arena, where does this leave us in regard to social policy? Can we do something to stem the flow of jobs out of this country that has resulted from outsourcing and the shift of some information technologies from the United States to India, China, and other emerging nations?

Friedman suggests that if the United States is to remain competitive in the global economy, we must not continue to do things as they have previously been done. He believes we should have a thoughtful national discussion about what globalization means in all of our lives. As Friedman (2005a) states, "When it comes to responding to the challenges of the flat world, there is no help line we can call. We have to dig into ourselves. We in America have all the basic economic and educational tools to do that. But we have not been improving those tools as much as we should."

Because there has been a shift from large-scale corporate players in the global economy to individual entrepreneurs and freelancers, we must look to each *individual* in our country to see how we can best play the economic game in the twenty-first century. In addition to nationwide policies, Friedman (2005a) believes that our social policy regarding globalization must begin at home; children and young adults must be encouraged to rise to the economic challenge that faces them:

> We need to get going immediately. It takes 15 years to train a good engineer, because, ladies and gentlemen, this really is rocket science. So parents, throw away the Game Boy, turn off the television and get your kids to work. There is no sugar-coating this: in a flat world, every individual is going to have to run a little faster if he or she wants to advance his or her standard of living. When I was growing up, my parents used to say to me, "Tom, finish your dinner—people in China are starving." But after sailing to the edges of the flat world for a year, I am now telling my own daughters, "Girls, finish your homework—people in China and India are starving for your jobs."

Reflect & Analyze

When social policy becomes personal (as Friedman believes), are we willing to engage in the changes it requires? Are Friedman's assumptions about the changing world order accurate? What do you think? What other arguments might be presented?

Chapter Review

• What is the primary function of the economy?

The economy is the social institution that ensures the maintenance of society through the production, distribution, and consumption of goods and services.

• What are the three sectors of economic production?

Economic production is divided into primary, secondary, and tertiary sectors. In primary sector production, workers extract raw materials and natural resources from the environment and use them without much processing. Industrial societies engage in secondary sector production, which is based on the processing of raw materials (from the primary sector) into finished goods. Postindustrial societies engage in tertiary sector production by providing services rather than goods.

• How do the major contemporary economic systems differ?

Throughout the twentieth century, capitalism, socialism, and mixed economies were the main systems in industrialized countries. Capitalism is characterized by ownership of the means of production, pursuit of personal profit, competition, and limited government intervention. Socialism is characterized by public ownership of the means of production, the pursuit of collective goals, and centralized decision making. In mixed economies, elements of a capitalist, market economy are combined with elements of a command, socialist economy. These mixed economies are often referred to as democratic socialism.

• What are the functionalist, conflict, and symbolic interactionist perspectives on the economy and work?

According to functionalists, the economy is a vital social institution because it is the means by which needed goods and services are produced and distributed. Conflict theorists suggest that the capitalist economy is based on greed. In order to maximize profits, capitalists suppress the wages of workers, who, in turn, cannot purchase products, making it necessary for capitalists to reduce production, close factories, lay off workers, and adopt other remedies that are detrimental to workers and society. Symbolic interactionists focus on the microlevel of the economic system, particularly on the social organization of work and its effects on workers' attitudes and behavior. Many workers experience job satisfaction when they like their job responsibilities and the organizational structure in which they work and when their individual needs and values are met. Alienation occurs when workers do not gain a sense of self-identity from their jobs and when their work is done completely for material gain and not for personal satisfaction.

• What are occupations, and how do they differ in the primary and secondary labor markets?

Occupations are categories of jobs that involve similar activities at different work sites. The primary labor market consists of well-paying jobs with good benefits that have some degree of security and the possibility for future advancement. The secondary labor market consists of low-paying jobs with few benefits and very little job security or possibility for future advancement.

• What are the characteristics of professions?

Professions are high-status, knowledge-based occupations characterized by abstract, specialized knowledge, autonomy, authority over clients and subordinate occupational groups, and a degree of altruism.

• What are marginal jobs?

Marginal jobs differ in some manner from mainstream employment norms: Jobs should be legal, be covered by government regulations, be relatively permanent, and provide adequate hours and pay in order to make a living. Marginal jobs fall below some or all of these norms.

• What is contingent work?

Contingent work is part-time work, temporary work, or subcontracted work that offers advantages to employers but may be detrimental to workers. Through the use of contingent workers, employers are able to cut costs and maximize profits, but workers have little or no job security.

www.cengage.com/login

Want to maximize your online study time? Take this easy-to-use study system's diagnostic pre-test, and it will create a personalized study plan for you. By helping you to identify the topics that you need to understand better and then directing you to valuable online resources, it can speed up your chapter review. CengageNOW even provides a post-test so you can confirm that you are ready for an exam.

Key Terms

capitalism 416
conglomerates 420
contingent work 432
corporations 417
democratic socialism 423
economy 410
interlocking corporate directorates 420
labor union 417
marginal jobs 431
mixed economy 423
occupations 426
oligopoly 420
primary labor market 427
primary sector production 411
professions 428
secondary labor market 427
secondary sector production 413
shared monopoly 420
socialism 422
subcontracting 433
tertiary sector production 414
transnational corporations 417
unemployment rate 435

Questions for Critical Thinking

1. If you were the manager of a computer software division, how might you encourage innovation among your technical employees? How might you encourage efficiency? If you were the manager of a fast-food restaurant, how might you increase job satisfaction and decrease job alienation among your employees?
2. Using Chapter 2 as a guide, design a study to determine the degree of altruism in certain professions. What might be your hypothesis? What variables would you study? What research methods would provide the best data for analysis?
3. What types of occupations will have the highest prestige and income in 2020? The lowest prestige and income? What, if anything, does your answer reflect about the future of the U.S. economy?

The Kendall Companion Website

www.cengage.com/sociology/kendall

Visit this book's companion website, where you'll find more resources to help you study and successfully complete course projects. Resources include quizzes and flash cards, as well as special features such as an interactive sociology timeline, maps, General Social Survey (GSS) data, and Census 2000 data. The site also provides links to useful websites that have been selected for their relevance to the topics in this chapter and include those listed below. (*Note:* Visit the book's website for updated URLs.)

National Institute for Occupational Safety and Health (NIOSH)

http://www.cdc.gov/niosh

NIOSH is the federal agency responsible for conducting research and making recommendations for the prevention of work-related disease and injury. The institute's website features extensive information, publications, and databases on policies and laws regarding worker safety and health, as well as alerts about occupational illnesses, injuries, and deaths.

National Bureau of Economic Research (NBER)

http://www.nber.org

NBER is a private, nonprofit, nonpartisan research organization that strives to promote a greater understanding of how the economy works. In addition to interesting links, publications, and research, this website contains a detailed data section with information and statistics from the macrolevel perspective to the microlevel perspective.

U.S. Department of Labor

http://www.dol.gov

This site is continually updated with new initiatives and reports, and it contains a vast amount of information to inform the work force on topics such as wages, health and safety, job training, job discrimination, and unemployment.

chapter

14 Politics and Government in Global Perspective

Chapter Focus Question

What effect does the intertwining of politics and the media have on the United States and other nations?

Do you think that mainstream TV news is boring? Does the idea of sitting down to watch *NewsHour with Jim Lehrer* make you yawn . . . just thinking about it? If so, you're not alone—in fact, this is now the norm among the college-aged demographic. A recent article [notes] that many young adults eschew traditional nightly news for [Jon Stewart's] *The Daily Show,* . . . which proudly bills itself as "the most trusted name in fake news."

Wow. While I'm a huge fan of Comedy Central, I usually tune in to watch something with an inherently comedic purpose—like South Park or the Colbert Report. While Colbert is billed as a blatant mockery of conservative TV pundits, the lines are more blurred when it comes to *The Daily Show*. Yes, it is largely satirical, but I

Jon Stewart, host of the extremely popular Comedy Central production *The Daily Show,* parodies mainstream news programming. Many people now watch Stewart's reports instead of watching network news programs. How might comedy news shows affect our perceptions of politics?

IN THIS CHAPTER

- Politics, Power, and Authority
- Political Systems in Global Perspective
- Perspectives on Power and Political Systems
- The U.S. Political System
- Governmental Bureaucracy
- The Military and Militarism
- Terrorism and War
- Politics and Government in the Future

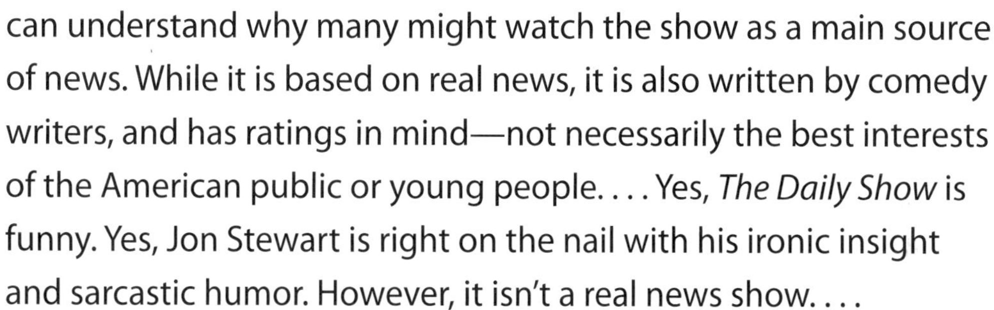

can understand why many might watch the show as a main source of news. While it is based on real news, it is also written by comedy writers, and has ratings in mind—not necessarily the best interests of the American public or young people. . . . Yes, *The Daily Show* is funny. Yes, Jon Stewart is right on the nail with his ironic insight and sarcastic humor. However, it isn't a real news show. . . .

—Katie Stapleton-Paff (2007), a writer for the *Daily of the University of Washington,* describing her concerns about the fact that many people between the ages of eighteen and twenty-five use "comedy" news programs as a prime source of information

I love the mock interviews, but I never go into the show thinking I'm watching real news. What I see is what I get and in the case of *The Daily Show* I see a funny show that makes fun of the day's news. . . . Fair and balanced news is extremely hard to come by these days and if students can't get fair and balanced as well as entertaining, they'll just stick with what's entertaining.

—Allen, a blogger, responding to the article by Stapleton-Paff

However comedic the show might be, it is important to understand that fiction and comedy have largely been a critique of mainstream

(Continued)

Sharpening Your Focus

- *What are the major political systems around the world?*
- *How does the center of power differ in the pluralist and the elite models of the U.S. power structure?*
- *How is government shaped by political parties and political attitudes?*
- *Why are government bureaucracies so powerful?*
- *What is the place of democracy in the future?*

industrialized societies, people do not share the same viewpoint on many issues and tend to openly question traditional authority. As the division of labor in a society becomes more complex, political and economic institutions become increasingly interdependent (Durkheim, 1933/1893).

Gender, race, and class relations are closely intertwined with traditional authority. Weber noted that traditional authority is often based on a system of patriarchy in which men are assumed to have authority in the household and in other small groups. Political scientist Zillah R. Eisenstein (1994) suggests that *racialized patriarchy*—the continual interplay of race and gender—reinforces traditional structures of power in contemporary societies. According to Eisenstein (1994: 2), "Patriarchy differentiates women from men while privileging men. Racism simultaneously differentiates people of color from whites and privileges whiteness. These processes are distinct but intertwined." Although racialized patriarchy has been increasingly challenged, many believe that it remains a reality in both preindustrial and industrialized nations. Class relations may also be linked to traditional authority in industrial nations such as the United States. In some upper-class families, for example, holding political office is considered to be a birthright and a family tradition. In families such as the Kennedys, Rockefellers, and Du Ponts, capitalism and some degree of traditional authority in politics appear to have been mutually reinforcing (see Baltzell, 1958; Domhoff, 1983).

Charismatic Authority ***Charismatic authority* is power legitimized on the basis of a leader's exceptional personal qualities** or the demonstration of extraordinary insight and accomplishment that inspire loyalty and obedience from followers. According to Weber, charismatic individuals are able to identify themselves with the central facts of people's lives and—through the force of their personalities—communicate their inspirations to others and lead them in new directions (Turner, Beeghley, and Powers, 2002). Charismatic leaders may be politicians, soldiers, and entertainers, among others.

From Weber's perspective, a charismatic leader may be either a tyrant or a hero. Thus, charismatic authority has been attributed to such diverse historical figures as Jesus Christ, Napoleon, Julius Caesar, Adolf Hitler, Winston Churchill, Franklin Roosevelt, Martin Luther King, Jr., and César Chávez. Few women other than Joan of Arc, Mother Teresa, Indira Gandhi of India, Eva Perón of Argentina, and Margaret Thatcher of the United Kingdom have had the opportunity to become charismatic leaders due to the predominantly patriarchal social structures in recorded history. Because women are seldom permitted to assume positions of leadership, they are much less likely to become charismatic leaders.

Charismatic authority tends to be temporary and relatively unstable; it derives primarily from individual leaders (who may change their minds, leave, or die) and from an administrative structure usually limited to a small number of faithful followers. For this reason, charismatic authority often becomes routinized. The ***routinization of charisma* occurs when charismatic authority is succeeded by a bureaucracy controlled by a rationally established authority or by a combination of traditional and bureaucratic authority** (Turner, Beeghley, and Powers, 2002). According to Weber (1968/1922: 1148), "It is the fate of charisma to recede . . . after it has entered the permanent structures of social action."

Rational–Legal Authority According to Weber, ***rational–legal authority* is power legitimized by law or written rules and regulations.** Rational–legal authority—also known as *bureaucratic authority*—is based on an organizational structure that includes a clearly defined division of labor, hierarchy of authority, formal rules, and impersonality. Power is legitimized by procedures; if leaders obtain their positions in a procedurally correct manner (such as by election or appointment), they have the right to act.

Rational–legal authority is held by elected or appointed government officials and by officers in a formal organization. However, authority is invested in the *office,* not in the *person* who holds the office. For example, although the U.S. Constitution grants rational–legal authority to the office of the presidency, a president who fails to uphold the public trust may be removed from office. In contemporary society, the media may play an important role in bringing to light allegations about presidents or other elected officials. Examples include the media blitzes surrounding the Watergate investigation of the 1970s that led to the resignation of President Richard M. Nixon and the late-1990s political firestorm over campaign fund-raising and the sex scandal involving President Bill Clinton.

In a rational–legal system, the governmental bureaucracy is the apparatus responsible for creating and enforcing rules in the public interest. Weber believed that rational–legal authority was the only means to attain efficient, flexible, and competent regulation under a rule of law (Turner, Beeghley, and Powers, 2002). Weber's three types of authority are summarized in the Concept Quick Review.

Political Systems in Global Perspective

Political systems as we know them today have evolved slowly. In the earliest societies, politics was not an entity separate from other aspects of life. Political in-

CONCEPT QUICK REVIEW

Weber's Three Types of Authority

Type	Description	Examples
Traditional	Legitimized by long-standing custom	Patrimony (authority resides in traditional leader supported by larger social structures, as in old British monarchy)
	Subject to erosion as traditions weaken	Patriarchy (rule by men occupying traditional positions of authority, as in the family)
Charismatic	Based on leader's personal qualities Temporary and unstable	Napoleon Adolf Hitler Martin Luther King, Jr. César Chávez Mother Teresa
Rational–Legal	Legitimized by rationally established rules and procedures	Modern British Parliament
	Authority resides in the office, not the person	U.S. presidency, Congress, federal bureaucracy

stitutions first emerged in agrarian societies as they acquired surpluses and developed greater social inequality. Elites took control of politics and used custom or traditional authority to justify their position. When cities developed circa 3500–3000 B.C.E., the *city-state*—a city whose power extended to adjacent areas—became the center of political power.

Nation-states as we know them began to develop in Europe between the twelfth and fifteenth centuries (see Tilly, 1975). A *nation-state* is a unit of political organization that has recognizable national boundaries and whose citizens possess specific legal rights and obligations. Nation-states emerge as countries develop specific geographic territories and acquire greater ability to defend their borders. Improvements in communication and transportation make it possible for people in a larger geographic area to share a common language and culture. As charismatic and traditional authority are superseded by rational–legal authority, legal standards come to prevail in all areas of life, and the nation-state claims a monopoly over the legitimate use of force (Kennedy, 1993).

Approximately 190 nation-states currently exist throughout the world; today, everyone is born, lives, and dies under the auspices of a nation-state (see Skocpol and Amenta, 1986). Four main types of political systems are found in nation-states: monarchy, authoritarianism, totalitarianism, and democracy.

Monarchy

***Monarchy* is a political system in which power resides in one person or family and is passed from generation to generation through lines of inheritance.** Monarchies are most common in agrarian societies and are associated with traditional authority patterns. However, the relative power of monarchs has varied across nations, depending on religious, political, and economic conditions.

Absolute monarchs claim a hereditary right to rule (based on membership in a noble family) or a divine right to rule (a God-given right to rule that legitimizes the exercise of power). In limited monarchies, rulers depend on powerful members of the nobility to retain their thrones. Unlike absolute monarchs, *limited monarchs* are not considered to be above the law. In *constitutional monarchies,* the royalty serve as symbolic rulers or heads of state while actual authority is held by elected officials in national parliaments. In present-day monarchies such as the United Kingdom, Sweden, Spain, and the Netherlands, members of royal families primarily perform ceremonial functions. In the

charismatic authority power legitimized on the basis of a leader's exceptional personal qualities.

routinization of charisma the process by which charismatic authority is succeeded by a bureaucracy controlled by a rationally established authority or by a combination of traditional and bureaucratic authority.

rational–legal authority power legitimized by law or written rules and procedures. Also referred to as *bureaucratic authority.*

monarchy a political system in which power resides in one person or family and is passed from generation to generation through lines of inheritance.

Box 14.2 Sociology in Global Perspective

The European Union: Transcending National Borders and Governments

What do people living in the following nations have in common?

Austria	Estonia	Hungary	Luxembourg	Slovakia
Belgium	Finland	Ireland	Malta	Slovenia
Cyprus	France	Italy	Netherlands	Spain
Czech Rep.	Germany	Latvia	Poland	Sweden
Denmark	Greece	Lithuania	Portugal	UK

If you answered that all of these countries are part of the European Union (EU), you are correct! The European Union is a treaty-based framework that defines and manages economic and political cooperation among its member nations. Based on treaties beginning after World War II, the EU became a sustained effort to create communities that share sovereignty in matters of coal and steel production, trade, and nuclear energy.

Although the EU has produced cooperation among the many governmental bodies in these nations, two forms of cooperation are most visible to the outsider who travels in EU countries. First, the euro has been established as the official currency in all of the nations except Denmark (where the krone is used), Sweden (where the krona is used), and the United Kingdom (where the British pound is used), which makes international commerce and tourism easier. The euro banknotes (similar to dollar bills) are identical in all of the other countries, but each country produces its own coins, with one common side and one national side. However, both the notes and coins can be used anywhere in the euro zone. Second, travel within the EU countries is easier and faster because citizens of any of the EU nations do not have to present a visa upon crossing a national border.

© plainpicture GmbH/Alamy

This dramatic sculpture representing the Euro—the European currency—is located near the European Centralbank in Frankfurt, Germany, and is symbolic of the spirit of cooperation emerging among a number of European countries.

Sociologically speaking, it will be interesting to see whether the EU can bring about changes in cultural and media practices in the various nations in this union. Recently, an effort has been made to enact legislation to reduce all forms of discrimination beyond the usual boundaries of work, trade, and labor, which the European Commission (a governing body within the EU) has previously addressed. The 1997 Treaty of Amsterdam gave the EU a mandate to combat sexual discrimination, as well as discrimination based on age, race, religion, and disability, among other things.

The European Union offers many challenges and opportunities for its member nations and for other countries around the globe. In 2005 the union experienced a series of political setbacks, including the rejection of a proposed constitution (which would need to be approved by all EU members to become effective) by voters in France and The Netherlands. Dissension among EU member nations continues to grow over many other issues, including the union's budget and whether or not to accept new members. Despite such disagreements, some analysts note that the EU has produced more than fifty years of stability, peace, and economic prosperity, and that this alliance could be a model for international governmental cooperation. Other analysts believe that the alliance has become unwieldy, contributing to problems of unemployment, declining benefits, and diminished working conditions in many of the member nations.

Reflect & Analyze

What does the future hold for the European Union? Will the EU ultimately transcend national borders and produce a stronger, more unified voice for Europe? Or will the union as it now exists disappear amid the voices of criticism and dissension that appear to be growing stronger?

Sources: Based on Alvarez, 2003; and Europa, 2003.

United Kingdom, for example, the media often focus large amounts of time and attention on the royal family, especially the personal lives of its members. Recently, the European Union (of which the United Kingdom, Spain, Sweden, and the Netherlands are all members) has also received media attention as a form of governmental cooperation across national boundaries but not one that weakens the powers of the present-day monarchies (see Box 14.2, "Sociology in Global Perspective").

Through its many ups and downs, the British royal family has remained a symbol of Great Britain's monarchy, today headed by Queen Elizabeth. Monarchies typically pass power from generation to generation, and the two young men on the left, Princes William and Harry, represent the future of the royal family's rule.

Authoritarianism

***Authoritarianism* is a political system controlled by rulers who deny popular participation in government.** A few authoritarian regimes have been absolute monarchies whose rulers claimed a hereditary right to their position. Today, Saudi Arabia and Kuwait are examples of authoritarian absolute monarchies. In *dictatorships,* power is gained and held by a single individual. Pure dictatorships are rare; all rulers need the support of the military and the backing of business elites to maintain their position. *Military juntas* result when military officers seize power from the government, as has happened in recent decades in Argentina, Chile, and Haiti. Today, authoritarian regimes exist in Fidel Castro's Cuba and in the People's Republic of China. Authoritarian regimes seek to control the media and to suppress coverage of any topics or information that does not reflect upon the regime in a favorable light.

Totalitarianism

***Totalitarianism* is a political system in which the state seeks to regulate all aspects of people's public and private lives.** Totalitarianism relies on modern technology to monitor and control people; mass propaganda and electronic surveillance are widely used to influence people's thinking and control their actions. One example of a totalitarian regime was the National Socialist (Nazi) Party in Germany during World War II; military leaders there sought to control all aspects of national life, not just government operations. Other examples include the former Soviet Union and contemporary Iraq before the fall of Saddam Hussein's regime.

To keep people from rebelling, totalitarian governments enforce conformity: People are denied the right to assemble for political purposes, access to information is strictly controlled, and secret police enforce compliance, creating an environment of constant fear and suspicion.

Many nations do not recognize totalitarian regimes as being the legitimate government for a particular country. Afghanistan in the year 2001 was an example. As the war on terrorism began in the aftermath of the September 11 terrorist attacks on the United States, many people developed a heightened awareness of the Taliban regime, which ruled most of Afghanistan and was engaged in fierce fighting to capture the rest of the country. The Taliban regime maintained absolute control over the Afghan people in most of that country. For example, it required that all Muslims take part in prayer five times each day and that men attend prayer at mosques, where women were forbidden (Marquis, 2001). Taliban leaders claimed that their actions were based on Muslim law and espoused a belief in never-ending *jihad*—a struggle against one's perceived enemies. Although the totalitarian nature of the Taliban regime was difficult for many people, it was particularly oppressive for women, who were viewed by this group as being "biologically, religiously and prophetically" inferior to men (McGeary, 2001: 41). Consequently, this regime made the veil obligatory and banned women from public life. U.S. government officials believed that the Taliban regime was protecting Osama bin Laden, the man thought to have been the mastermind behind numerous terrorist attacks on U.S. citizens and facilities, both on the mainland and abroad. As a totalitarian regime, the Taliban leadership was recognized by only three other governments, despite controlling most of Afghanistan.

Once the military action commenced in Afghanistan, most of what U.S. residents knew about the Taliban and about the war on terrorism was based on media accounts and "expert opinions" that were voiced on television. According to the political analyst Michael Parenti (1998), the media play a significant role in framing the information we receive about the political systems of other countries. As discussed in previous chapters, *framing* refers to how news is packaged, including the amount of exposure given to a story, its placement, the positive or negative tone of the story, the headlines and photographs, and the accompanying

authoritarianism a political system controlled by rulers who deny popular participation in government.

totalitarianism a political system in which the state seeks to regulate all aspects of people's public and private lives.

visual and auditory effects if the story is being broadcast. In politics and government, framing is not limited to information we receive about other countries; it can also be used to frame a political agenda in this country (see Box 14.3).

Democracy

***Democracy* is a political system in which the people hold the ruling power either directly or through elected representatives.** The literal meaning of *democracy* is "rule by the people" (from the Greek words *demos,* meaning "the people," and *kratein,* meaning "to rule"). In an ideal-type democracy, people would actively and directly rule themselves. *Direct participatory democracy* requires that citizens be able to meet together regularly to debate and decide the issues of the day. However, if all 302 million people in the United States came together in one place for a meeting, they would occupy an area of more than seventy square miles, and a single round of five-minute speeches would require more than five thousand years (based on Schattschneider, 1969).

In countries such as the United States, Canada, Australia, and the United Kingdom, people have a voice in the government through *representative democracy,* whereby citizens elect representatives to serve as bridges between themselves and the government. The U.S. Constitution requires that each state have two senators and a minimum of one member in the House of Representatives. The current size of the House (435 seats) has not changed since the apportionment following the 1910 census. Therefore, based on Census 2000, those 435 seats were reapportioned based on the increase or decrease in a state's population between 1990 and 2000.

In a representative democracy, elected representatives are supposed to convey the concerns and interests of those they represent, and the government is expected to be responsive to the wishes of the people. Elected officials are held accountable to the people through elections. However, representative democracy is not always equally accessible to all people in a nation. Throughout U.S. history, members of subordinate racial–ethnic groups have been denied full participation in the democratic process. Gender and social class have also limited some people's democratic participation. For example, women have not always had the same rights as men. Full voting rights were not gained by women until the ratification of the Nineteenth Amendment in 1920.

Even representative democracies are not all alike. As compared to the winner-takes-all elections in the United States, which are usually decided by who wins the most votes, the majority of European elections are based on a system of proportional representation, meaning that each party is represented in the national legislature according to the proportion of votes that party received. For example, a party that won 40 percent of the vote would receive 40 seats in a 100-seat legislative body, and a party receiving 20 percent of the votes would receive 20 seats.

Perspectives on Power and Political Systems

Is political power in the United States concentrated in the hands of the few or distributed among the many? Sociologists and political scientists have suggested many different answers to this question; however, two prevalent models of power have emerged: pluralist and elite.

Functionalist Perspectives: The Pluralist Model

The pluralist model is rooted in a functionalist perspective which assumes that people share a consensus on central concerns, such as freedom and protection from harm, and that the government serves important functions no other institution can fulfill (see ▶ Figure 14.1). According to Emile Durkheim (1933/1893), the purpose of government is to socialize people to be good citizens, to regulate the economy so that it operates effectively, and to provide necessary services for citizens. Contemporary functionalists state the four main functions as follows: (1) maintaining law and order, (2) planning and directing society, (3) meeting social needs, and (4) handling international relations, including warfare.

But what happens when people do not agree on specific issues or concerns? Functionalists suggest that divergent viewpoints lead to a system of political pluralism in which the government functions as an arbiter between competing interests and viewpoints. According to the ***pluralist model,* power in political systems is widely dispersed throughout many competing interest groups** (Dahl, 1961).

Key Elements Political scientists Thomas R. Dye and Harmon Zeigler (2008) have summarized the key elements of pluralism as follows:

- The diverse needs of women and men, people of all religions and racial–ethnic backgrounds, and the wealthy, middle class, and poor are met by political leaders who engage in a process of bargaining, accommodation, and compromise.
- Competition among leadership groups in government, business, labor, education, law, medicine,

Box 14.3 Framing Politics in the Media

Hero Framing and the Selling of an Agenda

> You have to do more than just go and have a little press conference. So the spectacle, showmanship, selling, promoting, marketing, publicizing, all of these things are extremely important. . . . We did all of this to make the people pay attention. . . . You also have to think about how you can sell the policy, how can we get it so that everyone in California at home starts paying attention to it.
>
> —California Governor Arnold Schwarzenegger, whose second term ends in 2010, explaining why he uses the same media tactics to sell his political agenda that he once used to plug films that he starred in, such as *The Terminator* (Murphy, 2005: ST1, ST2)

During his terms as governor, Arnold Schwarzenegger has actively promoted his state and his political agenda in much the same way that he pitched himself and his films when he was an actor. For example, Schwarzenegger has appeared in tourism ads for the state of California, and when he went on a trade mission to Japan, he hired a Terminator look-alike to ride through the crowd on a motorcycle throwing spectators promotional T-shirts while he stood in front of an oversize poster of himself that read, "See California. Buy California," and described for the audience the virtues of California beaches, wines, and almonds to encourage trade and tourism (Murphy, 2005). According to one journalist, "As in his Hollywood career Mr. Schwarzenegger relies on friendly media outlets, uses flamboyant public stunts to attract attention and self-deprecatingly jokes about his relentless selling in a way few career politicians would" (Murphy, 2005: ST2).

What at first glance might appear to be egregious self-promotion is actually Governor Schwarzenegger using media framing to develop a favorable response to his political initiatives. He has been able to blend his star power with his governorship so that they appear to be one and the same. It is easy for media audiences to believe that an action hero such as Schwarzenegger has the power to get things done that mere mortals (ordinary politicians) might never accomplish. This type of hero framing is aided by the fact that entertainment and reality are blurred in contemporary society (Murphy, 2005). And, apparently, this approach to selling policies is working for the governor. California voters have accepted a massive borrowing plan to at least temporarily help solve the state's budget deficit, and many of the statewide ballot initiatives he has campaigned for have passed. However, critics believe that media framing that blends real people and important social issues with imaginary heroes may be bad for the general public because the blending of the real and the imaginary may provide a smoke screen behind which a politician can hide what is really happening in his or her administration.

Although Governor Schwarzenegger is framing his story in such a manner as to make his policies appealing to media audiences, and particularly to California voters, this political leader is not alone in how he sells himself and frames his ideas in the most interesting light in the media. Many other political leaders have used media images (such as former President George W. Bush as the cowboy-boot-wearing Texan who is so down-home that he clears brush at his ranch himself) to further their political agendas.

Reflect & Analyze

How can we become more aware of policies that are being proposed or promoted by our political leaders? Do we need to look behind the façade to see what is actually going on, or can we trust our leaders to provide us with valid information? What do you think?

and consumer organizations, among others, helps prevent abuse of power by any one group. These groups often operate as *veto groups* that attempt to protect their own interests by keeping others from taking actions that would threaten those interests.

- Power is widely dispersed in society. Leadership groups that wield influence on some decisions are not the same groups that may be influential in other decisions.
- Public policy is not always based on majority preference; rather, it reflects a balance among competing interest groups.
- Everyday people can influence public policy by voting in elections, participating in existing special

democracy a political system in which the people hold the ruling power either directly or through elected representatives.

pluralist model an analysis of political systems that views power as widely dispersed throughout many competing interest groups.

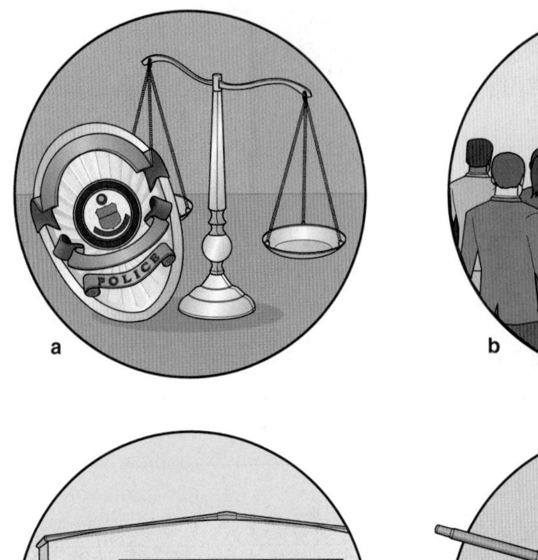

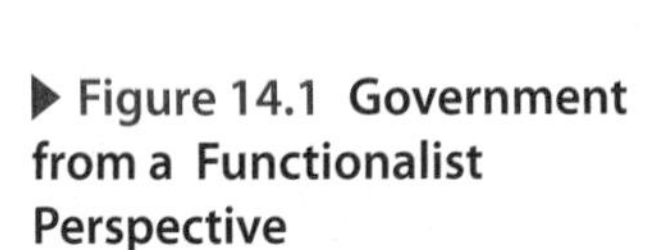

▶ **Figure 14.1 Government from a Functionalist Perspective**
From the functionalist perspective, government serves important functions no other institution can fulfill. Contemporary functionalists identify four main functions:
(a) maintaining law and order,
(b) planning and directing society,
(c) meeting social needs, and
(d) handling international relations, including warfare.

interest groups, or forming new ones to gain access to the political system.

Special Interest Groups ***Special interest groups*** **are political coalitions made up of individuals or groups that share a specific interest they wish to protect or advance with the help of the political system** (Greenberg and Page, 2002). Examples of special interest groups include the AFL-CIO (representing the majority of labor unions) and public interest or citizens' groups such as the American Conservative Union and Zero Population Growth.

What purpose do special interest groups serve in the political process? According to some analysts, special interest groups help people advocate their own interests and further their causes. Broad categories of special interest groups include banking, business, education, energy, the environment, health, labor, persons with a disability, religious groups, retired persons, women, and those espousing a specific ideological viewpoint; obviously, many groups overlap in interests and membership. Special interest groups are also referred to as *pressure groups* (because they put pressure on political leaders) or *lobbies*. Lobbies are often referred to in terms of the organization they represent or the single issue on which they focus—for example, the "gun lobby" and the "dairy lobby." The people who are paid to influence legislation on behalf of specific clients are referred to as *lobbyists*.

Over the past four decades, special interest groups have become more involved in "single-issue politics," in which political candidates are often supported or rejected solely on the basis of their views on a specific issue—such as abortion, gun control, gay and lesbian rights, or the environment. Single-issue groups derive their strength from the intensity of their beliefs; leaders have little room to compromise on issues.

Political Action Committees The funding of lobbying efforts has become more complex in recent years. Reforms in campaign finance laws in the 1970s set limits on direct contributions to political candidates and led to the creation of ***political action committees*** **(PACs)—organizations of special interest groups that solicit contributions from donors and fund campaigns to help elect (or defeat) candidates based on their stances on specific issues.**

As the cost of running for political office has skyrocketed, candidates have relied more on PACs for financial assistance. PACs contributed more than $316,800,000 to candidates for the U.S. House and Senate between January 1, 2005, and October 18, 2006 (the most recent period for which data are available), representing 28 percent of all contributions received (Federal Election Commission, 2006). Advertising, staff, direct-mail operations, telephone banks, computers, consultants, travel expenses, office rentals, and other expenses incurred in political campaigns make PAC money vital to candidates.

© AP Images/Stephen J. Boitano

Special interest groups help people advocate their interests and further their causes. Advocates may run for public office and gain a wider voice in the political process. An example is Ben Nighthorse Campbell of Colorado. Campbell, who was a U.S. senator from 1993 to 2005, is also a Cheyenne chief. He is shown here walking to the Senate floor after participating in a ground-breaking ceremony for the National Museum of the American Indian in Washington, D.C.

Some PACs represent the "public interest" and ideological interest groups such as gay rights or the National Rifle Association. Other PACs represent the capitalistic interests of large corporations. Realistically, members of the least-privileged sectors of society are not represented by PACs. As one senator pointed out, "There aren't any Poor PACs or Food Stamp PACs or Nutrition PACs or Medicare PACs" (qtd. in Greenberg and Page, 1993: 240). Critics of pluralism argue that "Big Business" wields such disproportionate power in U.S. politics that it undermines the democratic process (see Lindblom, 1977; Domhoff, 1978).

As an outgrowth of record-setting campaign spending in the 1996 national election, campaign financing abuses were alleged by both Republicans and Democrats in Washington. At the center of the controversy was the issue of "soft money" contributions, which are made outside the limits imposed by federal election laws. In 2002, Congress passed the McCain–Feingold campaign finance law, prohibiting soft money contributions in federal elections, and the U.S. Supreme Court upheld the soft money provisions of that law in 2003. However, the McCain–Feingold law applies only to federal elections and does not bar soft money contributions in state and local elections.

Conflict Perspectives: Elite Models

Although conflict theorists acknowledge that the government serves a number of important purposes in society, they assert that government exists for the benefit of wealthy or politically powerful elites who use the government to impose their will on the masses. According to the ***elite model,*** **power in political systems is concentrated in the hands of a small group of elites, and the masses are relatively powerless.** Early Italian sociologist Vilfredo Pareto (1848–1923) first used the term *elite* to refer to "the few who rule the many" (Marshall, 1998). Similarly, Karl Marx claimed that under capitalism, the government serves the interests of the ruling (or capitalist) class that controls the means of production.

Key Elements Contemporary elite models are based on the following assumptions (Dye and Zeigler, 2008):

- Decisions are made by the elite, which possesses greater wealth, education, status, and other resources than do the "masses" it governs.
- Consensus exists among the elite on the basic values and goals of society; however, consensus does not exist among most people in society on these important social concerns.

special interest groups political coalitions composed of individuals or groups that share a specific interest that they wish to protect or advance with the help of the political system.

political action committees (PACs) organizations of special interest groups that solicit contributions from donors and fund campaigns to help elect (or defeat) candidates based on their stances on specific issues.

elite model a view of society that sees power in political systems as being concentrated in the hands of a small group of elites whereas the masses are relatively powerless.

- Power is highly concentrated at the top of a pyramid-shaped social hierarchy; those at the top of the power structure come together to set policy for everyone.
- Public policy reflects the values and preferences of the elite, not the preferences of the people.

The pluralist and elite models are compared in ▶ Figure 14.2.

C. Wright Mills and the Power Elite Who makes up the U.S. power elite? According to the sociologist C. Wright Mills (1959a), the ***power elite* is made up of leaders at the top of business, the executive branch of the federal government, and the military.** Of these three, Mills speculated that the "corporate rich" (the highest-paid officers of the biggest corporations) were the most powerful because of their unique ability to parlay the vast economic resources at their disposal into political power (see "Sociology Works!"). At the middle level of the pyramid, Mills placed the legislative branch of government, special interest groups, and local opinion leaders. The bottom (and widest layer) of the pyramid is occupied by the unorganized masses, who are relatively powerless and are vulnerable to economic and political exploitation.

Mills emphasized that individuals who make up the power elite have similar class backgrounds and interests; many of them also interact on a regular basis. In addition, many frequently shift back and forth among the business, government, and military sectors. For example, it is not unusual for corporate executives to assume positions in a president's cabinet and then return to the business world. Similarly, a "revolving door" exists between the military and the executive branch, as well as between the military and corporations. Members of the power elite are able to influence many important decisions, including federal spending.

G. William Domhoff and the Ruling Class Sociologist G. William Domhoff (2002) asserts that this nation in fact has a *ruling class*—the corporate rich, who constitute less than 1 percent of the U.S. population. Domhoff uses the term *ruling class* to signify a relatively fixed group of privileged people who wield power sufficient to constrain political processes and serve underlying capitalist interests. Although the power elite controls the everyday operation of the political system, who *governs* is less important than who *rules.*

Like Mills, Domhoff asserts that individuals in the upper echelon are members of a business class based on the ownership and control of large corporations.

PLURALIST MODEL

- Decisions are made on behalf of the people by leaders who engage in bargaining, accommodation, and compromise.
- Competition among leadership groups makes abuse of power by any one group difficult.
- Power is widely dispersed, and people can influence public policy by voting.
- Public policy reflects a balance among competing interest groups.

ELITE MODEL

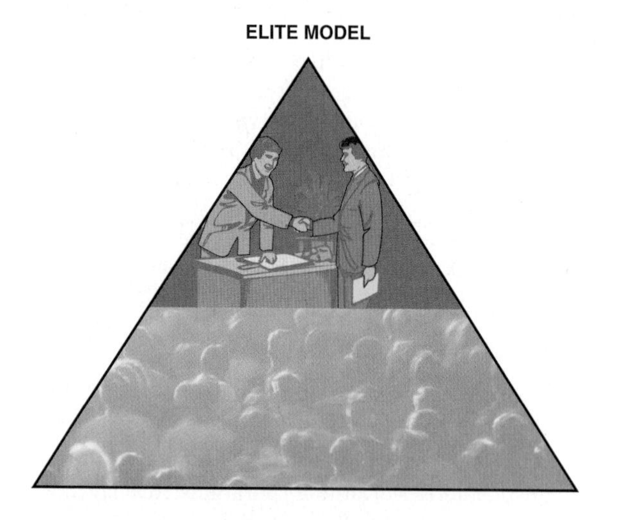

- Decisions are made by a small group of elite people.
- Consensus exists among the elite on the basic values and goals of society.
- Power is highly concentrated at the top of a pyramid-shaped social hierachy.
- Public policy reflects the values and preferences of the elite.

▶ **Figure 14.2 Pluralist and Elite Models**

The intertwining of the upper class and the corporate community produces economic and social cohesion. Economic interdependence among members of the ruling class is rooted in common stock ownership and is visible in interlocking corporate directorates that serve as a communications network (Domhoff, 2002). Members of the ruling class are also socially linked through exclusive clubs, expensive private schools and debutante parties for their children, and listings in the Social Register (an address book for upper-class families in major metropolitan areas).

According to Domhoff (2002), the corporate rich influence the political process in three ways. First, they affect the candidate selection process by helping to finance campaigns and providing favors to political candidates. Second, through participation in the special interest process, the corporate rich are able to obtain favors, tax breaks, and favorable regulatory rulings. Finally, the corporate rich gain access to the policy-making process through their appointments to governmental advisory committees, presidential commissions, and other governmental positions. Today, some members of the ruling class influence international politics through their involvement in banking, business services, and law firms that have a strong interest in overseas sales, investments, or raw materials extraction.

Clearly, owners of some media conglomerates would be classified in Domhoff's power elite. Consider, for example, journalist Ken Aulette's discussion with Rupert Murdock, head of News Corporation—the conglomerate that owns over a hundred newspapers worldwide, major movie studios, publishing interests, the Fox TV network, and numerous cable channels—about the enormous growth of Murdock's media empire:

> Does [he] tire of the constant competition to get bigger, to win each war? When does Murdock say *enough*? I asked him this question at the end of a long night that started with a drink and a stroll over his six acre property in Beverly Hills. . . .
>
> After dinner in the dining room of his comfortable Spanish-style home, Murdock sipped a glass of California Chardonnay and treated the question as something so alien as to be incomprehensible. "You go on," he responded, opaquely. "There is a global village in some sense. You are competing everywhere. . . . And I just enjoy it." (Aulette, 1998: 287)

Class Conflict Perspectives Most contemporary elite models are based on the work of Karl Marx; however, there are divergent viewpoints about the role of the state within the Marxist (or class conflict) perspective. *Instrumental Marxists* argue that the state invariably acts to perpetuate the capitalist class. From this perspective, capitalists control the government through special interest groups, lobbying, campaign financing, and other types of "influence peddling" to get legislatures and the courts to make decisions favorable to their class (Miliband, 1969; Domhoff, 2002). In other words, the state exists only to support the interests of the dominant class (Marger, 1987).

By contrast, *structural Marxists* contend that the state is not simply a passive instrument of the capitalist class. Because the state must simultaneously preserve order and maintain a positive climate for the accumulation of capital, not all decisions can favor the immediate wishes of the dominant class (Quadagno, 1984; Marger, 1987). For example, at various points in the history of this country, the state has had to institute social welfare programs, regulate business, and enact policies that favor unions in order to placate people and maintain "law and order." Ultimately, however, such actions serve the long-range interests of the capitalists by keeping members of subordinate groups from rebelling against the dominant group (O'Connor, 1973).

Critique of Pluralist and Elite Models

Pluralist and elite models each make a unique contribution to our understanding of power. The pluralist model emphasizes that many different groups, not just elected officials, compete for power and advantage in society. This model also shows how coalitions may shift over time and how elected officials may be highly responsive to public opinion on some occasions. However, critics counter that extensive research evidence contradicts the notion that the U.S. political system is pluralistic. They note instead that our system has the "veneer" of pluralism when, in fact, it is remarkably elitist for a society that claims to value ordinary people's input (Domhoff, 2002). For example, a wide disparity exists between the resources and political clout of "Big Business" when compared with those of interest groups that represent infants and children or persons with a disability. According to critics, consensus is difficult, if not impossible, in populations composed of people from different classes, religions, and racial–ethnic and age groups.

power elite C. Wright Mills's term for the group made up of leaders at the top of business, the executive branch of the federal government, and the military.

Sociology *Works!*

C. Wright Mills's Ideas About the Media: Ahead of His Time?

In a mass society the dominant type of communication is the formal media, and the [various audiences] become mere *media markets:* all those exposed to the contents of a given mass media.

—In *The Power Elite,* the influential sociologist C. Wright Mills (1959a: 299) provided this early critique of the U.S. mass media and its effect on the general public.

During the mid-twentieth century, when Mills wrote these words, the press in the United States was often described as a "watchdog for the public." It was widely believed that journalists would reliably report what the public needed to know to stay informed, including any government and/or corporate dishonesty or ineptitude (Headley, 1985: 333). However, C. Wright Mills rejected this notion because he believed the media industry lulled individuals into complacency and persuaded them to accept the status quo rather than encouraging the audience member to become a "public" person who formulates his or her own opinion on a given topic. As one contemporary scholar explained, "To recover Mills's 'public man' who is capable of formulating opinions, we would have to salvage whatever is left of him after he is stripped of MTV, CNN, and http://www., whose Italian translation, *ragnatele mondiale,* is useful in conveying the sense 'spider-web' and therefore the suspicion that we may be flies" (Cinquemani, 1997: 89).

Do the contemporary media inform people or trap them in a "spiderweb"? Mills hoped that individuals would develop the capacity for critical judgment based on information they received from the media industry; however, he was doubtful that this would occur for two important sociological reasons: (1) media communication involves a limited number of people who communicate *to* a great number of others ("the masses"), and (2) audiences have no effective way of answering back, making mass communication a one-way process.

As part of his legacy for making sociology work in the real world, Mills made an important prediction. According to Mills (1956: 303–304), critical thinking is more likely to occur among individuals under the following circumstances: "If public communications are so organized that there is a chance immediately and effectively to answer back any opinion expressed in public. Opinion formed by such discussion readily finds an outlet in effective action, even against—if necessary—the prevailing system of authority." If media audiences are able to respond immediately and provide their own opinions, media communications may become a two-way street. Technologies and methods of communication—including cell phones, text

Mills's power-elite model highlights the interrelationships of the economic, political, and military sectors of society and makes us aware that the elite may be a relatively cohesive group. Similarly, Domhoff's ruling-class model emphasizes the role of elites in setting and implementing policies that benefit the capitalist class. Power-elite models call our attention to a central concern in contemporary U.S. society: the ability of democracy and its ideals to survive in the context of the increasingly concentrated power held by capitalist oligarchies (see Chapter 13).

The U.S. Political System

The U.S. political system is made up of formal elements, such as the legislative process and the duties of the president, and informal elements, such as the role of political parties in the election process. We now turn to an examination of these informal elements, including political parties, political socialization, and voter participation.

Political Parties and Elections

A ***political party*** **is an organization whose purpose is to gain and hold legitimate control of government;** it is usually composed of people with similar attitudes, interests, and socioeconomic status. A political party (1) develops and articulates policy positions, (2) educates voters about issues and simplifies the choices for them, and (3) recruits candidates who agree with those policies, helps those candidates win office, and holds the candidates responsible for implementing the party's policy positions. In carrying out these functions, a party may try to modify the demands of special interests, build a consensus that could win majority support, and provide simple and identifiable choices for the voters on election day. Political parties create a *platform,* a formal statement of the party's political positions on various social and economic issues.

© Petrified Collection/Getty Images

© HANS DERYK/Reuters/Landov

C. Wright Mills believed that mass communication was a one-way street: The audience had no way to react to it and tended to become passive recipients of its content. However, today's digital technology allows people to respond quickly and, through developments such as blogs, to provide news content and opinions of their own. What are the benefits and drawbacks of interactive technologies?

messaging, e-mail, and websites—now provide us with opportunities for media interactivity through which we can respond to media discourse and make our thoughts known.

Sociology works today in Mills's ideas because he encourages us to think for ourselves and to express our ideas and opinions rather than being passive recipients of information from the media. Mills challenges us to think about the extent to which we rely on *mediated* experiences of others, such as the ideas presented by television commentators or online bloggers, as compared with becoming our own "public person" and responding with thoughts and judgments of our own.

Reflect & Analyze

How might we relate Mills's ideas to the contemporary role of the media industries in shaping political and economic "realities" around the world?

Since the Civil War, the Democratic and Republican parties have dominated the U.S. political system. Although one party may control the presidency for several terms, at some point the voters elect the other party's nominee, and control shifts. See ▶ Figure 14.3 for a look at the major political parties in U.S. history.

Ideal Type Versus Reality How well do the parties measure up to the ideal-type characteristics of a political party as described here? Although both parties have been quite successful at getting their candidates elected at various times, they generally do not meet the ideal characteristics of a political party. Thomas Dye and Harmon Zeigler (2008) suggest several reasons for that failure. First, the two parties do not offer voters clear policy alternatives. Most voters view themselves as being close to the center of the political spectrum (extremely liberal being the far left of that spectrum and extremely conservative being the far right). Although the definitions of *liberal* and *conservative* vary over time, *liberals* tend to focus on equality of opportunity and the need for government regulation and social safety nets. By contrast, *conservatives* are more likely to emphasize economic liberty and freedom from government interference (Greenberg and Page, 2002). Because most voters consider themselves to be moderates, neither party has much incentive to veer very far from the middle.

Second, the two parties are oligarchies, dominated by active elites. The active party elites hold views that are further from the center of the political spectrum (Democrats to the left, Republicans to the right, usually) than are those of a majority of members of their party. Thus, platforms state the policy goals of the oligarchies, whereas the goal of winning elections necessitates nominating more-centrist candidates.

Third, primary elections (in which the nominees of political parties for most offices other than president and vice president are selected) determine nominees. Thus, voters in the primaries may select nominees whose

political party an organization whose purpose is to gain and hold legitimate control of government.

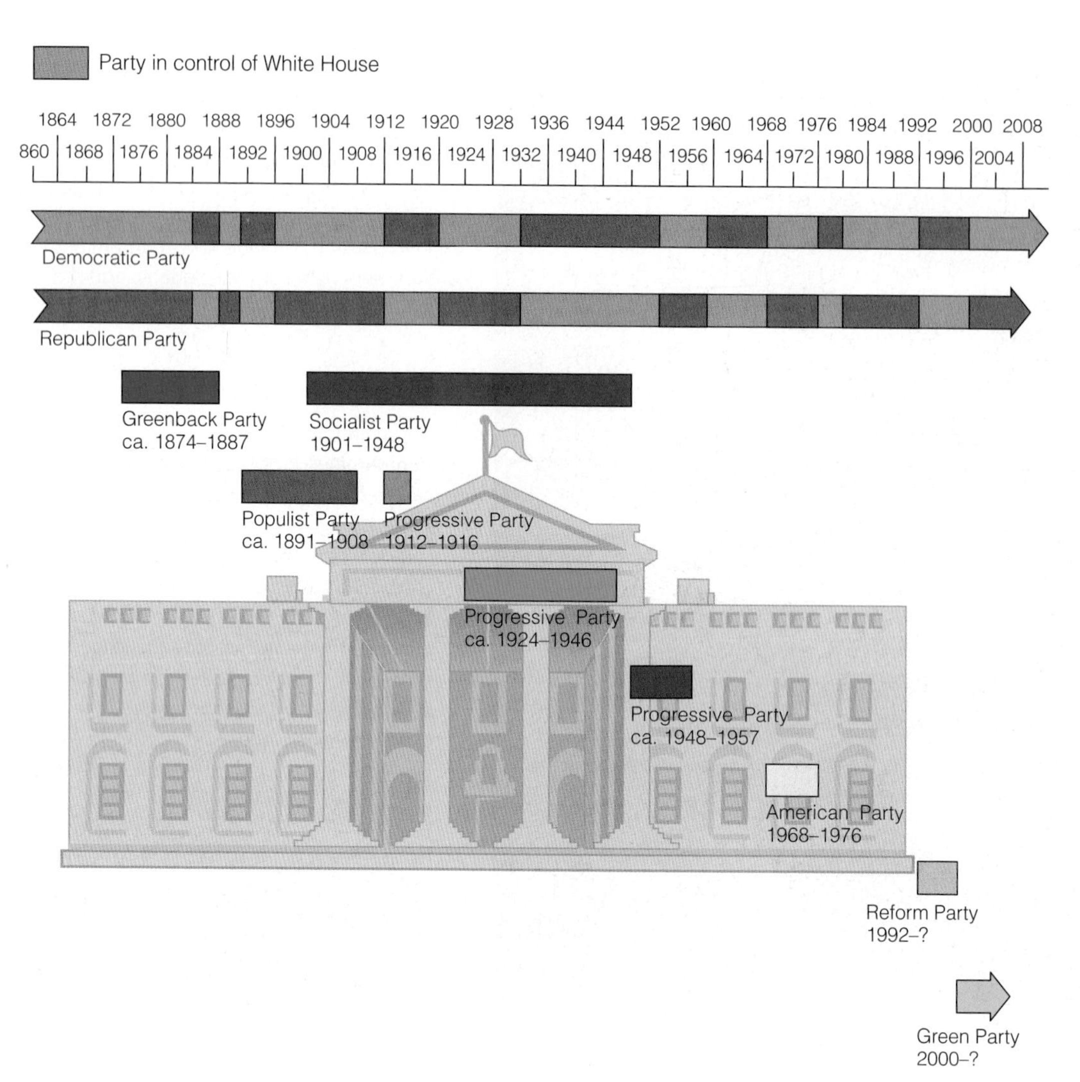

▶ **Figure 14.3 Major U.S. Political Parties**
Despite recurring attempts by other groups to organize third parties, the Democratic and Republican parties have dominated national politics in the United States. Control of the presidency has alternated between these two parties since the Civil War.

Note: Three different "third parties" have gone by the name Progressive Party.

views are closer to the center of the political spectrum and further away from the party's own platform. The nominee may establish a personal organization (committee) to work for the nominee's election *outside* the party organization. That committee may contain both Democratic and Republican members, thus being more centrist than the party platform or organization.

Fourth, party loyalties are declining. Many people today vote in one party's primary but then cast their ballot in general elections without total loyalty to that party. They may cast a "split-ticket" ballot (voting for one party's candidate in one race and another party's candidate in another). As a result, it is harder for the party to meet the "ideal" of holding the candidate accountable for implementing the party's platform.

Finally, the media have replaced the party as a means of political communication. Often, the candidate who wins does so as a result of media presentation, not the political party's platform. Candidates no longer need political parties to carry their message to the people.

Despite the fact that the Republican and Democratic parties do not have all of the ideal characteristics of political parties, they both flourish. On the local and state levels, other political parties—or even candidates who identify themselves as "independents"—may win office, but the Republican and Democratic parties have basically controlled the U.S. political scene for almost 150 years. In general, voters tend to select candidates and political parties based on social and economic issues that they consider to be important in their lives.

Social Issues Social issues are those relating to moral judgments or civil rights, ranging from abortion rights to gun control. Other social issues include the rights of people of color and persons with a disability, school prayer, and the environment. For example, persons with a liberal perspective on social issues tend to believe that women have the right to an abortion (at least under certain circumstances), that criminals should be rehabilitated instead of punished, and that the government has an obligation to protect the rights of subordinate groups. Conservatives tend to believe in limiting the expansion of individual rights on social issues and tend to oppose social programs that they see as promoting individuals on the basis of minority status rather than merit. Based on these distinctions, Democrats are more likely to seek passage of social programs that make the government a more active participant in society, promoting social welfare and equality. By contrast, Republicans are more likely to believe that government should not be responsible for financial equality.

Economic Issues Economic issues fall into two broad categories: (1) the amount that should be spent on government programs and (2) the extent to which these programs should encourage a redistribution of income and assets. Liberals believe that without governmental intervention, income and assets would become concentrated in the hands of even fewer people and that the government must act to redistribute wealth, thus ensuring that everyone gets a "fair slice" of the economic "pie." In order to accomplish this, liberals envision that larger sums of money must be raised and spent by the government on such programs. Conservatives contend that such programs are not only unnecessary but also counterproductive. That is, programs financed by tax increases lower people's incentive to work and to be innovative, and make people dependent upon the government. However, conservatives do believe in taxes that help maintain the status quo—funds for education, for the criminal justice system to maintain law and order, and for a strong military establishment.

© AP Images/Lynne Sladky

When people move from one nation to another, they learn new political attitudes, values, and behavior. Immigrants to the United States are required to undergo a process of political socialization before being "naturalized." Ironically, newcomers who have experienced poverty and repression in other lands are often stalwart in their devotion to democratic ideals.

Political Participation and Voter Apathy

Why do some people vote whereas others do not? How do people come to think of themselves as being conservative, moderate, or liberal? Key factors include individuals' political socialization, attitudes, and participation.

Political Socialization ***Political socialization* is the process by which people learn political attitudes, values, and behavior.** For young children, the family is the primary agent of political socialization, and children tend to learn and hold many of the same opinions as their parents. By the time children reach school age, they typically identify with the political party (if any) of their parents (Burnham, 1983). As children grow older, other agents of socialization begin to affect their political beliefs, including peers, teachers, and the media. Over time, these other agents may cause people's political attitudes and values to change, and individuals may cease to identify with the political party of their parents. Even for adults, political socialization continues through the media, friends, neighbors, and colleagues in the workplace.

Political Attitudes In addition to the socialization process, people's socioeconomic status affects their political attitudes, values, and beliefs. For example, individuals who are very poor or are unable to find employment tend to believe that society has failed them and therefore tend to be indifferent toward the political system (Zipp, 1985; Pinderhughes, 1986). Believing that casting a ballot would make no difference to their own circumstances, they do not vote.

People in the upper classes tend to be more conservative on economic issues and more liberal on social issues. Based on a philosophy of *noblesse oblige,* which asserts that those who are well-off have a responsibility for the welfare of the poor and disadvantaged, upper-class conservatives generally favor equality of opportunity but do not want their own income and assets taxed heavily to abolish poverty or other societal problems that they believe some people bring upon themselves. By contrast, people in the lower classes tend to be conservative on

political socialization the process by which people learn political attitudes, values, and behavior.

social issues, such as school prayer or abortion rights, but liberal on economic issues, such as increasing the minimum wage.

Political Participation and Election Results Democracy in the United States has been defined as a government "of the people, by the people, and for the people." Accordingly, it would stand to reason that "the people" would actively participate in their government at any or all of four levels: (1) voting, (2) attending and taking part in political meetings, (3) actively participating in political campaigns, and (4) running for and/or holding political office. At most, about 10 percent of the voting-age population in this country participates at a level higher than simply voting, and over the past 40 years, less than half of the voting-age population has voted in nonpresidential elections. Even in presidential elections, voter turnout often is relatively low. In the 2008 presidential election, about 62 percent of the 208.3 million eligible voters cast ballots, compared with 60.6 percent in the 2004 presidential election. The number of ballots cast in 2008 was the highest in history because about 6.5 million more people were registered to vote in 2008. The larger turnout in 2008 was partly a result of significant increases in voting by younger people, Latinos/as, and African American voters. Women's votes were also a significant factor in the election of Barack Obama because women strongly preferred Obama (56 percent) to John McCain (43 percent), whereas men split their votes almost evenly between Obama (49 percent) and McCain (48 percent). State-by-state differences in voting preferences are also highly visible in what political analysts refer to as the "red states" and the "blue states." See ▶ Map 14.1.

Why is it that so many eligible voters in this country stay away from the polls? During any election, millions of voting-age persons do not go to the polls due to illness, disability, lack of transportation, nonregistration, or absenteeism. However, these explanations do not account for why many other people do not vote. According to some conservative analysts, people may not vote because they are satisfied with the status quo or because they are apathetic and uninformed—they lack an understanding of both public issues and the basic processes of government.

By contrast, liberals argue that people stay away from the polls because they feel alienated from politics at all levels of government—federal, state, and local—due to political corruption and influence peddling by special interests and large corporations. Participation in politics is influenced by gender, age, race/ethnicity, and, especially, socioeconomic status (SES). One explanation for the higher rates of political participation at higher SES levels is that advanced levels of education may give people a better understanding of government processes, a belief that they have more at stake in the political process, and greater economic resources to contribute to the process. Some studies suggest that during their college years, many people develop assumptions about political participation that continue throughout their lives.

*One electoral vote in Nebraska remains undecided. The state allocates its electoral votes on the basis of the results in each Congressional district. Only 569 votes separate John McCain and Barack Obama in unofficial returns from the 2nd District.

▶ **Map 14.1 2008 Presidential Election: State by State**

Source: *New York Times*, 2008.

Governmental Bureaucracy

When most people think of political power, they overlook one of its major sources—the governmental bureaucracy. As previously discussed, Weber's rational–legal authority finds its contemporary embodiment in bureaucratic organizations. Negative feelings about bureaucracy are perhaps strongest when people are describing the "faceless bureaucrats" and "red tape" with which they must deal in government. But who are these "faceless bureaucrats," and what do they do?

Characteristics of the Federal Bureaucracy

Bureaucratic power tends to take on a life of its own. During the nineteenth century, the government had a relatively limited role in everyday life. In the 1930s, however, the scope of government was extended greatly during the Great Depression to deal with labor–management relations, public welfare, and the regulation of the securities markets. With dramatic increases in technology and increasing demands from the public that the government "do something" about problems facing society, the government has grown still more in recent decades. Today, even with slight reductions in size, the federal bureaucracy employs more than two million people in civilian positions.

Much of the actual functioning of the government is carried on by its bureaucracy. As strange as it may seem, even the president, the White House staff, and cabinet officials have difficulty establishing control over the bureaucracy (Dye and Zeigler, 2008). Many employees in the federal bureaucracy have seen a number of presidents "come and go." For example, when President Clinton promised to make his administration "look like America," analysts watched to see who would be appointed to his cabinet but did not watch for changes in the *permanent government* in Washington, made up of top-tier civil service bureaucrats who have built a major power base. ▶ Figure 14.4 shows characteristics of the "typical" federal civilian employee.

The governmental bureaucracy has been able to perpetuate itself and expand because many of its employees have highly specialized knowledge and skills and

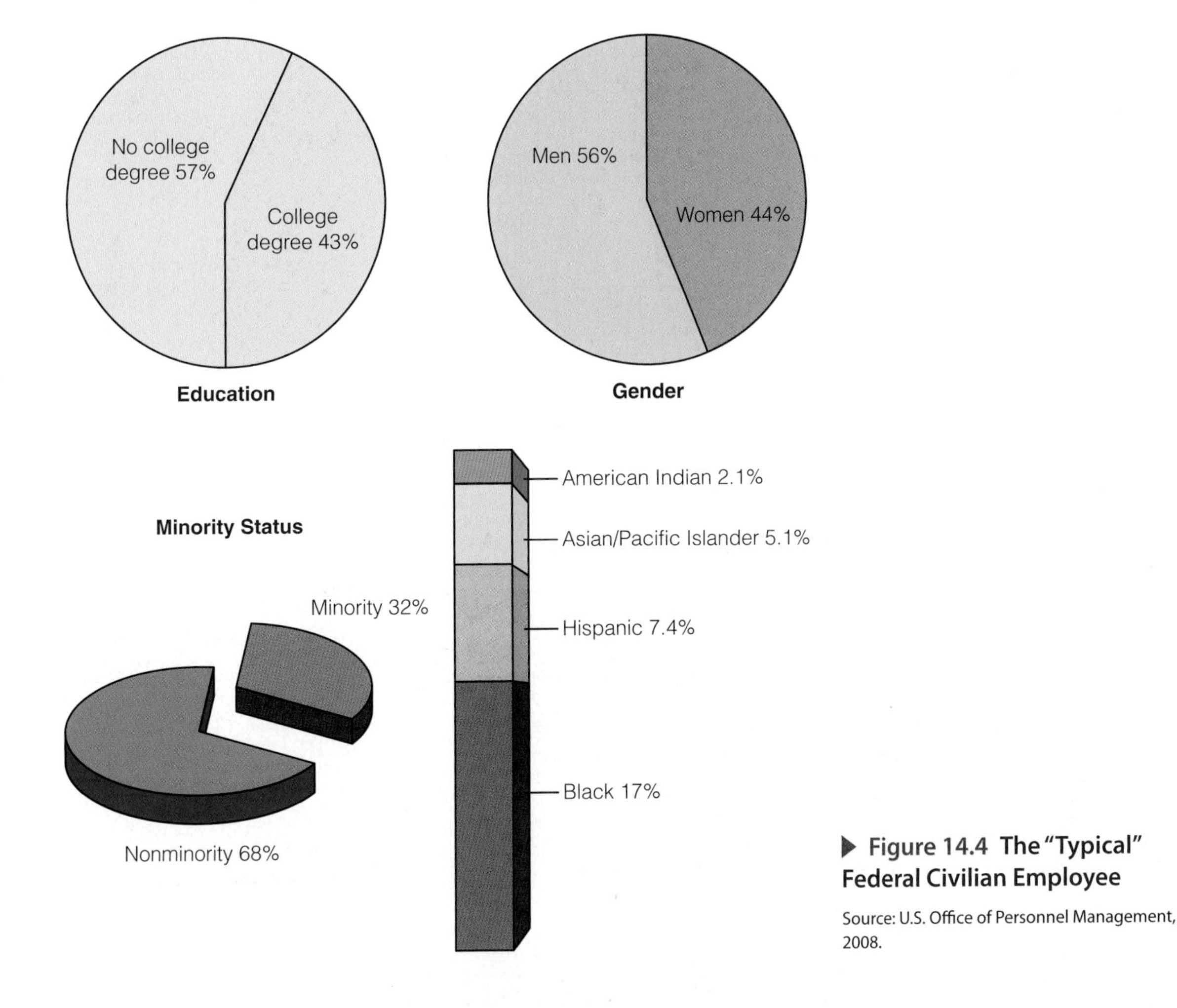

▶ **Figure 14.4 The "Typical" Federal Civilian Employee**

Source: U.S. Office of Personnel Management, 2008.

cannot be replaced easily by "outsiders." In addition, as the United States has grown in size and complexity, public policy is increasingly made by bureaucrats rather than by elected officials. For example, offices and agencies have been established to create rules, policies, and procedures for dealing with complex issues such as nuclear power, environmental protection, and drug safety; bureaucracies announce an estimated twenty rules or regulations for every one law passed by Congress (Dye and Zeigler, 2008). Typically, these bureaucracies receive little, if any, direction from Congress or the president. The actions of these agencies are subject to challenge in the courts, but most agencies still operate in a highly autonomous manner.

The executive branch is also highly bureaucratized, as shown in ▶ Figure 14.5. Cabinet-level secretaries, who are appointed by the president and approved by the Senate, head fifteen departments that carry out governmental functions. Like all other areas of governmental bureaucracy, these departments have grown in number and size over the years. Adding to the layers of bureaucracy are the bureaus and agencies that are subdivisions within cabinet departments, as well as government corporations—agencies organized like private companies and operating in a market setting. For example, the U.S. Postal Service competes with United Parcel Service, FedEx, DHL, and other delivery services (Greenberg and Page, 2002).

The federal budget is the central ingredient in the bureaucracy. Preparing the annual federal budget is a major undertaking for the president and the Office of Management and Budget, one of the most important agencies in Washington. Getting the budget approved by Congress is an even more monumental task; however, as Dye and Zeigler (2008) point out, even with the highly publicized wrangling over the budget by the president and Congress, the final congressional appropriations are usually within 2–3 percent of the budget originally proposed by the president.

What role do special interest groups play in influencing the federal bureaucracy? Interest groups have an effect on the budgets received by various agencies and departments. Even though the president has budgetary authority over the bureaucracy, any agency that feels it did not get its "fair share" can raise a public outcry by contacting friendly interest groups and congressional subcommittees. This outcry may force the president to restore funding to the agency or prod Congress

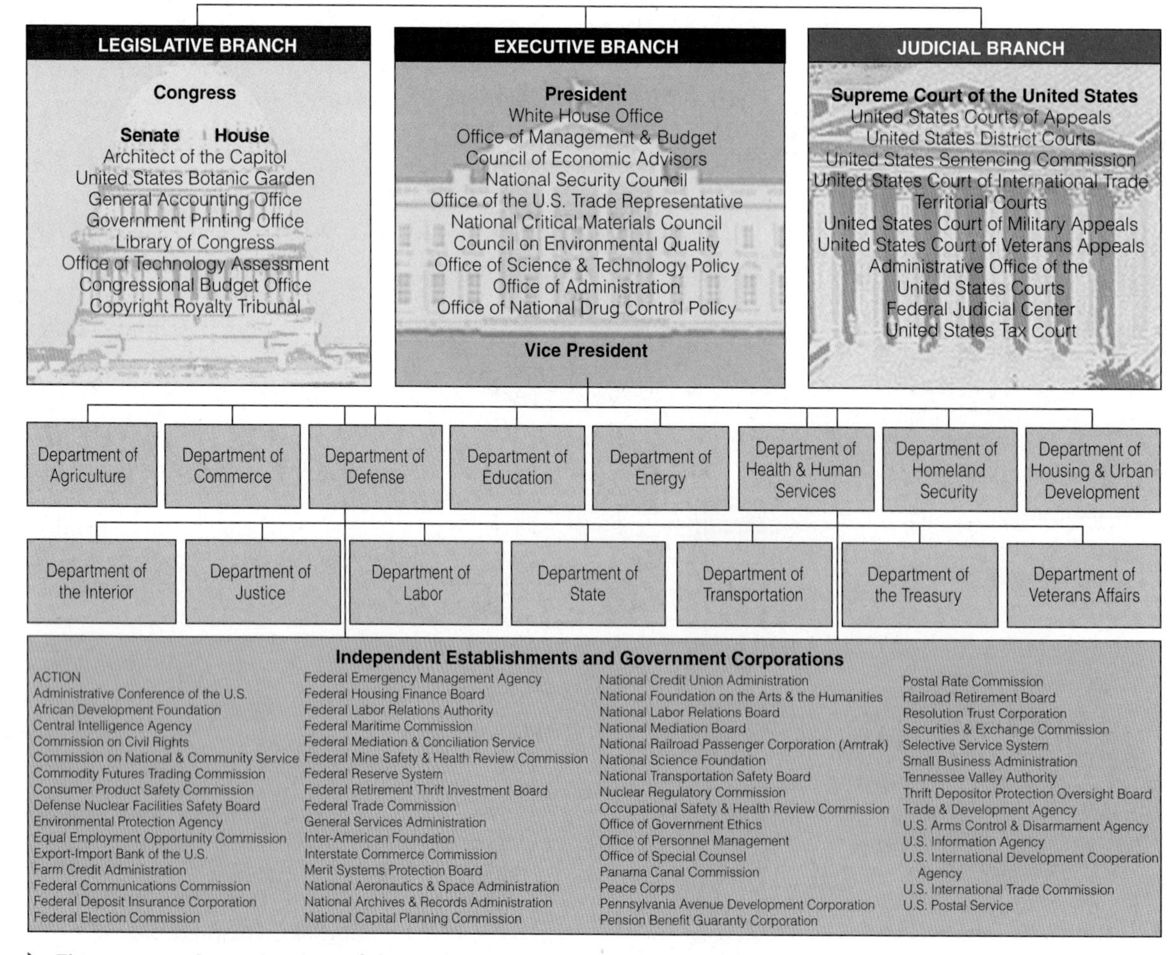

▶ **Figure 14.5 Organization of the U.S. Government**

to appropriate money not requested by the president, who then may cooperate to avoid a confrontation.

Another way in which special interest groups exert a powerful influence on the bureaucracy is the *iron triangle of power*—a three-way arrangement in which a private interest group (usually a corporation), a congressional committee or subcommittee, and a bureaucratic agency make the final decision on a political issue that is to be decided by that agency. ▶ Figure 14.6 illustrates the alliance among the Defense Department (Pentagon), private military (or defense) contractors, and members of Congress. We will now examine this relationship more closely.

The Iron Triangle and the Military–Industrial Complex

What exactly is the iron triangle, and how does it work? According to the sociologist Joe Feagin,

> The Iron Triangle has a revolving door of money, influence, and jobs among these three sets of actors, involving trillions of dollars. Military contractors who receive contracts from the Defense Department serve on the advisory committees that recommend what weapons they believe are needed. Many people move around the triangle from job to job, serving in the military, then in the Defense Department, then in military industries. (Feagin and Feagin, 1994: 405)

The long-term relationships found in this arrangement are part of what is referred to as the ***military–industrial complex*—the mutual interdependence of the military establishment and private military contractors.** The term was used by President Dwight D. Eisenhower, a retired five-star general, in his presidential farewell address in 1961, when he warned against the potential power of a huge military establishment working with a large arms industry:

> The conjecture of an immense military establishment and a huge arms industry is new in the American experience. The total influence—economic, political, and even spiritual—is felt in every city, every state house, and every office of the federal government. . . . In the councils of government, we must guard against the acquisition of unwarranted influence, whether sought or unsought, by the military–industrial complex. (Eisenhower, 1961, qtd. in Hartung, 1999)

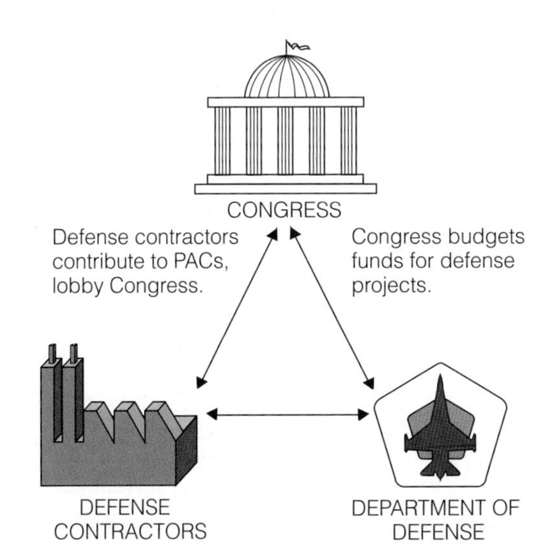

▶ **Figure 14.5 Example of the Iron Triangle of Power**

Sociologists such as C. Wright Mills (1976) have stated that an alliance of economic, military, and political power could result in a "permanent war economy" or "military economy." However, the economist John Kenneth Galbraith (1985) argued that war and the threat of war can benefit the economy. If the nation is seen as having dangerous enemies, government money will be spent on weapons; in turn, these expenditures will stimulate the private sector of the economy, create jobs, and encourage spending. For example, at the beginning of the war on terrorism, the Pentagon awarded what was expected to become the largest military contract in U.S. history to Lockheed Martin to build a new generation of supersonic stealth fighter jets. The contract was expected to be worth more than $200 billion, creating thousands of new jobs in an economy that was struggling at the time of the announcement (Dao, 2000).

Background Until World War I, most U.S. military actions took place within this country (such as the conquest of Native Americans and the Civil War) or close to home (as in the war with Mexico). After proving its military might at the international level in World War I, the United States took on the role of leader of the free world at the end of World War II.

During World War II, a massive expansion of the industrial infrastructure of the economy occurred as factories were built to produce uniforms, tanks, airplanes, ships, and so on. After the war, a major issue was what to do with the large military establishment

military–industrial complex the mutual interdependence of the military establishment and private military contractors.

that had helped the nation become a superpower. If the country wanted to remain the world leader, many argued, then it had to maintain this military establishment and make defense spending a national priority. The nation did just that. Starting in the 1950s, the United States and the Soviet Union engaged in a *cold war*—a conflict between nations based on the threat of war rather than on actual armed conflict. Because the Soviet Union wanted to be a superpower, as well as maintain control over the nations on its periphery in Eastern and Southern Europe, it entered into an arms race with the United States (Kennedy, 1987).

Defense Contracts and Arms Sales Between 1945 (the end of World War II) and the early 1990s, the U.S. government spent an estimated $10.2 trillion on national defense (in constant 1987 dollars). Later in the 1990s and continuing until immediately prior to the war on terrorism, the United States spent more than $270 billion per year on the military budget, a figure that is dwarfed by the costs of waging a war. The military budgets for 2003 and 2004 were almost $400 billion per year. Most defense spending on weapons was directed toward the "Big Three" weapons manufacturers—Lockheed Martin, Boeing, and Raytheon—which were created in a rash of military-industry mergers (Hartung, 1999). For example, Lockheed Martin was created by the merger of Lockheed with Martin Marietta, Loral Defense, the General Dynamics combat aircraft division, and many other military companies. The result is a $35-billion megacorporation that receives $18 billion annually in Pentagon contracts (Hartung, 1999).

Some of the largest defense contractors have a virtual monopoly over defense contracts because they function in a market with only one buyer (the U.S. government) and few (if any) competitors who can meet Pentagon specifications for a particular item. This monopoly is especially strong when *grease men* (independent consultants who act as a liaison between the Pentagon and defense contractors) provide "insider" information to the corporations. Thus, defense contractors have nothing to lose: They do not have to compete on costs, and their profits are guaranteed. Frequently, these corporations do not have to prove the efficacy of their products because the Pentagon is traditionally anxious to publicize, and occasionally deploy, new weapons systems before rivals can do so.

Some members of Congress have been willing to support measures that expand the military–industrial complex because doing so provides a unique opportunity for them to assist their local constituencies by authorizing funding for defense-related industries, military bases, and space centers in their home districts. These activities constitute a long-standing political practice known as *pork* or *pork barrel*—projects designed to bring jobs and public monies to the home state of members of Congress, for which they can then take credit (Greenberg and Page, 2002). Another reason for congressional support of the defense industry is that the politically powerful military industry employs millions and spends extravagantly on contributions to candidates and political parties.

In 2009 President Barack Obama acknowledged that costly defense contracts and other government spending were excessive and ordered a reform of the way the federal government does business. According to Obama, "The days of giving defense contractors a blank check are over" (qtd. in Colvin, 2009). Of the many excessive defense contracts that Obama believed existed, he highlighted the ballooning costs of a Lockheed Martin project to build a new presidential helicopter fleet and stated that it showed how the procurement process had "gone amok" (qtd. in Colvin, 2009). Obama believed that his plan would save taxpayers $40 billion a year, but defense companies argued that they used oversight and accountability in their work and had not been running wild with taxpayers' money. It remains to be seen whether or not the Obama administration or future presidents will be able to control the billions of dollars that go into defense contracts and military spending, to which we now turn our attention.

U.S. government expenditures for weapons and fighter jets such as the F-22 Fighter Jet shown here have contributed to the expansion of the military–industrial complex. In 2009, the military wanted to buy seven more jets of this type, which President Obama opposed, and the Senate backed the president and withdrew the requisite funding from the authorization bill.

The Military and Militarism

***Militarism* is a societal focus on military ideals and an aggressive preparedness for war.** Core U.S. values such as patriotism, courage, reverence, loyalty, obedience, and faith in authority help to support militarism, as the sociologist Cynthia H. Enloe (1987: 542–543) notes:

> Military expenditures, militaristic values, and military authority now influence the flow of foreign trade and determine which countries will or will not receive agricultural assistance. They shape the design and marketing of children's toys and games and of adult fashions and entertainment. Military definitions of progress and security dominate the economic fate of entire geographic regions. The military's ways of doing business open or shut access to information and technology for entire social groups. Finally, military mythologies of valor and safety influence the sense of self-esteem and well-being of millions of people.

However, note that *militarism* is not synonymous with *war;* it involves an ideology that perpetuates societal activities directed toward military preparedness and enormous spending on weapons production and "defense" as compared with other national priorities (Cock, 1994).

Explanations for Militarism

Sociologists have proposed several reasons for militarism. One is the economy. According to Enloe (1987: 527), people who see capitalism as the moving force behind the military's influence "believe that government officials enhance the status, resources, and authority of the military in order to protect the interests of private enterprises at home and overseas." From this perspective, the origin of militarism is found in the boardroom, not the war room. Because government and military funding is also a prime source of support for research in the physical, biological, and social sciences, administrators and faculty in some institutions of higher learning feel the need to support war-related spending to obtain grants for research. In addition, workers come to rely on military spending for jobs; labor unions have supported defense spending because it has provided well-paid, stable employment for union members.

A second explanation focuses on the role of the nation and its inclination toward coercion in response to perceived threats. From this perspective, nations will inevitably use force to ensure compliance within their society and to protect themselves from outside attacks. A third explanation is based primarily on patriarchy and the relationship between militarism and masculinity. Across cultures and over time, the military has been a male institution, and the "meanings attached to masculinity appear to be so firmly linked to compliance with military roles that it is often impossible to disentangle the two" (Enloe, 1987: 531). Symbolic interactionists might view this perspective as the *social construction of masculinity.* In other words, certain assumptions, teachings, and expectations that serve as the standard for appropriate male behavior—in this case, values of dominance, power, aggression, and violence—are created and re-created. Presumably, such qualities may be learned through gender socialization, including that which is received in military training. Historically, the development of manhood and male superiority has been linked to militarism and combat—the ultimate test of a man's masculinity (Enloe, 1987; Cock, 1994).

Gender, Race, and the Military

In recent years, some significant changes have occurred in the policies and composition of the U.S. armed forces. Especially with the introduction of the all-volunteer force and the end of the draft in the early 1970s, the focus of the military shifted from training "good citizen-soldiers" to recruiting individuals who would enlist in the military in the same way that another person might take a job in the private sector. Now, in return for joining the military, a person is rewarded with a reasonable salary, medical and dental care, the chance to travel, educational opportunities, and other benefits (Stiehm, 1989).

Since the introduction of the volunteer forces, considerable pressure has been placed on the military to recruit women. In 2009, for example, women made up approximately 20 percent of the total uniformed force. However, in spite of some changes in the gender composition of the military, certain analysts argue that it remains essentially a male-dominated world. According to one military analyst, women must break through the "brass ceiling" if they hope to rise to the top of "a masculine institution" so that they can become a general or a flag officer (Women's Research & Education Institute, 2009). In addition to facing a brass ceiling if they attempt to rise through the ranks of the military, some women have also faced sexual

militarism a term used to describe a societal focus on military ideals and an aggressive preparedness for war.

assault. During 2008, for example, the U.S. military services reported an 8 percent increase in the number of incidents of sexual assault. About 63 percent of the reported incidents involved rape or aggravated assault; 251 incidents occurred in combat areas such as Iraq and Afghanistan (Kruzel and Carden, 2009). Although the U.S. Defense Department believes that this increase may be attributed to more people reporting incidents of sexual assault, government officials indicate that new policies have been established to reduce these offenses.

Terrorism and War

In the first decade of this century, people in the United States became acutely aware of the problem of terrorism on their own soil, a concern that people in nations around the world have faced for many years. Today, the terrorist attacks of 2001 are remembered by some people only when they face heightened security at airports or when the nation commemorates the anniversary of that horrific event on September 11 of each year. However, many political leaders and social analysts suggest that we should not forget that constant vigilance is necessary to combat the possibility of future terrorism.

© AP Images/Gloria Wright

Regarding homosexuality, the U.S. military has a "don't ask/don't tell" policy. Shown here is Lt. Dan Choi, a West Point graduate and Iraq War veteran, who publicly revealed that he is gay. Choi is scheduled to be expelled from the service. However, the policy is not favored by President Obama, who may try to eliminate or modify it during his administration.

Types of Terrorism

***Terrorism* is the use of calculated, unlawful physical force or threats of violence against a government, organization, or individual to gain some political, religious, economic, or social objective.** Terrorist tactics include bombing, kidnapping, hostage taking, hijacking, assassination, and extortion (Vetter and Perlstein, 1991). Although terrorists sometimes attack government officials and members of the military, they more often target civilians as a way of pressuring the government. Until 2001, most international acts of terrorism that targeted U.S. interests took place outside of this country. The events of September 2001 were a frightening and shocking experience for people in the United States, who realized that the terrorists—whose objectives were not clearly known and whose linkage to any one government or political regime was vague—had wreaked immense destruction within our nation's borders, leaving thousands dead and the entire nation fearful for its long-term safety.

Around the world, acts of terrorism extract a massive toll by producing rampant fear, widespread loss of human life, and extensive destruction of property. In 2008, for example, 173 people were killed and more than 300 were wounded in Mumbai, India, when members of the Lashkar-e-Taiba, a Pakistan-based terrorist organization, instigated a series of eight attacks at locations in Mumbai, including the Taj Mahal Palace, a hospital, and a local college.

One form of terrorism—political terrorism—is actually considered a form of unconventional warfare. *Political terrorism* uses intimidation, coercion, threats of harm, and other violence that attempts to bring about a significant change in or overthrow an existing government. There are at least three types of political terrorism: revolutionary terrorism, state-sponsored terrorism, and repressive terrorism. *Revolutionary terrorism* refers to acts of violence against civilians that are carried out by enemies of the government who want to bring about political change. Some groups believe that if they perpetrate enough random acts of terrorism, they will achieve a political goal. Members of radical religious or revolutionary movements may engage in terrorist activities such as bombings, kidnappings, and assassinations of leading officials. The goal of the terrorists is to traumatize a civilian population so that people will put their government under such pressure that political officers and departments of the government can no longer work effectively and can be brought down (Jacquard, 2001). Although exact linkages may be difficult to pinpoint, some revolutionary terrorists may receive assistance from governments that support their objectives (Vetter and Perlstein, 1991).

Unlike revolutionary terrorism, *state-sponsored terrorism* occurs when a government provides financial resources, weapons, and training for terrorists who conduct their activities in other nations. In Libya, for example, Colonel Muammar Qaddafi has provided money and training for terrorist groups such as the Arab National Youth Organization, which was responsible for skyjacking a Lufthansa airplane over Turkey and forcing the German government to free the surviving members of Black September. Black September was the terrorist group responsible for killing Israeli Olympic athletes in the 1970s (Parry, 1976).

The third type of political terrorism, *repressive terrorism,* is conducted by a government against its own citizens for the purpose of protecting an existing political order. Repressive terrorism has taken place in many countries around the world, including Haiti, the People's Republic of China, and Cambodia, where the Pol Pot regime killed more than a million people in the five years between 1975 and 1979 (Mydans, 1997a).

Terrorism in the United States

The years 1995 and 2001 stand out in recent U.S. history as the periods when this nation suffered from the two worst terrorist attacks that had ever occurred in the continental United States. The April 1995 bombing of the Federal Building in Oklahoma City took the lives of 168 adults and children, and injured 850 others. Many other people were left with psychological scars. The Oklahoma City bombing was described as an act of domestic terrorism because it was attributed to two men who had no apparent connection to external enemies of the United States.

By contrast, external enemies (who had planted perpetrators in the United States) were believed to be responsible for the attacks on New York's World Trade Center, first in 1993 in the form of a bombing that killed six people and injured more than a thousand, and then in the 2001 destruction of the Trade Center's twin towers (and damage to the Pentagon, in Washington, D.C., that same morning) by suicide hijackers using passenger airplanes as lethal weapons, resulting in the deaths of almost three thousand people.

Prior to the events of September 11, 2001, terrorism in the form of biological and chemical attacks was considered by most analysts to be a remote possibility in the United States. However, with the deaths of postal workers and private citizens from the delivery of anthrax, suspicions deepened that, for the first time, terrorists—whether homegrown or international—might use germ warfare to achieve their subversive political goals. In the twenty-first century, people in the United States are confronted with threats of terrorism and problems associated with war that have been facts of life in some other countries for many years.

War

Generally speaking, ***war* is organized, armed conflict between nations or distinct political factions.** Social scientists define war to include both *declared wars* between nations or parties and *undeclared wars:* civil and guerrilla wars, covert operations, and some forms of terrorism. War is an institution that involves *violence*—behavior intended to bring pain, physical injury, and/or psychological stress to people or to harm or destroy property (Sullivan, 2000). As such, war is a form of *collective violence* by people who are seeking to promote their cause or resist social policies or practices that they consider oppressive (Sullivan, 2000).

As previously discussed, early U.S. military action took place at home or close to home, but the location and the nature of U.S. military action changed dramatically in the twentieth century. Wars were now fought on foreign soil, and larger numbers of U.S. military personnel were involved. In World War I, for example, about 5 million people served in the U.S. armed forces; more than 16 million men and women served in World War II (Ehrenreich, 1997). It remains to be seen what types of military response will be required in the war against terrorism in the twenty-first century.

The direct effects of war are loss of human life and serious physical and psychological effects on some survivors. Although it is impossible to determine how many human lives have been lost in wars throughout human history, social analyst Ruth Sivard (1991, 1993) estimated that 589 wars have been fought by 142 countries since 1500 and that approximately 142 million lives have been lost. However, according to Sivard, more lives were lost in wars during the twentieth century than in all of previous history. World War I took the lives of approximately 8 million combatants and 1 million civilians. In World War II, more than 50 million people (17 million combatants and 35 million civilians) lost their lives. During World War II, U.S. casualties alone totaled almost 300,000, and more

terrorism the calculated unlawful use of physical force or threats of violence against persons or property in order to intimidate or coerce a government, organization, or individual for the purpose of gaining some political, religious, economic, or social objective.

war organized, armed conflict between nations or distinct political factions.

© AP Images/Evan Vucci

For eighteen years the U.S. military banned the media from taking and publishing photographs of the caskets of military casualties after they were shipped back to the United States. Critics believed that the military was trying to "sanitize" the war and avoid facing its human costs. However, the policy has recently been reversed—such images now may be publicized with the permission of the deceased's family. Shown here in the first such photo published since the policy change is the casket of Air Force Staff Sergeant Phillip Myers, who was killed in Afghanistan on April 4, 2009.

than 600,000 were wounded. The number of casualties in conventional warfare pales when compared to the potential loss of human life if all-out nuclear, biological, or chemical warfare were to occur. The devastation would be beyond description.

Media watch groups have asked why the major news networks in the United States have made few efforts to accurately tally the number of deaths in recent wars. Some cite the fact that they believe the media are more likely to report the spin control of elected and appointed federal officials in the White House and the Pentagon rather than engage in investigative reporting on the true nature and extent of damage and deaths in these war-torn regions (see FAIR, 2003, for example).

While the conflicts in Afghanistan and Iraq continue (as of this writing, in 2009), U.S. armed forces are also engaged in limited operations in a number of other nations. No doubt, civilian and military casualties will continue to increase in various regions around the world as we have moved from a brief era of peacetime in the United States into what appears to be a protracted era of wartime in various regions of the world.

Politics and Government in the Future

Thinking about U.S. politics and government in the future is very much like the old story about optimists and pessimists. According to the story, an eight-ounce cup containing exactly four ounces of water is placed on a table. The optimist comes in, sees the cup, and says, "The cup is half full." The pessimist comes in, sees the cup, and says, "The cup is half empty." Clearly, both the pessimist and the optimist are looking at the same cup containing the same amount of water, but their perspective on what they see is quite different. For some analysts, looking at the future of the U.S. government and its political structure is very much like this.

Views of the future of politics and government relate to specific concerns about the United States:

- What will be the future of political parties? What did the 2008 presidential election tell us about the nature of these institutions?
- Are global corporate interests and the concerns of the wealthy in this nation and elsewhere overshadowing the needs and interests of everyday people?
- Is it possible to prevent future terrorist attacks in the United States (and other nations as well) through tightening organizational intelligence and reorganizing some governmental bureaucracies?
- Will we be able to balance the need for national security with the individual's right to privacy and freedom of movement within this country?
- How will elected politicians and appointed government officials handle the challenges regarding the changing demographics of the United States? For example, how will the aging population influence their decisions? Will increasing racial, ethnic, and religious diversity influence their decisions?
- Will immigration and employment policies be based on the best interests of the largest number of people, or will these policies be based on the best interests of a small elite who are major contributors to political campaigns?
- Do the media accurately report what is going on at all levels of government? To what extent can individuals and grassroots organizations influence the media and the political process? (See Box 14.4, "You Can Make a Difference.")

Box 14.4 You Can Make a Difference

Keeping an Eye on the Media

Do we get all of the news that we should about how our government operates and about the pressing social problems of our nation? Consider this list (shown by number) of the five top news stories that some media analysts believe were *not* adequately covered by the U.S. media in recent years:

1. Over One Million Iraqi Deaths Caused by U.S. Occupation
2. Security and Prosperity Partnership: Militarized NAFTA
3. InfraGard: The FBI Deputizes Business
4. ILEA: Is the US Restarting Dirty Wars in Latin America?
5. Seizing War Protesters' Assets (Project Censored, 2009)

According to the group that compiles this list, Project Censored, which is managed through the Department of Sociology at Sonoma State University, many important stories are either missing from the news altogether or do not receive the attention they deserve. Through the work of more than 200 people, Project Censored keeps an eye on the news media and publishes an annual list of significant stories that its researchers believe were neglected. (To view the entire list, visit Project Censored's website at **http://www.projectcensored.org.**) According to a statement by Project Censored, "Corporate media seem to have abdicated their First Amendment responsibility to keep the public informed. The traditional journalist values of supporting democracy by maintaining an educated electorate now take second place to profits and ratings" (Phillips and Project Censored, 2002).

What should be the role of the media in keeping us informed? The media are often referred to as the "fourth estate" or the "fourth branch of the government" because they are supposed to provide people relevant information on important topics regarding how the government operates in a democratic society. This information can then be used by citizens to decide how they will vote on candidates and issues presented for their approval or disapproval on the election ballot. However, some analysts believe that much media coverage, especially the political talk program genre that has proliferated on network and cable TV channels, may actually *reduce* public understanding of issues rather than add to it. Why? Because the media "hammer" home the message that "issues" do not matter except as items for politicians to squabble about (Fallows, 1997).

The first step in keeping an eye on the news is to become more analytical about the "news" that we do receive. How can we evaluate the information we receive from the media? In *How to Watch TV News,* the media analysts Neil Postman and Steve Powers (1992: 160–168) suggest the following:

1. We should keep in mind that television news shows are called "shows" for a reason. They are not a public service or a public utility.
2. We should never underestimate the power of commercials, which tell us much about our society.
3. We should learn about the economic and political interests of those who run television stations or own a controlling interest in a media conglomerate.
4. We should pay attention to the *language* of newscasts, not just the visual imagery. For example, a *question* may reveal as much about the *questioner* as the person answering the question.

Becoming aware of the media's role in influencing people's opinions about how our government is run is the first step toward becoming an informed participant in the democratic political process.

The second step in keeping an eye on the news is becoming aware of national and international events that should receive more coverage than they do or that might not be reported in a fair and unbiased manner. To form your own opinion about media coverage of the news, you may wish to also visit these sites, which examine media practices:

- FAIR (Fairness and Accuracy in Reporting): **http://www.fair.org**
- MediaChannel.org, which states, "As the media watch the world, we watch the media": **http://www.mediachannel.org**

As we enter the second decade of the twenty-first century, these are a few of the many questions regarding politics and the government that face us and people in other nations. How we (and our elected officials) answer these (and related) questions will in large measure determine the future of politics and government in the United States. Our answers will also have a profound influence on people and governments in other countries, whether they be high-, middle-, or low-income nations, around the world.

Chapter Review

- **What is power?**

Power is the ability of persons or groups to carry out their will even when opposed by others.

- **What is the relationship between power and politics?**

Politics is the social institution through which power is acquired and exercised by some people or groups. A strong relationship between politics and power exists in all countries, and the military is closely tied to the political system.

- **What are the three types of authority?**

Max Weber identified these types of authority: traditional, charismatic, and rational–legal. Traditional authority is based on long-standing custom. Charismatic authority is power based on a leader's personal qualities. Rational–legal authority is based on law or written rules and regulations, as found in contemporary bureaucracies.

- **What are the main types of political systems?**

The main types of political systems are monarchies, authoritarian systems, totalitarian systems, and democratic systems. In a monarchy, one person is the hereditary ruler of the nation. In authoritarian systems, rulers tolerate little or no public opposition and generally cannot be removed from office by legal means. In totalitarian systems, the state seeks to regulate all aspects of society and to monopolize all societal resources in order to exert complete control over both public and private life. In democratic systems, the powers of government are derived from the consent of all the people.

- **How do pluralist and power elite perspectives view power in the United States?**

According to the pluralist (functionalist) model, power is widely dispersed throughout many competing interest groups. People influence policy by voting, joining special interest groups and political action campaigns, and forming new groups. According to the elite (conflict) model, power is concentrated in a small group of elites, whereas the masses are relatively powerless.

- **Who makes up the power elite, and why are they important?**

According to C. Wright Mills, the power elite is composed of influential business leaders, key government leaders, and the military. The elites possess greater resources than the masses, and public policy reflects their preferences.

- **What is the military–industrial complex?**

The military–industrial complex refers to a mutual interdependence of the government military establishment and private military contractors.

- **What is terrorism?**

Terrorism is the use of calculated, unlawful physical force or threats of violence against a government, organization, or individual to gain some political, religious, economic, or social objective.

- **What are the major types of political terrorism?**

There are three types of political terrorism: revolutionary terrorism, state-sponsored terrorism, and repressive terrorism. Revolutionary terrorism refers to acts of violence against civilians that are carried out by enemies of the government who want to bring about political change. State-sponsored terrorism occurs when a government provides financial resources, weapons, and training for terrorists who conduct their activities in other nations. Repressive terrorism is conducted by a government against its own citizens for the purpose of protecting an existing political order.

www.cengage.com/login

Want to maximize your online study time? Take this easy-to-use study system's diagnostic pre-test, and it will create a personalized study plan for you. By helping you to identify the topics that you need to understand better and then directing you to valuable online resources, it can speed up your chapter review. CengageNOW even provides a post-test so you can confirm that you are ready for an exam.

Key Terms

authoritarianism 453
authority 449
charismatic authority 450
democracy 454
elite model 457
government 446
militarism 469
military–industrial complex 467
monarchy 451
pluralist model 454
political action committee 456
political party 460
political socialization 463
political sociology 447
politics 446
power 448
power elite 458
rational–legal authority 450
routinization of charisma 450
special interest group 456
state 447
terrorism 470
totalitarianism 453
traditional authority 449
war 471

Questions for Critical Thinking

1. Who is ultimately responsible for decisions and policies that are made in a democracy such as the United States—the people or their elected representatives?
2. How would you design a research project to study the relationship between campaign contributions to elected representatives and their subsequent voting records? What would be your hypothesis? What kinds of data would you need to gather? How would you gather accurate data?
3. How does your school (or workplace) reflect a pluralist or elite model of power and decision making?
4. Can democracy survive in a context of rising concentrated power in the capitalist oligarchies? Why or why not?

The Kendall Companion Website

www.cengage.com/sociology/kendall

Visit this book's companion website, where you'll find more resources to help you study and successfully complete course projects. Resources include quizzes and flash cards, as well as special features such as an interactive sociology timeline, maps, General Social Survey (GSS) data, and Census 2000 data. The site also provides links to useful websites that have been selected for their relevance to the topics in this chapter and include those listed below. (*Note:* Visit the book's website for updated URLs.)

Roll Call

http://www.rollcall.com

Roll Call is an organization that tracks political news and provides commentary on current political events. In addition to reporting on current events, the site features useful links, information on policy briefings, editorial cartoons, and editorial opinions.

Federal Election Commission (FEC)

http://www.fec.gov

An independent regulatory agency, the FEC was created in 1975 to administer and enforce the Federal Election Campaign Act (FECA), which governs the financing of federal elections. The FEC's website includes current news releases, information on elections and campaign finance, and various guides to local, state, and federal election systems.

Open Secrets

http://www.opensecrets.org

This comprehensive site provides information on campaign financing and profiles such current topics as nuclear energy, electricity deregulation, and managed care. The site allows you to view the contribution trends in specific industries, access specific campaign finance information for candidates, and conduct searches for information pertinent to your locality.

chapter

15 Families and Intimate Relationships

Chapter Focus Question

Why are many family-related concerns—such as divorce and child care—viewed primarily as personal problems rather than as social concerns requiring macrolevel solutions?

My son began commuting between his two homes at age 4. He traveled with a Hello Kitty suitcase with a pretend lock and key until it embarrassed him. . . . He graduated to a canvas backpack filled with a revolving arsenal of essential stuff: books and journals, plastic vampire teeth, "Star Trek" Micro Machines, a Walkman, CD's, a teddy bear.

The commuter flights between San Francisco and Los Angeles were the only times a parent wasn't lording over him, so he was able to order Coca-Cola, verboten at home. . . . But such benefits were insignificant when contrasted with his preflight nightmares about plane crashes. . . .

Like so many divorcing couples, we divided the china and art and our young son. First he was ferried back and forth

© Myrleen Ferguson Cate/PhotoEdit

Changing family patterns and relocations for employment opportunities have made sights such as this increasingly familiar, as children travel alone to various destinations to be reunited with a parent or other family member.

IN THIS CHAPTER

- Families in Global Perspective
- Theoretical Perspectives on Families
- Developing Intimate Relationships and Establishing Families
- Child-Related Family Issues and Parenting
- Transitions and Problems in Families
- Diversity in Families
- Family Issues in the Future

between our homes across town, and then, when his mother moved to Los Angeles, across the state. For the eight years since, he has been one of the thousands of American children with two homes, two beds, two sets of clothes and toys and two toothbrushes.

—David Sheff (1995: 64), several years ago, explaining how divorce and joint custody had affected his son, Nick Sheff

Four years later, Nick Sheff stated his own view on long-distance joint custody:

I am 16 now and I still travel back and forth, but it's mostly up to me to decide when. I've chosen to spend more time with my friends at the expense of visits with my mom. When I do go to L.A., it's like my stepdad put it: I have a cameo role in their lives. I say my lines and I'm off. It's painful.

What's the toll of this arrangement? I'm always missing somebody. When I'm in northern California, I miss my mom and stepdad. But when I'm in L.A., I miss hanging out with my friends, my other set of parents and little brothers and sisters. After all these back-and-forth flights, I've learned not to get too emotionally attached. I have to protect myself. . . . No child should be subjected to the hardship of

(Continued)

Sharpening Your Focus

- *Why is it difficult to define* family?
- *How do marriage patterns vary around the world?*
- *What are the key assumptions of functionalist, conflict/feminist, and symbolic interactionist perspectives on families?*
- *What significant trends affect many U.S. families today?*

long-distance joint custody. To prevent it, maybe there should be an addition to the marriage vows: Do you promise [that] if you ever have children and wind up divorced, [you will] stay within the same geographical area as your kids? . . . Or how about some common sense? If you move away from your children, *you* have to do the traveling to see them. (Sheff, 1999: 16)

Nick's family is one of millions around the globe experiencing major change as a result of divorce, death, or some other significant event. In this chapter, we examine the increasing complexity and diversity of contemporary families in the United States and other nations. Pressing social issues such as divorce, child care, and new reproductive technologies will be used as examples of how families and intimate relationships continue to change. Before reading on, test your knowledge about some contemporary trends in U.S. family life by taking the quiz in Box 15.1.

Families in Global Perspective

As the nature of family life and work has changed in high-, middle-, and low-income nations, the issue of what constitutes a "family" has been widely debated. For many years, the standard sociological definition of *family* has been a group of people who are related to one another by bonds of blood, marriage, or adoption and who live together, form an economic unit, and bear and raise children. Many people believe that this definition should not be expanded—that social approval should not be extended to other relationships simply because the persons in those relationships wish to consider themselves a family. However, others challenge this definition because it simply does not match the reality of family life in contemporary society (Lamanna and Riedmann, 2009). Today's families include many types of living arrangements and relationships, including single-parent households, unmarried couples, lesbian and gay couples, and multiple generations (such as grandparent, parent, and child) living in the same household. To accurately reflect these changes in family life, some sociologists believe that we need a more encompassing definition of what constitutes a family. Accordingly, we will define ***families* as relationships in which people live together with commitment, form an economic unit and care for any young, and consider their identity to be significantly attached to the group.** Sexual expression and parent–child relationships are a part of most, but not all, family relationships.

Although families differ widely around the world, they also share certain common concerns in their everyday lives. For example, women and men of all racial–ethnic categories, nationalities, and income levels face problems associated with child care. Various nongovernmental agencies of the United Nations have established international priorities to help families. Suggestions have included the development of "social infrastructures for the care and education of children of working parents in order to reduce their . . . burden" and the implementation of flexible working hours so that parents will have more opportunities to spend time with their children (Pietilä and Vickers, 1994: 115).

How do sociologists approach the study of families? In our study of families, we will use our sociological imagination to see how our personal experiences are related to the larger happenings in society. At the microlevel, each of us has a "biography," based on our experience within our family; at the macrolevel, our families are embedded in a specific social context that has a major effect on them. We will examine the institution of the family at both of these levels, starting with family structure and characteristics.

Family Structure and Characteristics

In preindustrial societies, the primary form of social organization is through kinship ties. ***Kinship* refers to a social network of people based on common ancestry, marriage, or adoption.** Through kinship networks, people cooperate so that they can acquire the basic necessities of life, including food and shelter. Kinship systems can also serve as a means by which property is transferred, goods are produced and distributed, and power is allocated.

In industrialized societies, other social institutions fulfill some of the functions previously taken care of

Box 15.1 Sociology and Everyday Life

How Much Do You Know About Contemporary Trends in U.S. Family Life?

True	False	
T	F	1. Today, people in the United States are more inclined to get married than at any other time in history.
T	F	2. Most U.S. family households are composed of a married couple with one or more children under age 18.
T	F	3. One in two U.S. preschoolers has a mother in the paid labor force.
T	F	4. Recent studies have found that sexual activity is more satisfying to people who are in sustained relationships such as marriage.
T	F	5. People under age 45 are more likely to cohabit than are people over age 45.
T	F	6. With the strong U.S. economy at the end of the 1990s, the number of children living in extreme poverty went down significantly.
T	F	7. Recent studies have found that adult children of divorced parents are more likely to dissolve their own marriages than they were two decades ago.
T	F	8. Teenage pregnancies have increased dramatically over the past 30 years.
T	F	9. Couples who already have children at the beginning of their marriage are less likely to divorce.
T	F	10. The number of children living with grandparents and with no parent in the home increased by more than 50 percent between 1990 and 2000.

Answers on page 480.

by the kinship network. For example, political systems provide structures of social control and authority, and economic systems are responsible for the production and distribution of goods and services. Consequently, families in industrialized societies serve fewer and more-specialized purposes than do families in preindustrial societies. Contemporary families are responsible primarily for regulating sexual activity, socializing children, and providing affection and companionship for family members.

Families of Orientation and Procreation During our lifetime, many of us will be members of two different types of families—a family of orientation and a family of procreation. The ***family of orientation*** **is the family into which a person is born and in which early socialization usually takes place.** Although most people are related to members of their family of orientation by blood ties, those who are adopted have a legal tie that is patterned after a blood relationship. The ***family of procreation*** **is the family that a person forms by having or adopting children.** Both legal and blood ties are found in most families of procreation. The relationship between a husband and wife is based on legal ties; however, the relationship between a parent and child may be based on either blood ties or legal ties, depending on whether the child has been adopted.

In the United States, although many young people leave their family of orientation as they reach adulthood, finish school, and/or get married, recent studies have found that many people maintain family ties across generations, particularly as older persons remain actively involved in relationships with their adult children.

Extended and Nuclear Families Sociologists distinguish between extended and nuclear families based on the number of generations that live within

families relationships in which people live together with commitment, form an economic unit and care for any young, and consider their identity to be significantly attached to the group.

kinship a social network of people based on common ancestry, marriage, or adoption.

family of orientation the family into which a person is born and in which early socialization usually takes place.

family of procreation the family that a person forms by having or adopting children.

Box 15.1 **Sociology and Everyday Life**

Answers to the Sociology Quiz on Contemporary Trends in U.S. Family Life

1. False. Census data show that the marriage rate has gone down by about one-third since 1960. In 1960 there were about 73 marriages per 1,000 unmarried women age 15 and up, whereas today the rate is about 49 per 1,000.

2. False. Less than 25 percent of all family households are composed of married couples with one or more children under age 18.

3. True. One in two preschoolers in the United States has a mother in the paid labor force, which makes affordable, high-quality child care an important issue for many families.

4. True. Recent research indicates that people who are married or are in a sustained relationship with their sexual partner express greater satisfaction with their sexual activity than do those who are not.

5. True. People under age 45 are more likely to cohabit than are older individuals.

6. False. Despite the strong economy prior to 2001, the number of children living in extreme poverty—with incomes below one-half of the poverty line—crept upward by almost 400,000.

7. False. Adult children of divorced parents are less likely to dissolve their own marriages than they were two decades ago.

8. False. Teenage pregnancies have actually decreased over the past 30 years. However, the percentage of *unmarried* teenage pregnancies has increased dramatically.

9. False. Couples who already have children at the beginning of their marriage are *more* likely to divorce than those who do not.

10. True. The number of children living with grandparents and with no parent in the home grew by 51.5 percent between 1990 and 2000. Today, more than 2.5 million grandparents are raising their grandchildren.

Sources: Based on U.S. Census Bureau, 2005; and Children's Defense Fund, 2008.

a household. An ***extended family*** **is a family unit composed of relatives in addition to parents and children who live in the same household.** These families often include grandparents, uncles, aunts, or other relatives who live close to the parents and children, making it possible for family members to share resources. In horticultural and agricultural societies, extended families are extremely important; having a large number of family members participate in food production may be essential for survival. Today, ex-

© Michael Newman/PhotoEdit

© Steven Rubin/The Image Works

Whereas the relationship between a husband and wife is based on legal ties, relationships between parents and children may be established by either blood ties or legal ties.

tended family patterns are found in Latin America, Africa, Asia, and some parts of Eastern and Southern Europe (Busch, 1990). With the advent of industrialization and urbanization, maintaining the extended family pattern becomes more difficult in societies. Increasingly, young people move from rural to urban areas in search of employment in the industrializing sector of the economy. At that time, the nuclear family typically becomes the predominant family form in the society.

A ***nuclear family*** **is a family composed of one or two parents and their dependent children, all of whom live apart from other relatives.** A traditional definition specifies that a nuclear family is made up of a "couple" and their dependent children; however, this definition became outdated when a significant shift occurred in the family structure. A comparison of Census Bureau data from 1970 and 2005 shows that there has been a significant decline in the percentage of U.S. households comprising a married couple with their own children under eighteen years of age (see "Census Profiles: Household Composition"). Conversely, there has been an increase in the percentage of households in which either a woman or a man lives alone.

Marriage Patterns

Across cultures, families are characterized by different forms of marriage. ***Marriage*** **is a legally recognized and/or socially approved arrangement between two or more individuals that carries certain rights and obligations and usually involves sexual activity.** In most societies, marriage involves a mutual commitment by each partner, and linkages between two individuals and families are publicly demonstrated.

In the United States, the only legally sanctioned form of marriage is ***monogamy*****—a marriage between two partners, usually a woman and a man.** For some people, marriage is a lifelong commitment that ends only with the death of a partner. For others, marriage is a commitment of indefinite duration. Through a pattern of marriage, divorce, and remarriage, some people practice *serial monogamy*—a succession of marriages in which a person has several spouses over a lifetime but is legally married to only one person at a time.

Polygamy **is the concurrent marriage of a person of one sex with two or more members of the opposite sex** (Marshall, 1998). The most prevalent form of polygamy is ***polygyny*****—the concurrent marriage of one man with two or more women.** Polygyny has been practiced in a number of Islamic societies, including some regions of contemporary Africa and southern Russia. For example, government officials in Africa estimate that 20 percent of Zambian marriages today are polygynous (Chipungu, 1999). How many wives and children might a polygynist have at one time? According to one report, Rodger Chilala of southern Zambia claimed to have 14 wives and more than 40 children; he stated that he previously had 24 wives but found that he could not afford the expenses associated with that many spouses (Chipungu, 1999). Although the cost of providing for multiple wives and many children makes the practice of polygyny implausible for all but the wealthiest men, such marriages still exist in both urban and rural areas of Africa and among people of divergent social status and age groups (Chipungu, 1999). Polygyny is also allowed in southern Russia, where efforts are under way to revive Islamic traditions (ITAR/TASS, 1999b). However, after legislation was passed in the Republic of Ingushetia that allowed men to have up to four wives, only three men took advantage of the new law (ITAR/TASS, 1999a). Some analysts believe that the practice of polygamy contributes to the likelihood that families will live in poverty (Chipungu, 1999).

The second type of polygamy is ***polyandry*****—the concurrent marriage of one woman with two or more men.** Polyandry is very rare; when it does occur, it is typically found in societies where men greatly outnumber women because of high rates of female infanticide or where marriages are arranged between two brothers and one woman ("fraternal polyandry"). According to recent research, polyandry is never the only form of marriage in a society: Whenever polyandry occurs, polygyny co-occurs (Trevithick, 1997).

extended family a family unit composed of relatives in addition to parents and children who live in the same household.

nuclear family a family composed of one or two parents and their dependent children, all of whom live apart from other relatives.

marriage a legally recognized and/or socially approved arrangement between two or more individuals that carries certain rights and obligations and usually involves sexual activity.

monogamy a marriage between two partners, usually a woman and a man.

polygamy the concurrent marriage of a person of one sex with two or more members of the opposite sex.

polygyny the concurrent marriage of one man with two or more women.

polyandry the concurrent marriage of one woman with two or more men.

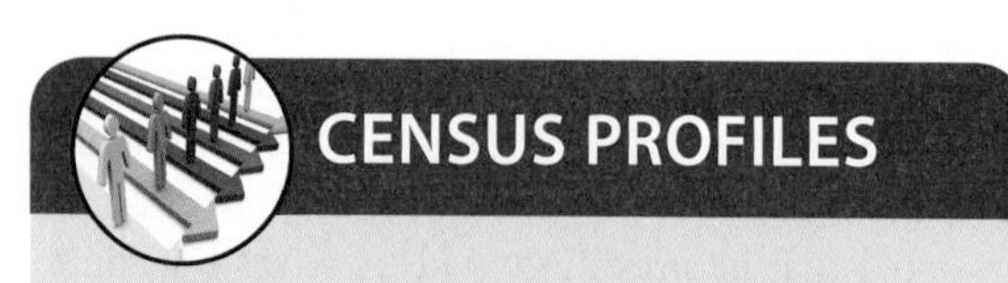
CENSUS PROFILES

Household Composition, 1970 and 2005

The Census Bureau asks a representative sample of the U.S. population about their marital status and also asks those individuals questions about other persons residing in their household. Based on the most recent data, the Census Bureau reports that the percentage distribution of nonfamily and family households has changed substantially during the past three decades. The most noticeable trend is the decline in the number of married-couple households with their own children living with them, which decreased from about 40 percent of all households in 1970 to 23.1 percent in 2005. The distribution shown below reflects this significant change in family structure.

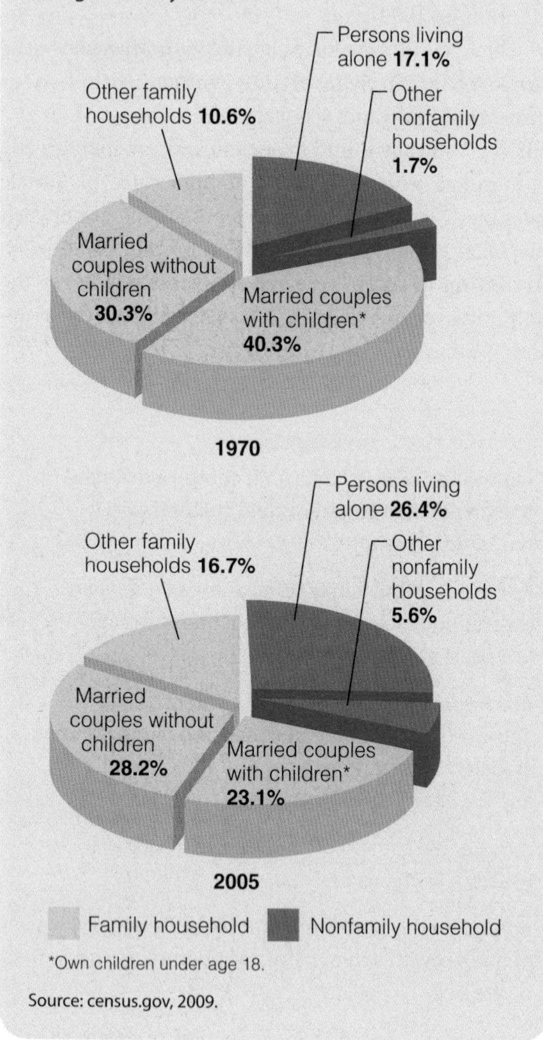

*Own children under age 18.

Source: census.gov, 2009.

Although Tibetans are the most frequently studied population where polyandry exists, anthropologists have also identified the Sherpas, Paharis, Sinhalese, and various African groups as sometimes practicing polyandry (Trevithick, 1997). An anthropological study of the Nyinba, an ethnically Tibetan population living in northwestern Nepal, found that fraternal polyandry is the normative form of marriage and that the practice continues to be highly valued culturally (Levine and Silk, 1997). Despite the fact that some polyandrous marriages fail, societies such as this pass on the practice from one generation to the next (Levine and Silk, 1997).

Patterns of Descent and Inheritance

Even though a variety of marital patterns exist across cultures, virtually all forms of marriage establish a system of descent so that kinship can be determined and inheritance rights established. In preindustrial societies, kinship is usually traced through one parent (unilineally). The most common pattern of unilineal descent is ***patrilineal descent*—a system of tracing descent through the father's side of the family.** Patrilineal systems are set up so that a legitimate son inherits his father's property and sometimes his position upon the father's death. In nations such as India, where boys are seen as permanent patrilineal family members but girls are seen as only temporary family members, girls tend to be considered more expendable than boys (O'Connell, 1994). Recently, some scholars have concluded that cultural and racial nationalism in China is linked to the idea of patrilineal descent being crucial to the modern Chinese national identity (Dikotter, 1996).

Even with the less common pattern of ***matrilineal descent*—a system of tracing descent through the mother's side of the family**—women may not control property. However, inheritance of property and position is usually traced from the maternal uncle (mother's brother) to his nephew (mother's son). In some cases, mothers may pass on their property to daughters.

By contrast, kinship in industrial societies is usually traced through both parents (bilineally). The most common form is ***bilateral descent*—a system of tracing descent through both the mother's and father's sides of the family.** This pattern is used in the United States for the purpose of determining kinship and inheritance rights; however, children typically take the father's last name.

Power and Authority in Families

Descent and inheritance rights are intricately linked with patterns of power and authority in families. The most prevalent forms of familial power and authority are patriarchy, matriarchy, and egalitarianism. A ***patriarchal family* is a family structure in which author-**

Polygamy is the concurrent marriage of a person of one sex with two or more persons of the opposite sex. Although most people in Iran do not practice this pattern of marriage, some men are married to more than one wife.

ity is held by the eldest male (usually the father). The male authority figure acts as head of the household and holds power and authority over the women and children, as well as over other males. A ***matriarchal family*** **is a family structure in which authority is held by the eldest female (usually the mother).** In this case, the female authority figure acts as head of the household.

The most prevalent pattern of power and authority in families is patriarchy. Across cultures, men are the primary (and often sole) decision makers regarding domestic, economic, and social concerns facing the family. The existence of patriarchy may give men a sense of power over their own lives, but it can also create an atmosphere in which some men feel greater freedom to abuse women and children.

An ***egalitarian family*** **is a family structure in which both partners share power and authority equally.** Recently, a trend toward more-egalitarian relationships has been evident in a number of countries as women have sought changes in their legal status and increased educational and employment opportunities. Some degree of economic independence makes it possible for women to delay marriage or to terminate a problematic marriage (O'Connell, 1994). However, one study of the effects of egalitarian values on the allocation and performance of domestic tasks in the family found that changes were relatively slow in coming. According to the study, fathers were more likely to share domestic tasks in nonconventional families where members held more-egalitarian values. Similarly, children's gender-role typing was more closely linked to their parents' egalitarian values and nonconventional lifestyles than to the domestic tasks they were assigned (Weisner, Garnier, and Loucky, 1994).

Residential Patterns

Residential patterns are interrelated with the authority structure and method of tracing descent in families. ***Patrilocal residence*** **refers to the custom of a married couple living in the same household (or community) as the husband's family.** Across cultures, patrilocal residency is most common. One example of contemporary patrilocal residency can be found in al-Barba, a lower-middle-class neighborhood in the Jordanian city of Irbid (McCann, 1997). According to researchers, the high cost of renting an apartment or building a new home has resulted in many sons building their own living quarters onto their parents' home, resulting in multifamily households consisting of an older married couple, their unmarried children, their married sons, and their sons' wives and children.

patrilineal descent a system of tracing descent through the father's side of the family.

matrilineal descent a system of tracing descent through the mother's side of the family.

bilateral descent a system of tracing descent through both the mother's and father's sides of the family.

patriarchal family a family structure in which authority is held by the eldest male (usually the father).

matriarchal family a family structure in which authority is held by the eldest female (usually the mother).

egalitarian family a family structure in which both partners share power and authority equally.

patrilocal residence the custom of a married couple living in the same household (or community) as the husband's family.

Few societies have residential patterns known as ***matrilocal residence*—the custom of a married couple living in the same household (or community) as the wife's parents.** In industrialized nations such as the United States, most couples hope to live in a ***neolocal residence*—the custom of a married couple living in their own residence apart from both the husband's and the wife's parents.**

To this point, we have examined a variety of marriage and family patterns found around the world. Even with the diversity of these patterns, most people's behavior is shaped by cultural rules pertaining to endogamy and exogamy. ***Endogamy* is the practice of marrying within one's own group.** In the United States, for example, most people practice endogamy: They marry people who come from the same social class, racial–ethnic group, religious affiliation, and other categories considered important within their own social group. Social scientists refer to this practice as ***homogamy*—the pattern of individuals marrying those who have similar characteristics, such as race/ethnicity, religious background, age, education, and/or social class.** Homogamy is similar to endogamy; however, issues pertaining to *why* people marry within their own group (endogamy) are somewhat different from the issues associated with why people choose to marry persons with similar characteristics and social status (homogamy). Various reasons have been given to explain why endogamy is so prevalent. One reason may be the proximity of other individuals in one's own group as contrasted with those who are geographically separated from it. Another reason may be that a person's marriage choice is often influenced by the opinions of parents, friends, and other people with whom the person associates (Kalmijn, 1998).

Although endogamy is the strongest marital pattern in the United States, more people now marry outside their own group. ***Exogamy* is the practice of marrying outside one's own social group or category.** Depending on the circumstances, exogamy may not be noticed at all, or it may result in a person being ridiculed or ostracized by other members of the "in" group. The three most important sources of positive or negative sanctions for intermarriage are the family, the church, and the state. Participants in these social institutions may look unfavorably on the marriage of an in-group member to an "outsider" because of the belief that it diminishes social cohesion in the group (Kalmijn, 1998). However, educational attainment is also a strong indicator of marital choices. Higher education emphasizes individual achievement, and college-educated people may be less likely than others to identify themselves with their social or cultural roots and thus more willing to marry outside their own social group or category if their potential partner shares a similar level of educational attainment (Hwang, Saenz, and Aguirre, 1995; Kalmijn, 1998).

Theoretical Perspectives on Families

The ***sociology of family* is the subdiscipline of sociology that attempts to describe and explain patterns of family life and variations in family structure.** Functionalist perspectives emphasize the functions that families perform at the macrolevel of society, whereas conflict and feminist perspectives focus on families as a primary source of social inequality. Symbolic interactionists examine microlevel interactions that are integral to the roles of different family members.

Functionalist Perspectives

Functionalists emphasize the importance of the family in maintaining the stability of society and the well-being of individuals. According to Emile Durkheim, marriage is a microcosmic replica of the larger society; both marriage and society involve a mental and moral fusion of physically distinct individuals (Lehmann, 1994). Durkheim also believed that a division of labor contributes to greater efficiency in all areas of life—including marriages and families—even though he acknowledged that this division imposes significant limitations on some people.

In the United States, Talcott Parsons was a key figure in developing a functionalist model of the family. According to Parsons (1955), the husband/father fulfills the *instrumental role* (meeting the family's economic needs, making important decisions, and providing leadership), whereas the wife/mother fulfills the *expressive role* (running the household, caring for children, and meeting the emotional needs of family members).

Contemporary functionalist perspectives on families derive their foundation from Durkheim. Division of labor makes it possible for families to fulfill a number of functions that no other institution can perform as effectively. In advanced industrial societies, families serve four key functions:

1. *Sexual regulation.* Families are expected to regulate the sexual activity of their members and thus control reproduction so that it occurs within specific boundaries. At the macrolevel, incest taboos prohibit sexual contact or marriage between certain relatives. For example, virtually all societies prohibit sexual relations between parents and their children and between brothers and sisters.

© Thinkstock/Corbis

Functionalist theorists believe that families serve a variety of functions that no other social institution can adequately fulfill. In contrast, conflict and feminist theorists believe that families may be a source of conflict over values, goals, and access to resources and power. Children in upper-class families have many advantages and opportunities that are not available to other children.

2. *Socialization.* Parents and other relatives are responsible for teaching children the necessary knowledge and skills to survive. The smallness and intimacy of families make them best suited for providing children with the initial learning experiences they need.
3. *Economic and psychological support.* Families are responsible for providing economic and psychological support for members. In preindustrial societies, families are economic production units; in industrial societies, the economic security of families is tied to the workplace and to macrolevel economic systems. In recent years, psychological support and emotional security have been increasingly important functions of the family.
4. *Provision of social status.* Families confer social status and reputation on their members. These statuses include the ascribed statuses with which individuals are born, such as race/ethnicity, nationality, social class, and sometimes religious affiliation. One of the most significant and compelling forms of social placement is the family's class position and the opportunities (or lack thereof) resulting from that position. Examples of class-related opportunities include access to quality health care, higher education, and a safe place to live.

Functionalist explanations of family problems examine the relationship between family troubles and a decline in other social institutions. Changes in the economy, in religion, in the educational system, and in the law or government programs can all contribute to family problems.

Conflict and Feminist Perspectives

Conflict and feminist analysts view functionalist perspectives on the role of the family in society as idealized and inadequate. Rather than operating harmoniously and for the benefit of all members, families are sources of social inequality and conflict over values, goals, and access to resources and power.

According to some conflict theorists, families in capitalist economies are similar to workers in a factory. Women are dominated by men in the home in the same manner that workers are dominated by capitalists and managers in factories (Engels, 1970/1884). Although childbearing and care for family members in the home contribute to capitalism, these activities also reinforce the subordination of women through unpaid (and often devalued) labor. Other conflict analysts are concerned with the effect that class conflict has on the family. The exploitation of the lower classes by the upper classes contributes to family problems such as high rates of divorce and overall family instability.

Some feminist perspectives on inequality in families focus on patriarchy rather than class. From this viewpoint, men's domination over women existed long before capitalism and private ownership of property (Mann, 1994). Women's subordination is rooted in patriarchy and men's control over women's labor power (Hartmann, 1981). According to one scholar, "Male

matrilocal residence the custom of a married couple living in the same household (or community) as the wife's parents.

neolocal residence the custom of a married couple living in their own residence apart from both the husband's and the wife's parents.

endogamy cultural norms prescribing that people marry within their social group or category.

homogamy the pattern of individuals marrying those who have similar characteristics, such as race/ethnicity, religious background, age, education, or social class.

exogamy cultural norms prescribing that people marry outside their social group or category.

sociology of family the subdiscipline of sociology that attempts to describe and explain patterns of family life and variations in family structure.

power in our society is expressed in economic terms even if it does not originate in property relations; women's activities in the home have been undervalued at the same time as their labor has been controlled by men" (Mann, 1994: 42). In addition, men have benefited from the privileges they derive from their status as family breadwinners.

Symbolic Interactionist Perspectives

Early symbolic interactionists such as Charles Horton Cooley and George Herbert Mead provided key insights on the roles we play as family members and how we modify or adapt our roles to the expectations of others—especially significant others such as parents, grandparents, siblings, and other relatives. How does the family influence the individual's self-concept and identity? Contemporary symbolic interactionist perspectives examine the roles of husbands, wives, and children as they act out their own part and react to the actions of others. From such a perspective, what people think, as well as what they say and do, is very important in understanding family dynamics.

According to the sociologists Peter Berger and Hansfried Kellner (1964), interaction between marital partners contributes to a shared reality. Although newlyweds bring separate identities to a marriage, over time they construct a shared reality as a couple. In the process, the partners redefine their past identities to be consistent with new realities. Development of a shared reality is a continuous process, taking place not only in the family but in any group in which the couple participates together. Divorce is the reverse of this process; couples may start with a shared reality and, in the process of uncoupling, gradually develop separate realities (Vaughan, 1985).

Symbolic interactionists explain family relationships in terms of the subjective meanings and everyday interpretations that people give to their lives. As the sociologist Jessie Bernard (1982/1973) pointed out, women and men experience marriage differently. Although the husband may see *his* marriage very positively, the wife may feel less positive about *her* marriage, and vice versa. Researchers have found that husbands and wives may give very different accounts of the same event and that their "two realities" frequently do not coincide (Safilios-Rothschild, 1969).

How do symbolic interactionists view problems within the family? Some focus on the terminology used to describe these problems, examining the extent to which words convey assumptions or "realities" about the nature of the problem. For example, violence between men and women in the home is often referred to as "spouse abuse" or "domestic violence." However, these terms imply that women and men play equal roles in perpetrating violence in families, overlooking the more active part that men usually play in such aggression. In addition, the term *domestic violence* suggests that this is the "kind of violence that women volunteer for, or inspire, or provoke" (Jacobs, 1994: 56). Some scholars and activists use terms such as *wife battering* or *wife abuse* to highlight the gendered nature of such behavior (see Bograd, 1988). However, others argue that *battered woman* suggests a "woman who is more or less permanently black and blue and helpless" (Jacobs, 1994: 56).

Other symbolic interactionists have examined ways in which individuals communicate with one another and interpret these interactions. According to Lenore Walker (1979), females are socialized to be passive and males are socialized to be aggressive long before they take on the adult roles of battered and batterer. However, even women who have not been socialized by their parents to be helpless and passive may be socialized into this behavior by abusive husbands. Three factors contribute to the acceptance of the roles of batterer and battered: (1) low self-esteem on the part of both people involved, (2) a limited range of behaviors (he only knows how to be jealous and possessive/she only knows how to be dependent and anxious to make everyone happy), and (3) a belief by both in stereotypic gender roles (she should be feminine and pampered/he should be aggressive and dominant). Other analysts suggest that this pattern is changing as more women are gaining paid employment and becoming less dependent on their husbands or male companions for economic support.

Postmodernist Perspectives

Although postmodern theorists disparage the idea that a universal theory can be developed to explain social life, a postmodernist perspective might provide insights on questions such as this: How is family life different in the "information age"? Social scientist David Elkind (1995) describes the postmodern family as *permeable*—capable of being diffused or invaded in such a manner that an entity's original purpose is modified or changed. According to Elkind (1995), if the nuclear family is a reflection of the age of modernity, the permeable family reflects the postmodern assumptions of difference, particularity, and irregularity. Difference is evident in the fact that the nuclear family is now only one of many family forms. Similarly, the idea of romantic love under modernity has given way to the idea of consensual love: Individuals agree to have sexual relations with others whom they have no intention of marrying or, if they marry, do not necessarily see the marriage as having permanence. Mater-

© Bob Barkany/Getty Images

© Corbis Super RF/Alamy

© Michael Newman/PhotoEdit

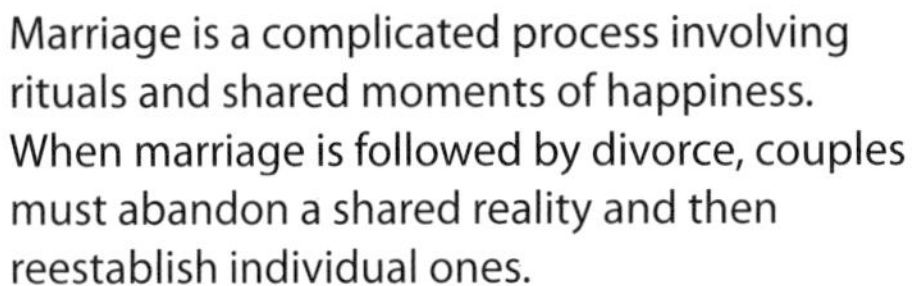
Marriage is a complicated process involving rituals and shared moments of happiness. When marriage is followed by divorce, couples must abandon a shared reality and then reestablish individual ones.

nal love has also been transformed into shared parenting, which includes not only mothers and fathers but also caregivers who may either be relatives or nonrelatives (Elkind, 1995).

Urbanity is another characteristic of the postmodern family. The boundaries between the public sphere (the workplace) and the private sphere (the home) are becoming much more open and flexible. In fact, family life may be negatively affected by the decreasing distinction between what is work time and what is family time. As more people are becoming connected "24/7" (twenty-four hours a day, seven days a week), the boss who would not call at 11:30 P.M. or when an employee is on vacation may send an e-mail asking for an immediate response to some question that has arisen while the person is away with family members (Leonard, 1999). According to some postmodern analysts, this is an example of the "power of the new communications technologies to integrate and control labour despite extensive dispersion and decentralization" (Haraway, 1994: 439).

Social theorist Jean Baudrillard's idea that the simulation of reality may come to be viewed by some people as "reality" can be applied to family interactions in the "Information Age." Does the ability to contact someone anywhere and any time of the day or night provide greater happiness and stability in families? Or is "reach out and touch someone" merely an ideology promulgated by the consumer society? Journalists have written about the experience of watching a family gathering at an amusement park, restaurant, mall, or other location only to see family members pick up their cell phones to receive or make calls to individuals not present, rather than spending "face time" with those family members who are present.

Even as postmodern perspectives call our attention to cyberspace, consumerism, and the hyperreal, it is important to recall that there is a growing "digital divide" and a "new kind of cyber class warfare," as some journalists refer to it, going on in the United States and around the world (Alter, 1999). New economic trends that are making the richest 2.7 million Americans even richer are making the poorest fifth of Americans even poorer. Although in 2000 more than 50 percent of all U.S. households owned computers and 41 percent of all households had Internet access, many families were left out as the gap in Internet access between those at the highest and the lowest income levels increased.

The Concept Quick Review summarizes sociological perspectives on the family. Taken together, these perspectives on the social institution of families reflect various ways in which familial relationships may be viewed in contemporary societies. Now we shift our

CONCEPT QUICK REVIEW

Theoretical Perspectives on Families

	Focus	Key Points	Perspective on Family Problems
Functionalist	Role of families in maintaining stability of society and individuals' well-being.	In modern societies, families serve the functions of sexual regulation, socialization, economic and psychological support, and provision of social status.	Family problems are related to changes in social institutions such as the economy, religion, education, and law/government.
Conflict/Feminist	Families as sources of conflict and social inequality.	Families both mirror and help perpetuate social inequalities based on class and gender.	Family problems reflect social patterns of dominance and subordination.
Symbolic Interactionist	Family dynamics, including communication patterns and the subjective meanings that people assign to events.	Interactions within families create a shared reality.	How family problems are perceived and defined depends on patterns of communication, the meanings that people give to roles and events, and individuals' interpretations of family interactions.
Postmodernist	Permeability of families.	In postmodern societies, families are diverse and fragmented. Boundaries between workplace and home are blurred.	Family problems are related to cyberspace, consumerism, and the hyperreal in an age increasingly characterized by high-tech "haves" and "have-nots."

focus to love, marriage, intimate relationships, and family issues in the United States.

Developing Intimate Relationships and Establishing Families

The United States has been described as a "nation of lovers"; it has been said that we are "in love with love." Why is this so? Perhaps the answer lies in the fact that our ideal culture emphasizes *romantic love*, which refers to a deep emotion, the satisfaction of significant needs, a caring for and acceptance of the person we love, and involvement in an intimate relationship (Lamanna and Riedmann, 2009).

Love and Intimacy

In the late nineteenth century, during the Industrial Revolution, people came to view work and home as separate spheres in which different feelings and emotions were appropriate (Coontz, 1992). The public sphere of work—men's sphere—emphasized self-reliance and independence. By contrast, the private sphere of the home—women's sphere—emphasized the giving of services, the exchange of gifts, and love. Accordingly, love and emotions became the domain of women, and work and rationality became the domain of men (Lamanna and Riedmann, 2009). Although the roles of women and men changed dramatically in the twentieth century, women and men still may not share the same perceptions about romantic love today. According to the sociologist Francesca Cancian (1990), women tend to express their feelings verbally whereas men tend to express their love through nonverbal actions, such as running an errand for someone or repairing a child's broken toy.

Love and intimacy are closely intertwined. Intimacy may be psychic ("the sharing of minds"), sexual, or both. Although sexuality is an integral part of many intimate relationships, perceptions about sexual activities vary from one culture to the next and from one time period to another. For example, kissing is found primarily in Western cultures; many African and Asian cultures view kissing negatively (Reinisch, 1990).

For more than forty years, the work of the biologist Alfred C. Kinsey was considered to be the definitive

In the United States, the notion of romantic love is deeply intertwined with our beliefs about how and why people develop intimate relationships and establish families. Not all societies share this concern with romantic love. However, in this country the number of opportunities for romance is sometimes increased by online matching services, such as this one, which caters to farmers and people who love nature.

research on human sexuality, even though some of his methodology had serious limitations. More recently, the work of Kinsey and his associates has been superseded by the National Health and Social Life Survey conducted by the National Opinion Research Center at the University of Chicago (see Laumann et al., 1994; Michael et al., 1994). Based on interviews with more than 3,400 men and women aged 18 to 59, this random survey tended to reaffirm the significance of the dominant sexual ideologies. Most respondents reported that they engaged in heterosexual relationships, although 9 percent of the men said they had at least one homosexual encounter resulting in orgasm. Although 6.2 percent of men and 4.4 percent of women said that they were at least somewhat attracted to others of the same gender, only 2.8 percent of men and 1.4 percent of women identified themselves as gay or lesbian. According to the study, persons who engaged in extramarital sex found their activities to be more thrilling than those with a marital partner, but they also felt more guilt. Persons in sustained relationships such as marriage or cohabitation found sexual activity to be the most satisfying emotionally and physically.

Cohabitation and Domestic Partnerships

Attitudes about cohabitation have changed in the past four decades. Until recently, the Census Bureau defined cohabitation as the sharing of a household by one man and one woman who are not related to each other by kinship or marriage, but now the Census Bureau is using a more inclusive definition. For our purposes, we will define ***cohabitation* as referring to two people who live together, and think of themselves as a couple, without being legally married.** It is not known how many people actually cohabit because the Census Bureau does not ask about emotional or sexual involvement between unmarried individuals sharing living quarters or between gay and lesbian couples.

Based on Census Bureau data, the people who are most likely to cohabit are under age 45, have been married before, or are older individuals who do not want to lose financial benefits (such as retirement benefits) that are contingent upon not remarrying. Among younger people, employed couples are more likely to cohabit than college students.

For couples who plan to eventually get married, cohabitation somewhat follows the *two-stage marriage* pattern set out by the anthropologist Margaret Mead, who argued that dating patterns in the United States are not adequate preparation for marriage and parenting responsibilities. Instead, Mead suggested that marriage should occur in two stages, each with its own ceremony and responsibilities. In the first stage, the *individual marriage,* two people would make a serious commitment to each other but agree not to have children during this stage. In the second stage, the *parental marriage,* the couple would decide to have children and to share responsibility for the children's upbringing (Lamanna and Riedmann, 2009).

Today, some people view cohabitation as a form of "trial marriage." Some people who have cohabited do eventually marry the person with whom they have been living, whereas others do not. A recent study of 11,000 women found that there was a 70 percent marriage rate for women who remained in a cohabiting relationship for at least 5 years. However, of the women in that study who cohabited and then married their partner, 40 percent became divorced within a 10-year

cohabitation a situation in which two people live together, and think of themselves as a couple, without being legally married.

period (Bramlett and Mosher, 2001). Whether these findings will be supported by subsequent research remains to be seen. But we do know that studies over the past decade have supported the proposition that couples who cohabit before marriage do not necessarily have a stable relationship following marriage (Bumpass, Sweet, and Cherlin, 1991; London, 1991).

Among heterosexual couples, many reasons exist for cohabitation; for gay and lesbian couples, however, no alternative to cohabitation exists in most U.S. states. For that reason, some lesbians and gays seek recognition of their ***domestic partnerships*—household partnerships in which an unmarried couple lives together in a committed, sexually intimate relationship and is granted the same rights and benefits as those accorded to married heterosexual couples** (Aulette, 1994; Gerstel and Gross, 1995). Benefits such as health and life insurance coverage are extremely important to *all* couples, as Gayle, a lesbian, points out: "It makes me angry that [heterosexuals] get insurance benefits and all the privileges, and Frances [her partner] and I take a beating financially. We both pay our insurance policies, but we don't get the discounts that other people get and that's not fair" (qtd. in Sherman, 1992: 197).

Over the past few years, much controversy has arisen over the legal status of gay and lesbian couples, particularly those who seek to make their relationship a legally binding commitment through marriage. Some states have allowed marriage licenses to be issued to same-sex couples only to have voters decide or courts rule that such marriages will not be allowed. Other states, such as Vermont, Iowa, Connecticut, Maine, New Hampshire, and Massachusetts, have legalized same-sex marriages, and similar decisions are pending in several other states. Clearly, opponents of same-sex marriage will continue to press for state and national constitutional amendments to restrict marriage to the union of a man and a woman (see Box 15.2).

Marriage

Why do people get married? Couples get married for a variety of reasons. Some do so because they are "in love," desire companionship and sex, want to have children, feel social pressure, are attempting to escape from a bad situation in their parents' home, or believe that they will have more money or other resources if they get married. These factors notwithstanding, the selection of a marital partner is actually fairly predictable. As previously discussed, most people in the United States tend to choose marriage partners who are similar to themselves. As previously discussed, *homogamy* refers to the pattern of individuals marrying those who have similar characteristics, such as race/ethnicity, religious background, age, education, or social class. However, homogamy provides only the general framework within which people select their partners; people are also influenced by other factors. For example, some researchers claim that people want partners whose personalities match their own in significant ways. Thus, people who are outgoing and friendly may be attracted to other people with those same traits. However, other researchers claim that

© AP Images/Hector Mata

© AP Images/Christopher Gannon

The debate over who should be allowed to get married is extremely divisive. At left are demonstrators against California's Proposition 8, a successful referendum that outlawed gay marriage in California and that was upheld by the California Supreme Court in May 2009. In April 2009, however, the Iowa Supreme Court ruled that gay marriage is legal in that state; at right are some supporters of this decision.

Box 15.2 Sociology and Social Policy

Should the U.S. Constitution Be Amended to Define "Marriage"?

Hello. My name is Michael and I am 15 years old and a sophomore at West Springfield High School. I live . . . with my mother, who is a lesbian. I wanted to thank you for being here because a lot of people have been talking about "gay marriage" and the proposed amendment banning it. All too often, this discussion focuses on abstract ideas, and I think more people . . . need to understand that there are people, *real* people—that this amendment would affect.

The people who wrote this amendment are basically saying that families that have gay and lesbian parents—like mine—aren't real families. Not only are they not real families . . . but the supporters of the amendment want to take away the few rights and protections that our families might obtain through local governments. That makes no sense, and it isn't fair. It isn't right. It isn't what living in America is about. . . .

I am here today because I believe in America and the freedoms with which the country was founded. The Constitution was written by the people, for the people and some of those people are gay. Writing discrimination into the Constitution is simply wrong and goes against the spirit of the document. That's not what America is about. It's about freedom, and I for one, think every family should have the freedom and the right to be a family.

—high school student Michael Cooper stating why he opposes a federal marriage amendment to the U.S. Constitution (Cooper, 2004)

Should two people of the same sex be allowed to marry each other? As a result of concerns that some states might legalize such a marriage, Congress in 1996 enacted the Defense of Marriage Act ("DOMA"), which provides that no state is required to recognize a same-sex marriage conducted under the laws of another state. However, as officials in some states and cities—based on court decisions or their own beliefs—began issuing permits and conducting such marriages, many people started to argue that a constitutional amendment banning same-sex marriages throughout the nation is needed. One proposed federal marriage amendment states that marriage consists only of the union between one man and one woman, and that no other union shall be construed as a marriage.

Proponents of this amendment argued that because the Constitution requires that each state give "full faith and credit" to actions lawfully taken under the laws of another state, the courts might hold that a statute such as DOMA could not override that constitutional requirement and that same-sex partners who became married in a state that permitted such unions would have to be accorded the rights of a married couple in every other state. Some opponents of the amendment—such as Michael—argued that gays and lesbians should have the right to marry each other if they so desire.

However, some opponents of the amendment also opposed same-sex marriages. They argued that the definition of what constitutes a marriage is an issue best left to the people of each state, rather than to the federal government, and that the Constitution should promote individual freedoms rather than imposing moral judgments of what is right and wrong (Barr, 2004).

Article V of the Constitution requires that two-thirds of the Congress and three-fourths of the states must approve any change in order for it to become an amendment. As a result, in the past 200 years only 27 amendments have been adopted out of the more than 14,000 that have been proposed (Rios, 2004). Although it seems unlikely that such an amendment regarding the definition of marriage will be passed, proponents of this measure continue to fight for its passage, particularly as more states legalize same-sex marriages.

Reflect & Analyze

Do you have a traditional view of marriage—that is, between a man and a woman—or do you have an alternative view? In either case, what in your upbringing and experience might have contributed to your opinion?

people look for partners whose personality traits differ from but complement their own.

Regardless of the individual traits of marriage partners, research indicates that communication and emotional support are crucial to the success of marriages. Common marital problems include a lack of emotional intimacy, poor communication, and lack

domestic partnerships household partnerships in which an unmarried couple lives together in a committed, sexually intimate relationship and is granted the same rights and benefits as those accorded to married heterosexual couples.

of companionship. One study concluded that for many middle- and upper-income couples, women's paid work is critical to the success of their marriages. People who have a strong commitment to their work have two distinct sources of pleasure—work and family. For members of the working class, however, work may not be a source of pleasure. For all women and men, balancing work and family life is a challenge.

Housework and Child-Care Responsibilities

Today, over 50 percent of all marriages in the United States are ***dual-earner marriages*—marriages in which both spouses are in the labor force.** More than half of all employed women hold full-time, year-round jobs. Even when their children are very young, most working mothers work full time. For example, in 2005 more than 60 percent of employed mothers with children under age 6 worked full time (U.S. Census Bureau, 2007). Moreover, as Chapter 11 points out, many married women leave their paid employment at the end of the day and then go home to perform hours of housework and child care. Sociologist Arlie Hochschild (1989, 2003) refers to this as the ***second shift*—the domestic work that employed women perform at home after they complete their workday on the job.** Thus, many married women today contribute to the economic well-being of their families and also meet many, if not all, of the domestic needs of family members by cooking, cleaning, shopping, taking care of children, and managing household routines. According to Hochschild, the unpaid housework that women do on the second shift amounts to an extra month of work each year. In households with small children or many children, the amount of housework increases. Across race and class, numerous studies have confirmed that domestic work remains primarily women's work (Gerstel and Gross, 1995). Hochschild (2003: 28) states that continuing problems regarding the second shift in many families are a sign that the gender revolution has stalled:

> The move of masses of women into the paid workforce has constituted a revolution. But the slower shift in ideas of "manhood," the resistance of sharing work at home, the rigid schedules at work make for a "stall" in this gender revolution. It is a stall in the change of institutional arrangements of which men are the principal keepers.

As Hochschild points out, the second shift remains a problem for many women in dual-earner marriages.

In recent years, more husbands have attempted to share some of the household and child-care responsibilities, especially in families in which the wife's earnings are essential to family finances. Overall, when husbands share some of the household responsibilities, they typically spend much less time in these activities than do their wives. Women and men perform different household tasks, and the deadlines for their work vary widely. Recurring tasks that have specific times for completion (such as bathing a child or cooking a meal) tend to be the women's responsibility; by contrast, men are more likely to do the periodic tasks that have no highly structured schedule (such as mowing the lawn or changing the oil in the car) (Hochschild, 1989). Men are also more reluctant to perform undesirable tasks such as scrubbing the toilet or diapering a baby, or to give up leisure pursuits.

Couples with more-egalitarian ideas about women's and men's roles tend to share more equally in food

Juggling housework, child care, and a job in the paid work force is all part of the average day for many women. Why does sociologist Arlie Hochschild believe that many women work a "second shift"?

preparation, housework, and child care (Wright et al., 1992). For some men, the shift to a more-egalitarian household occurs gradually, as Wesley, whose wife works full time, explains:

> It was me taking the initiative, and also Connie pushing, saying, "Gee, there's so much that has to be done." At first I said, "But I'm supposed to be the breadwinner," not realizing she's also the breadwinner. I was being a little blind to what was going on, but I got tired of waiting for my wife to come home to start cooking, so one day I surprised the hell out [of] her and myself and the kids, and I had supper waiting on the table for her. (qtd. in Gerson, 1993: 170)

In the United States, millions of parents rely on child care so that they can work and so that their young children can benefit from early educational experiences that will help in their future school endeavors. For millions more parents, after-school care for school-age children is an urgent concern. Nearly five million children are home alone after school each week in this country. The children need productive and safe activities to engage in while their parents are working. Although child care is often unavailable or unaffordable for many parents, those children who are in day care for extended hours often come to think of child-care workers and other caregivers as members of their extended families because they may spend nearly as many hours with them as they do with their own parents. For children of divorced parents and other young people living in single-parent households, the issue of child care is often a pressing concern because of the limited number of available adults and lack of financial resources.

Child-Related Family Issues and Parenting

Not all couples become parents. Those who decide not to have children often consider themselves to be "child-free," whereas those who do not produce children through no choice of their own may consider themselves "childless."

Deciding to Have Children

Cultural attitudes about having children and about the ideal family size began to change in the United States in the late 1950s. Women, on average, are now having 2.1 children each (see "Sociology Works!"). However, rates of fertility differ across racial and ethnic categories. In 2006, for example, Latinas (Hispanic women) had a total fertility rate of 2.9, which was 50 percent above that of white (non-Hispanic) women (National Center for Health Statistics, 2007). Among Latinas, the highest rate of fertility was found among Mexican American women, whereas Puerto Rican and Cuban American women had relatively lower rates.

Advances in birth control techniques over the past four decades—including the birth control pill and contraceptive patches and shots—now make it possible for people to decide whether or not they want to have children, how many they wish to have, and to determine (at least somewhat) the spacing of their births. However, sociologists suggest that fertility is linked not only to reproductive technologies but also to women's beliefs that they do or do not have other opportunities in society that are viable alternatives to childbearing (Lamanna and Riedmann, 2009).

Today, the concept of reproductive freedom includes both the desire *to have* or *not to have* one or more children. According to the sociologists Leslie King and Madonna Harrington Meyer (1997), many U.S. women spend up to one-half of their life attempting to control their reproductivity. Other analysts have found that women, more often than men, are the first to choose a child-free lifestyle (Seccombe, 1991). However, the desire not to have children often comes in conflict with our society's *pronatalist bias,* which assumes that having children is the norm and can be taken for granted, whereas those who choose not to have children believe they must justify their decision to others (Lamanna and Riedmann, 2009).

However, some couples experience the condition of *involuntary infertility,* whereby they want to have a child but find that they are physically unable to do so. *Infertility* is defined as an inability to conceive after a year of unprotected sexual relations. Today, infertility affects nearly five million U.S. couples, or one in twelve couples in which the wife is between the ages of fifteen and forty-four (Gabriel, 1996). Research suggests that fertility problems originate in females in approximately 30–40 percent of the cases and with males in about 40 percent of the cases; in the other 20 percent of the cases, the cause is impossible to determine (Gabriel, 1996). A leading cause of infertility is sexually transmitted diseases, especially those cases

dual-earner marriages marriages in which both spouses are in the labor force.

second shift Arlie Hochschild's term for the domestic work that employed women perform at home after they complete their workday on the job.

Sociology *Works!*

Social Factors Influencing Parenting: From the Housing Market to the Baby Nursery

> One reason there are so few children in Italy is that housing is so hard to come by. Houses are bigger in the U.S. and generally more available. That may help explain why Americans have more babies.
>
> —Robert Engelman, vice president for programs at the Worldwatch Institute, an environmental research organization, and author of *More: Population, Nature, and What Women Want* (qtd. in Leland, 2008: A12)

> Social scientists have long traced a connection between housing and fertility. When homes are scarce or beyond the means of young couples, as in the 1930s, couples delay marriage or have fewer children. This tendency helps account for the relatively dismal birth rates of many developed nations. . . . (Leland, 2008: A12)

For many years, demographers and other sociologists who specialize in population trends have sought to identify how biological and social factors affect fertility rates in various nations. Although biological factors such as general health and levels of nutrition in a region clearly affect the number of children a couple may produce, social factors are also important in determining the estimated number of children a woman will have in her lifetime (see Chapter 19 for additional information). A key social factor is the housing market where a couple lives: The ability to buy a house and having a relatively large home may influence a couple's decision about how many children to have. Early in this decade the housing market in some areas of the United States provided an opportune time for more young couples to purchase their own home. Mortgages were readily available, and it typically did not take much cash (in the way of a down payment) up front to sign the contract and move into one's dream home. The availability of such loans to people who really couldn't afford the home they were buying has since come back to haunt many underfunded home buyers. By 2006, however, the babies were arriving, and the fertility rate in the United States had grown to an estimated 2.1 children for every woman of childbearing age, reaching the highest level since the 1970s (Leland, 2008).

Sociological insights on the social aspects of fertility, which at first might appear to be primarily a biological phenomenon, have provided us with new information on why people decide to have children and how many children they might have. However, much remains unknown about the relationship between the housing market and the maternity ward, including how income and feelings of optimism or pessimism about the local and national economy might affect a couple's decisions regarding parenting.

Will the baby boomlet of 2006 continue in the future? According to some social analysts, the boomlet may be short-lived because of a major downturn in the economy and the crisis in the housing market, which has brought about many foreclosures—factors that often discourage couples from having children or producing larger families.

Reflect & Analyze

How might sociological findings about factors that influence a couple's decision to have children be useful in your community? For example, why is information about the availability of housing and local fertility trends important to school board members and administrators when they are making enrollment projections or deciding where to build a new school in the future?

that develop into pelvic inflammatory disease (Gold and Richards, 1994). It is estimated that about half of infertile couples who seek treatments such as fertility drugs, artificial insemination, and surgery to unblock fallopian tubes can be helped; however, some are unable to conceive despite expensive treatments such as *in vitro fertilization,* which costs as much as $11,000 per attempt (Gabriel, 1996). According to the sociologist Charlene Miall (1986), women who are involuntarily childless engage in "information management" to combat the social stigma associated with childlessness. Their tactics range from avoiding people who make them uncomfortable to revealing their infertility so that others will not think of them as "selfish" for being childless. Some people who are involuntarily childless may choose surrogacy or adoption as an alternative way of becoming a parent (see Box 15.3).

Adoption

Adoption is a legal process through which the rights and duties of parenting are transferred from a child's biological and/or legal parents to new legal parents. This procedure gives the adopted child all the rights of a biological child. In most adoptions, a new birth certificate is issued, and the child has no future contact

© James Devaney/WireImage/Getty Images

Megastars Angelina Jolie and Brad Pitt have produced their own children and adopted others, sometimes in controversial situations.

with the biological parents; however, some states have "right-to-know" laws under which adoptive parents must grant the biological parents visitation rights.

Matching children who are available for adoption with prospective adoptive parents can be difficult. The available children have specific needs, and the prospective parents often set specifications on the type of child they want to adopt. Some adoptions are by relatives of the child; others are by infertile couples who cannot produce a child of their own (although many fertile couples also adopt). Although thousands of children are available for adoption each year in the United States, many prospective parents seek out children in developing nations such as Romania, South Korea, and India. The primary reason is that the available children in the United States are thought to be "unsuitable." They may have disabilities, or they may be sick, nonwhite (most of the prospective parents are white), or too old (Zelizer, 1985). In addition, fewer infants are available for adoption today than in the past because better means of contraception exist, abortion is more readily available, and more unmarried teenage parents decide to keep their babies.

Ironically, although many couples who would like to have a child are unable to do so, other couples conceive a child without conscious intent. Consider the fact that for the approximately 6.4 million women who become pregnant each year in the United States, about 2.8 million (44 percent) pregnancies are intended whereas about 3.6 million (56 percent) are unintended (Gold and Richards, 1994). As with women and motherhood, some men feel that their fatherhood was planned; others feel that it was thrust upon them (Gerson, 1993). Unplanned pregnancies usually result from failure to use contraceptives or from using contraceptives that do not work. Even with "planned pregnancies," it is difficult to plan exactly when conception will occur and a child will be born.

Teenage Pregnancies

Teenage pregnancies are a popular topic in the media and political discourse, and the United States has the highest rate of teen pregnancy in the Western industrialized world (National Campaign to Prevent Teen Pregnancy, 1997). In 2006 the total number of live births per 1,000 women aged 15 to 17 was 22.0, and for women who were 18 and 19 the number was 73.0, in both instances the first increase since 1991 (National Center for Health Statistics, 2007).

What are the primary reasons for the high rates of teenage pregnancy? At the microlevel, several issues are most important: (1) many sexually active teenagers do not use contraceptives; (2) teenagers—especially those from some low-income families and/or subordinate racial and ethnic groups—may receive little accurate information about the use of, and problems associated with, contraception; (3) some teenage males (due to a double standard based on the myth that sexual promiscuity is acceptable among males but not females) believe that females should be responsible for contraception; and (4) some teenagers view pregnancy as a sign of male prowess or as a way to gain adult status. At the macrolevel, structural factors also contribute to teenage pregnancy rates. Lack of education and employment opportunities in some central-city and rural areas may discourage young people's thoughts of upward mobility that might make early parenting

© Fox Searchlight/Photofest

In the 2007 film *Juno*, the title character, played by Ellen Page, becomes pregnant while still in high school. Although this film is a comedy, it takes a surprisingly straightforward look at the reality of teenage pregnancy.

Box 15.3 Sociology in Global Perspective

Wombs-for-Rent: Outsourcing Births to India

Picture four people—three adults and one infant—as they might be shown on *CBS News:* One person in the photo is Karen Kim, a lovely young woman from California, who is cuddling her infant son, Brady. Another person is Karen's husband, Thomas, who lovingly looks on at the mother and child. Brady, the Kims' new son, is the third person in the photo, but who is the fourth person—a woman with long black hair, a red dress, and pearl earrings? Her name is Dr. Nayna Patel, and she is the physician who made it possible for the Kims to become parents because she runs Akanksha Fertility Clinic in Anand, India, where surrogate mothers give birth so that infertile couples such as the Kims can have children. (*CBS News,* 2007b)

The notion of extracting resources from the Third World in order to enrich the First World is hardly new. It harks back to imperialism in its most literal form: The nineteenth-century extraction of gold, ivory, and rubber from the Third World. . . . Today, as love and care become the "new gold," the female part of the story has grown in prominence.

—Sociologist Arlie Russell Hochschild (2003: 194) describes what she believes is happening as young women in low-income nations such as the Philippines leave their own children behind to work abroad for long periods of time, taking care of other people's children and households; however, these words can certainly also be applied to nations such as India, where women are serving as surrogate mothers for infertile couples in the United States, Great Britain, Taiwan, and beyond. (qtd. in Dolnick, 2008)

The pregnant women at Akanksha Fertility Clinic (where the Kims' son was born) are professional surrogate mothers. Each surrogate must have at least one child of her own before she is allowed to become a surrogate mother. The clinic established this rule based on the assumption that having a child shows potential clients that the surrogate can successfully carry and deliver a baby and that she has other children at home to love and will not be resistant to giving up a newborn that she gestated and birthed (Kohl, 2007).

Why do some infertile couples in the United States, Britain, and elsewhere want to "hire" a woman in India to have their child? Most couples that engage in this practice have made numerous attempts to have a child through in vitro fertilization and other assisted reproductive technologies. If they have been unsuccessful in their efforts, the couple may first attempt to find a surrogate in the United States, but they quickly learn that a gestational surrogate costs more than $50,000, whereas the cost of an Indian surrogate ranges from about $2,250 to $5,000 plus medical expenses (Kohl, 2007; United Press International, 2007). According to Dr. Patel, earning money through surrogacy helps uplift Indian women: It provides money for their household and makes them more independent. For example, the typical woman might earn more for one surrogate

appear less appealing. Likewise, religious and political opposition has resulted in issues relating to reproductive responsibility not being dealt with as openly in the United States as in some other nations. Finally, advertising, films, television programming, magazines, music, and other forms of media often flaunt the idea of being sexually active without showing the possible consequences of such behavior.

Teen pregnancies have been of concern to analysts who suggest that teenage mothers may be less skilled at parenting, are less likely to complete high school than their counterparts without children, and possess few economic and social supports other than their relatives (Maynard, 1996; Moore, Driscoll, and Lindberg, 1998). In addition, the increase in births among unmarried teenagers may have negative long-term consequences for mothers and their children, who could have severely limited educational and employment opportunities and a high likelihood of living in poverty. Moreover, the Children's Defense Fund estimates that among those who first gave birth between the ages of fifteen and nineteen, 43 percent will have a second child within three years.

Teenage fathers have largely been left out of the picture. According to the sociologist Brian Robinson (1988), a number of myths exist regarding teenage fathers: (1) they are worldly wise "superstuds" who engage in sexual activity early and often, (2) they are "Don Juans" who sexually exploit unsuspecting females, (3) they have "macho" tendencies because they are psychologically inadequate and need to prove their masculinity, (4) they have few emotional feelings for the women they impregnate, and (5) they are "phantom fathers" who are rarely involved in caring for and rearing their children. However, these assumptions overlook the fact that some teenage males try to be good fathers.

© AP Images/Ajit Solanki

A surrogate mother (left) has delivered a baby for Karen Kim (center), with the help of infertility specialist Dr. Nayna Patel (right). This practice, sometimes called "rent-a-womb," remains controversial.

pregnancy than she would earn in fifteen years from other kinds of employment (*CBS News*, 2007b).

Is there any problem with global "rent-a-womb"? If there is an agreement between a surrogate mother and a couple who badly wants a child, some analysts believe that "offer and acceptance" is nothing more than capitalism at work—where there is a demand (for infants by infertile couples), there will be a supply (from low-income surrogate mothers). However, some ethicists raise troubling questions about the practice of commercial surrogacy: A mother should give birth to her child because it is hers and she loves it, not because she is being paid to give birth to someone else's baby. Other social critics are concerned about the potential mistreatment of low-income women who may be exploited or may suffer long-term emotional damage from functioning as a surrogate (Dunbar, 2007). For the time being, in clinics such as the one in Anand, India, hopeful parents just provide the egg, the sperm, and the money, and all the rest is done for them by the clinic and the surrogates, who live in a spacious house where they are taken care of by maids, cooks, and doctors who want them to remain healthy and happy throughout their pregnancies—after all, it's good business!

Reflect & Analyze

What are your thoughts on surrogacy? Is there any difference between surrogacy when it occurs in high-income nations such as the United States and Britain as compared to situations in which the parents live in a high-income nation and the surrogate mother lives in a lower-income nation? How might we relate the specific issue of outsourced surrogacy to some larger concerns about families and intimate relationships that we have discussed in this chapter?

Single-Parent Households

In recent years, there has been a significant increase in single- or one-parent households due to divorce and to births outside of marriage. Even for a person with a stable income and a network of friends and family to help with child care, raising a child alone can be an emotional and financial burden. Single-parent households headed by women have been stereotyped by some journalists, politicians, and analysts as being problematic for children. About 42 percent of all white children and 86 percent of all African American children will spend part of their childhood living in a household headed by a single mother who is divorced, separated, never married, or widowed (Garfinkel and McLanahan, 1986). According to sociologists Sara McLanahan and Karen Booth (1991), children from mother-only families are more likely than children in two-parent families to have poor academic achievement, higher school absentee and dropout rates, early marriage and parenthood, higher rates of divorce, and more drug and alcohol abuse. Does living in a one-parent family *cause* all of this? Certainly not! Many other factors—including poverty, discrimination, unsafe neighborhoods, and high crime rates—contribute to these problems.

Lesbian mothers and gay fathers are counted in some studies as single parents; however, they often share parenting responsibilities with a same-sex partner. Due to homophobia (hatred and fear of homosexuals and lesbians), lesbian mothers and gay fathers are more likely to lose custody to a heterosexual parent in divorce disputes (Falk, 1989; Robson, 1992). In any case, between one million and three million gay men in the United States and Canada are fathers. Some gay men are married natural fathers, others are single gay men, and still others are part of gay couples who have adopted children. Very little research exists on gay fathers; what research does exist tends to show that noncustodial gay fathers try to maintain good relationships with their children (Bozett, 1988).

Single fathers who do not have custody of their children may play a relatively limited role in the lives of those children. Although many remain actively involved in their children's lives, others may become "Disneyland daddies" who take their children to recreational activities and buy them presents for special occasions but have a very small part in their children's day-to-day life. Sometimes, this limited role is by choice, but more often it is caused by workplace demands on time and energy, location of the ex-wife's residence, and limitations placed on visitation by custody arrangements.

Two-Parent Households

In recent decades, the percentage of children living in two-parent households has dropped, while the percentage living with a single parent has increased (see ▶ Figure 15.1). In 2005, the latest year for which comprehensive statistics are available, 67 percent of children lived with two parents, while 23 percent lived with only their mother and 5 percent lived with only their father. In computing these statistics, parents include not only biological parents but also stepparents who adopt their children. However, foster parents are considered nonrelatives.

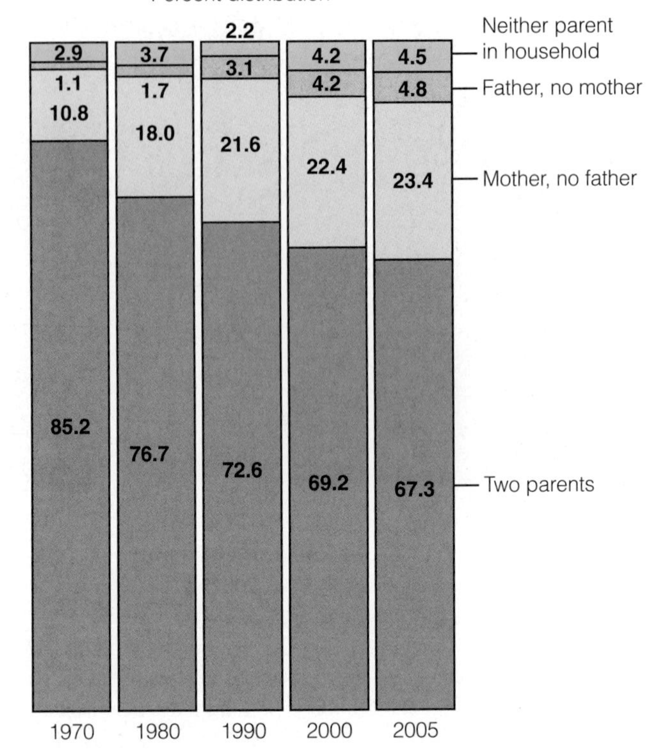

Note: Numbers may not add to 100 percent due to rounding.
Source: U.S. Census Bureau, 2006.

▶ Figure 15.1 Living Arrangements of Children for Selected Years: 1970 to 2005

Parenthood in the United States is idealized, especially for women. According to the sociologist Alice Rossi (1992), maternity is the mark of adulthood for women, whether or not they are employed. By contrast, men secure their status as adults by their employment and other activities outside the family (Hoffnung, 1995).

For families in which a couple truly shares parenting, children have two primary caregivers. Some parents share parenting responsibilities by choice; others share out of necessity because both hold full-time jobs. Some studies have found that men's taking an active part in raising the children is beneficial not only for mothers (who then have a little more time for other activities) but also for the men and the children. The men benefit through increased access to children and greater opportunity to be nurturing parents (Coltrane, 1989).

Transitions and Problems in Families

Families go through many transitions and experience a wide variety of problems, ranging from high rates of divorce and teen pregnancy to domestic abuse and family violence. These all-too-common experiences highlight two important facts about families: (1) for good or ill, families are central to our existence, and (2) the reality of family life is far more complicated than the idealized image of families found in the media and in many political discussions. Although some families provide their members with love, warmth, and satisfying emotional experiences, other families may be hazardous to the individual's physical and mental well-being. Because of this dichotomy in family life, sociologists have described families as both a "haven in a heartless world" (Lasch, 1977) and a "cradle of violence" (Gelles and Straus, 1988).

Family Violence

Violence between men and women in the home is often called spouse abuse or domestic violence. *Spouse abuse* refers to any intentional act or series of acts—whether physical, emotional, or sexual—that causes injury to a female or male spouse (Wallace, 2002). According to sociologists, the term *spouse abuse* refers not only to people who are married but also to those who are cohabiting or involved in a serious relationship, as well as those individuals who are separated or living apart from their former spouse (Wallace, 2002).

How much do we know about family violence? Women, as compared with men, are more likely to be the victim of violence perpetrated by intimate partners. Recent statistics indicate that women are five times more likely than men to experience such violence and that many of these women live in households with children younger than twelve. However, we cannot know the true extent of family violence because much of it is not reported to police. For example, it is estimated that only about one-half of the intimate-partner violence against women was reported to police in the 1990s (U.S. Department of Justice, 2000). African American women were more likely than other women to report such violence, which may further skew data about who is most likely to be victimized by a domestic partner (U.S. Department of Justice, 2000).

Although everyone in a household where family violence occurs is harmed psychologically, whether or not they are the victims of violence, children are especially affected by household violence. It is estimated that between three million and ten million children witness some form of domestic violence in their homes each year, and there is evidence to suggest that domestic violence and child maltreatment often take place in the same household (Children's Defense Fund, 2002). According to some experts, domestic violence is an important indicator that child abuse and neglect are also taking place in the household.

In some situations, family violence can be reduced or eliminated through counseling, the removal of one parent from the household, or other steps that are taken either by the family or by social service agencies or law enforcement officials. However, children who witness violence in the home may display certain emotional and behavioral problems that adversely affect their school life and communication with other people. In some families, the problems of family violence are great enough that the children are removed from the household and placed in foster care.

Children in Foster Care

Not all of the children in foster care have come from violent homes, but many foster children have been in dysfunctional homes where parents or other relatives lacked the ability to meet the children's daily needs. *Foster care* refers to institutional settings or residences where adults other than a child's own parents or biological relatives serve as caregivers. States provide financial aid to foster parents, and the intent of such programs is that the children will either return to their own families or be adopted by other families. However, this is often not the case for "difficult to place" children, particularly those who are over ten years of age, have illnesses or disabilities, or are perceived as suffering from "behavioral problems." More than 568,000 children are in foster care at any given time (Barovick, 2001). About 60 percent of children in foster care are children of color, with about 42 percent of them being African American—almost three times the percentage of African American children in the total U.S. child population (Children's Defense Fund, 2002). Even when the number of children entering foster care for the first time remains relatively stable, the total number of children in foster care continues to increase because fewer children are leaving foster care and being adopted or placed in permanent homes (Children's Defense Fund, 2002). Many children in foster care have limited prospects for finding a permanent home; however, a few innovative programs offer hope for children who previously had been moved from one foster care setting to another (see Box 15.4).

Problems in the family contribute to the large numbers of children who are in foster care. Such factors include parents' illness, unemployment, or death; violence or abuse in the family; and high rates of divorce.

Divorce

Divorce is the legal process of dissolving a marriage that allows former spouses to remarry if they so choose. Most divorces today are granted on the grounds of *irreconcilable differences,* meaning that there has been a breakdown of the marital relationship for which neither partner is specifically blamed. Prior to the passage of more-lenient divorce laws, many states required that the partner seeking the divorce prove misconduct on the part of the other spouse. Under *no-fault divorce laws,* however, proof of "blameworthiness" is generally no longer necessary.

Over the past 100 years, the U.S. divorce rate (number of divorces per 1,000 population) has varied from a low of 0.7 in 1900 to an all-time high of 5.3 in 1981; by 2004, it had decreased to 3.7 (U.S. Census Bureau, 2007). Although many people believe that marriage should last for a lifetime, others believe that marriage is a commitment that may change over time.

Studies have shown that 43 percent of first marriages end in separation or divorce within 15 years (National Centers for Disease Control, 2001). (▶ Figure 15.2 shows U.S. divorce rates for each state.) Many first marriages do not last even fifteen years: One in three first marriages ends within ten years, and one in five ends within five years. The likelihood of divorce goes up with each subsequent marriage in the serial monogamy pattern. When divorces that terminate second or subsequent marriages are taken into account, there are about half as many divorces each year in the United States as there are marriages. These data concern researchers in agencies such as the Centers for

	Rates per 1,000 population[a, b]			
	1990	2000	2004	2005
United States[c]	**4.7**	**4.2**	**3.7**	**3.6**
Alabama	6.1	5.4	4.7	4.9
Alaska	5.5	4.4	4.8	5.8
Arizona	6.9	4.4	4.2	4.1
Arkansas	6.9	6.9	6.3	6.0
California	4.3	n.a.	n.a.	n.a.
Colorado	5.5	n.a.	4.4	4.4
Connecticut	3.2	2.0	2.9	2.7
Delaware	4.4	4.2	3.7	3.9
District of Columbia	4.5	3.0	1.7	2.0
Florida	6.3	5.3	4.8	4.6
Georgia	5.5	3.9	n.a.	n.a.
Hawaii	4.6	3.9	n.a.	n.a.
Idaho	6.5	5.4	5.1	4.9
Illinois	3.8	3.2	2.6	2.5
Indiana	n.a.	n.a.	n.a.	n.a.
Iowa	3.9	3.3	2.8	2.7
Kansas	5.0	4.0	3.3	3.1
Kentucky	5.8	5.4	4.9	4.5
Louisiana	n.a.	n.a.	n.a.	n.a.
Maine	4.3	4.6	3.6	3.5
Maryland	3.4	3.3	3.1	3.1
Massachusetts	2.8	3.0	2.2	2.2
Michigan	4.3	4.0	3.5	3.4
Minnesota	3.5	3.1	2.8	n.a.
Mississippi	5.5	5.2	4.5	4.5

	Rates per 1,000 population[a, b]			
	1990	2000	2004	2005
Missouri	5.1	4.8	3.8	3.6
Montana	5.1	2.4	3.8	3.8
Nebraska	4.0	3.8	3.6	3.4
Nevada	11.4	9.6	6.4	7.7
New Hampshire	4.7	5.8	3.9	3.3
New Jersey	3.0	3.1	3.0	2.9
New Mexico	4.9	5.3	4.6	4.6
New York	3.2	3.4	3.0	2.8
North Carolina	5.1	4.8	4.4	3.8
North Dakota	3.6	3.2	2.8	2.4
Ohio	4.7	4.4	3.7	3.6
Oklahoma	7.7	3.7	n.a.	5.6
Oregon	5.5	5.0	4.1	4.3
Pennsylvania	3.3	3.2	2.5	2.3
Rhode Island	3.7	3.1	3.0	2.9
South Carolina	4.5	3.7	3.2	2.9
South Dakota	3.7	3.6	3.2	3.0
Tennessee	6.5	6.1	5.0	4.6
Texas	5.5	4.2	3.6	3.2
Utah	5.1	4.5	3.9	4.0
Vermont	4.5	8.6	3.9	3.3
Virginia	4.4	4.3	4.0	3.9
Washington	5.9	4.7	4.1	4.0
West Virginia	5.3	5.2	4.7	5.1
Wisconsin	3.6	3.3	3.1	3.0
Wyoming	6.6	5.9	5.3	5.3

▶ **Figure 15.2 U.S. Divorce Rates by State, 1990–2005**

[a]Based on total population residing in area; population enumerated as of April 1 for 1990: estimated as of July 1 for all other years.
[b]Includes annulments.
[c]U.S. totals for the number of divorces are an estimate that includes states not reporting (California, Colorado, Indiana, and Louisiana).
Source: U.S. Census Bureau, 2008.

Disease Control because separation and divorce often have adverse effects on the health and well-being of both children and adults (National Centers for Disease Control, 2001).

Causes of Divorce Why do divorces occur? As you will recall from Chapter 2, sociologists look for correlations (relationships between two variables) in attempting to answer questions such as this. Existing research has identified a number of factors at both the macrolevel and microlevel that make some couples more or less likely to divorce. At the macrolevel, societal factors contributing to higher rates of divorce include changes in social institutions, such as religion and family. Some religions have taken a more lenient attitude toward divorce, and the social stigma associated with divorce has lessened. Further, as we have seen in this chapter, the family institution has undergone a

Box 15.4 You Can Make a Difference

Providing Hope and Help for Children

> I take it personally when I see kids mistreated. I just think they need an advocate to fight for them. . . . For me, it's very simple: The kids' needs come first. That's the bottom line at Hope Meadows. We make decisions as if these are our own children, and when you think that way, your decisions are different than if you are just trying to work within a bureaucratic system.
>
> —sociologist Brenda Eheart, describing why she founded Hope Meadows (qtd. in Smith, 2001: 22)

After five years of research into the adoptions of older children, the sociologist Brenda Eheart realized that foster families faced many problems when they tried to help children who had been moved from home to home. Thinking that she might be able to make a difference, Eheart developed the plan for Hope Meadows, a community established in 1994 on an abandoned Air Force base in Illinois. Hope Meadows is made up of a three-block-long series of ranch houses that provide multigenerational and multiracial housing for foster children, their temporary families, and older adults who live and work with the children. Older adults who interact with the children receive reduced rent in exchange for at least six hours per week of volunteer work with the children. Foster families that reside at Hope Meadows gain a feeling of community as they work together to help children who have experienced severe abuse or neglect, have been exposed to drugs and numerous foster homes, and often have physical, emotional, and behavioral problems.

Since its commencement, Hope Meadows has been largely successful in helping children to get adopted. However, children are not the only beneficiaries of this community: Older residents gain the benefit of interacting with children and feeling that they can *make a positive contribution* to the lives of others (Barovick, 2001). Debbie Calhoun, a foster parent at Hope Meadows, has suggested things that children need the most when they come there, and we can make a difference by providing the children in our lives with these same things (based on Smith, 2001):

- *Understanding.* We need to gain an awareness of how children feel and why they say and do certain things.
- *Trust.* We need to help children to see us as people they can rely on and believe in.
- *Love.* We must show children that they are loved and that they will still be loved even when they make mistakes.
- *Compassion.* We must show children compassion because they must experience compassion in order to be able to show it to others.
- *Time.* We must give children time to be a part of our lives, and we must also give them time to adjust and to start over when they need to do so.
- *Security.* We must help children to feel secure in their surroundings and to believe that there is stability or permanency in their living arrangements.
- *Praise.* We must tell children when they are doing well and not always be critical of them.
- *Discipline.* We must let children know what behavior is acceptable and what behavior is not, all the while showing them that we love them, even when discipline is necessary.
- *Self-Esteem.* We must help children feel good about themselves.
- *Pride.* We should provide opportunities for children to learn to take pride in their accomplishments and in themselves.

If these suggestions are beneficial for children in foster care settings, then they are certainly useful ideas for each of us to implement in our own families and communities, as well. What other ideas would you add to the list? Why?

---

major change that has resulted in less economic and emotional dependency among family members—and thus reduced a barrier to divorce.

At the microlevel, a number of factors contribute to a couple's "statistical" likelihood of divorcing. Here are some of the primary social characteristics of those most likely to get divorced:

- Marriage at an early age (59 percent of marriages to brides under 18 end in separation or divorce within 15 years) (National Centers for Disease Control, 2001)
- A short acquaintanceship before marriage
- Disapproval of the marriage by relatives and friends
- Limited economic resources and low wages
- A high school education or less (although deferring marriage to attend college may be more of a factor than education per se)
- Parents who are divorced or have unhappy marriages
- The presence of children (depending on their gender and age) at the beginning of the marriage

The interrelationship of these and other factors is complicated. For example, the effect of age is intertwined with economic resources; persons from families at the low end of the income scale tend to marry earlier than those at more affluent income levels. Thus, the question becomes whether age itself is a factor or whether economic resources are more closely associated with divorce.

The relationship between divorce and factors such as race, class, and religion is another complex issue. Although African Americans are more likely than whites of European ancestry to get a divorce, other factors—such as income level and discrimination in society—must also be taken into account. Latinos/as share some of the problems faced by African Americans, but their divorce rate is only slightly higher than that of whites of European ancestry. As the sociologist Demie Kurz (1995: 21) notes,

> A person's socioeconomic position has a strong influence on the likelihood that they will divorce. Despite the stereotype that divorce is a middle-class phenomenon, and despite increases in the divorce rate in all socioeconomic levels, divorce rates have always been higher among lower-income people, and among those with less education. These patterns are related to higher rates of unemployment and job insecurity among workers from lower socioeconomic backgrounds.

Religion may affect the divorce rate of some people, including many Latinos/as who are Roman Catholic. However, despite the Catholic doctrine that discourages divorce, the rate of Catholic divorces is now approximately equal to that of Protestant divorces.

Consequences of Divorce Divorce may have a dramatic economic and emotional impact on family members. An estimated 60 percent of divorcing couples have one or more children. By age 16, about one out of every three white children and two out of every three African American children will experience divorce within their families. As a result, most of them will remain with their mothers and live in a single-parent household for a period of time (Thornton and Freedman, 1983). In recent years, there has been a debate over whether children who live with their same-sex parent after divorce are better off than their peers who live with an opposite-sex parent. However, sociologists have found virtually no evidence to support the belief that children are better off living with a same-sex parent (Powell and Downey, 1997).

Although divorce decrees provide for parental joint custody of approximately 100,000–200,000 children annually, this arrangement may create unique problems for some children, as the personal narratives of David and Nick Sheff at the beginning of this chapter demonstrate. Furthermore, some children experience more than one divorce during their childhood because one or both of their parents may remarry and subsequently divorce again.

But divorce does not have to be always negative. For some people, divorce may be an opportunity to terminate destructive relationships (Lund, 1990; Kurz, 1995). For others, it may represent a means to achieve personal growth by managing their lives and social relationships and establishing their own social identity (see Reissman, 1991). Still others choose to remarry one or more times.

Remarriage

Most people who divorce get remarried. In recent years, more than 40 percent of all marriages were between previously married brides and/or grooms. Among individuals who divorce before age thirty-five, about half will remarry within three years of their first divorce (Bramlett and Mosher, 2001). Most divorced people remarry others who have been divorced. However, remarriage rates vary by gender and age. At all ages, a greater proportion of men than women remarry, often relatively soon after the divorce. Among women, the older a woman is at the time of divorce, the lower her likelihood of remarrying. Women who have not graduated from high school and who have young children tend to remarry relatively quickly; by contrast, women with a college degree and without children are less likely to remarry.

As a result of divorce and remarriage, complex family relationships are often created. Some people become part of stepfamilies or *blended families,* which consist of a husband and wife, children from previous marriages, and children (if any) from the new marriage. At least initially, levels of family stress may be fairly high because of rivalry among the children and hostilities directed toward stepparents or babies born into the family. In spite of these problems, however, many blended families succeed. The family that results from divorce and remarriage is typically a complex, binuclear family in which children may have a biological parent and a stepparent, biological siblings and stepsiblings, and an array of other relatives, including aunts, uncles, and cousins.

According to the sociologist Andrew Cherlin (1992), the norms governing divorce and remarriage are ambiguous. Because there are no clear-cut guidelines, people must make decisions about family life (such as whom to invite for a birthday celebration or wedding) based on their beliefs and feelings about the people involved.

Diversity in Families

Although marriage at increasingly younger ages was the trend in the United States during the first half of the twentieth century, by the 1960s the trend had reversed, and many more adults were remaining single.

Remarriage and blended families create new opportunities and challenges for parents and children alike.

Currently, almost 80 million single adults reside in the United States (U.S. Census Bureau, 2007). Some people remain single by choice, whereas others are single because they are not in the age brackets most likely to be married.

Diversity Among Singles

Never-married singles who remain single by choice may be more interested in opportunities for a career (especially for women), find readily available sexual partners without marriage, believe that the single lifestyle is full of excitement, or have a desire for self-sufficiency and freedom to change and experiment (Stein, 1976, 1981). According to some marriage and family analysts, individuals who prefer to remain single typically hold more-individualistic values and are less family oriented than those who choose to marry. They also tend to value friends and personal growth more highly than getting married and having children (Cargan and Melko, 1982; Alwin, Converse, and Martin, 1985).

Other never-married singles remain single out of necessity. For some people, being single is an economic necessity: They simply cannot afford to marry and set up their own household. Structural changes in the economy limit the options of young people from working-class and low-income family backgrounds. Even some college graduates have found that they cannot earn enough money to set up a separate household away from their parents. Consequently, a growing proportion of young adults are living with one or both parents. Approximately 13 percent of young adults between the ages of 25 and 34 reside with their parents; 16 percent of men live with their parents, as compared with 9 percent of women. According to one young man who lived at home,

> The rent is low and utilities are free. There is hot food on the table and clean socks in the drawer. Mom nags a little and dad scowls a lot, but mostly they don't get in the way. And there's money left at the end of the month for a car payment. (qtd. in Gross, 1991: A1)

The proportion of singles varies significantly by racial and ethnic group, as shown in ▶ Figure 15.3. Among persons age 15 and over, 40.8 percent of African Americans have never married, compared with 30.1 percent of Latinos/as, 26.8 percent of Asian and Pacific Islander Americans, and 22.6 percent of whites. Among women age 20 and over, the difference is even more pronounced; almost twice as many African American women in this age category have never married, compared with U.S. women of the same age in general (U.S. Census Bureau, 2008).

Social scientists have found that five primary factors contribute to the lower marriage rate of African American women:

1. There are more African American women than men in the United States as a result of the high rate of mortality among young African American men (Staples, 1994). In addition, college-educated African American women significantly outnumber

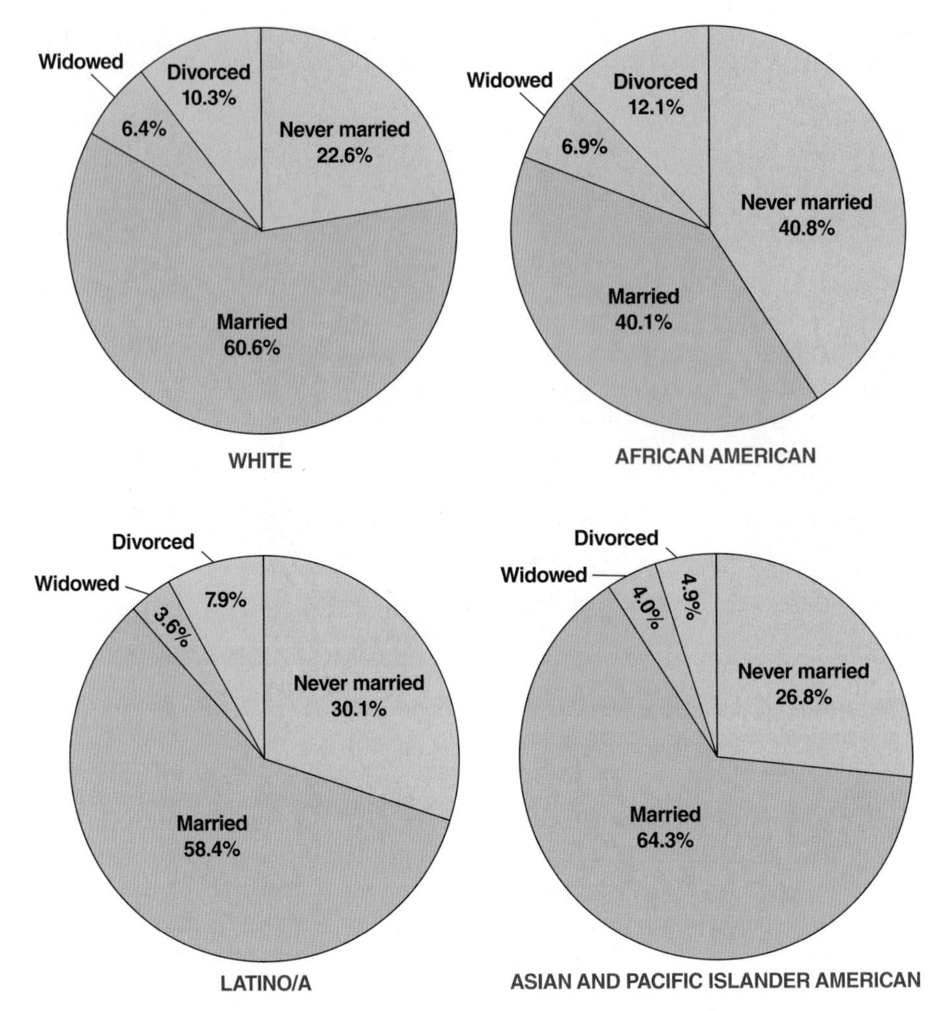

▶ **Figure 15.3 Marital Status of U.S. Population Age 15 and Over by Race/Ethnicity**

Source: U.S. Census Bureau, 2008.

college-educated African American men and tend to earn more money than the men (Lichter, LeClere, and McLaughlin, 1991; Roberts, 1994; Staples, 1994).

2. African American males who have been subjected to discriminatory practices and limited opportunities may perceive that their only economic options are to serve in the military or participate in criminal activity, thus making them poor candidates for marriage (Staples, 1994).
3. A higher rate of homosexuality exists among African American men than among women (Staples, 1994).
4. More African American men than women marry members of other racial–ethnic groups (Staples, 1994).
5. Working-class African American families often stress education for their children and encourage their daughters to choose education over marriage (Higginbotham, 1991; Higginbotham and Weber, 1995).

Although a number of studies have examined why African Americans remain single, few studies have focused on Latinos/as. Some analysts cite the diversity of experiences among Mexican Americans, Cuban Americans, and Puerto Ricans as the reason for this lack of research. Existing research attributes increased rates of singlehood among Latinas/os to several factors, including the youthful age of the Latino/a population and the economic conditions experienced by many young Latinas/os (see Mindel, Habenstein, and Wright, 1988).

African American Families

As with other racial–ethnic groups, there is no such thing as *the* African American family (McAdoo, 1990). Although many African Americans live in nuclear families, a higher proportion of African Americans than whites live in extended family households (Hofferth, 1984). The extended family often provides emotional and financial support not otherwise available. When an emergency arises, three or more generations may work together to support and care for one another (Taylor, Chatters, and Mays, 1988). Intergenerational care and concern by family members contribute to the well-being of adults and children alike, as Dawn March acknowledges:

> They have been there more so during my adult years than a lot of other families that I know about. My mother kept all of my children until they were old enough to go to day care. And she not only kept them, she'd give them a bath for me during the daytime and feed them before I got home from work. Very, very supportive people. So, I really would say I owe them for that. (qtd. in Higginbotham and Weber, 1995: 141)

As in March's case, mutually beneficial relationships often exist between older African American women and their grandchildren, nieces, and nephews. However, among middle- and upper-middle-class African American families, nuclear family patterns are more prevalent than extended family ties. In these nuclear families, visits with relatives outside the immediate family are limited to special occasions such as births, weddings, and funerals even when family members live close to one another (Willie, 1991).

Latina/o Families

Family support systems are also found in many Latina/o families. Referred to as *la familia,* this network spans a wide array of relatives, including a married couple and their children, parents, aunts and uncles, cousins, brothers and sisters, and their children. *a familia* may also include *compadres* (co-parents), *padrinos* (godfathers), and *madrinas* (godmothers) who are not biologically related to the family but become members of the extended family. *Compadres* participate in the major religious ceremonies in the children's lives, including baptism, confirmation, first communion, and marriage (see Griswold del Castillo, 1984). Poet Cherrié Moraga (1994: 41) describes *la familia* as "cross-generational bonding, deep emotional ties. . . . It is finding familia among friends where blood ties are formed through suffering and celebration shared." However, sociologists question the extent to which extended families and *familialism* (a strong sense of the importance of family held by all its members) exist among Latinos/as across social classes. According to the sociologist Norma Williams (1990), extended family networks have been disappearing, especially among economically advantaged Mexican Americans in urban centers. Similar patterns have been observed among Puerto Rican families, a trend that is not always pleasing to older family members such as Gloria Santos:

> [My son and daughter] both have good salaries but call me only once or twice a month. I hardly know my grandchildren.... They only visit me once a year and only for one or two days. I've told my daughter that instead of sending me money she could call me more often. I was a good mother and worked hard in order for them to get a good education and have everything. All I expected from them was to show me they care, that they love me. (qtd. in Sanchez-Ayendez, 1995: 265)

Asian American Families

Many Asian Americans live in nuclear families; however, others (especially those residing in Chinatowns) have extended family networks. The family of Tony Hom, a twenty-five-year-old Chinese American, is an example of what is known as a *semi-extended family* because other relatives live close by but not necessarily in the same household:

> My grandparents were always close by. When we lived in an apartment, they were always a few floors up. Now, my parents have a two family house, and my grandparents live next door. So this was always pointed out to me. Maybe they are fearful that one day we would become too Americanized and not take care of them. You always hear them talking about how the "lo-fans" (referring to Caucasians) don't take care of the elders, while the Chinese community prides itself on how it takes care of its old family members. (qtd. in Lee, 1992: 153)

Unlike Tony's family, extended family networks of some Vietnamese Americans are limited because some family members died in the war in Vietnam and others did not emigrate to the United States (Tran, 1988).

Native American Families

Family ties are also strong among many Native Americans, as Mary Crow Dog states:

> Our people have always been known for their strong family ties, for people within one family group caring for each other, for the "helpless ones," the old folks and especially the children, the coming generation.... At the center of the old Sioux society was the tiyospaye, the extended family group, the basic hunting band, which included grandparents, uncles, aunts, in-laws, and cousins. The tiyospaye was like a warm womb cradling all within it. Kids were never alone, always fussed over by not one but

© Davis Barber/PhotoEdit

The increasing diversity of the United States can be experienced by visiting a family-oriented attraction such as a theme park, as this photo from Disneyland indicates.

> several mothers, watched and taught by several fathers. (Crow Dog and Erdoes, 1991: 12–13)

Today, extended family patterns are common among lower-income Native Americans living on reservations; most others live in nuclear families (Sandefur and Sakamoto, 1988). Many Native Americans still share a sense of their extended family ties even when they do not live close to one another.

Biracial Families

Since the 1970s, there has been a 300 percent increase in marriages between people of different races. When these couples produce offspring, their children are considered to be biracial. Interracial marriage—which is most often thought of as marriage between whites and blacks—was illegal in sixteen states until the U.S. Supreme Court overturned miscegenation laws in 1967.

Today, the term *interracial marriage,* or racially mixed marriage, has taken on a much broader interpretation as people from nations around the globe, representing a wide diversity of racial–ethnic and cultural backgrounds, form couples, marry, and produce children of mixed racial parentage. Skin color and place of birth have ceased to be reliable indicators of a person's identity or origin.

What effect does interracial parenting have on children and families? "Penny Yang," a Hmong woman whose family immigrated to the United States from Laos, describes how her mixed racial–ethnic background affects her family life:

> My mom is Japanese and my dad is Hmong and my stepmom is American. It feels kind of different because I never met anyone who is Hmong and Japanese and American before, but I'm proud of it. I kind of get background from all sides, exposure from all different cultures. I wish, though, that I knew how to speak Hmong or Japanese really, really well. . . . When I'm with my dad's relatives, I wish I could understand more of what's going on. I wish I could communicate with them because it's kind of put a barrier between me and them that I don't speak Hmong. Even though I love them and know that we're a family and everything, it's holding me back because I don't know the language. So in a lot of ways I feel left out, and I know I would feel different if I could speak Hmong. . . . If my Hmong or Japanese relatives don't understand me, then I really can't do anything about it. That's just the way I am. I'm an American with a mixed background, just like a lot of Americans. (qtd. in Faderman, 1998: 238–239)

Many biracial and bicultural children find through the lens of their personal experiences that they have an enhanced ability to understand the meaning of race and culture as these factors influence daily life. They also have opportunities to map a new ethnic terrain for the United States and other nations that transcends traditional racial and cultural divisions (O'Hearn, 1998). Journalist Lise Funderburg (1994), who identifies herself as biracial, sums up the significance of the increasing number of biracial marriages and biracial and bicultural children in this way:

> To some extent, as the number of biracial people increases in this country, the choices they make about how to raise their children and how to influence their children's attitudes toward race will help determine not just how future generations of biracial children will be welcomed or shunned by society at large but also how all Americans will view and value race. . . . (Funderburg, 1994: 348)

Family Issues in the Future

As we have seen, families and intimate relationships changed dramatically during the twentieth century. Some people believe that the family as we know it is doomed. Others believe that a return to traditional family values will save this important social institution and create greater stability in society. However, the sociologist Lillian Rubin (1986: 89) suggests that clinging to a traditional image of families is hypocritical in light of our society's failure to support families: "We are after all, the only advanced industrial nation that has no public policy of support for the family whether with family allowances or decent publicly-sponsored child-care facilities." Some laws even have the effect of hurting children whose families do not fit the traditional model. For example, cutting back on government programs that provide food and medical care for pregnant women and infants will result in seriously ill children rather than model families (Aulette, 1994).

According to the psychologist Bernice Lott (1994: 155), people's perceptions about what constitutes a family will continue to change in the future:

> Persons on whom one can depend for emotional support, who are available in crises and emergencies, or who provide continuing affections, concern, and companionship can be said to make up a family. Members of such a group may live together in the same household or in separate households, alone or with others. They may be related by birth, marriage, or a chosen commitment to one another that has not been legally formalized.

Some of these changes are already becoming evident. For example, many men are attempting to take an active role in raising their children and helping with household chores. Many couples terminate abusive relationships and marriages.

Regardless of problems facing families today, many people still demonstrate their faith in the future by choosing to have children. It will be interesting to see what people today, some of whom are faced with various social and economic challenges, decide to do about having children. How many children will people have? What will family life be like in 2020? What will your own family be like?

Chapter Review

• What is the family?

Today, families may be defined as relationships in which people live together with commitment, form an economic unit and care for any young, and consider their identity to be significantly attached to the group.

• How does the family of orientation differ from the family of procreation?

The family of orientation is the family into which a person is born; the family of procreation is the family that a person forms by having or adopting children.

• What pattern of marriage is legally sanctioned in the United States?

In the United States, monogamy is the only form of marriage sanctioned by law. Monogamy is a marriage between two partners, usually a woman and a man.

• What are the functionalist, conflict, and symbolic interactionist perspectives on families?

Functionalists emphasize the importance of the family in maintaining the stability of society and the well-being of individuals. Functions of the family include sexual regulation, socialization, economic and psychological support, and provision of social status. Conflict and feminist perspectives view the family as a source of social inequality and an arena for conflict. Symbolic interactionists explain family relationships in terms of the subjective meanings and everyday interpretations that people give to their lives. Reflecting the individualism, particularity, and irregularity of social life in the Information Age, postmodern analysts view families as being permeable, capable of being diffused or invaded so that their original purpose is modified.

- **How are families in the United States changing?**

Families are changing dramatically in the United States. Cohabitation has increased significantly in the past three decades. With the increase in dual-earner marriages, women have increasingly been burdened by the second shift—the domestic work that employed women perform at home after they complete their workday on the job. Many single-parent families also exist today.

- **What is divorce, and what are some of its causes?**

Divorce is the legal process of dissolving a marriage. At the macrolevel, changes in social institutions may contribute to an increase in divorce rates; at the microlevel, factors contributing to divorce include age at marriage, length of acquaintanceship, economic resources, education level, and parental marital happiness.

www.cengage.com/login

Want to maximize your online study time? Take this easy-to-use study system's diagnostic pre-test, and it will create a personalized study plan for you. By helping you to identify the topics that you need to understand better and then directing you to valuable online resources, it can speed up your chapter review. CengageNOW even provides a post-test so you can confirm that you are ready for an exam.

Key Terms

bilateral descent 482
cohabitation 489
domestic partnerships 490
dual-earner marriages 492
egalitarian family 483
endogamy 484
exogamy 484
extended family 480
families 478
family of orientation 479
family of procreation 479
homogamy 484
kinship 478
marriage 481
matriarchal family 483
matrilineal descent 482
matrilocal residence 484
monogamy 481
neolocal residence 484
nuclear family 481
patriarchal family 482
patrilineal descent 482
patrilocal residence 483
polyandry 481
polygamy 481
polygyny 481
second shift 492
sociology of family 484

Questions for Critical Thinking

1. In your opinion, what constitutes an ideal family? How might functionalist, conflict, feminist, and symbolic interactionist perspectives describe the ideal family?
2. Suppose that you wanted to find out about women's and men's perceptions about love and marriage. What specific issues might you examine? What would be the best way to conduct your research?

3. You have been appointed to a presidential commission on child-care problems in the United States. How to provide high-quality child care at affordable prices is a key issue for the first meeting. What kinds of suggestions would you take to the meeting? How do you think your suggestions should be funded? How does the future look for children in high-, middle-, and low-income families in the United States?

The Kendall Companion Website

www.cengage.com/sociology/kendall

Visit this book's companion website, where you'll find more resources to help you study and successfully complete course projects. Resources include quizzes and flash cards, as well as special features such as an interactive sociology timeline, maps, General Social Survey (GSS) data, and Census 2000 data. The site also provides links to useful websites that have been selected for their relevance to the topics in this chapter and include those listed below. (*Note:* Visit the book's website for updated URLs.)

American Association for Marriage and Family Therapy (AAMFT)

http://www.aamft.org

AAMFT is a professional association of marriage and family therapists that provides resources for therapists and clients. Its website offers information and resources on a variety of topics, including bereavement and loss, infertility, adolescent behavior problems, alcohol, ADHD, children and divorce, and post-traumatic stress disorder.

Family Violence Prevention Fund (FVPF)

http://endabuse.org

The FVPF is a national, nonprofit organization that focuses on policy and education to prevent and to eventually end domestic violence. Its website includes the latest research, news reports, state and national statistics, celebrity watch, personal stories, advice on how to get help, and additional resources.

National Council on Family Relations (NCFR)

http://www.ncfr.org

In addition to promoting family well-being through the formulation of family-friendly policies, the NCFR provides the opportunity for family researchers, educators, and practitioners to share in the development and distribution of information about families and family relationships. This website offers family policy information, research, links, news, and a searchable database of information provided by NCFR members on topics such as relationships for couples, parenting, divorce, and families and technology

chapter 16 Education

Chapter Focus Question

How do race, class, and gender affect people's access to and opportunities in education?

During my first semester at Santa Barbara City College . . . my classes included aerobics, typing and remedial math. Who would have guessed that after spending three years there, I would be attending Harvard University?

Without the opportunity to study at my local community college, I probably wouldn't have gone to college at all. My high school grades would have sufficed to get me into a decent university, but I didn't consider myself college material. After all, none of my six brothers and sisters had attended college; most didn't even finish high school. My parents have the equivalent of a second-grade education. Mexican immigrants who do not speak English, my mother worked as a maid and my father as a dishwasher. Considering

© Sky Bonillo/PhotoEdit

In U.S. society, education is an extremely important institution, requiring enormous outlays of time, labor, and money. Education is an extremely controversial subject as well, raising questions of equality on the issues of race/ethnicity, gender, and social class.

IN THIS CHAPTER

- An Overview of Education
- Education in Historical–Global Perspective
- Sociological Perspectives on Education
- Inequalities Among Elementary and Secondary Schools
- Problems Within Elementary and Secondary Schools
- School Safety and School Violence
- Opportunities and Challenges in Colleges and Universities
- Continuing Issues and Future Trends in Education

my resources, I thought the convenience and limited cost of a community college made it my most viable option.

But enrolling in a community college was also one of the smartest decisions I ever made. Despite my slow start, I learned the skills I needed to move ahead academically. . . . I also began to explore educational alternatives. And transferring to a reputable four-year university became my most important goal. I specifically wanted to transfer to an Ivy League university because I thought that would open more doors for me.

—Cynthia G. Inda (1997: 31) describing how a community college gave her the opportunity to pursue her educational and career goals

Today, I am announcing the most significant down payment yet on reaching [the goal of having the highest proportion of college graduates in the world] in the next ten years. It's called the American Graduation Initiative. It will reform and strengthen community colleges from coast to coast so that they get the resources students and schools need—and the results workers and businesses demand. Through this plan, we seek to help an

(Continued)

Sharpening Your Focus

- *How do educational goals differ in various nations?*
- *What are the key assumptions of functionalist, conflict, and symbolic interactionist perspectives on education?*
- *What major problems are being faced by U.S. schools today?*
- *How are the issues in higher education linked to the problems of the larger society?*

additional five million Americans earn degrees and certificates in the next decade.

—President Barack Obama, addressing an audience at Macomb Community College in Michigan and announcing the beginning of an initiative to strengthen community colleges in the twenty-first century (whitehouse.gov, 2009a)

In acknowledging the importance of community colleges, President Obama described the importance of higher education for people like Cynthia, who is one of millions of people who have attended community colleges to gain knowledge and skills that will benefit them and enhance their opportunities in the future. In fact, a substantial proportion of college students start their higher education at community colleges.

From pre-kindergarten through postgraduate studies, education is one of the most significant social institutions in the United States and other high-income nations. Although most social scientists agree that schools are supposed to be places where people acquire knowledge and skills, not all of them agree on how a large number of factors—including class, race, gender, age, religion, and family background—affect individuals' access to educational opportunities or to the differential rewards that accrue at various levels of academic achievement. Consider this question, for example: Is higher education (at the college and university level) stratified by social class? According to the sociologist Dennis Gilbert (2008), the answer is both yes and no:

> (1) The American system of higher education is so big and so open that it provides major opportunities for talented youths from lower and middle-class families to prepare themselves for successful careers that raise them above the level of their parents, and (2) the American system of higher education is sufficiently stratified that its main function is to reproduce for each generation of children the status positions held by their parents. Does the American system of higher education promote mobility or reproduction? The answer appears to be both.

In this chapter, we will look further into this issue as we discuss education as a key social institution and analyze some of the problems that affect contemporary elementary, secondary, and higher education. Before reading further, take the quiz in Box 16.1 to see what you know about U.S. education.

An Overview of Education

***Education* is the social institution responsible for the systematic transmission of knowledge, skills, and cultural values within a formally organized structure.** Education is a powerful and influential force in contemporary societies. As a social institution, education imparts values, beliefs, and knowledge considered essential to the social reproduction of individual personalities and entire cultures (Bourdieu and Passeron, 1990). Education grapples with issues of societal stability and social change, reflecting society even as it attempts to shape it. Early socialization is primarily informal and takes place within our families and friendship networks; socialization then passes to the schools and other, more-formalized organizations created for the specific purpose of educating people. Today, education is such a significant social institution that an entire subfield of sociology—the *sociology of education*—is devoted to its study.

How did education emerge as such an important social institution in contemporary industrialized nations? To answer this question, we begin with a brief examination of education in historical–global perspective.

Education in Historical–Global Perspective

Education serves an important purpose in all societies. At the microlevel, people must acquire the basic knowledge and skills they need to survive in society. At the macrolevel, the social institution of education is an essential component in maintaining and perpetuating the culture of a society across generations. ***Cultural transmission*—the process by which children and recent immigrants become acquainted with the dominant cultural beliefs, values, norms, and accumulated knowledge of a society**—occurs through

Box 16.1 Sociology and Everyday Life

How Much Do You Know About U.S. Education?

True	False	
T	F	1. Public education in the United States dates back more than 160 years.
T	F	2. Equality of opportunity is a vital belief in the U.S. educational system.
T	F	3. Women and men earn about the same number of doctoral degrees in the United States each year.
T	F	4. Although there is much discussion about gangs in schools, only 10 percent of students in recent surveys reported that there were gangs at their school.
T	F	5. Young people who bully are more likely to smoke, drink alcohol, and get into fights.
T	F	6. Every school day more than 7,000 U.S. students drop out of high school or leave, never to return.
T	F	7. In public schools in the United States, core classes such as history and mathematics were never taught in any language other than English before the 1960s civil rights movement.
T	F	8. The federal government has limited control over how funds are spent by individual school districts because most of the money comes from the state and local levels.

Answers on page 514.

informal and formal education. However, the process of cultural transmission differs in preliterate, preindustrial, and industrial nations.

Informal Education in Preliterate Societies

Preliterate societies existed before the invention of reading and writing. These societies have no written language and are characterized by very basic technology and a simple division of labor. Daily activity often centers around the struggle to survive against natural forces, and the earliest forms of education are survival oriented. People in these societies acquire knowledge and skills through ***informal education*—learning that occurs in a spontaneous, unplanned way.** Through direct informal education, parents and other members of the group provide information about how to gather food, find shelter, make weapons and tools, and get along with others. For example, a boy might learn skills such as hunting, gathering, fishing, and farming from his father, whereas a girl might learn from her mother how to plant, gather, and prepare food or how to take care of her younger sisters and brothers. Such informal education often occurs through storytelling or ritual ceremonies that convey cultural messages and provide behavioral norms. Over time, the knowledge shared through informal education may become the moral code of the group.

Formal Education in Preindustrial, Industrial, and Postindustrial Societies

Although *preindustrial societies* have a written language, few people know how to read and write, and formal education is often reserved for the privileged. Education becomes more formalized in preindustrial and industrial societies. ***Formal education* is learning that takes place within an academic setting such as a school, which has a planned instructional process and teachers who convey specific knowledge, skills, and thinking processes to students.** Perhaps the earliest formal

education the social institution responsible for the systematic transmission of knowledge, skills, and cultural values within a formally organized structure.

cultural transmission the process by which children and recent immigrants become acquainted with the dominant cultural beliefs, values, norms, and accumulated knowledge of a society.

informal education learning that occurs in a spontaneous, unplanned way.

formal education learning that takes place within an academic setting such as a school, which has a planned instructional process and teachers who convey specific knowledge, skills, and thinking processes to students.

Box 16.1 Sociology and Everyday Life

Answers to the Sociology Quiz on U.S. Education

1. True. As far back as 1848, free public education was believed to be important in the United States because of the high rates of immigration and the demand for literacy so that the country would have an informed citizenry that could function in a democracy.

2. True. Despite the fact that equality of educational opportunity has not been achieved, it remains a goal of many people in this country. A large number subscribe to the belief that this country's educational system provides equal educational opportunities and that it is up to each individual to make the most of it.

3. False. Although in some years, more women earn doctoral degrees than men, in many other years, including 2006, more men (54.8 percent) than women (45.0 percent) earned doctorate degrees. However, in engineering and the physical sciences, men earned 80 percent and 72 percent, respectively, of the doctorates (*Chronicle of Higher Education,* 2008).

4. False. Approximately 24 percent of students reported gangs at their schools. Students in urban schools were more likely to report gang activity than students in suburban and rural schools.

5. True. Government data show that young people who bully *are* more likely to engage in other problematic behavior, such as smoking, drinking alcohol, and getting into fights.

6. True. At least 7,000 students leave high school and do not return. This means that as many as four million 16–24-year-olds are not enrolled in high school and have not earned a high school diploma or GED.

7. False. Late in the nineteenth century and early in the twentieth century, some classes were conducted in the language of recent immigrants, such as the Italians, the Polish, and the Germans.

8. True. Most funding for public education comes from state and local property taxes, and similar sources of revenue.

education occurred in ancient Greece and Rome, where philosophers such as Socrates, Plato, and Aristotle taught elite males the skills required to become thinkers and orators who could engage in the art of persuasion (Ballantine and Hammack, 2009). During the Middle Ages, the first colleges and universities were developed under the auspices of the Catholic Church.

The Renaissance and the Industrial Revolution had a profound effect on education. During the Renaissance, the focus of education shifted to the importance of developing well-rounded and liberally educated people. With the rapid growth of industrial capitalism and factories during the Industrial Revolution, it became necessary for workers to have basic skills in reading, writing, and arithmetic. However, from the Middle Ages until the end of World War I, only the sons of the privileged classes were able to attend European universities. Agriculture was the economic base of society, and literacy for people in the lower classes was not deemed important.

As societies industrialize, the need for formal education of the masses increases significantly. In the United States, the free public school movement was started in 1848 by Horace Mann, who stated that education should be the "great equalizer." By the mid-1850s, the process of mass education had begun in the United States as all states established free, tax-supported elementary schools that were readily available to children throughout the country. ***Mass education*** **refers to providing free, public schooling for wide segments of a nation's population.**

As industrialization and bureaucratization intensified, managers and business owners demanded that schools educate students beyond the third or fourth grade so that well-qualified workers would be available for rapidly emerging "white-collar" jobs in management and clerical work (Bailyn, 1960). By the middle of the nineteenth century, the idea of universal free public schooling was fairly well established in the United States. In addition to educating the next generation of children for the workplace, public schools were also supposed to serve as the primary agents of socialization for millions of European immigrants arriving in the United States seeking economic opportunities and a better life. By the 1920s, educators had introduced the "core" curriculum: courses such as mathematics,

© Kayte M. Deioma/PhotoEdit

Some early forms of mass education took place in one-room schoolhouses such as the one shown here, where children in various grades were all taught by the same teacher. How do changes in the larger society bring about changes in education?

social sciences, natural sciences, and English. This core is reflected in the contemporary "back to basics" movement, which calls for teaching the "three R's" (reading, 'riting, and 'rithmetic) and enforcing stricter discipline in schools.

Contemporary U.S. education attempts to meet the needs of the industrial and postindustrial society by teaching a wide diversity of students a myriad of topics, including history and science, computer skills, how to balance a checkbook, and how to avoid contracting AIDS. According to sociologists, many functions performed by other social institutions in the past are now under the auspices of the public schools. For example, full-day kindergartens and extended-day programs for school-age children are provided by many U.S. school districts because of the growing number of working parents who need high-quality, affordable care for their children. Within the regular classroom setting, many teachers feel that their job description encompasses too many divergent tasks. According to Ruth Prale, an elementary reading specialist,

> A teacher today is a social worker, surrogate parent, a bit of disciplinarian, a counselor, and someone who has to see to it that they eat. Many teachers are mandated to teach sex education, drug awareness, gang awareness. And we're supposed to be benevolent. And we haven't come to teaching yet! (qtd. in Collins and Frantz, 1993: 83)

At all levels of U.S. education, from kindergarten through graduate school, controversy exists over *what* should be taught, *how* it should be taught, and *who* should teach it. Do other nations have similar questions regarding their educational systems? Let's take a brief look at how educational systems are organized in Japan and Germany.

Contemporary Education in Other Nations

In this section, we will examine schools in two nations that frequently show up in the U.S. media. Because the United States and Japan are often compared to each other in the global marketplace, social analysts frequently compare the educational systems of these high-income countries. Similarly, schooling in Germany is often discussed in the United States because the science and math scores of students in these two nations are compared to assess the quality of education in each country.

Education in Japan Like other countries, Japan did not make public education mandatory for children until the country underwent industrialization. During the Meiji period (1868–1912), feudalism was eliminated, and Japan embarked on a new focus on youth and "bureaucratic" universal education (White, 1994). In hopes of catching up with the West, Japanese officials created an educational system and national educational goals. Education was viewed as a form of economic and national development and as a means of identifying talent for a new technological elite (White, 1994).

Today, Japanese educators, parents, students, and employers all view education as a crucial link to their children's future and to their nation's economic success. Japanese schools not only emphasize conformity and nationalism, but also highlight the importance of obligation to one's family and of learning the skills necessary for employment. Beginning at about three years of age, many Japanese toddlers are sent to cram schools (*jukus*) to help them qualify for good preschools.

In both cram schools and public schools, students learn discipline and thinking skills, along with physical activities such as karate and gymnastics to improve agility. By the time children reach elementary school, they are expected to engage in cooperative activities with their classmates. In some schools, children are responsible for preparing, serving, and cleaning up after the midday meal. At the end of the day, children may be seen cleaning the chalkboards and even mopping the floors, all as a part of the spirit of cooperation they

mass education the practice of providing free, public schooling for wide segments of a nation's population.

Schools in Japan emphasize conformity even at an early age. Many Japanese people believe that high-quality education is a crucial factor in their country's economic success.

are being taught. Virtually all of Japanese children attend public elementary schools; less than one percent attend a private school.

In Japan, the middle-school years are especially crucial: Students' futures are based on the academic status track on which they are placed while in middle school. Moreover, the kind of jobs they will hold in the future depends on the university or vocational-technical school they attend, which in turn depends on the high school they attend. By the third year of middle school, most students are acutely aware of their academic future and of how their lives and friendships change as they go through the educational process, as one former student recalled:

> Middle school was fun; I was with my friends and I could throw myself into sports and I didn't have to study very hard. I couldn't be sure which of my friends would be with me in high school. Now that I am in high school, I know these will be my friends for life—they've been with me through very tough times. (qtd. in White, 1994: 77)

Up until high school, the student population of a school typically reflects the neighborhoods in which children live. However, at the high school level, entrance to a particular school is based on ability: Some Japanese students enter vocational schools that teach them skills for the workplace; others enter schools that are exclusively for the college-bound. Many analysts have noted the extent to which most Japanese vocational schools provide state-of-the-art equipment and instruction in fields with wide employment opportunities. Consequently, graduates of most vocational high schools do not have problems finding relatively well-paid employment upon graduation.

Typically, the instruction in Japanese high schools for the college-bound is highly structured, and all students are expected to respond in unison to questions posed by the teacher. Students are expected to be fluent in more than one language; many Japanese high schools teach English, particularly vocabulary and sentence construction. Science and math courses are challenging, and Japanese students often take courses such as algebra and calculus several years before their U.S. counterparts. Students must be prepared for a variety of college entrance examinations because each college and university gives its own test, and all tests occur within a few weeks of one another.

Girls and young women in the United States would likely feel stifled by the lack of educational opportunities experienced by their counterparts in Japan. Although there have been some changes, many parents and educators still believe that a good junior college education is all that young women need in order to be employable and marriageable (White, 1994). At the college and university level, the absence of women as students and professors is especially profound. Although a woman in the 1990s became the first-ever female president of a state-run university (Nara Women's University) in Japan, women account for fewer than 5 percent of all presidents of colleges and universities (Findlay-Kaneko, 1997). Moreover, the lack of child-care facilities within the universities remains a pressing problem for women students and faculty in Japanese higher education.

Young men also experience extreme pressures in the Japanese system. In fact, high rates of school truancy occur as tens of thousands of students balk at going to school, and still others experience school-related health problems such as stomach ulcers, allergy disorders, and high blood pressure (White, 1994).

Education in Germany Residents of Germany have long believed that education is important for the future of individual citizens and for their country. A belief in compulsory education in Germany can be traced back to the theologian and professor Martin Luther, who, in the sixteenth century, strongly urged that all children needed schooling not only in basics—such as reading, writing, and arithmetic—but also in ethics, obedience, and duty. With the formation of the German Empire in the late 1800s, four distinct types of secondary schools were introduced that remain central to how a student is educated in Germany in the twenty-first century.

Today, about 82.4 million people live in Germany, a nation that is slightly smaller than the state of Montana. Because each state has its own unique political, cultural, and religious traditions, it also has its own ministry of education, and state and local authorities typically control primary and secondary education. At the national level, a Conference of Ministers of Education (KMK) coordinates the states' educational efforts and issues resolutions on topics such as textbooks, curriculum, and initiatives to improve math and science instruction. One of the unique challenges faced by the states and the KMK has been the greater diversity of students who have entered German schools from Southern European countries (such as Greece, Italy, and Turkey) and from nations of the former Soviet Union (such as Russia and Ukraine) as their parents have arrived as immigrant workers. More than 16 million people living in Germany are of non-German descent.

Although *kindergarten* is both a German word and a German creation that has been widely adopted around the world, it is generally not part of the state-supported school system in that country. Optional kindergarten education is provided for all children between three and six years old. However, about 85 percent of German children between the ages of three and six attend community- or church-supported kindergartens. Beginning at age six, all children are required to attend elementary school (*grundschule*), which lasts either four or six years, depending on the state. School attendance is compulsory for ten to thirteen years of each person's life. The stated goal of *grundschule* is to foster each child's full potential, and students are not separated or tracked by ability group. Children often remain with the same class and the same teacher for four years in an effort to promote bonding among them. In fact, students typically remain in the same classroom throughout the day, and teachers move from room to room teaching their specific subject each day. A student's performance in the final year of *grundschule* is very important because it strongly influences where he or she will attend school the next year.

The type of secondary (middle) school a student attends is determined by factors such as admissions tests, grade point average, teachers' recommendations, and parents' wishes (National Center for Education Statistics, 2005a). Students with higher grades and good teacher recommendations will most likely transfer to the *gymnasium* or the *realschule,* whereas those with lower performance evaluations will enter the *hauptschule,* which provides students with a vocational education. By contrast, the *gymnasium* specializes in gifted students, preparing them for the university. As one *gymnasium* teacher described the process,

> If the child is not suited to the *Gymnasium,* if he cannot learn foreign languages and is not good at abstract thought, then this child will continually fail and will continually experience frustration, and such children are deformed in the development of their personality. They have continual experiences of failure and no successful experience. . . . As a result, we say, do not send children to the *Gymnasium* who are not suited for it. Such children belong as soon as possible in the correct educational path. (qtd. in U.S. Department of Education, 1999)

As this teacher's statement suggests, the German education system provides a variety of schooling options after elementary school. *Hauptschule,* a lower-level general secondary school (until grade nine), is less academically oriented, has courses taught at a slower pace, and offers more basic instruction. A certificate of completion from this type of school typically leads to vocational or on-the-job training. Some, but not all, German states offer *gesamtschule,* an integrated secondary school where students of all ability groupings are educated under one roof. In this type of school, students are not divided into general education (vocational) and academic tracks, and they have a chance to take courses appropriate to their ability level, even if these vary widely. By contrast, the *realschule* offers students more enhanced academic studies (from grades five through ten) in both liberal and practical education, but the primary emphasis is on liberal education.

The learning environment at *realschule* is less competitive than it is at *gymnasium,* the highest academic secondary school, which continues until grade twelve or thirteen. *Gymnasium* provides students with a liberal education, including classical language, modern language, and mathematics/natural sciences. Students who successfully complete the *gymnasium* and pass their comprehensive exams receive the *abitur,* which automatically qualifies them to study at a university or a top-tier vocational school. In recent decades, attendance at *realschule* or *gymnasium* has been much more popular with students and parents than the other types of secondary schools because these schools offer students a chance to gain higher status and better-paying jobs.

Beyond secondary education, some students in Germany will attend either a vocational school or a university. Vocational schools include the *berufsschule,* a three- to four-year vocational program that typically includes an apprenticeship, and *fachoberscule,* a one- to two-year advanced vocational school. Universities in Germany are free, but students must pay for their books and housing, for there is no university-provided student housing on campus. Students may apply for financial aid, half of which is a grant and the other half a loan that must be paid back when the student is working following graduation. There are very few private universities in the country. Although there is a set curriculum in the liberal arts and sciences for the first few years at the university, students select which courses they want to take, and they take exams at the end of their period of study. For this reason, many students in

Germany take more years to graduate from the university than their U.S. counterparts, and German students may change schools three or four times in the course of their studies.

As is true in the United States, educational reform is a subject of much debate in Germany. School discipline has become a problem in urban areas, and more students desire to attend universities—especially in certain fields of studies—than the crowded universities can accept.

Sociological Perspectives on Education

Sociologists have divergent perspectives on the purpose of education in contemporary society. Functionalists believe that education contributes to the maintenance of society and provides people with an opportunity for self-enhancement and upward social mobility. Conflict theorists argue that education perpetuates social inequality and benefits the dominant class at the expense of all others. Symbolic interactionists focus on classroom dynamics and the effect of self-concept on grades and aspirations. Each of the three perspectives can provide valuable insights.

Functionalist Perspectives

Functionalists view education as one of the most important components of society. According to Emile Durkheim, education is crucial for promoting social solidarity and stability in society: Education is the "influence exercised by adult generations on those that are not yet ready for social life" (Durkheim, 1956: 28) and helps young people travel the great distance that it has taken people many centuries to cover. In other words, we can learn from what others have already experienced. Durkheim also asserted that *moral education* is very important because it conveys moral values—the foundation of a cohesive social order. He believed that schools are responsible for teaching a commitment to the common morality.

What values are these schoolchildren being taught? Is there a consensus about what today's schools should teach? Why or why not?

From this perspective, students must be taught to put the group's needs ahead of their individual desires and aspirations. Contemporary functionalists suggest that education is responsible for teaching U.S. values. According to the sociologist Amitai Etzioni (1994: 258–259),

> We ought to teach those values Americans share, for example, that the dignity of all persons ought to be respected, that tolerance is a virtue and discrimination abhorrent, that peaceful resolution of conflicts is superior to violence, that . . . truth telling is morally superior to lying, that democratic government is morally superior to totalitarianism and authoritarianism, that one ought to give a day's work for a day's pay, that saving for one's own and one's country's future is better than squandering one's income and relying on others to attend to one's future needs.

Etzioni suggests that "shared" values should be transmitted by schools from kindergarten through college. However, not all analysts agree on what those shared values should be or what functions education should serve in contemporary societies. In analyzing the values and functions of education, sociologists using a functionalist framework distinguish between manifest functions and latent functions, which are compared in ▶ Figure 16.1.

Manifest Functions of Education Some functions of education are *manifest functions*—previously defined as open, stated, and intended goals or consequences of activities within an organization or institution. Education serves five major manifest functions in society:

1. *Socialization.* From kindergarten through college, schools teach students the student role, specific academic subjects, and political socialization. In kindergarten, children learn the appropriate attitudes and behavior for the student role (Ballantine and Hammack, 2009). In primary and secondary schools, students are taught specific subject matter appropriate to their age, skill level, and previous ed-

Manifest functions—open, stated, and intended goals or consequences of activities within an organization or institution. In education, these are:

- socialization
- transmission of culture
- social control
- social placement
- change and innovation

Latent functions—hidden, unstated, and sometimes unintended consequences of activities within an organization. In education, these include:

- matchmaking and production of social networks
- restricting some activities
- creating a generation gap

Figure 16.1 Manifest and Latent Functions of Education

ucational experience. At the college level, students focus on more detailed knowledge of subjects they have previously studied and are exposed to new areas of study and research. Throughout their schooling, students receive political socialization in the form of history and civics lessons.

2. *Transmission of culture.* Schools transmit cultural norms and values to each new generation and play an active part in the process of assimilation, whereby recent immigrants learn dominant cultural values, attitudes, and behavior so that they can be productive members of society. However, questions remain as to *whose* culture is being transmitted. Because of the great diversity in the United States today, it is virtually impossible to define a single culture.
3. *Social control.* Schools are responsible for teaching values such as discipline, respect, obedience, punctuality, and perseverance. Schools teach conformity by encouraging young people to be good students, conscientious future workers, and law-abiding citizens. The teaching of conformity rests primarily with classroom teachers (see Sociology Works!).
4. *Social placement.* Schools are responsible for identifying the most-qualified people to fill the positions available in society. As a result, students are channeled into programs based on individual ability and academic achievement. Graduates receive the appropriate credentials to enter the paid labor force.
5. *Change and innovation.* Schools are a source of change and innovation. As student populations change over time, new programs are introduced to meet societal needs; for example, sex education, drug education, and multicultural studies have been implemented in some schools to help students learn about pressing social issues. Innovation in the form of new knowledge is required in colleges and universities. Faculty members are encouraged—and sometimes required—to engage in research and to share the results with students, colleagues, and others. In medical schools, for example, innovative technologies (such as new drugs) and new techniques (such as gene splicing) are developed and tested.

Latent Functions of Education All social institutions, including education, have *latent functions*—previously defined as hidden, unstated, and sometimes unintended consequences of activities within an organization or institution. Education serves at least three latent functions:

1. *Restricting some activities.* Early in the twentieth century, all states passed *mandatory education laws* that require children to attend school until they reach a specified age (usually age sixteen) or complete a minimum level of formal education (generally the eighth grade). The assumption was that an educated citizenry and work force are necessary for the smooth functioning of democracy and capitalism. Out of these laws grew one latent function of

Sociology *Works!*

Stopping Bullying: Character Building and Social Norms

Two Viewpoints:

> Bullying, particularly in adolescence, is epidemic, not just in the USA but around the world.
>
> —Marvin Berkowitz, professor of character education at the University of Missouri–St. Louis (qtd. in mindoh.com, 2007)

> What we've seen consistently is that risk behaviors [and] problems are overestimated, which [means] much of the bullying or violence or substance abuse can continue because the people engaged in that think everybody else is doing it.
>
> —H. Wesley Perkins, a sociology professor and director of the Alcohol Education Project at Hobart and William Smith Colleges in Geneva, New York (qtd. in Teicher, 2006)

What is bullying? By definition, bullying is aggressive behavior that is intentional and that involves an imbalance of power or strength. Bullying is often repeated, and the child who is routinely the object of the bullying usually has a difficult time defending himself or herself (U.S. Department of Health and Human Services, 2008).

How great a problem is bullying in today's schools? What have sociologists learned that might be useful in combating this problem? In regard to the frequency of bullying at school, some studies have found that as many as eight out of ten U.S. students in upper elementary, middle, and high schools who have access to an anonymous online messaging service that provides them with the ability to communicate the problem to school counselors and administrators report that they have been the victims of bullying (Gary, 2007). Other studies have shown that between 15 to 25 percent of U.S. students are bullied with some frequency and that 15 to 20 percent admit that they bully others with some frequency during the school year (U.S. Department of Health and Human Services, 2008). Although the nature and extent of this problem are unknown, experts are in agreement on one thing: Bullying can and must be addressed and reduced in schools. What methods work best for eliminating negative behaviors and maintaining social control in schools? One method of reducing bullying is rooted in character education, which focuses on the reinforcement of acceptable character traits, such as respect and responsibility. From character education courses and online studies, students learn problem-solving techniques, conflict-resolution approaches, and communication skills to help them interact with one another in a more positive manner.

However, an alternative approach has been suggested by such sociologists as H. Wesley Perkins, who emphasizes that the best practice in bullying prevention and intervention is to focus on the *social environment* of the school. According to Perkins, to reduce bullying it is important to change the climate of the school and social norms regarding bullying. As you will recall from Chapter 3, norms are established rules of behavior or standards of conduct; therefore, a school's social norms must establish standards of conduct that make it "uncool" for one person to bully another but "cool" for individuals to help out those who are bullied. Based on his previous research on alcohol use among college students, Perkins developed a social norms approach to preventing bullying which assumes that a

education, which is to keep students off the streets and out of the full-time job market for a number of years, thus helping keep unemployment within reasonable bounds (Braverman, 1974).

2. *Matchmaking and production of social networks.* Because schools bring together people of similar ages, social class, and race/ethnicity, young people often meet future marriage partners and develop social networks that may last for many years.
3. *Creation of a generation gap.* Students may learn information in school that contradicts beliefs held by their parents or their religion. Debates over the content of textbooks and library books typically center on information that parents deem unacceptable for their children. When education conflicts with parental attitudes and beliefs, a generation gap is created if students embrace the newly acquired perspective.

Dysfunctions of Education Functionalists acknowledge that education has certain dysfunctions. Some analysts argue that U.S. education is not promoting the high-level skills in reading, writing, science, and mathematics that are needed in the workplace and the global economy. For example, mathematics and science education in the United States does not compare favorably with that found in many other industrialized countries (see ◆ Table 16.1). In the latest available (2007) Trends in International Mathematics and Science Study (TIMSS), which compares the mathematics and science performance of U.S. students with that of their peers in other nations, U.S. eighth-grade students scored lower than did students in

© Radius Images/Jupiterimages

Sociologist H. Wesley Perkins sees the problem of bullying as an issue involving norms: If bullying becomes uncool, young people will be less likely to take part in it and more likely to help someone who is a victim of this very harmful activity.

person's behavior may be strongly influenced by the *incorrect* perceptions the individual holds about how other members of his or her social group think and act. From this approach, the first step to reducing the problem of bullying is to correct students' misperceptions about how prevalent this type of behavior really is: If students think that bullying is very common in their school and that "everybody is doing it," they may believe that the social norms support such behavior. If this misperception can be corrected, bullying will decrease as students come to see that the "nonbully status" is the social norm (Teicher, 2006).

Emphasizing the social environment of education and its norms is an important sociological contribution to our understanding of both positive and negative actions that take place in schools. It also makes us aware that social institutions fulfill a variety of important functions in society and that one of those functions for education is to maintain social control and to encourage civility among participants, whether or not they subscribe to moral values that teach them to "Love thy neighbor. . . ." Although it is also important to focus on character education for students and to teach them to respect others and have a sense of responsibility for their own actions, all of us have a responsibility for maintaining a social environment in which all individuals are treated with kindness and respect. Bullying isn't just a matter of "kids being kids": It is a serious sociological problem and must be dealt with as such.

Reflect & Analyze

What do you think about the problem of bullying? How can sociology help us to reduce problems such as this?

a number of other nations, including Chinese Taipei, Singapore, the Republic of Korea, and England (National Center for Education Statistics, 2009).

Are U.S. schools dysfunctional as a result of scores such as these? Analysts do not agree on what these score differences mean. For example, Japanese students may outperform their U.S. counterparts because their schools are more structured and teachers focus on drill and practice (Celis, 1994). Educators David C. Berliner and Bruce J. Biddle (1995) believe that data on cross-cultural differences in educational attainment actually involve comparisons of "apples" and "oranges": They found that studies such as this often compare the achievement of eighth-grade Japanese students who have already taken algebra with the achievement of U.S. students who typically take such courses a year or two later. Among U.S. students who had already completed an algebra course, most did at least as well as their Japanese counterparts on the mathematics exam (Berliner and Biddle, 1995).

Clearly, test scores are subject to a variety of interpretations; however, for most functionalist analysts, lagging test scores are a sign that dysfunctions exist in the nation's educational system. According to this approach, improvements will occur only when more stringent academic requirements are implemented for students. Dysfunctions may also be reduced by more thorough teacher training and consistent testing of instructors. Overall, functionalists typically advocate the importance of establishing a more rigorous academic environment in which students are required to learn the basics that will make them competitive in

◆ **Table 16.1 Trends in International Mathematics and Science Study (TIMSS)**
Results of the 2007 TIMSS (selected nations)

Country	Math Score	Science Score
Singapore	593	567
Korea, Republic of	597	553
Hong Kong SAR	572	530
Chinese Taipei	598	561
Japan	570	554
England	513	542
Hungary	517	539
Russian Federation	512	530
United States	508	520
Sweden	491	511
Israel	463	468
Turkey	432	454
Indonesia	397	427
Columbia	380	417
Saudi Arabia	329	403
El Salvador	340	387
Ghana	309	303

Source: National Center for Education Statistics, 2009.

school and job markets. An underlying assumption of functionalist thinking is that education can be an effective agent in the fight to reduce social inequality in society.

Conflict Perspectives

Conflict theorists do not believe that public schools reduce social inequality in society; rather, they believe that schools often perpetuate class, racial–ethnic, and gender inequalities as some groups seek to maintain their privileged position at the expense of others (Ballantine and Hammack, 2009).

Cultural Capital and Class Reproduction Although many factors—including intelligence, family income, motivation, and previous achievement—are important in determining how much education a person will attain, conflict theorists argue that access to quality education is closely related to social class. From this approach, education is a vehicle for reproducing existing class relationships. According to the French sociologist Pierre Bourdieu, the school legitimates and reinforces the social elites by engaging in specific practices that uphold the patterns of behavior and the attitudes of the dominant class. Bourdieu asserts that students from diverse class backgrounds come to school with different amounts of ***cultural capital*****—social assets that include values, beliefs, attitudes, and competencies in language and culture** (Bourdieu and Passeron, 1990). Cultural capital involves "proper" attitudes toward education, socially approved dress and manners, and knowledge about books, art, music, and other forms of high and popular culture. Middle- and upper-income parents endow their children with more cultural capital than do working-class and poverty-level parents. Because cultural capital is essential for acquiring an education, children with less cultural capital have fewer opportunities to succeed in school. For example, standardized tests that are used to group students by ability and to assign them to classes often measure students' cultural capital rather than their "natural" intelligence or aptitude. Thus, a circular effect occurs: Students with dominant cultural values are more highly rewarded by the educational system. In turn, the educational system teaches and reinforces those values that sustain the elite's position in society.

In her study of working-class women who returned to school after dropping out, the sociologist Wendy Luttrell (1997: 113–114) concluded that schools play a critical role in class reproduction and determining people's self-concept and identity:

> School denied but at the same time protected certain students' unearned advantages related to class, gender, and skin color in ways that made the women doubt their own value, voice, and abilities. . . . [When] the women were degraded by teachers and school officials for their speech, styles of dress, deportment, physical appearance, skin color, and forms of knowledge, they learned to recognize as "intelligent" or "valuable" only the styles, traits, and knowledge possessed by the economically advantaged students[:] . . . white, middle-class . . . behaviors and [appearance] . . . and urban or suburban mannerisms and styles of speech. Most important, [it was assumed that] those who possessed such cultural capital [were] entitled to their superior positions.

Tracking and Social Inequality Closely linked to the issue of cultural capital is how tracking in schools is related to social inequality. ***Tracking*** **refers to the practice of assigning students to specific curriculum groups and courses on the basis of their test scores, previous grades, or other criteria.** Conflict theorists believe that tracking seriously affects many students' educational performance and their overall academic accomplishments. In elementary schools, tracking is often referred to *ability grouping* and is based on the assumption that it is easier to teach a group of students who have similar abilities. However, class-based factors also affect which children are most likely to be placed

© Cleve Bryant/PhotoEdit

Children who are able to visit museums, libraries, and musical events may gain cultural capital that other children do not possess. What is cultural capital? Why is it important in the process of class reproduction?

in "high," "middle," or "low" groups, often referred to by such innocuous terms as "Blue Birds," "Red Birds," and "Yellow Birds." This practice is described by Ruben Navarrette, Jr. (1997: 274–275), who tells us about his own experience with tracking:

> One fateful day, in the second grade, my teacher decided to teach her class more efficiently by dividing it into six groups of five students each. Each group was assigned a geometric symbol to differentiate it from the others. There were the Circles. There were the Squares. There were the Triangles and Rectangles.
>
> I remember being a Hexagon. . . . I remember something else, an odd coincidence. The Hexagons were the smartest kids in the class. These distinctions are not lost on a child of seven. . . . Even in the second grade, my classmates and I knew who was smarter than whom. And on the day on which we were assigned our respective shapes, we knew that our teacher knew, too. As Hexagons, we would wait for her to call on us, then answer by hurrying to her with books and pencils in hand. We sat around a table in our "reading group," chattering excitedly to one another and basking in the intoxication of positive learning. We did not notice, did not care to notice, over our shoulders, the frustrated looks on the faces of Circles and Squares and Triangles who sat quietly at their desks, doodling on scratch paper or mumbling to one another. We knew also that, along with our geometric shapes, our books were different and that each group had different amounts of work to do. . . . The Circles had the easiest books and were assigned to read only a few pages at a time. . . . Not surprisingly, the Hexagons had the most difficult books of all, those with the biggest words and the fewest pictures, and we were expected to read the most pages.
>
> The result of all of this education by separation was exactly what the teacher had imagined that it would be: Students could, and did, learn at their own pace without being encumbered by one another. Some learned faster than others. Some, I realized only [later], did not learn at all.

As Navarrette suggests, tracking does make it possible for students to work together based on their perceived abilities and at their own pace; however, it also extracts a serious toll for students who are labeled as "underachievers" or "slow learners." Race, class, language, gender, and many other social categories may determine the placement of children in elementary tracking systems as much as or more than their actual academic abilities and interests.

The practice of tracking continues in middle school/junior high and high school. Although schools in some communities bring together students from diverse economic and racial/ethnic backgrounds, the students do not necessarily take the same courses or move on the same academic career paths (Gilbert, 2008). Numerous studies over the past three decades have found that ability grouping and tracking affect students' academic achievements and career choices (Oakes, 1985; Welner and Oakes, 2000). Education scholar Jeannie Oakes found that tracking affects students' perceptions of classroom goals and achievements, as the

cultural capital Pierre Bourdieu's term for people's social assets, including values, beliefs, attitudes, and competencies in language and culture.

tracking the assignment of students to specific curriculum groups and courses on the basis of their test scores, previous grades, or other criteria.

As Ruben Navarette, Jr., so powerfully describes, school is extremely tedious for underachieving students, who may find themselves "tracked" in such a way as to deny them upward mobility in the future.

following statements from high- and low-track students suggest:

- I want to be a lawyer and debate has taught me to dig for answers and get involved. I can express myself. (High Track English)
- To understand concepts and ideas and experiment with them. Also to work independently. (High Track Science)
- To behave in class. (Low Track English)
- To be a better listener in class. (Low Track English)
- I have learned that I should do my questions for the book when he asks me to. (Low Track Science) (Oakes, 1985: 86–89)

Perceptions of the students on the "low tracks" reflect the impact that years of tracking and lowered expectations can have on people's educational and career aspirations. Often, the educational track—vocational or college-bound—on which high school students are placed has a significant influence on their future educational and employment opportunities. Although the stated purpose of tracking systems is to permit students to study subjects that are suitable to their skills and interests, most research reveals that this purpose has not been achieved (Oakes, 1985; Welner and Oakes, 2000). Moreover, some social scientists believe that tracking is one of the most obvious mechanisms through which students of color and those from low-income families receive a diluted academic program, making it much more likely that they will fall even further behind their white, middle-class counterparts (see Miller, 1995). For example, a recent study of Latinas concluded that school practices such as tracking impose low expectations that create self-fulfilling prophecies for many of these young women (Ginorio and Huston, 2000). Instead of enhancing school performance, tracking systems may result in students dropping out of school or ending up in "dead-end" situations because they have not taken the courses required to go to college if they choose to do so.

The Hidden Curriculum According to conflict theorists, the ***hidden curriculum* is the transmission of cultural values and attitudes, such as conformity and obedience to authority, through implied demands found in the rules, routines, and regulations of schools** (Snyder, 1971). As compared to the published curriculum of schools, which includes information such as what courses a student is required to take, the hidden curriculum includes "the often unarticulated and unacknowledged things that students are taught in school" (Johnson, 2000: 143). For students from dominant groups in society, the way they are treated and what they learn in school tend to enhance their self-esteem and expectations that they will attain success. By contrast, students of color, students from lower-income families or families that have recently migrated to this country, and many young women may be treated in such a manner that they acquire inferior self-images and hold less optimistic views of their education options and future prospects. Let's look more closely at this argument.

Social Class and the Hidden Curriculum Although students from all social classes are subjected to the hidden curriculum, working-class and poverty-level students may be affected the most adversely (Polakow, 1993; Ballantine and Hammack, 2009; Oakes and Lipton, 2003). When teachers from middle- and upper-middle-class backgrounds instruct students from working- and lower-income families, the teachers often have a more structured classroom and a more controlling environment for students. These teachers may also have lower expectations for students' academic achievements. For example, one study of five elementary schools in different communities found significant differences in how knowledge was transmitted to students even though the general curriculum of the school was organized similarly (Anyon, 1980, 1997). Schools for working-class students emphasize procedures and rote memorization without much decision making, choice, or explanation of why something is done a particular way. Schools for middle-class students stress the processes (such as figuring and decision making) involved in getting the right answer. Schools for affluent students focus on creative activities in which students express their own ideas and apply them to the subject under consideration. Schools for students from elite families work to develop students' analytical powers and critical-thinking skills, applying abstract principles to problem solving.

Through the hidden curriculum, schools make working-class and poverty-level students aware that they will be expected to take orders from others, arrive at work punctually, follow bureaucratic rules, and experience high levels of boredom without complaining (Ballantine and Hammack, 2009). Over time, these students may be disqualified from higher education and barred from obtaining the credentials necessary for well-paid occupations and professions (Bowles and Gintis, 1976). Educational credentials are extremely important in societies that emphasize ***credentialism*—a process of social selection in which class advantage and social status are linked to the possession of academic qualifications** (Collins, 1979; Marshall, 1998). Credentialism is closely related to *meritocracy*—previously defined as a social system in which status is assumed to be acquired through individual ability and effort (Young, 1994/1958). Persons who acquire the appropriate credentials for a job are assumed to have gained the position through what they know, not who they are or whom they know. According to conflict theorists, the hidden curriculum determines in advance that the most valued credentials will primarily stay in the hands of the elites, so the United States is not actually as meritocratic as some might claim.

© Spencer Grant/PhotoEdit

Signs in this elementary classroom list the rules, rewards, and consequences of different types of student behavior. According to conflict theorists, schools impose rules on working-class and poverty-level students so that they will learn to follow orders and to be good employees in the workplace. How would functionalists and symbolic interactionists interpret these same signs?

Gender Bias and the Hidden Curriculum According to conflict theorists, gender bias is embedded in both the formal and the hidden curricula of schools. Although most girls and young women in the United States have a greater opportunity for education than those living in developing nations, their educational opportunities are not equal to those of boys and young men in their social class (see AAUW, 1995; Orenstein, 1995). For many years, reading materials, classroom activities, and treatment by teachers and peers contributed to a feeling among many girls and young women that they were less important than male students. Over time, this kind of differential treatment undermined females' self-esteem and discouraged them from taking certain courses, such as math and science, which were usually dominated by male teachers and students (Raffalli, 1994).

In recent years, some improvements have taken place in girls' education, as more females have enrolled in advanced placement or honors courses and in academic areas, such as math and science, where they had previously lagged (AAUW, 1998). However, girls are still not enrolled in higher-level science (such as physics) and computer sciences courses in the same numbers as boys. Girls and young women continue to make up only a small percentage of students in computer science and computer design classes, and the gender gap grows even wider from grade eight to grade eleven. Whereas male students are more likely to enroll in courses in which they learn how to develop computer programs, female students are more likely to enroll in clerical and data-entry classes (AAUW, 1998). Some researchers find that the hidden curriculum works against young women in that some educators do not provide females with as

hidden curriculum the transmission of cultural values and attitudes, such as conformity and obedience to authority, through implied demands found in rules, routines, and regulations of schools.

credentialism a process of social selection in which class advantage and social status are linked to the possession of academic qualifications.

much information about economic trends and the relationship among curriculum, course-taking choices, and career options as they provide to male students from middle- and upper-income families. Latinas are particularly affected by the hidden curriculum as it relates to gender, race/ethnicity, and language (AAUW, 1998).

Ethnicity, Language, and the Hidden Curriculum How are ethnicity and language related to the hidden curriculum? Teachers who are responsible for English as a Second Language (ESL) courses frequently find that they are not only teaching children English but also teaching them social skills, such as how to talk to other people in a polite manner that is acceptable to people in the prevalent culture. One ESL instructor described how the hidden curriculum works in her classes:

> I believe that teaching our ESL students what is considered good manners in the United States is very important. I spend a lot of time teaching students how to give and receive compliments, to thank someone for something, to answer the telephone, to ask directions, to say "thank you" and to make small talk. . . . Many educators believe that second language learners will acquire socially appropriate behavior and language simply by being with English-speaking natives. I don't think this should be left to chance. (Haynes, 1999)

Through the process of teaching a new language to students, some teachers also instill certain values and behaviors that they believe will be beneficial for the children as they adapt to their new cultural environment. However, this socialization process is not always viewed in the same manner by their parents. ESL teacher Judie Haynes (1999) describes what happened when a third-grade ESL student uttered a swear word in the classroom:

> An excited ESL student in my school recently used an "x-rated" expression in his third grade classroom. The teacher was understandably distressed and required that the student write an apology for homework. Even more upsetting to the teacher was that the student's parents did not take the infraction of school rules seriously. I explained to her that swearing does not have the same "shock" value in a person's second language as it does in their first. So the parents were not "shocked" by their child's use of this language. What is considered "shocking" or inappropriate language for the classroom must sometimes be directly taught.

As in other manifestations of the hidden curriculum, this example shows that students learn far more—both positively and negatively—than just the subject matter that is being taught in classrooms. They are exposed to a wide range of beliefs, values, attitudes, and behavioral expectations that are not directly related to subjects such as English, algebra, or history. Conflict theorists use these examples to show how inequality is structurally produced and reproduced by formal and informal socialization processes in schools and other educational settings.

Symbolic Interactionist Perspectives

Unlike functionalist analysts, who focus on the functions and dysfunctions of education, and conflict theorists, who focus on the relationship between education and inequality, symbolic interactionists focus on classroom communication patterns and educational practices, such as labeling, that affect students' self-concept and aspirations.

Labeling and the Self-Fulfilling Prophecy Chapter 7 explains that *labeling* is the process whereby a person is identified by others as possessing a specific characteristic or exhibiting a certain pattern of behavior (such as being deviant). According to symbolic interactionists, the process of labeling is directly related to the power and status of those persons who do the labeling and those who are being labeled. In schools, teachers and administrators are empowered to label children in various ways, including grades, written comments on deportment (classroom behavior), and placement in classes. For example, based on standardized test scores or classroom performance, educators label some children as "special ed" or low achievers, whereas others are labeled as average or "gifted and talented." For some students, labeling amounts to a *self-fulfilling prophecy*—previously defined as an unsubstantiated belief or prediction resulting in behavior that makes the originally false belief come true (Merton, 1968). A classic form of labeling and the self-fulfilling prophecy occurs through the use of IQ (intelligence quotient) tests, which claim to measure a person's inherent intelligence, apart from any family or school influences on the individual. In many school systems, IQ tests are used as one criterion in determining student placement in classes and ability groups.

Using Labeling Theory to Examine the IQ Debate The relationship between IQ testing and labeling theory has been of special interest to sociologists. In the 1960s, two social scientists conducted an experiment in an elementary school during which they intentionally misinformed teachers about the intelligence test scores of students in their classes (Rosenthal and Jacobson, 1968). Despite the fact that the students were randomly selected for the study and had no measurable differences in intelligence, the researchers informed the teachers that some of the students had extremely high IQ test scores, whereas others had average to below-average scores. As the researchers observed, the teachers began to teach "exceptional" students in a different manner from other students. In turn, the "exceptional" students began to outperform their "average" peers and to excel in their classwork. This study called attention to the labeling effect of IQ scores.

However, experiments such as this also raise other important issues: What if a teacher (as a result of stereotypes based on the relationship between IQ and race) believes that some students of color are less capable of learning? Will that teacher (often without realizing it) treat such students as if they are incapable of learning? In their controversial book *The Bell Curve: Intelligence and Class Structure in American Life,* Richard J. Herrnstein and Charles Murray (1994) argue that intelligence is genetically inherited and that people cannot be "smarter" than they are born to be, regardless of their environment or education. According to Herrnstein and Murray, certain racial–ethnic groups differ in average IQ and are likely to differ in "intelligence genes" as well. For example, they point out that on average, people living in Asia score higher on IQ tests than white Americans and that African Americans score 15 points lower on average than white Americans. Based on an all-white sample, the authors also concluded that low intelligence leads to social pathology, such as high rates of crime, dropping out of school, and winding up poor. In contrast, high intelligence typically leads to success, and family background plays only a secondary role.

Many scholars disagree with Herrnstein and Murray's research methods and conclusions. Two major flaws found in their approach were as follows: (1) the authors used biased statistics that underestimate the impact of hard-to-measure factors such as family background, and (2) they used scores from the Armed Forces Qualification Test, an exam that depends on the amount of schooling that people have completed. Thus, what the authors claim is immutable intelligence is actually acquired skills (Weinstein, 1997). Despite this refutation, the idea of inherited mental inferiority tends to acquire a life of its own when people want to believe that such differences exist (Duster, 1995; Hauser, 1995; Taylor, 1995). According to researchers, many African American and Mexican American children are placed in special education classes on the basis of IQ scores when the students are not fluent in English and thus cannot understand the directions given for the test. Moreover, when children are labeled as "special ed" students or as being *learning disabled,* these terms are social constructions that may lead to stigmatization and become a self-fulfilling prophecy (Carrier, 1986; Coles, 1987).

Labeling students based on IQ scores has been an issue for many decades. Immigrants from Southern and Eastern Europe—particularly from Italy, Poland, and Russia—who arrived in this country at the beginning of the twentieth century had lower IQ scores on average than did Northern European immigrants who had arrived earlier from nations such as Great Britain. For many of the white ethnic students, IQ testing became a self-fulfilling prophecy: Teachers did not expect them to do as well as children from a Northern European (WASP) family background and thus did not encourage them or give them an opportunity to overcome language barriers or other educational obstacles. Although many students persisted and achieved an education, the possibility that differences in IQ scores could be attributed to linguistic, cultural, and educational biases in the tests was largely ignored. Debates over the possible intellectual inferiority of white ethnic groups are unthinkable today, but arguments pertaining to African Americans and IQ continue to surface.

A self-fulfilling prophecy can also result from labeling students as gifted. Gifted students are considered to be those with above-average intellectual ability, academic aptitude, creative or productive thinking, or leadership skills (Ballantine and Hammack, 2009). When some students are labeled as better than others, they may achieve at a higher level because of the label. Ironically, such labeling may also result in discrimination against these students. For example, according to law professor Margaret Chon (1995: 238), the "myth of the superhuman Asian" creates a self-fulfilling prophecy for some Asian American students:

> When I was in college, I applied to the Air Force ROTC program. . . . I was given the most complete physical of my life [, and] I took an intelligence test. When I reported back to the ROTC staff, they looked glum. What is it? I thought. Did the physical

Question 2: Consider the following two statements: all farmers who are also ranchers cannot come near town; and most of the ranchers who are also farmers cannot surf. Which of the following statements MUST be true?

- ❍ Most of the farmers who cannot come near town can surf
- ❍ Only some farmers who ranch can surf near town
- ❍ A surfer who ranches and farms cannot surf near town
- ❍ Some ranchers who farm can come to town to learn to surf
- ❍ Any farmer who cannot surf also ranches

IQ tests containing items such as this are often used to place students in ability groups. Such placement can set the course of a person's entire education.

> turn up some life-threatening defect? . . . It turned out I had gotten the highest test score ever at my school. . . . Rather than feeling pleased and flattered, I felt like a sideshow freak. The recruiters were not happy either. I think our reactions had a lot to do with the fact that I did not resemble a typical recruit. I am a woman of East Asian, specifically Korean, descent. . . . They did not want me in ROTC no matter how "intelligent" I was.

According to Chon, painting Asian Americans as superintelligent makes it possible for others to pretend that they do not exist: "Governments ignore us because we've already made it. Schools won't recruit us because we do so well on the SATs. . . . Asian Americans seem almost invisible, except when there is a grocery store boycott—or when we're touted as the model minority" (Chon, 1995: 239–240).

Labeling and the self-fulfilling prophecy are not unique to U.S. schools. Around the globe, students are labeled by batteries of tests and teachers' evaluations of their attitudes, academic performance, and classroom behavior. Other problems in education, ranging from illiteracy and school discipline to unequal school financing and educational opportunities for students with disabilities, are concerns in elementary and secondary education in many nations.

Postmodernist Perspectives

As discussed in Chapter 15, postmodern theories often highlight *difference* and *irregularity* in society. From this perspective, education—like the family—is a social institution characterized by its permeability. In contemporary schools, a wide diversity of family kinship systems is recognized, and educators attempt to be substitute parents and promulgators of self-esteem in students. Urbanity is reflected in multicultural and anti-bias curricula that are initially introduced in early-childhood education. Similarly, autonomy is evidenced in policies such as voucher systems, under which parents have a choice about which schools their children will attend. Because the values of individual achievement and competition have so permeated contemporary home and school life, social adjustment (how to deal with others) has become of little importance to many people (Elkind, 1995).

How might a postmodern approach describe higher education? Postmodern views of higher education might incorporate the ideas of the sociologist George Ritzer (1998), who believes that "McUniversity" can be thought of as a means of educational consumption that allows students to consume educational services

Wellness/Recreation Center, University of Northern Iowa. Photo provided by University Marketing & Public Relations.

Many colleges and universities now offer students a wide array of new amenities and recreational facilities, such as this rock-climbing wall located in the Wellness/Recreation Center at the University of Northern Iowa.

and eventually obtain "goods" such as degrees and credentials:

> Students (and often, more importantly, their parents) are increasingly approaching the university as consumers; the university is fast becoming little more than another component of the consumer society. . . . Parents are, if anything, likely to be even more adept as consumers than their children and because of the burgeoning cost of higher education more apt to bring a consumerist mentality to it. (Ritzer, 1998: 151–152)

Savvy college and university administrators are aware of the permeability of higher education and the "students-as-consumers" model:

> [Students] want education to be nearby and to operate during convenient hours—preferably around the clock. They want to avoid traffic jams, to have easy, accessible and low cost parking, short lines, and polite and efficient personnel and services. They also want high-quality products but are eager for low costs. They are willing to shop—placing a premium on time and money. (Levine, 1993: 4)

To attract new students and enhance current students' opportunities for consumption, many campuses have student centers equipped with amenities such as food courts, ATMs, video games, Olympic-sized swimming pools, and massive rock-climbing walls. "High-tech" or "wired" campuses are also a major attraction for student consumers, and virtual classrooms make it possible for some students to earn college credit without having to look for a parking place at the traditional brick-and-mortar campus.

The permeability of contemporary universities may be so great that eventually it will be impossible to distinguish higher education from other means of consumption. For example, Ritzer (1998) believes that officials of "McUniversity" will start to emphasize the same kinds of production values as CNN or MTV, resulting in a simulated world of education somewhat like postmodernist views of Disneyland. Based on Baudrillard's fractal stage, where everything interpenetrates, Ritzer (1998: 160) predicts that we may enter a "transeducational" era: "Since education will be everywhere, since everything will be educational, in a sense nothing will be educational." Based on a postmodern approach, what do you believe will be the dominant means by which future students will consume educational services and goods at your college or university?

The Concept Quick Review summarizes the major theoretical perspectives on education.

Inequalities Among Elementary and Secondary Schools

As in colleges and universities, education in kindergarten through high school is a microcosm of many of the issues and problems facing the United States and other nations. Today, there are almost 15,000 school districts in the United States in what is probably the most decentralized system of public education in any high-income, developed nation of the world. Countries such as France have an education ministry that is officially responsible for every elementary school in the nation. Countries such as Japan have a centrally controlled curriculum; in nations such as England, national achievement tests are administered by the government, and students must pass them in order to advance to the next level of education (Lemann, 1997).

Inequality in Public Schools Versus Private Schools

Often, there is a perceived competition between public schools and private schools for students and financial resources. However, far more students and their parents are dependent on public schools than private ones for providing a high-quality education. Enrollment in U.S. elementary and secondary education (grades K through 12) totals 53 million, and these numbers are projected to increase to over 55 million by 2020 (Children's Defense Fund, 2002). Almost 90 percent of U.S. elementary and secondary students are educated in public schools. About 9.5 percent of all students are educated in low-tuition private schools, primarily Catholic (parochial) schools; only 1.5 percent of all students attend private schools with tuition of more than $5,000 per year (Lemann, 1997).

Private secondary boarding schools tend to be reserved for students from high-income families and for a few lower-income or subordinate-group students who are able to acquire academic or athletic scholarships that cover their tuition, room and board, and other expenses. The cost for seven-day tuition and room and board at

CONCEPT QUICK REVIEW

Sociological Perspectives on Education

	Key Points
Functionalist Perspectives	Education is one of the most important components of society: Schools teach students not only content but also to put group needs ahead of the individual's.
Conflict Perspectives	Schools perpetuate class, racial–ethnic, and gender inequalities through what they teach to whom.
Symbolic Interactionist Perspectives	Labeling and the self-fulfilling prophecy are an example of how students and teachers affect each other as they interpret their interactions.
Postmodernist Perspectives	In contemporary schools, educators attempt to become substitute parents and promulgators of self-esteem in students; students and their parents become the consumers of education.

secondary boarding schools is between $16,000 and $25,000 per year (American Universities Admission Program, 2003).

Among parents whose children attend private secondary schools, an important factor for a majority (51 percent) of the parents is the emphasis on academics that they believe exists in private (as opposed to public) schools. The most important factor for 17 percent of the parents is the moral and ethical standards that they believe private secondary schools instill in students. Overall, many families believe that private schools are a better choice for their children because they feel that these schools are more academically demanding, motivate students to learn, provide more-stringent discipline, and do not have many of the inadequacies found in public schools. However, according to some social analysts, there is little evidence to substantiate the claim that private schools (other than elite academies attended by the children of the wealthiest and most-influential families) are inherently better than public schools (Berliner and Biddle, 1995).

"Rich" schools and "poor" schools are readily identifiable by their buildings and equipment. What are the long-term social consequences of unequal funding for schools?

Unequal Funding of Public Schools

One of the biggest problems in public education today results from unequal funding. Why does unequal funding exist? Most educational funds come from state legislative appropriations and local property taxes (see ▶ Figure 16.2). As shown in Figure 16.2, state sources contribute 46.6 percent of public elementary–secondary school system revenue, 44.4 percent comes from local sources, and only 9 percent comes from federal sources. Much of the money from federal sources is earmarked for special programs for students who are disadvantaged (e.g., the Head Start program) or who have a disability. Although public school spending increases in some regions of the country each year, an even greater increase occurs in student enrollment.

Per-capita spending on public and secondary education varies widely from state to state (see ▶ Map 16.1). In part, this is because the local property-tax base has been eroding in central cities as major industries have relocated or gone out of business. Many middle- and upper-income families have moved to suburban areas with their own property-tax base so their children can attend relatively new schools equipped with the latest

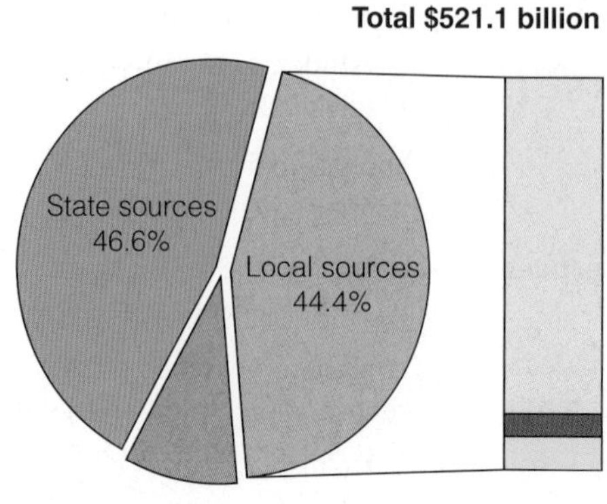

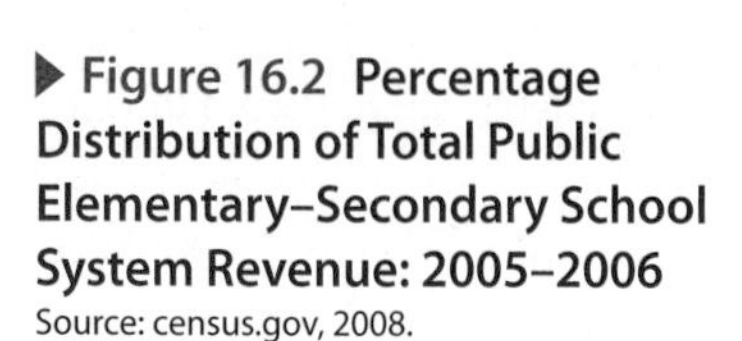
▶ Figure 16.2 Percentage Distribution of Total Public Elementary–Secondary School System Revenue: 2005–2006
Source: census.gov, 2008.

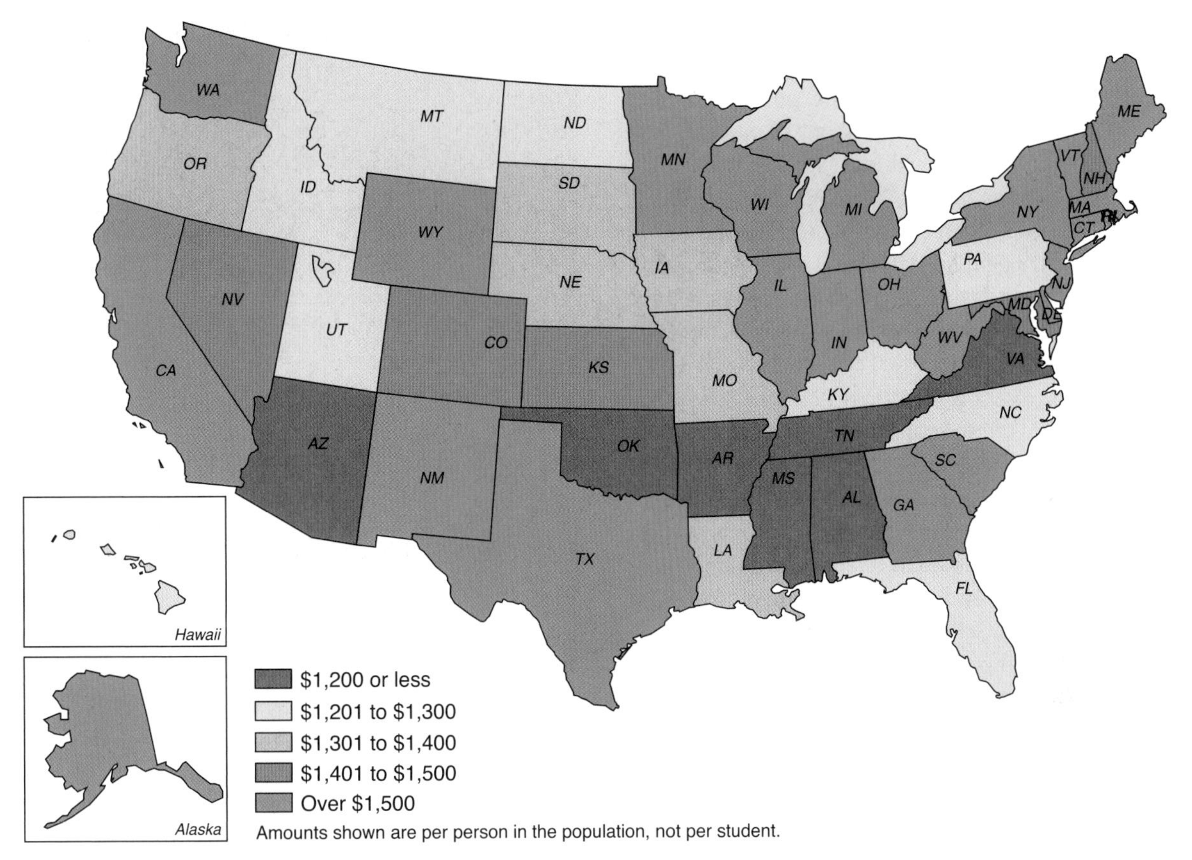

▶ **Map 16.1 Per-Capita Public Elementary and Secondary Spending by State**

Source: U.S. Census Bureau 2008.

textbooks and state-of-the-art computers—advantages that schools in central cities and poverty-ridden rural areas lack (see Kozol, 1991; Ballantine and Hammack, 2009).

Inequality Within Public School Systems

Another major problem in public school systems is the inequality that exists *within* local public school systems. Although the problem described in this section could be found in any major school district in the United States, we will focus on a case study conducted by education scholar John Devine (1996) regarding the New York City public schools. According to Devine, public high schools in New York City are stratified into higher-, middle-, and lower-tier schools. Although the New York school system does not officially designate high schools on this basis, the higher-tier schools are at the pinnacle of the educational structure. These schools provide students with a completely different educational environment and learning experience than the lower-tier schools. Higher-tier schools are highly specialized, admission is limited to students who pass a highly competitive entrance examination, and dropout rates are low. These schools turn out winners of academic achievement awards, and a large number of graduates attend prestigious universities. At the next level are middle-stratum schools that are academically comprehensive and located in fairly well-integrated neighborhoods. Referred to as "ed op" (educational option) schools, these high schools offer training for specific careers, such as health or computer science, and college prep courses. Some "ed op" schools serve as *magnet schools,* which offer a specialized curriculum that focuses on areas such as science, music, or art, and enroll high-achieving students from other neighborhoods.

According to Devine (1996: 23), in New York City it is not only the private schools of Manhattan and the Catholic schools that "siphon the best and the brightest—and the wealthiest—away from the public system . . . rather, the system itself does the most thorough job of drawing 'the best' to the top." In sharp contrast, students who attend schools in the lower tier tend to have low test scores and poor attendance. Most lower-tier schools are very large, overcrowded neighborhood schools located in highly segregated, deteriorating, and violent neighborhoods. Among New York City public schools, the lower-tier high schools have the highest dropout rates, the lowest graduation rates, the worst scores on standardized tests, the poorest attendance patterns, and the worst statistics on assault and possession of weapons.

Recent studies have found that the official dropout rate in New York City schools is about 20 percent, but if students who are "pushed out" are included, that number could be as high as 25 to 30 percent (Lewin and Medina, 2003). So that a school will not have low pass rates on stringent Regents exams, which a student must pass before receiving a high school diploma, some schools have allegedly been pushing out low-achieving students, getting them to leave school of their own accord so that the schools' statistics will not be tarnished if these students fail to graduate on time (Lewin and Medina, 2003). (To see how the media tend to treat the subject of standardized tests, see Box 16.2.) Push-out rates and other problems exist in schools across the nation because schools are a reflection of the larger political, economic, and social environment in which they exist. As such, schools also reflect the racial and ethnic divisions within the larger society.

Although many people believe that the United States is a racially integrated nation, a look at schools throughout the country reveals that many of them remain segregated or have become largely resegregated in recent decades.

Racial Segregation and Resegregation

In many areas of the United States, schools remain racially segregated or have become resegregated after earlier attempts at integration failed. In 1954 the U.S. Supreme Court ruled (in *Brown v. The Board of Education of Topeka, Kansas*) that "separate but equal" segregated schools are unconstitutional because they are inherently unequal. However, five decades later, racial segregation remains a fact of life in education.

Efforts to bring about *desegregation*—the abolition of legally sanctioned racial–ethnic segregation—or *integration*—the implementation of specific action to change the racial–ethnic and/or class composition of the student body—have failed in many districts throughout the country. Some school districts have bused students across town to achieve racial integration. Others have changed school attendance boundaries or introduced magnet schools with specialized programs such as science or the fine arts to change the racial–ethnic composition of schools. But school segregation does not exist in isolation. Racially segregated housing patterns are associated with the high rate of school segregation experienced by African American and Latina/o students. Although most racial segregation has affected African Americans, Latinas/os in the Southwest are even more likely than African American or white (non-Latina/o) students to attend segregated schools (Brooks and South, 1995).

Racial segregation and class segregation are apparent in many urban school districts where students of color constitute the vast majority of the student body. In contrast, private urban schools or upscale public schools in the suburbs are attended primarily by middle-class or upper-middle-class white students. In some school districts, more than 90 percent of the students are African American or Latina/o, and many live in low-income homes where English is not spoken (Judson, 1995).

Why are racially segregated schools a problem? Most research shows that such schools have serious negative consequences for minority students. According to recent studies, predominantly subordinate-group schools have lower student-retention rates, have higher teacher–student ratios, and employ less-qualified teachers who typically have lower expectations for their students (Feagin and Feagin, 2003).

Even in more-integrated schools, resegregation often occurs at the classroom level (Mickelson and Smith, 1995). Because of past racial discrimination and current socioeconomic inequalities, many children of color are placed in lower-level courses and special education classes. At the same time, non-Latina/o white and Asian American students are more likely to be enrolled in high-achievement courses and programs for the gifted and talented. However, racial segregation occurs even in required courses such as physical education (McLarin, 1994). Thus, the achievement gap between African American and white first-graders is much less than the gap that exists between African American and white twelfth-graders. According to researchers, schools tend to reinforce—rather than eliminate—the disadvantages of race and class during the years of a child's educational experience (Mickelson and Smith, 1995).

Box 16.2 Framing Education in the Media

"A Bad Report Card"—How the Media Frame Stories About U.S. Schools

- Test Scores Lag Behind Rising Grades (MSNBC.com, 2007)
- Failing Science (Futter, 2006)
- Most Students in Big Cities Lag Badly in Basic Science (Schemo, 2006)
- Students Ace State Tests, but Earn D's from U.S. (Dillon, 2005)

Headlines such as these have been used by the media in reporting the latest results of national exams given to students in the United States. Newspapers, Internet news sites, and television newscasts often use this type of framing—*report card framing*—in articles and news segments that describe students' academic achievement based on the National Assessment of Educational Progress (NAEP), the tests in reading, science, and mathematics that are given to fourth-, eighth-, and twelfth-grade students nationwide. The media tend to use the NAEP scores to "grade" students and schools in much the same way that educators use report cards to tell students and their parents how the students are doing.

What is report card framing? As discussed in previous chapters, media framing generally refers to the manner in which reporters and other journalists package information before it is presented to an audience. In regard to education reporting, journalists decide which key features or points to highlight in their coverage of education and which to minimize or leave out. For reporters on the "education beat," the results of nationwide student testing are among the most important things happening in schooling—and among the easiest subjects for an article—because the journalists receive numerous press releases about the NAEP and may be invited to press conferences by the Department of Education or other federal agencies during which the latest trends in test scores are discussed. Recent examples of report card framing include articles and news accounts in which journalists have puzzled over the fact that although high school seniors are taking more challenging courses and earning higher grades in school, their scores on national math and reading exams are lagging. According to media reports, trends that reveal no improvement in students' test scores over a period of years are particularly bad news for schools because educators are under pressure to place more emphasis on reading, math, and science in order to avoid funding cuts or school closings (MSNBC.com, 2007).

In sum, *report card framing* (when that concept is applied to education) refers to how the media typically give teachers, students, and schools "passing" or "failing" grades based on standardized test scores. This type of framing assumes that these scores are a good measure of the academic achievements of children, regardless of their family's income, their racial or ethnic heritage, or their urban/suburban/rural residential location. Report card framing also assumes that the tests are a good way to determine how well teachers teach and how well the U.S. educational system functions overall. However, the media use report card framing in other contexts as well, such as in "grading" whether the local police force is doing a good job or how well the various members of some legislative body are performing.

Are other framing devices also used by the media in reporting on education? Very definitely. Although report card framing is easily identifiable in media stories, the most frequently reported news on education typically relates to public school funding, staffing, and wage-and-benefit disputes. If we borrow the title of a popular ABBA song, "Money, Money, Money," to describe this type of framing, we see that the central focus of *"Money, Money, Money" framing* is on how much more money is needed for education than is now being spent, where the money might come from, and who is most likely to benefit from revenues that are expended. Most reporters' sources for this type of framing are government officials or public-school-affiliated sources who often have a vested interest in how the media report on their issues because such reports often influence policy makers and public opinion regarding the appropriate allocation of money for public schools.

One of the less frequent media approaches to reporting on education is an inside-the-classroom view of what is actually going on in schools. About the closest that most reporters get to this type of coverage is *students-as-human-interest-story framing,* which refers to situations such as when the media highlight students who have overcome great obstacles to achieve an education. One recent example is the *NBC Nightly News* (2007) account of students in a rural North Georgia school who were at high risk of dropping out of school but did not quit because their parents, neighbors, friends, coaches, and tutors kept praising them and telling them that they had the ability to finish high school. In the end, this approach paid off: The graduation rate was over 75 percent at that school, whereas other schools in the same area had much lower rates.

Reflect & Analyze

Should the media expand their reporting on education in the United States? Do some reporters miss out on what's really happening in classrooms across the nation because they primarily rely on press releases from government or school officials to set the agenda for reporting on education? What do you think?

Problems Within Elementary and Secondary Schools

The inequalities in educational opportunities that we have just discussed focus primarily on macrolevel issues in grades K through 12 and frequently involve decisions made by state legislatures, school boards, and school administrators. The problems we examine in this section, including school discipline, bullying and sexual harassment, and school violence, typically take place within the microcosm of individual schools, although the effect of these problems is much more widespread than at the original locations where they take place.

School Discipline and Teaching Styles

Is it possible for teachers and school administrators to handle the wide array of discipline problems that occur in many contemporary schools? Over the past three decades, this question has been asked by sociologists and education scholars; however, a lack of consensus exists regarding whether it is possible to make changes within the schools without reducing or eliminating larger structural problems in society that are intertwined with school problems. For example, one study conducted by the California Department of Education compared school discipline problems over the past sixty years. In the 1940s, some of the leading discipline problems included getting out of place in line, talking without permission during class, or not putting paper in the trash can (Collins and Frantz, 1993). By contrast, in the 1990s and into the twenty-first century, the leading discipline problems include violence, drug abuse, suicide, robbery, assault, and other forms of aggressive behavior.

Although schools in the past used corporal punishment (usually in the form of spankings or "paddling" of students) to discipline errant students, most states now prohibit the use of corporal punishment. Some school districts also prohibit disciplinary actions such as keeping children in during recess or expelling them from class. Moreover, many teachers and school administrators do not believe that corporal punishment or any other form of negative sanctions will deter the kinds of behavior that are occurring in classrooms and on the school grounds, as Evelyn Campbell, who has been teaching for more than thirty years, explains:

> Not in this day and time, I don't think corporal punishment will work. We've seen violence among students escalate in this district—kids carrying knives and weapons, kids with Uzis in their cars. A teacher dare not take that risk to hit a kid. For that reason alone, we need to stay away from corporal punishment. It could endanger a teacher's life. When they gave kids rights and parents rights, it meant hands off. We still have detention, expulsion. That's all we do. (qtd. in Collins and Frantz, 1993: 143)

Maintaining classroom order and discipline is often a difficult task for teachers. Changing views on how teachers and students should interact has created a perception among some students that they can control their schools and behave in an inappropriate manner in the classroom with few negative consequences.

Teaching and disciplinary styles may differ across racial and ethnic lines. Most studies on this issue have focused on the styles of white and African American teachers. According to one study, many white teachers try to ignore racial and ethnic differences in their students. As one teacher stated, "I don't see color, I only see children" (qtd. in Delpit, 1995: 177). However, other analysts believe that this approach suggests that there is something wrong with being black or brown and that color *should* be noticed. According to the education scholar Lisa Delpit (1995: 177), "If one does not see color, then one does not really see children. Children made 'invisible' in this manner become hard-pressed to see themselves as worthy of notice."

Can educational reforms reduce or eliminate school discipline problems? According to the education scholar Jean Anyon (1997), educational reform has a long way to go in bringing about better discipline—and other needed changes—because of structural inequalities in the larger society. Thus, when "blackness and whiteness, extreme poverty and relative affluence, cultural marginalization and social legitimacy come together—and conflict—within a school," the prospects of better discipline and other educational reform, including a reduction in bullying, teasing, sexual harassment, and school violence, are greatly diminished (based on Anyon, 1997: 36–37).

Bullying and Sexual Harassment

Although some people think of harassment as a "women's issue," in schools throughout the United States, both boys and girls report that they have experienced harassment, bullying, and teasing from their classmates. *Sexual harassment* is defined as "unwanted and unwelcome sexual behavior that interferes with your life. Sexual harassment is not behaviors that you like or want (for example, wanted kissing, touching, or flirting)" (AAUW, 2001). Based on this definition, many of the students believed that their academic experience had been diminished by the actions of other students, even as early as elementary school. And, as a spokesperson for the National Education Agency commented, "For children who are constantly picked on, ridiculed or harassed, school becomes torture" (qtd. in AAUW, 2001).

The National Centers for Disease Control (2008) reported the following findings regarding nonfatal victimization in U.S. schools:

- Students between the ages of 12 and 18 were the victims of about 628,200 violent crimes at school in 2005. These included rape, sexual and aggravated assault, and robbery.
- Thirty percent of students reported moderate ("sometimes") or frequent ("once a week or more") bullying. This included 13 percent as bully, 10.6 percent as a victim, and 6.3 percent as both.
- Young people who bully are more likely to get into fights, vandalize property, skip school, drop out of school, smoke, and drink alcohol.

It should be noted that not all injuries students experience at school are visible to other people. Some students experience depression, anxiety, and other psychological problems because of school harassment and violence. In 2005, for example, 6 percent of high school students who participated in a nationwide survey reported that they did not go to school on one or more of the previous thirty days because they feared for their safety (National Centers for Disease Control, 2008).

Dropping Out

Although there has been a slight decrease in the overall school dropout rate during this decade, each year almost one-third of public high school students fail to graduate from high school (see ▶ Figure 16.3 for graduation rates). Overall, every school day, 7,000 U.S. students leave high school never to return. Gender differences are evident in the dropout rate: Male students are consistently 8 percent less likely to graduate than female students, and the gap is as large as 14 percent between male and female African American students (National High School Center, 2008). Latinos/as (Hispanics) have the highest dropout rate (18.2 percent), followed by African Americans (9.1 percent) and non-Hispanic whites (7.8 percent) (U.S. Census Bureau, 2009).

Why are Latinos/as more likely to drop out of high school than are other racial and ethnic groups? First, the category of "Hispanic" or "Latino/a" incorporates a wide diversity of young people—including those who trace their origins to Mexico, Puerto Rico, Haiti, and countries in Central and South America—who may leave school for a variety of reasons. Second, some students may drop out of school partly because their teachers have labeled them as "troublemakers." Some students have been repeatedly expelled from school before they actually become "dropouts." In a study by the sociologist Howard Pinderhughes, one of the respondents—Rocco—discusses how he was expelled several times before he finally dropped out:

> I never liked school. The teachers all hated my guts because I didn't just accept what they had to say. I was kind of a wiseass, and that got me in trouble. But that's not why they kept kicking me out of school. Either I didn't come to school or I got in fights when I did come. It got so I got a reputation that people heard about and there was always some kid challengin' me. (qtd. in Pinderhughes, 1997: 58)

Some students may view school as a waste of time. Another respondent in Pinderhughes's study—Ronnie—explains why he dropped out:

> I thought [school] was a big waste of time. I knew older guys who had dropped out of the tenth grade and within a couple of years were makin' cash money as a mechanic or a carpenter. My brothers did the same thing. I didn't think I was learnin' no skill that I could get a job with in school. I wasn't very good in school. It was pretty damn boring. So I left. (qtd. in Pinderhughes, 1997: 58)

Students who drop out of school may be skeptical about the value of school even while they are still attending

Graduation Rates Among 50 Largest U.S. School Districts

BEST		WORST	
Fairfax County, Va.	82.5%	Detroit	21.7%
Wake County, N.C.	82.2	City of Baltimore, Md.	38.5
Baltimore County, Md.	81.9	New York City	38.9
Montgomery County, Md.	81.5	Milwaukee	43.1
Cypress-Fairbanks, Texas	81.3	Cleveland	43.8

▶ Figure 16.3 Top Five and Bottom Five Graduation Rankings Among the Fifty Largest Public School Districts in the United States

Source: Reproduced with permission from the June 21, 2006, issue of *The Christian Science Monitor* (www.csmonitor.com).

because they believe that school will not increase their job opportunities. Upon leaving school, many dropouts have high hopes of making some money and enjoying their newfound freedom. However, these feelings often turn to disappointment when they find that few jobs are available and that they do not meet the minimum educational requirements for any "good" jobs that exist (Pinderhughes, 1997).

In a study of at-risk Latino/a high school students, the sociologists Harriett Romo and Toni Falbo (1996) found that a growing number of students were taking the GED (General Equivalency Diploma) exam rather than graduating from high school. The most common reasons given for dropping out of school included lack of interest in school, serious personal problems, serious family problems, poor grades, and alcohol and/or drug problems (Romo and Falbo, 1996). Other studies have found that Latina/o students are disadvantaged in school districts that tend to view bilingualism and some cultural values as a liability rather than an asset. A study conducted by the AAUW found that the high school graduation rate for Latinas was lower than for girls in any other racial or ethnic group, although it was still higher than the graduation rate for Latinos (Ginorio and Huston, 2000). Compared to their white or Asian counterparts, Latinas were also less likely to take the SAT exam and were the least likely of any group of women to complete a bachelor's degree.

Violent rampages have become all too common at schools across the United States, and college and university campuses have not been immune to the problem. In 2007 Virginia Tech students, faculty, and administrators rose up with hope and courage to overcome the recent violence at their campus and to move forward.

School Safety and School Violence

Problems such as bullying, harassment, and high dropout rates are a major concern in education because schools should provide a safe and supportive environment where students and teachers alike can feel secure. Today, officials in schools ranging from the elementary years to two-year colleges and four-year universities are focusing on how to reduce or eliminate bullying, harassment, and violence. In many schools, teachers and counselors are instructed in anger management and peer mediation, and they are encouraged to develop classroom instruction that teaches values such as respect and responsibility (National Education Association, 2007). Some schools create partnerships with local law enforcement agencies and social service organizations to link issues of school safety to larger concerns about safety in the community and the nation. For example, the National Education Association's Safe Schools Program works with other national organizations to advocate for safe schools and communities and to create a positive learning environment for all students.

Clearly, some efforts to make schools a safe haven for students and teachers are paying off. Statistics related to school safety continue to show that U.S. schools are among the safest places for young people. According to "Indicators of School Crime and Safety," jointly released by the National Center for Education Statistics and the U.S. Department of Justice's Bureau of Justice Statistics, young people are more likely to be victims of violent crime at or near their home, on the streets, at commercial establishments, or at parks than they are at school (National Education Association, 2007). However, these statistics do not keep many people from believing that schools are becoming more dangerous with each passing year and that all schools should have high-tech surveillance equipment to help maintain a safe environment. As a result, many students attend classes in an academic environment that is somewhat similar to a prison. In some schools, all students, faculty, and staff are required to wear a photo ID around their necks or present a school-issued identity card upon entering the school grounds or buildings. Individuals entering school facilities are frequently required to pass through a weapons-scanning metal detector and have their backpacks or other personal items inspected or sent through an X-ray machine. Many schools are equipped with magnetic door locks and have a large staff of security guards who carry walkie-talkies and weapons. Faculty and security guards often use criminal justice vernacular such as "scanning," "holding areas," and "corridor sweeps" to describe their daily activities at the schools (Devine, 1996).

Even with all of these safety measures in place, violence and fear of violence continue to be pressing problems in schools throughout the United States. Each horrendous killing on a school campus further intensifies our fright and heightens our concern about how safe our schools really are. This concern extends from kindergarten through grade twelve because violent acts have resulted in numerous elementary and high school deaths in communities such as Jonesboro, Arkansas; Springfield, Oregon; Littleton, Colorado; Santee, California; Red Lake, Minnesota; and an Amish schoolhouse in rural Pennsylvania.

College and university campuses are also not immune to violence and multiple deaths, as deranged individuals have engaged in acts of personal terrorism at the expense of students, professors, and other victims. In 1966 Charles Whitman, a University of Texas at Austin student, went to the top of the university's twenty-seven-story tower and began firing shots at people walking around the campus below him. By the time that he was shot and killed by an Austin police officer, Whitman had taken the lives of fifteen people and seriously wounded thirty-one others. People who remember when Whitman went on his rampage sadly relived that experience of grief four decades later on April 16, 2007 (almost exactly eight years after the Columbine High school massacre in Littleton, Colorado, left fifteen people dead at the hands of two high school student gunmen), when Cho Seung-Hui, a Virginia Tech student, shot and killed thirty-two victims and himself on the college campus in Blacksburg, Virginia (Dewan and Broder, 2007). This latest act of violence on a school or university campus heightened concerns about how vulnerable people are even in supposedly safe settings such as a college dormitory or a classroom.

In the aftermath of the Virginia Tech tragedy, there was a massive outpouring of public sympathy and a call for greater campus security. As had been true following previous shootings, activist groups offered suggestions on how to make schools safer: Gun control advocates called for greater control over the licensing and ownership of firearms and for heightened police security on college campuses, whereas pro-gun advocates argued that students should be allowed to carry firearms on campus for their own protection (Buchholz, 2007). Neither of these approaches seems sufficient to deal with the real issue of school violence today. Given the problematic nature of discussions about safety and violence in elementary and secondary schools, as well as on college and university campuses, we must reassess at each level of schooling what we as a nation hope to accomplish through the education process and how this can best be achieved. Now let's look more closely at other opportunities and challenges in today's colleges and universities.

Opportunities and Challenges in Colleges and Universities

Who attends college? What sort of college or university do they attend? For students who complete high school, access to colleges and universities is determined not only by prior academic record but also by the ability to pay. One of the most remarkable success stories over the last fifty years has been the development and rapid growth of community colleges in the United States.

Opportunities and Challenges in Community Colleges

One of the fastest growing areas of U.S. higher education today is the community college; however, the history of two-year colleges goes back more than a century. Today, the oldest existing public two-year college is Joliet Junior College in Illinois. Like other community colleges, Joliet offers pre-baccalaureate programs for students planning to transfer to a four-year university, occupational education leading directly to employment, adult education and literacy programs, work-force and workplace development services, and support services to help students succeed (Joliet Junior College, 2005). Originally, two-year colleges focused on general liberal arts courses, but during the 1930s these institutions began offering job-training programs. Following World War II, the GI Bill of Rights provided the opportunity for more people to attend college, and, in 1948, a presidential commission report called for the establishment of a network of public community colleges that would charge little or no tuition, serve as cultural centers, be comprehensive in their program offerings, and serve the area in which they were located (Vaughan, 2000).

Hundreds of community colleges were opened across the nation during the 1960s, and the number of such institutions has steadily increased since that time as community colleges have responded to the needs of their students and local communities. Community colleges offer a variety of courses, some of which are referred to as "transfer courses" in which students earn credits that are fully transferable to a four-year college or university. Other courses are in technical/occupational programs, which provide formal instruction in fields such as nursing, emergency medical technology, plumbing, carpentry, and computer information technology. Community colleges also pride themselves on offering remedial education for students who need to gain additional background or competence in a subject, as well as courses to benefit those international students who need assistance in learning a new language or developing other skills. Finally, community colleges offer continuing education or lifelong learning courses and work-force development activities in which the schools partner with business and industry to provide skilled workers in their communities.

Did you know that community colleges educate about half of the nation's undergraduates? According to the American Association of Community Colleges (2005), there are a total of 1,166 community colleges (including public and private colleges) in the United States, and these institutions enroll almost 12 million students in credit and noncredit courses. Community college enrollment accounts for 46 percent of all U.S. undergraduates.

Who benefits most from community colleges? Community colleges provide significant educational opportunities to students across lines of income, gender, and race/ethnicity. Because community colleges are more affordable, with an average of about one-half the tuition and fees of the typical four-year college, more students are able to take advantage of the educational opportunities provided in their community. According to the American Association of Community Colleges (2005), almost 40 percent of all community college students receive financial aid to help meet the $1,076 average annual tuition. Women make up a slight majority (58 percent) of community college students, and for working women and mothers of young children, these schools provide a unique opportunity to attend classes on a part-time basis as their schedule permits. Men also benefit from flexible scheduling because they can work part time or full time while enrolled in school. About 62 percent of all community college students are enrolled part time, while 36 percent are full-time students (taking 12 or more credit hours each semester).

Although students of color remain underrepresented in many U.S. colleges and universities, African Americans, Latinos/Latinas, Asian Americans and Pacific Islanders, and Native Americans have found opportunities in community colleges that are not available elsewhere. Did you know that 47 percent of African American undergraduate students attend a community college? How about 56 percent of Hispanic students? Likewise, 48 percent of Asian/Pacific Islander and 57 percent of Native American students attend community colleges. For students across all racial/ethnic categories, more years of formal education translate into higher earnings and employment opportunities than might otherwise be available to them. These are important statistics when we think of how good an investment community colleges are for students and communities.

One of the greatest challenges facing community colleges today is money. Across the nation, state and local governments struggling to balance their budgets have slashed funding for community colleges. In a number of regions, these cuts have been so severe that schools have been seriously limited in their ability to meet the needs of their students. In some cases, colleges have terminated programs, slashed course offerings, reduced the number of faculty, and eliminated essential student services. Many people are hopeful that the Obama administration's "American Graduation Initiative," discussed earlier in this chapter, will strengthen community colleges and offer greater opportunities for students to achieve their educational goals. Of course, limited resources are a problem not only for community colleges but also for many four-year colleges and universities.

Opportunities and Challenges in Four-Year Colleges and Universities

More than ten million students attend public or private four-year colleges or universities in the United States. Whereas community colleges award certificates and associate degrees, four-year institutions offer a variety of degrees, including the bachelor's degree, master's degree, and the doctorate, the highest degree awarded. Some also award professional degrees in fields such as law or medicine. According to the Association of American Colleges and Universities, providing a liberal education is the goal of many institutions of higher education. Liberal education is a philosophy of education that aims to empower individuals, liberate the mind from ignorance, and cultivate social responsibility. For this reason, four-year schools typically offer a general education curriculum that gives students exposure to multiple disciplines and ways of knowing, along with more in-depth study (known as a "major") in at least one area of concentration. Having a liberal education provides opportunities for students in that it offers them a diversity of ideas and experiences that will help them not only in a career but in their interpersonal relationships and civic engagements as well. Today, it is increasingly important for people to acquire education beyond the high school level because the demand for college-educated workers has risen faster than the supply. Average earnings of college graduates are much higher than those of persons with only a high school diploma. However, many challenges are faced by four-year institutions, ranging from the cost of higher education and problems with students timely completing a degree program to racial and ethnic differences in enrollment and lack of faculty diversity. We now turn to the key challenges in higher education.

The Soaring Cost of a College Education

What does a college education cost? Studies by the College Board (a nonprofit organization that provides tests and many other educational services for students, schools, and colleges) have found that a college education is quite expensive and that increases in average yearly tuition for four-year colleges are higher than the overall rate of inflation (Bagby, 1997). Although public institutions such as community colleges and state colleges and universities typically have lower tuition and overall costs—because they are funded primarily by tax dollars—than private colleges have, the cost of attending public institutions has increased dramatically over the past decade. The total number of low-income students has dropped since the 1980s as a result of declining scholarship funds and also because many students must work full time or part time to pay for their education.

According to some social analysts, a college education is a bargain—even at about $90 a day for private schools or $35 for public schools—because for their money students receive instruction, room, board, and other amenities such as athletic facilities and job placement services. However, other analysts believe that the high cost of a college education reproduces the existing class system: Students who lack money may be denied access to higher education, and those who are able to attend college tend to receive different types of education based on their ability to pay. For example, a community college student who receives an associate degree or completes a certificate program may be prepared for a position in the middle of the occupational status range, such as a dental assistant, computer programmer, or auto mechanic (Gilbert, 2008). In contrast, university graduates with four-year degrees are more likely to find initial employment with firms where they stand a chance of being promoted to high-level management and executive positions. Although higher education may be a source of upward mobility for talented young people from poor families, the U.S. system of higher education is sufficiently stratified that it may also reproduce the existing class structure (Gilbert, 2008).

© BRIAN SNYDER/REUTERS/Landov

Soaring costs of both public and private institutions of higher education are a pressing problem for today's college students and their parents. What factors have contributed to the higher overall costs of obtaining a college degree?

Racial and Ethnic Differences in Enrollment

How does college enrollment differ by race and ethnicity? People of color (who are more likely than the average white student to be from lower-income families) are underrepresented in higher education. However, some increases in minority enrollment have occurred over the past three decades. For example, the total college enrollment for all institutions of higher education in fall 2006 was 17,758,900, with Latinas/os accounting for 1,964,300 students (11 percent), up from 1,166,100 in fall 1996 (*Chronicle of Higher Education*, 2008).

African American enrollment increased from 1,505,600 in fall 1996 to 2,279,600 (13 percent) in fall 2006. Gender differences are evident in African American enrollment: Women accounted for 1,484,200, or about 65 percent, of all African American college students in 2006 (*Chronicle of Higher Education*, 2008). There has been some increase in Native American enrollment rates, moving up from 137,600 in 1996 to 181,100 in 2006; however, this still amounts to only .01 percent of all student enrollment. About 45 percent of Native Americans attend two-year community colleges. A number of these colleges are referred to as tribal colleges because they are located on reservations, where they were originally founded to overcome racism experienced by Native American students in traditional colleges and to shrink the high dropout rate among Native American college students. There are now about thirty-five two-year and four-year colleges chartered and run by the Native American nations. The proportionately low number of people of color enrolled in colleges and universities is reflected in the educational achievement of people age twenty-five and over, as shown in the "Census Profiles" feature.

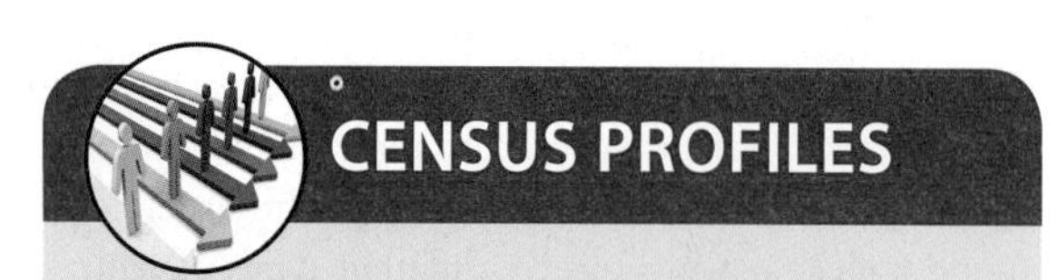

Educational Achievement of Persons Age 25 and Over

The Census Bureau asks people to indicate the highest degree or level of schooling they have completed. Sixteen categories, ranging from "no schooling completed" to "doctorate degree," are set forth as responses on the form that is used; however, we are looking only at the categories of high school graduate and above.

As shown below, census data reflect that the highest levels of educational attainment are held by Asian Americans, followed by non-Hispanic white respondents. For these statistics to change significantly, greater educational opportunities and more-affordable higher education would need to be readily available to African Americans and Latinos/as, who historically have experienced racial discrimination and inadequately funded public schools with high dropout rates and low high school graduation rates.

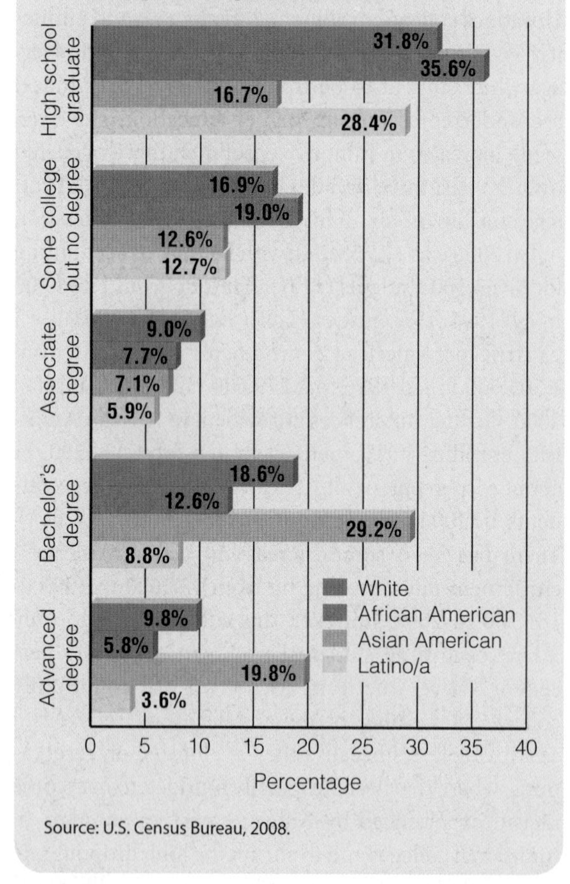

Source: U.S. Census Bureau, 2008.

Despite some recent gains in higher education, students of color continue to experience problems of prejudice and discrimination at predominantly white colleges and universities. According to the sociologists Joe R. Feagin, Hernán Vera, and Nikitah Imani (1996), racial discrimination presents major barriers for African American students in many colleges and universities. The barriers are erected by white students, professors, administrators, police officers, and other staff members. The authors concluded the following:

> [T]he black students and parents we interviewed call attention to African American students not being recognized as full members of the campus community. From university publications to the daily rhythm of life on campus, many symbols, comments, and actions suggest to black students that they really do not belong. The strength of their perceptions and feelings on these matters underscores the need for making majority-white educational institutions more welcoming to students whose cultures are not those of the dominant group. (Feagin, Vera, and Imani, 1996: 173)

Similarly, the sociologist Felix M. Padilla (1997) found that Latino/a students and professors in a major university sought to overcome isolation and discrimination by forging their own alliances and developing specialized curricula. According to Padilla (1997), Latino/a professors must speak out against injustices against Latino/a students, but this is difficult on campuses where Latina/o scholars are underrepresented in tenured faculty positions.

The Lack of Faculty Diversity

Despite the widely held assumption that there has been a significant increase in the number of minority professors, the latest figures available (fall 2005) indicate that African Americans account for only 35,458 (.05 percent) of the 675,624 full-time faculty members holding ranks ranging from professor to instructor/lecturer. By contrast, white (non-Hispanic) faculty account for 527,900 (78 percent) of the 675,624 full-time educators. Asian Americans totaled 48,457 (.07 percent), Hispanics numbered 22,818 (.03 percent), and Native Americans totaled 3,231 (.005 percent). Although there has been a slight increase in their overall representation, minority scholars are in the lowest tiers of the academic profession. In 2005, 28 percent of the white faculty members and 22 percent of the Asian American faculty members surveyed were full professors (with tenure), compared with only 15 percent of the African American and 16 percent of Latino/a faculty members (*Chronicle of Higher Education,* 2008). Gender is also a factor in faculty diversity. In all ranks and racial and ethnic categories, men made up 59.4 percent of the full-time faculty in 2005, while women accounted for 40.6 percent. Across racial and ethnic categories, women are underrepresented at the level of full professor and overrepresented at the assistant professor and instructor levels. Although assistant professors may be on a "tenure track," neither of these last two ranks provides the security of tenured positions.

The Continuing Debate Over Affirmative Action

For some periods of time, affirmative action drops out of the headlines and campus debates. However, some topic eventually arises, such as the confirmation of a new U.S. Supreme Court Justice, and the issue once again becomes a "hot topic" for debate on campuses and on radio and cable television talk shows. Why does the issue of affirmative action continue to generate such a controversy among people? And what is affirmative action anyway?

Affirmative action is a term that describes policies or procedures that are intended to promote equal opportunity for categories of people deemed to have been previously excluded from equality in education, employment, and other fields on the basis of characteristics such as race or ethnicity.

Education was one of the earliest targets of social policy pertaining to civil rights in the United States. Increased educational opportunity has been a goal of many subordinate-group members because of the widely held belief that education is the key to economic and social advancement. Beginning in the 1970s, most public and private colleges developed guidelines for admissions, financial aid, scholarships, and faculty hiring that took race, ethnicity, and gender into account. These affirmative action policies were challenged in a number of lawsuits, especially when the policies involved public colleges. Critics of affirmative action often assert that these policies amount to *reverse discrimination,* a term that describes a situation in which a person who is better qualified is denied enrollment in an educational program or employment in a specific position as a result of another person receiving preferential treatment as a result of affirmative action.

In 2003 the U.S. Supreme Court ruled in *Grutter v. Bollinger* (involving admissions policies of the University of Michigan's law school) and *Gratz v. Bollinger* (involving the undergraduate admissions policies of the same university) that race can be a factor for universities in shaping their admissions programs, but only within carefully defined limits. Some analysts believe that these cases were a victory for affirmative action; other analysts believe that the effect of these decisions will be limited in scope. One thing remains clear in either event: Our discussions regarding access to higher education—particularly regarding the way that access is influenced by income, race/ethnicity, gender, nationality, and other characteristics and attributes—are far from over as our country grows increasingly diverse in its population.

Continuing Issues and Future Trends in Education

Many books have been written about the issues facing education in the United States and other nations throughout the world. In this section, we will examine only a few of the concerns that are currently being discussed.

Academic Standards and Functional Illiteracy

Some social analysts believe that academic standards are not high enough in U.S. schools. As part of their evidence for this concern, they cite the rate of functional illiteracy in this country. ***Functional illiteracy* is the inability to read and/or write at the skill level necessary for carrying out everyday tasks.** It is estimated that 56 percent of adult Latinas/os are functionally illiterate in English, compared with 44 percent of adult African Americans and 16 percent of adult (non-Latino/a) whites. How can the United States have such a high rate of functional illiteracy when we have a relatively high rate of high school graduation? According to one study conducted by the U.S. Department of Education, many people who are functionally illiterate have graduated from high school. In fact, half of the people who scored in the lowest 20 percent on a literacy test were high school graduates. Today, 15 to 20 percent of people who have graduated from high school cannot read at the sixth-grade level (Kaplan, 1993).

However, we must distinguish between *functional* illiteracy and illiteracy. The National Literacy Act of 1991 defines *literacy* as "an individual's ability to read, write, and speak English and compute and solve problems at levels of proficiency necessary to function on the job and in society, to achieve one's goals, and to develop one's knowledge and potential" (U.S. Department of Education, 1993: 3). Based on this definition, the U.S. Department of Education concluded in its research that nearly half (95 million people) of the adult U.S. population was illiterate when it comes to tasks such as understanding a bus schedule, filling out

functional illiteracy the inability to read and/or write at the skill level necessary for carrying out everyday tasks.

a bank deposit slip, or computing the cost of having some work done (Kaplan, 1993). However, other social analysts note that this report classified people as illiterate on the basis of *one* reading comprehension test and did not take into account that English is not the native language of many respondents or that some respondents were visually impaired or had physical or mental disabilities (Berliner and Biddle, 1995).

Most suggestions for reducing both illiteracy and functional illiteracy have focused on school reforms such as more testing of students and teachers, increasing the requirements for high school graduation, and increasing the number of school days per year. Although some critics believe that the illiteracy problem is largely tied to high rates of immigration, the author Jonathan Kozol (1986), who has extensively studied the problem, argues that illiteracy is a homegrown problem that is closely linked to problems of poverty and segregated schools, both of which exist even without immigration. However, illiteracy is a global problem as well as a national one (see ◆ Table 16.2).

◆ **Table 16.2 Nations with the Lowest Literacy Rates**

Country	Literacy Rate
Burkina Faso	23.6%
Mali	24.0%
Chad	25.7%
Niger	29%
Guinea	30%
Benin	35%
Ethiopia	36%
Mozambique	39%

Source: United Nations, 2008.

The Debate Over Bilingual Education

As the United States becomes more racially, ethnically, and culturally diverse, schools become the focus of debate on how to best educate children who do not speak English as their primary language. The issue of bilingual and bicultural education is not a new one. During the large waves of immigration to the United States in the late nineteenth and early twentieth centuries, children often received classroom instruction from teachers who spoke the children's native language. For example, Italian American students studied Italian in school, and some students were instructed in the core subjects in their native language until they achieved proficiency in English. During World War I, however, members of white ethnic groups such as German Americans believed that they should show their loyalty to the United States and—as a consequence—that their children should be taught in English. As a result, there was little bilingual education until the years following World War II, when a large influx of Mexican and Asian immigrants arrived in the United States, and many schools enrolled large numbers of students who did not speak English.

© Don Smetzer/PhotoEdit

Obtaining a quality education is a problem across all age categories. Today, many adults are enrolled in classes that will help them learn to read, write, and speak English.

The issue of bilingual education resurfaced in the 1960s with the civil rights movement. In 1968, Congress mandated that public schools must provide non-English-speaking students with *bilingual education*—instruction in both a non-English language and in English. However, few school districts had the necessary money or zeal to hire teachers who could provide instruction in two or more languages. Only after a U.S. Supreme Court decision did bilingual education become a reality in this nation. In *Lau v. Nichols* (1974), the Court affirmed the responsibility of the states and local districts for providing appropriate education for minority-language students. As officials in school districts implemented bilingual programs, however, they were confronted by opposition from some members of the general public and from some political leaders, even as the number of non-English-speaking people living in the United States increased dramatically.

How successful are bilingual education programs? Not all social analysts and education scholars agree on the answer to this question. Critics point out that many children remain in the programs far too long. One report criticized schools in New York for allowing tens of thousands of students to remain in transitional bilingual education (TBE) programs for up to six years (Ravitch and Viteritti, 1997). These critics believe that TBE programs are much less efficient in bringing students into the English-speaking mainstream than are classes conducted primarily in English (Ravitch and Viteritti, 1997). But other social analysts disagree because they believe that TBE programs are highly effective in helping children build competency both in their native language and in English. Moreover, advocates of TBE believe that all children in U.S. public schools should have exposure to more than one language (Berliner and Biddle, 1995).

Equalizing Opportunities for Students with Disabilities

Another recent concern in education has been how to provide better educational opportunities for students with disabilities. As discussed in Chapter 18, the term *disability* has a wide range of definitions. For the purposes of this chapter, disability is regarded as any physical and/or mental condition that limits students' access to, or full involvement in, school life. Along with other provisions, the Americans with Disabilities Act of 1990 requires schools to make their facilities, services, activities, and programs accessible to people with disabilities. The law that specifically covers the treatment of children with disabilities is the Individuals with Disabilities Education Act (IDEA), which mandates that students with disabilities must receive free and appropriate education. Many schools have attempted to *mainstream* children with disabilities by providing *inclusion programs,* under which the special education curriculum is integrated with the regular education program and each child receives an *individualized education plan* that provides annual educational goals (Weinhouse and Weinhouse, 1994). *Inclusion* means that children with disabilities work with a wide variety of people; over the course of a day, children may interact with their regular education teacher, the special education teacher, a speech therapist, an occupational therapist, a physical therapist, and a resource teacher, depending on the child's individual needs.

Although much remains to be done, recent measures to enhance education for children with disabilities have increased the inclusion of many young people who were formerly excluded or marginalized in the educational system. But the problem of equal educational opportunities does not end at the elementary and high school levels for students with disabilities. If these students complete high school and continue on to college, they find new sets of physical and academic barriers that limit their access to higher education. However, many colleges and universities have provided relatively inexpensive accommodations to make facilities more accessible to students with disabilities. For example, special computers and wheelchair-accessible showers can make a major difference in whether or not the students can complete a college education.

Despite the difficulties associated with U.S. public schools, many people have not lost hope that schools can be improved. Some individuals have begun their own initiatives to make a difference in education (see Box 16.3). Others have sought alternatives for their children's education, including school vouchers, charter schools, and home schooling.

School Vouchers

School vouchers have been a topic of controversy for some time in public education. During the George W. Bush administration, some elected officials, business leaders, and educators advocated school voucher programs, which give parents the choice of what school their child will attend. Although vouchers have been widely discussed for many years, only a few school systems currently have programs under which public funds (tax dollars) are provided to students so that they can pay tuition to attend either public or private schools of their choice. According to researchers who assessed the voucher program in Cleveland, Ohio (one of the districts in which such a program exists, although vouchers are provided only to low-income students there), two-thirds of the parents whose children received vouchers were "very satisfied" with the academic quality of the private or parochial school their child attended, whereas less than 30 percent of the parents who applied for vouchers but whose children remained in the public schools were satisfied with

Computers make it possible for students with a disability to gain educational opportunities previously unavailable to them. Specialized software allows this hearing-impaired student to complete her assignment.

Box 16.3 You Can Make a Difference

Reaching Out to Youth: College Student Tutors

Tutoring is the highlight of my week. While going to a university, it is easy to lose a sense of curiosity and enthusiasm for learning because we get caught up in homework and grades. But, when I am at Martin Luther King Elementary, I feel refreshed and invigorated because of the excitement and curiosity with which young students approach their learning. . . . For this reason, I think that I get more out of this tutoring experience at times than the kids do. I have so much fun working with students and feel lucky to have the chance to help them learn.

—Zahra, a University of Washington pre-med student, describing how it feels to be a tutor at a local public school (qtd. in Stickler, 2004)

Across the nation, increasing numbers of college students such as Zahra are actively participating in tutoring and mentoring programs in kindergarten through grade twelve. Some college student tutors are education majors, but many come from diverse undergraduate majors including pre-med, engineering, the sciences and math, humanities, and social sciences such as sociology, psychology, and women's studies. At the University of Washington, for example, only about 40 percent of the tutors are planning to go into teaching as a career (Stickler, 2004).

College students tutoring students in local public schools is a win–win situation for everyone. Public school teachers benefit from the presence of college tutors in their classroom because it provides the teachers with a chance to work individually with more students or to have the tutor assist a child who needs special help in reading, writing, spelling, math, science, or some other subject. Students in kindergarten through grade twelve benefit by having a college tutor because they have the opportunity to learn in a one-on-one situation and to ask questions about the subject matter that they otherwise might not ask their teacher. With the assistance of tutors, top students are able to achieve even more than they otherwise would, while middle- and lower-tier students are able to gain not only academic skills but also a greater feeling of confidence as they succeed in school.

For college students, tutoring provides an opportunity to make a difference in the lives of children and young people who are looking for role models closer to their own age with whom they can identify. Gaining knowledge of the real world through field experience and service activities is an important component of a college education. For individuals who plan to go into the education field, tutoring in the classroom provides hand-on experience that cannot be gained in the standard college classroom (Musick, 2004). For college students who are planning to go into other careers, tutoring in public schools opens up new worlds of communication and social interaction with young people from diverse backgrounds. Even as college students provide a service for kids and local communities, the tutors also gain valuable experience and insights that may be useful throughout life. As Jill, another University of Washington student tutor, stated,

> I would, without hesitation, encourage anyone [who is] up to a meaningful and worthwhile challenge: to get involved in [the tutoring program]. Not only does the organization provide an incredible service to the kids and communities it holds classes in, but it offers real life learning unlike anything found in a textbook or university classroom. (qtd. in Stickler, 2004)

Are you interested in making a difference in education? Would you like to learn more about children and young people and help them meet their educational needs? If so, tutoring might be a good avenue for you to begin to develop this interest, and you may wish to find out what tutoring programs are available at your college or university.

their children's education (Lewis, 1997). Critics of the school voucher system believe that such programs undermine public education and might cause the public school system to collapse. Supporters of the voucher system believe that fairness is a central issue: The poor should have the same chance for a quality education as the more affluent (Lacayo, 1997). According to this approach, vouchers make it possible for poor children to leave behind the problems of inner-city public schools and find better educational opportunities elsewhere. Early in his administration, President Obama indicated that he believed that vouchers might hurt the public school system. One Washington, D.C., program was curtailed that included about 1,700 children a year, and this event produced a new controversy about school vouchers. The history of the debate over school vouchers is further discussed in Box 16.4.

Charter Schools and "For Profit" Schools

As compared to the voucher system, the charter school movement creates public schools that are free from many of the bureaucratic rules that often limit classroom performance. These schools operate under a charter contract negotiated by the school's organizers (often parents or teachers) and a sponsor (usually a local school board, a state board of education, or a university) that oversees the provisions of the contract. A charter school is freed from the day-to-day bureaucracy of a larger school district and may provide more autonomy for individual students and teachers. Critics of the charter school movement argue that it takes money away from conventional public schools. But, according to some education scholars, charter schools serve an important function in education: A large number of minority students receive a higher-quality education than they would in the public schools (Bierlein, 1997).

Other recent approaches for improving education include "contracting out"—hiring for-profit companies on a contract basis to operate public schools. Most contracting is done by three private companies: Whittle Communications' Edison Project, Sabis International Education Alternatives Incorporated, and Alternative Public Schools (Hill, 1997). Although many have applauded the innovative nature of this approach, critics are concerned about the privatization of education, which they believe undermines the public school system.

Home Schooling

A final alternative, home schooling, has been chosen by some parents who hope to avoid the problems of public schools while providing a quality education for their children. It is estimated that about 1.5 million children are home-schooled in grades K through 12 (National Center for Education Statistics, 2008). This is a significant increase from the estimated 1.1 million students who were home-schooled in 2003. The primary reasons that parents indicated for preferring to home-school their children were (1) concern about the school environment, (2) the desire to provide religious or moral instruction, and (3) dissatisfaction with the academic instruction available at traditional schools. Typically, home-schoolers' parents are better educated, on average, than other parents; however, their income is about the same. Researchers have found that boys and girls are equally likely to be home-schooled.

Parents who educate their children at home believe that their children are receiving a better education at home because instruction can be individualized to the needs and interests of their children. Some parents also indicate religious reasons for their decision to home-school their children. An association of home-schoolers now provides communication links for parents and children, and technological advances in computers and the Internet have made it possible for home-schoolers to gain information and communicate with one another. According to home-schooling advocates, home-school students typically have high academic achievements and high rates of employment.

Critics of home schooling question the knowledge of most parents and their competence to educate their own children at home, particularly in rapidly changing academic subjects such as science and computer technology. Some states have passed accountability laws that must be met by parents who teach their children at home. For example, Florida requires that parents register their children with their school districts or with umbrella schools that maintain and submit records to the state. Parents must maintain portfolios of their children's work and submit annual evaluations, including grades on standardized exams, to school officials (Miller, 2003). However, many people believe that the home-schooling movement will grow rapidly if problems in public education are not alleviated in the near future.

Concluding Thoughts

What will be the future of education in the United States? The answer to this question, particularly in regard to public elementary and secondary schools in this country, is linked to the No Child Left Behind Act of 2001. The purpose of this act is to close the achievement gap between rich and poor students. No Child Left Behind represents the most far-reaching educational reform to be implemented since compulsory education laws were passed early in the twentieth century. Under this law, schools are to be held accountable for students' learning, and specific steps are set forth toward producing a more accountable education system. The law requires states to test every student's progress toward meeting specific standards that were

Home schooling has grown in popularity in recent decades as parents have sought to have more control over their children's education. Although some home-school settings may resemble a regular classroom, other children learn in more informal settings such as the family kitchen.

Box 16.4 Sociology and Social Policy

The Ongoing Debate Over School Vouchers

As discussed in Chapter 3 ("Culture"), one of the core American values is equality—at least equality of *opportunity,* an assumed equal opportunity to achieve success. However, with problems in public education such as are discussed in this chapter, does every child really have that equal opportunity? If it is at least possible that a child could get a better education at a private school than at an inadequately funded public school, should the government provide that child's family with the resources to send the child to a private school in order to have the same opportunity as a child from a wealthier family? Should the government provide that funding even if the private school teaches specific religious beliefs as part of the academic curriculum? Questions such as these have been of concern to social policy makers, educators, parents, and others in the United States for decades, and it appears that the controversy is far from being resolved. To understand the issues involved in the school voucher controversy, let's briefly look at the intended purpose of vouchers and at the arguments for and against voucher programs.

The original idea, first proposed in 1955, was to provide a voucher (equal in value to the average amount spent per student by the local school district) to the family of any student who left the public schools to attend a private school (Henderson, 1997). The private school could exchange the voucher for that amount of money and apply it to the student's tuition at the private school; the school district would save an equivalent amount of money as a result of the student not enrolling in the public schools. Over the years, variations on the original idea have been proposed, such as allowing vouchers to be used for transfers between public schools in the same school district, transfers from one public school district to another, transfers only to schools with no religious connections, and vouchers for use only by students from low-income families.

Although the legislatures in at least twenty-six states have rejected school voucher programs (*New York Times,* 2002), school districts in a number of cities have adopted various types of school choice programs. Some of those plans authorize only transfers between public schools in the same district; other plans authorize the use of vouchers for transfers to private schools. For example, the Cleveland, Ohio, school district allows students to attend participating public or private schools of their parents' choosing—even public schools in adjacent districts—and provides tuition aid to the parents according to financial need. The tuition aid goes to the public or private school in which the student is enrolled. In 1999–2000, 96 percent of the students participating in this program were enrolled in religiously affiliated schools. The program was challenged in court on the basis that it violated the constitutional separation of church and state.

established for the end of each grade. School districts must report students' results to demonstrate that they are making progress toward meeting these standards. Schools that close the education gap will receive additional federal dollars, but schools and districts that do not show adequate progress may lose funding and pupils: Parents may be allowed to move their children from low-performing schools to other schools that meet or exceed their district's educational standards.

In 2007, on the sixth anniversary of the No Child Left Behind Act, President George W. Bush and Education Secretary Margaret Spellings issued a statement indicating that across-the-board improvements had occurred in fourth- and eighth-grade reading and math scores nationwide. The report concluded that African American and Hispanic students had made significant gains in closing the achievement gap in areas such as reading and mathematics. However, the Bush administration acknowledged that No Child Left Behind had also produced a series of unanticipated problems and that much remained to be done to improve education in this country. More than seventy-five specific suggestions on how to improve teaching, learning, and student performance were made in the report (*New York Times,* 2007a). Among the needed changes was the development of rigorous, voluntary national standards that might more effectively prepare students for success in college and the workplace (whitehouse.gov, 2008). Current problems also exist regarding how states collect data and assess both student and school performance. Other problems are related to the quality of teachers and how effectively they instruct their students. For schools to be able to attract outstanding teachers, the best teachers will need to be more adequately rewarded, and incentives must be offered to encourage good instructors to teach in underperforming schools.

At the bottom line, many critics believe that No Child Left Behind has fundamentally misdefined the problems facing education in this country. They state that schools need more money and more incentives, not more testing of students or teachers. One major criticism of this law is that it does not adequately address the profound educational inequalities that are so prevalent in the United States, particularly with high-spending schools outspending low-spending schools by at least three to one in most states (Darling-Hammond, 2007).

In *Zelman v. Simmons-Harris*, a divided U.S. Supreme Court upheld the Cleveland program against the constitutional challenge, with the majority's opinion holding that the plan was adopted for the purpose of providing educational assistance to poor children and was neutral toward religion. Would the same result have been reached if the plan was for *all* children instead of only those from low-income families?

According to some analysts, the Court's decision shifted the battle regarding voucher programs from a legal argument based on a constitutional question to a political debate involving social policy and politics (Nagourney, 2002). Advocates of the Cleveland program argue that the Court's decision was good because children from lower-income families can now have a wider range of educational choices, a range similar to that currently available to children from middle-income families. They assert that competition for students' vouchers will improve public education, forcing school administrators and teachers to perform at a higher level and thereby producing greater competencies in students. By contrast, opponents of school vouchers argue that the Court's decision could be the beginning of the end of public schools, that public education may be unable to withstand the loss of funds and the transfer of gifted students to private schools.

What are the major social policy issues surrounding voucher systems? Many people still strongly believe that using public tax dollars (even though they are technically given to the students and their parents rather than the schools) for funding private, religiously based schools violates the separation of church and state. Despite the Court's decision, they will continue to make that argument, whether on constitutional grounds or on the basis of public policy. Ultimately, from a social policy standpoint, the future of school vouchers may depend on the willingness (or unwillingness) of political leaders to authorize vouchers and the willingness of voters to pay for them.

How the social issues are framed may be significant in shaping the future of vouchers. The education reform outlined by President Obama does not provide for a growth in school voucher programs. His plan emphasizes personal and parental responsibility, but it also relies on states to enact their own measures of educational reform. It remains to be seen whether this will mark the end of the school voucher debate or provide more fuel for future debates regarding this ongoing topic.

Reflect & Analyze

In regard to the questions we raised at the beginning of this feature, what do you think? If it is at least possible that a child could get a better education at a private school than at an inadequately funded public school, should the government provide that child's family with the resources to send the child to a private school in order to have the same opportunity as a child from a wealthier family? Should the government provide that funding even if the private school teaches specific religious beliefs as part of the academic curriculum?

Similarly, colleges and universities face many challenges today, particularly given the major financial constraints that many institutions face. President Obama's American Graduation Initiative, discussed earlier in this chapter, was established to create a community-college challenge fund because these colleges are often underfunded and lack the basic resources they need to improve instruction, build ties with businesses, and adopt other reforms. Under the president's plan, community colleges will receive new competitive grants that will help them develop partnerships with businesses and create career pathways for workers. It will also help schools expand course offerings, provide dual enrollment at high schools and universities, and improve remedial and adult education programs (whitehouse.gov, 2009a). If these measures are implemented, many more people will have an opportunity to acquire a two-year degree, complete a certificate program lasting six months to a year that provides specific credentials for middle-skill jobs that require more than a high school diploma but less than a four-year degree, or acquire sufficient college credits to transfer to a four-year college or university. Making college more affordable and significantly increasing the number of students enrolled in community colleges by 2020 are commendable goals for enhancing education in the United States. What is crucial is the approximately $12-billion price tag, which the president suggests it will cost, to bring about necessary changes in the current system.

At the university level, institutions continue to expand their focus while undergoing strenuous budget cuts coupled with increasing demands to meet the needs of widely diverse student populations. In recent years, budget cuts for higher education were especially harsh in the South and West; however, colleges and universities in the Northeast were not immune to the effects of the nation's economic downturn and the resulting drop in the state tax revenues that fund public higher education (*Chronicle of Higher Education*, 2008). Florida is among the states that have been hardest hit in regard to funding for higher education: Explosive population growth, limited tax revenues, a struggling housing industry, and other types of financial woes have contributed to layoffs and hiring freezes at public institutions. Similarly, California has been undergoing financial turmoil, rapid growth, and demands that colleges and universities strenuously curb hiring and spending.

Tightening of the financial resources available to colleges and universities will lead to even more schools seeking alternative ways to fund their operations. Some will further increase tuition paid by students; others will seek different sources of funding. Some will move beyond the United States to find ways in which they can expand their base of operation. For example, some U.S. universities are expanding their educational operations to emerging nations where demand is high for certain kinds of curricula, such as advanced business and petroleum engineering courses in Qatar and other Middle Eastern countries. One of the major problems, present and future, in colleges and universities is the increasing cost of education for students and the part that higher education plays in maintaining and perpetuating social inequality in the larger society.

In conclusion, education is an important social institution that must be maintained and enhanced in a variety of ways because we have learned that simply spending larger sums of money on education does not guarantee that many of the problems facing this social institution will be resolved.

Chapter Review

- **What is education?**

Education is the social institution responsible for the systematic transmission of knowledge, skills, and cultural values within a formally organized structure.

- **What is the functionalist perspective on education?**

According to functionalists, education has both manifest functions (socialization, transmission of culture, social control, social placement, and change and innovation) and latent functions (keeping young people off the streets and out of the job market, matchmaking and producing social networks, and creating a generation gap).

- **What is the conflict perspective on education?**

From a conflict perspective, education is used to perpetuate class, racial–ethnic, and gender inequalities through tracking, ability grouping, and a hidden curriculum that teaches subordinate groups conformity and obedience.

- **What is the symbolic interactionist perspective on education?**

Symbolic interactionists examine classroom dynamics and study ways in which practices such as labeling may become a self-fulfilling prophecy for some students, such that these students come to perform up—or down—to the expectations held for them by teachers.

- **How are U.S. public schools funded?**

Most educational funds come from state legislative appropriations and local property taxes. State governments contribute about 47 percent of total educational expenses, local sources contribute about 44 percent, and the federal government pays the remaining 9 percent.

- **What are some of the leading discipline problems in today's schools?**

Schools have been plagued by the same array of problems that are occurring in the larger society, including violence, drug abuse, suicide, robbery, assault, and other forms of aggressive behavior.

- **How is the increasing racial and ethnic diversity of the United States affecting public education?**

Nearly 40 percent of U.S. schoolchildren are African American, Latino/a, Asian American, or Native American, whereas most teachers are middle-class white females.

- **What is functional illiteracy, and how does it differ from illiteracy?**

Functional illiteracy is the inability to read and/or write at the skill level necessary for carrying out everyday tasks. In contrast, illiteracy is the inability to read and/or write at the most basic level. Both forms of illiteracy are a problem in the United States.

- **What controversies persist in education?**

Racial segregation and integration, bilingual education, and how to equalize educational opportunities for students with disabilities are among the pressing issues facing U.S. public education today.

- **What are the major problems in higher education?**

Among the most pressing problems are the high cost of a college education, the underrepresentation of minorities as students and faculty in many schools and degree programs, and the continuing debate over affirmative action.

www.cengage.com/login

Want to maximize your online study time? Take this easy-to-use study system's diagnostic pre-test, and it will create a personalized study plan for you. By helping you to identify the topics that you need to understand better and then directing you to valuable online resources, it can speed up your chapter review. CengageNOW even provides a post-test so you can confirm that you are ready for an exam.

Key Terms

credentialism 525
cultural capital 522
cultural transmission 512
education 512
formal education 513
functional illiteracy 541
hidden curriculum 524
informal education 513
mass education 514
tracking 522

Questions for Critical Thinking

1. What are the major functions of education for individuals and for societies?
2. Why do some theorists believe that education is a vehicle for decreasing social inequality whereas others believe that education reproduces existing class relationships?
3. Why does so much controversy exist over what should be taught in U.S. public schools?
4. How are the values and attitudes that you learned from your family reflected in your beliefs about education?

The Kendall Companion Website

www.cengage.com/sociology/kendall

Visit this book's companion website, where you'll find more resources to help you study and successfully complete course projects. Resources include quizzes and flash cards, as well as special features such as an interactive sociology timeline, maps, General Social Survey (GSS) data, and Census 2000 data. The site also provides links to useful websites that have been selected for their relevance to the topics in this chapter and include those listed below. (*Note:* Visit the book's website for updated URLs.)

National Center for Education Statistics (NCES)

http://nces.ed.gov

The NCES is the federal organization primarily responsible for the collection and distribution of data on topics related to education in the United States and around the world. The NCES website features information on literacy, education costs, selecting colleges, student assessment, students with disabilities, and much more.

American Association of University Women (AAUW)

http://www.aauw.org

The AAUW is a national organization that advocates gender equity in education and promotes awareness of issues related to gender and education. The AAUW website features research reports, public policy initiatives, news updates, informative Web links, and information on how to take action against gender inequities in education.

International Bureau of Education (IBE)

http://www.ibe.unesco.org

The IBE is a private, nongovernmental organization that was created in 1925 to centralize information related to public and private education and to promote scientific research in the education field. In addition to profiles of national education systems and reports on the development of education, the IBE website provides access to extensive data banks, including the World Data on Education collection.

chapter 17 Religion

Chapter Focus Question

What is the relationship between society and religion, and what role does religion play in people's everyday lives?

We are teaching our children a theory [evolution] that most of us don't believe in. I don't think God creates everything on a day-to-day basis, like the color of the sky. But I do believe he created Adam and Eve—instantly.

—Steve Farrell, a resident of Dover, Pennsylvania, explaining why he approves of the Dover school board's decision to require eighth-grade biology teachers to teach "intelligent design"—an assertion that the universe is so complex that an intelligent, supernatural power must have created it—as an alterative to the theory of evolution (Powell, 2004: A1)

I *definitely* would prefer to believe that God created me than that I'm 50th cousin to a silverback ape. What's wrong with wanting our children to hear about all the holes in the theory of evolution?

—Lark Myers, another resident of Dover, who also wants her child to learn about intelligent design at school (Powell, 2004: A1)

For many years, people have argued about what should (or should not) be taught in U.S. public schools, including the teaching of creationism or intelligent design as contrasted with evolution. Shown here is Dr. Kenneth Miller, a biology professor, during a discussion of the pros and cons of incorporating the teaching of intelligent design into the Ohio state science curriculum.

IN THIS CHAPTER

- The Sociological Study of Religion
- Sociological Perspectives on Religion
- World Religions
- Types of Religious Organization
- Trends in Religion in the United States
- Religion in the Future

I believe it is wrong to introduce a non-scientific "explanation" of the origins of life into the science curriculum. This policy was not endorsed by the Dover High School science department. I think this policy was approved for religious reasons, not to improve science education for my child.

—Tammy Kitzmiller, one of the eleven parents who filed a lawsuit (*Kitzmiller v. Dover*) against the school board, challenging its controversial decision (ACLU, 2005)

People have an impatience about science. They think it's this practical process that explains how everything works, but that's the least interesting part. We understand a lot of the mechanisms of evolution but it's what we *don't* understand that makes it exciting. . . . It's very clear that intelligent design has become a stalking-horse. If these school boards had their druthers, they would teach Noah's flood and the 6,000-year-old design of Earth. My fear is that they are making real headway in the popular imagination.

—Kenneth R. Miller, a university biologist and author of the biology textbook used in Dover before the school board's decision, explaining why he believes that the teaching of intelligent design in public classrooms is a very bad idea (Powell, 2004: A1)

Sharpening Your Focus

- *What are the key components of religion?*
- *How do functionalist, conflict, and symbolic interactionist perspectives on religion differ?*
- *What are the central beliefs of the world's religions?*
- *How do religious bodies differ in organizational structure?*
- *What is the future of religion in the United States?*

What is all the controversy about? How did a small school district draw so much attention to itself and end up with a district judge ruling that the school board's decision to introduce intelligent design as an alternative to evolution violated the First Amendment to the United States Constitution?

The argument over intelligent design is the latest debate in a lengthy battle over the teaching of creationism versus evolutionism in public schools, and it is only one of many arguments that will continue to take place regarding the appropriate relationship between public education and religion in the United States. More than seventy years ago, for example, evolutionism versus creationism was hotly debated in the famous "Scopes monkey trial," so named because of Charles Darwin's assertion that human beings had evolved from lower primates. In this case, John Thomas Scopes, a substitute high school biology teacher in Tennessee, was found guilty of teaching evolution, which denied the "divine creation of man as taught in the Bible." Although an appeals court later overturned Scopes's conviction and $100 fine (on the grounds that the fine was excessive), teaching evolution in Tennessee's public schools remained illegal until 1967 (Chalfant, Beckley, and Palmer, 1994). By contrast, recent U.S. Supreme Court rulings have looked unfavorably on the teaching of creationism in public schools, based on a provision in the Constitution that requires a "wall of separation" between church (religion) and state (government). Initially, this wall of separation was erected to protect religion from the state, not the state from religion.

As these examples show, religion can be a highly controversial topic. One group's deeply held beliefs or cherished religious practices may be a source of irritation to another. Today, religion is a source of both stability and conflict not only in the United States but throughout the world (Kurtz, 1995). In this chapter, we examine how religion influences life in the United States and in other areas of the world. Before reading on, test your knowledge about how religion affects public education in this country by taking the quiz in Box 17.1.

The Sociological Study of Religion

What is religion? ***Religion* is a system of beliefs and practices (rituals)—based on some sacred or supernatural realm—that guides human behavior, gives meaning to life, and unites believers into a single moral community** (Durkheim, 1995/1912). Religion is one of the most significant social institutions in society. As such, it consists of a variety of elements, including beliefs about the sacred or supernatural, rituals, and a social organization of believers drawn together by their common religious tradition (Kurtz, 1995). This system of beliefs seeks to bridge the gap between the known and the unknown, the seen and the unseen, and the sacred ("holy, set apart, or forbidden") and the secular (things of this world). Most religions attempt to answer fundamental questions such as those regarding the meaning of life and how the world was created. Most religions also provide comfort to persons facing emotional traumas such as illness, suffering, grief, and death. According to the sociologist Lester Kurtz (1995: 9), religious beliefs are typically woven into a series of narratives, including stories about how ancestors and other significant figures had meaningful experiences with supernatural powers. Moreover, religious beliefs are linked to practices that bind people together and to rites of passage such as birth, marriage, and death. People with similar religious beliefs and practices often gather themselves together in a moral community (such as a church, mosque, temple, or synagogue) where they can engage in religious beliefs and practices with similarly minded people (Kurtz, 1995).

Given the diversity and complexity of religion, how is it possible for sociologists to study this social institution? Most sociologists studying religion are committed to the pursuit of "disinterested scholarship," meaning that they do not seek to make value judgments about religious beliefs or to determine whether particular religious bodies are "right" or "wrong." However, many acknowledge that it is impossible to completely rid themselves of those values and beliefs into which they were socialized (Bruce, 1996). Therefore, for the most part, sociologists study religion by using sociological methods such as historical analysis, experimentation, participant observation, survey research, and content analysis that can be verified and replicated (Roberts, 2004). As a result, most studies in the sociology of religion focus on tangible elements that can be *seen,* such as written texts, patterns of behavior, or individuals' opinions about religious matters, and that can be studied using standard sociological research tools. According to the sociologist Keith A. Roberts (2004: 28), beliefs constitute only a small part of a sociological examination of religion:

> The sociological approach focuses on religious groups and institutions (their formation, maintenance, and demise), on the behavior of individuals within these groups (e.g., social processes that affect conversion, ritual behavior, or decision to defect to another group), and on conflicts between religious groups (such as Catholic vs. Protestant, Christian vs. [Muslim], mainline denomination vs.

Box 17.1 Sociology and Everyday Life

How Much Do You Know About the Effect of Religion on U.S. Education?

True	False	
T	F	1. The Constitution of the United States originally specified that religion should be taught in the public schools.
T	F	2. Virtually all sociologists have advocated the separation of moral teaching from academic subject matter.
T	F	3. Parochial schools have decreased in enrollment as interest in religion has waned in the United States.
T	F	4. The number of children from religious backgrounds other than Christianity and Judaism has grown steadily in public schools over the past three decades.
T	F	5. Debates over the content of textbooks focus only on elementary education because of the vulnerability of young children.
T	F	6. Prayer or a moment of silence in public schools will cease to be a political issue in the United States during this century.

Answers on page 554.

cult). For the sociologist, beliefs are only one small part of religion.

Recently, more U.S. scholars have started examining religion from a global perspective to determine "ways in which religious ideas are performed on the world stage" (Kurtz, 1995: 16). As Kurtz (1995: 211) points out, conflicts in the global village are often deeply intertwined with religious differences: "In the twentieth century, the twin crises of modernism and multiculturalism . . . added a religious dimension to many ethnic, economic, and political battles, providing cosmic justifications for the most violent struggles." Of course, conflict is not always inherently bad. It can be the source of constructive change in communities and societies.

How does the sociological study of religion differ from the theological approach? Unlike the sociological approach, which primarily focuses on the visible aspects of religion, *theologians* study specific religious doctrines or belief systems, including answers to questions such as what the nature of God or the gods is and what the relationship is among supernatural power, human beings, and the universe. Many theologians primarily study the religious beliefs of a specific religion (such as Christianity, Judaism, Buddhism, or Hinduism), denomination (such as the Baptists, Catholics, Methodists, or Episcopalians), or religious leader (such as the Reverend Sun Myung Moon or L. Ron Hubbard, founder of Scientology) so that they can interpret this information for laypersons who seek answers for seemingly unanswerable questions about the meaning of life and death.

Religion and the Meaning of Life

Religion seeks to answer important questions such as why we exist, why people suffer and die, and what happens when we die. Sociologist Peter Berger (1967) referred to religion as a *sacred canopy*—a sheltering fabric hanging over people that gives them security and provides answers for the questions of life (see "Sociology Works!"). However, this sacred canopy requires that people have ***faith*—unquestioning belief that does not require proof or scientific evidence.** Science and medicine typically rely on existing scientific evidence to respond to questions of suffering, death, and injustice, whereas religion seeks to explain such phenomena by referring to the sacred. According to Emile Durkheim (1995/1912), ***sacred* refers to those aspects of life that are extraordinary or supernatural**—in other words, those things that are set apart as "holy." People feel a sense of awe, reverence, deep respect, or fear for that which is considered sacred. Across cultures and in

religion a system of beliefs and practices (rituals)—based on some sacred or supernatural realm—that guides human behavior, gives meaning to life, and unites believers into a single moral community.

faith an unquestioning belief that does not require proof or scientific evidence.

sacred those aspects of life that are extraordinary or supernatural.

Box 17.1 Sociology and Everyday Life

Answers to the Sociology Quiz on Religion and U.S. Education

1. False. Due to the diversity of religious backgrounds of the early settlers, no mention of religion was made in the original Constitution. Even the sole provision that currently exists (the establishment clause of the First Amendment) does not speak directly about the issue of religious learning in public education.

2. False. Obviously, contemporary sociologists hold strong beliefs and opinions on many subjects; however, most of them do not think that it is their role to advocate specific stances on a topic. Early sociologists were less inclined to believe that they had to be "value free." For example, Durkheim strongly advocated that education should have a moral component and that schools had a responsibility to perpetuate society by teaching a commitment to the common morality.

3. False. Just the opposite has happened. As parents have felt that their children were not receiving the type of education they desired in public schools, parochial schools have flourished. Christian schools have grown to over five thousand; Jewish parochial schools have also grown rapidly over the past decade.

4. True. Although about 86 percent of those age 18 and over in the 48 contiguous states of the United States describe their religion as some Christian denomination, there has still been a significant increase in those who adhere either to no religion (7.5 percent) or who are Jewish, Muslim/Islamic, Unitarian–Universalist, Buddhist, or Hindu.

5. False. Attempts to remove textbooks occur at all levels of schooling. A recent case involved the removal of Chaucer's "The Miller's Tale" and Aristophanes's *Lysistrata* from a high school curriculum.

6. False. Regardless of whether the current proposed school prayer amendment becomes part of the Constitution, school prayer will remain an issue due to the diversity of the population in this country.

Sources: Based on Ballantine, 2001; Gibbs, 1994; Greenberg and Page, 2002; C. Johnson, 1994; Kosmin and Lachman, 1993; and Roof, 1993.

different eras, many things have been considered sacred, including invisible gods, spirits, specific animals or trees, altars, crosses, holy books, and special words or songs that only the initiated could speak or sing (Collins, 1982). Those things that people do not set apart as sacred are referred to as ***profane*—the everyday, secular ("worldly") aspects of life** (Collins, 1982). Sacred beliefs are rooted in the holy or supernatural, whereas secular beliefs have their foundation in scientific knowledge or everyday explanations. In the educational debate over creationism and evolutionism, for example, advocates of creationism view their belief as founded in sacred (Biblical) teachings, but advocates of evolutionism argue that their beliefs are based on provable scientific facts.

In addition to beliefs, religion also comprises symbols and rituals. According to the anthropologist Clifford Geertz (1966), religion is a set of cultural symbols that establishes powerful and pervasive moods and motivations to help people interpret the meaning of life and establish a direction for their behavior. People often act out their religious beliefs in the form of ***rituals*—regularly repeated and carefully prescribed forms of behaviors that symbolize a cherished value or belief** (Kurtz, 1995). Rituals range from songs and prayers to offerings and sacrifices that worship or praise a supernatural being, an ideal, or a set of supernatural principles. For example, Muslims bow toward Mecca, the holy city of Islam, five times a day at fixed times to pray to God, whereas Christians participate in the celebration of communion (or the "Lord's Supper") to commemorate the life, death, and resurrection of Jesus. Rituals differ from everyday actions in that they involve very strictly determined behavior. The rituals involved in praying or in observing communion are carefully orchestrated and must be followed with precision. According to the sociologist Randall Collins (1982: 34), "In rituals, it is the forms that count. Saying prayers, singing a hymn, performing a primitive sacrifice or a dance, marching in a procession, kneeling before an idol or making the sign of the cross—in these, the action must be done the right way."

Not all sociologists believe that the "sacred canopy" metaphor suggested by Berger accurately describes contemporary religion. Some analysts believe that a more accurate metaphor for religion in the global vil-

Sociology *Works!*

Religion as a Sacred Canopy: Sacredness and Everyday Life?

Religion is the establishment, through human activity, of an all-embracing sacred order, that is, of a sacred cosmos that will be capable of maintaining itself in the ever-present face of chaos.

—Peter Berger, *The Sacred Canopy* (1967: 51)

In *The Sacred Canopy,* Peter Berger describes people as "world-builders" who create man-made worlds that are made precarious by the crises of human life, including illness, disaster, death, and various tragedies that demonstrate the brevity of life and the rapid passage of time. Consequently, Berger believes there is a need for answers that religion can provide to important "why" questions, such as "Why do people die?" and "Why do horrible disasters occur?" By seeking to provide answers to difficult questions such as these, religion serves an important function for both individuals and societies. Religion serves as a sacred canopy in that it provides individuals with answers and some peace of mind; it also offers societies much-needed norms and stability. As the sociologist Robert Wuthnow (1992: 5) explains, "Experiences of grief—or even experiences of extreme joy or ecstasy—shatter the willing suspension of our doubts and raise questions about the deeper meanings of life. And the compartmentalized spheres of relevance in which we perform our routine tasks require some means of integration if we are to function as whole persons. In short, there seems to be a requirement for meaning that goes beyond the confines of everyday life."

We can see that sociology works if we apply Peter Berger's idea of the "sacred canopy" to contemporary life. His work has been widely used not only in sociology but also in other fields such as cultural anthropology and religion to bring about a greater understanding about the relation of religion to everyday life. Although sociologists of religion have produced newer theories and empirical research since Berger's day, his assertion that religion is a canopy that protects people from the roughest edges and greatest tragedies of human existence remains important. Wuthnow (1992: 6) describes the contribution of Berger's sacred canopy in this way:

> Religious teachings characteristically serve to shelter the individual from chaos—from a reality that seems to make no sense—by providing explanations for suffering, death, tragedy, and injustice. They integrate the individual's biography by providing an overarching frame of reference that applies to all of life, that locates the individual ultimately in space and time, that specifies an ultimate purpose for the individual's life and thus permits daily activities to be organized around the fulfillment of this purpose.

Berger also makes us aware that religion provides rituals that set a framework for containing the potentially disruptive experience of mourning (rituals such as memorial services and funerals) or of great joy (rituals such as weddings and christenings). In Berger's words (1967: 33), "Religion legitimates social institutions by bestowing upon them an ultimately valid ontological status, that is, by *locating* them within a sacred and cosmic frame of reference."

The Sacred Canopy, written more than forty years ago, shows the lasting influence of key sociological concepts in how we frame our own thinking about religion and everyday life.

Reflect & Analyze

Does "the sacred canopy" provide us with a view of religion that more closely fits the functionalist, conflict, or symbolic interactionist perspective on religion? Do you believe that social institutions other than religion might provide people and societies with a sacred canopy in the twenty-first century?

lage is that of the *religious marketplace,* in which religious institutions and traditions compete for adherents, and worshippers shop for a religion in much the same way that consumers decide what goods and services they will purchase in the marketplace (Moore, 1995). However, other analysts do not believe that moral and ethical beliefs are bought and sold like groceries, shoes, or other commodities, and they note that many of the world's religions have persisted from earlier eras to advanced technological societies. But this poses another question: When did the earliest religions begin?

Although it is difficult to establish exactly when religious rituals first began, anthropologists have concluded that all known groups over the past hundred

profane the everyday, secular, or "worldly" aspects of life.

rituals regularly repeated and carefully prescribed forms of behaviors that symbolize a cherished value or belief.

© Jeff Greenberg/Alamy

College students—like people in all walks of life—may turn to religion for answers to important questions for which there are no easy answers. Rituals help individuals outwardly express their beliefs and provide a sense of cohesion and belonging.

thousand years have had some form of religion (Haviland, 1999). Religions have been classified into four main categories based on their dominant belief: simple supernaturalism, animism, theism, and transcendent idealism. In very simple preindustrial societies, religion often takes the form of ***simple supernaturalism*—the belief that supernatural forces affect people's lives either positively or negatively.** This type of religion does not acknowledge specific gods or supernatural spirits but focuses instead on impersonal forces that may exist in people or natural objects. For example, simple supernaturalism has been used to explain mystifying events of nature, such as sunrises and thunderstorms, and ways that some objects may bring a person good or bad luck. By contrast, ***animism* is the belief that plants, animals, or other elements of the natural world are endowed with spirits or life forces that have an impact on events in society.** Animism is associated with early hunting and gathering societies and with many Native American societies, in which everyday life is not separated from the elements of the natural world (Albanese, 2007).

The third category of religion is ***theism*—a belief in a god or gods.** Horticultural societies were among the first to practice ***monotheism*—a belief in a single, supreme being or god who is responsible for significant events such as the creation of the world.** Three of the major world religions—Christianity, Judaism, and Islam—are monotheistic. By contrast, Hinduism, Shinto, and a number of the indigenous religions of Africa are forms of ***polytheism*—a belief in more than one god.** The fourth category of religion, transcendent idealism, is a ***nontheistic religion*—a religion based on a belief in divine spiritual forces such as sacred principles of thought and conduct, rather than a god or gods.** Transcendent idealism focuses on principles such as truth, justice, affirmation of life, and tolerance for others, and its adherents seek an elevated state of consciousness in which they can fulfill their true potential.

Religion and Scientific Explanations

During the Industrial Revolution, scientific explanations began to compete with religious views of life. Rapid growth in scientific and technological knowledge gave rise to the idea that science would ultimately answer questions that had previously been in the realm of religion. Many scholars believed that increases in scientific knowledge would result in ***secularization*—the process by which religious beliefs, practices, and institutions lose their significance in sectors of society and culture** (Berger, 1967). Secularization involves a decline of religion in everyday life and a corresponding increase in organizations that are highly bureaucratized, fragmented, and impersonal (Chalfant, Beckley, and Palmer, 1994).

As previously discussed, some people argue that science and technology have overshadowed religion in the United States, but others point to the resurgence of religious beliefs in recent years and an unprecedented development of alternative religions (Kosmin and Lachman, 1993; Roof, 1993; Singer with Lalich, 1995). The issue of whether religious or scientific explanations best explain various aspects of social life (such as when a person becomes a "human being" or when a person dies) is not limited to the teachings of schools and religious organizations: Members of the contemporary media are also key players in the framing of religious and scientific debates, and journalists may influence how we view a number of key issues (see Box 17.2).

Sociological Perspectives on Religion

Religion as a social institution is a powerful, deeply felt, and influential force in human society. Sociologists study the social institution of religion because of the importance that religion holds for many people; they also want to know more about the influence of religion on society, and vice versa. For example, some people believe that the introduction of prayer or religious instruction in public schools would have a positive effect on the teaching of values such as honesty, compassion, courage, and tolerance because these values could be given a moral foundation. However, society has strongly influenced the practice of religion in the United States as a result of court rulings and laws that have limited religious activities in public settings, including schools.

CONCEPT QUICK REVIEW

Theoretical Perspectives on Religion

Key Points	
Functionalist Perspectives	Sacred beliefs and rituals bind people together and help maintain social control.
Conflict Perspectives	Religion may be used to justify the status quo (Marx) or to promote social change (Weber).
Symbolic Interactionist Perspectives	Religion may serve as a reference group for many people, but because of race, class, and gender, people may experience it differently.

The major sociological perspectives (which are summarized in the Concept Quick Review) have different outlooks on the relationship between religion and society. Functionalists typically emphasize the ways in which religious beliefs and rituals can bind people together. Conflict explanations suggest that religion can be a source of false consciousness in society. Symbolic interactionists focus on the meanings that people give to religion in their everyday lives.

Functionalist Perspectives on Religion

Emile Durkheim was one of the first sociologists to emphasize that religion is essential to the maintenance of society. He suggested that religion is a cultural universal found in all societies because it meets basic human needs and serves important societal functions.

Durkheim on Religion In *The Elementary Forms of Religious Life* (1995/1912: 44), Durkheim defined *religion* as "a unified system of beliefs and practices relative to sacred things, that is to say, things set apart and forbidden—beliefs and practices which unite into one single moral community . . . all those who adhere to them." According to Durkheim, all religions share three elements: (1) beliefs held by adherents, (2) practices (rituals) engaged in collectively by believers, and (3) a moral community based on the group's shared beliefs and practices pertaining to the sacred.

For Durkheim, the central feature of all religions is the presence of sacred beliefs and rituals that bind people together in a collectivity. In his studies of the religion of the Australian aborigines, for example, Durkheim found that each clan had established its own sacred totem, which included kangaroos, trees, rivers, rock formations, and other animals or natural creations. To clan members, their totem was sacred; it symbolized some unique quality of their clan. People developed a feeling of unity by performing ritual

Throughout recorded history, churches and other religious bodies have provided people with a sense of belonging and of being part of something larger than themselves. Members of this congregation show their unity as they visit one another.

simple supernaturalism the belief that supernatural forces affect people's lives either positively or negatively.

animism the belief that plants, animals, or other elements of the natural world are endowed with spirits or life forces having an effect on events in society.

theism a belief in a god or gods.

monotheism belief in a single, supreme being or god who is responsible for significant events such as the creation of the world.

polytheism a belief in more than one god.

nontheism a religion based on a belief in divine spiritual forces such as sacred principles of thought and conduct, rather than a god or gods.

secularization the process by which religious beliefs, practices, and institutions lose their significance in sectors of society and culture.

Box 17.2 Framing Religion in the Media

Shaping the Intersections of Science and Religion

After religious teachers accomplish the refining process indicated, they will surely recognize with joy that true religion has been ennobled and made more profound by scientific knowledge.

—Albert Einstein, a theoretical physicist known for formulating the theory of relativity (religioustolerance.org, 2005)

Science is almost totally incompatible with religion.

—Peter Atkins, a chemist at the University of Oxford (religioustolerance.org, 2005)

As the statements by these two well-known scientists show, a lack of consensus exists regarding the relationship between science and religion. However, members of the media are often called upon to write about situations in which science and religion apparently intersect in societies. When these events occur, many print and electronic journalists use specific types of framing to shape their discussion of the intersections of science and religion. Theology professor James Schaefer (2005) has identified several ways that the media might approach this topic, and the following discussion is a modification of three of those approaches.

Conflict Framing

According to Schaefer (2005), "The image of conflict between science and religion is conventional in the media today. Coverage of a story is more dramatic when extreme views are highlighted while more subtle positions go unreported." When this type of framing is used by the media, experts are often carefully chosen from widely divergent viewpoints so that extreme viewpoints will be highlighted for readers and viewers. The result is often a form of framing that emphasizes hostility toward religion or toward science and technology. An example of conflict framing is stories about the teaching of evolution versus the teaching of creationism in the classroom.

Conflation Framing

When conflation framing is employed by the media, it becomes difficult to distinguish between religion and science as distinct human endeavors (Schaefer, 2005). Conflation exists when two things are combined into one. An example that Schaefer (2005: 220) provides for this type of framing is the "practice of attributing to God's activity natural phenomena that cannot be explained scientifically." In other words, if there is not a ready scientific explanation for a natural or social phenomenon that occurs, it may be described as supernatural or within the realm of God's doing, not that of human beings. When the media rush to a Catholic church to describe people who are watching a portrait of the Virgin Mary that

dances around their totem, causing them to abandon individual self-interest. Durkheim suggested that the correct performance of the ritual gives rise to religious conviction. Religious beliefs and rituals are *collective representations*—group-held meanings that express something important about the group itself (McGuire, 2002). Because of the intertwining of group consciousness and society, functionalists suggest that religion is functional because it meets basic human needs.

Functions of Religion From a functionalist perspective, religion has three important functions in any society: (1) providing meaning and purpose to life, (2) promoting social cohesion and a sense of belonging, and (3) providing social control and support for the government.

Meaning and Purpose Religion offers meaning for the human experience. Some events create a profound sense of loss on both an individual basis (such as injustice, suffering, or the death of a loved one) and a group basis (such as famine, earthquake, economic depression, or subjugation by an enemy). Inequality may cause people to wonder why their own situation is no better than it is. Most religions offer explanations for these concerns. Explanations may differ from one religion to another, yet each tells the individual or group that life is part of a larger system of order in the universe (McGuire, 2002). Some (but not all) religions even offer hope of an afterlife for persons who follow the religion's tenets of morality in this life. Such beliefs help make injustices easier to endure.

In a study of religious beliefs among Baby Boomers (persons born between 1946 and 1964) in the United States, religion and society scholar Wade Clark Roof (1993) found that a number of people had returned to organized religion as part of a personal quest for meaning. Roof noted that they were looking "for something to believe in, for answers to questions about life," as reflected in these comments by a woman in North Carolina:

> Something was missing. You turn around and you go, is this it? I have a nice husband, I have a nice

appears to be weeping or bleeding, some journalists explain this phenomenon with a combination of religious and scientific explanations. For example, the occurrence may be described by onlookers as an act of God, but the weeping or bleeding may also be attributed to environmental factors such as stains on the artwork or the way in which light and shadows at different times of the year affect paintings in the church. Journalists who combine religious and scientific explanations of phenomenon such as the "bleeding Virgin" tend to conflate religion and science for media audiences.

Contrast Framing

Although media framing based on conflict and media framing based on conflation both suggest that religion and science are interrelated, contrast framing is just the opposite. In the words of Schaefer (2005: 221), the contrast approach views religion and science as "totally independent and autonomous ways of knowing. Each is valid only within its clearly defined sphere of inquiry." Contrast framing in the media is based on the assumption that science and religion are too different to be interrelated:

1. Religion and science deal with different types of questions.
2. Religion and science use different languages to explain *reality.*
3. Religion and science tackle different tasks.
4. Religion and science follow different authorities (see also Gilkey, 1993).

According to Schaefer (2005: 221), "Science examines the natural world empirically, while religion addresses the ultimate reality that transcends the empirically known world" (see also Haught, 1995). Today, some people view religion and science as conflicting world views, whereas other individuals view religion and science as indistinguishable.

How the media frame stories about the intersection of religion and science may influence our views on a number of topics, ranging from the origins of the Earth and human beings to how and when life on Earth might end. Media stories about religion and science may also influence our thinking about moral issues such as what constitutes "right" and "wrong" in our society and who should be allowed to determine policies on key social issues such as abortion, the death penalty, and end-of-life decisions for individuals with a critical illness.

Reflect & Analyze

Does it make a difference how the media frame stories about the intersection of the social institutions of religion and science? Do the media affect your views on issues such as abortion and the death penalty? If so, how? If not, why not?

house; I was just about to finish graduate school. I knew I was going to have a very marketable degree. I wanted to do it. And you turn and you go, here I am. This is it. And there were just things that were missing. I just didn't have stimulation. I didn't have the motivation. And I guess when you mentioned faith, I guess that's what was gone. (qtd. in Roof, 1993: 158)

Social Cohesion and a Sense of Belonging By emphasizing shared symbolism, religious teachings and practices help promote social cohesion. An example is the Christian ritual of communion, which not only commemorates a historical event but also allows followers to participate in the unity ("communion") of themselves with other believers (McGuire, 2002). All religions have some form of shared experience that rekindles the group's consciousness of its own unity.

Religion has played an important part in helping members of subordinate groups develop a sense of social cohesion and belonging even when they are the objects of prejudice and discrimination by dominant-group members. For example, some scholars suggest that African Americans initially brought into the United States as slaves found cohesion and stability in religion:

> Common religious beliefs and practices provided a new form of social cohesion in place of the old forms that had been destroyed when the slaves were seized in Africa and forcibly brought to America. Black churches brought about a distinctive culture and worldview that paralleled rather than replicated the culture of the land in which blacks resided involuntarily. The terms of their faith—salvation, freedom, and the Kingdom of God—were rooted in the black experience and expressed themselves in joy and jubilation tinged with mournfulness. (Kosmin and Lachman, 1993: 31)

Religion has also been important to those who voluntarily migrated to the United States. For example, Irish Catholic and Italian Catholic churches helped

Irish Americans and Italian Americans preserve a sense of identity and belonging (Greeley, 1972; Roberts, 2004). Since the 1960s, Korean Americans have found religious and ethnic fellowship in more than two thousand Korean American churches, mostly Presbyterian, Southern Baptist, and United Methodist (Kosmin and Lachman, 1993). Since the late 1980s and early 1990s, Russian Jewish immigrants have found a sense of belonging in some congregations, and even though they did not initially speak the language of their new country, they still shared established religious rituals and a sense of history. Shared experiences such as these strengthen not only the group but also the individual's commitment to the group's expectations and goals (McGuire, 2002).

Social Control and Support for the Government How does religion help bind society together and maintain social control? All societies attempt to maintain social control through systems of rewards and punishments. Sacred symbols and beliefs establish powerful, pervasive, long-lasting motivations based on the concept of a general order of existence. In other words, if individuals consider themselves to be part of a larger order that holds the ultimate meaning in life, they will feel bound to one another (and to past and future generations) in a way that might not be possible otherwise (McGuire, 2002).

Religion also helps maintain social control in society by conferring supernatural legitimacy on the norms and laws of a society. In some societies, social control occurs as a result of direct collusion between the dominant classes and the dominant religious organizations. Niccolo Machiavelli, an influential sixteenth-century statesman and author, wrote that it was "the duty of princes and heads of republics to uphold the foundations of religion in their countries, for then it is easy to keep their people religious, and consequently well conducted and united" (qtd. in McGuire, 2002: 242). And, as discussed in Chapter 14, absolute monarchs have often claimed a divine right to rule.

In the United States, the separation of church and state reduces religious legitimation of political power (see Box 17.3). Nevertheless, political leaders often use religion to justify their decisions, stating that they have prayed for guidance in deciding what to do. This informal relationship between religion and the state has been referred to as ***civil religion*—the set of beliefs, rituals, and symbols that makes sacred the values of the society and places the nation in the context of the ultimate system of meaning.** Civil religion is not tied to any one denomination or religious group; it has an identity all its own. For example, many civil ceremonies in the United States have a marked religious quality. National values are celebrated on "high holy days" such as Memorial Day and the Fourth of July. Political inaugurations and courtroom trials often require people to place their hand on a Bible while swearing to do their duty or tell the truth, as the case may be. The U.S. flag is the primary sacred object of our civil religion, and the Pledge of Allegiance includes the phrase "one nation under God." U.S. currency bears the inscription "In God We Trust."

Some critics have attempted to eliminate all vestiges of civil religion from public life. However, the sociologist Robert Bellah (1967), who has studied civil religion extensively, argues that civil religion is not the same thing as Christianity; rather, it is limited to affirmations that members of any denomination can accept. As the sociologist Meredith McGuire (2002: 203) explains,

> Civil religion is appropriate to actions in the official public sphere, and Christianity and other religions are granted full liberty in the sphere of personal piety and voluntary social action. This division of spheres of relevance is particularly important for countries such as the United States, where religious pluralism is both a valued feature of sociopolitical life and a barrier to achieving a unified perspective for decision making.

However, this assertion may not resolve the problem for those who do not believe in the existence of God or for those who believe that *true* religion is trivialized by civil religion.

Civil religion may serve either a *priestly function* by celebrating the greatness of the United States or a *prophetic function* by pointing out discrepancies between the nation's ideals and the realities of its actions. For example, in his famous "I Have a Dream" speech, Martin Luther King, Jr., appealed to people of all religions to end racial discrimination in the United States based on patriotic values (Roberts, 2004). However, civil religion was also used to justify two opposing stances on U.S. involvement in the war in Vietnam; some bumper stickers read "America—Love It or Leave It!" while others demanded "America—Change It or Lose It!" (McGuire, 2002).

Conflict Perspectives on Religion

Although many functionalists view religion, including civil religion, as serving positive functions in society, some conflict theorists view religion negatively.

Karl Marx on Religion For Marx, *ideologies*—systematic views of the way the world ought to be—are embodied in religious doctrines and political values (Turner, Beeghley, and Powers, 2002). These ideologies serve to justify the status quo and retard social change. The capitalist class uses religious ideology as a tool of domination to mislead the workers about their true interests. For this reason, Marx wrote his now-famous statement that religion is the "opiate of the masses." People become complacent because they have been taught to believe in an afterlife in which they will be rewarded for their suffering and misery in this life. Although these religious teachings soothe the masses' distress, any relief is illusory. Religion unites people in a "false consciousness" that they share common interests with members of the dominant class (Roberts, 2004).

From a conflict perspective, religion tends to promote strife between groups and societies. For example, the new religious right in the United States has incorporated both the priestly and prophetic functions into its agenda. While calling for moral reform, it also calls the nation back to a covenant with God (Roberts, 2004). According to McGuire (2002), Weber's distinction between people's "class situation" (stratification based on economic factors) and "status situation" (stratification based on lifestyle, honor, and prestige) is useful in understanding how the new religious right can press for change while at the same time demanding a "return" to traditional family values and prayer in public schools. McGuire (2002: 241) suggests that members of new-right religious organizations who may feel that their status (prestige or honor) is "threatened by changing cultural norms assert their values politically in order to re-establish the ideological basis of their status."

Separation of church and state is a constitutional requirement that is often contested by people who believe that religion should be a part of public life. These workers are complying with a federal court order to remove a monument bearing the Ten Commandments from the rotunda of the Alabama State Judicial Building.

According to conflict theorists, conflict may be *between* religious groups (for example, anti-Semitism), *within* a religious group (for example, when a splinter group leaves an existing denomination), or between a religious group and the *larger society* (for example, the conflict over religion in the classroom). Conflict theorists assert that in attempting to provide meaning and purpose in life while at the same time promoting the status quo, religion is used by the dominant classes to impose their own control over society and its resources. Many feminist scholars object to the patriarchal nature of most religions, some advocate a break from traditional religions, and others seek to reform religious language, symbols, and rituals in order to eliminate the elements of patriarchy.

Max Weber's Response to Marx Whereas Marx believed that religion retards social change, Weber argued just the opposite. For Weber, religion could be a catalyst to produce social change. In *The Protestant Ethic and the Spirit of Capitalism* (1976/1904–1905), Weber asserted that the religious teachings of John Calvin are directly related to the rise of capitalism. Calvin emphasized the doctrine of *predestination*—the belief that even before they are born, all people are divided into two groups, the saved and the damned, and only God knows who will go to heaven (the elect) and who will go to hell. Because people cannot know whether they will be saved, they tend to look for earthly signs that they are among the elect. According to the Protestant ethic, those who have faith, perform good works, and achieve economic success are more likely to be among the chosen of God. As a result, people work hard, save their money, and do not spend it on worldly frivolity; instead, they reinvest it in their land, equipment, and labor (Chalfant, Beckley, and Palmer, 1994).

The spirit of capitalism grew in the fertile soil of the Protestant ethic. Even as people worked ever harder to prove their religious piety, structural conditions in Europe led to the Industrial Revolution, free markets, and the commercialization of the economy—developments that worked hand in hand with Calvinist religious teachings. From this viewpoint, wealth is an unintended consequence of religious piety and hard work. However, Weber (1976/1904–1905: 182) recognized that for some people, "the pursuit of wealth, stripped of its religious and ethical meaning, tends

civil religion the set of beliefs, rituals, and symbols that makes sacred the values of the society and places the nation in the context of the ultimate system of meaning.

Box 17.3 Sociology and Social Policy

Should Prayer Be Allowed in Public Schools? Issues of Separation of Church and State

What we want is actual prayer [as opposed to a moment of silence for contemplation]. It happened to have been around on Sept. 11. The next day at some . . . schools, there was open prayer all through the schools. Even the [U.S.] president is asking for prayer. But [in] the very institutions that we need to have prayer the most, it has been outlawed. So why not [have it] where it is needed the most and where it can have a lasting effect?

—Ronald J. Waters, a city alderman in Harvey, Illinois, explaining why he believes students should be able to pray in school, particularly in the aftermath of a national disaster such as the 9/11 terrorist attacks (qtd. in Fountain, 2001: A18)

In U.S. public schools, including Thornton Township High School in Harvey (a suburb south of Chicago), some students meet outside of regular school hours in classrooms or at other school facilities to have prayer and Bible study, although such activities are believed by many critics to violate the separation of church and state mandated by law. Known as Prayer Warriors for Christ, the group of Harvey students and some city officials in their community are waging a war on the wall that separates church and state. They want group prayer during the school day (Fountain, 2001). Many people in school districts across the land are waging that same war.

Why is the issue of prayer and other religious observances in public schools such a concern? As we think about this question, it is necessary to understand how social policy has been historically used to establish a division between church and state. When the U.S. Constitution was

Should prayer be permitted in the classroom? On the school grounds? At school athletic events? Given the diversity of beliefs that U.S. people hold, arguments and court cases over activities such as prayer around the school flagpole will no doubt continue in the future.

to become associated with purely mundane passions, which actually give it the character of sport."

With the contemporary secularizing influence of wealth, people often think of wealth and material possessions as the major (or only) reason to work. Although the "Protestant ethic" is rarely invoked today, many people still refer to the "work ethic" in somewhat the same manner that Weber did. For example, political and business leaders in the United States often claim that "the work ethic is dead."

Like Marx, Weber was acutely aware that religion could reinforce existing social arrangements, especially the stratification system. The wealthy can use religion to justify their power and privilege: It is a sign of God's approval of their hard work and morality. As for the poor, if they work hard and live a moral life, they will be richly rewarded in another life. The Hindu belief in reincarnation is an example of religion reinforcing the stratification system. Because a person's social position in the current life is the result of behavior in a former life, the privileges of the upper class must be protected so that each person may enjoy those privileges in another incarnation.

Does Weber's thesis about the relationship between religion and the economy withstand the test of time? Recently, the sociologist Randall Collins reexamined Weber's assertion that the capitalist breakthrough occurred just in Christian Europe and concluded that this belief is only partially accurate. According to Collins, Weber was correct that religious institutions are among the most likely places within agrarian societies for capitalism to begin. However, Collins believes that the foundations for capitalism in Asia, particularly in Japan, were laid in the Buddhist monastic economy of late medieval Japan. Collins (1997: 855) states that "The temples were the first entrepreneurial organizations in Japan: the first to combine control of the factors of labor, capital, and land so as to allocate them for enhancing production." Because of an ethic of self-discipline and restraint on consumption, high levels of accumulation and investment took place in medieval Japanese Buddhism. Gradually, secular capitalism emerged from temple capitalism as

ratified in 1789, the colonists who made up the majority of the population of the original states were of many different faiths. Due to this diversity, there was no mention of religion in the original Constitution. However, in 1791 the First Amendment added the following provision (to this day, the only reference to religion in that document): "Congress shall make no law respecting an establishment of religion, or prohibiting the free exercise thereof." The ban was binding only on the federal government; however, in 1947 the U.S. Supreme Court held that the Fourteenth Amendment ("No State shall make or enforce any law which shall abridge the privileges or immunities of citizens") had the effect of making the First Amendment's separation of church and state applicable to state governments as well.

Historically, the Supreme Court has been called on to define the boundary between permissible and impermissible governmental action with regard to religion. In 1947 the Court held that the ban on establishing a religion made it unconstitutional for a state to use tax revenues to support an institution that taught religion. In 1962 and 1963, the Court expanded this ruling to include many types of religious activities at schools or in connection with school activities, including group prayer, invocations at sporting events, and distribution of religious materials at school. However, in 1990 the Court ruled that religious groups (such as the Prayer Warriors) could meet on school property if certain conditions were met, including that attendance must be voluntary, the meetings must be organized and run by students, and the activities must occur outside of regular class hours.

Compromises regarding prayer in schools have been attempted in some states as legislators passed laws that either permit or mandate a daily moment of silence in public schools; however, the effect of such compromises has often been that neither side in the debate is appeased. For example, those who favor school prayer have proposed constitutional amendments such as "Nothing in this Constitution shall be construed to prohibit individual or group prayer in public schools or other public institutions. No person shall be required by the United States or by any state to participate in prayer. Neither the United States nor any state shall compose the words of any prayer to be said in public schools."

The church–state division continues to be a topic of extensive debate among school officials and political leaders. Advocates of allowing more religious activities in public education during school hours as well as after school believe that the constitutional dictate prohibiting "establishment of religion" was not intended to keep religion out of the public schools and that students need greater access to religious and moral training. Opponents believe that any entry of religious training and religious observances into public education and taxpayer-supported school facilities or events violates the Constitution and might be used by some people to promote their religion over others, or over a person's right to have no religion at all.

Reflect & Analyze

Where should the line be drawn? What do you think?

Sources: Based on Albanese, 2007; Fountain, 2001; Greenberg and Page, 2002; and Morse, 2001.

new guilds arose that were independent of the temples, and the gap between the clergy and everyday people narrowed through property transformation brought about by uprisings of the common people and wars with outside entities. Moreover, the capitalist dynamic in the monasteries was eventually transferred to the secular economy, opening the way to the Industrial Revolution in Japan (Collins, 1997). From the works of Weber and Collins, we can conclude that the emergence of capitalism through a religious economy happened in several parts of the world, not just one, and that it occurred in both Christian and Buddhist forms (Collins, 1997).

Symbolic Interactionist Perspectives on Religion

Thus far, we have been looking at religion primarily from a macrolevel perspective. Symbolic interactionists focus their attention on a microlevel analysis that examines the meanings people give to religion in their everyday lives.

Religion as a Reference Group For many people, religion serves as a reference group to help them define themselves. For example, religious symbols have meaning for large bodies of people. The Star of David holds special significance for Jews, just as the crescent moon and star do for Muslims and the cross does for Christians. For individuals as well, a symbol may have a certain meaning beyond that shared by the group. For instance, a symbolic gift given to a child may have special meaning when he or she grows up and faces war or other crises. It may not only remind the adult of a religious belief but also create a feeling of closeness with a relative who is now deceased. It has been said that the symbolism of religion is so very powerful because it "expresses the essential facts of our human existence" (Collins, 1982: 37).

***Her* Religion and *His* Religion** Not all people interpret religion in the same way. In virtually all religions, women have much less influence in establishing social definitions of appropriate gender roles both within the religious community and in the larger community. Therefore, women and men may belong to the same religious group, but their individual religion will not necessarily be a carbon copy of the group's entire system of beliefs. According to McGuire (2002), women's versions of a certain religion probably differ markedly from men's versions. For example, although an Orthodox Jewish man may focus on his public ritual roles and his discussion of sacred texts, women have few ritual duties and are more likely to focus on their responsibilities in the home. Consequently, the meaning of being Jewish may be different for women than for men.

Religious symbolism and language typically create a social definition of the roles of men and women. For example, religious symbolism may depict the higher deities as male and the lower deities as female. Females are sometimes depicted as negative or evil spiritual forces. For example, the Hindu goddess Kali represents men's eternal battle against the evils of materialism (Daly, 1973). Historically, language has defined women as being nonexistent in the world's major religions. Phrases such as *for all men* in Catholic and Episcopal services have gradually been changed to *for all;* however, some churches retain the traditional liturgy. And although there has been resistance, especially by women, to some traditional terms, inclusive language is less common, overall, than older male terms for God (Briggs, 1987).

Many women resist the subordination that they have experienced in organized religion. They have worked to change the existing rules that have excluded them or placed them in a clearly subordinate position.

© James Marshall/Getty Images

These Jews at the Western Wall in Jerusalem—a wall that holds special significance for all Jews—express their faith in God and in the traditions of their ancestors.

World Religions

Although there are many localized religions throughout the world, those religions classified as *world religions* cover vast expanses of the Earth and have millions of followers. Six world religions—Hinduism, Buddhism, Confucianism, Judaism, Islam, and Christianity—have more than four billion adherents, almost 75 percent of the world's population. These six religions are compared in ◆ Table 17.1.

Hinduism

We begin with Hinduism because it is believed to be one of the world's oldest current religions, having originated along the banks of the Indus River in Pakistan between 3,500 and 4,500 years ago. Hinduism began before written records were kept, so modern scholars have only limited information about its earliest leaders and their teachings (Kurtz, 1995). Hindu beliefs and practices have been preserved through an oral tradition and expressed in texts and hymns known as the *Vedas* (meaning "knowledge" or "wisdom"); however, this religion does not have a "sacred" book, such as the Judeo-Christian Bible or the Islamic Qur'an (Sharma, 1995). Consequently, Hindu beliefs and practices emerged over the centuries across the subcontinent of India in a variety of forms, reflecting the influence of the various regional cultures (Kurtz, 1995). Because Hinduism has no scriptures that are thought to be inspired by a god or gods and is not based on the teachings of any one person, religion scholars refer it as to as an *ethical religion*—a system of beliefs that calls upon adherents to follow an ideal way of life. For most Hindus, this is partly achieved by adhering to the expectations of the caste system (see Chapter 9).

Central to Hindu teachings is the belief that individual souls (*jivas*) enter the world and roam the universe until they break free into the limitless atmosphere of illumination (*moska*) by discovering their own *dharma*—duties or responsibilities. According to Hinduism, individual *jivas* pass through a sequence of bodies over time as they undergo a process known as reincarnation (*samsara*)—an endless passage through cycles of life, death, and rebirth until the soul earns liberation (Kurtz, 1995). The soul's acquisition of each new body is tied to the law of *karma* (deed or act), which is a doctrine of the moral law of cause and effect. The present condition of the soul—how happy or unhappy it is, for example—is directly related to what it has done in the past, and its present thoughts and decisions are the ultimate determinants of what its future will be. The final goal of Hindu existence is enter-

◆ **Table 17.1 Major World Religions**

	Current Followers	Founder/Date	Beliefs
Christianity	1.7 billion	Jesus; 1st century C.E.	Jesus is the son of God. Through good moral and religious behavior (and/or God's grace), people achieve eternal life with God.
Islam	1 billion	Muhammad; ca. 600 C.E.	Muhammad received the Qur'an (scriptures) from God. On Judgment Day, believers who have submitted to God's will, as revealed in the Qur'an, will go to an eternal Garden of Eden.
Hinduism	719 million	No specific founder; ca. 1500 B.C.E.	Brahma (creator), Vishnu (preserver), and Shiva (destroyer) are divine. Union with ultimate reality and escape from eternal reincarnation are achieved through yoga, adherence to scripture, and devotion.
Buddhism	309 million	Siddhartha Gautama; 500 to 600 B.C.E.	Through meditation and adherence to the Eightfold Path (correct thought and behavior), people can free themselves from desire and suffering, escape the cycle of eternal rebirth, and achieve nirvana (enlightenment).
Judaism	18 million	Abraham, Isaac, and Jacob; ca. 2000 B.C.E.	God's nature and will are revealed in the Torah (Hebrew scripture) and in His intervention in history. God has established a covenant with the people of Israel, who are called to a life of holiness, justice, mercy, and fidelity to God's law.
Confucianism	5.9 million	K'ung Fu-Tzu (Confucius); ca. 500 B.C.E.	The sayings of Confucius (collected in the *Analects*) stress the role of virtue and order in the relationships among individuals, their families, and society.

ing the state of *nirvana*—becoming liberated from the world by uniting the individual soul with the universal soul (*Brahma*).

Hinduism has been devoid of some of the social conflict experienced by other religions. Because Hinduism is based on the assumption that there are many paths to the "Truth" and that the world's religions are alternate paths to that goal, Hindus typically have not engaged in religious debates or "holy wars" with those holding differing beliefs. One of the best-known Hindu leaders of modern times was Mohandas ("Mahatma") Gandhi, the champion of India's independence movement, who was devoted to the Hindu ideals of nonviolence, honesty, and courage (Sharma, 1995). However, some social analysts note that Hinduism has been closely associated with the perpetuation of the caste system in India. Although people in the lower castes are taught to live out their lives with dignity, even in the face of poverty and despair, they may also come to believe that their lowly position is the acceptable and appropriate place for them to be—which allows the upper castes to exploit them (Kurtz, 1995).

The Hindu religion is almost as diverse as the wide array of people who adhere to its teachings. It is estimated that there are more than 700 million Hindus in the world today, with 95 percent of them residing in India, over 80 percent of whose population is Hindu (Sharma, 1995).

Since changes in U.S. immigration laws in the 1960s brought a wave of immigrants from India, the number of Hindus in this country has increased significantly (Albanese, 2007). Most Asian Indian immigrants to the United States have been well-educated professionals who have joined the ranks of the U.S. middle and upper-middle classes and have been active in supporting the more than forty Hindu temples in the nation. For most people of Asian Indian descent in the United States, these temples are sites of worship and gathering places where they can maintain a sense of community. They are also ritual centers where language, arts, and practices from their ethnic past can be preserved (Albanese, 2007).

How have Hindus fared in the United States? Intolerance has been an ongoing problem for many members of the Asian Indian community in this country. In view of numerous violent hate crimes perpetrated against "dot heads" (so-called by adversaries because some Asian Indian women wear a dot in the middle of their forehead), many have bound together to fight intolerance across lines of race, class, and religion.

Despite discrimination and initial confrontations with members of other racial–ethnic groups, the influence

According to Marx and Weber, religion serves to reinforce social stratification in a society. For example, according to Hindu belief, a person's social position in his or her current life is a result of behavior in a former life.

of Hindus in the United States is likely to be profound as their number increases. It is estimated that there are about 800,000 adherents in the United States today—a number eight times as large as it was twenty years ago (Shorto, 1997).

Buddhism

When Buddhism first emerged in India some twenty-five hundred years ago, it was thought of as a "new religious movement," arising as it did around the sixth century B.C.E., after many earlier religions had become virtually defunct. Buddhism's founder, Siddhartha Gautama of the Sakyas (also known as Gautama Buddha), was born about 563 B.C.E. into the privileged caste. His father was King Suddhodana (who was more like a feudal lord than a king because many kingdoms existed on the Indian subcontinent during that era). According to historians, the king attempted to keep his young son in the palace at all times so that he would neither see how poor people lived nor experience the suffering present in the outside world. As a result, Siddhartha was oblivious to social inequality until he began to make forays beyond the palace walls and into the "real" world, where he observed how other people lived. On one excursion, he saw a monk with a shaven head and became aware that some people withdraw from the secular world and live a life of strict asceticism. Later, Siddhartha engaged in intense meditation underneath a bodhi tree in what is now Nepal, eventually declaring that he had obtained Enlightenment—an awakening to the true nature of reality (Kurtz, 1995). From that day forward, Siddhartha was referred to as *Buddha,* meaning "the Enlightened One" or the "Awakened One," and spent his life teaching others how to reach nirvana (Smith, 1991).

Because of the efforts of a series of invaders, Buddhism had ceased to exist in India by the thirteenth century but had already expanded into other nations in various forms. *Theravadin Buddhism,* which focuses on the life of the Buddha and seeks to follow his teachings, gained its strongest toehold in Southeast Asia. *Mahayana Buddhism* is centered in Japan, China, and Korea, and primarily focuses on meditation and the Four Noble Truths:

1. Life is *dukkha*—physical and mental suffering, pain, or anguish that pervades all human existence.
2. The cause of life's suffering is rooted in *tanha*—grasping, craving, and coveting.
3. One can overcome *tanha* and be released into Ultimate Freedom in Perfect Existence (nirvana).
4. Overcoming desire can be accomplished through the Eightfold Path to Nirvana. This path is a way of living that avoids extremes of indulgence and suggests that a person can live in the world but not be worldly. The path's eight steps are *right view* (proper belief), *right intent* (renouncing attachment to the world), *right speech* (not lying, slandering, or using abusive talk), *right action* (avoiding sexual in-

Worshippers at this Buddhist temple in Los Angeles celebrate the Thai New Year.

dulgence), *right livelihood* (avoiding occupations that do not enhance spiritual advancement), *right effort* (preventing potential evil from arising), *right mindfulness,* and *right concentration* (overcoming sensuous appetites and evil desires). (Smith, 1991; Matthews, 2004)

The third major branch of Buddhism—*Vajrayana*—incorporates the first two branches along with some aspects of Hinduism; it emerged in Tibet in the seventh century (Albanese, 2007). Like Hinduism, the teachings of this type of Buddhism—and specifically those of the Dalai Lama, the Tibetan Buddhist leader—emphasize the doctrine of *ahimsa,* or nonharmfulness, and discourage violence and warfare. It is believed that Buddhism may have suppressed caste-related tensions resulting from the vast economic inequality found in early India (Kurtz, 1995).

When did Buddhism first arrive in this country? According to most scholars, some branches appeared as early as the 1840s, when Chinese immigrants arrived on the West Coast. Shortly thereafter, temples were erected in San Francisco's Chinatown; however, ethnic Buddhism in the United States expanded after the Civil War, when Japanese immigrants arrived first in Hawaii (then a U.S. possession) and a decade later in California. The branch of Pure Land Buddhism brought by the Japanese probably had the greatest chance of succeeding in the United States because it most closely resembled Christianity. Pure Land Buddhism included teachings about God and His son Jesus, Who brought human beings to an eternal life in paradise (Albanese, 2007). Buddhism went through the process of Americanization, which is reflected in the contemporary use of terms such as *church, bishop,* and *Sunday school*—terms previously unknown in this religion.

Today, Buddhism is one of the fastest-growing Eastern religions in the United States. Zen and Tibetan Buddhism are extremely popular forms. Zen sprang up among white, middle-class young people, many of whom were well-educated and lived in California or New York. Buddhist monks fleeing the takeover of Tibet by the Communist Chinese in the 1960s brought Tibetan Buddhism to this country. Recently, Tibetan Buddhism has received extensive media coverage, at least partly due to the conversion of some U.S. celebrities, including Richard Gere, Steven Seagal, and Adam Yauch (Van Biema, 1997).

Confucianism

Confucianism—which means the "family of scholars"—started as a school of thought or a tradition of learning before its eventual leader, Confucius, was born (Wei-ming, 1995). Confucius (the Latinized form of K'ung Fu-tzu) lived in China between 551 and 479 B.C.E. and emerged as a teacher at about the same time that the Buddha became a significant figure in India.

Confucius—whose sayings are collected in the *Analects*—taught that people must learn the importance of *order* in human relationships and must follow a strict code of moral conduct, including respect for others, benevolence, and reciprocity (Kurtz, 1995). A central teaching of Confucius was that humans are by nature good and that they learn best by having an example or a role model. As a result, he created a *junzi* ("chun-tzu"), or model person, who has such attributes as being upright regardless of outward circumstances, being magnanimous by expressing forgiveness toward others, being directed by internal principles rather than external laws, being sincere in speech and action, and being earnest and benevolent. Confucius wanted to demonstrate these traits, and he believed that he should be a role model for his students. The *junzi's* behavior is to be based on the Confucian principle of *Li,* meaning righteousness or propriety, which refers both to ritual and to correct conduct in public. One of the central attributes exhibited by the *junzi* is *ren* (*jen*), which means having deep empathy or compassion for other humans (Matthews, 2004).

Confucius established the foundation for social hierarchy—and potential conflict—when he set forth his Five Constant Relationships: *ruler–subject, husband–wife, elder brother–younger brother, elder friend–junior friend,* and *father–son.* In each of these pairs, one person is unequal to the other, but each is expected to carry out specific responsibilities to the other (Matthews, 2004).

Confucianism is based on the ethical teachings formulated by Confucius, shown here in a portrait created by a Manchu prince in 1735.

Confucius taught that due authority is not automatic; it must be earned. The subject does not owe loyalty to the ruler or authority figure if that individual does not fulfill his or her end of the bargain.

Confucianism is based on the belief that Heaven and Earth are not separate places but rather a continuum in which both realms are constantly in touch with each other. According to this approach, those who inhabit Heaven are ruled over by a supreme ancestor and are the ancestors of those persons who are on Earth. These forefathers are eventually joined by those who are currently on Earth; therefore, death is nothing more than the promotion to a more honorable estate.

Until the Communist takeover and the establishment of the People's Republic, Confucianism was the official religion of China. After the takeover, political leaders sought to replace Confucianism, Taoism, and Buddhism with Maoism—belief in the teachings of the Chinese Communist leader—but the Confucian influence remains in East Asian countries such as Japan, Korea, Taiwan, and Singapore.

How prominent is Confucianism in the United States? It is difficult to provide an accurate answer to this question because some people view Confucianism as a set of ethical teachings rather than as a religion (Smith, 1991). However, many recent immigrants from China and Southeast Asia adhere to the teachings of Confucius, although perhaps mixed with those of other great Eastern religious philosophers and teachers. Neo-Confucianism, which has emerged on the West Coast, is heavily influenced by Buddhism and Taoism (Matthews, 2004).

We now move from the Eastern religions (Hinduism, Buddhism, and Confucianism), which tend to be based on ethics or values more than on a deity or Supreme Being, to the Western religions (Judaism, Christianity, and Islam), which are founded on the Abrahamic tradition and place an emphasis on God and a relationship between human beings and a Supreme Being. The original locations of all six of these religions are shown in ▶ Map 17.1.

Judaism

Although Judaism has fewer adherents worldwide than some other major religions, its influence is deeply felt in Western culture. Today, there are an estimated 18 million Jews residing in about 134 countries worldwide; however, the majority reside in the United States or Israel (J. Wright, 1997).

Central to contemporary Jewish belief is monotheism, the idea of a single god, called Yahweh, the God of Abraham, Isaac, and Jacob. The Hebrew tradition emerged out of the relationship of Abraham and Sarah, a husband and wife, with Yahweh. According to Jewish tradition, the God of the Jews made a covenant with Abraham and Sarah—His chosen people—that He would protect and provide for them if they swore Him love and obedience. When God appeared to Abraham in about the eighteenth century B.C.E., He encouraged Abraham to emigrate to the area near the Sea of Galilee and the Dead Sea (what is now Israel), leaving behind the ancient fertile crescent of the Middle East (present-day Iraq).

The descendants of Abraham and Sarah migrated to Egypt, where they became slaves of the Egyptians. In a vision, God's chosen leader, Moses, was instructed to liberate His chosen people from the bondage and slavery imposed upon them by the pharaoh. After experiencing a series of ten plagues, the pharaoh decided to free the slaves. The tenth and final plague had involved killing all of the firstborn in the land of Egypt—human beings and lower animals, as well—except the firstborn children of the Hebrews, who had put the blood of a lamb on the doorposts of their houses so that they would be passed over. This practice inspired the Jewish holiday Passover, which commemorates God's deliverance of the Hebrews from slavery in Egypt during the time of Moses (Matthews, 2004). It is believed that the first Passover took place in Egypt in about 1300–1200 B.C.E. (Matthews, 2004).

Wandering in the desert after their release, the Hebrews established a covenant with God, Who promised that if they would serve Him exclusively, He would give the Hebrews a promised land and make them a great

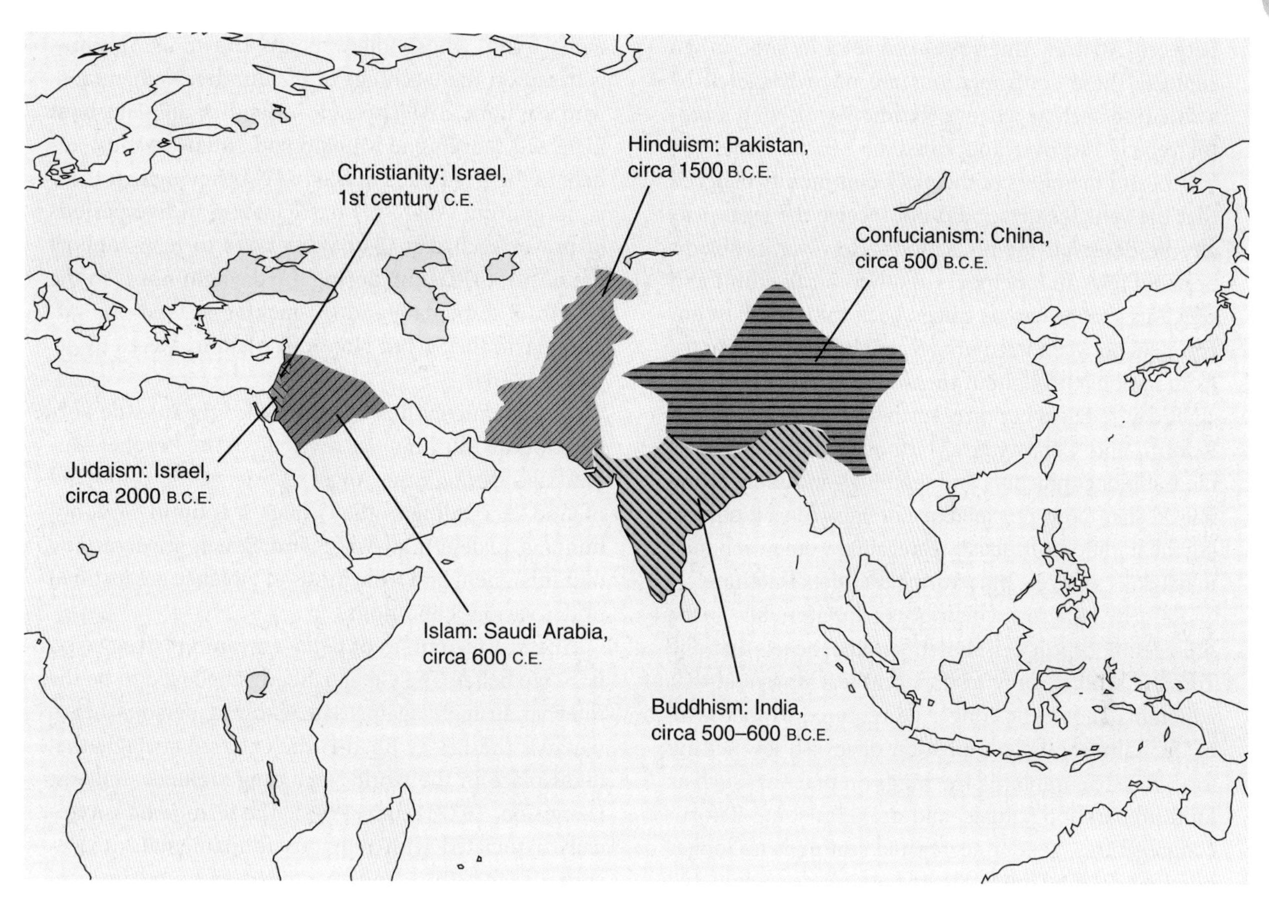

▶ Map 17.1 Original Locations of the World's Major Religions

nation. Known as the Ten Commandments (or *Decalogue*), the covenant between God and human beings was given to Moses on top of Mount Sinai. The Ten Commandments and discussions of moral, ceremonial, and cultural laws are contained in four books of the Torah: Exodus, Leviticus, Numbers, and Deuteronomy. The Torah, also known as the Pentateuch, is the sacred book of contemporary Judaism.

The Jewish people believe that they have a unique relationship with God, affirmed on the one hand by His covenant and on the other by His law. Judaism has three key components: God (the deity), Torah (God's teachings), and Israel (the community or holy nation). Although God guides human destiny, people are responsible for making their own ethical choices in keeping with His law; when they fail to act according to the law, they have committed a sin. Also fundamental to Judaism is the belief that one day the Messiah will come to Earth, ushering in an age of peace and justice for all.

Today, Jews worship in synagogues in congregations led by a *rabbi*—a teacher or ordained interpreter and leader of Judaism. The Sabbath is observed from sunset Friday to sunset Saturday, based on the story of Creation in Genesis, especially the belief that God rested on the seventh day after He had created the world. Worship services consist of readings from scripture, prayer, and singing. Jews celebrate a set of holidays distinct from U.S. dominant cultural religious celebrations. The most important holidays in the Jewish calendar are Rosh Hashana (New Year), Yom Kippur (Day of Atonement), Hanukkah (Festival of Lights), and Pesach (Passover).

Throughout their history, Jews have been the object of prejudice and discrimination. The Holocaust, which took place in Nazi Germany (and several other nations that the Germans occupied) between 1933 and 1945, remains one of the saddest eras in history. After the rise of Hitler in Germany in 1933 and the Nazi invasion of Poland, Jews were singled out with special registrations, passports, and clothing. Many of their families were separated by force, and some family members were sent to slave labor camps while others were sent to "resettlement." Eventually, many Jews were imprisoned in death camps, where six million lost their lives.

Anti-Semitism has been a continuing problem in the United States since the late nineteenth century. Like other forms of prejudice and discrimination, anti-Semitism has extracted a heavy toll on multiple generations of Jewish Americans. In the 1980s and 1990s, for example, there was an increase in interethnic violence between African Americans and Jews in large urban centers. These confrontations are often triggered by a situation such as when a Hasidic Jew driving a motor vehicle ran over and killed an African American youth, and members of the black community believed that the vehicle's driver did not receive the legal penalty he deserved for his actions. However, problems between Jews and members of other racial–ethnic and religious groups can be traced back to the earliest encounters between immigrants arriving in this country from nations throughout the world.

Today, Judaism has three main branches—Orthodox, Reform, and Conservative. Orthodox Judaism follows the traditional practices and teachings, including eating only kosher foods prepared in a designated way, observing the traditional Sabbath, segregating women and men in religious services, and wearing traditional clothing.

Reform Judaism, which began in Germany in the nineteenth century, is based on the belief that the Torah is binding only in its moral teachings and that adherents should no longer be required to follow all of the Talmud, the compilation of Jewish law setting forth the strict rabbinic teachings on practices such as food preparation, rituals, and dress. In some Reform congregations, gender-segregated seating is no longer required. In the United States, services are conducted almost entirely in English, a Sunday Sabbath is observed, and less emphasis is placed on traditional Jewish holidays (Albanese, 2007).

Conservative Judaism emerged between 1880 and 1914 with the arrival of many Jewish immigrants in the United States from countries such as Russia, Poland, Rumania, and Austria. Seeking freedom and an escape from persecution, these new arrivals settled in major cities such as New York and Chicago, where they primarily became factory workers, artisans, and small shopkeepers. Conservative Judaism, which became a middle ground between Orthodox and Reform Judaism, teaches that the Torah and Talmud must be followed and that *Zionism*—the movement to establish and maintain a Jewish homeland in Israel—is crucial to the future of Judaism. In Conservative synagogues, worship services are typically performed in Hebrew. Men are expected to wear head coverings, and women have roles of leadership in the congregation; some may become ordained rabbis (Matthews, 2004). Despite centuries of religious hatred and discrimination, Judaism persists as one of the world's most influential religions.

Islam

Like Judaism, Islam is a religion in the Abrahamic tradition; both religions arise through sons of Abraham—Judaism through Isaac and Islam through Ishmael. Islam, whose followers are known as Muslims, is based on the teachings of its founder, Muhammad, who was born in Mecca (now in Saudi Arabia) in about 570 C.E. According to Muhammad, followers must adhere to the five Pillars of Islam: (1) believing that there is no god but Allah, (2) participating in five periods of prayer each day, (3) paying taxes to help support the needy, (4) fasting during the daylight hours in the month of Ramadan, and (5) making at least one pilgrimage to the Sacred House of Allah in Mecca (Matthews, 2004).

The Islamic faith is based on the Qur'an—the holy book of the Muslims—as revealed to the Prophet Muhammad through the Angel Gabriel at the command of God. According to the Qur'an, it is up to God, not humans, to determine which individuals are deserving of punishment and what kinds of violence are justified under various conditions.

The Islamic notion of *jihad*—meaning "struggle"—is a core belief. The Greater Jihad is believed to be the internal struggle against sin within a person's heart, whereas the Lesser Jihad is the external struggle that takes place in the world, including violence and war (Ferguson, 1977; Kurtz, 1995). The term *jihad* is typically associated with religious fundamentalism. Despite the fact that fundamentalism is found in most of the world's religions, some social analysts believe that Islamic fundamentalism is uniquely linked to the armed struggles of groups such as Hamas, an alleged terrorist organization, and the militant Islamic Jihad, which is believed to engage in continual conflict (see Barber, 1996).

The Muslim women shown here pray at a mosque courtyard in Bangladesh during the fasting month of Ramadan. According to Muslim teaching, Ramadan marks God's revelation of the Qur'an to the Prophet Muhammad.

Today, more than 19 percent of the world's population considers itself to be Muslim. Most of the more than one billion adherents of this religion reside in the Middle East, but the majority of people residing in northern Africa and western Asia also consider themselves to be Muslim. Other large populations of Muslims are located in Pakistan, India, Bangladesh, Indonesia, and the southern regions of the former Soviet Union.

Islam is one of the fastest-growing religions in the United States. Driven by recent waves of migration and a relatively high rate of conversion, there has been a significant increase in the number of Muslims in this country. At about six million, they outnumber members of several major Protestant denominations, including the Presbyterians (estimated at 3.5 million) and the Episcopalians (estimated at 2.3 million). Recent Muslim arrivals in the United States typically have come from countries such as Pakistan, Iran, and Saudi Arabia. Most have settled in the Midwest or on the East Coast. Approximately 30 percent of Muslims are from southern Asia, 25 percent are African American, and 20 percent are Arab Americans (Shorto, 1997). The oldest U.S. Islamic group is the Federation of Islamic Associations, and many university campuses now have Muslim student associations. The largest "umbrella" organization of U.S. Muslims is the Islamic Society of North America (Matthews, 2004).

Muslims in the United States have not always felt welcome. During the Carter administration, the Iranian hostage crisis triggered negative stereotypes, including movies and television programs depicting Muslims as terrorists. The bombing of the World Trade Center in New York in 1993 and the terrorist attacks on the United States in 2001, which claimed the lives of about 3,000 people, intensified stereotypes regarding people appearing to be Muslim or from countries associated with Islam. According to some analysts, media distortions of "Islam" and "Muslims" intensify prejudice against Arab Americans.

Despite these negative stereotypes and some acts of violence against Muslims, the more than six hundred mosques and Islamic centers in this country today continue to serve as centers of worship and as social and educational centers, where people can maintain cultural ties and hold ceremonial rites of passage such as weddings and funerals (Matthews, 2004).

Christianity

Along with Judaism and Islam, Christianity follows the Abrahamic tradition, tracing its roots to Abraham and Sarah. Although Jews and Christians share common scriptures in the portions of the Bible known to Christians as the "Old Testament," they interpret them differently. The Christian teachings in the "New Testament" present a world view in which the old covenant between God and humans, as found in the Old Testament, is obsolete in light of God's offer of a new covenant to the followers of Jesus, whom Christians believe to be God's only son (Matthews, 2004).

As described in the New Testament, Jesus was born to the virgin Mary and her husband, Joseph. After a period of youth in which He prepared himself for the ministry, Jesus appeared in public and went about teaching and preaching, including performing a series of miracles—events believed to be brought about by divine intervention—such as raising people from the dead.

The central themes in the teachings of Jesus are the kingdom of God and standards of personal conduct for adherents of Christianity. Jesus emphasized the importance of righteousness before God and of praying to the Supreme Being for guidance in the daily affairs of life (Matthews, 2004).

One of the central teachings of Christianity is linked to the unique circumstances surrounding the death of Jesus. Just prior to His death, Jesus and His disciples held a special supper, now referred to as "the last supper," which is commemorated in contemporary Christianity in the sacrament of Holy Communion. Afterward, Jesus was arrested by a group sent by the priests and scribes for claiming to be king of the Jews. After being condemned to death by political leaders, Jesus was executed by crucifixion, which made the cross a central symbol of the Christian religion. According to the New Testament, Jesus died, was placed in a tomb, and on the third day was resurrected—restored to life—establishing that He is the son of God. Jesus then remained on Earth for forty days, after which He ascended into heaven on a cloud. Two thousand years later, many Christian churches teach that one day Jesus

Christians around the world have been drawn to cathedrals such as the Basilica of Sacré Coeur in Paris (built between 1875 and 1914) to worship God and celebrate their religious beliefs.

will "come again in glory" and that His second coming will mark the end of the world as we know it.

Whereas Judaism is basically an inherited religion and most adherents are born into the community, Christianity has universalistic criteria for membership—meaning that it does not have ethnic or tribal qualifications—based on acceptance of a set of beliefs. Becoming a Christian requires personal belief that Jesus is the son of God; that He died, was buried, and on the third day rose from the dead; and that the Supreme Being is a sacred *trinity*—the Christian belief of "God the Three in One," comprising God the Father, Jesus the Son, and the Holy Spirit—a presence that lives within those who have accepted Jesus as savior (Barna, 1996). To become members of the religious community, believers must affirm their faith and go through a rite of passage of baptism (Matthews, 2004). According to the teachings of Christianity, those who believe in Jesus as their savior will be resurrected from death and live eternally in the presence of God, whereas those who are wicked will endure an eternity in hell (Matthews, 2004).

For centuries, Christians, Jews, and Muslims have lived together in peace and harmony in some areas of the world; in others, however, they have engaged in strife and fighting. The wounds and animosities between Muslims and Christians have remained since the early Christian era; nevertheless, some religion scholars believe that there is hope for a genuine Christian–Muslim dialogue in the future (Cox, 1995).

Today, almost one-third of the world's population (between 1.5 billion and 2 billion) refer to themselves as Christians. The majority of Christians live in North or South America or in Europe. According to Kurtz (1995), a sociological analysis of Christianity would reveal that it became the dominant religion not necessarily due to its theology but because of its alliance with the power structures of Western civilization, beginning with the fourth-century conversion of Roman Emperor Constantine and following with its expansion into Western Europe during the Middle Ages. Missionary movements helped spread Christianity outward from Europe to other regions of the world in the nineteenth century, as missionaries also conquered land, cultures, and the economic and political resources of indigenous populations (Kurtz, 1995). As Christianity moved across cultures, it underwent dramatic transformations and became, in actuality, a tremendous variety of "Christianities" rather than just one highly integrated body of religious beliefs and practices (Kurtz, 1995).

Roman Catholics sent the first missionaries to North America, where they established 40 missions and converted about 26,000 Native Americans to Christianity before the early Protestant settlers arrived on this continent. From the earliest days of the British colonies in this country, a variety of religions were represented, including Anglicans (the forerunners of the Protestant Episcopal church), Baptists, Quakers, Presbyterians, Methodists, and Lutherans. Freedom of religion provided people with the opportunity to establish other denominations and sects and generally to worship as they pleased.

The African American church was the center of community life first for slaves and freed slaves, and then for generations of blacks who experienced ongoing prejudice and discrimination based on their race. The theory of nonviolent protest and civil disobedience used in the civil rights movement in the 1960s, and largely orchestrated by the Rev. Dr. Martin Luther King, Jr., emerged from the African American church in the South. Over the years, members of other minority groups, including Latinos/as and Asian Americans, have benefited from religious and social ties to various Christian denominations, including the Roman Catholic church. Today, about 160 million people in the United States are associated with Christian churches, of whom about 67 million consider themselves to be Catholics.

Types of Religious Organization

Religious groups vary widely in their organizational structure. Although some groups are large and somewhat bureaucratically organized, others are small and have a relatively informal authority structure. Some require total commitment of their members; others expect members to have only a partial commitment. Sociologists have developed typologies or ideal types of religious organization to enable them to study a wide variety of religious groups. The most common categorization sets forth four types: ecclesia, church, sect, and cult.

Ecclesia

Some countries have an official or state religion known as the ***ecclesia*—a religious organization that is so integrated into the dominant culture that it claims as its membership all members of a society.** Membership in the ecclesia occurs as a result of being born into the society, rather than by any conscious decision on the part of individual members. The linkages between the social institutions of religion and government are often very strong in such societies. Although no true ecclesia exists in the contemporary world, the Anglican church (the official church of England), the

◆ Table 17.2 Characteristics of Churches and Sects

Characteristic	Church	Sect
Organization	Large, bureaucratic organization, led by a professional clergy	Small, faithful group, with high degree of lay participation
Membership	Open to all; members usually from upper and middle classes	Closely guarded membership, usually from lower classes
Type of Worship	Formal, orderly	Informal, spontaneous
Salvation	Granted by God, as administered by the church	Achieved by moral purity
Attitude Toward Other Institutions and Religions	Tolerant	Intolerant

Lutheran church in Sweden and Denmark, the Roman Catholic church in Italy and Spain, and the Islamic mosques in Iran and Pakistan come fairly close.

The Church–Sect Typology

To help explain the different types of religious organizations found in societies, Ernst Troeltsch (1960/1931) and his teacher, Max Weber (1963/1922), developed a typology that distinguishes between the characteristics of churches and sects (see ◆ Table 17.2). Unlike an ecclesia, a church is not considered to be a state religion; however, it may still have a powerful influence on political and economic arrangements in society. A ***church*** **is a large, bureaucratically organized religious organization that tends to seek accommodation with the larger society in order to maintain some degree of control over it.** Church membership is largely based on birth; typically, children of church members are baptized as infants and become lifelong members of the church. Older children and adults may choose to join the church, but they are required to go through an extensive training program that culminates in a ceremony similar to the one that infants go through. Churches have a bureaucratic structure, and leadership is hierarchically arranged. Usually, the clergy have many years of formal education. Churches have very restrained services that appeal to the intellect rather than the emotions. Religious services are highly ritualized; they are led by clergy who wear robes, enter and exit in a formal processional, administer sacraments, and read services from a prayer book or other standardized liturgical format. The Lutheran church and the Episcopal church are two examples.

Midway between the church and the sect is the ***denomination*****—a large organized religion characterized by accommodation to society but frequently lacking in ability or intention to dominate society** (Niebuhr, 1929). Denominations have a trained ministry, and although involvement by lay members is encouraged more than in the church, their participation is usually limited to particular activities, such as readings or prayers. Denominations tend to be more tolerant and are less likely than churches to expel or excommunicate members. This form of organization is most likely to thrive in societies characterized by *religious pluralism*—a situation in which many religious groups exist because they have a special appeal to specific segments of the population. Perhaps because of its diversity, the United States has more denominations than any other nation. ◆ Table 17.3 shows this diversity in Christian denominations. Today, denominations range from Baptists and members of the Church of Christ to Unitarians and Congregationalists.

A ***sect*** **is a relatively small religious group that has broken away from another religious organization to renew what it views as the original version of the faith.** Unlike churches, sects offer members a more personal religion and an intimate relationship with a supreme being, depicted as taking an active interest in the individual's everyday life. Sects have informal prayers composed at the time they are given, whereas churches use formalized prayers, often from a prayer book. Typically,

ecclesia a religious organization that is so integrated into the dominant culture that it claims as its membership all members of a society.

church a large, bureaucratically organized religious organization that tends to seek accommodation with the larger society in order to maintain some degree of control over it.

denomination a large organized religion characterized by accommodation to society but frequently lacking in ability or intention to dominate society.

sect a relatively small religious group that has broken away from another religious organization to renew what it views as the original version of the faith.

◆ **Table 17.3 Major U.S. Denominations That Self-Identify as Christian**

Religious Body	Members	Churches
Roman Catholic Church	69,135,000	18,992
Southern Baptist Convention	16,270,000	43,669
United Methodist Church	8,075,000	34,660
Church of Jesus Christ of Latter Day Saints	5,691,000	12,753
Church of God in Christ[a]	5,500,000	15,300
National Baptist Convention, U.S.A.[a]	5,000,000	9,000
Evangelical Lutheran Church in America	4,851,000	10,519
National Baptist Convention of America[a]	3,500,000	(N/A)
Presbyterian Church (U.S.A.)	3,099,000	10,960
Assemblies of God	2,831,000	12,298
African Methodist Episcopal Church[a]	2,500,000	4,174
National Missionary Baptist Convention of America[a]	2,500,000	(N/A)
Progressive National Baptist Convention[a]	2,500,000	2,000
Lutheran Church–Missouri Synod	2,441,000	6,144
Episcopal Church	2,248,000	7,200
Churches of Christ[a]	1,639,000	15,000
Greek Orthodox Church	1,500,000	566
Pentecostal Assemblies of the World[a]	1,500,000	1,750
African Methodist Episcopal Zion Church	1,440,000	3,260
American Baptist Churches in U.S.A.	1,397,000	5,740
United Church of Christ	1,244,000	5,567
Baptist Bible Fellowship, International[a]	1,200,000	4,500
Christian Churches and Churches of Christ[a]	1,072,000	5,579
Jehovah's Witnesses	1,046,000	12,384

[a]Current data not available; prior data used may no longer be comparable.
Source: U.S. Census Bureau, 2008.

religious sects appeal to those who might be characterized as lower class, whereas denominations primarily appeal to the middle and upper-middle classes, and churches focus on the upper classes.

According to the church–sect typology, as members of a sect become more successful economically and socially, they tend to focus more on this world and less on the next. However, sect members who do not achieve financial success often believe that they are being left behind as the other members, and sometimes the minister, shift their priorities to things of this world. Eventually, this process weakens some religious

This mass wedding ceremony of thousands of brides and grooms brought widespread media attention to the Reverend Sun Myung Moon and the Unification Church, which many people view as a religious cult.

organizations, and the dissatisfied or downwardly mobile split off to create new, less worldly versions of the original group that will be more committed to "keeping the faith." Those who defect to form a new religious organization may start another sect or form a cult (Stark and Bainbridge, 1981).

Cults

A ***cult*** **is a loosely organized religious group with practices and teachings outside the dominant cultural and religious traditions of a society.** Although many people view cults negatively, some major religions (including Judaism, Islam, and Christianity) and some denominations (such as the Mormons) started as cults. Cult leadership is based on *charismatic* characteristics (personal magnetism or mystical leadership) of the individual leader, including an unusual ability to communicate and to form attachments with other people. An example is the religious movement started by Reverend Sun Myung Moon, a Korean electrical engineer who believed that God had revealed to him that the Judgment Day was rapidly approaching. Out of this movement, the Unification church, or "Moonies," grew and flourished, recruiting new members through their personal attachments to present members. Some cult leaders have not fared well, including Jim Jones, whose ill-fated cult ended up committing mass suicide in Guyana in 1978, and more recently Marshall Herff Applewhite ("Do"), who led his 38 Heaven's Gate followers to commit mass suicide in 1997 at their mansion in Rancho Santa Fe, California. Applewhite's followers were convinced that the comet Hale-Bopp, which swung by Earth in late March of that year, would be their celestial chariot taking them to a higher level (*Newsweek*, 1997).

Are all cults short-lived? Over time, some cults disappear; however, others undergo transformation into sects or denominations. For example, cult leader Mary Baker Eddy's Christian Science church has become an established denomination with mainstream methods of outreach, such as a Christian Science Reading Room strategically placed in an office building or shopping mall, where persons who otherwise might know nothing of the organization learn of its beliefs during their routine activities.

Trends in Religion in the United States

As we have seen throughout this chapter, religion in the United States is very diverse. Pluralism and religious freedom are among the cultural values most widely espoused, and no state church or single denomination predominates. As shown in ◆ Table 17.4, Protestants constitute the largest religious body in the United States, followed by Roman Catholics, Muslims, Jews, Eastern churches, and others.

◆ **Table 17.4 U.S. Religious Traditions' Membership**

Religious Tradition	Percentage of All U.S. Adults
Protestants	51.3
Roman Catholics	23.9
Mormons	1.7
Jews	1.7
Jehovah's Witnesses	0.7
Buddhists	0.7
Orthodox Christians	0.6
Muslims	0.6
Hindus	0.4
Other faiths	1.5
Unaffiliated	16.1
Don't know/no answer	0.6

Source: Pew Forum on Religion and Public Life, 2008.

Religion and Social Inequality

Social science research continues to identify significant patterns between social class and religious denominations. Historically, those denominations with the most-affluent members and the highest social status have been the mainline liberal churches, such as the Episcopalians, the Presbyterians, and the Congregationalists. The earliest members of some of these churches in the United States were immigrants from Britain, who were also instrumental in establishing the U.S. government and the capitalist economy. Since that time, people with high levels of economic and political power have been closely linked to these denominations; however, members of these denominations also come from the middle and working classes and from minority groups. Although the Presbyterian church is typically characterized as middle class and the Baptist church as working class and lower income, both of these denominations also have members from a wide diversity of income levels (Kosmin and Lachman, 1993). According to some

cult a religious group with practices and teachings outside the dominant cultural and religious traditions of a society.

analysts, the socioeconomic position of families that adhere to conservative religious beliefs may sometimes be lower because traditional gender roles assign women to the home and men to the workplace, thus limiting the family's income, as contrasted with religions that do not discourage women from working outside the home (Kosmin and Lachman, 1993). A comparison of different denominations yields some information on class and income differences, but a significant difference in race and class is seen in many central-city and suburban churches.

Race and Class in Central-City and Suburban Churches

As more middle- and upper-income individuals and families moved to the suburbs during the twentieth century, some churches followed the members of their congregations to the suburbs. This pattern is known as "upgrading" and results in the church having newer facilities and more members who are in the upper-middle and upper classes. Meanwhile, the church's former building site is often taken over by a minority congregation or by a denomination that appeals primarily to members of the working class, older members living on a fixed income, or recent immigrant groups such as Vietnamese Americans (Hudnut-Beumler, 1994).

The sociologist Nancy Ammerman (1997) and her associates examined churches struggling to survive in central cities and concluded that the efforts of many churches that remain in the central city and engage in community outreach have had mixed results. When church leaders seek to include people from the immediate neighborhood in the life of the church, some longtime members find the experience too disruptive and discourage future endeavors to reach out to the surrounding community (Ammerman, 1997). As a result, the church may stagnate as older members die or move away.

On the other hand, racial and cultural minorities who feel overpowered by their lack of economic resources may be drawn to churches that help them establish a sense of dignity and personal integrity that is otherwise missing in their daily social interactions. For example, African American churches have provided members with a sense of personal dignity and worth in the face of persistent racial prejudice and discrimination while serving as a family life and social service center, a community organization, and a means of providing hope for the next generation. As the sociologist Andrew Billingsley (1992: 349) explains, "Over the centuries, the church has become the strongest institution in [the African American] community. It is prevalent, independent, and has extensive outreach."

Consider how one pastor in a contemporary African American church described his role:

> As a pastor . . . I have watched families struggle with an assortment of devastating problems. I have shared the pain of families in which members have been accused or convicted of theft, drug addiction, prostitution, rape, and murder. I have been involved with homeless families who have been so desperate for a place to live that squatting in abandoned houses was their only recourse. I have witnessed elderly persons lose all sense of autonomy because of homelessness, illness, and loneliness. I have heard the cries of children, parents, and the elderly as they faced conditions of hopelessness. Through it all I have witnessed a remarkable fact—for these persons the church has been the central authenticating reality in their lives. When the world has so often been willing to say only "no" to these people, the church has said "yes." For black people the church has been the one place where they have been able to experience unconditional positive regard. (qtd. in Smith, 1985: 14)

As previously noted, members of other subordinate racial and ethnic groups have also found support and hope in their churches.

© Rachel Epstein/The Image Works

Churches in converted buildings such as this seek to win new religious followers and to offer solace and hope to people in low-income areas.

Secularization and the Rise of Religious Fundamentalism

How strong is the influence of religion in contemporary life? Some analysts believe that *secularization*—the decline in the significance of the sacred in daily life—has occurred; consequently, there has been a resurgence of religious fundamentalism among some groups. ***Fundamentalism* is a traditional religious doctrine that is conservative, is typically opposed to modernity, and rejects "worldly pleasures" in favor of otherworldly spirituality.** In the United States, traditional fundamentalism primarily appealed to people from lower-income, rural, southern backgrounds; however, the "new" fundamentalist movement of the 1980s and 1990s has had a much wider appeal to people from all socioeconomic levels, geographical areas, and occupations. One reason for the rise of fundamentalism has been a reaction against modernization (Shupe and Hadden, 1989). Around the world, those who adhere to fundamentalism—whether they are Muslims, Christians, or followers of one of the other world religions—believe that sacred traditions must be revitalized. In the United States, public education has been the focus of some who follow the tenets of Christian fundamentalism. For example, various religious and political leaders have vowed to bring the Christian religion "back" into the public life of this country. They have been especially critical of educators who teach what they perceive to be *secular humanism*—the belief that human beings can become better through their own efforts rather than through belief in God and a religious conversion. According to Christian fundamentalists, elementary schoolchildren do not receive a fair and balanced picture of the Christian religion, but instead are taught that their parents' religion is inferior and perhaps irrational (Carter, 1994).

But how might students and teachers who come from the diverse religious heritages that we have examined in this chapter feel about Christian religious instruction or organized prayer in public schools? Many social analysts believe that such practices would cause conflict and perhaps discrimination on the basis of religion. For example, Rick Nelson—a teacher in the Fairfax County, Virginia, public school system—describes the potential effect of religious teaching and group prayer on students in his classroom:

> I think it really trivializes religion when you try to take such a serious topic with so many different viewpoints and cover it in the public schools. At my school we have teachers and students who are Hindu. They are really devout, but they are not monotheistic. . . . I am not opposed to individual prayer by students. I expect students to pray when I give them a test. They need to do that for my tests. . . . But when there is group prayer . . . who's going to lead the group? And if I had my Hindu students lead the prayer, I will tell you it will disrupt many of my students and their parents. It will disrupt the mission of my school, unfortunately . . . if my students are caused to participate in a group Hindu prayer. (CNN, 1994)

Religion in the Future

Debates over religion, particularly issues such as secularization and fundamentalism, will no doubt continue; however, many social scientists believe that religion is alive and well in the United States and in other nations of the world. Not only are we seeing the creation of new religious forms, but we are also seeing a dramatic revitalization of traditional forms of religious life (Kurtz, 1995).

One example of this change is ***liberation theology*—the Christian movement that advocates freedom from political subjugation within a traditional perspective and the need for social transformation to benefit the poor and downtrodden** (Kurtz, 1995). Although liberation theology initially emerged in Latin America as people sought to free themselves from the historical oppression of that area, this perspective has been embraced by a wide variety of people, ranging from African and African American Christians to German theologians and some feminists.

Another example of changes in the nature of theology is found in some feminist movements that have turned to pagan religions and witchcraft as a means of countering what they consider to be the patriarchal structure and content of the world's religions. For instance, the *Goddess movement* encompasses a variety of countercultural beliefs based on paganism and feminism, and is rooted in acknowledgment of the legitimacy of female power as a "beneficent and independent power" (Christ, 1987: 121).

fundamentalism a traditional religious doctrine that is conservative, is typically opposed to modernity, and rejects "worldly pleasures" in favor of otherworldly spirituality.

liberation theology the Christian movement that advocates freedom from political subjugation within a traditional perspective and the need for social transformation to benefit the poor and downtrodden.

Box 17.4 You Can Make a Difference

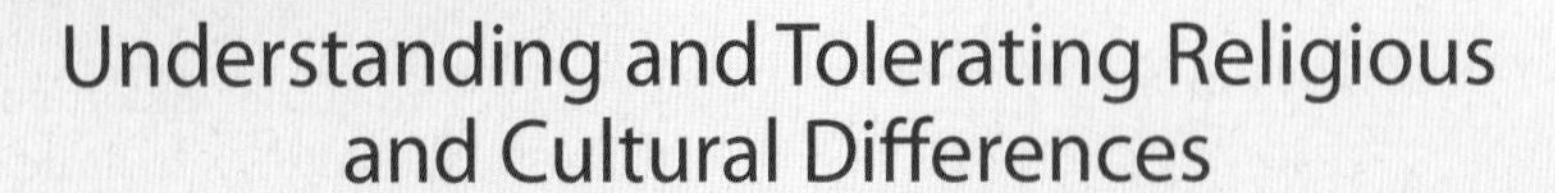

Understanding and Tolerating Religious and Cultural Differences

> [The small strip of kente cloth that hangs on the marble altar and the flags of a dozen West Indies nations, including Trinidad, Jamaica, and Barbados,] represent all of the people in this parish. It's an opportunity for us to celebrate everyone's culture. You won't find your typical Episcopal church looking like that.
>
> —Rev. Cannon Peter P. Q. Golden, rector of St. Paul's Episcopal Church in Brooklyn, New York (qtd. in Pierre-Pierre, 1997: A11)

One part of the history of St. Paul's Episcopal Church in Flatbush, Brooklyn, is etched into its huge stained-glass windows, which tell the story of the prominent descendants of English and Dutch settlers who founded the church. However, the church's more recent history is told in the faces of its parishioners—most of whom are Caribbean immigrants from the West Indies who labor as New York City's taxi drivers, factory workers, accountants, and medical professionals (Pierre-Pierre, 1997).

Traveling only a short distance, we find a growing enclave of Muslims in Brooklyn. It is their wish that people will recognize that Islam shares a great deal with Christianity and Judaism, including the fact that all three believe in one God, are rooted in the same part of the world, and share some holy sites (Sengupta, 1997). Some Muslim adherents also hope that more people in this country will learn greater tolerance toward those who have a different religion and celebrate different holidays. For example, in some years the Islamic holy season, Ramadan, falls at about the same time as Christmas and Hanukkah, but many Muslim children and adults are disparaged by their neighbors because they do not celebrate the same holidays as others. As one person observed, "As Muslims, we have to respect all religions" (Sengupta, 1997: A12). Left unsaid was the belief that other people should do likewise.

How can each of us—regardless of our race, color, creed, or national origin—help to bring about greater tolerance of religious diversity in this country? Here is one response to this question, from Huston Smith (1991: 389–390), a historian of religion:

> Whether religion is, for us, a good word or bad; whether (if on balance it is a good word) we side with a single religious tradition or to some degree open our arms to all: How do we comport ourselves in a pluralistic world that is riven by ideologies, some sacred, some profane? . . . We listen. . . . If one of the [world's religions] claims us, we begin by listening to it. Not uncritically, for new occasions teach new duties and everything finite is flawed in some respects. Still, we listen to it expectantly, knowing that it houses more truth than can be encompassed in a single lifetime.
>
> But we also listen to the faith of others, including the secularists. We listen first because . . . our times require it. The community today can be no single tradition; it is the planet. Daily, the world grows smaller, leaving understanding the only place where peace can find a home. . . . Those who listen work for peace, a peace built not on ecclesiastical or political hegemonies but on understanding and mutual concern.

Perhaps this is how each of us can make a difference—by *learning* more about our own beliefs and about the diverse denominations and world religions represented in the United States and around the globe, and *listening* to what other people have to say about their own beliefs and religious experiences.

Will religious tolerance increase in the United States? In the world? What steps can you take to help make a difference?

What significance will religion have in the future? Religion will continue to be important in the lives of many people. Moreover, the influence of religion may be felt even by those who claim no religious beliefs of their own. In many nations, the rise of *religious nationalism* has led to the blending of strongly held religious and political beliefs. The rise of religious nationalism is especially strong in the Middle East, where Islamic nationalism has spread rapidly and where the daily lives of people, particularly women and children, have been greatly affected (Juergensmeyer, 1993). Similarly, in the United States the influence of religion will be evident in ongoing political battles over social issues such as school prayer, abortion, gay and lesbian rights, and family issues. On the one hand, religion may unify people; on the other, it may result in tensions and confrontations among individuals and groups. However, according to the legal scholar Stephen L. Carter (1994), the tension between religion and other social institutions is not always negative. Carter believes that one of the most cherished freedoms in the United States is religious liberty. Maintaining an appropriate balance between religion and other aspects of social life will be an important challenge for the future. In contrast, totalitarian states find all conflict threatening and often quickly remove religious liberty when a statist dictator takes control.

As we have seen in this chapter, the debate continues over what religion is, what it should do, and what its relationship to other social institutions such as education should be. It will be up to your generation to understand other religions and to work for greater understanding among the diverse people who make up our nation and the world (see Box 17.4). But there is reason for hope, as one scholar explains:

> [People] know that religion, for all its institutional limitations, holds a vision of life's unity and meaningfulness, and for that reason it will continue to have a place in their narrative. In a very basic sense, religion itself was never the problem, only social forms of religion that stifle the human spirit. The sacred lives on and is real to those who can access it. (Roof, 1993: 261)

Chapter Review

- **What is religion, and what purpose does it serve in society?**

Religion is a system of beliefs, symbols, and rituals, based on some sacred or supernatural realm, that guides human behavior, gives meaning to life, and unites believers into a community.

- **What is the functionalist perspective on religion?**

According to functionalists, religion has three important functions in any society: (1) providing meaning and purpose to life, (2) promoting social cohesion and a sense of belonging, and (3) providing social control and support for the government.

- **What is the conflict perspective on religion?**

From a conflict perspective, religion can have negative consequences in that the capitalist class uses religion as a tool of domination to mislead workers about their true interests. However, Max Weber believed that religion could be a catalyst for social change.

- **What is the symbolic interactionist perspective on religion?**

Symbolic interactionists focus on a microlevel analysis of religion, examining the meanings that people give to religion and the meanings that they attach to religious symbols in their everyday life.

- **What are the major types of religious organization?**

Religious organizations can be categorized as ecclesia, churches, denominations, sects, and cults.

- **What are the major world religions?**

The major world religions are Buddhism, Hinduism, Judaism, Islam, and Christianity. More than 75 percent of the world's population is represented in one of these religions.

- **How has modernization contributed to the growth of religious orthodoxy (fundamentalism)?**

Fundamentalism has emerged in many religions because people do not like to see social changes taking place that affect their most treasured beliefs and values. In some situations, people have viewed modernity, especially science and new technologies, as a threat to their traditional beliefs and practices, which are important components of self-identity and group cohesion. As a result, fundamentalism opposes religious accommodation to the things of this world and demands higher standards of adherents, including a code of conduct and acceptance of specific beliefs.

- **Will religion continue as a major social institution?**

Although a few scholars of religion have "written off" this social institution, most believe that it is not in decline but rather is experiencing changes and some new definitions. One thing appears certain: With the influx of recent immigrants and increasing cultural diversity, religious congregations in the United States in the future will be much more diverse and will hold a wider variety of beliefs than did the traditional mainline denominations of the past.

www.cengage.com/login

Want to maximize your online study time? Take this easy-to-use study system's diagnostic pre-test, and it will create a personalized study plan for you. By helping you to identify the topics that you need to understand better and then directing you to valuable online resources, it can speed up your chapter review. CengageNOW even provides a post-test so you can confirm that you are ready for an exam.

Key Terms

animism 556
church 573
civil religion 560
cult 575
denomination 573
ecclesia 572
faith 553
fundamentalism 577
liberation theology 577
monotheism 556
nontheism 556
polytheism 556
profane 554
religion 552
rituals 554
sacred 553
sect 573
secularization 556
simple supernaturalism 556
theism 556

Questions for Critical Thinking

1. Why do people who believe they have "no religion" subscribe to civil religion? How would you design a research project to study the effects of civil religion on everyday life?
2. How is religion a force for social stability? How is it a force for social change?
3. What is the relationship among race, class, gender, and religious beliefs in contemporary society?
4. If Durkheim, Marx, and Weber were engaged in a discussion about religion, on what topics might they agree? On what topics would they disagree?

The Kendall Companion Website

www.cengage.com/sociology/kendall

Visit this book's companion website, where you'll find more resources to help you study and successfully complete course projects. Resources include quizzes and flash cards, as well as special features such as an interactive sociology timeline, maps, General Social Survey (GSS) data, and Census 2000 data. The site also provides links to useful websites that have been selected for their relevance to the topics in this chapter and include those listed below. (*Note:* Visit the book's website for updated URLs.)

The Association of Religion Data Archives (ARDA)

http://www.thearda.com

Funded by the Lilly Endowment, the ARDA is a project dedicated to collecting quantitative data for the study of American religion. The site offers extensive data on churches, church memberships, religious professionals, and religious groups, as well as maps and useful Web links.

The Religious Movements Homepage

http://religiousmovements.lib.virginia.edu

Developed at the University of Virginia, this website seeks to provide an understanding of the processes and changes that religious groups undergo and to promote the tolerance and appreciation of all faiths. In addition to profiles of religious movements, the site provides information on cult controversies, world religions, religious freedom, and many other topics.

The American Religious Experience

http://are.as.wvu.edu

Created by Brian Turley at West Virginia University, the American Religious Experience is a project devoted to scholarship about American religious history. The site features a site-specific search engine, a film archive, student projects, articles, and links to information on topics such as women and ethnic groups in religion.

chapter 18 Health, Health Care, and Disability

Chapter Focus Question

Why are health, health care, and disability significant concerns not only for individuals but also for entire societies?

Medicine is, I have found, a strange and in many ways disturbing business. The stakes are high, the liberties taken tremendous. We drug people, put needles and tubes into them, manipulate their chemistry, biology, and physics, lay them unconscious and open their bodies up to the world. We do so out of an abiding confidence in our know-how as a profession. What you find when you get in close, however—close enough to see the furrowed brows, the doubts and missteps, the failures as well as the successes—is how messy, uncertain, and also surprising medicine turns out to be.

The thing that still startles me is how fundamentally human an endeavor it is. Usually, when we think about medicine and its remarkable abilities, what comes to mind is the

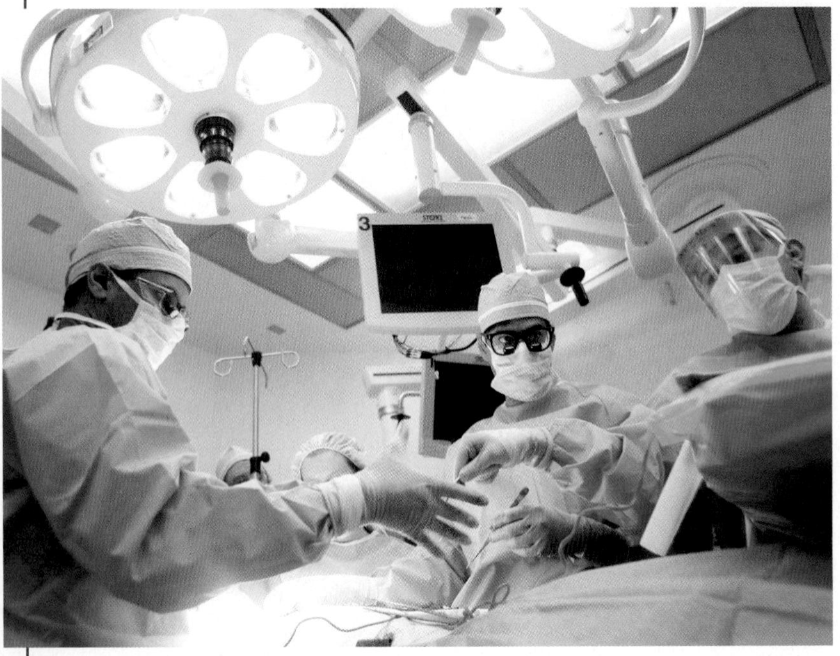

Dr. Atul Gawande (center) has written movingly about the differences between people's expectations of physicians and the medical establishment and the realities that they find in health care today. Sociologists study these contradictions to better understand a very complex and important part of U.S. social life.

IN THIS CHAPTER

science and all it has given us to fight sickness and misery: the tests, the machines, the drugs, the procedures. And without question, these are at the center of virtually everything medicine achieves. But we rarely see how it all actually works. You have a cough that won't go away—and then? It's not science you call upon but a doctor. A doctor with good days and bad days. A doctor with a weird laugh and a bad haircut. A doctor with three other patients to see and, inevitably, gaps in what he knows and skills he's still trying to learn. . . . We look for medicine to be an orderly field of knowledge and procedure. But it is not. It is an imperfect science, an enterprise of constantly changing knowledge, uncertain information, fallible individuals, and at the same time lives on the line. There is science in what we do, yes, but also habit, intuition, and sometimes plain old guessing. The gap between what we know and what we aim for persists. And this gap complicates everything we do.

—Atul Gawande, M.D. (2002: 4, 5, 7), a surgeon at Brigham and Women's Hospital in Boston, was a surgical resident when he wrote these words describing how he feels about the power and the limits of medicine.

Sharpening Your Focus

- *What is the relationship between the social environment and health and illness?*
- *What are the major issues in U.S. health care?*
- *How do functionalist, conflict, and symbolic interactionist approaches differ in their analysis of health and health care?*
- *What is mental illness, and why is it a difficult topic for sociological research?*
- *What are some of the consequences of disability?*

The everyday life of a doctor like Atul Gawande is filled with its high and low points: Some patients benefit from medical treatments that they receive from physicians, whereas others have sustained injuries or developed illnesses that are too severe or are beyond the scope of current knowledge and practice in the health care system to be successfully resolved. Physicians are human beings just like the patients they treat; however, much more is expected of them because of the availability of health care in the United States and other high-income nations and also because the dominant role of doctors in modern high-tech medicine has led many individuals to believe that virtually anything should be possible when it comes to one's health and longevity. However, this assumption is often not an accurate reflection of how health, illness, and health care actually work.

In this chapter, we will explore the dynamics of health, health care, and disability from a sociological perspective, as well as look at health-related issues through the eyes of those who have experienced medical problems. Before reading on, test your knowledge about health, illness, and health care by taking the quiz in Box 18.1.

What does the concept of health mean to you? At one time, health was considered to be simply the absence of disease. However, the World Health Organization (2003: 7) defines ***health*** as **a state of complete physical, mental, and social well-being.** According to this definition, health involves not only the absence of disease but also a positive sense of wellness. In other words, health is a multidimensional phenomenon: It includes physical, social, and psychological factors.

What is illness? Illness refers to an interference with health; like health, illness is socially defined and may change over time and between cultures. For example, in the United States and Canada, obesity is viewed as unhealthy, whereas in other times and places, obesity indicated that a person was prosperous and healthy.

What happens when a person is perceived to have an illness or disease? Healing involves both personal and institutional responses to perceived illness and disease. One aspect of institutional healing is health care and the health care delivery system in a society. ***Health care*** **is any activity intended to improve health.** When people experience illness, they often seek medical attention in hopes of having their health restored. A vital part of health care is ***medicine*****—an institutionalized system for the scientific diagnosis, treatment, and prevention of illness.**

Health in Global Perspective

Studying health and health care issues around the world offers insights on illness and how political and economic forces shape health care in nations. Disparities in health are glaringly apparent between high-income and low-income nations when we examine factors such as the prevalence of life-threatening diseases, rates of life expectancy and infant mortality, and access to health services. In regard to global health, for example, the number of people infected with HIV/AIDS more than doubled between 1990 and 2000 (from fewer than 15 million to more than 34 million). AIDS has cut life expectancy by 5 years in Nigeria, 18 years in Kenya, and 33 years in Zimbabwe (U.S. Census Bureau, 2008). ***Life expectancy*** **refers to an estimate of the average lifetime of people born in a specific year.** AIDS results in higher mortality rates in childhood and young adulthood, stages in the life course when mortality is otherwise low. However, AIDS is not the only disease reducing life expectancy in some nations. Most deaths in low- and middle-income nations are linked to infectious and parasitic diseases that are now rare in high-income, industrialized nations. Among these diseases are tuberculosis, polio, measles, diphtheria, meningitis, hepatitis, malaria, and leprosy. Although it is estimated that only 13 percent of U.S. citizens and 9 percent of Canadians will die prior to age 60, health experts estimate that more than 1.5 billion people around the world will die prior to age 60. This is particularly true in low-income nations such as Zambia, where 80 percent of the people are not expected to see their sixtieth birthday.

The ***infant mortality rate*** **is the number of deaths of infants under 1 year of age per 1,000 live births in a given year.** The infant mortality rate in some low-income nations is staggering: 261 infants under 1 year of age die per 1,000 live births in Angola, 257 die in Sierra Leone, and 239 die in Niger (World Health Organization, 2004a). In fact, almost 14 percent of all children born in low-income nations die before they reach their first birthday. The World Health Organization (2004a) estimates that two-thirds of those infants die during the *first month* of life. A child born in Latin America or Asia can expect to live between 7 and 13 fewer years, on average, than one born in North America or Western Europe (Epidemiological Network for Latin America and the Caribbean, 2000). There are many reasons for these differences in life expectancy and infant mortality. Many people in low-income countries have insufficient or contaminated

Box 18.1 Sociology and Everyday Life

How Much Do You Know About Health, Illness, and Health Care?

True	False	
T	F	1. Some social scientists view sickness as a special form of deviant behavior.
T	F	2. The field of epidemiology focuses primarily on how individuals acquire disease and bodily injury.
T	F	3. The primary reason that African Americans have shorter life expectancies than whites is the high rate of violence in central cities and the rural South.
T	F	4. Native Americans have shown dramatic improvement in their overall health level since the 1950s.
T	F	5. Health care in most high-income, developed nations is organized on a fee-for-service basis as it is in the United States.
T	F	6. The medical–industrial complex has operated in the United States with virtually no regulation, and allegations of health care fraud have largely been overlooked by federal and state governments.
T	F	7. Media coverage of chronic depression and other mental conditions focuses primarily on these problems as "women's illnesses."
T	F	8. It is extremely costly for employers to "mainstream" persons with disabilities in the workplace.

Answers on page 586.

food; lack access to pure, safe water; and do not have adequate sewage and refuse disposal. Added to these hazards is a lack of information about how to maintain good health. Many of these nations also lack qualified physicians and health care facilities with up-to-date equipment and medical procedures.

Nevertheless, tremendous progress has been made in saving the lives of children and adults over the past 15 years. Life expectancy at birth has risen to more than 70 years in 84 countries, up from only 55 countries in 1990. Life expectancy in low-income nations increased on average from 53 to 62 years, and mortality of children under 5 years of age dropped from 149 to 85 per 1,000 live births. Although this increase has been attributed to a number of factors, an especially important advance has been the development of a safe water supply. The percentage of the world's population with access to safe water nearly doubled between 1990 and 2000 (United Nations Development Programme, 2003).

Will improvements in health around the world continue to occur? Organizations such as the United Nations argue that both public-sector and private-sector initiatives will be required to improve global health conditions. For example, a United Nations report states that in the era of globalization and dominance by transnational corporations, "money talks louder than need" when "cosmetic drugs and slow-ripening tomatoes come higher on the list [of priorities] than a vaccine against malaria or drought-resistant crops for marginal lands" (United Nations Development Programme, 1999: 68).

Recently, pressing questions have arisen about the availability of new technologies and life-saving drugs around the world. An example is the problem of providing access to drugs in countries with high rates of HIV/AIDS. Many people cannot afford to pay for drugs, such as the three-drug combination therapy that

health a state of complete physical, mental, and social well-being.

health care any activity intended to improve health.

medicine an institutionalized system for the scientific diagnosis, treatment, and prevention of illness.

life expectancy an estimate of the average lifetime of people born in a specific year.

infant mortality rate the number of deaths of infants under 1 year of age per 1,000 live births in a given year.

Box 18.1 Sociology and Everyday Life

Answers to the Sociology Quiz on Health, Illness, and Health Care

1. True. Some social scientists view sickness as a special form of deviant behavior. However, it is not equivalent to other forms of deviance such as crime or violent behavior. Unlike many who are defined as criminal, the sick are provided with therapeutic care so that their health will be restored and they can fulfill their roles in society (Weiss and Lonnquist, 2009).

2. False. The primary focus of the epidemiologist is on the health problems of social aggregates or large groups of people, not on individuals as such (Cockerham, 2010).

3. False. The lower life expectancy of African Americans as a category is due to a higher prevalence of life-threatening illnesses, such as cancer, heart disease, hypertension, and AIDS. However, it should be noted that African American males do have the highest death rates from homicide of any racial–ethnic category in the United States (Cockerham, 2010).

4. True. Native Americans (including American Indians and native Alaskans) as a category have had significant improvement in health in recent decades. Some analysts attribute this change to better nutrition and health care services. However, other analysts point out that Native Americans continue to have high rates of mortality from diabetes, alcohol-related illnesses, and suicide (Cockerham, 2010).

5. False. The United States is one of the few high-income, developed nations that does not have some form of universal health coverage. In the United States, health care has traditionally been purchased by the patient. In most other high-income nations, health care is provided by or purchased by the government (Cockerham, 2010).

6. False. A number of government investigations have focused on rising health care payments and allegations of fraud in the health care delivery system. Billing frauds have been found in Medicare and Medicaid payments to physicians, hospitals, nursing homes, home health agencies, medical labs, and medical equipment manufacturers.

7. False. Until recently, chronic depression and other mental conditions were most often depicted as "female" problems. However, male depression has become more widely publicized through documentaries on individuals' lives and through advertising for antidepressants.

8. False. Although disability expenditures nationwide may be costly, individual employers often find that they can accommodate the workplace needs of a worker with a disability for costs ranging from zero to several thousand dollars, thus opening up new opportunities for people previously excluded from certain types of jobs and careers.

prolongs the life of many AIDS patients. Transnational pharmaceutical companies fear that if they provide their name-brand drugs at a lower price in low-income countries, that might undercut their major sales base in high-income countries if those drugs become available as generic products (which are less costly and can be made by more than one manufacturer) or are reimported into the high-income countries at a reduced price. The companies claim that they need the money generated from sales of their name-brand drugs in order to fund research on other products that will reduce suffering and sometimes prolong human life. For this reason, pharmaceutical companies have increasingly marketed their prescription drugs to patients through the media, particularly television and print advertisements (see Box 18.2). Pharmaceutical companies that hold the patents on various drugs see their products as something that needs to be protected by law, whereas people in human relief agencies around the world are concerned about the fact that one-third of the world's population does not have access to essential medicines and that—even worse—this figure rises to one-half in the poorest parts of Africa and Asia (United Nations Development Programme, 2003). If we are to see a significant improvement in life expectancy and health among people in all of the nations of the world, improvements are needed in the availability of new medical technologies and life-saving drugs.

How about improvements in health and health care within one nation? Is there a positive relationship between the amount of money that a society spends on health care and the overall physical, mental, and social

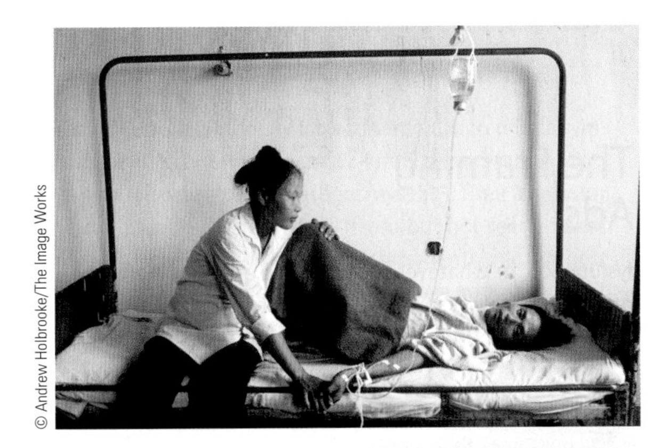

© Andrew Holbrooke/The Image Works

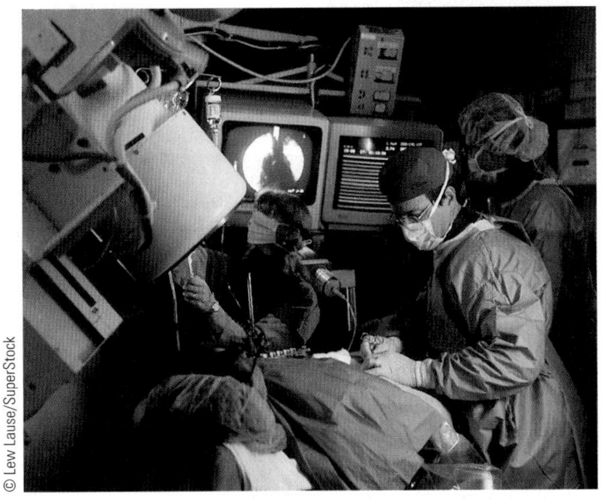

© Lew Lause/SuperStock

Access to quality health care is much greater for some people than for others. The factors that are involved vary not only for people within one nation but also across the nations of the world.

well-being of its people? Not necessarily. If there were such a relationship, people in the United States would be among the healthiest and most physically fit people in the world. In 2007 we spent more than $2.2 trillion—the equivalent of $7,421 per person—on health care (HHS.gov, 2008). By 2012, health care spending is projected to reach $3.1 trillion, and $4.3 trillion by 2016 if drastic changes are not made by the Obama administration. Health care spending accounts for about 20 percent of the gross domestic product (GDP) in the United States, but it accounts for only 10.9 percent of the GDP in Switzerland, 10.7 percent in Germany, 9.7 percent in Canada, and 9.5 percent in France (National Coalition on Health Care, 2009).

Health in the United States

Even if we limit our discussion (for the moment) to people in the United States, why are some of us healthier than others? Is it biology—our genes—that accounts for this difference? Does the environment within which we live have an effect? How about our own individual lifestyle?

Social Epidemiology

The field of social epidemiology attempts to answer questions such as these. ***Social epidemiology* is the study of the causes and distribution of health, disease, and impairment throughout a population** (Weiss and Lonnquist, 2009). Typically, the target of the investigation is disease agents, the environment, and the human host. *Disease agents* include biological agents such as insects, bacteria, and viruses that carry or cause disease; nutrient agents such as fats and carbohydrates; chemical agents such as gases and pollutants in the air; and physical agents such as temperature, humidity, and radiation. The *environment* includes the physical (geography and climate), biological (presence or absence of known disease agents), and social (socioeconomic status, occupation, and location of home) environments. The human *host* takes into account demographic factors (age, sex, and race/ethnicity), physical condition, habits and customs, and lifestyle (Weiss and Lonnquist, 2009). Let's look briefly at some of these factors.

Age Rates of illness and death are highest among the old and the young. Mortality rates drop shortly after birth and begin to rise significantly during middle age. After age 65, rates of chronic diseases and mortality increase rapidly. ***Chronic diseases* are illnesses that are long term or lifelong and that develop gradually or are present from birth;** in contrast, ***acute diseases* are illnesses that strike suddenly and cause dramatic incapacitation and sometimes death** (Weitz, 2004). An older woman once described how she coped with chronic conditions associated with aging, particularly at the start of each day:

> Every morning, I wake up in pain. I wiggle my toes. Good. They still obey. I open my eyes. Good. I can

social epidemiology the study of the causes and distribution of health, disease, and impairment throughout a population.

chronic diseases illnesses that are long term or lifelong and that develop gradually or are present from birth.

acute diseases illnesses that strike suddenly and cause dramatic incapacitation and sometimes death.

Because women on average live longer than men, it is easy to jump to the conclusion that they are healthier than men. However, although men at all ages have higher rates of fatal diseases, women have higher rates of chronic illness.

Race/Ethnicity and Social Class Racial/ethnic differences are also visible in statistics pertaining to life expectancy. Projections of life expectancy for infants born in 2010 reveal this sobering fact: Life expectancy for African American males is estimated at 70.2 years as compared to 77.2 for African American females and 75.7 years for white males.

Although race/ethnicity and social class are related to issues of health and mortality, research continues to show that income and the neighborhood in which a person lives may be equally or more significant than race or ethnicity with respect to these issues. How is it possible that the neighborhood you live in may significantly affect your risk of dying during the next year? According to a study by the Stanford Center for Research in Disease Prevention (Winkleby and Cubbin, 2003), people have a higher survival rate if they live in better-educated or wealthier neighborhoods than if the neighborhood is low income and has low levels of education. Among the reasons researchers believe that neighborhoods make a difference are the availability (or lack thereof) of safe areas to exercise and grocery stores with nutritious foods, and access to transportation, education, and good jobs. Many low-income neighborhoods are characterized by fast-food restaurants, liquor stores, and other facilities that do not afford residents healthy options.

As discussed in prior chapters, people of color are more likely to have incomes below the poverty line, and the poorest people typically receive less preventive care and less optimal management of chronic diseases than do other people (Leary, 1996). People living in central cities, where there are high levels of poverty and crime, or in remote rural areas generally have greater difficulty in getting health care because most doctors prefer to locate their practice in a "safe" area, particularly one with a patient base that will produce a high income. Although rural Americans make up 20 percent of the U.S. population, only about 10 percent of the nation's physicians practice in rural areas, and fewer specialists such as cardiologists are available in these areas.

Another factor is occupation. People with lower incomes are more likely to be employed in jobs that expose them to danger and illness—working in the construction industry or around heavy equipment in a factory, for example, or holding a job as a convenience store clerk or other position that exposes a person to the risk of armed robbery. Finally, people of color and poor people are more likely to live in areas that contain environmental hazards.

However, although Latinas/os are more likely than non-Latino/a whites to live below the poverty line, they have lower death rates from heart disease, cancer, accidents, and suicide, and an overall lower death rate. One explanation may be dietary factors and the strong family life and support networks found in many Latina/o families (Weiss and Lonnquist, 2009). Obviously, more research is needed on this point, for the answer might be beneficial to all people.

Can your neighborhood be bad for your health? According to recent research, it can indeed, especially if it predominantly contains fast-food restaurants, liquor stores, and similarly unhealthy lifestyle options.

Lifestyle Factors

As noted previously, social epidemiologists also examine lifestyle choices as a factor in health, disease, and impairment. We will examine three lifestyle factors as they relate to health: drugs, sexually transmitted diseases, and diet and exercise.

Drug Use and Abuse What is a drug? There are many different definitions, but for our purposes, a ***drug* is any substance—other than food and water—that, when taken into the body, alters its functioning in some way.** Drugs are used for either therapeutic or recreational purposes. *Therapeutic* use occurs when a person takes a drug for a specific purpose such as reducing a fever or controlling a cough. In contrast, *recreational* drug use occurs when a person takes a drug for no purpose other than achieving a pleasurable feeling or psychological state. Alcohol and tobacco are examples of drugs that are primarily used for recreational purposes; their use by people over a fixed age (which

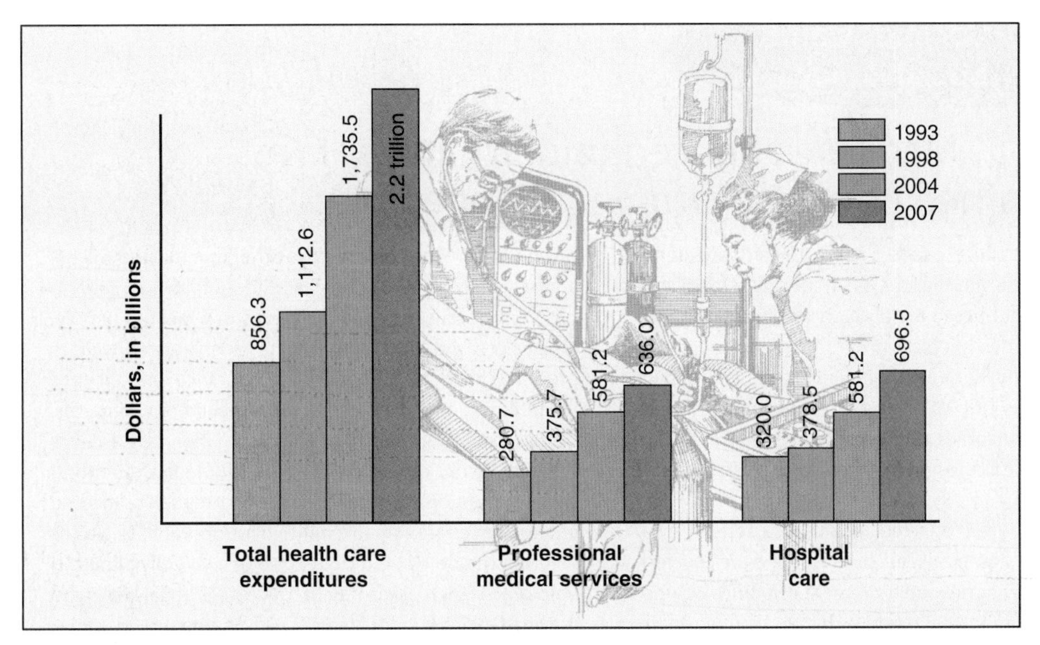

▶ **Figure 18.2 Increase in Cost of Health Care, 1993–2007**

Source: HHS.gov, 2008; and National Coalition on Health Care, 2009.

Medicare and Medicaid (see Box 18.3). Medicare is a program for persons age 65 or older who are covered by Social Security or who are eligible and "buy into" the program by paying a monthly premium. Medicare pays part of the health care costs of these people. Medicaid, a jointly funded federal–state–local program, was established to make health care more available to the poor. However, both the Medicaid program and the Medicare program are in financial difficulty. These two programs cost $760.6 billion annually and account for one-fourth of all federal spending. Medicare and Medicaid are growing more rapidly than the U.S. economy and the revenues that are used to finance them. Medicare spending grew 7.2 percent in 2007, for example, to $431.2 billion, which followed a growth of 18.5 percent in 2006, because of the one-time implementation of Medicare Part D (prescription coverage). Similarly, Medicaid spending grew 6.4 percent in 2007 to $329.4 billion (HHS.gov, 2008). As a result, the Congressional Budget Office estimates that the costs of these programs will more than double in the next decade unless significant changes are made. Medicare is particularly endangered because the number of older Americans who rely on the program to pay for their health care is rising dramatically. At the same time, health care spending is also growing rapidly because of the increased use of new medical technologies.

Health Maintenance Organizations (HMOs) Created in an effort to provide workers with health coverage by keeping costs down, ***health maintenance organizations (HMOs)*** **provide, for a set monthly fee, total care with an emphasis on prevention to avoid costly treatment later.** The doctors do not work on a fee-for-service basis, and patients are encouraged to get regular checkups and to practice good health practices (e.g., exercise and eat right). However, research shows that preventive care is good for the individual's health but does not necessarily lower total costs. As long as patients use only the doctors and hospitals that are affiliated with their HMO, they pay no fees, or only small co-payments, beyond their insurance premiums.

Recent concerns about physicians being used as gatekeepers who might prevent some patients from obtaining referrals to specialists or from getting needed treatment have resulted in changes in the policies of some HMOs, which now allow patients to visit health care providers outside an HMO's network or to receive other previously unauthorized services by paying a higher co-payment. However, critics charge that those HMOs whose primary-care physicians are paid on a capitation basis—meaning that they receive only a fixed amount per patient that they see, regardless of how long they spend with that patient—in effect encourage doctors to undertreat patients.

health maintenance organizations (HMOs) companies that provide, for a set monthly fee, total care with an emphasis on prevention to avoid costly treatment later.

Box 18.3 Sociology and Social Policy

Medicare and Medicaid: What We Have Learned About Government-Funded Medical Care

Today, as our political leaders work toward providing medical coverage for millions of uninsured Americans, the question continues to rise about how effective current government-funded medical programs are. Many of us hear Medicare and Medicaid discussed in the media and in political elections, but we often do not know much about the origins or outcomes of such programs. Let's look in more detail at Medicare and Medicaid from a social policy standpoint.

In 1965, Congress enacted Medicare, a federal program for people age sixty-five and over (who are covered by Social Security or railroad retirement insurance or who have been permanently and totally disabled for two years or more). This program was primarily funded through Social Security taxes paid by current workers. We refer to Medicare as an entitlement program because people who receive benefits under the plan must have paid something to be covered. Medicare Part A (hospital insurance) provides coverage for some inpatient hospital expenses, including critical access hospitals, and skilled nursing facilities. It also helps cover hospice care and limited home health care. Part B (medical insurance) helps cover doctors' services and outpatient care. It covers some of the services of physical and occupational therapists, and some home health care. Most people pay a monthly premium for Part B. Beginning in 2006, Medicare prescription drug coverage became available to everyone covered under Medicare. Private companies provide the coverage, and beneficiaries choose the drug plan and pay a monthly premium.

How successful has Medicare been as a social policy? Medicare has created greater access to health care services for many older people. However, it also has serious limitations, including the fact that it primarily provides for acute, short-term care and is limited to partial payment for ninety days of hospital care. With the aging population, it is problematic that this plan covers only a restricted amount of skilled nursing care and home health services. Patients must pay a deductible and co-payments (fees paid by people each time they see a health care provider), and Medicare does not include many of the things that some older people need the most, such as dental care, hearing aids, eyeglasses, long-term care in nursing homes, or custodial or nonmedical service, including adult day care or homemaker services. There are such serious gaps in Medicare coverage that many older people who can afford to do so buy medigap policies, which are supplementary private insurance policies that fill the gaps in the Medicare coverage.

As we have seen, one form of government-funded health care for older people—Medicare—is considered to be an entitlement program. The other one that we will examine—Medicaid—is more often thought of as a welfare program because it primarily serves low-income people. Medicaid is the joint federal–state means-tested welfare program that provides health care insurance for poor people of any age who meet specific eligibility requirements. Certain factors are taken into account when determining whether or not a person is eligible for Medicaid. Among these are age, disability, blindness, and pregnancy. Income and resources are also taken into consideration, as well as whether the person is a U.S. citizen or a lawfully admitted immigrant. Each state has its own rules regarding who may be covered under Medicaid, and some provide time-limited coverage for specific categories of individuals, such as uninsured women with breast or cervical cancer or those individuals diagnosed with TB (tuberculosis).

Unlike Medicare recipients, who are often seen as "worthy" of their health care benefits, some Medicaid recipients are stigmatized by some politicians and media outlets for their participation in this "welfare program." Today, many physicians refuse to take Medicaid patients because the administrative paperwork is burdensome and reimbursements are low—typically less than one-half of what private insurance companies pay for the same services.

When we analyze Medicare and Medicaid as social policies, we find that both programs provide broad coverage; however, both programs contain serious flaws: They are extremely expensive to administer, and they can become highly vulnerable to costly fraud by health care providers and some recipients. Finally, across lines of race/ethnicity and class, some studies have shown that these medical assistance programs typically provide higher-quality health care to older people who are white and more affluent than they do for people of color, very-low-income individuals, people with disabilities, and those who live in rural areas. These issues will remain important as we consider ways in which to overhaul the entire health care system now and in the future.

Reflect & Analyze

Is it possible to learn from present programs, such as Medicare and Medicaid, how we should structure the payment for and delivery of health care services in the future? Are there times when it is better to start over and create new social policy rather than trying to revise existing programs? How do you think the payment of health care might be revamped or reorganized to provide the best coverage for the largest number of people?

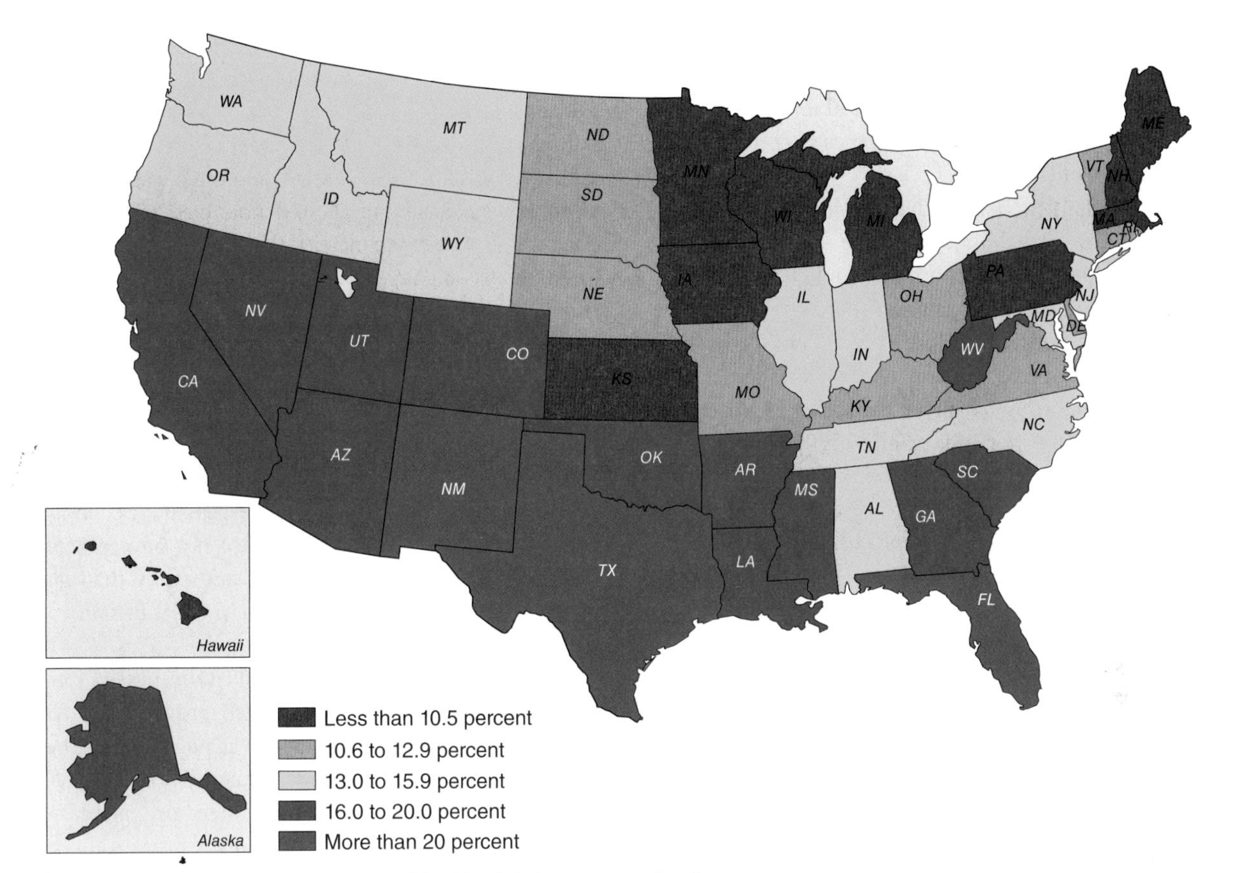

▶ **Map 18.3 Persons Not Covered by Health Insurance, by State**

Source: U.S. Census Bureau, 2008.

Managed Care Another approach to controlling health care costs in the United States is known as ***managed care*****—any system of cost containment that closely monitors and controls health care providers' decisions about medical procedures, diagnostic tests, and other services that should be provided to patients.** One type of managed care in the United States is a *preferred provider organization* ("PPO"), which is an organization of medical doctors, hospitals, and other health care providers who enter into a contract with an insurer or a third-party administrator to provide health care at a reduced rate to patients who are covered under specific insurance plans. In most managed-care programs, patients choose a primary-care physician from a list of participating doctors. Unlike many of the HMOs, when a patient covered under a PPO plan needs medical services, he or she may contact any one of a number of primary-care physicians or specialists who are "in-network" providers. Like HMOs, most PPO plans do contain a precertification requirement in which scheduled (nonemergency) hospital admissions and certain kinds of procedures must be approved in advance. Through measures such as this, these insurance plans have sought unsuccessfully to curb the rapidly increasing costs of medical care and to reduce the extensive paperwork and bureaucracy involved in the typical medical visit.

The Uninsured and the Underinsured Despite public and private insurance programs, about one-third of all U.S. citizens are without health insurance or had difficulty getting or paying for medical care at some time in the last year. As shown on ▶ Map 18.3, the number of people not covered by health insurance varies from state to state. Of the people not covered by health insurance, 8.7 million are children (DeNavas-Walt, Proctor, and Lee, 2007). An estimated 45.7 million people in the United States had no health insurance in 2007—more than 15 percent of the nation's population (Familiesusa.org, 2009). Although every racial and ethnic group is affected, Latinos/as and African Americans are more likely to be uninsured

managed care any system of cost containment that closely monitors and controls health care providers' decisions about medical procedures, diagnostic tests, and other services that should be provided to patients.

than whites (non-Hispanics). During 2007–2008, 55 percent of Latinos/as, 40.3 percent of African Americans, and 34.0 percent of other racial and ethnic minorities were without health insurance as compared to 25.8 percent of whites.

Social scientists believe that the number of uninsured will continue to increase as the economic downturn continues to take its toll on employment in this country. Some estimates suggest that as many as 6.9 million more people will lose their insurance by the end of 2010 (Familiesusa.org, 2009). Overall, one out of three people who are under the age of 65 were uninsured for some portion, or all, of 2007–2008. The working poor constitute a substantial portion of this category: It is estimated that four out of five individuals (79.2 percent) of those who were without health insurance were from working families. Furthermore, 69.7 percent were in families with a worker who was employed full time (Familiesusa.com, 2009). In a worst-case scenario, some who are uninsured (despite being employed full time) make too little to afford health insurance but too much to qualify for Medicaid, and their employers do not provide health insurance coverage. What happens when they need medical treatment? "They do without," explains Ray Hanley, medical services director for the Arkansas Department of Human Services (qtd. in Kilborn, 1997: A10).

Paying for Medical Care in Other Nations

Other industrialized and industrializing countries do not leave their citizens in the situation in which some people in the United States find themselves. Let's examine how other nations pay for health care.

Canada Prior to the 1960s, Canada's health care system was similar to that of the United States today. However, in 1962 the government of the province of Saskatchewan implemented a health insurance plan despite opposition from doctors, who went on strike to protest the program. The strike was not successful, as the vast majority of citizens supported the government, which maintained health services by importing doctors from Great Britain. The Saskatchewan program proved itself viable in the years following the strike, and by 1972 all Canadian provinces and territories had coverage for medical and hospital services (Kendall, Linden, and Murray, 2008). As a result, Canada has a ***universal health care* system—a health care system in which all citizens receive medical services paid for by tax revenues.** In Canada, these revenues are supplemented by insurance premiums paid by all taxpaying citizens.

One major advantage of the Canadian system over that in the United States is a significant reduction in administrative costs. Whereas more than 20 percent of the U.S. health care dollar represents administrative costs, in Canada the corresponding figure is 10 percent (Weiss and Lonnquist, 2009). However, the system is not without its critics, who claim that it is costly and often wasteful. For example, Canadians are allowed unlimited trips to the doctor, and doctors can increase their income by ordering extensive tests and repeat visits, just as in the United States (Kendall, Linden, and Murray, 2008).

The Canadian health care system does not constitute what is referred to as ***socialized medicine*—a health care system in which the government owns the medical care facilities and employs the physicians.** Rather, Canada has maintained the private nature of the medical profession. Although the government pays most health care costs, the physicians are not government employees and have much greater autonomy than do physicians in the health care system in Great Britain.

Great Britain In 1946, Great Britain passed the National Health Service Act, which provided for all health care services to be available at no charge to the entire population. Although physicians work out of offices or clinics—as in the United States or Canada—the government sets health care policies, raises funds and controls the medical care budget, owns health care facilities, and directly employs physicians and other health care personnel (Weiss and Lonnquist, 2009). Unlike the Canadian model, the health care system in Great Britain *does* constitute socialized medicine. Physicians receive capitation payments from the government: a fixed annual fee for each patient in their practice regardless of how many times they see the patient or how many procedures they perform. They also receive supplemental payments for each low-income or elderly patient in their practice, to compensate for the extra time such patients may require; bonus payments if they meet targets for providing preventive services, such as immunizations against disease; and financial incentives if they practice in medically underserved areas. Physicians may accept private patients, but such patients rarely constitute more than a small fraction of a physician's practice; hospitals reserve a small number of beds for private patients (Weiss and Lonnquist, 2009). Why would anyone want to be a private patient who pays for her or his own care or hospital bed? The answer is primarily found in the desire to avoid the long waits ("queues") that the general population encounters and the fact that private patients can enter the hospital for surgery at times convenient to the consumer rather than wait upon the convenience of the system (Gill, 1994).

China After a lengthy civil war, in 1949 the Communist Party won control of mainland China but found itself in charge of a vast nation with a population of one billion people, most of whom lived in poverty and misery. Malnutrition was prevalent, life expectancies were short, and infant and maternal mortality rates were high. In the cities, only the elite could afford medical care; in the rural areas where most of the population resided, Western-style health care barely existed (Weitz, 2004). With a lack of both financial resources and trained health care personnel, China needed to adopt innovative strategies in order to improve the health of its populace. One such policy was developing a large number of *physician extenders* and sending them out into the cities and rural areas to educate the public regarding health and health care and to treat illness and disease. Referred to as *street doctors* in urban areas and *barefoot doctors* in the countryside, these individuals had little formal training and worked under the supervision of trained physicians (Weitz, 2004).

Over the past four decades, medical training has become more rigorous, and supervision has increased. All doctors receive training in both Western and traditional Chinese medicine. Doctors who work in hospitals receive a salary; all other doctors now work on a fee-for-service basis. In urban areas, the cost of health care is paid by employers; however, 78 percent of the population lives in rural areas, where most work on family farms and are expected to pay for their own health care. The cost of health care generally remains low, but the cost of hospital care has risen; accordingly, many Chinese—if they can afford it—purchase health care insurance to cover the cost of hospitalization (Weitz, 2004). As a low-income country, China spends only 2 percent to 3 percent of its gross domestic product on health care, but the health of its citizens is only slightly below that of most industrialized nations.

Regardless of which system of delivering medical care that a nation may have, the health care providers and the general population of the nation are having to face new issues that arise as a result of new technology.

Social Implications of Advanced Medical Technology

Advances in medical technology are occurring at a speed that is almost unbelievable; however, sociologists and other social scientists have identified specific social implications of some of the new technologies (see Weiss and Lonnquist, 2009):

1. *The new technologies create options for people and for society, but options that alter human relationships.* An example is the ability of medical personnel to sustain a life that in earlier times would have ended as the result of disease or an accident. Although this can be beneficial, technologically advanced equipment that can sustain life after consciousness is lost and there is no likelihood that the person will recover can create a difficult decision for the family of that person if he or she has not left a *living will*—a document stating the person's wishes regarding the medical circumstances under which his or her life should be terminated. Federal law requires all hospitals and other medical facilities to honor the terms of a living will. Recent media coverage of individuals whose lives have been prolonged by new medical technologies has made more people aware of some end-of-life issues.
2. *The new technologies increase the cost of medical care.* For example, the computerized axial tomography (CT or CAT) scanner—which combines a computer with X-rays that are passed through the body at different angles—produces clear images of the interior of the body that are invaluable in investigating disease. However, the cost of such a scanner is around $1 million. Magnetic resonance imaging (MRI) equipment that allows pictures to be taken of internal organs ranges in cost from $1 million to $2.5 million. Can the United States afford such equipment in every hospital for every patient? The money available for health care is not unlimited, and when it is spent on high-tech equipment and treatment, it is being reallocated from other health care programs that might be of greater assistance to more people.
3. *The new technologies raise provocative questions about the very nature of life.* In Chapter 15, we briefly discuss in vitro fertilization—a form of assisted reproductive technology. But during 1997, Dr. Ian Williams and his associates in Scotland took in vitro fertilization a step further: They cloned a lamb (that they named Dolly) from the DNA of an adult sheep. Subsequently, scientists have cloned other animals in the same manner, raising a number of profound questions: If scientists can duplicate mammals from adult DNA, is it possible to clone a perfect (whatever that may be) human being instead of taking a chance on a child that is born to a couple? If it is possible, would it be ethical? For example, if—as discussed earlier in this text—most parents prefer a boy over a girl if they are going to

universal health care a health care system in which all citizens receive medical services paid for by tax revenues.

socialized medicine a health care system in which the government owns the medical care facilities and employs the physicians.

have only one child, would the world suddenly have substantially more boys than girls? Would everyone start to look alike, eliminating diversity? If a child were born other than through cloning and had some sort of biological defect, would the child have the right to sue his or her parents for negligence?

However, at the same time that high-tech medicine is becoming a major part of overall health care, many people are turning to holistic medicine and alternative healing practices.

Holistic Medicine and Alternative Medicine

When examining the subject of medicine, it is easy to think only in terms of conventional (or mainstream) medical treatment. By contrast, ***holistic medicine* is an approach to health care that focuses on prevention of illness and disease and is aimed at treating the whole person—body and mind—rather than just the part or parts in which symptoms occur.** Under this approach, it is important that people not look solely to medicine and doctors for their health, but rather that people engage in health-promoting behavior. Likewise, medical professionals must not only treat illness and disease but also work with the patient to promote a healthy lifestyle and self-image.

Many practitioners of *alternative medicine*—healing practices inconsistent with dominant medical practice—take a holistic approach, and today many people are turning to alternative medicine either in addition to or in lieu of traditional medicine. Jill Neimark tells how she feels about the two types of medicine:

> I'd like to make a confession: I believe. I believe in the brilliance of Western medicine, in the technological wonders of organ transplants, brain scans, and laparoscopic surgery—and I also believe that shamanic healers from indigenous cultures have discovered treatments for illnesses that traditional medicine can't touch, that the mind alone can trigger and then reverse illness, that dietary changes, along with vitamin and herbal supplements, can powerfully impact the course of disease. I believe in merging the best of both worlds—the orthodox and the alternative.
>
> I'm not alone in my belief. This is a season of unprecedented possibility in medicine. And yet in spite of the tremendous changes, there is still some resistance from mainstream medicine, as well as the government and the pharmaceutical industry. One can almost feel the tectonic plates of conventional and holistic medicine crashing up against each other and realigning the landscape of health care. (Neimark, 1997)

The use of herbal therapies is a form of alternative medicine that is increasing in popularity in the United States. How does this approach to health care differ from a more traditional medical approach?

Are medical doctors as opposed to alternative medicine as Neimark indicates? In understanding the medical establishment's reaction to alternative medicine, it is important to keep in mind the philosophy of scientific medicine—that medicine is a science, not an art. Thus, to the extent to which alternative medicine is "nonscientific," it must be quackery and therefore something that is undoubtedly worthless and possibly harmful. Undoubtedly, self-interest is also involved in mainstream medicine's reaction to alternative medicine: If the public can be persuaded that scientific medicine is the only legitimate healing practice, fewer health care dollars will be spent on a form of medical treatment that is (at least to some extent) in competition with the medical establishment (Weiss and Lonnquist, 2009). But if all forms of alternative medicine (including chiropractic, massage, and spiritual) are taken into account, people spend more money on unconventional therapies than they do for all hospitalizations (Weiss and Lonnquist, 2009).

Sociological Perspectives on Health and Medicine

Functionalist, conflict, symbolic interactionist, and postmodernist perspectives focus on different aspects of health and medicine; each provides us with significant insights on the problems associated with these pressing social concerns.

A Functionalist Perspective: The Sick Role

According to the functionalist approach, if society is to function as a stable system, it is important for people to be healthy and to contribute to their society. Consequently, sickness is viewed as a form of deviant behavior that must be controlled by society. This view was initially set forth by the sociologist Talcott Parsons (1951) in his concept of the ***sick role*—the set of patterned expectations that defines the norms and values appropriate for individuals who are sick and for those who interact with them.** According to Parsons, the sick role has four primary characteristics:

1. People who are sick are not responsible for their condition. It is assumed that being sick is not a deliberate and knowing choice of the sick person.
2. People who assume the sick role are temporarily exempt from their normal roles and obligations. For example, people with illnesses are typically not expected to go to school or work.
3. People who are sick must want to get well. The sick role is considered to be a temporary role that people must relinquish as soon as their condition improves sufficiently. Those who do not return to their regular activities in a timely fashion may be labeled as hypochondriacs or malingerers.
4. People who are sick must seek competent help from a medical professional to hasten their recovery.

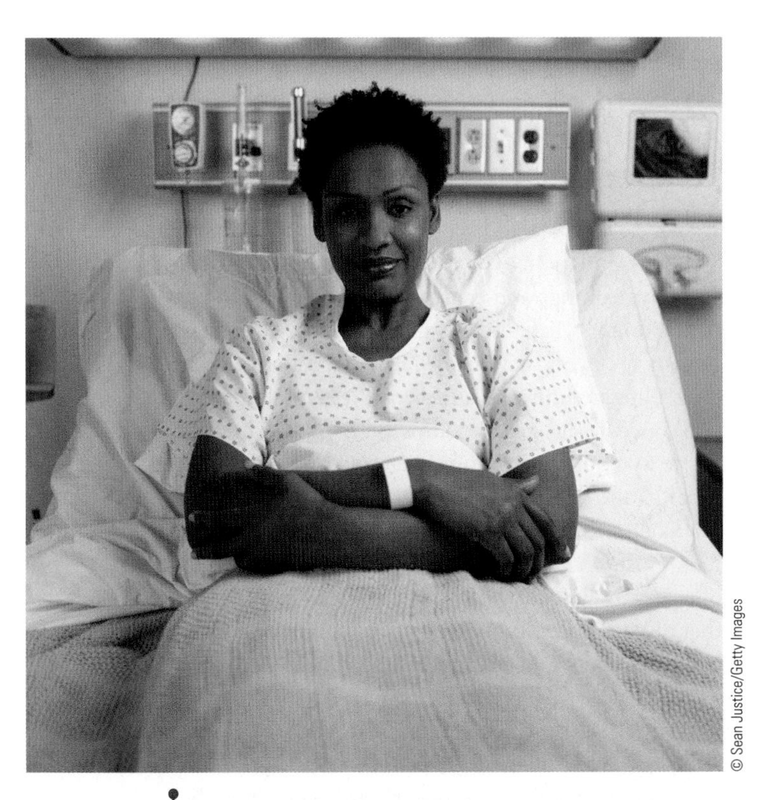

© Sean Justice/Getty Images

According to the functionalist perspective, the sick role exempts the patient from routine activities for a period of time but assumes that the individual will seek appropriate medical attention and get well as soon as possible.

As these characteristics show, Parsons believed that illness is dysfunctional for both individuals and the larger society. Those who assume the sick role are unable to fulfill their necessary social roles, such as being parents or employees. Similarly, people who are ill lose days from their productive roles in society, thus weakening the ability of groups and organizations to fulfill their functions.

According to Parsons, it is important for the society to maintain social control over people who enter the sick role. Physicians are empowered to determine who may enter this role and when patients are ready to exit it. Because physicians spend many years in training and have specialized knowledge about illness and its treatment, they are certified by the society to be "gatekeepers" of the sick role. When patients seek the advice of a physician, they enter into the patient–physician relationship, which does not contain equal power for both parties. The patient is expected to follow the "doctor's orders" by adhering to a treatment regime, recovering from the malady, and returning to a normal routine as soon as possible.

What are the major strengths and weaknesses of Parsons's model and, more generally, of the functionalist view of health and illness? Parsons's analysis of the sick role was pathbreaking when it was introduced. Some social analysts believe that Parsons made a major contribution to our knowledge of how society explains illness-related behavior and how physicians have attained their gatekeeper status. In contrast, other analysts believe that the sick-role model does not take into account racial–ethnic, class, and gender variations in the ways that people view illness and interpret this role. For example, this model does not take into account the fact that many individuals in the working class may choose not to accept the sick role unless they are seriously ill—because they cannot afford to miss

holistic medicine an approach to health care that focuses on prevention of illness and disease and is aimed at treating the whole person—body and mind—rather than just the part or parts in which symptoms occur.

sick role the set of patterned expectations that defines the norms and values appropriate for individuals who are sick and for those who interact with them.

time from work and lose a portion of their earnings. Moreover, people without health insurance may not have the option of assuming the sick role.

A Conflict Perspective: Inequalities in Health and Health Care

Unlike the functionalist approach, conflict theory emphasizes the political, economic, and social forces that affect health and the health care delivery system. Among the issues of concern to conflict theorists are the ability of all people to obtain health care; how race, class, and gender inequalities affect health and health care; power relationships between doctors and other health care workers; the dominance of the medical model of health care; and the role of profit in the health care system.

Who is responsible for problems in the U.S. health care system? According to many conflict theorists, problems in U.S. health care delivery are rooted in the capitalist economy, which views medicine as a commodity that is produced and sold by the medical–industrial complex. The ***medical–industrial complex*** **encompasses both local physicians and hospitals as well as global health-related industries such as insurance companies and pharmaceutical and medical supply companies** (Relman, 1992).

The United States is one of the few industrialized nations that rely almost exclusively on the medical–industrial complex for health care delivery and do not have universal health coverage, which provides some level of access to medical treatment for all people. Consequently, access to high-quality medical care is linked to people's ability to pay and to their position within the class structure. Those who are affluent or have good medical insurance may receive high-quality, state-of-the-art care in the medical–industrial complex because of its elaborate technologies and treatments. However, people below the poverty level and those just above it have greater difficulty gaining access to medical care. Referred to as the *medically indigent,* these individuals do not earn enough to afford private medical care but earn just enough money to keep them from qualifying for Medicaid (Weiss and Lonnquist, 2009). In the profit-oriented capitalist economy, these individuals are said to "fall between the cracks" in the health care system.

Who benefits from the existing structure of medicine? According to conflict theorists, physicians—who hold a legal monopoly over medicine—benefit from the existing structure because they can charge inflated fees. Similarly, clinics, pharmacies, laboratories, hospitals, supply manufacturers, insurance companies, and many other corporations derive excessive profits from the existing system of payment in medicine. In recent years, large drug companies and profit-making hospital corporations have come to occupy a larger and larger part of health care delivery. As a result, medical costs have risen rapidly, and the federal government and many insurance companies have placed pressure for cost containment on other players in the medical–industrial complex (Tilly and Tilly, 1998).

Theorists using a radical conflict framework for their analysis believe that the only way to reduce inequalities in the U.S. health care structure is to eliminate capitalism or curb the medical–industrial complex. From this approach, capitalism is implicated in both the rates of illness and how health care is delivered. An example of how rates of illness are related to capitalism is the predominance of workplace hazards and environmental pollution that are dangerous but are often not corrected because corrective measures would be too costly, reducing corporate profits.

Conflict theorists increase our awareness of inequalities of race, class, and gender as these statuses influence people's access to health care. They also inform us about the problems associated with health care becoming "big business." However, some analysts believe that the conflict approach is unduly pessimistic about the gains that have been made in health status and longevity—gains that are at least partially due to large investments in research and treatment by the medical–industrial complex.

A Symbolic Interactionist Perspective: The Social Construction of Illness

Symbolic interactionists attempt to understand the specific meanings and causes that we attribute to particular events. In studying health, symbolic interactionists focus on the meanings that social actors give their illness or disease and how these affect people's self-concept and relationships with others. According to symbolic interactionists, we socially construct "health" and "illness" and how both should be treated. For example, some people explain disease by blaming it on those who are ill. If we attribute cancer to the acts of a person, we can assume that we will be immune to that disease if we do not engage in the same behavior. Nonsmokers who learn that a lung cancer victim had a two-pack-a-day habit feel comforted that they are unlikely to suffer the same fate. Similarly, victims of AIDS are often blamed for promiscuous sexual conduct or intravenous drug use, regardless of how they contracted HIV. In this case, the social definition of the illness leads to the stigmatization of individuals who suffer from the disease.

Although biological characteristics provide objective criteria for determining medical conditions such as heart disease, tuberculosis, or cancer, there is also a subjective component to how illness is defined. This subjective component is very important when we look at conditions such as childhood hyperactivity, mental illness, alcoholism, drug abuse, cigarette smoking, and overeating, all of which have been medicalized. The term ***medicalization* refers to the process whereby nonmedical problems become defined and treated as illnesses or disorders.** Medicalization may occur on three levels: (1) the conceptual level (e.g., the use of medical terminology to define the problem), (2) the institutional level (e.g., physicians are supervisors of treatment and gatekeepers to applying for benefits), and (3) the interactional level (e.g., when physicians treat patients' conditions as medical problems). For example, the sociologists Deborah Findlay and Leslie Miller (1994: 277) explain how gambling has been medicalized:

> Habitual gambling . . . has been regarded by a minority as a sin, and by most as a leisure pursuit—perhaps wasteful but a pastime nevertheless. Lately, however, we have seen gambling described as a psychological illness—"compulsive gambling." It is in the process of being medicalized. The consequences of this shift in discourse (that is, in the way of thinking and talking) about gambling are considerable for doctors, who now have in gamblers a new market for their services or "treatment"; perhaps for gambling halls, which may find themselves subject to new regulations, insofar as they are deemed to contribute to the "disease"; and not least, for gamblers themselves, who are no longer treated as sinners or wastrels, but as patients, with claims on our sympathy, and to our medical insurance plans as well.

Is gambling a moral issue or a medical one? According to sociologists, the recent trend toward viewing compulsive gambling as a health care issue is an example of the medicalization of deviance.

Sociologists often refer to this form of medicalization as the *medicalization of deviance* because it gives physicians and other medical professionals greater authority to determine what should be considered "normal" and "acceptable" behavior and to establish the appropriate mechanisms for controlling "deviant behaviors."

According to symbolic interactionists, medicalization is a two-way process: Just as conditions can be medicalized, so can they be demedicalized. ***Demedicalization* refers to the process whereby a problem ceases to be defined as an illness or a disorder.** Examples include the removal of certain behaviors (such as homosexuality) from the list of mental disorders compiled by the American Psychiatric Association and the deinstitutionalization of mental health patients. The process of demedicalization also continues in women's health as advocates seek to redefine childbirth and menopause as natural processes rather than as illnesses (Conrad, 1996).

In addition to how health and illness are defined, symbolic interactionists examine how doctors and patients interact in health care settings (see "Sociology Works!"). Some physicians may hesitate to communicate certain kinds of medical information to patients, such as why they are prescribing certain medications or what side-effects or drug interactions may occur (Kendall, 2004).

Symbolic interactionist perspectives on health and health care provide us with new insights on the social construction of illness and how health and illness cannot be strictly determined by medical criteria. Symbolic interactionists also make us aware of the importance of communication between physicians and patients, including factors that may reduce effective medical treatment for some individuals. However, these approaches have been criticized for suggesting that few objective medical criteria exist for many illnesses and for overemphasizing microlevel issues without giving adequate recognition to macrolevel issues such as the effects on health care of managed care, health maintenance organizations, and for-profit hospital chains.

medical–industrial complex local physicians, local hospitals, and global health-related industries such as insurance companies and pharmaceutical and medical supply companies that deliver health care today.

medicalization the process whereby nonmedical problems become defined and treated as illnesses or disorders.

demedicalization the process whereby a problem ceases to be defined as an illness or a disorder.

Sociology *Works!*

Sociology Sheds Light on the Physician–Patient Relationship

Doctor: What's the problem?

(Chair noise).

Patient: . . . had since last Monday evening so it's a week of sore throat.

Doctor: . . . hm . . . hm . . .

Patient: . . . which turned into a cold . . . and then a cough.

Doctor: A cold you mean what? Stuffy nose?

Patient: uh stuffy nose yeah not a chest . . . cold . . .

Doctor: . . . hm . . . hm . . . And a cough.

Patient: And a cough . . . which is the most irritating aspect. . . .

Doctor: Okay. Uh, any fever?

Patient: Not that I know of . . . I took it a couple of times in the beginning but haven't felt like—

Doctor: How bout your ears? . . . (Mishler, 2005: 322)

In this brief excerpt from the transcript of a discussion between a doctor and patient, the patient responds to the doctor's request for information by telling him what her symptoms are and how they began to change when her illness "turned into a cold . . . and then a cough." According to sociologists who study the social organization of health care, this transcript indicates that the physician wants the patient to continue speaking when he makes sounds such as "hm . . . hm" but that he also wants to remain in control of the conversation. When the patient mentions that she has a "cold," for example, the doctor asks for further clarification of her specific symptoms so that he can determine, efficiently and in a short period of time, what kind of cold she has and what the treatment plan should be (Mishler, 2005). This is one brief passage from many pages of medical transcripts that researchers have used to study what they refer to as the struggle between the voice of medicine and the voice of the lifeworld (Mishler, 1984, 2005).

What are the voices of medicine and of the lifeworld? In this context, the voice of medicine refers to the technical, scientific attitude adopted by many doctors in their communication with patients. This type of discourse is generally abstract, neutral, and somewhat distant. By contrast, the voice of the lifeworld refers to the natural, everyday attitudes that are expressed by patients when they talk to their physician in the hope of gaining additional insight on their medical condition. Some sociologists believe that a constant struggle exists between these two "voices" in the doctor–patient relationship and that this struggle affects the outcome of each medical encounter. The voice of medicine makes it difficult for patients to believe that their concerns are being heard when the physician is visibly in a hurry, does not listen, interrupts frequently, and/or talks down to the patient.

To minimize the voice of medicine, some sociologists advocate *therapeutic communication* between doctors and patients. In therapeutic communication, (1) the physician engages in full and open communication with the patient and feels free to ask questions about psychosocial as well as physical conditions, (2) the patient provides full and open information to the physician and feels free to ask questions and seek clarifications, and (3) a genuine rapport develops between physician and patient (Weiss and Lonnquist, 2009). According to some social analysts, it takes a doctor no additional time to use a positive communication style that conveys friendliness, empathy, genuineness and candor, an openness to conversation, and a nonjudgmental attitude toward the patient. Such a communication style certainly helps physicians establish positive relationships with their patients. Of course, patients should also try to communicate in a positive manner with physicians, people whom the patients hope will be able to help them remain healthy or help them resolve an existing medical problem (Weiss and Lonnquist, 2009).

Reflect & Analyze

Sociologists who study the social organization of medicine will continue to look for new insights on the physician–patient relationship. What other sociological perspectives do you believe might be useful in explaining the dynamics of doctor–patient communications or other social interactions (such as between physicians and other health professionals) that routinely take place within the health care system?

CONCEPT QUICK REVIEW

Sociological Perspectives on Health and Medicine	
A Functionalist Perspective: The Sick Role	People who are sick are temporarily exempt from normal obligations but must want to get well and seek competent help.
A Conflict Perspective: Inequalities in Health and Health Care	Problems in health care are rooted in the capitalist system, exemplified by the medical–industrial complex.
A Symbolic Interactionist Perspective: The Social Construction of Illness	People socially construct both "health" and "illness," and how both should be treated.
A Postmodernist Perspective: The Clinical Gaze	Doctors gain power through observing patients to gather information, thus appearing to speak "wisely."

A Postmodernist Perspective: The Clinical Gaze

In *The Birth of the Clinic* (1994/1963), postmodern theorist Michel Foucault questioned existing assumptions about medical knowledge and the power that doctors have gained over other medical personnel and everyday people. Foucault asserted that truth in medicine—as in all other areas of life—is a social construction, in this instance one that doctors have created. Foucault believed that doctors gain power through the *clinical* (or "observing") *gaze,* which they use to gather information. Doctors develop the clinical gaze through their observation of patients; as the doctors begin to diagnose and treat medical conditions, they also start to speak "wisely" about everything. As a result, other people start to believe that doctors can "penetrate illusion and see . . . the hidden truth" (Shawver, 1998).

According to Foucault, the prestige of the medical establishment was further enhanced when it became possible to categorize all illnesses within a definitive network of disease classification under which physicians can claim that they know why patients are sick. Moreover, the invention of new tests made it necessary for physicians to gaze upon the naked body, to listen to the human heart with an instrument, and to run tests on the patient's body fluids. Patients who objected were criticized by the doctors for their "false modesty" and "excessive restraint" (Foucault, 1994/1963: 163). As the new rules allowed for the patient to be touched and prodded, the myth of the doctor's diagnostic wisdom was further enhanced, and "medical gestures, words, gazes took on a philosophical density that had formerly belonged only to mathematical thought" (Foucault, 1994/1963: 199). For Foucault, the formation of clinical medicine was merely one of the more-visible ways in which the fundamental structures of human experience have changed throughout history.

Foucault's work provides new insights on medical dominance, but it has been criticized for its lack of attention to alternative viewpoints. Among these is the possibility that medical breakthroughs and new technologies actually help physicians become wiser and more scientific in their endeavors. Another criticism is that Foucault's approach is based upon the false assumption that people are passive individuals who simply comply with doctors' orders—he does not take into account that people (either consciously or unconsciously) may resist the myth of the "wise doctor" and not follow "doctors' orders" (Lupton, 1997).

Foucault's analysis (1988/1961) was not limited to doctors who treat bodily illness; he also critiqued psychiatrists and the treatment of insanity.

The Concept Quick Review summarizes the major sociological perspectives on health and medicine.

Mental Illness

Mental illness affects many people; however, it is a difficult topic for sociological research. Social analysts such as Thomas Szasz (1984) have argued that mental illness is a myth. According to this approach, "mental illnesses" are actually individual traits or behaviors that society deems unacceptable, immoral, or deviant. According to Szasz, labeling individuals as "mentally ill" harms them because they often come to accept the label and are then treated accordingly by others.

Is mental illness a myth? After decades of debate on this issue, social analysts are no closer to reaching a consensus than they were when Szasz originally introduced his ideas. However, many scholars believe that mental illness is a reality that has biological, psychological, environmental, social, and other factors involved. Many medical professionals distinguish between a *mental disorder*—a condition that makes it difficult or impossible for a person to cope with everyday life—and *mental illness*—a condition in which a person has a severe mental disorder requiring extensive treatment with medication, psychotherapy, and sometimes hospitalization.

How many people are affected by mental illness? Some answers to this question have come from a systematic national study known as the National Comorbidity Survey. The term *comorbidity* refers to the physical and mental conditions—such as physical illness and depression—that compound each other and undermine the individual's overall well-being (Angel and Angel, 1993). Researchers in the study found that among respondents between the ages of 15 and 54, nearly 50 percent had been diagnosed with a mental disorder at some time in their lives (Kessler, 1994). However, severe mental illness—such as schizophrenia, bipolar affective disorder (manic-depression), and major depression—typically affects less than 15 percent of U.S. adults at some time in their lives (Bourdon et al., 1992; Kessler, 1994). According to the researchers, the prevalence of mental disorders in the United States is greater than most analysts had previously believed. The most widely accepted classification of mental disorders is the American Psychiatric Association's (1994) *Diagnostic and Statistical Manual of Mental Disorders (DSM),* which is illustrated in ◆ Table 18.1. When the next edition of the *DSM* is published in 2012, it will be interesting to see how the listings compare to those shown in this table.

Mental disorders are very costly to the nation. Direct costs associated with mental disorders include the price of medication, clinic visits, and hospital stays. However, many indirect costs are incurred as well. These include the loss of earnings by individuals, the costs associated with homelessness and incarceration, and other indirect costs that exist but are difficult to document.

The Treatment of Mental Illness

In *Madness and Civilization,* Michel Foucault (1988/1961) examined the "archaeology of madness" from 1500 to 1800 to determine how ideas of mental illness have changed over time and to describe the "birth of the asylum." According to Foucault, early in this period insanity was considered part of everyday life, and people with mental illnesses were free to walk the streets; however, beginning with the Renaissance and continuing into the seventeenth and eighteenth centuries, the mentally ill were viewed as a threat to others. During that time, asylums were built, and a clear distinction was drawn between the "insane" and the rest of humanity. According to Foucault (1988/1961: 252),

◆ **Table 18.1 Mental Disorders Identified by the American Psychiatric Association**

Disorder	Description
Anxiety Disorders	Disorders characterized by anxiety that is manifested in phobias, panic attacks, or obsessive–compulsive disorder
Dissociative Disorders	Problems involving a splitting or dissociation of normal consciousness such as amnesia and multiple personality
Disorders First Evident in Infancy, Childhood, or Adolescence	Including mental retardation, attention-deficit hyperactivity, anorexia nervosa, bulimia nervosa, and stuttering
Eating or Sleeping Disorders	Including such problems as anorexia and bulimia, or insomnia and other problems associated with sleep
Impulse Control Disorders	Including the inability to control undesirable impulses such as kleptomania, pyromania, and pathological gambling
Mood Disorders	Emotional disorders such as major depression and bipolar (manic-depressive) disorder
Organic Mental Disorders	Psychological or behavioral disorders associated with dysfunctions of the brain caused by aging, disease, or brain damage
Personality Disorders	Maladaptive personality traits that are generally resistant to treatment, such as paranoid and antisocial personality types
Schizophrenia and Other Psychotic Disorders	Disorders with symptoms such as delusions or hallucinations
Somatoform Disorders	Psychological problems that present themselves as symptoms of physical disease, such as hypochondria
Substance-Related Disorders	Disorders resulting from abuse of alcohol and/or drugs such as barbiturates, cocaine, or amphetamines

Source: Adapted from American Psychiatric Association, 1994.

people came to see "madness" as a minority status that does not have the right to autonomy:

> Madness is childhood. Everything at the [asylum] is organized so that the insane are transformed into minors. They are regarded as children who have an overabundance of strength and make dangerous use of it. They must be given immediate punishments and rewards; whatever is remote has no effect on them.

At the beginning of the twenty-first century, many people with mental disorders do not receive professional treatment. However, mental disorders—particularly substance-related ones (due to alcohol and other drug abuse)—are the leading cause of hospitalization for men between the ages of eighteen and forty-four, and the second leading cause (after childbirth) for women in that age group (Agency for Healthcare Research and Quality, 2000).

Many people seeking psychiatric assistance are treated with medications or psychotherapy—which is believed to help patients understand the underlying reasons for their problem—and sometimes treatment in psychiatric wards of local hospitals or in private psychiatric hospitals. However, the introduction of new psychoactive drugs to treat mental disorders and the deinstitutionalization movement in the 1960s have created dramatic changes in how people with mental disorders are treated. ***Deinstitutionalization*** **refers to the practice of rapidly discharging patients from mental hospitals into the community.** Originally devised as a *solution* for the problem of "warehousing" mentally ill patients in large, prison-like mental hospitals in the first half of the twentieth century, deinstitutionalization is now viewed as the *problem* by many social scientists. The theory behind this process was that patients' rights were being violated because many patients experienced involuntary commitment (i.e., without their consent) to the hospitals, where they remained for extended periods of time. Instead, some professionals believed that the patients' mental disorders could be controlled with proper medications and treatment from community-based mental health services. Advocates of deinstitutionalization also believed that this practice would relieve the stigma attached to mental illness and hospitalization. However, critics of deinstitutionalization argue that this process exacerbated long-term problems associated with inadequate care for people with mental illness.

Admitting people to mental hospitals on an involuntary basis ("involuntary commitment") has always been controversial; however, it remains the primary method by which police officers, judges, social workers, and other officials deal with people—particularly the homeless—whom they have reason to believe are mentally ill and imminently dangerous to others if not detained (Monahan, 1992). State mental hospitals continue to provide most of the chronic inpatient care for poor people with mental illnesses; these institutions tend to serve as a revolving door to poverty-level board-and-care homes, nursing homes, or homelessness, as contrasted with the situation of patients who pay their bills at private psychiatric facilities through private insurance coverage or Medicare (Brown, 1985).

According to the sociologist Erving Goffman (1961a), mental hospitals are a classic example of a *total institution*, previously defined as a place where people are isolated from the rest of society for a period of time and come under the complete control of the officials who run the institution.

Race, Class, Gender, and Mental Disorders

Most studies examining race/ethnicity and mental disorders have compared African Americans and white Americans and have uncovered no significant differences in diagnosable mental illness. However, researchers have found that African Americans report psychological distress or demoralization more often than white Americans do. In a qualitative study about the effects of racism on the everyday lives of middle-class African Americans, social scientists Joe R. Feagin and Melvin P. Sikes (1994) concluded that repeated personal encounters with racial hostility deeply affect the psychological well-being of most African Americans, regardless of their level of education or social class. In a subsequent study, Feagin and Hernán Vera (1995) found that white Americans also pay a high "psychic cost" for the prevalence of racism because it contradicts deeply held beliefs about the great American dream of equality under the law. In earlier work on the effects of discrimination on mental well-being, the social psychologist Thomas Pettigrew (1981) suggested that about 15 percent of whites have such high levels of racial prejudice that they tend to exhibit symptoms of serious mental illness. According to Pettigrew, racism in all its forms constitutes a mentally unhealthy situation in which people do not achieve their full potential.

A limited number of social science studies have focused on mental disorders among racial–ethnic groups such as Mexican immigrants and Mexican Americans, and some of these studies have yielded contradictory

deinstitutionalization the practice of rapidly discharging patients from mental hospitals into the community.

results. Researchers in one study found that strong extended (intergenerational) families—which are emphasized in Mexican culture—provide individuals with social support and sources of self-esteem, even if these individuals possess low levels of education and income (Mirowsky and Ross, 1980). Other analysts have challenged this assumption, noting that retention of Mexican culture, as compared to adopted Anglo culture, is not the primary consideration in whether or not Mexican Americans develop mental disorders (Burnham et al., 1987).

Although cultural factors may be important in determining rates of mental disorders, researchers have consistently agreed that social class has a significant effect on this issue. For example, one study examining the relationship among mental disorders, race, and class (as measured by socioeconomic status) found that rates of mental disorders decrease as social class increases for both white Americans and African Americans. For example, at the lowest socioeconomic level, researchers found few race differences (Williams, Takeuchi, and Adair, 1992). Despite agreement among many researchers about the relationship between class and mental disorders, not all of them agree on whether lower social class status causes mental illness or mental illness causes lower social class position (Weitz, 2004). Analysts using the *social stress framework* to examine schizophrenia—the disorder most consistently linked to class—believe that stresses associated with lower-class life lead to greater mental disorders. In contrast, analysts using the *social drift framework* argue that mental disorders cause people to drift downward in class position. To support their argument, they note that individuals diagnosed with schizophrenia typically hold lower-class jobs than would be expected based on their family backgrounds (Eaton, 1980; Weitz, 2004).

Researchers have also consistently found that the rate of diagnosable depression is about twice as high for women as for men. This gender difference typically emerges in puberty (Cleary, 1987) and increases in adulthood as women and men enter and live out their unequal adult statuses (Mirowsky, 1996). However, researchers have found no consistent relationship between gender and the rate of schizophrenia or other serious mental disorders in which the individual loses touch with reality. Although women have higher rates of minor depression and other disorders that cause psychological distress, men have higher rates of personality disorders such as compulsive gambling and drinking, as well as maladaptive personality traits such as antisocial behavior (Link and Dohrenwend, 1989; Weitz, 2004). Some analysts suggest that differences in mental disorders between women and men are linked to gender-role socialization that instills aggressiveness in men and learned helplessness in women. According to the *learned helplessness theory*, people become depressed when they believe they cannot control their lives (Seligman, 1975). Embedded within this theory is the assumption that women contribute to their own helplessness because of the *subjective* perception that they have no control over their lives. However, feminist analysts argue instead that powerlessness is an *objective* condition that may contribute to depression in many women's lives (Jack, 1993). Support for the latter assumption comes from numerous studies indicating that women in high-income, high-status jobs typically have higher levels of psychological well-being and fewer symptoms of mental disorders, regardless of their marital status (Horowitz, 1982; Angel and Angel, 1993).

Although there have been numerous studies about women with mild depression, women with serious mental illnesses have almost been ignored (Mowbray, Herman, and Hazel, 1992). Similarly, women's mental disorders have often been misdiagnosed, and on many occasions physical illnesses have been confused with psychiatric problems (see Busfield, 1996; Lerman, 1996; Klonoff, 1997). According to some analysts, additional studies focusing on women's diversity across lines of race, class, age, religion, sexual orientation, and other factors will be necessary before we can accurately assess the relationship between gender and mental illness and how women are treated in the mental health care industry (Gatz, 1995).

Disability

What is a disability? There are many different definitions. In business and government, disability is often defined in terms of work—for instance, "an inability to engage in gainful employment." Medical professionals tend to define it in terms of organically based impairments—the problem being entirely within the body (Albrecht, 1992). However, not all disabilities are visible to others or necessarily limit people physically. ***Disability* refers to a reduced ability to perform tasks one would normally do at a given stage of life and that may result in stigmatization or discrimination against the person with disabilities.** In other words, the notion of disability is based not only on physical conditions but also on social attitudes and the social and physical environments in which people live. In an elevator, for example, the buttons may be beyond the reach of persons using a wheelchair. In this context, disability derives from the fact that certain things have been made inaccessible to some people. According to disability rights advocates, disability must be thought of in terms of how society causes or contributes to the

© Michael Newman/PhotoEdit

One of the most compelling concerns for many people with a disability is access to buildings and other public accommodations. For persons using a wheelchair, steps such as these are a formidable obstacle in everyday life.

problem—not in terms of what is "wrong" with the person with a disability.

An estimated 49.7 million persons in the United States have one or more physical or mental disabilities. This number continues to increase for several reasons. First, with advances in medical technology, many people who once would have died from an accident or illness now survive, although with an impairment. Second, as more people live longer, they are more likely to experience diseases (such as arthritis) that may have disabling consequences. Third, persons born with serious disabilities are more likely to survive infancy because of medical technology. However, less than 15 percent of persons with a disability today were born with it; accidents, disease, and war account for most disabilities in this country.

Although anyone can become disabled, some people are more likely to be or to become disabled than others. African Americans have higher rates of disability than whites, especially more serious disabilities; persons with lower incomes also have higher rates of disability (Weitz, 2004). However, "disability knows no socioeconomic boundaries. You can become disabled from your mother's poor nutrition or from falling off your polo pony," says Patrisha Wright, a spokesperson for the Disability Rights Education and Defense Fund (qtd. in Shapiro, 1993: 10).

For persons with chronic illness and disability, life expectancy may take on a different meaning. Knowing that they will likely not live out the full life expectancy for persons in their age cohort, they may come to "treasure each moment," as did James Keller, a former baseball coach:

> In December 1992, I found out I have Lou Gehrig's disease—amyotrophic lateral sclerosis, or ALS. I learned that this disease destroys every muscle in the body, that there's no known cure or treatment and that the average life expectancy for people with ALS is two to five years after diagnosis.
>
> Those are hard facts to accept. Even today, nearly two years after my diagnosis, I see myself as a 42-year-old career athlete who has always been blessed with excellent health. Though not an hour goes by in which I don't see or hear in my mind that phrase "two to five years," I still can't quite believe it. Maybe my resistance to those words is exactly what gives me the strength to live with them and the will to make the best of every day in every way. (Keller, 1994)

As Keller's comments illustrate, disease and disability are intricately linked.

Environment, lifestyle, and working conditions may all contribute to either temporary or chronic disability. For example, air pollution in automobile-clogged cities leads to a higher incidence of chronic respiratory disease and lung damage, which may result in severe disability for some people. Eating certain types of food and smoking cigarettes increase the risk for coronary and cardiovascular diseases (Albrecht, 1992). In contemporary industrial societies, workers in the second tier of the labor market (primarily recent immigrants, white women, and people of color) are at the greatest risk for certain health hazards and disabilities. Employees in data processing and service-oriented jobs may also be affected by work-related disabilities. The extensive use of computers has been shown to harm some workers' vision; to produce joint problems such as arthritis, low-back pain, and carpal tunnel syndrome; and to place employees under high levels of stress that may result in neuroses and other mental health problems (Albrecht, 1992). As shown in ◆ Table 18.2, 20.8 percent of the

disability a reduced ability to perform tasks one would normally do at a given stage of life and that may result in stigmatization or discrimination against the person with disabilities.

◆ **Table 18.2 Percentage of U.S. Population with Disabilities**

Characteristic	Percentage[a]
With a disability	20.8
Severe	13.7
Not severe	7.0
Has difficulty or is unable to:	
See words and letters	3.5
Hear normal conversation	3.5
Have speech understood	1.2
Lift or carry ten pounds	6.9
Use stairs	9.2
Walk	9.4
Has difficulty or needs assistance with:	
Getting around inside the house	1.7
Getting in/out of bed or a chair	2.5
Taking a bath or shower	2.2
Dressing	1.7
Eating	0.8
Getting to or using the toilet	1.1
Has difficulty or needs assistance with:	
Going outside the home alone	4.0
Managing money and bills	2.2
Preparing meals	2.3
Doing light housework	3.1
Using the telephone	1.3

[a]Percentage of persons age 15 and older.
Source: Steinmetz, 2006.

people in the United States (more than one in five) have a "chronic health condition which, given the physical, attitudinal, and financial barriers built into the social system, makes it difficult to perform one or more activities generally considered appropriate for persons of their age" (Weitz, 2004).

Can a person in a wheelchair have equal access to education, employment, and housing? If public transportation is not accessible to those in wheelchairs, the answer is certainly no. As disability rights activist Mark Johnson put it, "Black people fought for the right to ride in the front of the bus. We're fighting for the right to get on the bus" (qtd. in Shapiro, 1993: 128).

Many disability rights advocates argue that persons with a disability have been kept out of the mainstream of society. They have been denied equal opportunities in education by being consigned to special education classes or special schools. For example, people who grow up deaf are often viewed as disabled; however, many members of the deaf community instead view themselves as a "linguistic minority" that is part of a unique culture (Lane, 1992; Cohen, 1994). They believe that they have been restricted from entry into schools and the work force not due to their own limitations, but by societal barriers.

Living with disabilities is a long-term process. For infants born with certain types of congenital (present at birth) problems, their disability first acquires social significance for their parents and caregivers. In a study of children with disabilities in Israel, the sociologist Meira Weiss (1994) challenged the assumption that parents automatically bond with infants, especially those born with visible disabilities. She found that an infant's appearance may determine how parents will view the child. Parents are more likely to be bothered by external, openly visible disabilities than by internal or disguised ones; some parents are more willing to consent to or even demand the death of an "appearance-impaired" child (Weiss, 1994). According to Weiss, children born with internal (concealed) disabilities are at least initially more acceptable to parents because they do not violate the parents' perceived body images of their children. Weiss's study provides insight into the social significance that people attach to congenital disabilities.

Among persons who acquire disabilities through disease or accidents later in life, the social significance of their disability can be seen in how they initially respond to their symptoms and diagnosis, how they view the immediate situation and their future, and how the illness and disability affect their lives. When confronted with a disability, most people adopt one of two strategies—avoidance or vigilance. Those who use the avoidance strategy deny their condition in order to maintain hopeful images of the future and elude depression; for example, some individuals refuse to participate in rehabilitation following a traumatic injury because they want to pretend that the disability does not exist. By contrast, those using the vigilance strategy actively seek knowledge and treatment so that they can respond appropriately to the changes in their bodies (Weitz, 2004).

Sociological Perspectives on Disability

How do sociologists view disability? Those using the functionalist framework often apply Parsons's sick role model, which is referred to as the *medical model* of disability. According to the medical model, people with disabilities become, in effect, chronic patients under the supervision of doctors and other medical personnel, subject to a doctor's orders or a program's rules, and not to their own judgment (Shapiro, 1993). From this perspective, disability is deviance.

The deviance framework is also apparent in some symbolic interactionist perspectives. According to symbolic interactionists, people with a disability experience *role ambiguity* because many people equate disability with deviance (Murphy et al., 1988). By labeling individuals with a disability as "deviant," other people can avoid them or treat them as outsiders. Society marginalizes people with a disability because they have lost old roles and statuses and are labeled as "disabled" persons. According to the sociologist Eliot Freidson (1965), how the people are labeled results from three factors: (1) their degree of responsibility for their impairment, (2) the apparent seriousness of their condition, and (3) the perceived legitimacy of the condition. Freidson concluded that the definitions of and expectations for people with a disability are socially constructed factors.

Finally, from a conflict perspective, persons with a disability are members of a subordinate group in conflict with persons in positions of power in the government, in the health care industry, and in the rehabilitation business, all of whom are trying to control their destinies (Albrecht, 1992). Those in positions of power have created policies and artificial barriers that keep people with disabilities in a subservient position (Asch, 1986; Hahn, 1987). Moreover, in a capitalist economy, disabilities are big business. When people with disabilities are defined as a social problem and public funds are spent to purchase goods and services for them, rehabilitation becomes a commodity that can be bought and sold by the medical–industrial complex (Albrecht, 1992). From this perspective, persons with a disability are objectified. They have an economic value as consumers of goods and services that will allegedly make them "better" people. Many persons with a disability endure the same struggle for resources faced by people of color, women, and older persons. Individuals who hold more than one of these ascribed statuses, combined with experiencing disability, are doubly or triply oppressed by capitalism.

Social Inequalities Based on Disability

People with visible disabilities are often the objects of prejudice and discrimination, which interfere with their everyday life. For example, Marylou Breslin, executive director of the Disability Rights Education and Defense Fund, was wearing a businesswoman's suit, sitting at the airport in her battery-powered wheelchair, and drinking a cup of coffee while waiting for a plane. A woman walked by and plunked a coin in the coffee cup that Breslin held in her hand, splashing coffee on Breslin's blouse (Shapiro, 1993). Why did the woman drop the coin in Breslin's cup? The answer to this question is found in stereotypes built on lack of knowledge about or exaggeration of the characteristics of people with a disability. Some stereotypes project the image that persons with disabilities are deformed individuals who may also be horrible deviants. For example, slasher movies such as the *Nightmare on Elm Street* series show a villain who was turned into a hateful, sadistic killer because of disfigurement resulting from a fire. Lighter fare such as the *Batman* movies depict villains as individuals with disabilities: the Joker, disfigured by a fall into a vat of acid, and the Penguin, born with flippers instead of arms (Shapiro, 1993). Other stereotypes show persons with disabilities as pathetic individuals to be pitied. Fund-raising activities by many charitable organizations—such as "poster child" campaigns showing a photograph of a friendly-looking child with a visible disability—sometimes contribute to this perception. Even apparently positive stereotypes become harmful to people with a disability. An example is what some disabled persons refer to as "supercrips"—people with severe disabilities who seem to excel despite the impairment and who receive widespread press coverage in the process. Disability rights advocates note that such stereotypes do not reflect the day-to-day reality of most persons with disabilities, who must struggle constantly with smaller challenges (Shapiro, 1993).

Today, many working-age persons with a disability in the United States are unemployed (see "Census Profiles: Disability and Employment Status"). Most of them believe that they could and would work if offered the opportunity. However, even when persons with a severe disability are able to find jobs, they typically earn less than persons without a disability (Yelin, 1992). On average, workers with a severe disability make 59 percent of what their co-workers without disabilities earn, and the gap is growing (U.S. Census Bureau, 2005). The problem has been particularly severe for African Americans and Latinos/as with disabilities. Among Latinos/as with a severe disability, only 26 percent are employed; those who work earn 80 percent of what white (non-Latino) persons with a severe disability earn.

Employment, poverty, and disability are related. On the one hand, people may become economically disadvantaged as a result of chronic illness or disability. On the other hand, poor people are less likely to be educated and more likely to be malnourished and have inadequate access to health care—all of which contribute to risk of chronic illness, physical and mental disability, and the inability to participate in the labor force. In addition, the type of employment available to people with limited resources increases their chances of becoming disabled. They may work in hazardous places such as mines, factory assembly lines, and chemical plants, or in the construction industry, where the chance of becoming seriously disabled is much higher (Albrecht, 1992).

Generalizations about the relationship between disability and income are difficult to make for at least three reasons. First, most research on disability is organized around specific conditions or impairments, making it problematic to reach conclusions about how much

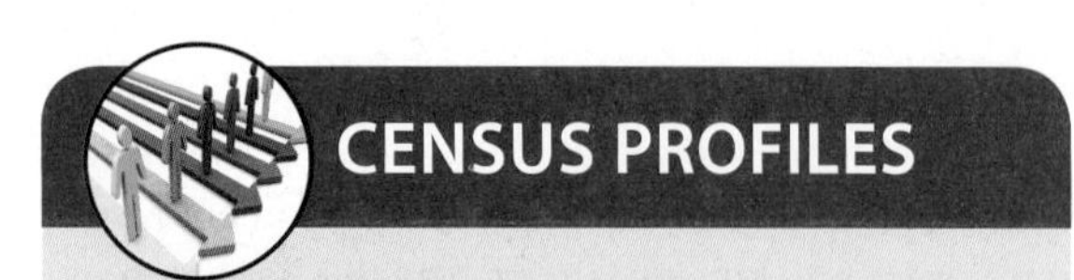

CENSUS PROFILES

Disability and Employment Status

One of the questions on Census 2000 asked respondents about long-lasting conditions such as a physical, mental, or emotional condition that substantially limited important basic activities. It also asked if they were working at a job or business. The answers allowed the Census Bureau to determine how many people with a disability and how many people without a disability were employed at the time the census was conducted. As you can see from the figure set forth below, for the population aged 21 to 64 years (the period during which people are most likely to be employed), less than 50 percent of persons with a disability were employed, compared with almost 80 percent of persons without a disability who were employed. Is employment among persons with a disability related primarily to their disability status, or do other facts—such as prejudice or lack of willingness to make the necessary accommodations that would allow such persons to hold a job—play a significant part in the high rate of unemployment of persons with a disability?

Source: U.S. Census Bureau, 2008.

income a person loses because of a disability (DeJong, Batavia, and Griss, 1989). Second, generalizations about entire categories of people (such as whites, African Americans, or Latinos/as) tend to be inaccurate because of differences *within* each group. For example, the sociologist Ronald Angel (1984) found significant differences in the effect of disabilities on Latinos depending on whether they were Mexican American, Puerto Rican, or Cuban American. Angel concluded that although both Mexican American and Puerto Rican men had lower rates of full-time employment and lower hourly wages relative to (non-Latino) whites, Puerto Ricans were worse off than Mexican Americans in terms of earnings and number of work hours. Third, there is the problem of determining which factor occurred first. For example, some of the problems for Puerto Ricans with disabilities are tied to their overall position of economic disadvantage (see Chapter 8). Although disability is associated with lower earnings and higher rates of unemployment for both males and females, a disability has a stronger negative effect on women's labor force participation than it does on men's. Compared to men, women with disabilities are overrepresented as clerical and service workers and underrepresented as managers and administrators. Women with disabilities are also much less likely to be covered by pension and health plans than are men (Russo and Jansen, 1988).

What does it cost to "mainstream" persons with disabilities? Disability expenditures for 1986 were estimated to be $169.4 billion for the working-age population, taking into account money spent by both the government and the private sector (Albrecht, 1992). However, between 1978 and 1992, Sears, Roebuck found that the average cost of an accommodation for a worker with a disability was only $126, with some accommodations involving little or no cost (such as flexible schedules, back-support belts, rest periods, and changes in employee work stations). Other accommodations cost about $1,000, with the largest single cost being $14,500 each for special Braille computer displays for visually impaired employees (Noble, 1995). Future opportunities for persons with disabilities are closely related to current expenditures that provide greater access to education and jobs.

Health Care in the Future

Central questions regarding the future of health care are how to provide coverage for the largest number of people and how to do this without bankrupting the entire nation. At the time of this writing, Congress is debating the possibilities of passing a health reform bill that would require virtually all Americans to carry health insurance or to pay a penalty. It would also require most businesses to provide health insurance for workers. The plan would expand Medicaid to cover many more poor people. It would make it possible for middle-income people to buy health insurance with government subsidies. A reform bill such as this would

Box 18.4 You Can Make a Difference

Joining the Fight Against Illness and Disease!

> Each day trained volunteers nurture female patients in hospitals and long-term care facilities by administering complimentary makeup and moisturizer. Among the benefits to patients receiving this caring social interaction and gentle touch are higher self-esteem and shorter recovery times. (Fiffer and Fiffer, 1994: 120)

Beverly Barnes did not intend to start the program now known as Patient Pride when she went to the hospital to visit a friend who was recuperating from surgery. However, by the time she left the hospital, having helped her friend put on makeup and freshen her appearance, Barnes realized how much seemingly small acts of kindness can mean to people who are ill or recuperating from surgery. Eventually, Patient Pride, Inc., became a nonprofit corporation that relies on volunteers to visit patients, and new chapters servicing hospitals and long-term health care facilities have opened throughout the country. Recently, the program has sought to include male patients as well as women.

Patient Pride is only one of many examples of how everyday people can make a difference in the lives of people experiencing illness or disability. Here are some Internet sources that you can check out to learn more about volunteering and participating in the fight against diseases such as cancer and cardiovascular disease:

- The American Cancer Society's home page (which has links to various volunteer activities): **http://www.cancer.org**
- The American Heart Association's home page: **http://www.amhrt.org**

constitute a form of universal health coverage. According to the Congressional Budget Office, about 97 percent of all legal residents in the United States would have health insurance by 2015. However, the price tag for this coverage has created great dissent, particularly among the more affluent (defined in the bill as families with adjusted gross incomes exceeding $350,000 and individuals making more than $280,000), who would bear a major share of the burden by paying $544 billion over the next decade through a graduated income surtax. The other part of the cost would be paid by greatly reducing current spending on Medicare. The bill would also expand prevention programs and wellness activities in hopes that these would cut down on the costs of illness and disability coverage in the future. The major sticking point on passage of the health reform bill is the surtax on the wealthy, and it will be interesting to see how this issue comes out both now and in the future. Overall, our best hope is that a payment system will be developed that pays for good patient results at a reasonable cost and thus brings about a transformation in the current U.S. health care system.

A key factor in contemporary health care is the role that advanced technologies play in the rising costs of medical care and their usefulness as major tools for diagnosis and treatment. Technology is a major stimulus for social change, and the health care systems in high-income nations such as the United States reflect the rapid rate of technological innovation that has occurred in the last few decades. In the future, advanced health care technologies will no doubt provide even more accurate and quicker diagnosis, effective treatment techniques, and increased life expectancy.

However, technology alone cannot solve many of the problems confronting us in health and health care delivery. In fact, some aspects of technological innovation may be dysfunctional for individuals and society. As we have seen, some technological "advances" raise new ethical concerns, such as the moral and legal issues surrounding the cloning of human life. Some "advances" also may fail: A new prescription drug may be found to cause side-effects that are more serious than the illness that it was supposed to remedy. Whether advanced technology succeeds or fails in some areas, it will probably continue to increase the cost of health care in the future. As a result, the gap between the rich and the poor in the United States will contribute to inequalities of access to vital medical services. On a global basis, new technologies may lower the death rate in some low-income countries, but it will primarily be the wealthy in those nations who will have access to the level of health care that many people in higher-income countries take for granted.

As previously discussed, the organization of U.S. health care will continue to change in the future. Finally, to a degree, health care in the future will be up to each of us. What measures will we take to safeguard ourselves against illness and disorders? How can we help others who are the victims of acute and chronic diseases or disabilities? Although we cannot change global or national health problems, there are some small (but not insignificant) things we can do to make a difference (see Box 18.4).

Chapter Review

- **What is health, and why are sociologists interested in studying health and medicine?**

According to the World Health Organization, health is a state of complete physical, mental, and social well-being. In other words, health is not only a biological issue but also a social issue. For this reason, sociologists are interested in studying health and medicine. As a social institution, medicine is one of the most important components of quality of life.

- **How do acute and chronic diseases differ, and which is most closely linked to spiraling health care costs?**

Acute diseases are illnesses that strike suddenly and cause dramatic incapacitation and sometimes death. Chronic diseases, such as arthritis, diabetes, and heart disease, are long-term or lifelong illnesses that develop gradually or are present at birth. Treatment of chronic diseases is typically more costly because of the duration of these problems.

- **How is health care paid for in the United States?**

Throughout most of the past hundred years, medical care in the United States has been paid for on a fee-for-service basis. The term *fee for service* means that patients are billed individually for each service they receive. This approach to paying for medical services is expensive because few restrictions are placed on the fees that doctors, hospitals, and other medical providers can charge patients. Recently, there have been efforts at cost containment, and HMOs and managed care have produced both positive and negative results in the contemporary practice of medicine. Health maintenance organizations (HMOs) provide, for a set monthly fee, total care with an emphasis on prevention to avoid costly treatment later. Managed care refers to any system of cost containment that closely monitors and controls health care providers' decisions about medical procedures, diagnostic tests, and other services that should be provided to patients.

- **What is the functionalist perspective on health and illness?**

According to the functionalist approach, if society is to function as a stable system, it is important for people to be healthy and to contribute to their society. Consequently, sickness is viewed as a form of deviant behavior that must be controlled by society. Sociologist Talcott Parsons (1951) described the sick role—the set of patterned expectations that defines the norms and values appropriate for individuals who are sick and for those who interact with them. Although individuals are given permission to not perform their usual activities for a period of time, they are expected to seek medical attention and get well as soon as possible so that they can go about their normal routine.

- **What is the conflict perspective on health and illness?**

Conflict theory tends to emphasize the political, economic, and social forces that affect health and the health care delivery system. Among these issues are the ability of all people to obtain health care; how race, class, and gender inequalities affect health and health care; power relations between doctors and other health care workers; the dominance of the medical model of health care; and the role of profit in the health care system.

- **What is the symbolic interactionist perspective on health and illness?**

In studying health, symbolic interactionists focus on the fact that the meaning that social actors give their illness or disease will affect their self-concept and their relationships with others. According to symbolic interactionists, we socially construct "health" and "illness" and how both should be treated. Symbolic interactionists also examine medicalization—the process whereby nonmedical problems become defined and treated as illnesses or disorders.

- **What is the postmodernist perspective on health and illness?**

Postmodern theorists such as Michel Foucault argue that doctors and the medical establishment have gained control over illness and patients at least partly because of the physicians' clinical gaze, which replaces all other systems of knowledge. The myth of the wise doctor has also been supported by the development of disease classification systems and new tests.

- **How did deinstitutionalization change the way that mental illness is treated?**

Deinstitutionalization shifted many mental patients from hospital treatment to community- or family-based care. This was possible because of newer drugs and treatments; however, it also created new issues because of social stereotypes about the "mentally ill" and differences of opinion about various treatment options.

- **What is a disability, and how prevalent are disabilities in the United States?**

Disability is a physical or health condition that stigmatizes or causes discrimination. An estimated 49.7 million persons in the United States have one or more physical or mental disabilities. This number continues to increase for several reasons. First, with advances in medical technology, many people who once would have died from an accident or illness now survive, although with an impairment. Second, as more people live longer, they are more likely to experience diseases (such as arthritis) that may have disabling consequences. Third, persons born with serious disabilities are more likely to survive infancy because of medical technology. However, less than 15 percent of persons with a disability today were born with it; accidents, disease, and war account for most disabilities in this country.

www.cengage.com/login

Want to maximize your online study time? Take this easy-to-use study system's diagnostic pre-test, and it will create a personalized study plan for you. By helping you to identify the topics that you need to understand better and then directing you to valuable online resources, it can speed up your chapter review. CengageNOW even provides a post-test so you can confirm that you are ready for an exam.

Key Terms

acute diseases 587
chronic diseases 587
deinstitutionalization 609
demedicalization 605
disability 610
drug 590
health 584
health care 584
health maintenance organizations (HMOs) 597
holistic medicine 602
infant mortality rate 584
life expectancy 584
managed care 599
medical–industrial complex 604
medicalization 605
medicine 584
sick role 603
social epidemiology 587
socialized medicine 600
universal health care 600

Questions for Critical Thinking

1. Why is it important to explain the social, as well as the biological, aspects of health and illness in societies?
2. In what ways are race, class, and gender intertwined with physical and mental disorders?
3. How would functionalists, conflict theorists, and symbolic interactionists suggest that health care delivery might be improved in the United States?
4. Based on this chapter, how do you think illness and disability will be handled in the United States in the near future? Are there things that we can learn from other nations regarding the delivery of health care? Why or why not?

The Kendall Companion Website

www.cengage.com/sociology/kendall

Visit this book's companion website, where you'll find more resources to help you study and successfully complete course projects. Resources include quizzes and flash cards, as well as special features such as an interactive sociology timeline, maps, General Social Survey (GSS) data, and Census 2000 data. The site also provides links to useful websites that have been selected for their relevance to the topics in this chapter and include those listed below. (*Note:* Visit the book's website for updated URLs.)

World Health Organization (WHO)
http://www.who.int

The website of the World Health Organization is an incredibly rich source of information on global health issues, covering diseases, environmental hazards, health systems, and much more.

National Institutes of Health (NIH)
http://www.nih.gov

The NIH is a federal organization dedicated to the discovery of new knowledge aimed at improving the health of individuals. In addition to news and events, scientific resources, and health information, the website provides links to all of the NIH institutes' home pages, each of which has its own databases, FAQs, quick facts, and publications.

Office of Minority Health (OMH)
http://www.omhrc.gov

The OMH was created in 1985 by the U.S. Department of Health and Human Services to address and improve the health of minority populations. Its website offers data and statistics, information on health disparities, a resource center, Web links, and publications.

chapter

19 Population and Urbanization

Chapter Focus Question

What effect does migration have on cities and on shifts in the global population?

The tomato-farming region of Immokalee, Florida:

It's a very bad thing because we're working very hard here and there's no support from the government. We're only working. We're not committing a sin.

—Rigoberto Morales expressing his frustration upon learning that Congress is doing little to help him become a U.S. citizen after he has spent eight long years working in the hot fields of Florida picking tomatoes and sending most of his earnings to his family in Mexico (qtd. in Goodnough and Steinhauer, 2006: A35)

We live here in fear. We fear Immigration will come, and many people just don't go out.

—Antonia Fuentes, a Mexican farmworker who has picked tomatoes for two years and would welcome even a guest worker program that would permit her to stay in the United States for a set period of time (qtd. in Goodnough and Steinhauer, 2006: A35)

A migrant shelter in Nogales, Mexico:

We want to try our luck [in the United States]. We

Many undocumented workers seek to enter the United States so that they can work and obtain a better quality of life for themselves and their families. However, the aspirations of these workers often run contrary to the mandate of law enforcement officials to patrol the border and keep out individuals who seek to enter the country illegally.

IN THIS CHAPTER

can't go back to Michoacan [a state in Mexico] because there is no future there. My brothers said there is plenty of work [in the United States], and that it looks like they will start giving [work] permits.

—Edith Mondragon explaining why she and her husband paid a smuggler in a failed attempt to get across the Mexican border into the United States (qtd. in CNN.com, 2006)

It's hard to cross. But it's harder to see your children have little to eat.

—Raul Gonzalez, who turned himself in to U.S. authorities after he was robbed and his feet started bleeding from walking for five days to get across the border into the United States, stating the common theme of most immigrants: We want a better life for ourselves and our children (qtd. in CNN.com, 2006)

Washington, D.C., immigration rally:

We want to be legal. We want to live without hiding, without fear. We have to speak so that our voices are listened to and we are taken into account.

—Ruben Arita, a construction worker from Honduras, explaining why he participated in a rally to push for legal status for undocumented workers residing in the United States (qtd. in Swarns, 2006: A1)

(Continued)

Sharpening Your Focus

- *How are people affected by population changes?*
- *How do ecological/functionalist models and political economy/conflict models differ in their explanations of urban growth?*
- *What is meant by the experience of urban life, and how do sociologists seek to explain this experience?*
- *What are the best-case and worst-case scenarios regarding population and urban growth in the twenty-first century, and how might some of the worst-case scenarios be averted?*

Immigrants are coming together in a way that we have never seen before, and it's going to keep going. . . . This is a movement. We're sending a strong message that we are people of dignity. All that we want is to have a shot at the American dream.

—Jaime Contreras, who entered the United States illegally from El Salvador but is now a U.S. citizen, discusses why he would like to see a major change in current immigration and citizenship laws (qtd. in Swarns, 2006: A1)

As economic conditions have worsened in the United States near the end of the first decade of the twenty-first century, the issue of immigration remains a source of contention among many people in this country. Some undocumented immigrants, such as the ones interviewed above, believe that they have earned the right to remain in the United States because they have worked hard, paid taxes, and supported the nation. On the other hand, some political leaders and conservative activists believe that immigration reform and "closing the borders" are long overdue in this country. According to these analysts, the Obama administration should "fix" this problem along with many others we have looked at, including economic stabilization, school reform, and containment of health care costs. Even as tens of thousands of demonstrators have taken to the streets in past years to proclaim "We are America" and to ask for legal status and citizenship for millions of illegal immigrants, critics argue that immigration will become the downfall of this nation.

In the United States, as well as in many other nations that have experienced high rates of immigration, questions of permanent residency and citizenship for immigrant workers have stirred up extensive debates and sometimes produced hostility or violence toward people who are labeled as "illegal immigrants" or "outsiders." According to various estimates, between 11.5 million and 12 million unauthorized migrants are living in the United States, and two-thirds (66 percent) of this population has been in the country for ten years or less. In fact, about 40 percent of the unauthorized population (about 4.4 million people) has been in the United States for five years or less (Passel, 2006). This upswing in unauthorized immigration has produced a complex debate involving many pressing economic, political, and social issues, and has divided people in this country. By contrast, other factors associated with population growth and urban change, such as the number of live births annually as compared with the number of deaths, have received very little recent attention.

In this chapter, we explore the dynamics of population and urbanization, with a focus on how migration, and particularly immigration, affect growth and change in societies such as ours. Before reading on, test your knowledge about current U.S. immigration issues by taking the quiz in Box 19.1.

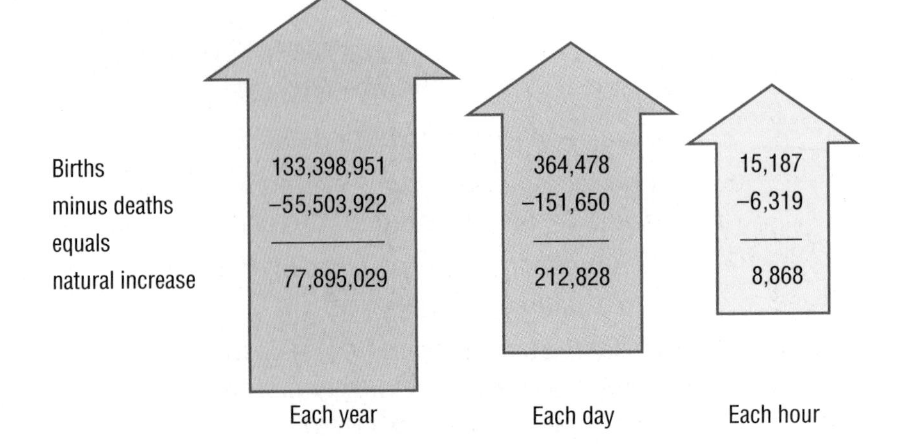

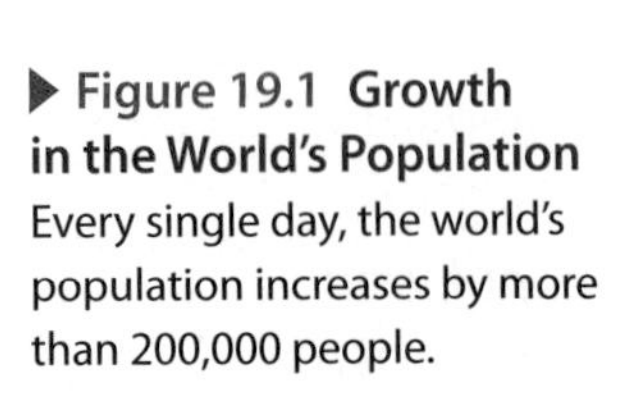
▶ Figure 19.1 Growth in the World's Population
Every single day, the world's population increases by more than 200,000 people.

Source: U.S. Census Bureau, 2008.

Box 19.1 Sociology and Everyday Life

How Much Do You Know About U.S. Immigration?

True	False	
T	F	1. All "unauthorized immigrants" in the United States entered the country illegally.
T	F	2. Nearly two-thirds of the children living in unauthorized immigrant families in the United States are U.S. citizens by birth.
T	F	3. Adult males make up a larger percentage of the unauthorized immigrant population than adult females.
T	F	4. Unauthorized immigrants from Mexico and Latin America represent slightly less than 50 percent of the unauthorized population in the United States.
T	F	5. During the past ten years, more unauthorized immigrants came to the United States from Europe and Canada than from Asia.
T	F	6. Unauthorized male immigrants are more likely to be employed than are males who are either legal immigrants or native-born.
T	F	7. The percentage of unauthorized immigrant workers in white-collar occupations has risen substantially in recent years.
T	F	8. Most undocumented immigrant workers settle in six states (California, New York, Florida, Texas, New Jersey, and Illinois).

Answers on page 622.

Demography: The Study of Population

Although population growth has slowed in the United States, the world's population of 6.8 billion in 2009 is increasing by more than 80.2 million people per year as a result of the larger number of births than deaths worldwide (see ▶ Figure 19.1). Between 2000 and 2030, almost all of the world's 1.4 percent annual population growth will occur in low-income countries in Africa, Asia, and Latin America (Population Reference Bureau, 2001).

What causes the population to grow rapidly in some nations? This question is of interest to scholars who specialize in the study of ***demography*—a subfield of sociology that examines population size, composition, and distribution.** Many sociological studies use demographic analysis as a component of the research design because all aspects of social life are affected by demography. For example, an important relationship exists between population size and the availability of food, water, energy, and housing. Population size, composition, and distribution are also connected to issues such as poverty, racial and ethnic diversity, shifts in the age structure of society, and concerns about environmental degradation (Weeks, 2008).

Increases or decreases in population can have a powerful impact on the social, economic, and political structures of societies. As used by demographers, a *population* is a group of people who live in a specified geographic area. Changes in populations occur as a result of three processes: fertility (births), mortality (deaths), and migration.

Fertility

***Fertility* is the actual level of childbearing for an individual or a population.** The level of fertility in a society is based on biological and social factors, the primary biological factor being the number of women of childbearing age (usually between ages 15 and 45). Other biological factors affecting fertility include the general health and level of nutrition of women of childbearing age. Social factors influencing the level of fertility include the roles available to women in a society and prevalent viewpoints regarding what constitutes the "ideal" family size.

Based on biological capability alone, most women could produce twenty or more children during their childbearing years. *Fecundity* is the potential number

demography a subfield of sociology that examines population size, composition, and distribution.

fertility the actual level of childbearing for an individual or a population.

Box 19.1 Sociology and Everyday Life

Answers to the Sociology Quiz on U.S. Immigration

1. False. Although the term "unauthorized immigrant" refers to a U.S. resident who is not a citizen of this country, who has not been admitted for permanent residence, or who does not have an authorized temporary status that permits longer-term residence and work, some "unauthorized immigrants" originally entered the country with valid visas but overstayed their visas' expiration or otherwise violated the terms of their admission.

2. True. Based on available estimates, about 3.1 million children, or 64 percent of all children living in unauthorized immigrant families, are U.S. citizens by birth.

3. True. It is estimated that about 58 percent of the unauthorized adult immigrant population is composed of males.

4. False. Unauthorized immigrants from Mexico and Latin America account for almost 80 percent of the unauthorized population in the United States. Most unauthorized immigrants come from Mexico, and the Pew Hispanic Center estimates that about 80 to 85 percent of all Mexican immigrants residing in the U.S. for fewer than 10 years are undocumented.

5. False. Unauthorized immigrants from South and East Asia made up about 13 percent of the undocumented population, whereas immigrants from Europe and Canada accounted for about 6 percent of all illegal U.S. immigration.

6. True. Undocumented male workers are often younger than legal immigrant or native-born males, and they are also less likely to attend college. As a result, a few years ago, 94 percent of unauthorized male immigrants were in the work force, as compared with 86 percent of male legal immigrants and 83 percent of native-born adult males. Today, these numbers may have changed due to economic conditions in the nation.

7. False. Unauthorized immigrant workers continue to be underrepresented in white-collar occupations such as management, business, and professional occupations and overrepresented in occupational categories (such as farming, cleaning, construction, and food preparations) that typically require less education and have no licensing requirement.

8. False. Although most immigrant workers previously settled in one of these six states, many more are moving to—or initially arriving at—destinations throughout various regions of the United States, including states such as North Carolina, Georgia, Nebraska, and Idaho.

Source: Based on Passel, 2006.

of children who could be born if every woman reproduced at her maximum biological capacity. Fertility rates are not as high as fecundity rates because people's biological capabilities are limited by social factors such as practicing voluntary abstinence and refraining from sexual intercourse until an older age, as well as by contraception, voluntary sterilization, abortion, and infanticide. Additional social factors affecting fertility include significant changes in the number of available partners for sex and/or marriage (as a result of war, for example), increases in the number of women of childbearing age in the work force, and high rates of unemployment. In some countries, governmental policies also affect the fertility rate. For example, China's two-decades-old policy of allowing only one child per family in order to limit population growth will result in that country's population starting to decline in 2042, according to United Nations projections (Beech, 2001).

The most basic measure of fertility is the ***crude birth rate*—the number of live births per 1,000 people in a population in a given year.** In 2008 the crude birth rate in the United States was almost 14 (13.82) per 1,000, as compared with an all-time high rate of 27 per 1,000 in 1947 (following World War II). This measure is referred to as a "crude" birth rate because it is based on the entire population and is not "refined" to incorporate significant variables affecting fertility, such as age, marital status, religion, and race/ethnicity.

In most areas of the world, women are having fewer children. Women who have six or seven children tend to live in agricultural regions of the world, where

◆ **Table 19.1 The Ten Leading Causes of Death in the United States, 1900 and 2006**

Cause of Death—1900	Rank	Cause of Death—2006
Influenza/pneumonia	1	Heart disease
Tuberculosis	2	Cancer
Stomach/intestinal disease	3	Stroke
Heart disease	4	Chronic lung disease
Cerebral hemorrhage	5	Accidents (unintentional injuries)
Kidney disease	6	Diabetes
Accidents	7	Alzheimer's disease
Cancer	8	Influenza and pneumonia
Diseases in early infancy	9	Nephritis, nephritic syndrome, and nephrosis
Diphtheria	10	Septicemia

Sources: Hostetler, 1994; Cockerham, 1995; and National Centers for Disease Control and Prevention, 2009.

children's labor is essential to the family's economic survival and child mortality rates are very high. For example, Uganda has a crude birth rate of almost 48 (47.8) per 1,000, as compared with 14 per 1,000 in the United States (U.S. Census Bureau, 2009). However, in Uganda and some other African nations, families need to have many children in order to ensure that one or two will live to adulthood due to high rates of poverty, malnutrition, and disease.

Mortality

The primary cause of world population growth in recent years has been a decline in ***mortality*—the incidence of death in a population.** The simplest measure of mortality is the ***crude death rate*—the number of deaths per 1,000 people in a population in a given year.** In 2005 the U.S. crude death rate was 8.3 per 1,000 (U.S. Census Bureau, 2009). In high-income, developed nations, mortality rates have declined dramatically as diseases such as malaria, polio, cholera, tetanus, typhoid, and measles have been virtually eliminated by vaccinations and improved sanitation and personal hygiene (Weeks, 2008). Just as smallpox appeared to be eradicated, however, HIV/AIDS rapidly rose to surpass the 30 percent fatality rate of smallpox. (The ten leading causes of death in the United States are shown in ◆ Table 19.1.) In low-income, less-developed nations, infectious diseases remain the leading cause of death; in some areas, mortality rates are rapidly increasing as a result of HIV/AIDS. Children under age 15 constitute a growing number of those who are infected with HIV.

But many children do not survive long enough to contract communicable diseases. On a global basis, large numbers of newborn infants do not live to see their first birthday. The measure of these deaths is referred to as the *infant mortality rate*, which is defined in Chapter 18 as the number of deaths of infants under 1 year of age per 1,000 live births in a given year. The infant mortality rate is an important reflection of a society's level of preventive (prenatal) medical care, maternal nutrition, childbirth procedures, and neonatal care for infants. Differential levels of access to these services are reflected in the divergent infant mortality rates for African Americans and whites. In 2005, for example, the U.S. mortality rate for white infants was 5.7 per 1,000 live births, as compared with 13.7 per 1,000 live births for African American infants (U.S. Census Bureau, 2009). This constituted a decrease in infant mortality rates for African Americans, which were at a high of 22.2 in 1980 and remained in the 14 to 15 category until 2004, when the rate dipped to 13.8 per 1,000 live births.

As discussed in Chapter 18, *life expectancy* is an estimate of the average lifetime in years of people born in a specific year. For persons born in the United States in 2008, for example, life expectancy at birth was 78.1 years, as compared with 82.1 years in Japan and 50 years or less in the African nations of Nigeria, Somalia, and South Africa (U.S. Census Bureau, 2009). Life expectancy varies by sex; for instance, females born in the United States in 2010 can expect to live almost 81 years as compared with 75.7 years for males. Life expectancy also varies by race. As we noted in Chapter

crude birth rate the number of live births per 1,000 people in a population in a given year.

mortality the incidence of death in a population.

crude death rate the number of deaths per 1,000 people in a population in a given year.

18, life expectancy for African American males born in 2010 is estimated at 70.2 years as compared to 77.2 for African American females and 75.7 years for white males (U.S. Census Bureau, 2009).

Migration

***Migration* is the movement of people from one geographic area to another for the purpose of changing residency.** Migration affects the size and distribution of the population in a given area. *Distribution* refers to the physical location of people throughout a geographic area. In the United States, people are not evenly distributed throughout the country; many of us live in densely populated areas. *Density* is the number of people living in a specific geographic area. In urbanized areas, density may be measured by the number of people who live per room, per block, or per square mile.

Migration may be either international (movement between two nations) or internal (movement within national boundaries). Internal migration has occurred throughout U.S. history and has significantly changed the distribution of the population over time.

Migration involves two types of movement: immigration and emigration. *Immigration* is the movement of people into a geographic area to take up residency. Each year, more than 1 million people enter the United States, primarily from Latin America and Asia. Over the last three decades, there has been a ten-fold increase in the number of adult Mexican immigrants living in the United States. Today, these immigrants alone account for more than 11 million to 12 million people in this country, and their children make up approximately 20 percent of the U.S. child population, with these rates being even higher in states such as Texas and California. Working with immigrant families and their children has become an important concern (see Box 19.2).

Immigration rates are not an accurate reflection of the actual number of immigrants who enter a country. The U.S. Immigration and Naturalization Service records only legal immigration based on entry visas and change-of-immigration-status forms. Similarly, few records are maintained regarding *emigration*—the movement of people out of a geographic area to take up residency elsewhere. To determine the net migration in a geographic area, the number of people leaving that area to take up permanent or semipermanent residence elsewhere (emigrants) is subtracted from the number of people entering that area to take up residence there (immigrants), unless more people are moving out of the area than into it, in which case the mathematical process is reversed.

People migrate either voluntarily or involuntarily. *Pull* factors at the international level, such as a democratic government, religious freedom, employment opportunities, or a more temperate climate, may draw voluntary immigrants into a nation. Within nations, people from large cities may be pulled to rural areas by lower crime rates, more space, and a lower cost of living. People such as Antonia Fuentes, whose decision to migrate to the United States is described at the beginning of this chapter, are drawn by pull factors such as greater economic opportunities at their destination and are pushed by factors such as low wages and few employment opportunities in their previous place of residence. *Push* factors at the international level, such as political unrest, violence, war, famine, plagues, and natural disasters, may encourage people to leave one area and relocate elsewhere. Push factors in regional U.S. migration include unemployment, harsh weather conditions, a high cost of living, inadequate school systems, and high crime rates.

Involuntary, or forced, migration usually occurs as a result of political oppression, such as when Jews fled Nazi Germany in the 1930s or when Afghans left their country to escape oppression there in the early 2000s. Slavery is the most striking example of involuntary migration; for example, the 10 million to 20 million

© Chris Hondros/Getty Images

© David Young-Wolff/PhotoEdit

Political unrest, violence, and war are "push" factors that encourage people to leave their country of origin. By contrast, job opportunities, such as construction work in the United States, are a major "pull" factor for people from low-income countries.

Box 19.2 You Can Make a Difference

Creating a Vital Link Between College Students and Immigrant Children

The program that I volunteered for is the Youth Tutoring Program. This program is designed to help refugee, immigrant, and underserved minority children get additional help on homework or class assignments. The program also offers activities to help children succeed in school. Its goal is to help the children develop effective problem-solving skills and create supportive peer networks.

My experience with the program and especially with the kids [has] been wonderful and truly rewarding. I love the children there, and I enjoy teaching them as well as building a relationship with them. . . . The students have taught me the importance of a strong and caring friendship.

—Sovanny That (2007), who was completing her doctor of pharmacy degree when she wrote these words, explaining why she feels so strongly about making a contribution through her volunteer work with immigrant and refugee children

Sovanny That's volunteer work took place with the Refugee Women's Alliance (ReWA), a nonprofit, multiethnic organization that provides services for newly arrived immigrant families in the state of Washington. Across the nation, volunteers provide much needed services for recent immigrants and their children, who may need assistance not only in learning about life in the United States but also in developing basic learning skills for school. Adjusting to a new environment is difficult for adults, but it can be even more challenging for young children who may not understand the reasons why their families have moved to faraway places.

Like Sovanny, thousands of other college students seek to make a difference in the lives of immigrant children through organized volunteer activities at their schools. At Northwestern University in Evanston, Illinois, for example, the Northwestern Community Development Corps (NCDC) is a student-run organization that engages students in community development in the greater Chicago area. According to the group's mission statement, "We promote civic engagement through direct service, social awareness, and advocacy" (NCDC, 2007). Among the projects offered by NCDC is the Centro Romero After School program, where college volunteers tutor seven- to twelve-year-old children from immigrant or refugee families in math, reading, and spelling. After the homework is completed, the volunteers enjoy playing games with the children. Tutoring programs for immigrant students exist nationwide in various public schools and colleges, and most college students who volunteer in these programs find their activities to be both rewarding and a learning experience as the children begin to thrive through their efforts.

Depending on the area of the country in which you live, children from immigrant and refugee families may trace their roots to Mexico, Central or South America, Somalia, Ethiopia, or Eastern Europe, among other places. By serving in such a program, volunteers may not only help the children gain a foothold in this country but may also help dispel a pervasive myth that immigrant students are a drain on the American educational system. Volunteers may help other people see that these students are a valuable resource in our schools because they provide firsthand knowledge of the larger world in which we live. As the sociologist Robert Crosnoe (2006) suggested in his recent book on how to help Mexican immigrant children succeed in school (and this would apply to other racial/ethnic groups as well), recent immigrants should be seen not as a *threat* but as a potential *resource* for our nation. Consequently, we should make every effort to see that the American Dream is a realistic goal, and not just an ideology, for new arrivals as well as for everyone else residing in this country.

Reflect & Analzye

Does your college or university have a volunteer coordinator who might provide you with the name of programs where you could put your skills to work helping immigrant students or other people who can benefit from your knowledge and expertise?

Africans forcibly transported to the Western Hemisphere prior to 1800 did not come by choice (see Chapter 10).

Population Composition

Changes in fertility, mortality, and migration affect the ***population composition*—the biological and social characteristics of a population, including age, sex, race, marital status, education, occupation, income, and size of household.**

migration the movement of people from one geographic area to another for the purpose of changing residency.

population composition the biological and social characteristics of a population, including age, sex, race, marital status, education, occupation, income, and size of household.

Photo Essay

Immigration and the Changing Face(s) of the United States

Throughout U.S. history, immigration has had a profound effect on our nation. Chances are very good that almost all of us can trace our heritage and our family roots to one or more other nations where our ancestors lived before coming to the United States. Immigration has also been a controversial topic at times throughout our nation's history, just as it is at the start of the twenty-first century.

When we look at the faces of the people around our country today, we see a wide diversity of human beings, most of whom are seeking to live their lives together positively and peacefully. Demographers and other social science researchers are interested in studying how people become part of the mainstream of a country to which they have immigrated while still maintaining their own unique cultural identity, and also why some of them do not want to become part of that mainstream.

As you view the pictures on these two pages, think about how you and other members of your family came to view yourselves as Americans—residents of the United States of America—and what this means to you and to them in terms of what you think and do on a daily basis. Doing so helps us gain a better understanding of some of the sociological issues relating to immigration.

At a march to show support for the Somali community in Lewiston, Maine, marchers urged people to "Love Thy Neighbor!" rather than to attempt to keep new arrivals from other countries from moving into the city.

© AP Images/Ann Heisenfelt

Immigrants and Educational Opportunities

For recent immigrants, education is a key that may open the door to greater involvement in one's new country. Some educational experiences may be long term (such as completing a college degree), whereas others serve more immediate needs, such as the acculturation and parenting class that these recent Hmong immigrants have completed to help them adjust to their new life in the United States.

Politics and Immigration

Voting and political participation help people have a voice in the U.S. democratic process. Here, four languages are used to encourage residents of South Boston to vote on "Super Tuesday" in 2008. How does politics influence our thinking about immigration? Does immigration influence our thinking about politics? Can our opinions change over time?

© AP Images/Charles Krupa

© Alexander Tamargo/Getty Images

Immigration and the Changing Face of the Media

As recent immigrants to the United States reach out to find media sources that reflect their culture and interests, executives at many media outlets strive to reach the large number of young people (typically between the ages of 18 and 24) who represent the future of the larger ethnic categories in this country. Latinos and Latinas, for example, are a key target audience for both Hispanic and so-called mainstream media. On the Spanish-language cable network program *Acceso Maximo*, for instance, viewers vote for their favorite videos and artists by text messaging or voting online.

© Tony Freeman/PhotoEdit

Immigrants and Employment

Many people often view the words *immigrants* and *workers* as being almost interchangeable because the primary purpose of much immigration is either to find work or for employers to have a larger pool of low-paid workers from which to hire new employees. Immigrant workers in the United States hold many jobs, ranging from agricultural and gardening positions to high-tech and health-care-related professions. As with many of our ancestors, these California landscape workers hope that their earnings will help their children have a more secure future in this country.

Reflect & Analyze

1. **When did your ancestors immigrate to the United States? Or, if you are a Native American, what is the history of your group? Do some research to find out about your ancestors.**
2. **Do you believe that non-English-speaking immigrants to this country should be expected—or required—to read and speak English at school, work, and other public places? Why or why not?**

Turning to Video

Watch the ABC video *Minutemen Patrol the Border* (run time 2:13), available on the Kendall Companion Website and through Cengage Learning eResources accounts. Following President George W. Bush's controversial efforts in 2005 to combat illegal immigration, this news report provides a look at the Minutemen, private citizens who usually live along U.S. borders and patrol for illegal immigrants, as well as communities where illegal immigration is an unexpected source of conflict. As you watch the video, think about the photographs, commentary, and questions you encountered in this photo essay. After you've watched the video, consider another question: To what degree is racism involved in U.S. citizens' anti-illegal-immigrant feelings and actions?

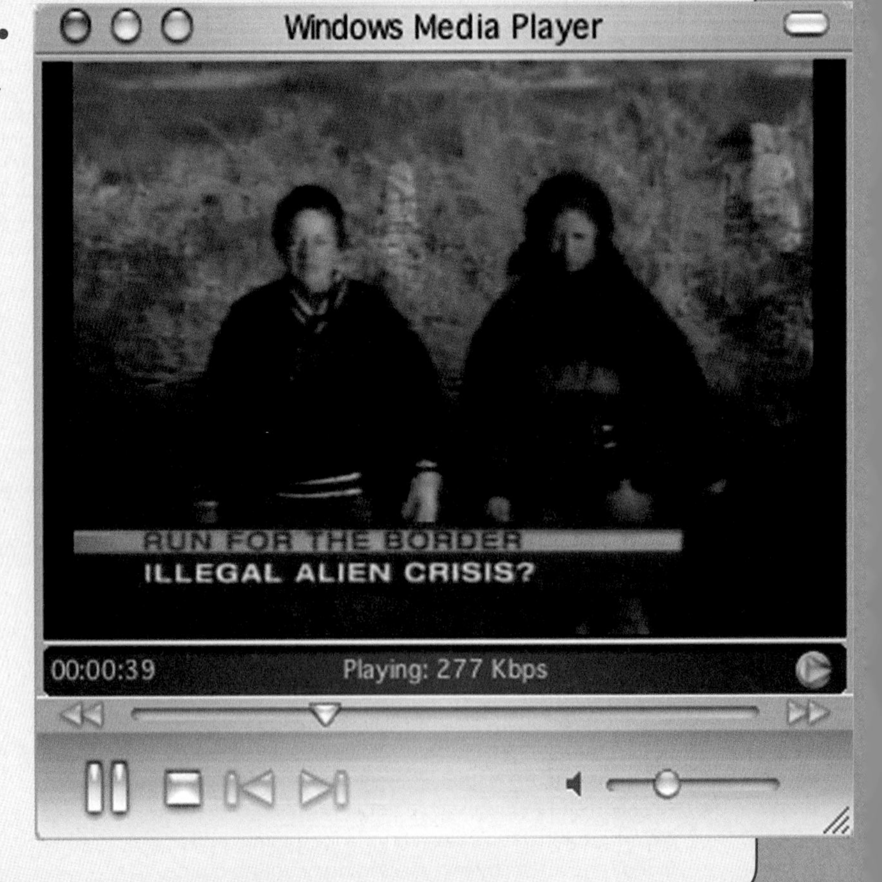

Sex Ratios of the U.S. Population Compared by Race and Ethnicity

According to the Census Bureau, the U.S. population in 2003 was made up of 147.7 million females (50.8 percent of the population) and 143.0 million males (49.2 percent of the population). Using this data, the Census Bureau computes the sex ratio (which it refers to as the male–female ratio) for various classifications of people by multiplying the number of males times 100, divided by the number of females.

When the sex ratios for various racial or ethnic categories of people are compared, some pronounced differences are evident, as shown at right. As you can see, the categories with the highest male–female ratios (more males than females) are Latino/a and Native Hawaiian or other Pacific Islander. To determine reasons for differences in the sex ratio, it would be important to take into account a number of factors, including the age of the people in the various categories. For example, up to age 24, the sex ratio of the U.S. population is about 105, reflecting the fact that more boys than girls are born every year and that boys continue to outnumber girls until the age 35 to 44 category, when the ratio slips to 98.9. More than 63 percent of the U.S. Latino/a population is younger than 35 years of age, compared to slightly less than 47 percent of the U.S. white population. However, more than 57 percent of the African American population in the United States is younger than age 35; if age were the only factor involved in these differences, the African American sex ratio would be between that for Latinos/as and whites. Because the African American male–female ratio is lower than that for the other categories shown, age is obviously not the only factor involved in these differences.

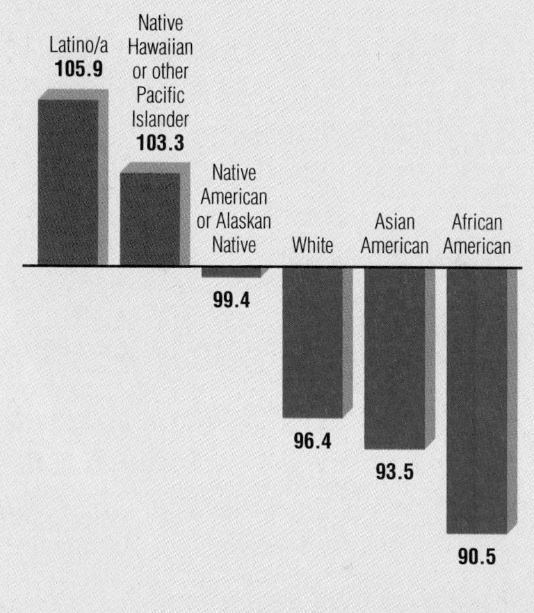

Source: National Centers for Disease Control, 2005.

One measure of population composition is the ***sex ratio*—the number of males for every hundred females in a given population.** A sex ratio of 100 indicates an equal number of males and females in the population. If the number is greater than 100, there are more males than females; if it is less than 100, there are more females than males. In the United States, the estimated sex ratio for 2009 was 97, which means there were 97 males per 100 females. Although approximately 124 males are conceived for every 100 females, male fetuses miscarry at a higher rate. From birth to age 14, the sex ratio is 1.04; in the age 15–64 category, however, the ratio shifts to 1.0, and from 65 upward, women outnumber men. By age 65, the sex ratio is about 75—that is, there are 75 men for every 100 women. As "Census Profiles: Sex Ratios of the U.S. Population Compared by Race and Ethnicity" demonstrates, the ratio of males to females varies among racial and ethnic categories in addition to varying by age.

For demographers, sex and age are significant population characteristics; they are key indicators of fertility and mortality rates. The age distribution of a population has a direct bearing on the demand for schooling, health, employment, housing, and pensions. The current distribution of a population can be depicted in a ***population pyramid*—a graphic representation of the distribution of a population by sex and age.** Population pyramids are a series of bar graphs divided into five-year age cohorts; the left side of the pyramid shows the number or percentage of males in each age bracket, and the right side provides the same information for females. The age/sex distribution in the United States and other high-income nations does not have the appearance of a classic pyramid, but rather is more rectangular or barrel-shaped. By contrast, low-income

sex ratio a term used by demographers to denote the number of males for every hundred females in a given population.

population pyramid a graphic representation of the distribution of a population by sex and age.

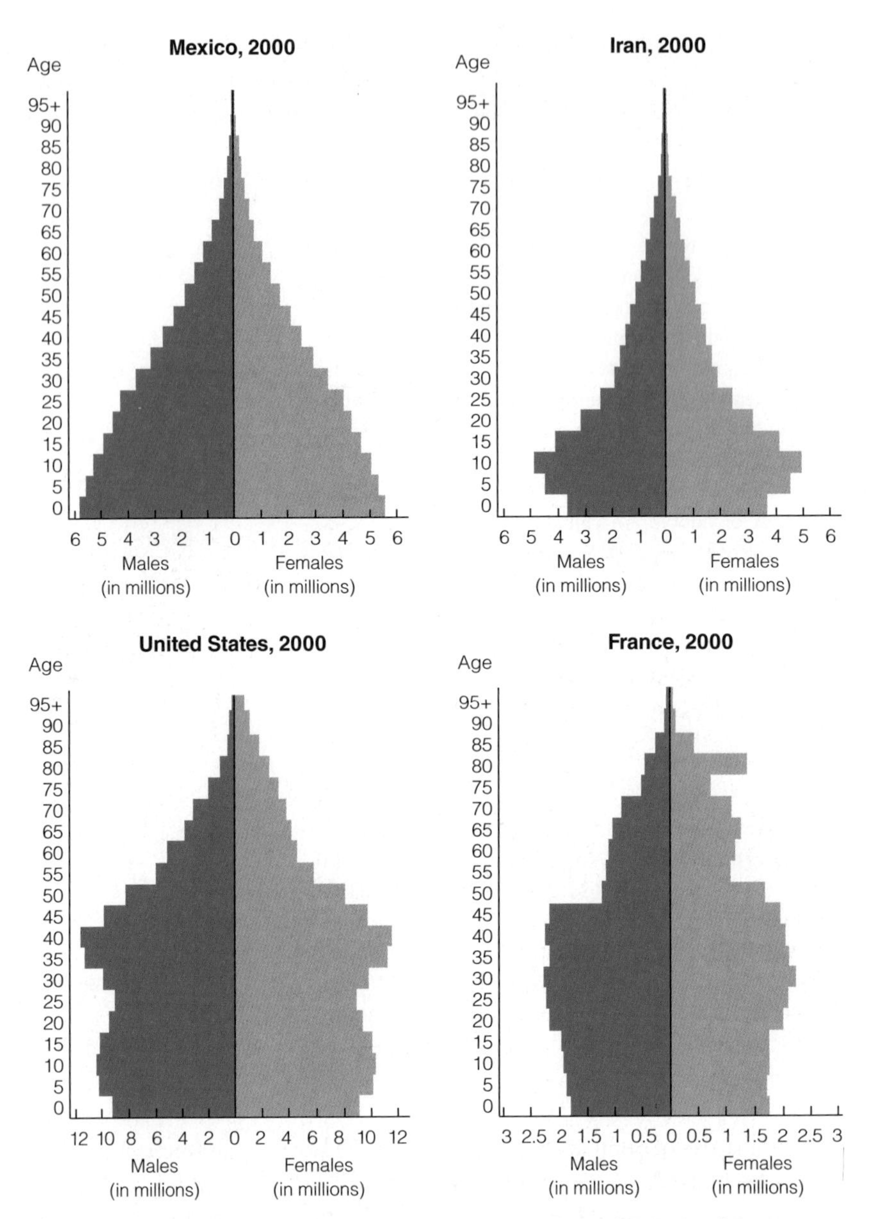

▶ **Figure 19.2 Population Pyramids for Mexico, Iran, the United States, and France**

Sources: Weeks, 2005; UNAIDS/WHO, 2000.

nations, such as Mexico and Iran, which have high fertility and mortality rates, do fit the classic population pyramid. ▶ Figure 19.2 compares the demographic composition of Mexico, Iran, the United States, and France.

Population Growth in Global Context

What are the consequences of global population growth? Scholars do not agree on the answer to this question. Some biologists have warned that Earth is a finite ecosystem that cannot support the 10 billion people predicted by 2050; however, some economists have emphasized that free-market capitalism is capable of developing innovative ways to solve such problems. The debate is not a new one; for several centuries, strong opinions have been voiced about the effects of population growth on human welfare.

The Malthusian Perspective

English clergyman and economist Thomas Robert Malthus (1766–1834) was one of the first scholars to systematically study the effects of population. Displeased with societal changes brought about by the Industrial Revolution in England, Malthus (1965/1798: 7) anonymously published *An Essay on the Principle of Popula-*

tion, As It Affects the Future Improvement of Society, in which he argued that "the power of population is infinitely greater than the power of the earth to produce subsistence [food] for man."

According to Malthus, the population, if left unchecked, would exceed the available food supply. He argued that the population would increase in a geometric (exponential) progression (2, 4, 8, 16 . . .) while the food supply would increase only by an arithmetic progression (1, 2, 3, 4 . . .). In other words, a *doubling effect* occurs: Two parents can have four children, sixteen grandchildren, and so on, but food production increases by only one acre at a time. Thus, population growth inevitably surpasses the food supply, and the lack of food ultimately ends population growth and perhaps eliminates the existing population (Weeks, 2008). Even in a best-case scenario, overpopulation results in poverty.

However, Malthus suggested that this disaster might be averted by either positive or preventive checks on population. *Positive checks* are mortality risks such as famine, disease, and war; *preventive checks* are limits to fertility. For Malthus, the only acceptable preventive check was *moral restraint;* people should practice sexual abstinence before marriage and postpone marriage as long as possible in order to have only a few children.

Malthus has had a lasting impact on the field of population studies. Most demographers refer to his dire predictions when they examine the relationship between fertility and subsistence needs. Overpopulation is still a daunting problem that capitalism and technological advances thus far have not solved, especially in middle- and low-income nations with rapidly growing populations and very limited resources.

The Marxist Perspective

Among those who attacked the ideas of Malthus were Karl Marx and Frederick Engels. According to Marx and Engels, the food supply is not threatened by overpopulation; technologically, it is possible to produce the food and other goods needed to meet the demands of a growing population. Marx and Engels viewed poverty as a consequence of the exploitation of workers by the owners of the means of production. For example, they argued that England had poverty because the capitalists skimmed off some of the workers' wages as profits. The labor of the working classes was used by capitalists to earn profits, which, in turn, were used to purchase machinery that could replace the workers rather than supply food for all.

From this perspective, overpopulation occurs because capitalists desire to have a surplus of workers (an industrial reserve army) so as to suppress wages and force workers concerned about losing their livelihoods to be more productive.

According to some contemporary economists, the greatest crisis today facing low-income nations is capital shortage, not food shortage. Through technological advances, agricultural production has reached the level at which it can meet the food needs of the world if food is distributed efficiently. Capital shortage refers to the lack of adequate money or property to maintain a business; it is a problem because the physical capital of the past no longer meets the needs of modern economic development. In the past, self-contained rural economies survived on local labor, using local materials to produce the capital needed for other laborers. For example, in a typical village a carpenter made the loom needed by the weaver to make cloth. Today, in the global economy, the one-to-one exchange between the carpenter and the weaver is lost. With an antiquated, locally made loom, the weaver cannot compete against electronically controlled, mass-produced looms. Therefore, the village must purchase capital from the outside, using its own meager financial resources. In the process, the complementary relationship between labor and capital is lost; modern technology brings with it steep costs and results in village noncompetitiveness and underemployment (see Keyfitz, 1994).

Marx and Engels made a significant contribution to the study of demography by suggesting that poverty, not overpopulation, is the most important issue with regard to food supply in a capitalist economy. Although Marx and Engels offer an interesting counterpoint to Malthus, some scholars argue that the Marxist perspective is self-limiting because it attributes the population problem solely to capitalism. In actuality, nations with socialist economies also have demographic trends similar to those of capitalist societies.

The Neo-Malthusian Perspective

More recently, *neo-Malthusians* (or "new Malthusians") have reemphasized the dangers of overpopulation. To neo-Malthusians, Earth is "a dying planet" with too many people and too little food, compounded by environmental degradation. Overpopulation and rapid population growth result in global environmental problems, ranging from global warming and rain-forest destruction to famine and vulnerability to epidemics (Ehrlich, Ehrlich, and Daily, 1995). Unless significant changes are made, including improving the status of women, reducing racism and religious prejudice, reforming the agriculture system, and shrinking the growing gap between rich and poor, the consequences will be dire (Ehrlich, Ehrlich, and Daily, 1995).

Early neo-Malthusians published birth control handbooks, and widespread acceptance of birth control eventually reduced the connection between people's sexual conduct and fertility (Weeks, 2008). Later neo-Malthusians have encouraged people to be part of the solution to the problem of overpopulation by having only one or two children in order to bring about ***zero population growth*—the point at which no population increase occurs from year to year** because the number of births plus immigrants is equal to the number of deaths plus emigrants (Weeks, 2008).

Demographic Transition Theory

Some scholars who disagree with the neo-Malthusian viewpoint suggest that the theory of demographic transition offers a more accurate picture of future population growth. ***Demographic transition* is the process by which some societies have moved from high birth and death rates to relatively low birth and death rates as a result of technological development.** Demographic transition is linked to four stages of economic development (see ▶ Figure 19.3):

- *Stage 1: Preindustrial societies.* Little population growth occurs because high birth rates are offset by high death rates. Food shortages, poor sanitation, and lack of adequate medical care contribute to high rates of infant and child mortality.
- *Stage 2: Early industrialization.* Significant population growth occurs because birth rates are relatively high whereas death rates decline. Improvements in health, sanitation, and nutrition produce a substantial decline in infant mortality rates. Overpopulation is likely to occur because more people are alive than the society has the ability to support.
- *Stage 3: Advanced industrialization and urbanization.* Very little population growth occurs because both birth rates and death rates are low. The birth rate declines as couples control their fertility through contraceptives and become less likely to adhere to religious directives against their use. Children are not viewed as an economic asset; they consume income rather than produce it. Societies in this stage attain zero population growth, but the actual number of births per year may still rise due to an increased number of women of childbearing age.
- *Stage 4: Postindustrialization.* Birth rates continue to decline as more women gain full-time employment and the cost of raising children continues to increase. The population grows very slowly, if at all, because the decrease in birth rates is coupled with a stable death rate.

Debate continues as to whether this evolutionary model accurately explains the stages of population growth in all societies. Advocates note that demographic transition theory highlights the relationship between technological development and population growth, thus making Malthus's predictions obsolete. Scholars also point out that demographic transitions occur at a faster rate in now-low-income nations than they previously did in the nations that are already developed. For example, nations in the process of development have higher birth rates and death rates than the now-developed societies did when they were going through the transition. The death rates declined in the now-developed nations as a result of internal economic development—not, as is the case today, through improved methods of disease control (Weeks, 2008). Critics suggest that this theory best explains development in Western societies.

Other Perspectives on Population Change

In recent decades, other scholars have continued to develop theories about how and why changes in population growth patterns occur. Some have studied the relationship between economic development and a decline in fertility; others have focused on the process of *secularization*—the decline in the significance of the sacred in daily life—and how a change from believing that otherworldly powers are responsible for one's life to a sense of responsibility for one's own well-being is linked to a

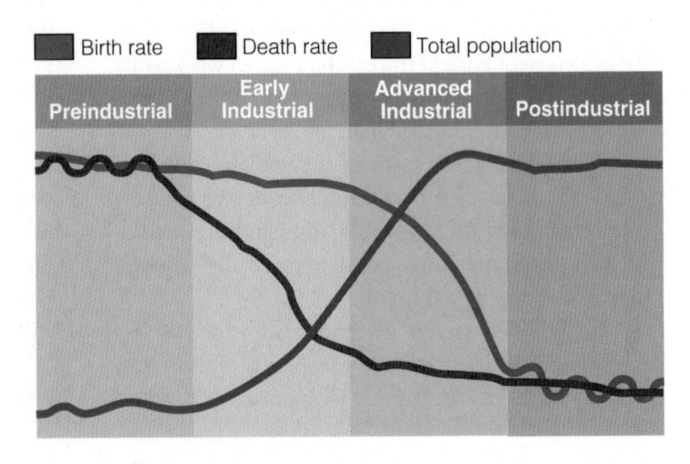

▶ **Figure 19.3 The Demographic Transition**

decline in fertility. Based on this premise, some analysts argue that the processes of industrialization and economic development are typically accompanied by secularization, but the relationship between these factors is complex when it comes to changes in fertility.

Shifting from the macrolevel to the microlevel, education and social psychological factors also play into the decisions that individuals make about how many children to have. Family planning information is more readily available to people with more years of formal education and may cause them to engage in decision making in accord with *rational choice theory,* which is based on the assumption that people make decisions based on a calculated cost–benefit analysis ("What do I gain and lose from a specific action?"). In low-income countries or other settings in which children are identified as an economic resource for their parents throughout life, fertility rates are higher than in higher-income countries. However, as modernization and urbanization occur in such societies, the positive economic effects of having more children may be offset by the cost of caring for those children and the lowered economic advantage gained from having children in an industrialized nation.

As demographers have reformulated demographic transition theory, they have highlighted additional factors that are likely to be causes of fertility decline, and they have suggested that demographic transition is not just one process, but rather a set of intertwined transitions. One is the epidemiological transition—the shift from deaths at younger ages due to acute, communicable diseases. Another is the fertility transition—the shift from natural fertility to controlled fertility, resulting in a decrease in the fertility rate. Other transitions include the migration transition, the urban transition, the age transition, and the family and household transition, which occur as a result of lower fertility, longer life, an older age structure, and predominantly urban residence.

As the demographer John R. Weeks (2008) points out, we can best understand demographic events and behavior by studying the context of global change to determine how factors such as political change, economic development, and perhaps the process of "westernization" may influence population growth and patterns of migration.

A Brief Glimpse at International Migration Theories

Why do people relocate from one nation to another? Several major theories have been developed in an attempt to explain international migration. The *neoclassical economic approach* assumes that migration patterns occur based on geographic differences in the supply of and demand for labor. The United States and other high-income countries that have had growing economies and a limited supply of workers for certain types of jobs have paid higher wages than are available in areas with a less-developed economy and a large labor force. As a result, people move to gain higher wages and sometimes better living conditions. They may also take jobs in other countries so that they can send money to their families in their country of origin (see Box 19.3). For example, it is estimated that Mexican workers in the United States send about half of what they earn to their families across the border, an amount that may total nearly $7 billion per year during good economic times (Ferriss, 2001).

Unlike the neoclassical explanation of migration, which focuses on individual decision making, the *new households economics of migration approach* emphasizes the part that entire families or households play in the migration process. From this approach, the previous example of Mexican workers' temporary migration to the United States would be examined not only from the perspective of the individual worker but also in terms of what the entire family gains from the process of having one or more migrant family members work in another country. By having a diversity of family income (originating from more than one source), the family is cushioned from the economic woes of the nation that most of the family members think of as "home."

Two conflict perspectives on migration add to our knowledge of why people migrate. Split-labor-market theory (as previously discussed in Chapter 10) suggests that immigrants from low-income countries are often recruited for secondary labor market positions: dead-end jobs with low wages, unstable employment, and sometimes hazardous working conditions. By contrast, migrants from higher-income countries may migrate for primary-sector employment—jobs in which well-educated workers are paid high wages and receive benefits such as health insurance and a retirement plan. The global migration of some high-tech workers is an example of this process, whereas the migration of farmworkers and construction helpers is an example of secondary labor market migration.

zero population growth the point at which no population increase occurs from year to year.

demographic transition the process by which some societies have moved from high birth rates and death rates to relatively low birth rates and death rates as a result of technological development.

Box 19.3 Framing Immigration in the Media

Media Framing and Public Opinion

Inside his little Western wear store tucked in a corner of East Riverside Drive, Francisco Javier Aceves can't help but feel a kinship with the angular young men who come in to buy jeans, cowboy boots, phone cards and cell phones. As sure as regular payday, they come in also to wire money to their families back home in Mexico, in places such as Veracruz, Tabasco, Chiapas and Oaxaca.

"Sometimes they come three or four in a car," Aceves said about his customers. "Sometimes they just start lining up to wire money. . . ." The flow of money repeats and repeats, affirming emotional bonds between people separated by a border. . . . It is part of a growing phenomenon. Every month, Mexican immigrants working in the United States in mostly low-paying jobs send more than $1 billion to their families back home, more than Mexico earns from tourism or foreign investment. (Castillo, 2003: J1)

This newspaper article about Mexican immigrants who send money to their families in Mexico is an example of sympathetic framing by the news media. *Sympathetic framing* refers to news writing that focuses on the human interest side of a story and shows that the individuals involved are caring people who are representative of a larger population.

In stark contrast to sympathetic framing of such stories are those news reports that employ negative framing to describe recent immigrants from countries such as Mexico. *Negative framing* describes immigrants as nothing more than cheap labor that benefits employers in this country who do not believe in paying a living wage. Articles focusing on the problematic aspects of illegal immigrants or of "guest worker" programs usually emphasize that such workers suppress the wages of other low-income employees because they are willing to work for less money. Negative framing also emphasizes that these workers bring in (or give birth to) millions of children who speak little English, contributing to the decline of public education.

Negative framing of the issue of immigrant labor is not new in the United States. For many years, "immigrant, foreign labor" has been described as a threat to the livelihood of other workers and as a menace to public safety. In 1904 the *San Francisco Chronicle* carried lengthy articles describing how Japanese laborers were taking jobs away from U.S. workers, reflecting a pattern of media reporting that continues to be a topic in the twenty-first century (Puette, 1992).

In the early 2000s, when former President George W. Bush introduced his "Guest Worker Proposal," both sympathetic and negative media framing of immigrant workers ensued. Some articles emphasized the importance of immigrant workers' contributions to the United States. Other reports highlighted the problematic aspects of guest worker programs, arguing that immigration undermines the future of the country.

How the media frame stories about immigrant workers may influence social policy. Will we close our borders to immigration? Will we develop new programs to allow limited entry of immigrant workers? Not only are these important legal and social policy questions, but they are also issues that journalists and other news analysts must face as they frame their stories on immigrants and the billions of dollars that are sent to other countries in exchange for the labor of these workers.

Perhaps the American Dream now transcends the borders of this country. Francisco Javier Aceves (the store proprietor) described to one reporter how the young men who came into his shop have their own dreams of making money, building a home in Mexico one piece at a time, and then going back: "They tell me this. They'll say, 'With this, I'm going to put the roof on my house.'" He paused, smiling slowly, "They're excited!" (Castillo, 2003: J4).

Reflect & Analyze

Since the recent decline in the U.S. housing market and the economic crisis in this country, the number of residential construction jobs has been drastically reduced. Would these problems have any effect on the "pull" factors that attract immigrants to this country? How might "push" factors still be important in determining if and when people immigrate to the United States?

Finally, world systems theory (discussed later in this chapter) views migration as linked to the problems caused by capitalist development around the world (Massey et al., 1993). As the natural resources, land, and work force in low-income countries with little or no industrialization have come under the influence of international markets, there has been a corresponding flow of migrants from those nations to the highly industrialized, high-income countries, especially those with which the poorer nations have had the most economic, political, or military contact.

After flows of migration commence, the pattern may continue because potential migrants have personal ties with relatives and friends who now live in the country of destination and can serve as a source of stability when the potential migrants relocate to the new country. Known as *network theory,* this approach suggests that once migration has commenced, it takes on a life of its own and that the migration pattern which ensues may be different from the original *push* or *pull* factors that produced the earlier migration. Another approach, *institutional theory,* suggests that migration may be fostered by groups—such as humanitarian aid organizations relocating refugees or smugglers bringing people into a country illegally—and that the actions of these groups may produce a larger stream of migrants than would otherwise be the case.

As you can see from these diverse approaches to explaining contemporary patterns of migration, the reasons that people migrate are numerous and complex, involving processes occurring at the individual, family, and societal levels.

Urbanization in Global Perspective

Urban sociology is a subfield of sociology that examines social relationships and political and economic structures in the city. According to urban sociologists, a *city* is a relatively dense and permanent settlement of people who secure their livelihood primarily through nonagricultural activities. Although cities have existed for thousands of years, only about 3 percent of the world's population lived in cities 200 years ago, as compared with almost 50 percent today. Current estimates suggest that two out of every three people around the world will live in urban areas by 2050 (United Nations, 2000). Thus, the process of urbanization continues on a global basis.

Emergence and Evolution of the City

Cities are a relatively recent innovation when compared with the length of human existence. The earliest humans are believed to have emerged anywhere from 40,000 to 1,000,000 years ago, and permanent human settlements are believed to have begun first about 8000 B.C.E. However, some scholars date the development of the first city between 3500 and 3100 B.C.E., depending largely on whether a formal writing system is considered as a requisite for city life (Sjoberg, 1965; Weeks, 2008; Flanagan, 2002).

According to the sociologist Gideon Sjoberg (1965), three preconditions must be present in order for a city to develop:

An increasing proportion of the world's population lives in cities. How is this scene in Lagos, Nigeria, similar to and different from major U.S. cities?

1. *A favorable physical environment,* including climate and soil conducive to the development of plant and animal life and an adequate water supply to sustain both.
2. *An advanced technology* (for that era) that could produce a social surplus in both agricultural and nonagricultural goods.
3. *A well-developed social organization,* including a power structure, in order to provide social stability to the economic system.

Based on these prerequisites, Sjoberg places the first cities in the Middle Eastern region of Mesopotamia or in areas immediately adjacent to it at about 3500 B.C.E. However, not all scholars concur; some place the earliest city in Jericho (located in present-day Jordan) at about 8000 B.C.E., with a population of about 600 people (see Kenyon, 1957).

The earliest cities were not large by today's standards. The population of the larger Mesopotamian centers was between 5,000 and 10,000 (Sjoberg, 1965). The population of ancient Babylon (probably founded around 2200 B.C.E.) may have grown as large as 50,000 people; Athens may have held 80,000 people (Weeks,

2008). Four to five thousand years ago, cities with at least 50,000 people existed in the Middle East (in what today is Iraq and Egypt) and Asia (in what today is Pakistan and China), as well as in Europe. About 3,500 years ago, cities began to reach this size in Central and South America.

Preindustrial Cities

The largest preindustrial city was Rome; by 100 C.E., it may have had a population of 650,000 (Chandler and Fox, 1974). With the fall of the Roman Empire in 476 C.E., the nature of European cities changed. Seeking protection and survival, those persons who lived in urban settings typically did so in walled cities containing no more than 25,000 people. For the next 600 years, the urban population continued to live in walled enclaves, as competing warlords battled for power and territory during the "dark ages." Slowly, as trade increased, cities began to tear down their walls.

Preindustrial cities were limited in size by a number of factors. For one thing, crowded housing conditions and a lack of adequate sewage facilities increased the hazards from plagues and fires, and death rates were high. For another, food supplies were limited. In order to generate food for each city resident, at least fifty farmers had to work in the fields (Davis, 1949), and animal power was the only means of bringing food to the city. Once foodstuffs arrived in the city, there was no effective way to preserve them. Finally, migration to the city was difficult. Many people were in serf, slave, and caste systems whereby they were bound to the land. Those able to escape such restrictions still faced several weeks of travel to reach the city, thus making it physically and financially impossible for many people to become city dwellers.

In spite of these problems, many preindustrial cities had a sense of *community*—a set of social relationships operating within given spatial boundaries or locations that provides people with a sense of identity and a feeling of belonging. The cities were full of people from all walks of life, both rich and poor, and they felt a high degree of social integration. You will recall that Ferdinand Tönnies (1940/1887) described such a community as *Gemeinschaft*—a society in which social relationships are based on personal bonds of friendship and kinship and on intergenerational stability, such that people have a commitment to the entire group and feel a sense of togetherness. By contrast, industrial cities were characterized by Tönnies as *Gesellschaft*—societies exhibiting impersonal and specialized relationships, with little long-term commitment to the group or consensus on values (see Chapter 5). In *Gesellschaft* societies, even neighbors are "strangers" who perceive that they have little in common with one another.

During the industrial era, people not only moved from the countryside into cities, but some people also moved from the cities to the suburbs after transportation became available to make getting from home to work and back again an easier process.

Industrial Cities

The Industrial Revolution changed the nature of the city. Factories sprang up rapidly as production shifted from the primary, agricultural sector to the secondary, manufacturing sector. With the advent of factories came many new employment opportunities not available to people in rural areas. Emergent technology, including new forms of transportation and agricultural production, made it easier for people to leave the countryside and move to the city. Between 1700 and 1900, the population of many European cities mushroomed. For example, the population of London increased from 550,000 to almost 6.5 million. Although the Industrial Revolution did not start in the United States until the mid-nineteenth century, the effect was similar. Between 1870 and 1910, for example, the population of New York City grew by 500 percent. In fact, New York City became the first U.S. *metropolis*—one or more central cities and their surrounding suburbs that dominate the economic and cultural life of a region. Nations, such as Japan and Russia, that became industrialized after England and the United States experienced a delayed pattern of urbanization, but this process moved quickly once it commenced in those countries.

Postindustrial Cities

Since the 1950s, postindustrial cities have emerged in nations such as the United States as their economies have gradually shifted from secondary (manufacturing) production to tertiary (service and information-processing) production. Postindustrial cities increasingly rely on an economic structure that is based on scientific knowledge rather than industrial production, and as a result, a class of professionals and technicians grows in size and influence. Postindustrial cities are dominated by "light" industry, such as software manufacturing; information-processing services, such as airline and hotel reservation services; educational complexes; medical centers; convention and entertainment centers; and retail trade centers and shopping malls. Most families do not live close to a central business district. Technological advances in communication and transportation make it possible for middle- and upper-income individuals and families to have more work options and to live greater distances from the workplace; however, these options are not often available to people of color and those at the lower end of the class structure.

On a global basis, cities such as New York, London, and Tokyo appear to fit the model of the postindustrial city (see Sassen, 2001). These cities have experienced a rapid growth in knowledge-based industries such as financial services. London, Tokyo, and New York have—at least until recently—experienced an increase in the number of highly paid professional jobs, and more workers have been in high-income categories. Many people have benefited for a number of years from these high incomes and have created a lifestyle that is based on materialism and the gentrification of urban spaces. Meanwhile, those persons outside the growing professional categories have seen their own quality of life further deteriorate and their job opportunities become increasingly restricted to secondary labor markets in their respective "global" cities.

Despite an increase in telecommuting and more diverse employment opportunities in the high-tech economy, our highways have grown increasingly congested. Can we implement measures to reduce the problems of urban congestion and environmental pollution, or will these problems grow worse with each passing year?

Perspectives on Urbanization and the Growth of Cities

Urban sociology follows in the tradition of early European sociological perspectives that compared social life with biological organisms or ecological processes. For example, Auguste Comte pointed out that cities are the "real organs" that make a society function. Emile Durkheim applied natural ecology to his analysis of *mechanical solidarity,* characterized by a simple division of labor and shared religious beliefs such as are found in small, agrarian societies, and *organic solidarity,* characterized by interdependence based on the elaborate division of labor found in large, urban societies (see Chapter 5). These early analyses became the foundation for ecological models/functionalist perspectives in urban sociology.

Functionalist Perspectives: Ecological Models

Functionalists examine the interrelations among the parts that make up the whole; therefore, in studying the growth of cities, they emphasize the life cycle of urban growth. Like the social philosophers and sociologists before him, the University of Chicago sociologist Robert Park (1915) based his analysis of the city on *human ecology*—the study of the relationship between people and their physical environment. According to Park (1936), economic competition produces certain regularities in land-use patterns and population distributions. Applying Park's idea to the study of urban land-use patterns, the sociologist Ernest W. Burgess (1925) developed the concentric zone model, an ideal construct that attempted to explain why some cities expand radially from a central business core.

The Concentric Zone Model Burgess's *concentric zone model* is a description of the process of urban growth that views the city as a series of circular areas or zones, each characterized by a different type of land use, that developed from a central core (see ▶ Figure 19.4a). *Zone 1* is the central business district and cultural center. In *Zone 2,* houses formerly occupied by wealthy families are divided into rooms and rented to

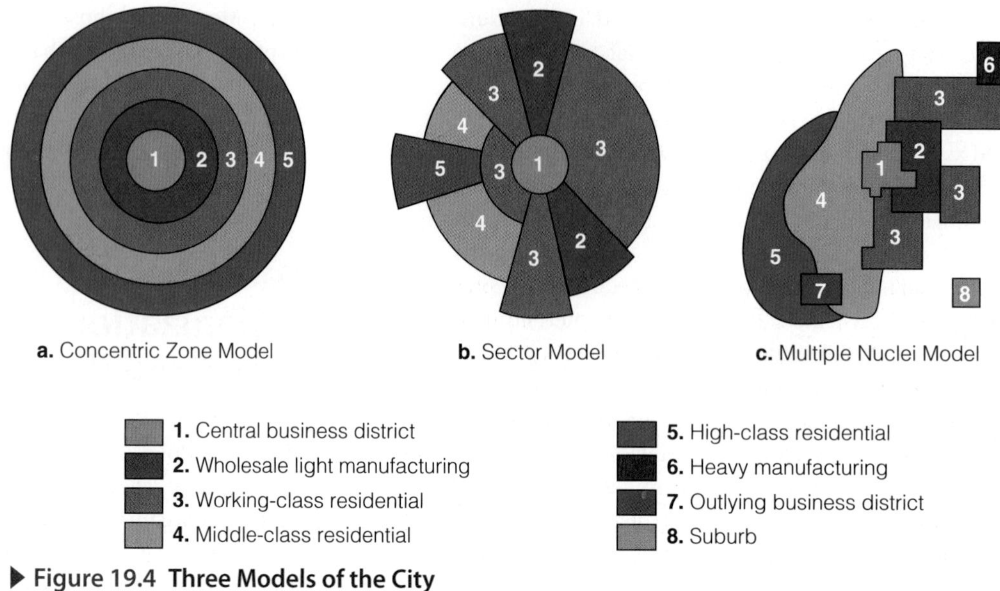

Figure 19.4 Three Models of the City

Source: Adapted from Harris and Ullman, 1945.

recent immigrants and poor persons; this zone also contains light manufacturing and marginal businesses (such as secondhand stores, pawnshops, and taverns). *Zone 3* contains working-class residences and shops and ethnic enclaves. *Zone 4* comprises homes for affluent families, single-family residences of white-collar workers, and shopping centers. *Zone 5* is a ring of small cities and towns populated by persons who commute to the central city to work and by wealthy people living on estates.

Two important ecological processes are involved in the concentric zone theory: invasion and succession. ***Invasion* is the process by which a new category of people or type of land use arrives in an area previously occupied by another group or type of land use** (McKenzie, 1925). For example, Burgess noted that recent immigrants and low-income individuals "invaded" Zone 2, formerly occupied by wealthy families. ***Succession* is the process by which a new category of people or type of land use gradually predominates in an area formerly dominated by another group or activity** (McKenzie, 1925). In Zone 2, for example, when some of the single-family residences were sold and subsequently divided into multiple housing units, the remaining single-family owners moved out because the "old" neighborhood had changed. As a result of their move, the process of invasion was complete and succession had occurred.

Invasion and succession theoretically operate in an outward movement: Those who are unable to "move out" of the inner rings are those without upward social mobility, so the central zone ends up being primarily occupied by the poorest residents—except when gentrification occurs. ***Gentrification* is the process by which members of the middle and upper-middle classes, especially whites, move into the central-city area and renovate existing properties.** Centrally located, naturally attractive areas are the most likely candidates for gentrification. To urban ecologists, gentrification is the solution to revitalizing the central city. To conflict theorists, however, gentrification creates additional hardships for the poor by depleting the amount of affordable housing available and by "pushing" them out of the area (Flanagan, 2002).

The concentric zone model demonstrates how economic and political forces play an important part in the location of groups and activities, and it shows how a large urban area can have internal differentiation (Gottdiener, 1985). However, the model is most applicable to older cities that experienced high levels of immigration early in the twentieth century and to a few midwestern cities such as St. Louis (Queen and Carpenter, 1953). No city, including Chicago (on which the model is based), entirely conforms to this model.

The Sector Model In an attempt to examine a wider range of settings, urban ecologist Homer Hoyt (1939) studied the configuration of 142 cities. Hoyt's *sector model* emphasizes the significance of terrain and the importance of transportation routes in the layout of cities. According to Hoyt, residences of a particular type and value tend to grow outward from the center of the city in wedge-shaped sectors, with the more-expensive residential neighborhoods located along the higher ground near lakes and rivers or along certain streets that stretch in one direction or another from the downtown area (see Figure 19.4b). By contrast, industrial areas tend to be located along river valleys and railroad lines. Middle-class residential zones exist on either side of the wealthier neighborhoods. Finally, lower-class residential areas occupy the remaining space, bordering the central business area and the industrial areas. Hoyt (1939) concluded that the sector model applied to cities such as Seattle, Minneapolis, San Francisco, Charleston (South Carolina), and Richmond (Virginia).

The Multiple Nuclei Model According to the *multiple nuclei model* developed by urban ecologists Chauncey Harris and Edward Ullman (1945), cities do not have one center from which all growth radiates, but rather have numerous centers of development based on specific urban needs or activities (see Figure 19.4c). As cities began to grow rapidly, they annexed formerly outlying and independent townships that had been communities in their own right. In addition to the central business district, other nuclei developed around entities such as an educational institution, a medical complex, or a government center. Residential neighborhoods may exist close to or far away from these nuclei. A wealthy residential enclave may be located near a high-priced shopping center, for instance, whereas less-expensive housing must locate closer to industrial and transitional areas of town. This model may be applicable to cities such as Boston. However, critics suggest that it does not provide insights about the uniformity of land-use patterns among cities and relies on an after-the-fact explanation of why certain entities are located where they are (Flanagan, 2002).

Contemporary Urban Ecology Urban ecologist Amos Hawley (1950) revitalized the ecological tradition by linking it more closely with functionalism. According to Hawley, urban areas are complex and expanding social systems in which growth patterns are based on advances in transportation and communication. For example, commuter railways and automobiles led to the decentralization of city life and the movement of industry from the central city to the suburbs (Hawley, 1981).

Other urban ecologists have continued to refine the methodology used to study the urban environment. *Social area analysis* examines urban populations in terms of economic status, family status, and ethnic classification (Shevky and Bell, 1966). For example, middle- and upper-middle-class parents with school-aged children tend to cluster together in "social areas" with a "good" school district; young single professionals may prefer to cluster in the central city for entertainment and nightlife.

The influence of human ecology on the field of urban sociology is still very strong today (see Frisbie and Kasarda, 1988). Contemporary research on European and North American urban patterns is often based on the assumption that spatial arrangements in cities conform to a common, most efficient design (Flanagan, 2002). However, some critics have noted that ecological models do not take into account the influence of powerful political and economic elites on the development process in urban areas (Feagin and Parker, 1990).

Conflict Perspectives: Political Economy Models

Conflict theorists argue that cities do not grow or decline by chance. Rather, they are the product of specific decisions made by members of the capitalist class and political elites. These far-reaching decisions regarding land use and urban development benefit the members of some groups at the expense of others (see Castells, 1977/1972). Karl Marx suggested that cities are the arenas in which the intertwined processes of class conflict and capital accumulation take place; class consciousness and worker revolt are more likely to develop when workers are concentrated in urban areas (Flanagan, 2002).

According to the sociologists Joe R. Feagin and Robert Parker (1990), three major themes prevail in political economy models of urban growth. First, both economic *and* political factors affect patterns of urban growth and decline. Economic factors include capitalistic investments in production, workers, workplaces, land, and buildings. Political factors include governmental protection of the right to own and dispose of privately held property as owners see fit and the role of government officials in promoting the interests of business elites and large corporations.

Second, urban space has both an exchange value and a use value. *Exchange value* refers to the profits that industrialists, developers, bankers, and others make from buying, selling, and developing land and buildings. By contrast, *use value* is the utility of space, land, and buildings for everyday life, family life, and neighborhood life. In other words, land has purposes

invasion the process by which a new category of people or type of land use arrives in an area previously occupied by another group or land use.

succession the process by which a new category of people or type of land use gradually predominates in an area formerly dominated by another group or activity.

gentrification the process by which members of the middle and upper-middle classes, especially whites, move into a central-city area and renovate existing properties.

▶ **Figure 19.5 The Value of Urban Space**

other than simply for generating profits—for example, for homes, open spaces, and recreational areas (see ▶ Figure 19.5) Today, class conflict exists over the use of urban space, as is evident in battles over rental costs, safety, and development of large-scale projects (see Tabb and Sawers, 1984).

Third, both structure and agency are important in understanding how urban development takes place. *Structure* refers to institutions such as state bureaucracies and capital investment circuits that are involved in the urban development process. *Agency* refers to human actors, including developers, business elites, and activists protesting development, who are involved in decisions about land use.

According to conflict theorists, exploitation by the capitalist class increasingly impoverishes poor whites and low-income subordinate-group members. Increasing rates of homelessness have made scenes such as this a recurring sight in many cities.

Capitalism and Urban Growth in the United States

According to political economy models, urban growth is influenced by capital investment decisions, power and resource inequality, class and class conflict, and government subsidy programs. Members of the capitalist class choose corporate locations, decide on sites for shopping centers and factories, and spread the population that can afford to purchase homes into sprawling suburbs located exactly where the capitalists think they should be located (Feagin and Parker, 1990).

Today, a few hundred financial institutions and developers finance and construct most major and many smaller urban development projects around the country, including skyscrapers, shopping malls, and suburban housing projects. These decision makers set limits on the individual choices of the ordinary citizen with regard to real estate, just as they do with regard to other choices (Feagin and Parker, 1990). They can make housing more affordable or totally unaffordable for many people. Ultimately, their motivation rests not in benefiting the community, but rather in making a profit; the cities that they produce reflect this mindset.

One of the major results of these urban development practices is *uneven development*—the tendency of some neighborhoods, cities, or regions to grow and prosper whereas others stagnate and decline (Perry and Watkins, 1977). Conflict theorists argue that uneven development reflects inequalities of wealth and power in society. The problem not only affects areas in a state of decline but also produces external costs, even in "boom" areas, that are paid by the entire community. Among these costs are increased pollution, increased traffic congestion, and rising rates of crime and violence. According to the sociologist Mark Gottdiener (1985: 214), these costs are "intrinsic to the very core of capitalism, and those who profit the most from development are not called upon to remedy its side effects."

The Gated Community in the Capitalist Economy The growth of *gated communities*—subdivisions or neighborhoods surrounded by barriers such as walls, fences, gates, or earth banks covered with bushes and shrubs, along with a secured entrance—is an example to many people of how developers, builders, and municipalities have encouraged an increasing division between public and private property in capitalist societies. Many gated communities are created by developers who hope to increase their profits by offering potential residents a semblance of safety, privacy, and luxury that they might not have in nongated residential areas. Other gated communities have been developed after the fact in established neighborhoods by adding walls, gates, and sometimes security guard stations. For example, a recent study noted situations in which residents of elite residential enclaves, such as the River Oaks area of Houston or the "Old Enfield" area of Austin, Texas, were able to gain approval from the city to close certain streets and create cul-de-sacs, or to erect other barriers to discourage or prevent outsiders from driving through the neighborhood (Kendall, 2002). Gated communities for upper-middle-class and upper-class residents convey the idea of exclusivity and privilege, whereas such communities for middle- and lower-income residents typically focus on such features as safety for children and the ability to share amenities such as a "community" swimming pool or recreational center with other residents.

Regardless of the social and economic reasons given for the development of gated communities, many analysts agree that these communities reflect a growing divide between public and private space in urban areas. According to a recent qualitative study by the anthropologist Setha Low (2003), gated communities do more than simply restrict access to the residents' homes: They also limit the use of public spaces, making it impossible for others to use the roads, parks, and open space contained within the enclosed community. Low (2003) refers to this phenomenon as the "fortressing of America."

Gender Regimes in Cities Feminist perspectives have only recently been incorporated in urban studies (Garber and Turner, 1995). From this perspective, urbanization reflects the workings not only of the political economy but also of patriarchy. According to the sociologist Lynn M. Appleton (1995), different kinds of cities have different *gender regimes*—prevailing ideologies of how women and men should think, feel, and act; how access to social positions and control of resources should be managed; and how relationships between men and women should be conducted. The higher density and greater diversity found in central cities such as New York City serve as a challenge to the private patriarchy found in the home and workplace in lower-density, homogeneous areas such as suburbs and rural areas. *Private patriarchy* is based on a strongly gendered division of labor in the home, gender-segregated paid employment, and women's dependence on men's income. At the same time, cities may foster *public patriarchy* in the form of women's increasing dependence on paid work and the state for income and their decreasing emotional interdependence with men. At this point, gender often intersects with class and race as a form of oppression because lower-income women of color often reside in central cities. Public patriarchy may be perpetuated by cities through policies that limit women's access to paid work and public transportation. However, such cities may also be a forum for challenging patriarchy; all residents who differ in marital status, paternity, sexual orientation, class, and/or race/ethnicity tend to live close to one another and may hold a common belief that both public and private patriarchy should be eliminated (Appleton, 1995).

Symbolic Interactionist Perspectives: The Experience of City Life

Symbolic interactionists examine the *experience* of urban life. How does city life affect the people who live in a city? Some analysts answer this question positively; others are cynical about the effects of urban living on the individual.

Simmel's View of City Life According to the German sociologist Georg Simmel (1950/1902–1917), urban life is highly stimulating, and it shapes people's thoughts and actions. Urban residents are influenced by the quick pace of the city and the pervasiveness of economic relations in everyday life. Due to the intensity of urban life, people have no choice but to become somewhat insensitive to events and individuals around them. Many urban residents avoid emotional involvement with one another and try to ignore events taking place around them. Urbanites feel wary toward other people because most interactions in the city are economic rather than social. Simmel suggests that attributes such as punctuality and exactness are rewarded but that friendliness and warmth in interpersonal relations are viewed as personal weaknesses. Some people act reserved to cloak their deeper feelings of distrust or dislike toward others. However, Simmel did not view city life as completely negative; he also pointed out that urban living could have a liberating effect on people because they had opportunities for individualism and autonomy (Flanagan, 2002).

Urbanism as a Way of Life Based on Simmel's observations on social relations in the city, the early Chicago School sociologist Louis Wirth (1938) suggested that urbanism is a "way of life." *Urbanism* refers to the distinctive social and psychological patterns of life typically found in the city. According to Wirth, the size, density, and heterogeneity of urban populations typically result in an elaborate division of labor and in spatial segregation of people by race/ethnicity, social class, religion, and/or lifestyle. In the city, primary-group ties are largely replaced by secondary relationships; social interaction is fragmented, impersonal, and often superficial. Even though people gain some degree of freedom and privacy by living in the city, they pay a price for their autonomy, losing the group support and reassurance that come from primary-group ties.

From Wirth's perspective, people who live in urban areas are alienated, powerless, and lonely. A sense of community is obliterated and replaced by the "mass society"—a large-scale, highly institutionalized society in which individuality is supplanted by mass messages, faceless bureaucrats, and corporate interests.

Festive occasions such as this street fair provide opportunities for urban villagers to mingle with others, enjoying entertainment and social interaction.

Gans's Urban Villagers In contrast to Wirth's gloomy assessment of urban life, the sociologist Herbert Gans (1982/1962) suggested that not everyone experiences the city in the same way. Based on research in the west end of Boston in the late 1950s, Gans concluded that many residents develop strong loyalties and a sense of community in central-city areas that outsiders may view negatively. According to Gans, there are five major categories of adaptation among urban dwellers. *Cosmopolites* are students, artists, writers, musicians, entertainers, and professionals who choose to live in the city because they want to be close to its cultural facilities. *Unmarried people and childless couples* live in the city because they want to be close to work and entertainment. *Ethnic villagers* live in ethnically segregated neighborhoods; some are recent immigrants who feel most comfortable within their own group. The *deprived* are poor individuals with dim future prospects; they have very limited education and few, if any, other resources. The *trapped* are urban dwellers who can find no escape from the city; this group includes persons left behind by the process of invasion and succession, downwardly mobile individuals who have lost their former position in society, older persons who have nowhere else to go, and individuals addicted to alcohol or other drugs. Gans concluded that the city is a pleasure and a challenge for some urban dwellers and an urban nightmare for others (see "Sociology Works!").

Gender and City Life In their everyday lives, do women and men experience city life differently? Ac-

Sociology *Works!*

Herbert Gans and Twenty-First-Century Urban Villagers

I moved to Austin because it's a high-tech city with a small-town feel. Kinda my own "urban village" where I can cycle around town when I want but still own a nice car to go out in. I chose my neighborhood because it's centrally located to downtown eating and live entertainment. Austin calls itself the "Live Music Capital of the World," and I have plenty of opportunities to hear the music I like. Overall, I'd say that I'm compatible with Austin, and Austin's compatible with me.

—"Brad," a twenty-four-year-old college graduate, explaining why he chose to become an "urban villager" in Austin, Texas (author's files, 2007)

In the more than five decades since the urban sociologist Herbert J. Gans examined life in Boston's west end and identified five major categories of adaptation among urban dwellers, we continue to find that many residents think of themselves as living in an "urban village" despite the differences in high-rise buildings and the smaller settings in which the Boston west enders lived. Today, many younger urban residents think of themselves as having strong loyalties to a specific segment of their community with which they share interests and common experiences. Although many contemporary studies of urban life have emphasized the problems of poverty, crime, racial and ethnic discrimination, inadequate health care, and poor schools in our nation's major cities, it is useful for us to examine positive aspects of urban life as well. We can also gain important insights from examining the experiences of middle- and upper-income residents of our cities.

Herbert Gans believed that the people he referred to as *cosmopolites* chose to live in the city so that they could be close to cultural facilities, while he thought that *unmarried people and childless couples* chose to live there because they wanted to be close to work and entertainment. Among some affluent residents in high-tech cities such as Austin, Texas, married people and families with children have increasingly joined the ranks of individuals who live in or near the city's downtown area. According to contemporary urban villagers such as "Brad," they can find other people who are like themselves, who participate in activities they enjoy, and who support one another much like the members of an extended family when they need friendship or assistance. From this perspective, Gans's ideas about the urban village work because they show us that an important way to understand city life is through the experiences of people who live there. In the final analysis, of course, all people—including lower-income individuals, who have been further disadvantaged or even displaced by gentrification, and the poor and homeless—must be included in any thorough sociological examination of city life in the twenty-first century. However, "urban villagers," as coined by Gans, has staying power as a concept because it encourages us to look at urban life as a kaleidoscope of diversity that includes the wealthy and the merely affluent, as well as those who are "just getting by" or who are poor, because they live near to one another as contemporary urban dwellers.

Reflect & Analyze

Can you identify categories of urban villagers in a city with which you are familiar? To what extent do people live in certain areas of the city based on personal choice? What factors appear to be beyond their control?

cording to the scholar Elizabeth Wilson (1991), some men view the city as *sexual space* in which women, based on their sexual desirability and accessibility, are categorized as prostitutes, lesbians, temptresses, or virtuous women in need of protection. Wilson suggests that more-affluent, dominant-group women are more likely to be viewed as virtuous women in need of protection by their own men or police officers. Cities offer a paradox for women: On the one hand, cities offer more freedom than is found in comparatively isolated rural, suburban, and domestic settings; on the other, women may be in greater physical danger in the city. For Wilson, the answer to women's vulnerability in the city is not found in offering protection for them, but rather in changing people's perceptions that they can treat women as sexual objects because of the impersonality of city life (Wilson, 1991).

Cities and Persons with a Disability Chapter 18 describes how disability rights advocates believe that structural barriers create a "disabling" environment for many people, particularly in large urban settings. Many cities have made their streets and sidewalks more user-friendly for persons in wheelchairs and for individuals with visual disability by constructing concrete ramps with slide-proof surfaces at intersections or installing traffic lights with sounds designating when to "Walk." However, both urban and rural areas have a long way to go before many persons with disabilities will have the access to the things they need to become productive members of the community: educational and employment opportunities. Some persons with disabilities cannot navigate the streets and sidewalks of their communities, and some face obstacles getting into buildings that marginally, at best, meet the accessibility standards of the Americans with Disabilities Act; thus, many persons with a disability are unemployed.

Political scientist Harlan Hahn (1997: 177–178) traces the problem of lack of access to the beginnings of industrialism:

> The rise of industrialism produced extensive changes in the lives of disabled as well as nondisabled people. As factories replaced private dwellings as the primary sites of production, routines and architectural configurations were standardized to suit nondisabled workers. Both the design of worksites and of the products that were manufactured gave virtually no attention to the needs of people with disabilities. As a result, patterns of aversion and avoidance toward disabled persons were embedded in the construction of commodities, landscapes, and buildings that would remain for centuries. . . .
>
> The social and economic changes fostered by industrialization may have been exacerbated by the accompanying process of urbanization. As workers increasingly moved from farms and rural villages to live near the institutions of mass production, the character of community life appeared to shift perceptibly. Deviant or atypical personal characteristics that may have gradually become familiar in a small community seemed bizarre or disturbing in an urban milieu.

As Hahn's statement suggests, historical patterns in the dynamics of industrial capitalism contributed to discrimination against persons with disabilities, and this legacy remains evident in contemporary cities. Structural barriers are further intensified when other people do not respond favorably toward persons with disabilities. Scholar and disability rights advocate Sally French

CONCEPT QUICK REVIEW

Perspectives on Urbanism and the Growth of Cities

Functionalist Perspectives: Ecological Models	Concentric zone model	Due to invasion, succession, and gentrification, cities are a series of circular zones, each characterized by a particular land use.
	Sector model	Cities consist of wedge-shaped sectors, based on terrain and transportation routes, with the most-expensive areas occupying the best terrain.
	Multiple nuclei model	Cities have more than one center of development, based on specific needs and activities.
Conflict Perspectives: Political Economy Models	Capitalism and urban growth	Members of the capitalist class choose locations for skyscrapers and housing projects, limiting individual choices by others.
	Gender regimes in cities	Different cities have different prevailing ideologies regarding access to social positions and resources for men and women.
	Global patterns of growth	Capital investment decisions by core nations result in uneven growth in peripheral and semiperipheral nations.
Symbolic Interactionist Perspectives: The Experience of City Life	Simmel's view of city life	Due to the intensity of city life, people become somewhat insensitive to individuals and events around them.
	Urbanism as a way of life	The size, density, and heterogeneity of urban population result in an elaborate division of labor and space.
	Gans's urban villagers	Five categories of adaptation occur among urban dwellers, ranging from cosmopolites to trapped city dwellers.
	Gender and city life	Cities offer women a paradox: more freedom than in more isolated areas, yet greater potential danger.

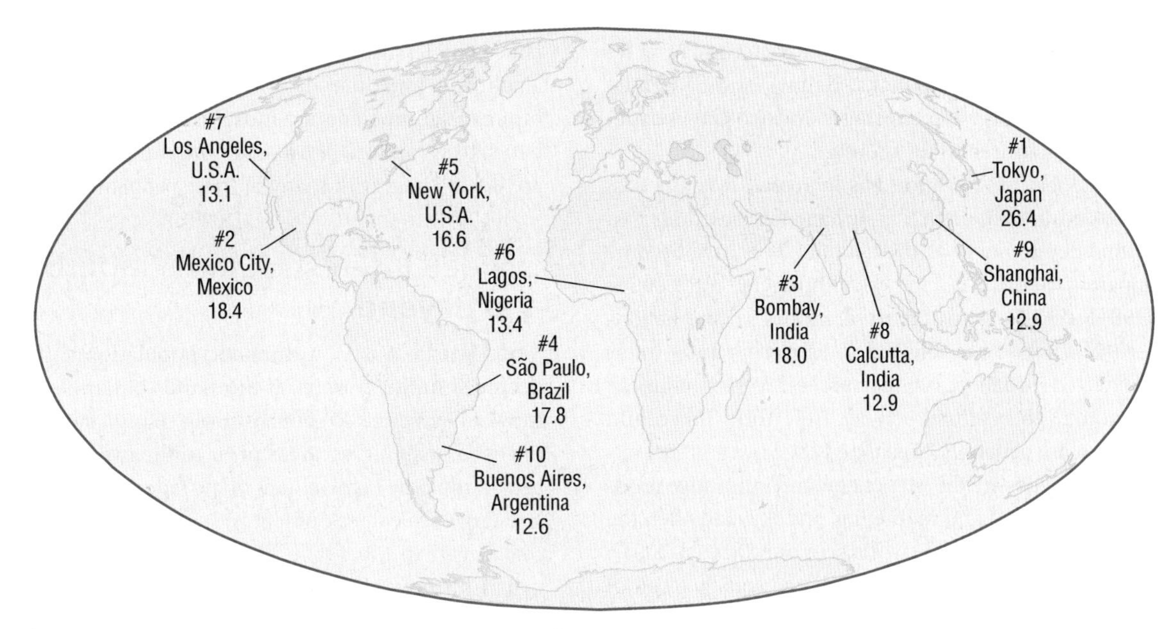

▶ **Figure 19.6 The World's Ten Largest Metropolises**

Note: 2000 population in millions.
Source: Population Reference Bureau, 2001.

(1999: 25–26), who is visually disabled, describes her own experience:

> I have lived in the same house for 16 years and yet I cannot recognize my neighbors. I know nothing about them at all; which children belong to whom, who has come and gone, who is old or young, ill or well, black or white. . . . On moving to my present house I informed several neighbors that, because of my inability to recognize them, I would doubtless pass them by in the street without greeting them. One neighbor, who had previously seen me striding confidently down the road, refused to believe me, but the others said they understood and would talk to me if our paths crossed. For the first couple of weeks it worked and I was surprised how often we met, but after that their greetings rapidly decreased and then ceased altogether. Why this happened I am not sure, but I suspect that my lack of recognition strained the interaction and limited the social reward they received from the encounter.

The Concept Quick Review examines the multiple perspectives on urban growth and urban living.

Problems in Global Cities

As we have seen, although people have lived in cities for thousands of years, the time is rapidly approaching when more people worldwide will live in or near a city than live in a rural area. In the middle-income and low-income regions of the world, Latin America is becoming the most urbanized: Four megacities—Mexico City, Buenos Aires, Lima, and Santiago—already contain more than half of the region's population and continue to grow rapidly. Within the next ten years, Rio de Janeiro and São Paulo are expected to have a combined population of about 40 million people living in a 350-mile-long megalopolis. By 2015, New York City will be the only U.S. city among the world's ten most populous. (▶ Figure 19.6 shows the populations in 2000 of the world's ten largest metropolises.)

Natural increases in population (higher birth rates than death rates) account for two-thirds of new urban growth, and rural-to-urban migration accounts for the rest. Some people move from rural areas to urban areas because they have been displaced from their land. Others move because they are looking for a better life. No matter what the reason, migration has caused rapid growth in cities in sub-Saharan Africa, India, Algeria, and Egypt. At the same time that the population is growing rapidly, the amount of farmland available for growing crops to feed people is decreasing. In Egypt, for example, land that was previously used for growing crops is now used for petroleum refineries, food-processing plants, and other factories (Kaplan, 1996).

Rapid global population growth in Latin America and other regions is producing a wide variety of urban problems, including overcrowding, environmental pollution, and the disappearance of farmland. In fact, many cities in middle- and low-income nations are quickly reaching the point at which food, housing, and basic public services are available to only a limited segment of the population (Crossette, 1996). With urban populations growing at a rate of 170,000 people per day in the late 1990s and even faster today, cities such as Cairo, Lagos, Dhaka, Beijing, and São Paulo are experiencing water shortages; Mexico City is experiencing a chronic water shortage.

As global urbanization has increased over the past three decades, differences in urban areas based on economic development at the national level have become apparent. Some cities in what Immanuel Wallerstein's (1984) world systems theory describes as core nations (see Chapter 9) are referred to as *global cities*—interconnected urban areas that are centers of political, economic, and cultural activity. New York, Tokyo, and London are generally considered the largest global cities. These cities are the sites of new and innovative product development and marketing, and they are often the "command posts" for the world economy (Sassen, 2001). But economic prosperity is not shared equally by all of the people in the core-nation global cities. Sometimes the living conditions of workers in low-wage service-sector jobs or in assembly production jobs more closely resemble the living conditions of workers in semiperipheral nations than they resemble the conditions of middle-class workers in their own country.

Most African countries and many countries in South America and the Caribbean are *peripheral* nations, previously defined as nations that depend on core nations for capital, have little or no industrialization (other than what may be brought in by core nations), and have uneven patterns of urbanization. According to Wallerstein (1984), the wealthy in peripheral nations support the exploitation of poor workers by core-nation capitalists in return for maintaining their own wealth and position. Poverty is thus perpetuated, and the problems worsen because of the unprecedented population growth in these countries.

In regard to the semiperipheral nations, only two cities are considered to be global cities: São Paulo, Brazil, the center of the Brazilian economy, and Singapore, the economic center of a multicountry region in Southeast Asia (Friedmann, 1995). Like peripheral nations, semiperipheral nations—such as India, Iran, and Mexico—are confronted with unprecedented population growth. In addition, a steady flow of rural migrants to large cities is creating enormous urban problems (see Box 19.4). What is the outlook for cities in the United States?

Urban Problems in the United States

Even the most optimistic of observers tends to agree that cities in the United States have problems brought on by years of neglect and deterioration. As we have seen in previous chapters, poverty, crime, racism, sexism, homelessness, inadequate public school systems, alcoholism and other drug abuse, gangs and guns, and other social problems are most visible and acute in urban settings. Issues of urban growth and development are intertwined with many of these problems.

Divided Interests: Cities, Suburbs, and Beyond

Since World War II, a dramatic population shift has occurred in this country as thousands of families have moved from cities to suburbs. Even though some people lived in suburban areas prior to the twentieth century, it took the involvement of the federal government and large-scale development to spur the dramatic shift that began in the 1950s (Palen, 1995). According to urban historian Kenneth T. Jackson (1985), postwar suburban growth was fueled by aggressive land developers, inexpensive real estate and construction methods, better transportation, abundant energy, government subsidies, and racial stress in the cities. However, the sociologist J. John Palen (1995) suggests that the Baby Boom following World War II and the liberalization of lending policies by federal agencies such as the Veterans Administration (VA) and the Federal Housing Authority (FHA) were significant factors in mass suburbanization.

Regardless of its causes, mass suburbanization has created a territorial division of interests between cities and suburban areas (Flanagan, 2002). Although many suburbanites rely on urban centers for their employment, entertainment, and other services, they pay their property taxes to suburban governments and school districts. Some affluent suburbs have state-of-the-art school districts, police and fire departments, libraries, and infrastructures (such as roads, sewers, and water treatment plants). By contrast, central-city services and school districts languish for lack of funds. Affluent families living in "gentrified" properties typically send their children to elite private schools, whereas the children of poor families living in racially segregated public housing projects attend underfunded (and often substandard) public schools.

Race, Class, and Suburbs The intertwining impact of race and class is visible in the division between central cities and suburbs. About 41 percent of central-city residents are persons of color, although they constitute a substantially smaller portion of the nation's population; just 27 percent of all African Americans live in suburbs. For most African American suburbanites, class is more important than race in determining one's neighbors. According to Vincent Lane, chairman of the Chicago Housing Authority, "Suburbanization

Box 19.4 Sociology in Global Perspective

Urban Migration and the "Garbage Problem"

Why do people around the world move from one location to another within their own country? In some low-income countries, people move to cities primarily because they have been displaced from their land. However, others move in hopes of finding new opportunities and a better quality of life. No matter what the reason, rural-to-urban migration has produced rapid growth in many cities, including ones located in Latin America, sub-Saharan Africa, India, and Egypt.

Although rapid global population growth and strong patterns of rural-to-urban migration have produced a wide variety of urban problems—including overcrowding, environmental pollution, and the disappearance of farmland—one pressing problem in many cities around the world is the collection of household garbage, something that people in high-income countries simply take for granted. We put our garbage in the apartment Dumpster or place our garbage can at curbside outside our residence, and it is picked up on the appointed day, never to be seen by us again. Obviously, we are not without garbage problems even in the United States and other high-income countries, as our landfills overflow and our city "dumps" become a source of environmental pollution and shame. However, our problems are not as large as those in urban areas in low-income countries, where some large cities have inadequate household-garbage-collection facilities. Consider the fact that less than 10 percent of the population in some cities benefits from the regular collection of household wastes. Estimates show that the proportion of garbage *not* collected by official means in Accra (Ghana) and Kampala (Uganda) is 90 percent and is 65 percent in Dar es Salaam (Tanzania) and 50 percent in Bogotá (Colombia) (see Middleton, 1999). Of course, these figures do not mean that garbage is not collected or "recycled" at all.

In the United States, we tend to think of recycling in terms of formal programs that encourage people to separate papers, cans, and bottles from other forms of trash so that these recyclables can be processed and used again. In low-income countries, however, the process is different. In Dar es Salaam, for example, informal scavenging at city dumps is a source of employment for many people. The scavenged resources are used as inputs to small-scale manufacturing industries that produce low-cost buckets, charcoal stoves, and lamps, which are then sold to residents of the city's squatter settlements for a lower cost than if the products had been made from imported raw materials (Middleton, 1999). This informal pattern of recycling and reuse takes place because of individual initiative and necessity rather than urban policies directed toward recycling. However, Dar es Salaam residents are not the only ones who deal with some aspects of the "garbage problem" in this manner. Mexico City and Cairo also have large squatter communities whose residents support themselves by living and working at official or unofficial rubbish dump sites.

Overall, the picture of the global garbage problem is not rosy: Uncollected garbage is a major problem in that it can be a serious fire hazard and a health hazard, attracting pests and becoming a breeding grounds for certain diseases. In Kampala, carnivorous Marabou storks live on the garbage and the rodents that are attracted to it. Today, demographers and other social analysts are concerned—just as analysts have been concerned since the days of Thomas Malthus—about rapid population growth and the patterns of migration to the largest cities of the world, many of which have already far exceeded their capacity to provide a safe urban environment for existing residents.

Reflect & Analyze

One way to reduce our "carbon footprint" is to recycle or reuse eligible materials and products. Do you regularly do either or both of these activities? Why or why not?

isn't about race now; it's about class. Nobody wants to be around poor people, because of all the problems that go along with poor people: poor schools, unsafe streets, gangs" (qtd. in De Witt, 1994: A12).

Nationally, most suburbs are predominantly white, and many upper-middle-class and upper-class suburbs remain virtually white. For example, only 5 percent of the population in northern Fulton County (adjoining Atlanta, Georgia) is African American. Likewise, in Plano (adjoining Dallas, Texas), nearly nine out of ten students in the public schools are white, whereas the majority of students in the Dallas Independent School District are African American, Latina/o, or Asian American. In the suburbs, people of color (especially African Americans) often become resegregated (see Feagin and Sikes, 1994). An example is Chicago, which remains one of the most-segregated metropolitan areas in the country in spite of its fair-housing ordinance. African Americans who have fled the high crime of Chicago's South Side primarily reside in nearby suburbs such as Country Club Hill and Chicago Heights, whereas suburban Asian Americans are most likely to live in Skokie and Naperville and suburban Latinos/as to reside in Maywood, Hillside, and Bellwood (De Witt, 1994). Similarly, suburban Latinas/os are highly concentrated in eight metropolitan areas in California, Texas, and Florida; by far, the largest such racial–ethnic concentration is found in the Los Angeles–Long Beach metropolitan area, with over 1.7 million Latinas/os. As with other groups, affluent Latinas/os live in affluent suburbs, whereas poorer Latinas/os remain segregated in less-desirable central-city areas (Palen, 1995).

Some analysts argue that the location of one's residence is a matter of personal choice. However, other analysts suggest that residential segregation reflects discriminatory practices by landlords, homeowners, and white realtors and their agents, who engage in *steering* people of color to different neighborhoods than those shown to their white counterparts. Lending practices of banks (including the *redlining* of certain properties so that acquiring a loan is virtually impossible) and the behavior of neighbors further intensify these problems (see Feagin and Sikes, 1994). In a study of suburban property taxes, the sociologist Andrew A. Beveridge found that African American homeowners are taxed more than whites on comparable homes in 58 percent of the suburban regions and 30 percent of the cities (cited in Schemo, 1994). Some analysts suggest that African Americans are more likely to move to suburbs with declining tax bases because they have limited finances, because they are steered there by real estate agents, or because white flight occurs as African American homeowners move in, leaving a heavier tax burden for the newcomers and those who remain behind. Longer-term residents may not see their property reassessed or their taxes go up for some period of time; in some cases, reassessment does not occur until the house is sold (Schemo, 1994).

Beyond the Suburbs: Edge Cities New urban fringes (referred to as *edge cities*) have been springing up beyond central cities and suburbs in recent years (Garreau, 1991). The Massachusetts Turnpike corridor west of Boston and the Perimeter area north of Atlanta are examples. Edge cities initially develop as residential areas; then retail establishments and office parks move into the area, creating the unincorporated edge city. Commuters from the edge city are able to travel around (rather than in and out of) the metropolitan region's center and can avoid its rush-hour traffic quagmires. Edge cities may not have a governing body or correspond to municipal boundaries; however, they drain taxes from central cities and older suburbs. Many businesses and industries have moved physical plants and tax dollars to these areas: Land is cheaper, and utility rates and property taxes are lower.

Lower taxes are a contributing factor to another recent development in the United States—the growth of Sunbelt cities in the southern and western states. In the 1970s, millions of people moved from the north-central

Affluent gated communities and enclaves of million-dollar homes stand in sharp contrast to low-income housing when we see them on the urban landscape. What sociological theories help us describe the disparity of lifestyles and life chances shown in these two settings?

and northeastern states to this area. Four reasons are generally given for this population shift: (1) more jobs and higher wages; (2) lower taxes; (3) pork-barrel programs that funneled federal money into projects in the Sunbelt, creating jobs and encouraging industry; and (4) easier transition to new industry because most industry in the northern states was based on heavy manufacturing rather than high technology.

The Continuing Fiscal Crises of the Cities

The largest cities in the United States have faced periodic fiscal crises for many years. In the twenty-first century, these crises have been intensified by higher employee health care and pension costs, declining revenues, and increased expenditures for public safety and homeland security. A 2004 survey of 328 cities found that 80 percent of those cities were less able to meet their financial needs than in the previous year (Fecht, 2004). As a result of declining revenues and increased costs, many cities have cut back on spending in areas other than public safety. According to John DeStefano, president of the National League of Cities, this financial crisis is hurting not only the cities themselves but also the people who live there: "Under-funded public schools, smaller police forces, deteriorating transportation systems, expensive health care, sprawl—these are [factors] that increasingly subvert our American ideal . . ." (qtd. in Fecht, 2004).

The current economic crisis in the United States has produced new problems for cities, and if a rebound occurs in the economy, the positive effects of such a rebound will not improve the budgetary problems of our cities and towns for a number of years.

Rural Community Issues in the United States

Although most people think of the United States as highly urbanized, about 20 percent of the U.S. population resides in rural areas, identified as communities of 2,500 people or less by the U.S. Census Bureau. Sociologists typically identify *rural communities* as small, sparsely settled areas that have a relatively homogeneous population of people who primarily engage in agriculture (Johnson, 2000). However, rural communities today are more diverse than this definition suggests.

Unlike the standard migration patterns from rural to urban places in the past, recently more people have moved from large urban areas and suburbs into rural areas. Many of those leaving urban areas today want to escape the high cost of living, crime, traffic congestion, and environmental pollution that make daily life difficult. Technological advances make it easier for people to move to outlying rural areas and still be connected to urban centers if they need to be. The proliferation of computers, cell phones, commuter airlines, and highway systems has made previously remote areas seem much more accessible to many people. However, many recent immigrants to rural areas do not face some traditional problems experienced by long-term rural residents, particularly farmers, small-business owners, teachers, doctors, and medical personnel in these rural communities.

For many people in rural areas who have made their livelihood through farming and other agricultural endeavors, recent decades have been very difficult, both financially and emotionally. Rural crises such as droughts, crop failures, and the loss of small businesses in the community have had a negative effect on many adults and their children. Like their urban counterparts, rural families have experienced problems of divorce, alcoholism, abuse, and other crises, but these issues have sometimes been exacerbated by such events as the loss of the family farm or business (Pitzer, 2003). Because home is also the center of work in farming families, the loss of the farm may also mean the loss of family and social life, and the loss of such things dear to children as their 4-H projects—often an animal that a child raises to show and sell (Pitzer, 2003). Some rural children and adolescents are also subject to injuries associated with farm work, such as livestock kicks or crushing, falling out of a tractor or pickup, and operating machinery designed for adults, that are not typically experienced by their urban counterparts (Schutske, 2002).

Economic opportunities are limited in many rural areas, and average salaries are typically lower than in urban areas, based on the assumption that a family can live on less money in rural communities than in cities. An example is rural teachers, who earn substantially less than their urban and suburban counterparts. Some rural areas have lost many teachers and administrators to higher-paying districts in other cities.

Although many of the problems we have examined in this book are intensified in rural areas, one of the most pressing is the availability of health services and doctors. Recently, some medical schools have established clinics and practices in outlying rural regions of the states in which they are located in an effort to increase the number of physicians available to rural residents. Typically, physicians who have just started to practice medicine have chosen to work in large urban centers with accessible high-tech medical facilities. Because of the pressing time constraints of tending to patients with life-threatening problems, the availability of community clinics and hospitals in rural areas may be a life-or-death matter for some residents. Losing these facilities can have a devastating effect on people's health and life chances.

In addition to the movement of some urban dwellers to rural areas, two other factors have changed the face of rural America in some regions. One is the proliferation of superstores, such as Wal-Mart, PetSmart, Lowe's, and Home Depot. In some cases, these superstores have effectively put small businesses such as hardware stores and pet shops out of business because local merchants cannot meet the prices established by these large-volume discount chains. The development of superstores and outlet malls along the rural highways of this country has raised new concerns about environmental issues such as air and water pollution, and has brought about new questions regarding whether these stores benefit the rural communities where they are located.

A second factor that has changed the face of some rural areas (and is sometimes related to the growth of megastores and outlet malls) is an increase in tourism in rural America (Brown, 2003). According to a recent study, about 87 million people (nearly two-thirds of all U.S. adults) have taken a trip to a rural destination, usually for leisure purposes, over the past few years. Tourism produces jobs; however, many of the positions are for food servers, retail clerks, and hospitality workers, which are often low-paying, seasonal jobs that have few benefits. Tourism may improve a community's tax base, but this does not occur when the outlet malls, hotels, and fast-food restaurants are located outside of the rural community's taxing authority, as frequently occurs when developers decide where to locate malls and other tourist amenities.

Population and Urbanization in the Future

In the future, rapid global population growth is inevitable. Although death rates have declined in many low-income nations, there has not been a corresponding decrease in birth rates. Between 1985 and 2025, 93 percent of all global population growth will have occurred in Africa, Asia, and Latin America; 83 percent of the world's population will live in those regions by 2025. Perhaps even more amazing is the fact that in the five-year span between 1995 and 2000, 21 percent of the entire world's population increase occurred in two countries: China and India. ▶ Figure 19.7a shows the net annual additions to the populations of six countries during that five-year period. Figure 19.7b shows the growth of the world's population from 1927 to 1999 and the expected growth to eight billion by the year 2025.

In the future, low-income countries will have an increasing number of poor people. While the world's population will *double,* the urban population will *triple* as people migrate from rural to urban areas in search of food, water, and jobs.

One of the many effects of urbanization is greater exposure of people to the media. Increasing numbers of poor people in less-developed nations will see images from the developed world that are beamed globally by news networks such as CNN. As futurist John L. Petersen (1994: 119) notes, "For the first time in his-

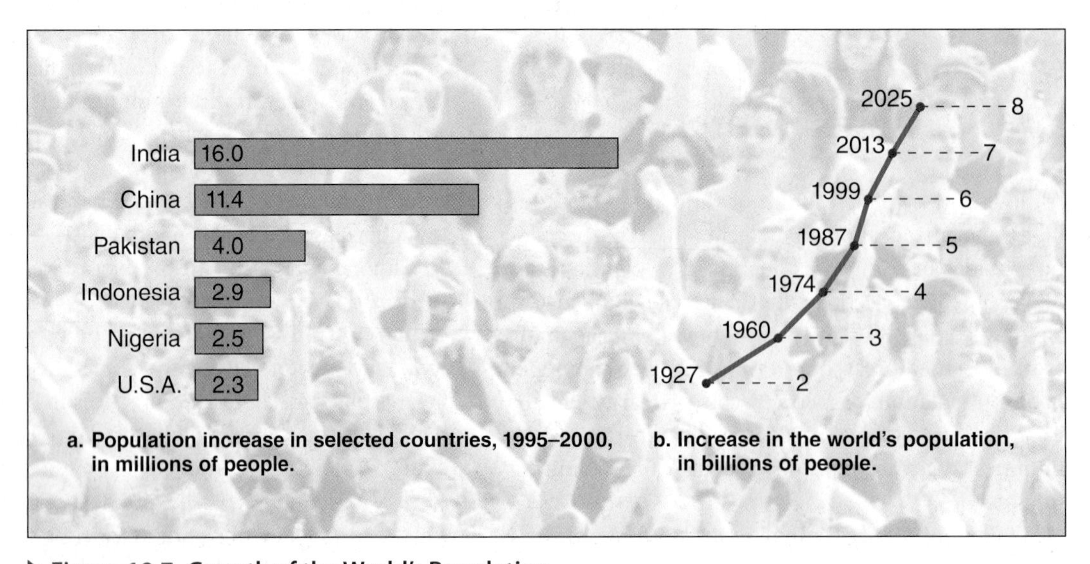

▶ **Figure 19.7 Growth of the World's Population**

Source: United Nations Population Division, 1999.

tory, the poor are beginning to understand how relatively poor they are compared to the rich nations. They see, in detail, how the rest of the world lives and feel their increasing disenfranchisement."

Infants born today will be teenagers in the year 2023. By then, in a worst-case scenario, central cities and nearby suburbs in the United States will have experienced bankruptcy exacerbated by sporadic race- and class-oriented violence. The infrastructure will be beyond the possibility of repair. Families and businesses with the ability to do so will have long since moved to "new cities," where they will inevitably diminish the quality of life that they originally sought there. Areas that we currently think of as being relatively free from such problems will be characterized by depletion of natural resources and by greater air and water pollution (see Ehrlich, Ehrlich, and Daily, 1995).

By contrast, in a best-case scenario, the problems brought about by rapid population growth in low-income nations will be remedied by new technologies that make goods readily available to people. International trade agreements are removing trade barriers and making it possible for all nations to fully engage in global trade. People in low-income nations will benefit by gaining jobs and opportunities to purchase goods at lower prices. Of course, the opposite may also occur: People may be exploited as inexpensive labor, and their country's natural resources may be depleted as transnational corporations buy up raw materials without contributing to the long-term economic stability of the nation.

In the United States, a best-case scenario for the future might include improvements in how taxes are collected and spent. Some analysts suggest that regional governments should be developed that would be responsible for water, wastewater (sewage), transportation, schools, and other public services over a wider area.

At the macrolevel, we may be able to do little about population and urbanization; however, at the microlevel, we may be able to exercise some degree of control over our communities and our lives. In both cases, futurists suggest that as we approach the future, we must "leave the old ways and invent new ones" (Petersen, 1994: 340). What aspects of our "old ways" do you think we should discard? Can you help invent new ones?

Chapter Review

- **What are the processes that produce population changes?**

Populations change as the result of fertility (births), mortality (deaths), and migration.

- **What is the Malthusian perspective?**

Over two hundred years ago, Thomas Malthus warned that overpopulation would result in poverty, starvation, and other major problems that would limit the size of the population. According to Malthus, the population would increase geometrically while the food supply would increase only arithmetically, resulting in a critical food shortage and poverty.

- **What are the views of Karl Marx and the neo-Malthusians on overpopulation?**

According to Karl Marx, poverty is the result of capitalist greed, not overpopulation. More recently, neo-Malthusians have reemphasized the dangers of overpopulation and encouraged zero population growth—the point at which no population increase occurs from year to year.

- **What are the stages in demographic transition theory?**

Demographic transition theory links population growth to four stages of economic development: (1) the preindustrial stage, with high birth rates and death rates; (2) early industrialization, with relatively high birth rates and a decline in death rates; (3) advanced industrialization and urbanization, with low birth rates and death rates; and (4) postindustrialization, with additional decreases in the birth rate coupled with a stable death rate.

- **What are the three functionalist models of urban growth?**

Functionalists view urban growth in terms of ecological models. The concentric zone model sees the city as a series of circular areas, each characterized by a different type of land use. The sector model describes urban growth in terms of terrain and transportation routes. The multiple nuclei model views cities as having numerous centers of development from which growth radiates.

chapter

20 Collective Behavior, Social Movements, and Social Change

Chapter Focus Question

Can collective behavior and social movements make people aware of important social issues such as environmental pollution?

One of the by-products of having grown up alongside the Houston Ship Channel was very nearly becoming desensitized to the vast amounts of pollutants the oil and chemical industry poured into East Houston's air and waterways. I once fell in the Ship Channel while working on a crew that built launching pads for a new supertanker. The resulting kidney infection took nine months to heal. I urinated blood for three weeks. No one can tell me that the current state of global consumerism does not impact the world's climate adversely. To [people] who pooh pooh the notion of global warming, I say this: Go take a swim in the Houston Ship Channel.

—Rodney Crowell, Grammy-winning songwriter and recording artist, explaining why he joined the virtual march against global warming (qtd. in StopGlobalWarming.org, 2006)

Environmental activism is a powerful type of social movement that seeks to call public attention to pressing social concerns such as air pollution. Activists often stage public events such as the one shown here in an attempt to gain the attention of political leaders and everyday citizens.

IN THIS CHAPTER

- Collective Behavior
- Social Movements
- Social Movement Theories
- Social Change in the Future

> When I circled the moon and looked back at Earth, my outlook on life and my viewpoint on Earth changed. You don't see Las Vegas, Boston or even New York. You don't see boundaries or people. No whites, blacks, French, Greeks, Christians or Jews. The Earth looks completely uninhabited, and you know that on Spaceship Earth, there live over six billion astronauts, all seeking the same things from life.
>
> When viewed in total, Earth is a spaceship just like Apollo. We are all the crew of Spaceship Earth; and just like Apollo, the crew must learn to live and work together. We must learn to manage the resources of this world with new imagination. The future is up to you.
>
> —Jim Lovell, a retired NASA astronaut, describing how his experience in space gave him a new perspective on environmental problems such as global warming (qtd. in StopGlobalWarming.org, 2006)

Along with many other well-known persons, Rodney Crowell and Jim Lovell are expressing their concerns about global warming and encouraging others to engage in collective behavior to try to save the planet from serious harm. The term *global warming* refers to a process that occurs when carbon dioxide (produced by cars, factories, and power plants, among other sources) stays in the atmosphere and acts like a blanket that holds in the heat. Over time, global warming results in higher temperatures, rises in sea levels,

Sharpening Your Focus

- *What causes people to engage in collective behavior?*
- *What are some common forms of collective behavior?*
- *How can different types of social movements be distinguished from one another?*
- *What draws people into social movements?*
- *What factors contribute to social change?*

and catastrophic weather such as powerful hurricanes (Revkin, 2006). However, global warming is not a major concern for most people, who, when asked on surveys to identify the most important problems facing the United States, put at the top of their list the war in the Middle East, the economy and jobs, immigration, terrorism, health care, and energy prices. Even when respondents mention environmental issues as a concern, global warming is placed far down the list after the pollution of rivers, lakes, reservoirs, and drinking water; air pollution; and toxic waste, according to a recent Gallup survey (*New York Times,* 2006: WK14). The message of the environmental movement is that we must act collectively and immediately to reduce global warming before havoc comes to the Earth: Social change is essential. Sociologists define ***social change*** **as the alteration, modification, or transformation of public policy, culture, or social institutions over time.** Social change is usually brought about by collective behavior and social movements.

In this chapter, we will examine collective behavior, social movements, and social change from a sociological perspective. We will use environmental activism as an example of how people may use social movements as a form of mass mobilization and social transformation (Buechler, 2000). Before reading on, test your knowledge about collective behavior and environmental issues by taking the quiz in Box 20.1.

Collective Behavior

Collective behavior **is voluntary, often spontaneous activity that is engaged in by a large number of people and typically violates dominant-group norms and values.** Unlike the *organizational behavior* found in corporations and voluntary associations (such as labor unions and environmental organizations), collective behavior lacks an official division of labor, hierarchy of authority, and established rules and procedures. Unlike *institutional behavior* (in education, religion, or politics, for example), it lacks institutionalized norms to govern behavior. Collective behavior can take various forms, including crowds, mobs, riots, panics, fads, fashions, and public opinion.

According to the sociologist Steven M. Buechler (2000), early sociologists studied collective behavior because they lived in a world that was responding to the processes of modernization, including urbanization, industrialization, and proletarianization of workers. Contemporary forms of collective behavior, particularly social protests, are variations on the themes that originated during the transition from feudalism to capitalism and the rise of modernity in Europe (Buechler, 2000). Today, some forms of collective behavior and social movements are directed toward public issues such as air pollution, water pollution, and the exploitation of workers in global sweatshops by transnational corporations (see Shaw, 1999). For example, Riverkeeper, the group of environmental activists who protect the Hudson River, mounts campaigns to lobby members of Congress. The members of this organization and its volunteers also focus more widely on a number of environmental concerns about waterways in many regions.

Conditions for Collective Behavior

Collective behavior occurs as a result of some common influence or stimulus that produces a response from a collectivity. A *collectivity* is a number of people who act together and may mutually transcend, bypass, or subvert established institutional patterns and structures. Three major factors contribute to the likelihood that collective behavior will occur: (1) structural factors that increase the chances of people responding in a particular way, (2) timing, and (3) a breakdown in social control mechanisms and a corresponding feeling of normlessness (McPhail, 1991; Turner and Killian, 1993).

A common stimulus is an important factor in collective behavior. For example, the publication of *Silent Spring* (1962) by former Fish and Wildlife Service biologist Rachel Carson is credited with triggering collective behavior directed at demanding a clean environment and questioning how much power large corporations should have. Carson described the dangers of pesticides such as DDT, which was then being promoted by the chemical industry as the miracle that could give the United States the unchallenged position as food supplier to the world (Cronin and Kennedy, 1999). Carson's activism has been described in this way:

> Carson was not a wild-eyed reformer intent on bringing the industrial age to a grinding halt. She wasn't even opposed to pesticides per se. She was a careful scientist and brilliant writer whose painstaking research on pesticides proved that the "miraculous" bursts of agricultural productivity had long-term costs undisclosed in the chemical industry's exaggerated puffery. Americans were losing things—their health, many birds and fishes, and the purity of their waterways—that they should value more than modest savings at the grocery store. (Cronin and Kennedy, 1999: 151)

Timing is another significant factor in bringing about collective behavior. For example, in the 1960s

Box 20.1 Sociology and Everyday Life

How Much Do You Know About Collective Behavior and Environmental Issues?

True	False	
T	F	1. Scientists are forecasting a global warming of between 2 and 11 degrees Fahrenheit over the next century.
T	F	2. The environmental movement in the United States started in the 1960s.
T	F	3. People who hold strong attitudes regarding the environment are very likely to be involved in social movements to protect the environment.
T	F	4. Environmental groups may engage in civil disobedience or use symbolic gestures to call attention to their issue.
T	F	5. People are most likely to believe rumors when no other information is readily available on a topic.
T	F	6. Influencing public opinion is a very important activity for many social movements.
T	F	7. Most social movements in the United States seek to improve society by changing some specific aspect of the social structure.
T	F	8. Sociologists have found that people in a community respond very similarly to natural disasters and to disasters caused by technological failures.

Answers on page 658.

smog had started staining the skies in this country; in Europe, birds and fish were dying from environmental pollution; and oil spills from tankers were provoking public outrage worldwide (Cronin and Kennedy, 1999). People in this country were ready to acknowledge that problems existed. By writing *Silent Spring,* Carson made people aware of the hazards of chemicals in their foods and the destruction of wildlife. However, that is not all she produced: As a consequence of her careful research and writing, she also produced anger in people at a time when they were beginning to wonder if they were being deceived by the very industries that they had entrusted with their lives and their resources. Once aroused to action, many people began demanding an honest, comprehensive accounting of where pollution was occurring and how it might be endangering public health and environmental resources. Public outcries also led to investigations in courts and legislatures throughout the United States as people began to demand legal recognition of the right to a clean environment (Cronin and Kennedy, 1999).

A breakdown in social control mechanisms has been a powerful force in triggering collective behavior regarding environmental protection and degradation. During the 1970s, people in the "Love Canal" area of Niagara Falls, New York, became aware that their neighborhood and their children's school had been built over a canal where tons of poisonous waste had been dumped by a chemical company between 1930 and 1950. After the company closed the site, covered it with soil, and sold it (for $1) to the city of Niagara Falls, homes and a school were built on the sixteen-acre site. Over the next two decades, an oily black substance began oozing into the homes in the area and killing the trees and grass on the lots; schoolchildren reported mysterious illnesses and feelings of malaise. Tests indicated that the dump site contained more than two hundred different chemicals, many of which could cause cancer or other serious health problems. Upon learning this information, Lois Gibbs, a mother of one of the schoolchildren, began a grassroots campaign to force government officials to relocate community members injured by seepages from the chemical dump. The collective behavior of neighborhood volunteers was not only successful in eventually bringing about social change but also inspired others to engage in collective behavior regarding environmental problems in their communities. After the Love Canal occurrence, Lois Gibbs founded Citizens' Clearinghouse for Hazardous Waste, which has assisted more

social change the alteration, modification, or transformation of public policy, culture, or social institutions over time.

collective behavior voluntary, often spontaneous activity that is engaged in by a large number of people and typically violates dominant-group norms and values.

Box 20.1 Sociology and Everyday Life

Answers to the Sociology Quiz on Collective Behavior and Environmental Issues

1. True. Global surface temperatures have increased about 0.4 degrees Fahrenheit during the past 25 years, and scientists believe that this trend will grow more pronounced during the next 100 years.

2. False. The environmental movement in the United States is the result of more than 100 years of collective action. The first environmental organization, the American Forestry Association (now American Forests), originated in 1875.

3. False. Since the 1980s, public opinion polls have shown that the majority of people in the United States have favorable attitudes regarding protection of the environment and banning nuclear weapons; however, far fewer individuals are actually involved in collective action to further these causes.

4. True. Environmental groups have held sit-ins, marches, boycotts, and strikes, which sometimes take the form of civil disobedience. Others have hanged political leaders in effigy or held officials hostage. Still others have dressed as grizzly bears to block traffic in Yellowstone National Park or created a symbolic "crack" (made of plastic) on the Glen Canyon Dam on the Colorado River to denounce development in the area.

5. True. Rumors are most likely to emerge and circulate when people have very little information on a topic that is important to them. For example, rumors abound in times of technological disasters, when people are fearful and often willing to believe a worst-case scenario.

6. True. Many social movements, including grassroots environmental activism, attempt to influence public opinion so that local decision makers will feel obliged to correct a specific problem through changes in public policy.

7. True. Most social movements are reform movements that focus on improving society by changing some specific aspect of the social structure. Examples include environmental movements and the disability rights movement.

8. False. Most sociological studies have found that people respond differently to natural disasters, which usually occur very suddenly, than to technological disasters, which may occur gradually. One of the major differences is the communal bonding that tends to occur following natural disasters, as compared with the extreme social conflict that may follow technological disasters.

Sources: Based on Adams, 1991; Gamson, 1990; Hynes, 1990; Worster, 1985; and Young, 1990.

than 1,700 community groups in fighting pollution in their neighborhoods (see Cable and Cable, 1995; Cronin and Kennedy, 1999). Similarly, the Hudson Valley Riverkeeper originated in a grassroots movement that was fueled by "people power and never lost touch with its power base" (Cronin and Kennedy, 1999: 277).

Dynamics of Collective Behavior

To better understand the dynamics of collective behavior, let's briefly examine several questions. First, how do people come to transcend, bypass, or subvert established institutional patterns and structures? Some environmental activists have found that they cannot get their point across unless they go outside established institutional patterns and organizations. For example, Lois Gibbs and other Love Canal residents initially tried to work within established means through the school administration and state health officials to clean up the problem. However, they quickly learned that their problems were not being solved through "official" channels. As the problem appeared to grow worse, organizational responses became more defensive and obscure. Accordingly, some residents began acting outside of established norms by holding protests and strikes (Gibbs, 1982). Some situations are more conducive to collective behavior than others. When people can communicate quickly and easily with one another, spontaneous behavior is more likely (Turner and Killian, 1993). When people are gathered together in one general location (whether lining the streets or assembled in a massive stadium), they are more likely to respond to a common stimulus.

Second, how do people's actions compare with their attitudes? People's attitudes (as expressed in public opinion surveys, for instance) are not always reflected in their political and social behavior. Issues pertaining to the environment are no exception. For example,

© William Campbell/Sygma/CORBIS

The Love Canal area of Niagara Falls, New York, has been the site of protests and other forms of collective behavior because of hazardous environmental pollution. Original protests in the 1970s, demanding a cleanup of the site, were followed in the 1990s by new protests, this time over the proposed resettlement of the area.

people may indicate in survey research that they believe the quality of the environment is very important, but the same people may not turn out on election day to support propositions that protect the environment or candidates who promise to focus on environmental issues. Likewise, individuals who indicate on a questionnaire that they are concerned about increases in ground-level ozone—the primary component of urban smog—often drive single-occupant, oversized vehicles that government studies have shown to be "gas guzzlers" that contribute to lowered air quality in urban areas. As a result, smog levels increase, contributing to human respiratory problems and dramatically reduced agricultural crop yields (Voynick, 1999).

Third, why do people act collectively rather than singly? As the sociologists Ralph H. Turner and Lewis M. Killian (1993: 12) note, people believe that there is strength in numbers: "the rhythmic stamping of feet by hundreds of concert-goers in unison is different from isolated, individual cries of 'bravo.'" Likewise, people may act as a collectivity when they believe it is the only way to fight those with greater power and resources. Collective behavior is not just the sum total of a large number of individuals acting at the same time; rather, it reflects people's joint response to some common influence or stimulus.

Distinctions Regarding Collective Behavior

People engaging in collective behavior may be divided into crowds and masses. A ***crowd*** **is a relatively large number of people who are in one another's immediate vicinity** (Lofland, 1993). Examples of crowds include the audience in a movie theater or people at a pep rally for a sporting event. By contrast, a ***mass*** **is a number of people who share an interest in a specific idea or issue but who are not in one another's immediate vicinity** (Lofland, 1993). An example is the popularity of blogging on the Internet. A *blog,* which is short for "web log," is an online journal maintained by an individual who frequently records entries that are maintained in a chronological order. People who self-publish blogs are widely diverse in their interests. Some may include poetry, diary entries, or discussions of such activities as body piercings. However, others express their beliefs about social issues such as the environment, terrorism, war, and their concerns about the future. Readers often share a common interest with the blogger on the topics the person is writing about, but these individuals have never met—and probably will never meet—each other in a face-to-face encounter. To distinguish between crowds and masses, think of the difference between a riot and a rumor: People who participate in a riot must be in the same general location; those who spread a rumor may be thousands of miles apart, communicating by telephone or the Internet.

Collective behavior may also be distinguished by the dominant emotion expressed. According to the sociologist John Lofland (1993: 72), the *dominant emotion* refers to the "publicly expressed feeling perceived by participants and observers as the most prominent in an episode of collective behavior." Lofland suggests that fear, hostility, and joy are three fundamental emotions found in collective behavior; however, grief, disgust, surprise, or shame may also predominate in some forms of collective behavior.

Types of Crowd Behavior

When we think of a crowd, many of us think of *aggregates,* previously defined as a collection of people who happen to be in the same place at the same time but who share little else in common. However, the presence of a relatively large number of people in the same location does not necessarily produce collective behavior. Sociologist Herbert Blumer (1946) developed a typology in which crowds are divided into four categories: casual, conventional, expressive, and

crowd a relatively large number of people who are in one another's immediate vicinity.

mass a number of people who share an interest in a specific idea or issue but who are not in one another's immediate vicinity.

acting. Other scholars have added a fifth category, protest crowds.

Casual and Conventional Crowds *Casual crowds* are relatively large gatherings of people who happen to be in the same place at the same time; if they interact at all, it is only briefly. People in a shopping mall or a subway car are examples of casual crowds. Other than sharing a momentary interest, such as a clown's performance or a small child's fall, a casual crowd has nothing in common. The casual crowd plays no active part in the event—such as the child's fall—which likely would have occurred whether or not the crowd was present; the crowd simply observes.

Conventional crowds are made up of people who come together for a scheduled event and thus share a common focus. Examples include religious services, graduation ceremonies, concerts, and college lectures. Each of these events has preestablished schedules and norms. Because these events occur regularly, interaction among participants is much more likely; in turn, the events would not occur without the crowd, which is essential to the event.

Expressive and Acting Crowds *Expressive crowds* provide opportunities for the expression of some strong emotion (such as joy, excitement, or grief). People release their pent-up emotions in conjunction with other persons experiencing similar emotions. Examples include worshippers at religious revival services; mourners lining the streets when a celebrity, public official, or religious leader has died; and revelers assembled at Mardi Gras or on New Year's Eve at Times Square in New York.

Acting crowds are collectivities so intensely focused on a specific purpose or object that they may erupt into violent or destructive behavior. Mobs, riots, and panics are examples of acting crowds, but casual and conventional crowds may become acting crowds under some circumstances. A ***mob*** **is a highly emotional crowd whose members engage in, or are ready to engage in, violence against a specific target—a person, a category of people, or physical property.** Mob behavior in this country has included lynchings, fire bombings, effigy hangings, and hate crimes. Mob violence tends to dissipate relatively quickly once a target has been injured, killed, or destroyed. Sometimes, actions such as an effigy hanging are used symbolically by groups that are not otherwise violent. For example, Lois Gibbs and other Love Canal residents called attention to their problems with the chemical dump site by staging a protest in which they "burned in effigy" the governor and the health commissioner to emphasize their displeasure with the lack of response from these public officials (A. Levine, 1982).

Compared with mob actions, riots may be of somewhat longer duration. A ***riot*** **is violent crowd behavior that is fueled by deep-seated emotions but not directed at one specific target.** Riots are often triggered by fear, anger, and hostility; however, not all riots are caused by deep-seated hostility and hatred—people may be expressing joy and exuberance when rioting occurs. Examples include celebrations after sports victories such as those that occurred in Montreal, Canada, following a Stanley Cup win and in Vancouver following a playoff victory (Kendall, Lothian Murray, and Linden, 2004).

A ***panic*** **is a form of crowd behavior that occurs when a large number of people react to a real or perceived threat with strong emotions and self-destructive behavior.** The most common type of panic occurs when people seek to escape from a perceived danger, fearing that few (if any) of them will be able to get away from that danger. Panics can also arise in response to events that people believe are beyond their control—such as a major disruption in the economy. Although panics are relatively rare, they receive massive media coverage because they provoke strong feelings of fear in readers and viewers, and the number of casualties may be large.

Crowds of people come together for a variety of reasons. The people pictured here wanted to be near the front of the line to purchase an iPhone on the first day the new device became available. How does a crowd such as this differ from other types of crowds?

Protest Crowds Sociologists Clark McPhail and Ronald T. Wohlstein (1983) added protest crowds to the four types of crowds identified by Blumer. *Protest crowds* engage in activities intended to achieve specific political goals. Examples include sit-ins, marches, boycotts, blockades, and strikes. In 1997 an International Day of Protest was staged against Nike's labor practices in countries such as Vietnam (Shaw, 1999). Some protests take the form of ***civil disobedience*****—nonviolent action that seeks to change a policy or law by refus-**

ing to comply with it. Acts of civil disobedience may become violent, as in a confrontation between protesters and police officers; in this case, a protest crowd becomes an *acting crowd.* In the 1960s, African American students and sympathetic whites used sit-ins to call attention to racial injustice and demand social change (see Morris, 1981; McAdam, 1982). Some of these protests can escalate into violent confrontations even when violence was not the intent of the organizers.

As you may recall, collective action often puts individuals into a situation where they engage in some activity as a group that they might not do on their own. Does this mean that people's actions are produced by some type of "herd mentality"? Some analysts have answered this question affirmatively; however, sociologists typically do not agree with that assessment.

Explanations of Crowd Behavior

What causes people to act collectively? How do they determine what types of action to take? One of the earliest theorists to provide an answer to these questions was Gustave Le Bon, a French scholar who focused on crowd psychology in his contagion theory.

Contagion Theory *Contagion theory* focuses on the social-psychological aspects of collective behavior; it attempts to explain how moods, attitudes, and behavior are communicated rapidly and why they are accepted by others (Turner and Killian, 1993). Le Bon (1841–1931) argued that people are more likely to engage in antisocial behavior in a crowd because they are anonymous and feel invulnerable. Le Bon (1960/1895) suggested that a crowd takes on a life of its own that is larger than the beliefs or actions of any one person. Because of its anonymity, the crowd transforms individuals from rational beings into a single organism with a collective mind. In essence, Le Bon asserted that emotions such as fear and hate are contagious in crowds because people experience a decline in personal responsibility; they will do things as a collectivity that they would never do when acting alone.

Le Bon's theory is still used by many people to explain crowd behavior. However, critics argue that the "collective mind" has not been documented by systematic studies.

© AP Images

Protest movements take place around the world when people believe that an injustice has occurred or that great inequalities exist in the distribution of societal resources. After the 2009 Iranian elections kept President Mahmoud Ahmadinejad in power, many observers disputed the results. The government then began a crackdown of the ensuing demonstrations. This woman is protesting the government's actions.

Social Unrest and Circular Reaction Sociologist Robert E. Park was the first U.S. sociologist to investigate crowd behavior. Park believed that Le Bon's analysis of collective behavior lacked several important elements. Intrigued that people could break away from the powerful hold of culture and their established routines to develop a new social order, Park added the concepts of social unrest and circular reaction to contagion theory. According to Park, social unrest is transmitted by a process of *circular reaction*—the interactive communication between persons such that the discontent of one person is communicated to another, who, in turn, reflects the discontent back to the first person (Park and Burgess, 1921).

Convergence Theory *Convergence theory* focuses on the shared emotions, goals, and beliefs that many people may bring to crowd behavior. Because of their

mob a highly emotional crowd whose members engage in, or are ready to engage in, violence against a specific target—a person, a category of people, or physical property.

riot violent crowd behavior that is fueled by deep-seated emotions but is not directed at one specific target.

panic a form of crowd behavior that occurs when a large number of people react to a real or perceived threat with strong emotions and self-destructive behavior.

civil disobedience nonviolent action that seeks to change a policy or law by refusing to comply with it.

individual characteristics, many people have a predisposition to participate in certain types of activities (Turner and Killian, 1993). From this perspective, people with similar attributes find a collectivity of like-minded persons with whom they can express their underlying personal tendencies. Although people may reveal their "true selves" in crowds, their behavior is not irrational; it is highly predictable to those who share similar emotions or beliefs.

Convergence theory has been applied to a wide array of conduct, from lynch mobs to environmental movements. In social psychologist Hadley Cantril's (1941) study of one lynching, he found that the participants shared certain common attributes: They were poor and working-class whites who felt that their status was threatened by the presence of successful African Americans. Consequently, the characteristics of these individuals made them susceptible to joining a lynch mob even if they did not know the target of the lynching.

Convergence theory adds to our understanding of certain types of collective behavior by pointing out how individuals may have certain attributes—such as racial hatred or fear of environmental problems that directly threaten them—that initially bring them together. However, this theory does not explain how the attitudes and characteristics of individuals who take some collective action differ from those who do not.

Convergence theory is based on the assumption that crowd behavior involves shared emotions, goals, and beliefs, such as the importance of protecting the environment. An example is the Earth Day events that brought together these children carrying this banner to foster environmental causes.

Emergent Norm Theory Unlike contagion and convergence theories, *emergent norm theory* emphasizes the importance of social norms in shaping crowd behavior. Drawing on the symbolic interactionist perspective, the sociologists Ralph Turner and Lewis Killian (1993: 12) asserted that crowds develop their own definition of a situation and establish norms for behavior that fit the occasion:

> Some shared redefinition of right and wrong in a situation supplies the justification and coordinates the action in collective behavior. People do what they would not otherwise have done when they panic collectively, when they riot, when they engage in civil disobedience, or when they launch terrorist campaigns, because they find social support for the view that what they are doing is the right thing to do in the situation.

According to Turner and Killian (1993: 13), emergent norms occur when people define a new situation as highly unusual or see a long-standing situation in a new light.

Sociologists using the emergent norm approach seek to determine how individuals in a given collectivity develop an understanding of what is going on, how they construe these activities, and what type of norms are involved. For example, in a study of audience participation, the sociologist Steven E. Clayman (1993) found that members of an audience listening to a speech applaud promptly and independently but wait to coordinate their booing with other people; they do not wish to "boo" alone.

Some emergent norms are permissive—that is, they give people a shared conviction that they may disregard ordinary rules such as waiting in line, taking turns, or treating a speaker courteously. Collective activity such as mass looting may be defined (by participants) as taking what rightfully belongs to them and punishing those who have been exploitative. For example, following the Los Angeles riots of 1992, some analysts argued that Korean Americans were targets of rioters because they were viewed by Latinos/as and African Americans as "callous and greedy invaders" who became wealthy at the expense of members of other racial–ethnic groups (Cho, 1993). Thus, rioters who used this rationalization could view looting and burning as a means of "paying back" Korean Americans or of gaining property (such as TV sets and microwave ovens) from those who had already taken from them. Once a crowd reaches some agreement on the norms, the collectivity is supposed to adhere to them. If crowd members develop a norm that condones looting or vandalizing property, they will proceed to cheer for those who conform and ridicule those who are unwilling to abide by the collectivity's new norms.

Emergent norm theory points out that crowds are not irrational. Rather, new norms are developed in a rational way to fit the immediate situation. However, critics note that proponents of this perspective fail to specify exactly what constitutes a norm, how new ones emerge, and how they are so quickly disseminated and accepted by a wide variety of participants. One variation of this theory suggests that no single dominant norm is accepted by everyone in a crowd; instead, norms are specific to the various categories of actors rather than to the collectivity as a whole (Snow, Zurcher, and Peters, 1981). For example, in a study of football victory celebrations, the sociologists David A. Snow, Louis A. Zurcher, and Robert Peters (1981) found that each week, behavioral patterns were changed in the postgame revelry, with some being modified, some added, and some deleted.

Mass Behavior

Not all collective behavior takes place in face-to-face collectivities. ***Mass behavior* is collective behavior that takes place when people (who often are geographically separated from one another) respond to the same event in much the same way.** For people to respond in the same way, they typically have common sources of information that provoke their collective behavior. The most frequent types of mass behavior are rumors, gossip, mass hysteria, public opinion, fashions, and fads. Under some circumstances, social movements constitute a form of mass behavior. However, we will examine social movements separately because they differ in some important ways from other types of dispersed collectivities.

Rumors and Gossip ***Rumors* are unsubstantiated reports on an issue or subject** (Rosnow and Fine, 1976). Whereas a rumor may spread through an assembled collectivity, rumors may also be transmitted among people who are dispersed geographically, including people posting messages on the Internet or talking by cell phone. Although rumors may initially contain a kernel of truth, they may be modified as they spread to serve the interests of those repeating them. Rumors thrive when tensions are high and when little authentic information is available on an issue of great concern. For example, when the blackout of August 2003 occurred, leaving 50 million people in 8 states and parts of Canada without electricity, the earliest rumors about the power failures reflected the turbulent times in which we live. One of the first rumors that began to spread was that the blackout was an act of terrorism. As one person stated in an e-mail (uncorrected by your author),

> I think its an act of terrorism. They are planning to strike. They need to know how the power system works and what cities and states will be affected so they can plan there strategy. In the meantime, USA is assuming that its natural. Typical of americans. Don't put your guards down. . . . (Cromwell, 2003)

Television broadcasters and public officials quickly tried to deflect this rumor for fear that people might panic and be injured as they sought to leave their workplaces in cities such as New York and make their way home. Other rumors that initially circulated ranged from the humorous to the accusatory: Some attributed the power failure to "UFOs," "the energy companies that want to gouge everyone," "President Bush's buddies at Enron," "homeland security crooks," and a wide variety of other explanations. Nearly a month after the massive blackout, investigators were still attempting to analyze the data to determine specific causes; however, the rumor mill had provided people with instant speculation about why this event had occurred. Fortunately, most people acted responsibly, and the riots and looting that took place during prior blackouts in New York City did not recur (*BBC News,* 2003; CNN.com, 2003b; Gibbs, 2003).

As the example of the blackout of 2003 shows, people are willing to give rumors credence when no opposing information is available. Environmental issues are similar. For example, in the past, when residents of Love Canal waited for information from health department officials about their exposure to the toxic chemicals and from the government about possible relocation at state expense to another area, new waves of rumors also spread through the community daily. By the time a meeting was called by health department officials to provide homeowners with the results of air-sample tests for hazardous chemicals (such as chloroform and benzene) performed on their homes, already fearful residents were ready to believe the worst, as Lois Gibbs (1982: 25) describes:

> Next to the names [of residents] were some numbers. But the numbers had no meaning. People stood there looking at the numbers, knowing nothing of what they meant but suspecting the worst.
>
> One woman, divorced and with three sick children, looked at the piece of paper with numbers and started crying hysterically: "No wonder my children are sick. Am I going to die? What's going to happen to my children?" No one could answer. . . .

mass behavior collective behavior that takes place when people (who often are geographically separated from one another) respond to the same event in much the same way.

rumor an unsubstantiated report on an issue or subject.

> The night was very warm and humid, and the air was stagnant. On a night like that, the smell of Love Canal is hard to describe. It's all around you. It's as though it were about to envelop you and smother you. By now, we were outside, standing in the parking lot. The woman's panic caught on, starting a chain reaction. Soon, many people there were hysterical.

Once a rumor begins to circulate, it seldom stops unless compelling information comes to the forefront that either proves the rumor false or makes it obsolete.

In industrialized societies with sophisticated technology, rumors come from a wide variety of sources and may be difficult to trace. Print media (newspapers and magazines) and electronic media (radio and television), fax machines, cellular networks, satellite systems, and the Internet aid the rapid movement of rumors around the globe. In addition, modern communications technology makes anonymity much easier. In a split second, messages (both factual and fictitious) can be disseminated to thousands of people through e-mail, computerized bulletin boards, and Internet newsgroups.

© Jonathan Fickies/Getty Images

When unexpected events such as the massive 2003 power outage in the United States and Canada occur, people frequently rely on rumors to help them know what is going on. Getting accurate information out quickly helped prevent people from panicking as tens of thousands of workers in Manhattan sought to get home any way they could while electricity was unavailable in the city.

Whereas rumors deal with an issue or a subject, ***gossip* refers to rumors about the personal lives of individuals.** Charles Horton Cooley (1963/1909) viewed gossip as something that spread among a small group of individuals who personally knew the person who was the object of the rumor. Today, this is frequently not the case; many people enjoy gossiping about people whom they have never met. Tabloid newspapers and magazines such as the *National Enquirer* and *People,* and television "news" programs that purport to provide "inside" information on the lives of celebrities, are sources of contemporary gossip, much of which has not been checked for authenticity.

Mass Hysteria and Panic *Mass hysteria* is a form of dispersed collective behavior that occurs when a large number of people react with strong emotions and self-destructive behavior to a real or perceived threat. Does mass hysteria actually occur? Although the term has been widely used, many sociologists believe that this behavior is best described as a panic with a dispersed audience.

An example of mass hysteria or a panic with a widely dispersed audience was actor Orson Welles's 1938 Halloween eve radio dramatization of H. G. Wells's science fiction classic *The War of the Worlds.* A CBS radio dance music program was interrupted suddenly by a news bulletin informing the audience that Martians had landed in New Jersey and were in the process of conquering Earth. Some listeners became extremely frightened even though an announcer had indicated before, during, and after the performance that the broadcast was a fictitious dramatization. According to some reports, as many as 1 million of the estimated 10 million listeners believed that this astonishing event had occurred. Thousands were reported to have hidden in their storm cellars or to have gotten in their cars so that they could flee from the Martians (see Brown, 1954). In actuality, the program probably did not generate mass hysteria, but rather a panic among gullible listeners. Others switched stations to determine if the same "news" was being broadcast elsewhere. When they discovered that it was not, they merely laughed at the joke being played on listeners by CBS. In 1988, on the fiftieth anniversary of the broadcast, a Portuguese radio station rebroadcast the program; once again, a panic ensued.

Fads and Fashions As you will recall from Chapter 3, a *fad* is a temporary but widely copied activity enthusiastically followed by large numbers of people. Fads can be embraced by widely dispersed collectivities; news networks such as CNN and Internet websites may bring the latest fad to the attention of audiences around the world. Recently, people in a number of countries have participated in a relatively new fad known as the "flash mob."

Unlike fads, fashions tend to be longer lasting. In Chapter 3, *fashion* is defined as a currently valued style of behavior, thinking, or appearance. Fashion also applies to art, music, drama, literature, architecture, interior design, and automobiles, among other things. However, most sociological research on fashion has focused on clothing, especially women's apparel (Davis, 1992).

In preindustrial societies, clothing styles remained relatively unchanged. With the advent of industrialization, items of apparel became readily available at low prices because of mass production. Fashion became more important as people embraced the "modern" way of life and as advertising encouraged "conspicuous consumption."

Georg Simmel, Thorstein Veblen, and Pierre Bourdieu have all viewed fashion as a means of status differentiation among members of different social classes. Simmel (1957/1904) suggested a classic "trickle-down" theory (although he did not use those exact words) to describe the process by which members of the lower classes emulate the fashions of the upper class. As the fashions descend through the status hierarchy, they are watered down and "vulgarized" so that they are no longer recognizable to members of the upper class, who then regard them as unfashionable and in bad taste (Davis, 1992). Veblen (1967/1899) asserted that fashion serves mainly to institutionalize conspicuous consumption among the wealthy. Eighty years later, Bourdieu (1984) similarly (but more subtly) suggested that "matters of taste," including fashion sensibility, constitute a large share of the "cultural capital" possessed by members of the dominant class.

Herbert Blumer (1969) disagreed with the trickle-down approach, arguing that "collective selection" best explains fashion. Blumer suggested that people in the middle and lower classes follow fashion because it is *fashion,* not because they desire to emulate members of the elite class. Blumer thus shifts the focus on fashion to collective mood, tastes, and choices: "Tastes are themselves a product of experience. . . . They are formed in the context of social interaction, responding to the definitions and affirmation given by others. People thrown into areas of common interaction and having similar runs of experience develop common tastes" (qtd. in Davis, 1992: 116). Perhaps one of the best refutations of the trickle-down approach is the way in which fashion today often originates among people in the lower social classes and is mimicked by the elites. In the mid-1990s, the so-called grunge look was a prime example of this.

Public Opinion ***Public opinion*** **consists of the attitudes and beliefs communicated by ordinary citizens to decision makers** (Greenberg and Page, 2002). It is measured through polls and surveys, which use research methods such as interviews and questionnaires, as described in Chapter 2. Many people are not interested in all aspects of public policy but are concerned about issues they believe are relevant to themselves. Even on a single topic, public opinion will vary widely based on race/ethnicity, religion, region, social class, education level, gender, age, and so on.

Scholars who examine public opinion are interested in the extent to which the public's attitudes are communicated to decision makers and the effect (if any) that public opinion has on policy making (Turner and Killian, 1993). Some political scientists argue that public opinion has a substantial effect on decisions at all levels of government (see Greenberg and Page, 2002); others strongly disagree. For example, Thomas Dye and Harmon Zeigler (2006: 439) argue that

> Public policy does not reflect demands of "the people," but rather the preferences, interests, and values of the very few who participate in the policy-making process. Changes or innovations in public policy come about when elites redefine their own interests or modify their own values. Policies decided by elites need not be oppressive or exploitative of the masses. Elites may be very public-regarding, and the welfare of the masses may be an important consideration in elite decision making. Yet it is the *elites* that make policy, not the *masses.*

From this perspective, polls may artificially create the appearance of a public opinion; pollsters may ask questions that those being interviewed had not even considered before the survey.

As the masses attempt to influence elites and vice versa, a two-way process occurs with the dissemination of ***propaganda*****—information provided by individuals or groups that have a vested interest in furthering their own cause or damaging an opposing one.** Although many of us think of propaganda in negative terms, the information provided can be correct and can have a positive effect on decision making.

gossip rumors about the personal lives of individuals.

public opinion the attitudes and beliefs communicated by ordinary citizens to decision makers.

propaganda information provided by individuals or groups that have a vested interest in furthering their own cause or damaging an opposing one.

In recent decades, grassroots environmental activists (including the Love Canal residents and the Riverkeepers) have attempted to influence public opinion. In a study of public opinion on environmental issues, the sociologist Riley E. Dunlap (1992) found that public awareness of the seriousness of environmental problems and support for environmental protection increased dramatically between the late 1960s and the 1990s. However, it is less clear that public opinion translates into action by either decision makers in government and industry or by individuals (such as a willingness to adopt a more ecologically sound lifestyle).

Initially, most grassroots environmental activists attempt to influence public opinion so that local decision makers will feel the necessity of correcting a specific problem through changes in public policy. Although activists usually do not start out seeking broader social change, they often move in that direction when they become aware of how widespread the problem is in the larger society or on a global basis. One of two types of social movements often develops at this point—one focuses on NIMBY ("not in my backyard"), whereas the other focuses on NIABY ("not in anyone's backyard") (Freudenberg and Steinsapir, 1992).

© AP Images

Martin Luther King, Jr., a leader of the civil rights movement in the 1950s and 1960s, advocated nonviolent protests that sometimes took the form of civil disobedience. Here he marches alongside his wife, Coretta Scott King, who for many years took over Dr. King's activities after he was assassinated.

Social Movements

Although collective behavior is short-lived and relatively unorganized, social movements are longer lasting, are more organized, and have specific goals or purposes. A ***social movement*** **is an organized group that acts consciously to promote or resist change through collective action** (Goldberg, 1991). Because social movements have not become institutionalized and are outside the political mainstream, they offer "outsiders" an opportunity to have their voices heard.

Social movements are more likely to develop in industrialized societies than in preindustrial societies, where acceptance of traditional beliefs and practices makes such movements unlikely. Diversity and a lack of consensus (hallmarks of industrialized nations) contribute to demands for social change, and people who participate in social movements typically lack power and other resources to bring about change without engaging in collective action. Social movements are most likely to spring up when people come to see their personal troubles as public issues that cannot be solved without a collective response.

Social movements make democracy more available to excluded groups (see Greenberg and Page, 2002). Historically, people in the United States have worked at the grassroots level to bring about changes even when elites sought to discourage activism (Adams, 1991). For example, the civil rights movement brought into its ranks African Americans in the South who had never been allowed to participate in politics (see Killian, 1984). The women's suffrage movement gave voice to women who had been denied the right to vote (Rosenthal et al., 1985). Similarly, a grassroots environmental movement gave the working-class residents of Love Canal a way to "fight city hall" and Hooker Chemicals (A. Levine, 1982), as Lois Gibbs (1982: 38–40) explains:

> People were pretty upset. They were talking and stirring each other up. I was afraid there would be violence. We had a meeting at my house to try to put everything together [and] decided to form a homeowners' association. We got out the word as best we could and told everyone to come to the Frontier Fire Hall on 102d Street. . . . The firehouse was packed with people, and more were outside. . . .
>
> I was elected president. . . . I took over the meeting but I was scared to death. It was only the second time in my life I had been in front of a microphone or a crowd. . . . We set four goals right at the beginning—(1) get all the residents within the Love Canal area who wanted to be evacuated, evacuated and relocated, especially during the construction and repair of the canal; (2) do something about propping up property values; (3) get the canal fixed properly; and (4) have air sampling and soil and water testing done throughout the whole area, so we could tell how far the contamination had spread. . . .

Most social movements rely on volunteers like Lois Gibbs to carry out the work. Traditionally, women have been strongly represented in both the membership and the leadership of many grassroots movements (A. Levine, 1982; Freudenberg and Steinsapir, 1992).

The Love Canal activists set the stage for other movements that have grappled with the kind of issues that the sociologist Kai Erikson (1994) refers to as a "new species of trouble." Erikson describes the "new species" as environmental problems that "contaminate rather than merely damage . . . they pollute, befoul, taint, rather than just create wreckage . . . they penetrate human tissue indirectly rather than just wound the surfaces by assaults of a more straightforward kind. . . . And the evidence is growing that they scare human beings in new and special ways, that they elicit an uncanny fear in us" (Erikson, 1991: 15). The chaos that Erikson (1994: 141) describes is the result of technological disasters—"meaning everything that can go wrong when systems fail, humans err, designs prove faulty, engines misfire, and so on." A recent example of such a disaster occurred in Japan, where more than 300,000 residents living within six miles of the nuclear plant at Tokaimura were told to stay indoors in the aftermath of three workers' mishandling of stainless-steel pails full of uranium, causing the worst nuclear accident in Japan's history (Larimer, 1999). Although no lives were immediately lost, workers in the plant soaked up potentially lethal doses of radiation, some of which also leaked from the plant into the community. The fifty-two nuclear power plants in Japan have been plagued by other accidents and radiation leaks, causing concern for many people around the globe (Larimer, 1999).

Social movements provide people who otherwise would not have the resources to enter the game of politics a chance to do so. We are most familiar with those movements that develop around public policy issues considered newsworthy by the media, ranging from abortion and women's rights to gun control and environmental justice. However, a number of other types of social movements exist as well.

Types of Social Movements

Social movements are difficult to classify; however, sociologists distinguish among movements on the basis of their *goals* and the *amount of change* they seek to produce (Aberle, 1966; Blumer, 1974). Some movements seek to change people, whereas others seek to change society.

Reform Movements Grassroots environmental movements are an example of *reform movements,* which seek to improve society by changing some specific aspect of the social structure. Members of reform movements usually work within the existing system to attempt to change existing public policy so that it more adequately reflects their own value system. Examples of reform movements (in addition to the environmental movement) include labor movements, animal rights movements, antinuclear movements, Mothers Against Drunk Driving, and the disability rights movement.

Sociologist Lory Britt (1993) suggested that some movements arise specifically to alter social responses to and definitions of stigmatized attributes. From this perspective, social movements may bring about changes in societal attitudes and practices while at the same time causing changes in participants' social emotions. For example, the civil rights and gay rights movements helped replace shame with pride (Britt, 1993). Such a sense of pride may extend beyond current members of a reform movement. The late César Chávez, organizer of a Mexican American farmworkers' movement that developed into the United Farm Workers Union, noted that the "consciousness and pride raised by our union is alive and thriving inside millions of young Hispanics who will never work on a farm!" (qtd. in Ayala, 1993: E4).

Revolutionary Movements Movements seeking to bring about a total change in society are referred to as *revolutionary movements.* These movements usually do not attempt to work within the existing system; rather, they aim to remake the system by replacing existing institutions with new ones. Revolutionary movements range from utopian groups seeking to establish an ideal society to radical terrorists who use fear tactics to intimidate those with whom they disagree ideologically (see Alexander and Gill, 1984; Berger, 1988; Vetter and Perlstein, 1991).

Movements based on terrorism often use tactics such as bombings, kidnappings, hostage taking, hijackings, and assassinations (Vetter and Perlstein, 1991). Over the past thirty years, a number of movements in the United States have engaged in terrorist activities or supported a policy of violence. However, the terrorist attacks in New York City and Washington, D.C., on September 11, 2001, and the events that followed those attacks proved to all of us that terrorism within this country can originate from the activities of revolutionary terrorists from outside the country as well.

Religious Movements Social movements that seek to produce radical change in individuals are typically

social movement an organized group that acts consciously to promote or resist change through collective action.

based on spiritual or supernatural belief systems. Also referred to as *expressive movements, religious movements* are concerned with renovating or renewing people through "inner change." Fundamentalist religious groups seeking to convert nonbelievers to their belief system are an example of this type of movement. Some religious movements are *millenarian*—that is, they forecast that "the end is near" and assert that an immediate change in behavior is imperative. Relatively new religious movements in industrialized Western societies have included the Hare Krishnas, the Unification Church, Scientology, and the Divine Light Mission, all of which tend to appeal to the psychological and social needs of young people seeking meaning in life that mainstream religions have not provided for them.

Alternative Movements Movements that seek limited change in some aspect of people's behavior are referred to as *alternative movements.* For example, early in the twentieth century the Women's Christian Temperance Union attempted to get people to abstain from drinking alcoholic beverages. Some analysts place "therapeutic social movements" such as Alcoholics Anonymous in this category; however, others do not, due to their belief that people must change their lives completely in order to overcome alcohol abuse (see Blumberg, 1977). More recently, a variety of "New Age" movements have directed people's behavior by emphasizing spiritual consciousness combined with a belief in reincarnation and astrology. Such practices as vegetarianism, meditation, and holistic medicine are often included in the self-improvement category. Beginning in the 1990s, some alternative movements have included the practice of yoga (usually without its traditional background in the Hindu religion) as a means by which the self can be liberated and union can be achieved with the supreme spirit or universal soul.

Resistance Movements Also referred to as *regressive movements, resistance movements* seek to prevent change or to undo change that has already occurred. Virtually all of the proactive social movements previously discussed face resistance from one or more reactive movements that hold opposing viewpoints and want to foster public policies that reflect their own beliefs. Examples of resistance movements are groups organized since the 1950s to oppose school integration, civil rights and affirmative action legislation, and domestic partnership initiatives. However, perhaps the most widely known resistance movement includes many who label themselves as "pro-life" advocates—such as Operation Rescue, which seeks to close abortion clinics and make abortion illegal under all circumstances (Gray, 1993; Van Biema, 1993). Protests by some radical anti-abortion groups have grown violent, resulting in the death of several doctors and clinic workers, and creating fear among health professionals and patients seeking abortions (Belkin, 1994).

Yoga has become an increasingly popular activity in recent years as many people have turned to alternative social movements derived from Asian traditions.

Stages in Social Movements

Do all social movements go through similar stages? Not necessarily, but there appear to be identifiable stages in virtually all movements that succeed beyond their initial phase of development.

In the *preliminary* (or *incipiency*) *stage,* widespread unrest is present as people begin to become aware of a problem. At this stage, leaders emerge to agitate others into taking action. In the *coalescence stage,* people begin to organize and to publicize the problem. At this stage, some movements become formally organized at local and regional levels. In the *institutionalization* (or *bureaucratization*) *stage,* an organizational structure develops, and a paid staff (rather than volunteers) begins to lead the group. When the movement reaches this stage, the initial zeal and idealism of members may diminish as administrators take over management of the organization. Early grassroots supporters may become disillusioned and drop out; they may also start another movement to address some as-yet-unsolved aspect of the original problem. For example, some national environmental organizations—such as the Sierra Club, the National Audubon Society, and the National Parks and Conservation Association—that started as grassroots conservation movements are currently viewed by many people as being unresponsive to local environmental problems (Cable and Cable, 1995). As a result, new movements have arisen.

Social Movement Theories

What conditions are most likely to produce social movements? Why are people drawn to these movements? Sociologists have developed several theories to answer these questions.

Relative Deprivation Theory

According to relative deprivation theory, people who are satisfied with their present condition are less likely to seek social change. Social movements arise as a response to people's perception that they have been deprived of their "fair share" (Rose, 1982). Thus, people who suffer relative deprivation are more likely to feel that change is necessary and to join a social movement in order to bring about that change. *Relative deprivation* refers to the discontent that people may feel when they compare their achievements with those of similarly situated persons and find that they have less than they think they deserve (Orum and Orum, 1968). Karl Marx captured the idea of relative deprivation in this description: "A house may be large or small; as long as the surrounding houses are small it satisfies all social demands for a dwelling. But let a palace arise beside the little house, and it shrinks from a little house to a hut" (qtd. in Ladd, 1966: 24). Movements based on relative deprivation are most likely to occur when an upswing in the standard of living is followed by a period of decline, such that people have *unfulfilled rising expectations*—newly raised hopes of a better lifestyle that are not fulfilled as rapidly as the people expected or are not realized at all.

Although most of us can relate to relative deprivation theory, it does not fully account for why people experience social discontent but fail to join a social movement. Even though discontent and feelings of deprivation may be necessary to produce certain types of social movements, they are not sufficient to bring movements into existence. In fact, the sociologist Anthony Orum (1974) found the best predictor of participation in a social movement to be prior organizational membership and involvement in other political activities.

Value-Added Theory

The value-added theory developed by sociologist Neil Smelser (1963) is based on the assumption that certain conditions are necessary for the development of a social movement. Smelser called his theory the "value-added" approach based on the concept (borrowed from the field of economics) that each step in the production process adds something to the finished product. For example, in the process of converting iron ore into automobiles, each stage "adds value" to the final product (Smelser, 1963). Similarly, Smelser asserted, six conditions are necessary and sufficient to produce social movements when they combine or interact in a particular situation:

1. *Structural conduciveness.* People must become aware of a significant problem and have the opportunity to engage in collective action. According to Smelser, movements are more likely to occur when a person, class, or agency can be singled out as the source of the problem; when channels for expressing grievances either are not available or fail; and when the aggrieved have a chance to communicate among themselves.
2. *Structural strain.* When a society or community is unable to meet people's expectations that something should be done about a problem, strain occurs in the system. The ensuing tension and conflict contribute to the development of a social movement based on people's belief that the problem would not exist if authorities had done what they were supposed to do.
3. *Spread of a generalized belief.* For a movement to develop, there must be a clear statement of the problem and a shared view of its cause, effects, and possible solution.
4. *Precipitating factors.* To reinforce the existing generalized belief, an inciting incident or dramatic event must occur. With regard to technological disasters, some (including Love Canal) gradually emerge from a long-standing environmental threat, whereas others (including the Three Mile Island nuclear power plant) involve a suddenly imposed problem.
5. *Mobilization for action.* At this stage, leaders emerge to organize others and give them a sense of direction.
6. *Social control factors.* If there is a high level of social control on the part of law enforcement officials, political leaders, and others, it becomes more difficult to develop a social movement or engage in certain types of collective action.

Value-added theory takes into account the complexity of social movements and makes it possible to test Smelser's assertions regarding the necessary and sufficient conditions that produce such movements. However, critics note that the approach is rooted in the functionalist tradition and views structural strains as disruptive to society. Smelser's theory has been described as a mere variant of convergence theory, which, you will recall, is based on the assumption that people with similar predispositions will be activated by a common event or object (Quarantelli and Hundley, 1993).

Resource Mobilization Theory

Smelser's value-added theory tends to underemphasize the importance of resources in social movements. By contrast, *resource mobilization theory* focuses on the ability of members of a social movement to acquire resources and mobilize people in order to advance their cause (Oberschall, 1973; McCarthy and Zald, 1977). Resources include money, people's time and skills, access to the media, and material goods such as property and equipment. Assistance from outsiders is essential for social movements. For example, reform movements are more likely to succeed when they gain the support of political and economic elites (Oberschall, 1973).

Resource mobilization theory is based on the assumption that participants in social movements are rational people. According to the sociologist Charles Tilly (1973, 1978), movements are formed and dissolved, mobilized and deactivated, based on rational decisions about the goals of the group, available resources, and the cost of mobilization and collective action. Resource mobilization theory also assumes that participants must have some degree of economic and political resources to make the movement a success. In other words, widespread discontent alone cannot produce a social movement; adequate resources and motivated people are essential to any concerted social action (Aminzade, 1973; Gamson, 1990). Based on an analysis of fifty-three U.S. social protest groups ranging from labor unions to peace movements between 1800 and 1945, the sociologist William Gamson (1990) concluded that the organization and tactics of a movement strongly influence its chances of success. However, critics note that this theory fails to account for social changes brought about by groups with limited resources.

Today, scholars continue to modify resource mobilization theory and to develop new approaches for investigating the diversity of movements at the beginning of the twenty-first century (see Buechler, 2000). For example, emerging perspectives based on resource mobilization theory emphasize the ideology and legitimacy of movements as well as material resources (see Zald and McCarthy, 1987; McAdam, McCarthy, and Zald, 1988). Additional perspectives are also needed on social movements in other nations to determine how activists in those countries acquire resources and mobilize people to advance causes such as environmental protection (see Box 20.2).

Social Constructionist Theory: Frame Analysis

Recent theories based on a symbolic interactionist perspective focus on the importance of the symbolic presentation of a problem to both participants and the general public (see Snow et al., 1986; Capek, 1993). Social constructionist theory is based on the assumption that a social movement is an interactive, symbolically defined, and negotiated process that involves participants, opponents, and bystanders (Buechler, 2000).

Research based on this perspective often investigates how problems are framed and what names they are given. This approach reflects the influence of the sociologist Erving Goffman's *Frame Analysis* (1974), in which he suggests that our interpretation of the particulars of events and activities is dependent on the framework from which we perceive them. According to Goffman (1974: 10), the purpose of frame analysis is "to try to isolate some of the basic frameworks of understanding available in our society for making sense out of events and to analyze the special vulnerabilities to which these frames of reference are subject." In other words, various "realities" may be simultaneously occurring among participants engaged in the same set of activities. Sociologist Steven M. Buechler (2000: 41) explains the relationship between frame analysis and social movement theory:

> Framing means focusing attention on some bounded phenomenon by imparting meaning and significance to elements within the frame and setting them apart from what is outside the frame. In the context of social movements, framing refers to the interactive, collective ways that movement actors assign meanings to their activities in the conduct of social movement activism. The concept of framing is designed for discussing the social construction of grievances as a fluid and variable process of social interaction—and hence a much more important explanatory tool than resource mobilization theory has maintained.

Sociologists have identified at least three ways in which grievances are framed. First, *diagnostic framing* identifies a problem and attributes blame or causality to some group or entity so that the social movement has a target for its actions. Second, *prognostic framing* pinpoints possible solutions or remedies, based on the target previously identified. Third, *motivational framing* provides a vocabulary of motives that compel people to take action (Benford, 1993; Snow and Benford, 1988). When successful framing occurs, the individual's vague dissatisfactions are turned into well-defined grievances, and people are compelled to join the movement in an effort to reduce or eliminate those grievances (Buechler, 2000).

Beyond motivational framing, additional frame alignment processes are necessary in order to supply a continuing sense of urgency to the movement. *Frame alignment* is the linking together of interpretive orientations of individuals and social movement organizations so that there is congruence between individuals' interests, beliefs, and values and the movement's ideologies, goals, and activities (Snow et al., 1986). Four distinct frame alignment processes occur in social movements: (1) *frame bridging* is the process by which movement organizations reach individuals who already share the same world view as the organization, (2) *frame amplification* occurs when movements appeal to deeply held values and beliefs in the general population and link those to movement issues so that people's preexisting value commitments serve as a "hook" that can be used to recruit them, (3) *frame extension* occurs when movements enlarge the boundaries of an initial frame to incorporate other issues that appear to be of importance to potential participants, and (4) *frame transformation* refers to the process whereby the creation and maintenance of new values, beliefs, and meanings induce movement participation by redefining activities and events in such a manner that people believe they must become involved in collective action (Buechler, 2000). Some or all of these frame alignment processes are used by social movements as they seek to define grievances and recruit participants.

Sociologist William Gamson (1995) believes that social movements borrow, modify, or create frames as they seek to advance their goals. For example, social activism regarding nuclear disasters involves differential framing over time. When the Fermi reactor had a serious partial meltdown in 1966, there was no sustained social opposition. However, when Pennsylvania's Three Mile Island nuclear power plant experienced a similar problem in 1979, the situation was defined as a technological disaster. At that time, social movements opposing nuclear power plants grew in number, and each intensified its efforts to reduce or eliminate such plants. This example shows that people react toward something on the basis of the meaning that thing has for them, and such meanings typically emerge from the political culture, including the social activism, of a specific era (Buechler, 2000).

Frame analysis provides new insights on how social movements emerge and grow when people are faced with problems such as technological disasters, about which greater ambiguity typically exists, and when people are attempting to "name" the problems associated with things such as nuclear or chemical contamination. However, frame analysis has been criticized for its "ideational biases" (McAdam, 1996). According to the sociologist Doug McAdam (1996), frame analyses of social movements have looked almost exclusively at ideas and their formal expression, whereas little attention has been paid to other significant factors such as movement tactics, mobilizing structures, and changing political opportunities that influence the signifying work of movements. In this context, "political opportunity" means government structure, public policy, and political conditions that set the boundaries for change and political action. These boundaries are crucial variables in explaining why various social movements have different outcomes (Meyer and Staggenborg, 1996; Gotham, 1999).

Political Opportunity Theory

Why do social protests occur? According to political opportunity theorists, the origins of social protests cannot be explained solely by the fact that people possess a variety of grievances or that they have resources available for mobilization. Instead, social protests are directly related to the political opportunities that potential protesters and movement organizers believe

© AP Images/Rich Pedroncelli

© Saul Loeb/epa/CORBIS

How is the issue of immigration framed in these photos? Research based on frame analysis often investigates how social issues are framed and what names they are given.

Box 20.2 Sociology in Global Perspective

China: A Nation of Environmental Woes and Emergent Social Activism

News Bulletin: Estimated number of premature deaths in China each year that are caused by pollution:

- Outdoor Air Pollution: 350,000 to 400,000 people
- Indoor Air Pollution: 300,000 people
- Water Pollution: 60,000 people

—World Bank data (Kahn and Yardley, 2008: A1)

China is frequently in the international news these days because of its rapid industrial growth and swift rise as a global economic power. However, accompanying this nation's double-digit growth rate has been an unprecedented pollution problem. According to some social analysts, "China is choking on its own success" (Kahn and Yardley, 2008: A1). Although the economy is on an upward swing, much of this growth is related to a vast expansion of industry and rapid patterns of urbanization. For this kind of growth to be possible, staggering amounts of energy are needed, and China derives almost all of its energy from coal, one of the dirtiest sources of energy.

If this is China's problem, why should those of us who live in the United States be concerned? For humanitarian reasons, we must be concerned about the effects of deadly pollution on the residents of China. But we must also be concerned about the effects of such environmental degradation because "What happens in China *does not* stay in China." China's pollution problems are not just national problems; they are global problems. According to the *Journal of Geophysical Research,* "Sulfur dioxide and nitrogen oxides spewed by China's coal-fired power plants fall as acid rain on Seoul, South Korea, and Tokyo. Much of the particulate pollution over Los Angeles originates in China" (qtd. in Kahn and Yardley, 2008: A6). Yes, that is correct: Some of the pollution found in Los Angeles, California, may be attributed to what happens in China!

Can anything be done about the problem? Are activists and environmental movements trying to bring about environmental conservation in China? Some environmental activists are indeed attempting to highlight the causes and consequences of the various forms of pollution that are assaulting their nation. For example, environmental activist Wu Lihong repeatedly warned public officials in Wuxi, China, that pollution was strangling Lake Tai, but little attention was paid to his concerns until after the city was forced to shut off its drinking water because a deadly algae bloom was growing rapidly on the lake. Environmental researchers partly attributed this algae bloom to heavy pollution in the area. However, rather than praising Wu Lihong for his efforts to mobilize people and raise awareness of

exist within the political system at any given point in time. Political opportunity theory is based on the assumption that social protests that take place *outside* of mainstream political institutions are deeply intertwined with more conventional political activities that take place *inside* these institutions. As used in this context, *opportunity* refers to "options for collective action, with chances and risks attached to them that depend on factors outside the mobilizing group" (Koopmans, 1999: 97). Political opportunity theory states that people will choose those options for collective action that are most readily available to them and those options that will produce the most favorable outcome for their cause.

What are some specific applications of political action theory? Urban sociologists and social movement analysts have found that those cities that provided opportunities for people's protests to be heard within urban governments were less likely to have extensive protests or riots in their communities because aggrieved people could use more conventional means to make their claims known. By contrast, urban riots were more likely to occur when activists believed that all conventional routes to protest were blocked (Eisinger, 1973). According to Doug McAdam (1982), changes in demography, migration, and the political economy in the United States (factors that were seemingly external to the civil rights movement) all contributed to a belief on the part of African Americans in the late 1960s and early 1970s that they could organize collective action and that their claims regarding the need for racial justice might be more readily heard by government officials. The study by McAdam was conducted over a period of time and looked at a single movement, and therefore was able to identify how certain aspects of the external world affect the development of social movements (Meyer and Minkoff, 2004).

Political opportunity theory has grown in popularity among sociologists who study social movements because this approach highlights the interplay of opportunity, mobilization, and political influence in determining when certain types of behavior may occur (Meyer and Minkoff, 2004). However, like other perspectives, this theory has certain limitations, includ-

the problem, public officials had him arrested on blackmail and extortion charges, claiming that he demanded money from businesses by threatening that he would expose them for illegal pollution (Bodeen, 2007). Other social movement organizers in China have also found that they risk arrest and prosecution if they publicize their concerns and try to gather resources and mobilize others for environmental causes. As a result, some organizers are hesitant to take action because they fear the consequences.

Indeed, it appears that "China is choking on its own success" and that social movements so far have made few, if any, inroads toward addressing the problem. According to the environmental researcher Wang Jinnan, "It is a very awkward situation for the country because our greatest achievement is also our biggest burden. There is pressure for change, but many people refuse to accept that we need a new approach so soon" (qtd. in Kahn and Yardley, 2008: A1, A6).

What will the future hold for environmental protection in China? According to resource mobilization theory, widespread discontent alone cannot produce a social movement: Adequate resources and motivated people are essential for any concerned social action. Some analysts believe that environmental leaders in China will eventually be able to mobilize people for change because more-affluent Chinese residents are becoming very concerned about quality-of-life issues and because cell phones, the Internet, and other methods of rapid communications are making it possible for people to organize quickly and demand governmental intervention on pressing problems such as this one.

Although some people believe that the Chinese government sends mixed messages about care for the environment, China has joined other countries in an effort to limit the use of plastic shopping bags. Other alternatives are shown here.

Reflect & Analyze

Do you believe that environmental movements in China might be organized like the most successful ones in the United States? Why or why not?

ing the fact that social movement organizations may not always be not completely distinct from, or external to, the existing political system. Social activists typically *create* their own opportunities rather than wait for them to emerge, and activists often are political entrepreneurs in their own right, much like the state and federal legislators and other governmental officials whom they seek to influence on behalf of their social cause. Overall, however, this theory calls our attention to how important the degree of openness of a political system is to the goals and tactics of social movements' organizers.

New Social Movement Theory

New social movement theory looks at a diverse array of collective actions and the manner in which those actions are based in politics, ideology, and culture. It also incorporates factors of identity, including race, class, gender, and sexuality, as sources of collective action and social movements. Examples of "new social movements" include ecofeminism and environmental justice movements. Ecofeminism emerged in the late 1970s and early 1980s out of the feminist, peace, and ecology movements. Prompted by the near-meltdown at the Three Mile Island nuclear power plant, ecofeminists established World Women in Defense of the Environment. *Ecofeminism* is based on the belief that patriarchy is a root cause of environmental problems. According to ecofeminists, patriarchy not only results in the domination of women by men but also contributes to a belief that nature is to be possessed and dominated, rather than treated as a partner (see Ortner, 1974; Merchant, 1983, 1992; Mies and Shiva, 1993).

Another "new social movement" focuses on environmental justice and the intersection of race and class in the environmental struggle (see "Sociology Works!"). Sociologist Stella M. Capek (1993) investigated a contaminated landfill in the Carver Terrace neighborhood of Texarkana, Texas, and found that residents were able to mobilize for change and win a federal buyout and relocation by symbolically linking their issue to a larger *environmental justice*

Sociology *Works!*

Fine-Tuning Theories and Data Gathering on Environmental Racism

Throughout *Sociology in Our Times,* we have examined social theories that help us understand the combined effect of factors such as race, class, and gender on the everyday lives of millions of people. In the "Sociology Works!" feature, we have focused on specific theories and how applications of those theories can help us understand the world and sometimes make it a better place in which to live.

In this chapter, we have looked at the work of new social movement theorists who have demonstrated the intersection of environmental justice with race and class: the belief that hazardous-waste treatment, storage, and disposal facilities are more likely to be located near low-income, nonwhite neighborhoods than to higher-income, predominantly white neighborhoods. This is an important issue because of the potential health risks that such sites may pose for people who live nearby. However, critics have scoffed at the suggestion that race- or class-based discrimination is involved in decisions about where hazardous-waste-materials facilities are located. Can more accurate data be gathered to help determine the nature and extent to which environmental racism exists?

During the 1980s and 1990s, the most frequently used method employed in national-level studies documenting the location of waste sites and other polluting industrial facilities was referred to as "unit-hazard coincidence" methodology. Based on this approach, researchers selected a predefined geographic unit (such as certain ZIP code areas or census tracts). Then they identified subsets of the units (areas located within a specific ZIP code or census tract) that had, or did not have, the hazard present. The researchers then compared the demographic characteristics of people living within each of the subsets to see if a larger minority population was present near the hazardous facility (see Mohai and Saha, 2007). Unit-hazard coincidence methodology assumes that the people who live within the predefined geographic units included in a study are located closer to the hazard than those individuals who do not live in those geographic units (Mohai and Saha, 2007). The problem with this approach is that the hazardous site is usually not located at the center of the ZIP code or census tract and that the geographic area being examined may be large or small, making it difficult to know for sure the racial and class characteristics of the people who live the closest to the waste facility.

In recent years, sociologists and other social scientists have begun to use other methods such as GIS (a computer system for capturing, storing, checking, integrating, manipulating, analyzing, and displaying data related to positions on the Earth's surface) to more adequately determine the distance between environmentally hazardous sites and nearby populations. By using distance-based methods to control for proximity around environmentally hazardous sites, those researchers have demonstrated that nonwhites, who made up about 25 percent of the nation's population in 1990, constituted over 40 percent of the population living within one mile of hazardous-waste facilities, meaning that racial disparities in the distribution of hazardous sites are much greater than what had been previously reported. According to social scientists Paul Mohai and Robin Saha (2007: 343), "We [find that] these disparities persist even when controlling for economic and sociopolitical variables, suggesting that factors uniquely associated with race, such as racial targeting, housing discrimination, or other race-related factors, are associated with the location of the nation's hazardous waste facilities."

Sociological theories and research pertaining to environmental racism have raised public awareness about the fact that the location of hazardous facilities is not purely coincidental in communities throughout our nation. Clearly, proximity to hazardous sites is related to the cost of the land on which the facilities are located, but the issue of proximity based on the racial/ethnic composition of residents raises an even more challenging social and ethical dilemma. However, it is also clear that the vast quantity of data available today—and the methods for obtaining that data—make it possible for us to fine-tune previous theories and obtain a better understanding of the social world in which we live.

Reflect & Analyze

New technology can cause problems for society, but it can also improve people's lives. Can you think of another way that current technology could be used to help correct a social problem in your community?

framework. Since the 1980s, the emerging environmental justice movement has focused on the issue of ***environmental racism*—the belief that a disproportionate number of hazardous facilities (including industries such as waste disposal/treatment and chemical plants) are placed in low-income areas populated primarily by people of color** (Bullard and Wright, 1992). These areas have been left out of most

Referred to as "Cancer Alley," this area of Baton Rouge, Louisiana, is home to a predominantly African American population and also to many refineries that heavily pollute the region. Sociologists suggest that environmental racism is a significant problem in the United States and other nations. What do you think?

of the environmental cleanup that has taken place in the last two decades (Schneider, 1993). Capek concludes that linking Carver Terrace with environmental justice led to it being designated as a cleanup site. She also views this as an important turning point in new social movements: "Carver Terrace is significant not only as a federal buyout and relocation of a minority community, but also as a marker of the emergence of environmental racism as a major new component of environmental social movements in the United States" (Capek, 1993: 21).

Sociologist Steven M. Buechler (2000) has argued that theories pertaining to twenty-first-century social movements should be oriented toward the structural, macrolevel contexts in which movements arise. These theories should incorporate both political and cultural dimensions of social activism:

> Social movements are historical products of the age of modernity. They arose as part of a sweeping social, political, and intellectual change that led a significant number of people to view society as a social construction that was susceptible to social reconstruction through concerted collective effort. Thus, from their inception, social movements have had a dual focus. Reflecting the political, they have always involved some form of challenge to prevailing forms of authority. Reflecting the cultural, they have always operated as symbolic laboratories in which reflexive actors pose questions of meaning, purpose, identity, and change. (Buechler, 2000: 211)

The Concept Quick Review summarizes the main theories of social movements.

As we have seen, social movements may be an important source of social change. Throughout this text, we have examined a variety of social problems that have been the focus of one or more social movements during the past hundred years. In the process of bringing about change, most movements initially develop innovative ways to get their ideas across to decision makers and the public. Some have been successful in achieving their goals; others have not. As the historian Robert A. Goldberg (1991) has suggested, gains made by social movements may be fragile, acceptance brief, and benefits minimal and easily lost. For this reason, many groups focus on preserving their gains while simultaneously fighting for those that they believe they still desire.

environmental racism the belief that a disproportionate number of hazardous facilities (including industries such as waste disposal/treatment and chemical plants) are placed in low-income areas populated primarily by people of color.

CONCEPT QUICK REVIEW

Social Movement Theories

	Key Components
Relative Deprivation	People who are discontent when they compare their achievements with those of others consider themselves relatively deprived and join social movements in order to get what they view as their "fair share," especially when there is an upswing in the economy followed by a decline.
Value-Added	Certain conditions are necessary for a social movement to develop: (1) structural conduciveness, such that people are aware of a problem and have the opportunity to engage in collective action; (2) structural strain, such that society or the community cannot meet people's expectations for taking care of the problem; (3) growth and spread of a generalized belief about causes and effects of and possible solutions to the problem; (4) precipitating factors, or events that reinforce the beliefs; (5) mobilization of participants for action; and (6) social control factors, such that society comes to allow the movement to take action.
Resource Mobilization	A variety of resources (money, members, access to media, and material goods such as equipment) are necessary for a social movement; people participate only when they feel the movement has access to these resources.
Social Construction Theory: Frame Analysis	Based on the assumption that social movements are an interactive, symbolically defined, and negotiated process involving participants, opponents, and bystanders, frame analysis is used to determine how people assign meaning to activities and processes in social movements.
Political Opportunity	People choose those options for collective action that are most readily available to them and that will produce the most favorable outcome for their cause. Social protests are directly related to the political opportunities that potential protesters and movement organizers believe exist within the political system at any given point in time.
New Social Movement	The focus is on sources of social movements, including politics, ideology, and culture. Race, class, gender, sexuality, and other sources of identity are also factors in movements such as ecofeminism and environmental justice.

Social Change in the Future

In this chapter, we have focused on collective behavior and social movements as potential forces for social change in contemporary societies. A number of other factors also contribute to social change, including the physical environment, population trends, technological development, and social institutions.

The Physical Environment and Change

Changes in the physical environment often produce changes in the lives of people; in turn, people can make dramatic changes in the physical environment, over which we have only limited control. Throughout history, natural disasters have taken their toll on individuals and societies. Major natural disasters—including hurricanes, floods, and tornados—can devastate an entire population. In September 2005, the United States experienced the worst natural disaster in this nation's history when Hurricane Katrina left a wide path of death and destruction through Louisiana, Mississippi, and Alabama. However, damage from the hurricane itself was just the beginning of how the physical environment abruptly changed, how this disaster altered the lives of millions of people, and how it raised serious questions about our national priorities and the future of the environment. Even people who did not lose family members or suffer extensive property loss in this disaster may have experienced trauma that will remain with them in the future. According to the sociologist Kai Erikson (1976, 1994), disaster-related trauma may outweigh the actual loss of physical property because memories of such events can haunt people for many years.

Some natural disasters are exacerbated by human decisions. For example, floods are viewed as natural disasters, but excessive development can contribute to a flood's severity. As office buildings, shopping malls, industrial plants, residential areas, and highways are developed, less land remains as groundcover to absorb rainfall. When heavier-than-usual rains occur, flooding becomes inevitable; some regions of the United States—such as in and around New Orleans—have remained underwater for days or even weeks in recent years. Clearly, humans cannot control the rain, but human decisions can worsen the consequences.

The destruction of large sections of New Orleans by Hurricane Katrina and the flooding that followed is an example of how human decisions may worsen the consequences of a natural disaster. If Hurricane Katrina's first wave was the storm itself, the second wave was a *man-made disaster* resulting in part from human decisions relating to planning and budgetary priorities, allocation of funds for maintaining infrastructure, and the importance of emergency preparedness. For many years it was widely known that New Orleans had a great risk of taking a direct hit from a major hurricane and that, in a worst-case scenario, the city might be flooded and badly damaged, with portions of the city rendered uninhabitable. However, despite the city's unique topography (some sections are located below sea level) and its vulnerability to harsh storms, these concerns simply were not a top priority in urban planning or resource allocation. Failure of components of the infrastructure also contributed to the city's problems when water pumps and several levees on Lake Pontchartrain failed, allowing millions of gallons of polluted water to pour out onto the city's already-flooded streets, thereby forcing residents to leave their homes. *Infrastructure* refers to a framework of support systems, such as transportation and utilities, that makes it possible to have specific land uses (commercial, residential, and recreational, for example) and a built environment (buildings, houses, highways, and such) in order to aid people's daily activities and the nation's economy. It takes money and commitment to make sure that the components of the infrastructure remain strong so that cities can withstand natural disasters and other catastrophes.

Hurricane Katrina was a massive lesson in sociology, bringing to the public's attention so many of the issues discussed in this text. By way of example, many of the people most affected by Hurricane Katrina were low-income African Americans whose residences were located in the low-lying areas of the city; the residences of many of the wealthier white residents of the city were located on higher ground and were spared the brunt of the flooding. Apparently, no adequate disaster evacuation plans had been developed for individuals without vehicles or sufficient money to leave on their own. In New Orleans, 28 percent of all African Americans have an income below the federal poverty level, a rate that is more than double the national rate.

Both the evacuation process before the hurricane and the relief efforts after the hurricane and flooding have been widely criticized because of the length of time it took for political leaders, the military, and officials in other governmental agencies to mobilize and actively begin to rescue victims, care for the ill and dying, and bring order to the city—again, a lesson in sociology. In essence, this natural disaster brought profound changes in the lives of many individuals, and it also revealed how much remains to be done if we are to overcome the deep divisions of race and class that affect everyone's identity, life chances, and opportunities. Finally, this dramatic change in the physical environment (the damage and destruction caused directly or indirectly by the hurricane) of a region revealed to our nation that it potentially lacks preparedness for massive disasters such as terrorist attacks, hurricanes, and floods.

Because flooding is one of the many problems that face the world today, it may seem strange that one of the major concerns in the twenty-first century is the availability of water. Many experts warn that water is a finite resource that is necessary for both human survival and the production of goods. However, water is being wasted and polluted, and the supply of *potable* (drinkable) water is limited. People are causing—or at least contributing to—that problem (see Box 20.3).

People also contribute to changes in the Earth's physical condition. Through soil erosion and other

Natural disasters such as Hurricane Katrina can not only produce flooding and devastation in cities such as New Orleans and in less populated areas, but they can also make us acutely aware of vast racial and economic inequalities that persist in the United States and other nations. A significant part of the recovery work in New Orleans was done by volunteers.

Box 20.3 Sociology and Social Policy

The Fight Over Water Rights

Who controls water rights—the rights to the world's finite supply of potable (drinkable) water? (▶ Figure 1 shows how finite that supply is.) Is the water supply something that individual people or governments should be able to *own*? Answers to that question may vary depending on whether an individual is one of the "haves" or the "have-nots" with regard to water. Court decisions speak of water rights in terms of *sovereignty* with regard to governmental actions and *riparian rights* with regard to individual or group water rights. *Riparian* refers to the rights of a person (or group) to water by virtue of owning or occupying the bank of a river or lake. Historically, if a river passed through your property, you had the right to take and use as much of its water as you wanted or needed, without

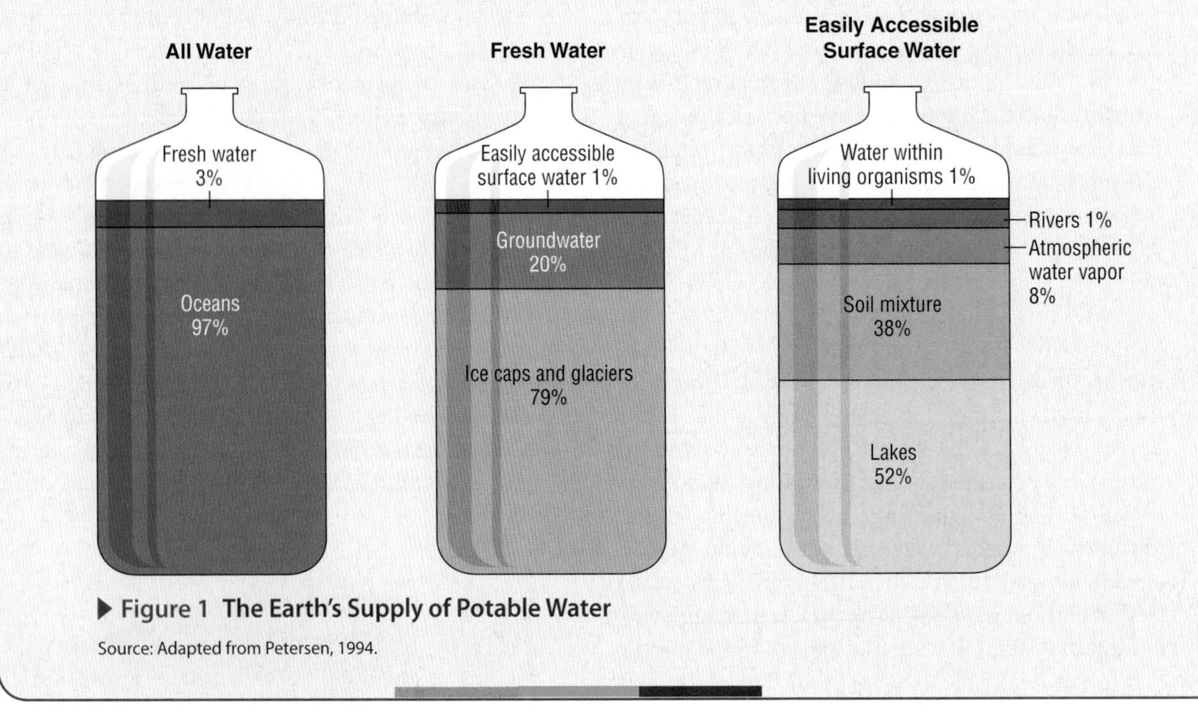

▶ **Figure 1 The Earth's Supply of Potable Water**

Source: Adapted from Petersen, 1994.

degradation of grazing land, often at the hands of people, an estimated 24 billion tons of topsoil is lost annually. As people clear forests to create farmland and pastures and to acquire lumber and firewood, the Earth's tree cover continues to diminish. As millions of people drive motor vehicles, the amount of carbon dioxide in the environment continues to rise each year, possibly resulting in global warming.

Just as people contribute to change in the physical environment, human activities must also be adapted to changes in the environment. For example, we are being warned to stay out of the sunlight because of increases in ultraviolet rays, a cause of skin cancer, as a result of the accelerating depletion of the ozone layer. If the ozone warnings are accurate, the change in the physical environment will dramatically affect those who work or spend their leisure time outside.

Population and Change

Changes in population size, distribution, and composition affect the culture and social structure of a society and change the relationships among nations. As discussed in Chapter 19, the countries experiencing the most rapid increases in population have a less-developed infrastructure to deal with those changes. How will nations of the world deal with population growth as the global population continues to move toward seven billion? Only time will provide a response to this question.

In the United States, a shift in population distribution from central cities to suburban and exurban areas has produced other dramatic changes. Central cities have experienced a shrinking tax base as middle- and upper-middle-income residents and businesses have moved to suburban and outlying areas. As a result, schools and

regard to the effect this had on people farther down the river. Accordingly, those who lived higher up a river could build a dam and divert water into lavish agricultural irrigation projects even when this resulted in water rationing for people farther down the river. Often, untreated wastewater was intentionally discharged back into the river, again without regard to the effect such action had on those farther down the river.

In the United States—as well as in the rest of the world—the assertion of riparian rights is being challenged by those whose water supply is threatened, whether by dwindling supplies or by pollution. A nation, a state, or a city may assert absolute sovereignty over the natural resources within its territory (such as its water supply), whereas nations, states, or cities farther downstream may assert another principle—governmental integrity, or the right to a supply that is adequate (in terms of both quantity and quality) to meet their own survival requirements.

One example of this conflict can be found in the Edwards Aquifer debate in Texas (an *aquifer* is an underground water supply). The Edwards Aquifer reaches from Austin to San Antonio (the nation's tenth largest city) and beyond. San Antonio relies on the aquifer for its potable water supply, but other—smaller—cities and many rural businesses also depend on the aquifer, including farmers who need the water to grow their crops. When the supply in the aquifer drops, the farmers must compete with the cities for water; attorneys representing both groups often go before the courts and regulatory agencies to argue over how much of the water each should be entitled to receive and use.

Referred to as the "western water wars," water policies have angered many people in Nevada and California. Over the past eighty years in Nevada, billions of tax dollars have been used to finance irrigation projects that siphoned off water from Pyramid Lake and diverted it to alfalfa farms and cattle ranches in the middle of the high desert east of Reno. However, the U.S. government recently started buying back much of that water and giving it to the Pyramid Lake Paiute Indians to restore fish runs and wetlands. The owners of the irrigated farms in the desert are extremely frustrated at this change in policy (Egan, 1997). In California, large volumes of water are being transferred from the big farms in the Central Valley to help depleted fish runs in the San Joaquin and Sacramento rivers, and an area "sucked dry" of its water by Los Angeles has demanded its water back (Brandon, 1997).

What will the future hold in regard to water? Many people agree that water policies are necessary; they just do not agree on what those policies should be. However, most of us believe that taking care of the environment—for today and for the future—is a worthy and necessary goal. Maintaining an adequate and nonpolluted supply of water is an integral part of that task; however, as you can see, social policy in this area often produces a great deal of conflict, and it is likely to produce even greater conflict in the future.

Reflect & Analyze

The average person in the United States uses sixty-five to seventy gallons of water per day, and this figure does not include the amount consumed outside the home—for example, water used in restaurants for preparing food and cleaning up after meals. Can you think of some easy, basic ways that individuals could conserve water during their daily lives?

public services have declined in many areas, leaving those people with the greatest needs with the fewest public resources and essential services. The changing composition of the U.S. population has resulted in children from more diverse cultural backgrounds entering school, producing a demand for new programs and changes in curricula. An increase in the birth rate has created a need for more child care; an increase in the older population has created a need for services such as medical care, placing greater demands on programs such as Social Security.

As we have seen in previous chapters, population growth and the movement of people to urban areas have brought profound changes to many regions and intensified existing social problems. Among other factors, growth in the global population is one of the most significant driving forces behind environmental concerns such as the availability and use of natural resources.

Technology and Change

Technology is an important force for change; in some ways, technological development has made our lives much easier. Advances in communication and transportation have made instantaneous worldwide communication possible but have also brought old belief systems and the status quo into question as never before. Today, we are increasingly moving information instead of people—and doing it almost instantly. Advances in science and medicine have made significant changes in people's lives in high-income countries. The light bulb, the automobile, the airplane, the assembly line, and recent high-tech developments have contributed to dramatic changes. Individuals in high-income nations have benefited from the use of the technology; those in low-income nations may have paid a disproportionate share of the cost of some of these inventions and discoveries.

Scientific advances will continue to affect our lives, from the foods we eat to our reproductive capabilities. Genetically engineered plants have been developed and marketed in recent years, and biochemists are creating potatoes, rice, and cassava with the same protein value as meat (Petersen, 1994). Advances in medicine have made it possible for those formerly unable to have children to procreate; women well beyond menopause are now able to become pregnant with the assistance of medical technology. Advances in medicine have also increased the human lifespan, especially for white and middle- or upper-class individuals in high-income nations; medical advances have also contributed to the declining death rate in low-income nations, where birth rates have not yet been curbed.

Just as technology has brought about improvements in the quality and length of life for many, it has also created the potential for new disasters, ranging from global warfare to localized technological disasters at toxic waste sites. As the sociologist William Ogburn (1966) suggested, when a change in the material culture occurs in society, a period of *cultural lag* follows in which the nonmaterial (ideological) culture has not caught up with material development. The rate of technological advance at the level of material culture today is mind-boggling. Many of us can never hope to understand technological advances in the areas of artificial intelligence, holography, virtual reality, biotechnology, cold fusion, and robotics.

One of the ironies of twenty-first-century high technology is the vulnerability that results from the increasing complexity of such systems. As futurist John L. Petersen (1994: 70) notes, "The more complex a system becomes, the more likely the chance of system failure. There are unknown secondary effects and particularly vulnerable nodes." He also asserts that most of the world's population will not participate in the technological revolution that is occurring in high-income nations (Petersen, 1994).

Pollution of lakes, rivers, and other bodies of water has an adverse effect on food supplies, air quality, and the entire environment. What influence does a "business as usual" approach have on environmental quality in your area?

Technological disasters may result in the deaths of tens of thousands, especially if we think of modern warfare as a technological disaster. Nuclear power, which can provide energy for millions, can also be the source of a nuclear war that could devastate the planet. As a government study on even limited nuclear war concluded,

> Natural resources would be destroyed; surviving equipment would be designed to use materials and skills that might no longer exist; and indeed some regions might be almost uninhabitable. Furthermore, pre-war patterns of behavior would surely change, though in unpredictable ways. (U.S. Congress, 1979, qtd. in Howard, 1990: 320)

Even when lives are not lost in technological disasters, families are uprooted and communities cease to exist as people are relocated. In many cases, the problem is not solved; people are moved either temporarily or permanently away from the site.

Social Institutions and Change

Many changes occurred in the family, religion, education, the economy, and the political system during the twentieth century and early in the twenty-first century. As we saw in Chapter 15, the size and composition of families in the United States changed with the dramatic increase in the number of single-person and single-parent households. Changes in families produced changes in the socialization of children, many of whom spend large amounts of time in front of a television set or in child-care facilities outside their own homes. Although some political and religious leaders advocate a return to "traditional" family life, numerous scholars argue that such families never worked quite as well as some might wish to believe.

Public education changed dramatically in the United States during the past century. This country was one of the first to provide "universal" education for students regardless of their ability to pay. As a result, at least until recently, the United States has had one of the most highly educated populations in the world. Today, the United States still has one of the best public education systems in the world for the top 15 percent of the students, but it badly fails the bottom 25 percent. As the nature of the economy changes, schools almost inevitably will have to change, if for no other reason than demands from leaders in business and industry for an educated work force that allows U.S. companies to compete in a global economic environment.

Political systems experienced tremendous change and upheaval in some parts of the world during the twentieth century. The United States participated in

Box 20.4 You Can Make a Difference

It's Now or Never: The Imperative of Taking Action Against Global Warming

It's pretty gut-wrenching. People say climate change is something for your kids to worry about. No. It's now!

—Canadian forestry scientist Allan Carroll emphasizing the importance of acting now to reduce global warming (qtd. in Struck, 2006)

A recurring theme of environmental social movements is that problems associated with the environment are *everyone's problem* and that change must take place before we run out of time. Clearly, we need massive initiatives implemented by elected officials and corporate leaders to reduce carbon emissions, reduce our dependency on fossil fuels, and "get serious" about dealing with the global warming problem. However, at the individual level, we each can take some steps to help reduce the amount of *greenhouse gases*—any gas that absorbs infrared radiation, such as carbon dioxide, methane, nitrous oxide, ozone, and hydrofluorocarbons—that we put into the atmosphere. Here are a few suggestions from experts:

- Walk, take mass transit, or ride a bike rather than driving alone in your car.
- Promote community carpooling and the creation of bike lanes.
- Buy a car that is fuel efficient (for every gallon of gas that is burned, twenty pounds of carbon dioxide go into the atmosphere).
- Purchase appliances, computers, televisions, and stereos that are energy efficient as shown by the ENERGY STAR® awarded by the Environmental Protection Agency.
- Have an energy audit performed by your local electric or gas utility, and make changes in your apartment or house.
- Turn off lights, televisions, and computers when you are not using them.
- Recycle cans, bottles, plastic bags, newspapers, and other items (recycling helps diminish waste disposal problems by reducing the amount of waste hauled off to landfills).
- Become actively involved in social movements and engage in political activism to put pressure on government officials to encourage industry to protect the health of our environment by reducing carbon emissions. (based on Union of Concerned Scientists, 2006, and EPA, 2006)

A famous statement by the author Mark Twain is often quoted: "Everybody talks about the weather, but nobody ever does anything about it." In the case of global warming, if nobody does anything to stop it, we may face dire consequences now, and future generations truly may be imperiled. For additional information on global warming and how you can get involved to bring about social change, visit these websites:

http://www.stopglobalwarming.org
http://www.ucsusa.org/global_warming

two world wars and numerous other "conflicts," the largest and most divisive being in Vietnam and the most recent being the war on terrorism. The cost of these wars and conflicts, and of increased security measures as a result of terrorism, is staggering.

A new concept of world security is emerging, requiring the cooperation of high-income countries in halting the proliferation of weapons of mass destruction and combating terrorism. That concept also requires the cooperation of *all* nations in reducing the plight of the poorest people in the low-income countries of the world.

Although we have examined changes in the physical environment, population, technology, and social institutions separately, they all operate together in a complex relationship, sometimes producing large, unanticipated consequences. As we move further into the twenty-first century, we need new ways of conceptualizing social life at both the macrolevel and the microlevel. The sociological imagination helps us think about how personal troubles—regardless of our race, class, gender, age, sexual orientation, or physical abilities and disabilities—are intertwined with the public issues of our society and the global community of which we are a part. As one analyst noted regarding Lois Gibbs and Love Canal,

> If Love Canal has taught Lois Gibbs—and the rest of us—anything, it is that ordinary people become very smart very quickly when their lives are threatened. They become adept at detecting absurdity, even when it is concealed in bureaucratese and scientific jargon. Lois Gibbs learned that one cannot always rely on government to act in the best interests of ordinary citizens—at least, not without considerable prodding. She determined that she would prod them until her objectives were attained. She led one of the most successful, single-purpose grass roots efforts of our time. (M. Levine, 1982: xv)

Taking care of the environment is an example of something that government and each of us as individuals can do to help (see Box 20.4). And it is vitally important that we all do everything that we can in order to protect the environment.

A Few Final Thoughts

In this text, we have covered a substantial amount of material, examined different perspectives on a wide variety of social issues, and have suggested different methods by which to deal with them. The purpose of this text is not to encourage you to take any particular point of view; rather, it is to allow you to understand different viewpoints and ways in which they may be helpful to you and to society in dealing with the issues of the twenty-first century. Possessing that understanding, we can hope that the future will be something we can all look forward to—producing a better way of life, not only in this country but worldwide as well.

Chapter Review

- **What is the relationship between social change and collective behavior?**

Social change—the alteration, modification, or transformation of public policy, culture, or social institutions over time—is usually brought about by collective behavior, which is defined as relatively spontaneous, unstructured activity that typically violates established social norms.

- **When is collective behavior likely to occur?**

Collective behavior occurs when some common influence or stimulus produces a response from a relatively large number of people.

- **What is a crowd?**

A crowd is a relatively large number of people in one another's immediate presence. Sociologist Howard Blumer divided crowds into four categories: (1) casual crowds, (2) conventional crowds, (3) expressive crowds, and (4) acting crowds (including mobs, riots, and panics). A fifth type of crowd is a protest crowd.

- **What causes crowd behavior?**

Social scientists have developed several theories to explain crowd behavior. Contagion theory asserts that a crowd takes on a life of its own as people are transformed from rational beings into part of an organism that acts on its own. A variation on this is social unrest and circular reaction—people express their discontent to others, who communicate back similar feelings, resulting in a conscious effort to engage in the crowd's behavior. Convergence theory asserts that people with similar attributes find other like-minded persons with whom they can release underlying personal tendencies. Emergent norm theory asserts that as a crowd develops, it comes up with its own norms that replace more conventional norms of behavior.

- **What are the primary forms of mass behavior?**

Mass behavior is collective behavior that occurs when people respond to the same event in the same way even if they are not in geographic proximity to one another. Rumors, gossip, mass hysteria, fads and fashions, and public opinion are forms of mass behavior.

- **What are the major types of social movements, and what are their goals?**

A social movement is an organized group that acts consciously to promote or resist change through collective action. Reform, revolutionary, religious, and alternative movements are the major types identified by sociologists. Reform movements seek to improve society by changing some specific aspect of the social structure. Revolutionary movements seek to bring about a total change in society—sometimes by the use of terrorism. Religious movements seek to produce radical change in individuals by way of spiritual or supernatural belief systems. Alternative movements seek limited change to some aspect of people's behavior. Resistance movements seek to prevent change or to undo change that has already occurred.

- **How do social movements develop?**

Social movements typically go through three stages: (1) a preliminary stage (unrest results from a perceived problem), (2) coalescence (people begin to organize), and (3) institutionalization (an organization is developed, and paid staff replaces volunteers in leadership positions).

- **How do relative deprivation theory, value-added theory, and resource mobilization theory explain social movements?**

Relative deprivation theory asserts that if people are discontented when they compare their accomplishments with those of others similarly situated, they are more likely to join a social movement than are people who are relatively content with their status. According to value-added theory, six conditions are required for a social movement: (1) a perceived problem, (2) a perception that the authorities are not resolving the problem, (3) a spread of the belief to an adequate number of people, (4) a precipitating incident, (5) mobilization of other people by leaders, and (6) a lack of social control. By contrast, resource mobilization theory asserts that successful social movements can occur only when they gain the support of political and economic elites, who provide access to the resources necessary to maintain the movement.

- **What is the primary focus of research based on frame analysis and new social movement theory?**

Research based on frame analysis often highlights the social construction of grievances through the process of social interaction. Various types of framing occur as problems are identified, remedies are sought, and people feel compelled to take action. Like frame analysis, new social movement theory has been used in research that looks at technological disasters and cases of environmental racism.

www.cengage.com/login

Want to maximize your online study time? Take this easy-to-use study system's diagnostic pre-test, and it will create a personalized study plan for you. By helping you to identify the topics that you need to understand better and then directing you to valuable online resources, it can speed up your chapter review. CengageNOW even provides a post-test so you can confirm that you are ready for an exam.

Key Terms

civil disobedience 660
collective behavior 656
crowd 659
environmental racism 674
gossip 664
mass 659
mass behavior 663
mob 660
panic 660
propaganda 665
public opinion 665
riot 660
rumors 663
social change 656
social movement 666

Questions for Critical Thinking

1. What types of collective behavior in the United States do you believe are influenced by inequalities based on race/ethnicity, class, gender, age, or disabilities? Why?
2. Which of the four explanations of crowd behavior (contagion theory, social unrest and circular reaction, convergence theory, and emergent norm theory) do you believe best explains crowd behavior? Why?
3. In the text, the Love Canal environmental movement is analyzed in terms of the value-added theory. How would you analyze that movement under the relative deprivation and resource mobilization theories?

The Kendall Companion Website

www.cengage.com/sociology/kendall

Visit this book's companion website, where you'll find more resources to help you study and successfully complete course projects. Resources include quizzes and flash cards, as well as special features such as an interactive sociology timeline, maps, General Social Survey (GSS) data, and Census 2000 data. The site also provides links to useful websites that have been selected for their relevance to the topics in this chapter and include those listed below. (*Note:* Visit the book's website for updated URLs.)

Terrorism Research Center (TRC)

http://www.terrorism.com

The mission of the TRC is to inform the public about terrorism, computer security, law enforcement, national security, and defense policy. The site provides information on current events, profiles of terrorist organizations and counterterrorism groups, original analyses, references, and links to numerous resources.

Envirolink

http://www.envirolink.org

Envirolink is an objective online information clearinghouse providing information about hundreds of environmental action groups. The website presents the latest environmental news, education resources, government links, career resources, articles, publications, and much more.

Environmental Protection Agency (EPA)

http://www.epa.gov

The mission of the EPA is to protect the natural environment and human health. Visit its website's newsroom to access the latest headlines, locate environmental information about where you live, search a comprehensive list of topics related to the environment, and access links and educational resources.

Glossary

absolute poverty a level of economic deprivation that exists when people do not have the means to secure the most basic necessities of life.

achieved status a social position that a person assumes voluntarily as a result of personal choice, merit, or direct effort.

activity theory the proposition that people tend to shift gears in late middle age and find substitutes for previous statuses, roles, and activities.

acute diseases illnesses that strike suddenly and cause dramatic incapacitation and sometimes death.

age stratification the inequalities, differences, segregation, or conflict between age groups.

agents of socialization the persons, groups, or institutions that teach us what we need to know in order to participate in society.

aggregate a collection of people who happen to be in the same place at the same time but share little else in common.

aging the physical, psychological, and social processes associated with growing older.

agrarian societies societies that use the technology of large-scale farming, including animal-drawn or energy-powered plows and equipment, to produce their food supply.

alienation a feeling of powerlessness and estrangement from other people and from oneself.

animism the belief that plants, animals, or other elements of the natural world are endowed with spirits or life forces having an effect on events in society.

anomie Emile Durkheim's designation for a condition in which social control becomes ineffective as a result of the loss of shared values and of a sense of purpose in society.

anticipatory socialization the process by which knowledge and skills are learned for future roles.

ascribed status a social position conferred at birth or received involuntarily later in life based on attributes over which the individual has little or no control, such as race/ethnicity, age, and gender.

assimilation a process by which members of subordinate racial and ethnic groups become absorbed into the dominant culture.

authoritarian leaders people who make all major group decisions and assign tasks to members.

authoritarian personality a personality type characterized by excessive conformity, submissiveness to authority, intolerance, insecurity, a high level of superstition, and rigid, stereotypic thinking.

authoritarianism a political system controlled by rulers who deny popular participation in government.

authority power that people accept as legitimate rather than coercive.

bilateral descent a system of tracing descent through both the mother's and father's sides of the family.

body consciousness a term that describes how a person perceives and feels about his or her body.

bureaucracy an organizational model characterized by a hierarchy of authority, a clear division of labor, explicit rules and procedures, and impersonality in personnel matters.

bureaucratic personality a psychological construct that describes those workers who are more concerned with following correct procedures than they are with getting the job done correctly.

capitalism an economic system characterized by private ownership of the means of production, from which personal profits can be derived through market competition and without government intervention.

capitalist class (or **bourgeoisie**) Karl Marx's term for the class that consists of those who own and control the means of production.

caste system a system of social inequality in which people's status is permanently determined at birth based on their parents' ascribed characteristics.

category a number of people who may never have met one another but share a similar characteristic, such as education level, age, race, or gender.

charismatic authority power legitimized on the basis of a leader's exceptional personal qualities.

chronic diseases illnesses that are long term or lifelong and that develop gradually or are present from birth.

chronological age a person's age based on date of birth.

church a large, bureaucratically organized religious organization that tends to seek accommodation with the larger society in order to maintain some degree of control over it.

civil disobedience nonviolent action that seeks to change a policy or law by refusing to comply with it.

civil religion the set of beliefs, rituals, and symbols that makes sacred the values of the society and places the nation in the context of the ultimate system of meaning.

class conflict Karl Marx's term for the struggle between the capitalist class and the working class.

class system a type of stratification based on the ownership and control of resources and on the type of work that people do.

cohabitation a situation in which two people live together, and think of themselves as a couple, without being legally married.

cohort a group of people born within a specified period in time.

collective behavior voluntary, often spontaneous activity that is engaged in by a large number of people and typically violates dominant-group norms and values.

comparable worth (or pay equity) the belief that wages ought to reflect the worth of a job, not the gender or race of the worker.

conflict perspectives the sociological approach that views groups in society as engaged in a continuous power struggle for control of scarce resources.

conformity the process of maintaining or changing behavior to comply with the norms established by a society, subculture, or other group.

conglomerate a combination of businesses in different commercial areas, all of which are owned by one holding company.

content analysis the systematic examination of cultural artifacts or various forms of communication to extract thematic data and draw conclusions about social life.

contingent work part-time work, temporary work, or subcontracted work that offers advantages to employers but that can be detrimental to the welfare of workers.

control group in an experiment, the group containing the subjects who are not exposed to the independent variable.

core nations according to world systems theory, dominant capitalist centers characterized by high levels of industrialization and urbanization.

corporate crime illegal acts committed by corporate employees on behalf of the corporation and with its support.

corporations large-scale organizations that have legal powers, such as the ability to enter into contracts and buy and sell property, separate from their individual owners.

correlation a relationship that exists when two variables are associated more frequently than could be expected by chance.

counterculture a group that strongly rejects dominant societal values and norms and seeks alternative lifestyles.

credentialism a process of social selection in which class advantage and social status are linked to the possession of academic qualifications.

crime behavior that violates criminal law and is punishable with fines, jail terms, and other sanctions.

criminology the systematic study of crime and the criminal justice system, including the police, courts, and prisons.

crowd a relatively large number of people who are in one another's immediate vicinity.

crude birth rate the number of live births per 1,000 people in a population in a given year.

crude death rate the number of deaths per 1,000 people in a population in a given year.

cult a religious group with practices and teachings outside the dominant cultural and religious traditions of a society.

cultural capital Pierre Bourdieu's term for people's social assets, including values, beliefs, attitudes, and competencies in language and culture.

cultural imperialism the extensive infusion of one nation's culture into other nations.

cultural lag William Ogburn's term for a gap between the technical development of a society (material culture) and its moral and legal institutions (nonmaterial culture).

cultural relativism the belief that the behaviors and customs of any culture must be viewed and analyzed by the culture's own standards.

cultural transmission the process by which children and recent immigrants become acquainted with the dominant cultural beliefs, values, norms, and accumulated knowledge of a society.

cultural universals customs and practices that occur across all societies.

culture the knowledge, language, values, customs, and material objects that are passed from person to person and from one generation to the next in a human group or society.

culture shock the disorientation that people feel when they encounter cultures radically different from their own and believe they cannot depend on their own taken-for-granted assumptions about life.

deinstitutionalization the practice of rapidly discharging patients from mental hospitals into the community.

demedicalization the process whereby a problem ceases to be defined as an illness or a disorder.

democracy a political system in which the people hold the ruling power either directly or through elected representatives.

democratic leaders leaders who encourage group discussion and decision making through consensus building.

democratic socialism an economic and political system that combines private ownership of some of the means of production, governmental distribution of some essential goods and services, and free elections.

demographic transition the process by which some societies have moved from high birth rates and death rates to relatively low birth rates and death rates as a result of technological development.

demography a subfield of sociology that examines population size, composition, and distribution.

denomination a large organized religion characterized by accommodation to society but frequently lacking in ability or intention to dominate society.

dependency theory the belief that global poverty can at least partially be attributed to the fact that the low-income countries have been exploited by the high-income countries.

dependent variable a variable that is assumed to depend on or be caused by one or more other (independent) variables.

deviance any behavior, belief, or condition that violates significant cultural norms in the society or group in which it occurs.

differential association theory the proposition that individuals have a greater tendency to deviate from societal norms when they frequently associate with persons who are more favorable toward deviance than conformity.

diffusion the transmission of cultural items or social practices from one group or society to another.

disability a reduced ability to perform tasks one would normally do at a given stage of life and that may result in stigmatization or discrimination against the person with disabilities.

discovery the process of learning about something previously unknown or unrecognized.

discrimination actions or practices of dominant-group members (or their representatives) that have a harmful effect on members of a subordinate group.

disengagement theory the proposition that older persons make a normal and healthy adjustment to aging when they detach themselves from their social roles and prepare for their eventual death.

domestic partnerships household partnerships in which an unmarried couple lives together in a committed, sexually intimate relationship and is granted the same rights and benefits as those accorded to married heterosexual couples.

dominant group a group that is advantaged and has superior resources and rights in a society.

dramaturgical analysis the study of social interaction that compares everyday life to a theatrical presentation.

drug any substance—other than food and water—that, when taken into the body, alters its functioning in some way.

dual-earner marriages marriages in which both spouses are in the labor force.

dyad a group composed of two members.

ecclesia a religious organization that is so integrated into the dominant culture that it claims as its membership all members of a society.

economy the social institution that ensures the maintenance of society through the production, distribution, and consumption of goods and services.

education the social institution responsible for the systematic transmission of knowledge, skills, and cultural values within a formally organized structure.

egalitarian family a family structure in which both partners share power and authority equally.

ego according to Sigmund Freud, the rational, reality-oriented component of personality that imposes restrictions on the innate pleasure-seeking drives of the id.

elder abuse a term used to describe physical abuse, psychological abuse, financial exploitation, and medical abuse or neglect of people age 65 or older.

elite model a view of society that sees power in political systems as being concentrated in the hands of a small group of elites whereas the masses are relatively powerless.

endogamy cultural norms prescribing that people marry within their social group or category.

entitlements certain benefit payments made by the government (Social Security, for example).

environmental racism the belief that a disproportionate number of hazardous facilities (including industries such as waste disposal/treatment and chemical plants) are placed in low-income areas populated primarily by people of color.

ethnic group a collection of people distinguished, by others or by themselves, primarily on the basis of cultural or nationality characteristics.

ethnic pluralism the coexistence of a variety of distinct racial and ethnic groups within one society.

ethnocentrism the practice of judging all other cultures by one's own culture.

ethnography a detailed study of the life and activities of a group of people by researchers who may live with that group over a period of years.

ethnomethodology the study of the commonsense knowledge that people use to understand the situations in which they find themselves.

exogamy cultural norms prescribing that people marry outside their social group or category.

experiment a research method involving a carefully designed situation in which the researcher studies the impact of certain variables on subjects' attitudes or behavior.

experimental group in an experiment, the group that contains the subjects who are exposed to an independent variable (the experimental condition) to study its effect on them.

expressive leadership an approach to leadership that provides emotional support for members.

extended family a family unit composed of relatives in addition to parents and children who live in the same household.

faith an unquestioning belief that does not require proof or scientific evidence.

families relationships in which people live together with commitment, form an economic unit and care for any young, and consider their identity to be significantly attached to the group.

family of orientation the family into which a person is born and in which early socialization usually takes place.

family of procreation the family that a person forms by having or adopting children.

feminism the belief that all people—both women and men—are equal and that they should be valued equally and have equal rights.

feminization of poverty the trend in which women are disproportionately represented among individuals living in poverty.

fertility the actual level of childbearing for an individual or a population.

field research the study of social life in its natural setting: observing and interviewing people where they live, work, and play.

folkways informal norms or everyday customs that may be violated without serious consequences within a particular culture.

formal education learning that takes place within an academic setting such as a school, which has a planned instructional process and teachers who convey specific knowledge, skills, and thinking processes to students.

formal organization a highly structured group formed for the purpose of completing certain tasks or achieving specific goals.

functional age a term used to describe observable individual attributes such as physical appearance, mobility, strength, coordination, and mental capacity that are used to assign people to age categories.

functional illiteracy the inability to read and/or write at the skill level necessary for carrying out everyday tasks.

functionalist perspectives the sociological approach that views society as a stable, orderly system.

fundamentalism a traditional religious doctrine that is conservative, is typically opposed to modernity, and rejects "worldly pleasures" in favor of otherworldly spirituality.

Gemeinschaft (guh-MINE-shoft) a traditional society in which social relationships are based on personal bonds of friendship and kinship and on intergenerational stability.

gender the culturally and socially constructed differences between females and males found in the meanings, beliefs, and practices associated with "femininity" and "masculinity."

gender bias behavior that shows favoritism toward one gender over the other.

gender identity a person's perception of the self as female or male.

gender role the attitudes, behavior, and activities that are socially defined as appropriate for each sex and are learned through the socialization process.

gender socialization the aspect of socialization that contains specific messages and practices concerning the nature of being female or male in a specific group or society.

generalized other George Herbert Mead's term for the child's awareness of the demands and expectations of the society as a whole or of the child's subculture.

genocide the deliberate, systematic killing of an entire people or nation.

gentrification the process by which members of the middle and upper-middle classes, especially whites, move into a central-city area and renovate existing properties.

gerontology the study of aging and older people.

Gesellschaft (guh-ZELL-shoft) a large, urban society in which social bonds are based on impersonal and specialized relationships, with little long-term commitment to the group or consensus on values.

goal displacement a process that occurs in organizations when the rules become an end in themselves rather than a means to an end, and organizational survival becomes more important than achievement of goals.

gossip rumors about the personal lives of individuals.

government the formal organization that has the legal and political authority to regulate the relationships among members of a society and between the society and those outside its borders.

groupthink the process by which members of a cohesive group arrive at a decision that many individual members privately believe is unwise.

Hawthorne effect a phenomenon in which changes in a subject's behavior are caused by the researcher's presence or by the subject's awareness of being studied.

health a state of complete physical, mental, and social well-being.

health care any activity intended to improve health.

health maintenance organizations (HMOs) companies that provide, for a set monthly fee, total care with an emphasis on prevention to avoid costly treatment later.

hermaphrodite a person in whom sexual differentiation is ambiguous or incomplete.

hidden curriculum the transmission of cultural values and attitudes, such as conformity and obedience to authority, through implied demands found in rules, routines, and regulations of schools.

high-income countries (sometimes referred to as **industrial countries**) nations with highly industrialized economies; technologically advanced industrial, administrative, and service occupations; and relatively high levels of national and personal income.

holistic medicine an approach to health care that focuses on prevention of illness and disease and is aimed at treating the whole person—body and mind—rather than just the part or parts in which symptoms occur.

homogamy the pattern of individuals marrying those who have similar characteristics, such as race/ethnicity, religious background, age, education, or social class.

homophobia extreme prejudice directed at gays, lesbians, bisexuals, and others who are perceived as not being heterosexual.

horticultural societies societies based on technology that supports the cultivation of plants to provide food.

hospice an organization that provides a homelike facility or home-based care (or both) for people who are terminally ill.

hunting and gathering societies societies that use simple technology for hunting animals and gathering vegetation.

hypothesis in research studies, a tentative statement of the relationship between two or more concepts.

id Sigmund Freud's term for the component of personality that includes all of the individual's basic biological drives and needs that demand immediate gratification.

ideal type an abstract model that describes the recurring characteristics of some phenomenon (such as bureaucracy).

illegitimate opportunity structures circumstances that provide an opportunity for people to acquire through illegitimate activities what they cannot achieve through legitimate channels.

impression management (presentation of self) Erving Goffman's term for people's efforts to present themselves to others in ways that are most favorable to their own interests or image.

income the economic gain derived from wages, salaries, income transfers (governmental aid), and ownership of property.

independent variable a variable that is presumed to cause or determine a dependent variable.

individual discrimination behavior consisting of one-on-one acts by members of the dominant group that harm members of the subordinate group or their property.

industrial societies societies based on technology that mechanizes production.

industrialization the process by which societies are transformed from dependence on agriculture and handmade products to an emphasis on manufacturing and related industries.

infant mortality rate the number of deaths of infants under 1 year of age per 1,000 live births in a given year.

informal education learning that occurs in a spontaneous, unplanned way.

informal side of a bureaucracy those aspects of participants' day-to-day activities and interactions that ignore, bypass, or do not correspond with the official rules and procedures of the bureaucracy.

ingroup a group to which a person belongs and with which the person feels a sense of identity.

institutional discrimination the day-to-day practices of organizations and institutions that have a harmful impact on members of subordinate groups.

instrumental leadership goal- or task-oriented leadership.

intergenerational mobility the social movement (upward or downward) experienced by family members from one generation to the next.

interlocking corporate directorates members of the board of directors of one corporation who also sit on the board(s) of other corporations.

internal colonialism according to conflict theorists, a practice that occurs when members of a racial or ethnic group are conquered or colonized and forcibly placed under the economic and political control of the dominant group.

interview a research method using a data-collection encounter in which an interviewer asks the respondent questions and records the answers.

intragenerational mobility the social movement (upward or downward) of individuals within their own lifetime.

invasion the process by which a new category of people or type of land use arrives in an area previously occupied by another group or land use.

invention the process of reshaping existing cultural items into a new form.

iron law of oligarchy according to Robert Michels, the tendency of bureaucracies to be ruled by a few people.

job deskilling a reduction in the proficiency needed to perform a specific job that leads to a corresponding reduction in the wages for that job.

juvenile delinquency a violation of law or the commission of a status offense by young people.

kinship a social network of people based on common ancestry, marriage, or adoption.

labeling theory the proposition that deviants are those people who have been successfully labeled as such by others.

labor union a group of employees who join together to bargain with an employer or a group of employers over wages, benefits, and working conditions.

laissez-faire leaders leaders who are only minimally involved in decision making and who encourage group members to make their own decisions.

language a set of symbols that expresses ideas and enables people to think and communicate with one another.

latent functions unintended functions that are hidden and remain unacknowledged by participants.

laws formal, standardized norms that have been enacted by legislatures and are enforced by formal sanctions.

liberation theology the Christian movement that advocates freedom from political subjugation within a traditional perspective and the need for social transformation to benefit the poor and downtrodden.

life chances Max Weber's term for the extent to which individuals have access to important societal resources such as food, clothing, shelter, education, and health care.

life expectancy an estimate of the average lifetime of people born in a specific year.

looking-glass self Charles Horton Cooley's term for the way in which a person's sense of self is derived from the perceptions of others.

low-income countries (sometimes referred to as **underdeveloped countries**) nations with little industrialization and low levels of national and personal income.

macrolevel analysis an approach that examines whole societies, large-scale social structures, and social systems.

managed care any system of cost containment that closely monitors and controls health care providers' decisions about medical procedures, diagnostic tests, and other services that should be provided to patients.

manifest functions functions that are intended and/or overtly recognized by the participants in a social unit.

marginal jobs jobs that differ from the employment norms of the society in which they are located.

marriage a legally recognized and/or socially approved arrangement between two or more individuals that carries certain rights and obligations and usually involves sexual activity.

mass a number of people who share an interest in a specific idea or issue but who are not in one another's immediate vicinity.

mass behavior collective behavior that takes place when people (who often are geographically separated from one another) respond to the same event in much the same way.

mass education the practice of providing free, public schooling for wide segments of a nation's population.

master status the most important status that a person occupies.

material culture a component of culture that consists of the physical or tangible creations (such as clothing, shelter, and art) that members of a society make, use, and share.

matriarchal family a family structure in which authority is held by the eldest female (usually the mother).

matriarchy a hierarchical system of social organization in which cultural, political, and economic structures are controlled by women.

matrilineal descent a system of tracing descent through the mother's side of the family.

matrilocal residence the custom of a married couple living in the same household (or community) as the wife's parents.

mechanical solidarity Emile Durkheim's term for the social cohesion of preindustrial societies, in which there is minimal division of labor and people feel united by shared values and common social bonds.

medical–industrial complex local physicians, local hospitals, and global health-related industries such as insurance companies and pharmaceutical and medical supply companies that deliver health care today.

medicalization the process whereby nonmedical problems become defined and treated as illnesses or disorders.

medicine an institutionalized system for the scientific diagnosis, treatment, and prevention of illness.

meritocracy a hierarchy in which all positions are rewarded based on people's ability and credentials.

microlevel analysis sociological theory and research that focus on small groups rather than on large-scale social structures.

middle-income countries (sometimes referred to as **developing countries**) nations with industrializing economies and moderate levels of national and personal income.

migration the movement of people from one geographic area to another for the purpose of changing residency.

militarism a term used to describe a societal focus on military ideals and an aggressive preparedness for war.

military–industrial complex the mutual interdependence of the military establishment and private military contractors.

mixed economy an economic system that combines elements of a market economy (capitalism) with elements of a command economy (socialism).

mob a highly emotional crowd whose members engage in, or are ready to engage in, violence against a specific target—a person, a category of people, or physical property.

modernization theory a perspective that links global inequality to different levels of economic development and suggests that low-income economies can move to middle- and high-income economies by achieving self-sustained economic growth.

monarchy a political system in which power resides in one person or family and is passed from generation to generation through lines of inheritance.

monogamy a marriage between two partners, usually a woman and a man.

monotheism belief in a single, supreme being or god who is responsible for significant events such as the creation of the world.

mores strongly held norms with moral and ethical connotations that may not be violated without serious consequences in a particular culture.

mortality the incidence of death in a population.

neolocal residence the custom of a married couple living in their own residence apart from both the husband's and the wife's parents.

network a web of social relationships that links one person with other people and, through them, with other people they know.

nonmaterial culture a component of culture that consists of the abstract or intangible human creations of society (such as attitudes, beliefs, and values) that influence people's behavior.

nontheism a religion based on a belief in divine spiritual forces such as sacred principles of thought and conduct, rather than a god or gods.

nonverbal communication the transfer of information between persons without the use of words.

norms established rules of behavior or standards of conduct.

nuclear family a family composed of one or two parents and their dependent children, all of whom live apart from other relatives.

occupational (white-collar) crime illegal activities committed by people in the course of their employment or financial affairs.

occupations categories of jobs that involve similar activities at different work sites.

oligopoly a condition existing when several companies overwhelmingly control an entire industry.

organic solidarity Emile Durkheim's term for the social cohesion found in industrial societies, in which people perform very specialized tasks and feel united by their mutual dependence.

organized crime a business operation that supplies illegal goods and services for profit.

outgroup a group to which a person does not belong and toward which the person may feel a sense of competitiveness or hostility.

panic a form of crowd behavior that occurs when a large number of people react to a real or perceived threat with strong emotions and self-destructive behavior.

participant observation a research method in which researchers collect data while being part of the activities of the group being studied.

pastoral societies societies based on technology that supports the domestication of large animals to provide food, typically emerging in mountainous regions and areas with low amounts of annual rainfall.

patriarchal family a family structure in which authority is held by the eldest male (usually the father).

patriarchy a hierarchical system of social organization in which cultural, political, and economic structures are controlled by men.

patrilineal descent a system of tracing descent through the father's side of the family.

patrilocal residence the custom of a married couple living in the same household (or community) as the husband's family.

peer group a group of people who are linked by common interests, equal social position, and (usually) similar age.

peripheral nations according to world systems theory, nations that are dependent on core nations for capital, have little or no industrialization (other than what may be brought in by core nations), and have uneven patterns of urbanization.

personal space the immediate area surrounding a person that the person claims as private.

pink-collar occupations relatively low-paying, nonmanual, semiskilled positions primarily held by women, such as day-care workers, checkout clerks, cashiers, and waitpersons.

pluralist model an analysis of political systems that views power as widely dispersed throughout many competing interest groups.

political action committees (PACs) organizations of special interest groups that solicit contributions from donors and fund campaigns to help elect (or defeat) candidates based on their stances on specific issues.

political crime illegal or unethical acts involving the usurpation of power by government officials, or illegal/unethical acts perpetrated against the government by outsiders seeking to make a political statement, undermine the government, or overthrow it.

political party an organization whose purpose is to gain and hold legitimate control of government.

political socialization the process by which people learn political attitudes, values, and behavior.

political sociology the area of sociology that examines the nature and consequences of power within or between societies, as well as the social and political conflicts that lead to changes in the allocation of power.

politics the social institution through which power is acquired and exercised by some people and groups.

polyandry the concurrent marriage of one woman with two or more men.

polygamy the concurrent marriage of a person of one sex with two or more members of the opposite sex.

polygyny the concurrent marriage of one man with two or more women.

polytheism a belief in more than one god.

popular culture the component of culture that consists of activities, products, and services that are assumed to appeal primarily to members of the middle and working classes.

population composition the biological and social characteristics of a population, including age, sex, race, marital status, education, occupation, income, and size of household.

population pyramid a graphic representation of the distribution of a population by sex and age.

positivism a term describing Auguste Comte's belief that the world can best be understood through scientific inquiry.

postindustrial societies societies in which technology supports a service- and information-based economy.

postmodern perspectives the sociological approach that attempts to explain social life in modern societies that are characterized by postindustrialization, consumerism, and global communications.

power according to Max Weber, the ability of people or groups to achieve their goals despite opposition from others.

power elite C. Wright Mills's term for the group made up of leaders at the top of business, the executive branch of the federal government, and the military.

prejudice a negative attitude based on faulty generalizations about members of selected racial and ethnic groups.

prestige the respect or regard with which a person or status position is regarded by others.

primary deviance the initial act of rule-breaking.

primary group Charles Horton Cooley's term for a small, less specialized group in which members engage in face-to-face, emotion-based interactions over an extended period of time.

primary labor market the sector of the labor market that consists of high-paying jobs with good benefits that have some degree of security and the possibility of future advancement.

primary sector production the sector of the economy that extracts raw materials and natural resources from the environment.

primary sex characteristics the genitalia used in the reproductive process.

probability sampling choosing participants for a study on the basis of specific characteristics, possibly including such factors as age, sex, race/ethnicity, and educational attainment.

profane the everyday, secular, or "worldly" aspects of life.

professions high-status, knowledge-based occupations.

propaganda information provided by individuals or groups that have a vested interest in furthering their own cause or damaging an opposing one.

property crime crimes including burglary (breaking into private property to commit a serious crime), motor vehicle theft, larceny-theft (theft of property worth $50 or more), and arson.

public opinion the attitudes and beliefs communicated by ordinary citizens to decision makers.

punishment any action designed to deprive a person of things of value (including liberty) because of some offense the person is thought to have committed.

questionnaire a printed research instrument containing a series of items to which subjects respond.

race a category of people who have been singled out as inferior or superior, often on the basis of physical characteristics such as skin color, hair texture, and eye shape.

racial socialization the aspect of socialization that contains specific messages and practices concerning the nature of one's racial or ethnic status.

racism a set of attitudes, beliefs, and practices that is used to justify the superior treatment of one racial or ethnic group and the inferior treatment of another racial or ethnic group.

random sampling a study approach in which every member of an entire population being studied has the same chance of being selected.

rational choice theory of deviance the belief that deviant behavior occurs when a person weighs the costs and benefits of nonconventional or criminal behavior and determines that the benefits will outweigh the risks involved in such actions.

rational–legal authority power legitimized by law or written rules and procedures. Also referred to as *bureaucratic authority.*

rationality the process by which traditional methods of social organization, characterized by informality and spontaneity, are gradually replaced by efficiently administered formal rules and procedures.

reference group a group that strongly influences a person's behavior and social attitudes, regardless of whether that individual is an actual member.

relative poverty a condition that exists when people may be able to afford basic necessities but are still unable to maintain an average standard of living.

reliability in sociological research, the extent to which a study or research instrument yields consistent results when applied to different individuals at one time or to the same individuals over time.

religion a system of beliefs and practices (rituals)—based on some sacred or supernatural realm—that guides human behavior, gives meaning to life, and unites believers into a single moral community.

research methods specific strategies or techniques for systematically conducting research.

resocialization the process of learning a new and different set of attitudes, values, and behaviors from those in one's background and previous experience.

respondents persons who provide data for analysis through interviews or questionnaires.

riot violent crowd behavior that is fueled by deep-seated emotions but is not directed at one specific target.

rituals regularly repeated and carefully prescribed forms of behaviors that symbolize a cherished value or belief.

role a set of behavioral expectations associated with a given status.

role conflict a situation in which incompatible role demands are placed on a person by two or more statuses held at the same time.

role exit a situation in which people disengage from social roles that have been central to their self-identity.

role expectation a group's or society's definition of the way that a specific role *ought* to be played.

role performance how a person *actually* plays a role.

role strain a condition that occurs when incompatible demands are built into a single status that a person occupies.

role-taking the process by which a person mentally assumes the role of another person in order to understand the world from that person's point of view.

routinization of charisma the process by which charismatic authority is succeeded by a bureaucracy controlled by a rationally established authority or by a combination of traditional and bureaucratic authority.

rumor an unsubstantiated report on an issue or subject.

sacred those aspects of life that are extraordinary or supernatural.

sanctions rewards for appropriate behavior or penalties for inappropriate behavior.

Sapir–Whorf hypothesis the proposition that language shapes the view of reality of its speakers.

scapegoat a person or group that is incapable of offering resistance to the hostility or aggression of others.

second shift Arlie Hochschild's term for the domestic work that employed women perform at home after they complete their workday on the job.

secondary analysis a research method in which researchers use existing material and analyze data that were originally collected by others.

secondary deviance the process that occurs when a person who has been labeled a deviant accepts that new identity and continues the deviant behavior.

secondary group a larger, more specialized group in which members engage in more-impersonal, goal-oriented relationships for a limited period of time.

secondary labor market the sector of the labor market that consists of low-paying jobs with few benefits and very little job security or possibility for future advancement.

secondary sector production the sector of the economy that processes raw materials (from the primary sector) into finished goods.

secondary sex characteristics the physical traits (other than reproductive organs) that identify an individual's sex.

sect a relatively small religious group that has broken away from another religious organization to renew what it views as the original version of the faith.

secularization the process by which religious beliefs, practices, and institutions lose their significance in sectors of society and culture.

segregation the spatial and social separation of categories of people by race, ethnicity, class, gender, and/or religion.

self-concept the totality of our beliefs and feelings about ourselves.

self-fulfilling prophecy the situation in which a false belief or prediction produces behavior that makes the originally false belief come true.

semiperipheral nations according to world systems theory, nations that are more developed than peripheral nations but less developed than core nations.

sex the biological and anatomical differences between females and males.

sex ratio a term used by demographers to denote the number of males for every hundred females in a given population.

sexism the subordination of one sex, usually female, based on the assumed superiority of the other sex.

sexual harassment unwanted sexual advances, requests for sexual favors, or other verbal or physical conduct of a sexual nature.

sexual orientation a person's preference for emotional–sexual relationships with members of the opposite sex (heterosexuality), the same sex (homosexuality), or both (bisexuality).

shared monopoly a condition that exists when four or fewer companies supply 50 percent or more of a particular market.

sick role the set of patterned expectations that defines the norms and values appropriate for individuals who are sick and for those who interact with them.

significant others those persons whose care, affection, and approval are especially desired and who are most important in the development of the self.

simple supernaturalism the belief that supernatural forces affect people's lives either positively or negatively.

slavery an extreme form of stratification in which some people are owned by others.

small group a collectivity small enough for all members to be acquainted with one another and to interact simultaneously.

social bond theory the proposition that the probability of deviant behavior increases when a person's ties to society are weakened or broken.

social change the alteration, modification, or transformation of public policy, culture, or social institutions over time.

social construction of reality the process by which our perception of reality is shaped largely by the subjective meaning that we give to an experience.

social control systematic practices developed by social groups to encourage conformity to norms, rules, and laws and to discourage deviance.

social Darwinism Herbert Spencer's belief that those species of animals, including human beings, best adapted to their environment survive and prosper, whereas those poorly adapted die out.

social devaluation a situation in which a person or group is considered to have less social value than other persons or groups.

social distance the extent to which people are willing to interact and establish relationships with members of racial and ethnic groups other than their own.

social epidemiology the study of the causes and distribution of health, disease, and impairment throughout a population.

social exclusion Manuel Castells's term for the process by which certain individuals and groups are systematically barred from access to positions that would enable them to have an autonomous livelihood in keeping with the social standards and values of a given social context.

social facts Emile Durkheim's term for patterned ways of acting, thinking, and feeling that exist *outside* any one individual but that exert social control over each person.

social group a group that consists of two or more people who interact frequently and share a common identity and a feeling of interdependence.

social institution a set of organized beliefs and rules that establishes how a society will attempt to meet its basic social needs.

social interaction the process by which people act toward or respond to other people; the foundation for all relationships and groups in society.

social mobility the movement of individuals or groups from one level in a stratification system to another.

social movement an organized group that acts consciously to promote or resist change through collective action.

social stratification the hierarchical arrangement of large social groups based on their control over basic resources.

social structure the complex framework of societal institutions (such as the economy, politics, and religion) and the social practices (such as rules and social roles) that make up a society and that organize and establish limits on people's behavior.

socialism an economic system characterized by public ownership of the means of production, the pursuit of collective goals, and centralized decision making.

socialization the lifelong process of social interaction through which individuals acquire a self-identity and the physical, mental, and social skills needed for survival in society.

socialized medicine a health care system in which the government owns the medical care facilities and employs the physicians.

society a large social grouping that shares the same geographical territory and is subject to the same political authority and dominant cultural expectations.

sociobiology the systematic study of how biology affects social behavior.

socioeconomic status (SES) a combined measure that, in order to determine class location, attempts to classify individuals, families, or households in terms of factors such as income, occupation, and education.

sociological imagination C. Wright Mills's term for the ability to see the relationship between individual experiences and the larger society.

sociology the systematic study of human society and social interaction.

sociology of family the subdiscipline of sociology that attempts to describe and explain patterns of family life and variations in family structure.

special interest groups political coalitions composed of individuals or groups that share a specific interest that they wish to protect or advance with the help of the political system.

split labor market a term used to describe the division of the economy into two areas of employment, a primary sector or upper tier, composed of higher-paid (usually dominant-group) workers in more-secure jobs, and a secondary sector or lower tier, composed of lower-paid (often subordinate-group) workers in jobs with little security and hazardous working conditions.

state the political entity that possesses a legitimate monopoly over the use of force within its territory to achieve its goals.

status a socially defined position in a group or society characterized by certain expectations, rights, and duties.

status symbol a material sign that informs others of a person's specific status.

stereotypes overgeneralizations about the appearance, behavior, or other characteristics of members of particular categories.

strain theory the proposition that people feel strain when they are exposed to cultural goals that they are unable to obtain because they do not have access to culturally approved means of achieving those goals.

subcontracting an agreement in which a corporation contracts with other (usually smaller) firms to provide specialized components, products, or services to the larger corporation.

subculture a group of people who share a distinctive set of cultural beliefs and behaviors that differs in some significant way from that of the larger society.

subordinate group a group whose members, because of physical or cultural characteristics, are disadvantaged and subjected to unequal treatment by the dominant group and who regard themselves as objects of collective discrimination.

succession the process by which a new category of people or type of land use gradually predominates in an area formerly dominated by another group or activity.

superego Sigmund Freud's term for the conscience, consisting of the moral and ethical aspects of personality.

survey a poll in which the researcher gathers facts or attempts to determine the relationships among facts.

symbol anything that meaningfully represents something else.

symbolic interactionist perspectives the sociological approach that views society as the sum of the interactions of individuals and groups.

taboos mores so strong that their violation is considered to be extremely offensive and even unmentionable.

technology the knowledge, techniques, and tools that allow people to transform resources into a usable form and the knowledge and skills required to use what is developed.

terrorism the calculated unlawful use of physical force or threats of violence against persons or property in order to intimidate or coerce a government, organization, or individual for the purpose of gaining some political, religious, economic, or social objective.

tertiary deviance deviance that occurs when a person who has been labeled a deviant seeks to normalize the behavior by relabeling it as nondeviant.

tertiary sector production the sector of the economy that is involved in the provision of services rather than goods.

theism a belief in a god or gods.

theory a set of logically interrelated statements that attempts to describe, explain, and (occasionally) predict social events.

total institution Erving Goffman's term for a place where people are isolated from the rest of society for a set period of time and come under the control of the officials who run the institution.

totalitarianism a political system in which the state seeks to regulate all aspects of people's public and private lives.

tracking the assignment of students to specific curriculum groups and courses on the basis of their test scores, previous grades, or other criteria.

traditional authority power that is legitimized on the basis of long-standing custom.

transnational corporations large corporations that are headquartered in one country but sell and produce goods and services in many countries.

transsexual a person in whom the sex-related structures of the brain that define gender identity are opposite from the physical sex organs of the person's body.

transvestite a male who lives as a woman or a female who lives as a man but does not alter the genitalia.

triad a group composed of three members.

unemployment rate the percentage of unemployed persons in the labor force actively seeking jobs.

universal health care a health care system in which all citizens receive medical services paid for by tax revenues.

unstructured interview an extended, open-ended interaction between an interviewer and an interviewee.

urbanization the process by which an increasing proportion of a population lives in cities rather than in rural areas.

validity in sociological research, the extent to which a study or research instrument accurately measures what it is supposed to measure.

values collective ideas about what is right or wrong, good or bad, and desirable or undesirable in a particular culture.

victimless crimes crimes that involve a willing exchange of illegal goods or services among adults.

violent crime actions—murder, forcible rape, robbery, and aggravated assault—involving force or the threat of force against others.

wage gap a term used to describe the disparity between women's and men's earnings.

war organized, armed conflict between nations or distinct political factions.

wealth the value of all of a person's or family's economic assets, including income, personal property, and income-producing property.

working class (or **proletariat**) those who must sell their labor to the owners in order to earn enough money to survive.

zero population growth the point at which no population increase occurs from year to year.

References

AARP. 2002. "Ten Facts to Remember About Social Security." Retrieved Apr. 16, 2005. Online: http://www.aarp.org/research/socialsecurity/benefits/aresearch-import-343-DD79.html#three

AAUW (American Association of University Women). 1995. *How Schools Shortchange Girls/The AAUW Report: A Study of Major Findings on Girls and Education.* New York: Marlowe.

———. 1998. *Gender Gap: Where Schools Still Fail Our Children.* Washington, DC: American Association of University Women. Retrieved Aug. 7, 2003. Online: http://aauw.org/research/girls_education/gg.cfm

———. 2001. *Hostile Hallways: Bullying, Teasing, and Sexual Harassment in School.* Washington, DC: American Association of University Women. Retrieved Aug. 6, 2003. Online: http://www.aauw.org/research/girls_education/hostile.cfm

ABA Banking Journal. 1990. "Easing Borrowers Off the Road to Bankruptcy," p. 42. Quoted in George Ritzer, *Expressing America: A Critique of the Global Credit Card Society.* Thousand Oaks, CA: Pine Forge, 1995, p. 82.

Aberle, D. F., A. K. Cohen, A. K. Davis, M. J. Leng, Jr., and F. N. Sutton. 1950. "The Functional Prerequisites of Society." *Ethics,* 60 (January): 100–111.

Aberle, David F. 1966. *The Peyote Religion Among the Navaho.* Chicago: Aldine.

Accuracy in Media. 2006. "Media Monitor." Retrieved Feb. 23, 2008. Online: http://www.aim.org/media-monitor/print/affirmative-action-for-conservatives

ACLU. 2005. "Pennsylvania Parents File First-Ever Challenge to 'Intelligent Design' Instruction in Public Schools." American Civil Liberties Union. Retrieved Mar. 18, 2006. Online: http://www.aclu.org//religion/schools/16372prs20041214.html

Adams, Tom. 1991. *Grass Roots: How Ordinary People Are Changing America.* New York: Citadel.

Adler, Jerry. 1999. "The Truth About High School." *Newsweek* (May 10): 56–58.

Adler, Patricia A., and Peter Adler. 1998. *Peer Power: Preadolescent Culture and Identity.* New Brunswick, NJ: Rutgers University Press.

———. 2003. *Constructions of Deviance: Social Power, Context, and Interaction* (4th ed.). Belmont, CA: Wadsworth.

Adorno, Theodor W., Else Frenkel-Brunswick, Daniel J. Levinson, and R. Nevitt Sanford. 1950. *The Authoritarian Personality.* New York: Harper & Row.

adrants. 2005. "Reality Is for Losers." Retrieved Apr. 9, 2005. Online: http://www.adrants.com/2005/02/reality-is-for-losers.php

Agency for Healthcare Research and Quality. 2000. "Hospitalization in the United States, 1997." Retrieved Aug. 7, 2003. Online: http://www.ahcpr.gov/data/hcup/factbk1/hcupfbk1.pdf

Agger, Ben. 1993. *Gender, Culture, and Power: Toward a Feminist Postmodern Critical Theory.* Westport, CT: Praeger.

Agnew, Robert. 1985. "Social Control Theory and Delinquency: A Longitudinal Test." *Criminology,* 23: 47–61.

Agonafir, Rebecca. 2002. "Workplace Surveillance of Internet and E-mail Usage." Retrieved July 6, 2002. Online: http://firstclass.wellesley.edu/~ragonafi/cs100.rp1.html

AGS Foundation for Health in Aging. 2007. "Aging in the Know: Nursing Home Care." Retrieved Mar. 29, 2007. Online: http://www.healthinaging.org/agingintheknow/chapters_ch_trial.asp?ch515

Aiello, John R., and S. E. Jones. 1971. "Field Study of Proxemic Behavior of Young School Children in Three Subcultural Groups." *Journal of Personality and Social Psychology,* 19: 351–356.

Akers, Ronald. 1992. *Drugs, Alcohol, and Society: Social Structure, Process, and Policy.* Belmont, CA: Wadsworth.

Akers, Ronald L. 1998. *Social Learning and Social Structure: A General Theory of Crime and Deviance.* Boston: Northeastern University Press.

Albanese, Catherine L. 2007. *America: Religions and Religion* (4th ed.). Belmont, CA: Wadsworth.

Albas, Daniel, and Cheryl Albas. 1988. "Aces and Bombers: The Post-Exam Impression Management Strategies of Students." *Symbolic Interaction,* 11 (Fall): 289–302.

Albrecht, Gary L., 1992. *The Disability Business: Rehabilitation in America.* Newbury Park, CA: Sage.

Alexander, Peter, and Roger Gill (Eds.). 1984. *Utopias.* London: Duckworth.

Alireza, Marianne. 1990. "Lifting the Veil of Tradition." *Austin American-Statesman* (Sept. 23): C1, C7.

Allport, Gordon. 1958. *The Nature of Prejudice* (abridged ed.). New York: Doubleday/Anchor.

Alonso, Alejandro A. 1998. "Urban Graffiti on the City Landscape." Paper presented at the Western Geography Graduate Conference, San Diego State University (Feb. 14). Retrieved Mar. 10, 2007. Online: http://www.streetgangs.com/academic/alonsograffiti.pdf

Altemeyer, Bob. 1981. *Right-Wing Authoritarianism.* Winnipeg: University of Manitoba Press.

———. 1988. *Enemies of Freedom: Understanding Right-Wing Authoritarianism.* San Francisco: Jossey-Bass.

Alter, Jonathan. 1999. "Bridging the Digital Divide." *Newsweek* (Sept. 20): 55.

Alvarez, Lizette. 2003. "A Push to Make la Différence Verboten in the New Europe." *New York Times* (July 27): YT8.

Alvarez, Lizette, and Michael Wilson. 2009. "Their Launching Pad: Like Sotomayor, Starting in Projects and Rising to Top of Their Fields." *New York Times* (May 31): A25.

Alwin, Duane, Philip Converse, and Steven Martin. 1985. "Living Arrangements and Social Integration." *Journal of Marriage and the Family,* 47: 319–334.

Alzheimer's Disease Education and Referral Center. 2005. "General Information." Retrieved Apr. 16, 2005. Online: http://www.alzheimers.org/generalinfo.htm

American Academy of Child and Adolescent Psychiatry. 1997. "Children and Watching TV." Retrieved June 22, 1999. Online: http://www.aacap.org/factsfam/tv.htm

American Anthropological Association. 2001. "What Is Anthropology?" Retrieved July 21, 2001. Online: http://www.aaanet.org/anthbroc.htm

American Association of Community Colleges. 2005. "Community College Facts." Retrieved Apr. 21, 2005. Online: http://www.aacc.nche.edu

American Association of Suicidology. 2006. "AAS Fact Sheets." Retrieved Jan. 28, 2007. Online: http://www.suicidology.org/displaycommon.cfm?an51&subarticlenbr5185

American Bar Association. 2003. "How Will You Maintain Your Privacy?" ABA Section of Business Law, Committee on Cyberspace Law, Subcommittee on Electronic Commerce. Retrieved June 8, 2003. Online: http://www.safeshopping.org/privacy/main.html

American Management Association. 2001. "Workplace Monitoring and Surveillance Survey, 2001." Retrieved July 8, 2002. Online: http://www.amanet.org/research/pdfs/emsfu_short.pdf

American Psychiatric Association. 1994. *Diagnostic and Statistical Manual of Mental Disorders IV.* Washington, DC: American Psychiatric Association.

American Society for Aesthetic Plastic Surgery. 2003. "Teenagers and Cosmetic (Aesthetic) Plastic Surgery." Retrieved Mar. 16, 2005. Online: http://www.surgery.org/press/news-release.php?iid51

American Sociological Association. 1997. *Code of Ethics.* Washington, DC: American Sociological Association (orig. pub. 1971).

American Universities Admission Program. 2003. "Prep School USA." Retrieved Aug. 9, 2003. Online: http://www.auap.com/prepschoolusa.html

Americans for Fairness in Lending. 2009. "Credit Cards." affil.org. Retrieved Mar. 23, 2009. Online: http://www.affil.org/consumer_rsc/credit_cards2.php

Aminzade, Ronald. 1973. "Revolution and Collective Political Violence: The Case of the Working Class of Marseille, France, 1830–1871." Working Paper #86, Center for Research on Social Organization. Ann Arbor: University of Michigan, October 1973.

Ammerman, Nancy Tatom. 1997. *Congregation and Community.* New Brunswick, NJ: Rutgers University Press.

Amott, Teresa. 1993. *Caught in the Crisis: Women and the U.S. Economy Today.* New York: Monthly Review Press.

Amott, Teresa, and Julie Matthaei. 1996. *Race, Gender, and Work: A Multicultural Economic History of Women in the United States* (rev. ed.). Boston: South End.

Ananova News Service. 2005. "Cosmetic Surgery Is Teenagers' Reward for Passing Exams." Retrieved Mar. 16, 2005. Online: http://www.ananova.com/news/story/sm_791290.html

Andersen, Margaret L., and Patricia Hill Collins (Eds.). 1998. *Race, Class, and Gender: An Anthology* (3rd ed.). Belmont, CA: Wadsworth.

Anderson, David C. 1993. "Ellen Baxter." *New York Times Magazine* (Dec. 19): 36–39.

Anderson, Elijah. 1990. *Streetwise: Race, Class, and Change in an Urban Community.* Chicago: University of Chicago Press.

———. 1999. *Code of the Street: Decency, Violence, and the Moral Life of the Inner City.* New York: Norton.

Anderson, Jenny. 2007. "Morgan Stanley to Settle Sex Bias Suit." *New York Times* (Apr. 25): C1, C18.

Angel, Ronald. 1984. "The Costs of Disability for Hispanic Males." *Social Science Quarterly,* 65: 426–443.

Angel, Ronald J., and Jacqueline L. Angel. 1993. *Painful Inheritance: Health and the New Generation of Fatherless Families.* Madison: University of Wisconsin Press.

Angelotti, Amanda. 2006. "Confessions of a Beauty Pageant Drop-Out." *Campus Progress: A Project of Center for American Progress* (Jan. 25). Retrieved Jan. 27, 2008. Online: http://www.campusprogress/org/features/727/confessions-of-a-beauty-pageant-drop-out

Angier, Natalie. 1993. "'Stopit!' She Said. 'Nomore!'" *New York Times Book Review* (Apr. 25): 12.

Anyon, Jean. 1980. "Social Class and the Hidden Curriculum of Work." *Journal of Education,* 162: 67–92.

———. 1997. *Ghetto Schooling: A Political Economy of Urban Educational Reform.* New York: Teachers College Press.

APA Online. 2000. "Psychiatric Effects of Violence." Public Information: APA Fact Sheet Series. Washington, DC: American Psychological Association. Retrieved Apr. 5, 2000. Online: http://www.psych.org/psych/htdocs/public_info/media_violence.html

Applebaum, Eileen R., and Ronald Schettkat. 1989. "Employment and Industrial Restructuring: A Comparison of the U.S. and West Germany." In E. Matzner (Ed.), *No Way to Full Employment.* Research Unit Labor Market and Employment, Discussion Paper FSI 89-16 (July): 394–448.

Appleton, Lynn M. 1995. "The Gender Regimes in American Cities." In Judith A. Garber and Robyne S. Turner (Eds.), *Gender in Urban Research.* Thousand Oaks, CA: Sage, pp. 44–59.

Arendt, Hannah. 1973. *On Revolution.* London: Penguin.

Arenson, Karen W. 2004. "Suicide of N.Y.U. Student, 19, Brings Sadness and Questions." *New York Times* (Mar. 10). Retrieved Jan. 8, 2006. Online: http://query.nytimes.com/gst/fullpage.html?sec5health&res59904EFDF153EF933A25750C0A9629C8B63

Argyris, Chris. 1960. *Understanding Organizational Behavior.* Homewood, IL: Dorsey.

———. 1962. *Interpersonal Competence and Organizational Effectiveness.* Homewood, IL: Dorsey.

Arnold, Regina A. 1990. "Processes of Victimization and Criminalization of Black Women." *Social Justice,* 17 (3): 153–166.

Asch, Adrienne. 1986. "Will Populism Empower Disabled People?" In Harry G. Boyle and Frank Reissman (Eds.), *The New Populism: The Power of Empowerment.* Philadelphia: Temple University Press, pp. 213–228.

Asch, Solomon E. 1955. "Opinions and Social Pressure." *Scientific American,* 193 (5): 31–35.

———. 1956. "Studies of Independence and Conformity: A Minority of One Against a Unanimous Majority." *Psychological Monographs,* 70 (9) (Whole No. 416).

Ashe, Arthur R., Jr. 1988. *A Hard Road to Glory: A History of the African-American Athlete.* New York: Warner.

Atchley, Robert C., and Amanda Barusch. 2004. *Social Forces and Aging: An Introduction to Social Gerontology* (10th ed.). Belmont: CA: Wadsworth.

Atkins, Rebecca. 2005. "Teenagers Opt for Cosmetic Surgery." *Manchester News.* Retrieved Mar. 16, 2005. Online: http://www.manchesteronline.co.uk/news/s/141/141788_teenagers_opt_for_cosmetic_surgery.html

Aulette, Judy Root. 1994. *Changing Families.* Belmont, CA: Wadsworth.

Aulette, Ken. 1998. *The Highwaymen: Warriors of the Information Superhighway.* San Diego, CA: Harvest/Harcourt.

Austin American-Statesman. 2001. "High-Tech Manufacturing in Mexico." (Sept. 24): D1.

Axinn, June. 1989. "Women and Aging: Issues of Adequacy and Equity." In J. D. Barner and Susan O. Mercer (Eds.), *Women As They Age: Challenges, Opportunity and Triumph.* Binghamton, NY: Haworth, pp. 339–362.

Axtell, Roger E. 1991. *Gestures: The Do's and Taboos of Body Language Around the World.* New York: Wiley.

Ayala, Elaine. 1993. "Unfinished Work: Cesar Chavez Is Dead But His Struggle Has Been Reborn, Followers Say." *Austin American-Statesman* (Sept. 6): E1, E4. Quoted from a speech by Cesar Chavez to the Commonwealth Club of San Francisco, Nov. 9, 1984.

Babbie, Earl. 2004. *The Practice of Social Research* (10th ed.). Belmont, CA: Wadsworth.

Bagby, Meredith (Ed.). 1997. *Annual Report of the United States of America 1997.* New York: McGraw-Hill.

Bahr, Howard M., and Theodore Caplow. 1991. "Middletown as an Urban Case Study." In Joe R. Feagin, Anthony M. Orum, and Gideon Sjoberg (Eds.), *A Case for the Case Study.* Chapel Hill: University of North Carolina Press, pp. 80–120.

Bailyn, Bernard. 1960. *Education in the Forming of American Society.* New York: Random House.

Baker, Robert. 1993. "'Pricks' and 'Chicks': A Plea for 'Persons.'" In Anne Minas (Ed.), *Gender Basics: Feminist Perspectives on Women and Men.* Belmont, CA: Wadsworth, pp. 66–68.

Ballantine, Jeanne H. 2001. *The Sociology of Education: A Systematic Analysis* (5th ed.). Englewood Cliffs, NJ: Prentice Hall.

Ballantine, Jeanne H., and Floyd M. Hammack. 2009. *The Sociology of Education: A Systematic Analysis* (6th ed.). Englewood Cliffs, NJ: Prentice Hall.

Ballara, Marcela. 1991. *Women and Literacy.* Prepared for the UN/NGO Group on Women and Development. Atlantic Highlands, NJ: Zed.

Baltzell, E. Digby. 1958. *Philadelphia Gentlemen: The Making of a National Upper Class.* New York: Free Press.

Bane, Mary Jo. 1986. "Household Composition and Poverty: Which Comes First?" In Sheldon H. Danziger and Daniel H. Weinberg (Eds.), *Fighting Poverty: What Works and What Doesn't.* Cambridge, MA: Harvard University Press.

Bane, Mary Jo, and David T. Ellwood. 1994. *Welfare Realities: From Rhetoric to Reform.* Cambridge, MA: Harvard University Press.

Banet-Weiser, Sarah. 1999. *The Most Beautiful Girl in the World: Beauty Pageants and National Identity.* Berkeley: University of California Press.

Banner, Lois W. 1993. *In Full Flower: Aging Women, Power, and Sexuality.* New York: Vintage.

Barakat, Matthew. 2000. "Survey: Women's Salaries Beat Men's in Some Fields." *Austin American-Statesman* (July 4): D1, D3.

Barber, Benjamin R. 1996. *Jihad vs. McWorld: How Globalism and Tribalism Are Reshaping the World.* New York: Ballantine.

Barboza, David. 2001. "From Golden Arches to Lightning Rod." *International Herald Tribune* (Oct. 15). Retrieved Oct. 22, 2001. Online: http://www.iht.com

———. 2005. "China, New Land of Shoppers, Builds Malls on Gigantic Scale." *New York Times* (May 25): A1.

Bardwell, Jill R., Samuel W. Cochran, and Sharon Walker. 1986. "Relationship of Parental Education, Race, and Gender to Sex Role Stereotyping in Five-Year-Old Kindergarteners." *Sex Roles,* 15: 275–281.

Barlow, Hugh D., and David Kauzlarich. 2002. *Introduction to Criminology* (8th ed.). Upper Saddle River, NJ: Prentice Hall.

Barna, George. 1996. *Index of Leading Spiritual Indicators.* Dallas: Word.

Barnard, Chester. 1938. *The Functions of the Executive.* Cambridge, MA: Harvard University Press.

Barnett, Harold. 1979. "Wealth, Crime and Capital Accumulation." *Contemporary Crises,* 3: 171–186.

Barovick, Harriet. 2001. "Hope in the Heartland." *Time* (special edition, July): G1–G3.

Barr, Bob. 2004. "Testimony Submitted by Bob Barr, Former Member of Congress, to the House Judiciary Committee, Subcommittee on the Federal Marriage Amendment." Retrieved Apr. 6, 2004. Online: http://www.aclu.org/LesbianGayRights/LesbianGayRights.cfm?ID515382&c5101

Barron, James. 1997. "A Life, Like a Race, May Be Long." *New York Times* (Nov. 1): A13.

Barrymore, Drew, with Todd Gold. 1994. *Little Girl Lost.* New York: Pocket. In Jay David (Ed.), *The Family Secret: An Anthology.* New York: Morrow, pp. 171–199.

Basow, Susan A. 1992. *Gender Stereotypes and Roles* (3rd ed.). Pacific Grove, CA: Brooks/Cole.

Baudrillard, Jean. 1983. *Simulations.* New York: Semiotext.

———. 1998. *The Consumer Society: Myths and Structures.* London: Sage (orig. pub. 1970).

Baxter, J. 1970. "Interpersonal Spacing in Natural Settings." *Sociology,* 36 (3): 444–456.

BBC News. 2003. "What Caused the Blackouts?" Retrieved Aug. 15, 2003. Online: http://news.bbc.co.uk/1/hi/business/3153237.stm

Becker, Howard S. 1963. *Outsiders: Studies in the Sociology of Deviance.* New York: Free Press.

Beech, Hannah. 2001. "China's Lifestyle Choice." *Time* (Aug. 6): 32.

Beeghley, Leonard. 2008. *The Structure of Social Stratification in the United States* (5th ed.). Boston: Allyn & Bacon.

Belkin, Lisa. 1994. "Kill for Life?" *New York Times Magazine* (Oct. 30): 47–51, 62–64, 76, 80.

———. 2006. "Life's Work: The Best Part Comes in the Third Act." *New York Times* (July 2). Retrieved Mar. 25, 2007. Online: http://select.nytimes.com/search/restricted/article?res5FB0E10F83E540C718CDDAE0894DE404482

Bell, Daniel. 1973. *The Coming of Post-Industrial Society.* New York: Basic.

Bell, Inge Powell. 1989. "The Double Standard: Age." In Jo Freeman (Ed.), *Women: A Feminist Perspective* (4th ed.). Mountain View, CA: Mayfield, pp. 236–244.

Bellah, Robert N. 1967. "Civil Religion." *Daedalus,* 96: 1–21.

Belsky, Janet K. 1999. *The Psychology of Aging: Theory, Research, and Interventions* (3rd ed.). Belmont, CA: Wadsworth.

Benford, Robert D. 1993. "'You Could Be the Hundredth Monkey': Collective Action Frames and Vocabularies of Motive Within the Nuclear Disarmament Movement." *Sociological Quarterly,* 34: 195–216.

Benjamin, Lois. 1991. *The Black Elite: Facing the Color Line in the Twilight of the Twentieth Century.* Chicago: Nelson-Hall.

Bennahum, David S. 1999. "For Kosovars, an On-Line Phone Directory of a People in Exile." *New York Times* (July 15): D7.

Benokraitis, Nijole V. 1999. *Marriages and Families: Changes, Choices, and Constraints* (3rd ed.). Upper Saddle River, NJ: Prentice Hall.

——. 2002. *Marriages and Families: Changes, Choices, and Constraints* (4th ed.). Upper Saddle River, NJ: Prentice-Hall.

Benokraitis, Nijole V., and Joe R. Feagin. 1995. *Modern Sexism: Blatant, Subtle, and Covert Discrimination* (2nd ed.). Englewood Cliffs, NJ: Prentice Hall.

Berg, Bruce L. 1998. *Qualitative Research Methods for the Social Sciences.* Boston: Allyn & Bacon.

Berger, Bennett M. 1988. "Utopia and Its Environment." *Society* (January/February): 37–41.

Berger, Peter. 1963. *Invitation to Sociology: A Humanistic Perspective.* New York: Anchor.

——. 1967. *The Sacred Canopy: Elements of a Sociological Theory of Religion.* New York: Doubleday.

Berger, Peter, and Hansfried Kellner. 1964. "Marriage and the Construction of Reality." *Diogenes,* 46: 1–32.

Berger, Peter, and Thomas Luckmann. 1967. *The Social Construction of Reality: A Treatise in the Sociology of Knowledge.* Garden City, NY: Anchor.

Berliner, David C., and Bruce J. Biddle. 1995. *The Manufactured Crisis: Myths, Fraud, and the Attack on America's Public Schools.* Reading, MA: Addison-Wesley.

Bernard, Jessie. 1982. *The Future of Marriage.* New Haven, CT: Yale University Press (orig. pub. 1973).

Better Health Channel. 2007. "Food and Celebrations." Retrieved Feb. 11, 2007. Online: http://www.betterhealth.vic.gov.au

Biblarz, Arturo, R. Michael Brown, Dolores Noonan Biblarz, Mary Pilgram, and Brent F. Baldree. 1991. "Media Influence on Attitudes Toward Suicide." *Suicide and Life-Threatening Behavior,* 21 (4): 374–385.

Bierlein, Louann A. 1997. "The Charter School Movement." In Diana Ravitch and Joseph P. Viteritti (Eds.), *New Schools for a New Century: The Redesign of Urban Education.* New Haven, CT: Yale University Press, pp. 37–60.

Billingsley, Andrew. 1992. *Climbing Jacob's Ladder: The Enduring Legacy of African-American Families.* New York: Touchstone.

Blanchard, Kendall. 1980. "Sport and Ritual in Choctaw Society: Structure and Perspective." In Helen Schwartzman (Ed.), *Play and Culture.* Champaign, IL: Leisure, pp. 83–91.

Blau, Peter. 1964. *Exchange and Power in Social Life.* New York: Wiley.

——. 1975. *Approaches to the Study of Social Structure.* New York: Free Press.

Blau, Peter M., and Otis Dudley Duncan. 1967. *The American Occupational Structure.* New York: Wiley.

Blau, Peter M., and Marshall W. Meyer. 1987. *Bureaucracy in Modern Society* (3rd ed.). New York: Random House.

Blauner, Bob. 1989. *Black Lives, White Lives.* Berkeley: University of California Press.

Blauner, Robert. 1964. *Alienation and Freedom.* Chicago: University of Chicago Press.

——. 1972. *Racial Oppression in America.* New York: Harper & Row.

Block, Fred, Anna C. Korteweg, and Kerry Woodward, with Zach Schiller and Imrul Mazid. 2008. "The Compassion Gap in American Poverty Policy." In Jeff Goodwin and James M. Jasper (Eds.), *The Contexts Reader.* New York: Norton, pp. 166–175. Originally appeared in *Contexts,* a journal published by the American Sociological Association (Spring 2006).

Bluestone, Barry, and Bennett Harrison. 1982. *The Deindustrialization of America.* New York: Basic.

Blumberg, Leonard. 1977. "The Ideology of a Therapeutic Social Movement: Alcoholics Anonymous." *Journal of Studies on Alcohol,* 38: 2122–2143.

Blumer, Herbert G. 1946. "Collective Behavior." In Alfred McClung Lee (Ed.), *A New Outline of the Principles of Sociology.* New York: Barnes & Noble, pp. 167–219.

——. 1969. *Symbolic Interactionism: Perspective and Method.* Englewood Cliffs, NJ: Prentice Hall.

——. 1974. "Social Movements." In R. Serge Denisoff (Ed.), *The Sociology of Dissent.* New York: Harcourt, pp. 74–90.

——. 1986. *Symbolic Interactionism: Perspective and Method.* Berkeley: University of California Press (orig. pub. 1969).

Bodeen, Christopher. 2007. "China's Woes Vindicate Whistle-Blower, But Court Case Against Him Continues." *International Business Times* (June 9). Retrieved Mar. 8, 2008. Online: http://www.ibtimes.com/articles/20070609/china-whistleblower-wu-lihong_all.htm

Bogardus, Emory S. 1925. "Measuring Social Distance." *Journal of Applied Sociology,* 9: 299–308.

——. 1968. "Comparing Racial Distance in Ethiopia, South Africa, and the United States." *Sociology and Social Research,* 52 (2): 149–156.

Bograd, Michele. 1988. "Feminist Perspectives on Wife Abuse: An Introduction." In Kersti Yllo and Michele Bograd (Eds.), *Feminist Perspectives on Wife Abuse.* Newbury Park, CA: Sage, pp. 11–26.

Bologh, Roslyn Wallach. 1992. "The Promise and Failure of Ethnomethodology from a Feminist Perspective: Comment on Rogers." *Gender & Society,* 6 (2): 199–206.

Bonacich, Edna. 1972. "A Theory of Ethnic Antagonism: The Split Labor Market." *American Sociological Review,* 37: 547–549.

——. 1976. "Advanced Capitalism and Black–White Relations in the United States: A Split Labor Market Interpretation." *American Sociological Review,* 41: 34–51.

Bonvillain, Nancy. 2001. *Women & Men: Cultural Constructs of Gender* (3rd ed.). Upper Saddle River, NJ: Prentice Hall.

Boodman, Sandra G. 2004. "For More Teenager Girls, Adult Plastic Surgery." *Washington Post.* Retrieved Mar. 16, 2005. Online: http://www.washingtonpost.com/ac2/wp-dyn/A61540-2004Oct25.html

Bordewich, Fergus M. 1996. *Killing the White Man's Indian.* New York: Anchor/Doubleday.

Bordo, Susan. 2004. *Unbearable Weight: Feminism, Western Culture, and the Body* (10th anniversary edition). Berkeley: University of California Press.

Bourdieu, Pierre. 1984. *Distinction: A Social Critique of the Judgement of Taste.* Trans. Richard Nice. Cambridge, MA: Harvard University Press.

Bourdieu, Pierre, and Jean-Claude Passeron. 1990. *Reproduction in Education, Society and Culture.* Newbury Park, CA: Sage.

Bourdon, Karen H., Donald S. Rae, Ben Z. Locke, William E. Narrow, and Darrel A. Regier. 1992. "Estimating the Prevalence of Mental Disorders in U.S. Adults from the Epidemiological Catchment Area Survey." *Public Health Reports,* 107: 663–668.

Bowles, Samuel, and Herbert Gintis. 1976. *Schooling in Capitalist America: Education and the Contradictions of Economic Life.* New York: Basic.

Boyes, William, and Michael Melvin. 2002. *Economics* (5th ed.). Boston: Houghton Mifflin.

Bozett, Frederick. 1988. "Gay Fatherhood." In Phyllis Bronstein and Carolyn Pape Cowan (Eds.), *Fatherhood Today: Men's Changing Role in the Family.* New York: Wiley, pp. 60–71.

Bramlett, Matthew D., and William D. Mosher. 2001. "First Marriage Dissolution, Divorce, and Remarriage: United States." DHHS publication no. 2001-1250 01-0384 (5/01). Hyattsville, MD: Department of Health and Human Services.

Brandl, Bonnie, and Loree Cook-Daniels. 2002. "Domestic Abuse in Later Life." Retrieved July 26, 2003. Online: http://www.elderabusecenter.org/pdf/research/abusers.pdf

Brandl, Steven, Meghan Stroshine, and James Frank. 2001. "Who Are the Complaint-Prone Officers? An Examination of the Relationship Between Police Officers' Attributes, Arrest Activity, Assignment, and Citizens' Complaints About Excessive Force." *Journal of Criminal Justice,* 29: 521–529.

Brandon, Karen. 1997. "Area Sucked Dry by L.A. Wants Its Water Back." *Austin American-Statesman* (Aug. 10): K3.

Braverman, Harry. 1974. *Labor and Monopoly Capital.* New York: Monthly Review Press.

Breault, K. D. 1986. "Suicide in America: A Test of Durkheim's Theory of Religious and Family Integration, 1933–1980." *American Journal of Sociology,* 92 (3): 628–656.

Bremner, Brian. 2000. "A Japanese Way of Death." *Business Week* (Aug. 22). Retrieved Aug. 25, 2001. Online: http://www.businessweek.com/bwdaily/dnflash/aug2000/nf20000822_176.htm

Briggs, Sheila. 1987. "Women and Religion." In Beth B. Hess and Myra Marx Ferree (Eds.), *Analyzing Gender: A Handbook of Social Science Research.* Newbury Park, CA: Sage, pp. 408–441.

Brint, Steven. 1994. *In an Age of Experts: The Changing Role of Professionals in Politics and Public Life.* Princeton, NJ: Princeton University Press.

Britt, Lory. 1993. "From Shame to Pride: Social Movements and Individual Affect." Paper presented at the 88th annual meeting of the American Sociological Association, Miami, August.

Broder, John M. 2003. "Debris Is Now Leading Suspect in Shuttle Catastrophe." *New York Times* (Feb. 4): A1–A25.

Bronfenbrenner, Urie. 1989. "Ecological Systems Theory." In Ross Vasta (Ed.), *Annals of Child Development: A Research Annual* (Vol. 6). Greenwich, CT: JAI.

——. 1990. "Five Critical Processes for Positive Development." From "Discovering What Families Do" in *Rebuilding the Nest: A New Commitment to the American Family.* Retrieved June 29, 1999. Online: http://www.montana.edu/wwwctf/process.html

Brooks, A. Phillips, and Jeff South. 1995. "School Choice Plans Worry Resegregation Critics." *Austin American-Statesman* (Apr. 9): A1, A18.

Brooks-Gunn, Jeanne. 1986. "The Relationship of Maternal Beliefs About Sex Typing to Maternal and Young Children's Behavior." *Sex Roles,* 14: 21–35.

Brown, Dennis M. 2003. "Rural Tourism: An Annotated Bibliography." United States Department of Agriculture. Retrieved Aug. 9, 2003. Online: http://www.nal.usda.gov/ric/ricpubs/rural_tourism.html#summary

Brown, E. Richard. 1979. *Rockefeller Medicine Men.* Berkeley: University of California Press.

Brown, Phil. 1985. *The Transfer of Care: Psychiatric Deinstitutionalization and Its Aftermath.* Boston: Routledge & Kegan Paul.

Brown, Robert W. 1954. "Mass Phenomena." In Gardner Lindzey (Ed.), *Handbook of Social Psychology* (vol. 2). Reading, MA: Addison-Wesley, pp. 833–873.

Browne, Colette, and Alice Broderick. 1994. "Asian and Pacific Island Elders: Issues for Social Work Practice and Education." *Social Work,* 39 (3): 252–260.

Bruce, Steve. 1996. *Religion in the Modern World.* New York: Oxford University Press.

Brustad, Robert J. 1996. "Attraction to Physical Activity in Urban Schoolchildren: Parental Socialization and Gender Influence." *Research Quarterly for Exercise and Sport,* 67: 316–324.

Buchholz, Brad. 2007. "Gunshots Shatter Serenity of Sacred Sanctuary." *Austin American-Statesman* (Apr. 22): G1, G4.

Bucks, Brian K., Arthur B. Kennickell, and Kevin B. Moore. 2006. "Recent Changes in U.S. Family Finances: Evidence from the 2001 and 2004 Survey of Consumer Finances." Federal Reserve Board. Retrieved Feb. 25, 2006. Online: http://www.federalreserve.gov/pubs/bulletin/2006/financesurvey.pdf

Buechler, Steven M. 2000. *Social Movements in Advanced Capitalism: The Political Economy and Cultural Construction of Social Activism.* New York: Oxford University Press.

Bullard, Robert B., and Beverly H. Wright. 1992. "The Quest for Environmental Equity: Mobilizing the African-American Community for Social Change." In Riley E. Dunlap and Angela G. Mertig (Eds.), *American Environmentalism: The U.S. Environmental Movement, 1970–1990.* New York: Taylor & Francis, pp. 39–49.

Bumpass, Larry, James E. Sweet, and Andrew J. Cherlin. 1991. "The Role of Cohabitation in Declining Rates of Marriage." *Journal of Marriage and the Family,* 53: 913–927.

Burawoy, Michael. 1991. "Introduction." In Michael Burawoy, Alice Burton, Ann Arnett Ferguson, and others, *Ethnography Unbounded: Power and Resistance in the Modern Metropolis.* Berkeley: University of California Press, pp. 1–7.

Burciaga, Jose Antonio. 1993. *Drink Cultura.* Santa Barbara, CA: Capra.

Burgess, Ernest W. 1925. "The Growth of the City." In Robert E. Park and Ernest W. Burgess (Eds.), *The City.* Chicago: University of Chicago Press, pp. 47–62.

Burkhauser, Richard V., Robert H. Haveman, and Barbara L. Wolfe. 1993. "How People with Disabilities Fare When Public Policies Change." *Journal of Policy Analysis and Management,* 12: 251–269.

Burnham, M. Audrey, Richard L. Hough, Marvin Karno, Javier I. Escobar, and Cynthia A. Telles. 1987. "Acculturation and Lifetime Prevalence of Psychiatric Disorders Among Mexican Americans in Los Angeles." *Journal of Health and Social Behavior,* 28: 89–102.

Burnham, Walter Dean. 1983. *Democracy in the Making: American Government and Politics.* Englewood Cliffs, NJ: Prentice Hall.

Burros, Marian. 1994. "Despite Awareness of Risks, More in U.S. Are Getting Fat." *New York Times* (July 17): 1, 8.

———. 1996. "Eating Well: A Law to Encourage Sharing in a Land of Plenty." *New York Times* (Dec. 11): B6.

Busch, Ruth C. 1990. *Family Systems: Comparative Study of the Family.* New York: Lang.

Busfield, Joan. 1996. *Men, Women and Madness: Understanding Gender and Mental Disorder.* Houndmills, Basingstoke, Hampshire: MacMillan.

Butler, John S. 1991. *Entrepreneurship and Self-Help Among Black Americans.* New York: SUNY Press.

Buvini_, Mayra. 1997. "Women in Poverty: A New Global Underclass." *Foreign Policy* (Fall): 38–53.

Cable, Sherry, and Charles Cable. 1995. *Environmental Problems, Grassroots Solutions: The Politics of Grassroots Environmental Conflict.* New York: St. Martin's.

Callaghan, Polly, and Heidi Hartmann. 1991. *Contingent Work.* Washington, DC: Economic Policy Institute.

Campbell, Anne. 1984. *The Girls in the Gang* (2nd ed.). Cambridge, MA: Basil Blackwell.

Campus Progress. 2006. "Five Minutes with: Morgan Spurlock." Retrieved Mar. 10, 2007. Online: http://www.campusprogress.org/features/336/five-minutes-with-morgan-spurlock

Cancian, Francesca M. 1990. "The Feminization of Love." In C. Carlson (Ed.), *Perspectives on the Family: History, Class, and Feminism.* Belmont, CA: Wadsworth, pp. 171–185.

———. 1992. "Feminist Science: Methodologies That Challenge Inequality." *Gender & Society,* 6 (4): 623–642.

Canetto, Silvia Sara. 1992. "She Died for Love and He for Glory: Gender Myths of Suicidal Behavior." *OMEGA,* 26 (1): 1–17.

Canter, R. J., and S. S. Ageton. 1984. "The Epidemiology of Adolescent Sex-Role Attitudes." *Sex Roles,* 11: 657–676.

Cantor, Muriel G. 1980. *Prime-Time Television: Content and Control.* Newbury Park, CA: Sage.

———. 1987. "Popular Culture and the Portrayal of Women: Content and Control." In Beth B. Hess and Myra Marx Ferree (Eds.), *Analyzing Gender: A Handbook of Social Science Research.* Newbury Park, CA: Sage, pp. 190–214.

Cantril, Hadley. 1941. *The Psychology of Social Movements.* New York: Wiley.

Capek, Stella M. 1993. "The 'Environmental Justice' Frame: A Conceptual Discussion and Application." *Social Problems,* 40 (1): 5–23.

Cargan, Leonard, and Matthew Melko. 1982. *Singles: Myths and Realities.* Newbury Park, CA: Sage.

Carnegie Council on Adolescent Development. 1995. *Great Transitions: Preparing Adolescents for a New Century.* New York: Carnegie Foundation.

Carrier, James G. 1986. *Social Class and the Construction of Inequality in American Education.* New York: Greenwood.

Carson, Rachel. 1962. *Silent Spring.* Boston: Houghton Mifflin.

Carter, Stephen L. 1994. *The Culture of Disbelief: How American Law and Politics Trivializes Religious Devotion.* New York: Anchor/Doubleday.

Cashmore, E. Ellis. 1996. *Dictionary of Race and Ethnic Relations* (4th ed.). London: Routledge.

Castells, Manuel. 1977. *The Urban Question.* London: Edward Arnold (orig. pub. 1972 as *La Question Urbaine,* Paris).

———. 1998. *End of Millennium.* Malden, MA: Blackwell.

Castillo, Juan. 2003. "U.S. Payday Is Something to Write Home About." *Austin American-Statesman* (Dec. 14): J1, J4.

Cavender, Gray. 1995. "Alternative Theory: Labeling and Critical Perspectives." In Joseph F. Sheley (Ed.), *Criminology: A Contemporary Handbook* (2nd ed.). Belmont, CA: Wadsworth, pp. 349–371.

CBS News. 2000. "Russian Mafia's Worldwide Grip." (July 21). Retrieved Apr. 23, 2005. Online: http://www.cbsnews.com/stories/2000/07/21/world/217683.shtml

———. 2004. "The Issues: Child Care." *CBS Evening News* (July 12). Retrieved Apr. 4, 2005. Online: http://www.cbsnews.com/stories/2004/07/12/eveningnews/main628891.shtml

———. 2007a. "How Old Is Too Old to Drive?" (Feb. 3). Retrieved Mar. 25, 2007. Online: http://www.cbsnews.com/stories/2007/02/03/health/main2428859.shtml

———. 2007b. "Outsourced 'Wombs-for-Rent' in India." Retrieved Feb. 9, 2008. Online: http://www.cbsnews.com/stories/2007/12/31/health/main3658750.shtml

Celis, William, III. 1994. "Nations Envied for Schools Share Americans' Worries." *New York Times* (July 13): B4.

census.gov. 2008. "Public Education Finances: 2006." Retrieved Aug. 26, 2009. Online: http://www2.census.gov/govs/school/06f33pub.pdf

———. 2009. "Current Population Survey." Retrieved July 20, 2009. Online: http://www.census.gov/population/www/socdemo/hh-fam.html

Center for Problem-Oriented Policing. 2002. "Graffiti." Retrieved Mar. 10, 2007. Online: http://www.popcenter.org/problems/problem-graffiti.htm

Chafetz, Janet Saltzman. 1984. *Sex and Advantage: A Comparative, Macro-Structural Theory of Sex Stratification.* Totowa, NJ: Rowman & Allanheld.

Chagnon, Napoleon A. 1992. *Yanomamo: The Last Days of Eden.* New York: Harcourt (rev. from 4th ed., *Yanomamo: The Fierce People,* published by Holt, Rinehart & Winston).

Chalfant, H. Paul, Robert E. Beckley, and C. Eddie Palmer. 1994. *Religion in Contemporary Society* (3rd ed.). Itasca, IL: Peacock.

Chambliss, William J. 1973. "The Saints and the Roughnecks." *Society,* 11: 24–31.

Chandler, Tertius, and Gerald Fox. 1974. *3000 Years of Urban History.* New York: Academic Press.

Charmaz, K., and V. Olesen. 1997. "Ethnographic Research in Medical Sociology: Its Foci and Distinctive Contributions." *Sociological Methods and Research,* 25: 452–494.

Chen, Hsiang-shui. 1992. *Chinatown No More: Taiwan Immigrants in Contemporary New York.* Ithaca, NY: Cornell University Press.

Cherlin, Andrew J. 1992. *Marriage, Divorce, Remarriage.* Cambridge, MA: Harvard University Press.

Chesney-Lind, Meda. 1989. "Girls' Crime and Woman's Place: Toward a Feminist Model of Female Delinquency." *Crime and Delinquency,* 35 (1): 5–29.

———. 1997. *The Female Offender.* Thousand Oaks, CA: Sage.

Children Now. 1999. "Children, Race and Advertising." Retrieved Mar. 27, 2005. Online: http://www.childrennow.org/media/medianow/mnwinter1999.html

Children's Defense Fund, 2001. *The State of America's Children: Yearbook 2001.* Boston: Beacon.

———. 2002. *The State of Children in America's Union: A 2002 Action Guide to Leave No Child Behind.* Retrieved June 29, 2003. Online: http://www.childrensdefense.org/pdf/minigreenbook.pdf

———. 2003. "Early Childhood Development: Frequently Asked Questions." Retrieved Apr. 23, 2005. Online: www.childrensdefense.org/earlychildhood/childcare/faq.aspx

———. 2008. *The State of America's Children: Yearbook 2008.* Retrieved July 28, 2009. Online: http://www.childrensdefense.org/child-research data-publications/data/state-of-americas-children-highlights.html

Chipungu, Joel. 1999. "Polygamy Is Alive and Well in Zambia." African News Service (July 22). Retrieved Sept. 11, 1999. Online: http://www.comtex.news.com

Chitale, Radha. 2009. "Job Loss Can Make You Sick." ABCNews.com (May 8). Retrieved June 24, 2009. Online: http://www.abcnews.go.com/id57530730

Cho, Sumi K. 1993. "Korean Americans vs. African Americans: Conflict and Construction." In Robert Gooding-Williams (Ed.), *Reading Rodney King, Reading Urban Uprising.* New York: Routledge, pp. 196–211.

Chon, Margaret. 1995. "The Truth About Asian Americans." In Russell Jacoby and Naomi Glauberman (Eds.), *The Bell Curve Debate: History, Documents, Opinions.* New York: Times Books, pp. 238–240.

Chow, Esther Ngan-Ling. 1994. "Asian American Women at Work." In Maxine Baca Zinn and Bonnie Thornton Dill (Eds.), *Women of Color in U.S. Society.* Philadelphia: Temple University Press, pp. 203–227.

Christ, Carol P. 1987. *Laughter of Aphrodite: Reflections on a Journey to the Goddess.* San Francisco: Harper & Row.

Chronicle of Higher Education. 2008. "Almanac of Higher Education: College Enrollment." Retrieved Aug. 26, 2009. Online: http://chronicle.com/section/Almanac-of-Higher-Education/141

Chu, Bryan. 2008. "Asian Americans Remain Rare in Men's College Basketball." SFGate.com. Retrieved June 13, 2009. Online: http://www.sfgate.com/cgi-bin/article.cgi?f5/c/a/2008/12/15/SPD213J9RD.DTL

Cinquemani, Anthony M. 1997. "C. Wright Mills, Media, and Mass: Only More So." *Educational Change* (Spring): 88–90.

City of Farmers Branch. 2006. "Resolution No. 2006-130." Retrieved Feb. 17, 2007. Online: http://www.ci.farmers-branch.tx.us/Communication/Resolution%202006-130.html

Clayman, Steven E. 1993. "Booing: The Anatomy of a Disaffiliative Response." *American Sociological Review,* 58 (1): 110–131.

Cleary, Paul D. 1987. "Gender Differences in Stress-Related Disorders." In Rosalind C. Barnett, Lois Biener, and Grace K. Baruch (Eds.), *Gender and Stress.* New York: Free Press, pp. 39–72.

Clinard, Marshall B., and Peter C. Yeager. 1980. *Corporate Crime.* New York: Free Press.

Clines, Francis X. 1996. "A Chef's Training Program That Feeds Hope as Well as Hunger." *New York Times* (Dec. 11): B1, B6.

Cloward, Richard A., and Lloyd E. Ohlin. 1960. *Delinquency and Opportunity: A Theory of Delinquent Gangs.* New York: Free Press.

CNN. 1994. "Both Sides: School Prayer." (Nov. 26).

CNN.com. 2003a. "Crash Witness: Rescue Was 'Collective Effort.'" Retrieved July 20, 2003. Online: http://www.cnn.com/2003/US/West/07/17/cnna.crisman/index.html

———. 2003b. "U.S. Energy Secretary Says Weeks Needed to Analyze Blackout Data." Retrieved Aug. 28, 2003. Online: http://www.cnn.com/2003/US/Northeast/08/27/blackout.investigation.ap/index.html

———. 2006. "Immigrant Hopefuls Gather Near U.S. Border." (Apr. 12). Retrieved Apr. 29, 2006. Online: http://www.cnn.com/2006/WORLD/americas/04/12/mexico.border.ap/index.html

Coakley, Jay J. 2004. *Sport in Society: Issues and Controversies* (8th ed.). New York: McGraw-Hill.

Coasports.org. 2009. "Estimated Probability of Competing in Athletics Beyond the High-School Interscholastic Level." Online: http://www.coasports.org/pdf/odds%20of%20becoming%20a%20pro.pdf

Coburn, Andrew F., and Elise J. Bolda. 1999. "The Rural Elderly and Long-Term Care." In Thomas C. Ricketts, III (Ed.), *Rural Health in the United States.* New York: Oxford University Press, pp. 179–189.

Cock, Jacklyn. 1994. "Women and the Military: Implications for Demilitarization in the 1990s in South Africa." *Gender & Society,* 8 (2): 152–169.

Cockerham, William C. 1995. *Medical Sociology* (6th ed.). Englewood Cliffs, NJ: Prentice Hall.

Cohen, Adam. 1999. "A Curse of Cliques." *Time* (May 3): 44–45.

Cohen, Leah Hager. 1994. *Train Go Sorry: Inside a Deaf World.* Boston: Houghton Mifflin.

Cole, David. 2000. *No Equal Justice: Race and Class in the American Criminal Justice System.* New York: New Press.

Cole, George F., and Christopher E. Smith. 2004. *The American System of Criminal Justice* (10th ed.). Belmont, CA: Wadsworth.

Coleman, James S. 1990. *Foundations of Social Theory.* Cambridge: Belknap Press of Harvard University Press.

Coleman, Richard P., and Lee Rainwater. 1978. *Social Standing in America: New Dimensions of Class.* New York: Basic.

Coles, Gerald. 1987. *The Learning Mystique: A Critical Look at "Learning Disabilities."* New York: Pantheon.

Collier, Peter, and David Horowitz. 1987. *The Fords: An American Epic.* New York: Summit.

Collins, Ann Marie. 1991. *How to Live in America: A Guide for the Japanese.* Tokyo: Yohan.

Collins, Catherine, and Douglas Frantz. 1993. *Teachers: Talking Out of School.* Boston: Little, Brown.

Collins, Patricia Hill. 1990. *Black Feminist Thought: Knowledge, Consciousness, and the Politics of Empowerment.* London: HarperCollins Academic.

———. 1991. "The Meaning of Motherhood in Black Culture." In Robert Staples (Ed.), *The Black Family: Essays and Studies.* Belmont, CA: Wadsworth, pp. 169–178. Orig. pub. in *SAGE: A Scholarly Journal on Black Women,* 4 (Fall 1987): 3–10.

———. 1998. *Fighting Words: Black Women and the Search for Justice.* Minneapolis: University of Minnesota Press.

Collins, Randall. 1971. "A Conflict Theory of Sexual Stratification." *Social Problems,* 19 (1): 3–21.

———. 1979. *The Credential Society: An Historical Sociology of Education.* New York: Academic Press.

———. 1982. *Sociological Insight: An Introduction to Non-Obvious Sociology.* New York: Oxford University Press.

———. 1987. "Interaction Ritual Chains, Power, and Property: The Micro-Macro Connection as an Empirically Based Theoretical Problem." In Jeffrey C. Alexander et al. (Eds.), *The Micro-Macro Link.* Berkeley: University of California Press, pp. 193–206.

———. 1994. *Four Sociological Traditions.* New York: Oxford University Press.

———. 1997. "An Asian Route to Capitalism: Religious Economy and the Origins of Self-Transforming Growth in Japan." *American Sociological Review,* 62 (December): 843–865.

Coltrane, Scott. 1989. "Household Labor and the Routine Production of Gender." *Social Problems,* 36: 473–490.

Colvin, Ross. 2009. "Obama Takes Aim at Costly Defense Contracts." Reuters News Service. Retrieved July 9, 2009. Online: http://www.reuters.com/articlePrint?articleID5USN0351345820090304

Comer, James P. 1988. *Maggie's American Dream: The Life and Times of a Black Family.* New York: New American Library/Penguin.

Condry, Sandra McConnell, John C. Condry, Jr., and Lee Wolfram Pogatshnik. 1983. "Sex Differences: A Study of the Ear of the Beholder." *Sex Roles,* 9: 697–704.

Conrad, Peter. 1996. "Medicalization and Social Control." In Phil Brown (Ed.), *Perspectives in Medical Sociology* (2nd ed.). Prospect Heights, IL: Waveland, pp. 137–162.

Cook, Sherburn F. 1973. "The Significance of Disease in the Extinction of the New England Indians." *Human Biology,* 45: 485–508.

Cookson, Peter W., Jr., and Caroline Hodges Persell. 1985. *Preparing for Power: America's Elite Boarding Schools.* New York: Basic.

Cooley, Charles Horton. 1963. *Social Organization: A Study of the Larger Mind.* New York: Schocken (orig. pub. 1909).

———. 1998. "The Social Self—the Meaning of 'I.'" In Hans-Joachim Schubert (Ed.), *On Self and Social Organization—Charles Horton Cooley.* Chicago: University of Chicago Press, pp. 155–175. Reprinted from Charles Horton Cooley, *Human Nature and the Social Order.* New York: Schocken, 1902.

Coontz, Stephanie. 1992. *The Way We Never Were: American Families and the Nostalgia Trap.* New York: Basic.

Cooper, Michael. 2004. "Statement of Michael Cooper, 15, Springfield, VA on the Federal Marriage Amendment." Retrieved Apr. 6, 2004. Online: http://www.aclu.org/LesbianGayRights/LesbianGayRights.cfm?ID515322&c5101

Corr, Charles A., Clyde M. Nabe, and Donald M. Corr. 2003. *Death and Dying, Life and Living* (4th ed.). Pacific Grove, CA: Brooks/Cole.

Corsaro, William A. 1985. *Friendship and Peer Culture in the Early Years.* Norwood, NJ: Ablex.

———. 1992. "Interpretive Reproduction in Children's Peer Cultures." *Social Psychology Quarterly,* 55 (2): 160–177.

———. 1997. *Sociology of Childhood.* Thousand Oaks, CA: Pine Forge.

Corsbie-Massay, Charisse L. 2005. "Beauty Pageants and Television: A Perfect Marriage." Retrieved Jan. 27, 2008. Online: http://alum.mit.edu/www/charisse

Cortese, Anthony J. 2004. *Provocateur: Images of Women and Minorities in Advertising* (2nd ed.). Latham, MD: Rowman and Littlefield.

Cose, Ellis. 1993. *The Rage of a Privileged Class.* New York: HarperCollins.

Coser, Lewis A. 1956. *The Functions of Social Conflict.* Glencoe, IL: Free Press.

Costello, Cynthia, and Anne J. Stone (Eds.), for the Women's Research and Education Institute. 1994. *The American Woman, 1994–95.* New York: Norton.

Coughlin, Ellen K. 1993. "Author of a Noted Study on Black Ghetto Life Returns with a Portrait of Homeless Women." *Chronicle of Higher Education* (Mar. 31): A7–A8.

Cowgill, Donald O. 1986. *Aging Around the World.* Belmont, CA: Wadsworth.

Cox, Harvey. 1995. "Christianity." In Arvind Sharma (Ed.), *Our Religions.* San Francisco: HarperCollins, pp. 359–423.
Cox, Oliver C. 1948. *Caste, Class, and Race.* Garden City, NY: Doubleday.
Craig, Steve. 1992. "Considering Men and the Media." In Steve Craig (Ed.), *Men, Masculinity, and the Media.* Newbury Park, CA: Sage, pp. 1–7.
Cressey, Donald. 1969. *Theft of the Nation.* New York: Harper & Row.
Creswell, John W. 1998. *Qualitative Inquiry and Research Design: Choosing Among Five Traditions.* Thousand Oaks, CA: Sage.
Crinnion, Walter J. 1995. "Are Organic Foods Really Healthier for You?" Retrieved Feb. 17, 2007. Online: http://lookwayup.com/free/organic.htm
Cromwell, Larcenia. 2003. "Terrorist." E-mail message posted Aug. 14, 2003.
Cronin, John, and Robert F. Kennedy, Jr. 1999. *The Riverkeepers: Two Activists Fight to Reclaim Our Environment as a Basic Human Right.* New York: Touchstone.
Crosnoe, Robert. 2006. *Mexican Roots, American Schools: Helping Mexican Immigrant Children Succeed.* Stanford, CA: Stanford University Press.
Crossette, Barbara. 1996. "Hope, and Pragmatism, for U.S. Cities Conference." *New York Times* (June 3): A3.
Crow Dog, Mary, and Richard Erdoes. 1991. *Lakota Woman.* New York: HarperPerennial.
Cumming, Elaine C., and William E. Henry. 1961. *Growing Old: The Process of Disengagement.* New York: Basic.
Currie, Elliott. 1998. *Crime and Punishment in America.* New York: Metropolitan.
Curtiss, Susan. 1977. *Genie: A Psycholinguistic Study of a Modern Day "Wild Child."* New York: Academic Press.
Cyrus, Virginia. 1993. *Experiencing Race, Class, and Gender in the United States.* Mountain View, CA: Mayfield.
Dahl, Robert A. 1961. *Who Governs?* New Haven, CT: Yale University Press.
Dahrendorf, Ralf. 1959. *Class and Class Conflict in an Industrial Society.* Stanford, CA: Stanford University Press.
Dailytroll.com. 2006. "Food as Bonding." Retrieved Dec. 29, 2007. Online: http://dailytroll.com/?p5890
Daly, Kathleen, and Meda Chesney-Lind. 1988. "Feminism and Criminology." *Justice Quarterly,* 5: 497–533.
Daly, Mary. 1973. *Beyond God the Father.* Boston: Beacon.
Daniels, Roger. 1993. *Prisoners Without Trial: Japanese-Americans in World War II.* New York: Hill & Wang.
Danziger, Sheldon, and Peter Gottschalk. 1995. *America Unequal.* Cambridge, MA: Harvard University Press.
Dao, James. 2000. "Lockheed Wins $200 Billion Deal for Fighter Jet." *New York Times* (Oct. 27): A1, A9.
Darley, John M., and Thomas R. Shultz. 1990. "Moral Rules: Their Content and Acquisition." *Annual Review of Psychology,* 41: 525–556.
Darling-Hammond, Linda. 2007. "Evaluating 'No Child Left Behind.'" *The Nation* (May 21). Retrieved Feb. 11, 2008. Online: http://www.thenation.com/doc/20070521/darling-hammond
Dart, Bob. 1999. "Kids Get More Screen Time Than School Time." *Austin American-Statesman* (June 28): A1, A5.
Davis, F. James. 1991. *Who Is Black?* University Park: Pennsylvania State University Press.
Davis, Fred. 1992. *Fashion, Culture, and Identity.* Chicago: University of Chicago Press.
Davis, Kingsley. 1940. "Extreme Social Isolation of a Child." *American Journal of Sociology,* 45 (4): 554–565.
———. 1949. *Human Society.* New York: Macmillan.
Davis, Kingsley, and Wilbert Moore. 1945. "Some Principles of Stratification." *American Sociological Review,* 7 (April): 242–249.
Dean, L. M., F. N. Willis, and J. N. la Rocco. 1976. "Invasion of Personal Space as a Function of Age, Sex and Race." *Psychological Reports,* 38 (3) (pt. 1): 959–965.
Death Penalty Information Center. 2007. "Facts About the Death Penalty, March 15, 2007." Retrieved Mar. 31, 2007. Online: http://www.deathpenaltyinfo.org/FactSheet.pdf
———. 2009. "Facts About the Death Penalty." Retrieved June 26, 2009. Online: http://www.deathpenaltyinfo.org/documents/FactSheet.pdf
Deegan, Mary Jo. 1988. *Jane Addams and the Men of the Chicago School, 1892–1918.* New Brunswick, NJ: Transaction.
Degher, Douglas, and Gerald Hughes. 1991. "The Identity Change Process: A Field Study of Obesity." *Deviant Behavior,* 12: 385–402.
DeJong, Gerben, Andrew I. Batavia, and Robert Griss. 1989. "America's Neglected Health Minority: Working-Age Persons with Disabilities." *Milbank Quarterly,* 67 (suppl. 2): 311–351.
Delgado, Richard. 1995. "Introduction." In Richard Delgado (Ed.), *Critical Race Theory: The Cutting Edge.* Philadelphia: Temple University Press, pp. xiii–xvi.
Delloff, Linda-Marie. 1987. "Distorted Images: The Elderly and the Media." *Christian Century* (Jan. 7–14): 12. Retrieved Apr. 10, 2005. Online: http://www.religion-online-org.showarticle.asp?title5977
Delpit, Lisa. 1995. *Other People's Children: Cultural Conflict in the Classroom.* New York: New Press.
DeNavas-Walt, Carmen, Robert W. Cleveland, and Bruce H. Webster, Jr. 2003. "Income in the United States: 2002." U.S. Census Bureau, Current Population Reports, P60–221. Washington, DC: U.S. Government Printing Office.
DeNavas-Walt, Carmen, Bernadette D. Proctor, and Cheryl Hill Lee. 2005. "Income, Poverty, and Health Insurance in the United States: 2004." U.S. Census Bureau, Current Population Reports, P60–229. Washington, DC: U.S. Government Printing Office.
DeNavas-Walt, Carmen, Bernadette D. Proctor, and Robert J. Mills. 2004. "Income, Poverty, and Health Insurance Coverage in the United States: 2003." U.S. Census Bureau, Current Population Reports, P60–226. Washington, DC: U.S. Government Printing Office.
DeNavas-Walt, Carmen, Bernadette D. Proctor, and Jessica Smith. 2007. "Income, Poverty, and Health Insurance Coverage in the United States: 2006." U.S. Census Bureau, Current Population Reports, P60-233. Retrieved Jan. 12, 2008. Online: http://www.census.gov/prod/2007pubs/p60-233.pdf
———. 2008. *Income, Poverty, and Health Insurance Coverage in the United States: 2007.* Washington, DC: U.S. Census Bureau.
Denzin, Norman K. 1989. *The Research Act* (3rd ed.). Englewood Cliffs, NJ: Prentice Hall.
DeParle, Jason. 2007. "A Global Trek to Poor Nations, from Poorer Ones." *New York Times* (Dec. 27): A1, A16.
Derber, Charles. 1983. *The Pursuit of Attention: Power and Individualism in Everyday Life.* New York: Oxford University Press.
Devine, John. 1996. *Maximum Security: The Culture of Violence in Inner-City Schools.* Chicago: University of Chicago Press.
Dewan, Shaila, and John M. Broder. 2007. "Rampage Gunman Was Student; 2-Hour Lag Tied to Early Chase." *New York Times* (Apr. 18): A1, A16.
De Witt, Karen. 1994. "Wave of Suburban Growth Is Being Fed by Minorities." *New York Times* (Aug. 15): A1, A12.
Dikotter, Frank. 1996. "Culture, 'Race' and Nation: The Formation of Identity in Twentieth Century China." *Journal of International Affairs* (Winter): 590–605.
Dill, Bonnie Thornton. 1988. "'Making Your Job Good Yourself': Domestic Service and the Construction of Personal Dignity." In Ann Bookman and Sandra Morgen (Eds.), *Women and the Politics of Empowerment.* Philadelphia: Temple University Press, pp. 33–52.
Dillon, Sam. 2005. "Students Ace State Tests, but Earn D's from U.S." *New York Times* (Nov. 26). Retrieved Mar. 4, 2007. Online: http://select.nytimes.com/search/restricted/article?res5F10F15F83C6550C758EDDA80994D
DiMicco, Joan Morris, and David R. Millen. 2007. "Identity Management: Multiple Presentations of Self in Facebook." Retrieved Jan. 5, 2008. Online: http://www.joandimicco.com/pubs/dimicco-millen-group07.pdf
Dobrzynski, Judith H. 1996. "When Directors Play Musical Chairs." *New York Times* (Nov. 17): F1, F8, F9.
Dohan, Daniel, and Martin Sanchez-Jankowski. 1998. "Using Computers to Analyze Ethnographic Field Data: Theoretical and Practical Considerations." *Annual Review of Sociology,* 24: 477–499.
Dollard, John, Neal E. Miller, Leonard W. Doob, O. H. Mowrer, and Robert R. Sears. 1939. *Frustration and Aggression.* New Haven, CT: Yale University Press.
Dolnick, Sam. 2008. "Surrogate Business Makes Birth the Latest Job Outsourced to India." *Austin American-Statesman* (Jan. 1): A10.
Domhoff, G. William. 1978. *The Powers That Be: Processes of Ruling Class Domination in America.* New York: Random House.
———. 1983. *Who Rules America Now? A View for the '80s.* Englewood Cliffs, NJ: Prentice Hall.
———. 2002. *Who Rules America? Power and Politics* (4th ed.). New York: McGraw-Hill.
Downey, A. 1984. "Relationship of Religiosity to Death Anxiety of Middle-Aged Males." *Psychological Reports,* 54: 811–822. Cited in N. Hooyman and H. A. Kiyak, 2002.
Driskell, Robyn Bateman, and Larry Lyon. 2002. "Are Virtual Communities True Communities? Examining the Environments and Elements of Community." *City & Community,* 1 (4): 1–18.
Du Bois, W. E. B. 1967. *The Philadelphia Negro: A Social Study.* New York: Schocken (orig. pub. 1899).
Dubowitz, Howard, Maureen Black, Raymond H. Starr, Jr., and Susan Zuravin. 1993. "A Conceptual Definition of Child Neglect." *Criminal Justice and Behavior,* 20 (1): 8–26.
Duffy, John. 1976. *The Healers.* New York: McGraw-Hill.
Dunbar, Polly. 2007. "Wombs to Rent: Childless British Couples Pay Indian Women to Carry Their Babies." *The Daily Mail* (Dec. 8). Retrieved Feb. 9, 2008. Online: http://www.dailymail.co.uk/pages/live/articles/news/worldnews.html?in_article_id5500601
Duncan, Otis Dudley. 1968. "Social Stratification and Mobility: Problems in Measurement of

Trend." In E. B. Sheldon and W. E. Moore (Eds.), *Indicators of Social Change.* New York: Russell Sage Foundation.
Dunlap, Riley E. 1992. "Trends in Public Opinion Toward Environmental Issues: 1965–1990." In Riley E. Dunlap and Angela G. Mertig (Eds.), *American Environmentalism: The U.S. Environmental Movement, 1970–1990.* New York: Taylor & Francis, pp. 89–113.
Dupre, Roslyn, and Paul Gains. 1997. "Fundamental Differences." *Women's Sports & Fitness* (October): 63–68.
Durkheim, Emile. 1933. *The Division of Labor in Society.* Trans. George Simpson. New York: Free Press (orig. pub. 1893).
———. 1956. *Education and Sociology.* Trans. Sherwood D. Fox. Glencoe, IL: Free Press.
———. 1964a. *The Rules of Sociological Method.* Trans. Sarah A. Solovay and John H. Mueller. New York: Free Press (orig. pub. 1895).
———. 1964b. *Suicide.* Trans. John A. Sparkling and George Simpson. New York: Free Press (orig. pub. 1897).
———. 1995. *The Elementary Forms of Religious Life.* Trans. Karen E. Fields. New York: Free Press (orig. pub. 1912).
Durling, Sharon. 2005. "Conquer the Compulsive Shopping Blues." Retrieved Feb. 10, 2007. Online: http://www.womenswallstreet.com/topics/topic.aspx?aid5460
Duster, Troy. 1995. "Symposium: The Bell Curve." *Contemporary Sociology: A Journal of Reviews,* 24 (2): 158–161.
Dye, Thomas R., and Harmon Zeigler. 2006. *The Irony of Democracy: An Uncommon Introduction to American Politics* (13th ed.). Belmont, CA: Wadsworth.
———. 2008. *The Irony of Democracy: An Uncommon Introduction to American Politics* (14th ed.). Belmont, CA: Wadsworth.
Early, Kevin E. 1992. *Religion and Suicide in the African-American Community.* Westport, CT: Greenwood.
Eaton, William W. 1980. "A Formal Theory of Selection for Schizophrenia." *American Journal of Sociology,* 86: 149–158.
Ebaugh, Helen Rose Fuchs. 1988. *Becoming an EX: The Process of Role Exit.* Chicago: University of Chicago Press.
The Economist. 2006. "Making Advances—Credit Cards." *The Economist.* Retrieved Jan. 28, 2007. Online: http://www.gecfmediawatch.com/europe/MAKINGADVANCES.shtml
Eder, Donna. 1995. *School Talk: Gender and Adolescent Culture* (with Catherine Colleen Evans and Stephen Parker). New Brunswick, NJ: Rutgers University Press.
Edwards, Richard. 1979. *Contested Terrain.* New York: Basic.
Egan, Timothy. 1997. "Where Water Is Power, the Balance Shifts." *New York Times* (Nov. 30): A1, A16.
Ehrenreich, Barbara. 1989. *Fear of Falling: The Inner Life of the Middle Class.* New York: HarperPerennial.
———. 1997. *Blood Rites: Origins and History of the Passions of War.* New York: Metropolitan.
———. 2001. *Nickel and Dimed: On (Not) Getting by in America.* New York: Metropolitan.
———. 2008. "The Communist Manifesto Hits 160." *Current Affairs* (Oct. 10). Retrieved July 4, 2009. Online: http://ehrenreich.blogs.com/barbaras_blog/2008/10/the-communist-manifesto-hits-160.html
Ehrlich, Paul R., Anne H. Ehrlich, and Gretchen C. Daily. 1995. *The Stork and the Plow: The Equity Answer to the Human Dilemma.* New Haven, CT: Yale University Press.
Eighner, Lars. 1993. *Travels with Lizbeth.* New York: St. Martin's.
Eisenhower, Dwight D. 1961. "Farewell Address to the Nation." Quoted in William D. Hartung, "Military–Industrial Complex Revisited: How Weapons Makers Are Shaping U.S. Foreign and Military Policies." Retrieved Sept. 11, 1999. Online: http://www.foreignpolicy-infocus.org/paper/micr/index.html
Eisenstein, Zillah R. 1994. *The Color of Gender: Reimaging Democracy.* Berkeley: University of California Press.
Eisinger, Peter. 1973. "The Conditions of Protest Behavior in American Cities." *American Political Science Review,* 81: 11–18.
Eisler, Benita. 1983. *Class Act: America's Last Dirty Secret.* New York: Franklin Watts.
Eitzen, D. Stanley, and George H. Sage. 1997. *The Sociology of North American Sport* (6th ed.). Dubuque, IA: Brown.
Elkin, Frederick, and Gerald Handel. 1989. *The Child and Society: The Process of Socialization* (5th ed.). New York: Random House.
Elkind, David. 1995. "School and Family in the Postmodern World." *Phi Delta Kappan* (September): 8–21.
Ellison, Christopher G., Jeffrey A. Burr, and Patricia L. McCall. 1997. "Religious Homogeneity and Metropolitan Suicide Rates." *Social Forces,* 76 (September): 273–300.
Elster, Jon. 1989. *Nuts and Bolts for the Social Sciences.* Cambridge, England: Cambridge University Press.
Emerson, Richard M. 1962. "Power–Dependency Relations." *American Sociological Review,* 27: 31–41.
Emling, Shelley. 1997a. "Haiti Held in Grip of Another Drought." *Austin American-Statesman* (Sept. 19): A17, A18.
———. 1997b. "In Haiti, It's Resort vs. Reality." *Austin American-Statesman* (Sept. 27): A17, A19.
Engels, Friedrich. 1970. *The Origins of the Family, Private Property, and the State.* New York: International (orig. pub. 1884).
Engerman, Stanley L. 1995. "The Extent of Slavery and Freedom Throughout the World as a Whole and in Major Subareas." In Julian L. Simon (Ed.), *The State of Humanity.* Cambridge, MA: Blackwell, pp. 171–177.
Enloe, Cynthia H. 1987. "Feminists Thinking About War, Militarism, and Peace." In Beth H. Hess and Myra Marx Ferree (Eds.), *Analyzing Gender: A Handbook of Social Science Research.* Newbury Park, CA: Sage, pp. 526–547.
EPA. 2006. "Global Warming: We Can Make a Difference!" United States Environmental Protection Agency. Retrieved May 6, 2006. Online: http://www.epa.gov/globalwarming/kids/difference.html
Epidemiological Network for Latin America and the Caribbean. 2000. "HIV and AIDS in the Americas: An Epidemic with Many Faces." Retrieved Nov. 23, 2001. Online: http://www.census.gov/ipc/www/hivaidinamerica.pdf
Epstein, Cynthia Fuchs. 1988. *Deceptive Distinctions: Sex, Gender, and the Social Order.* New Haven, CT: Yale University Press.
Erikson, Erik H. 1963. *Childhood and Society.* New York: Norton.
———. 1980. *Identities and the Life Cycle.* New York: Norton (orig. pub. 1959).
Erikson, Kai T. 1962. "Notes on the Sociology of Deviance." *Social Problems,* 9: 307–314.
———. 1964. "Notes on the Sociology of Deviance." In Howard S. Becker (Ed.), *The Other Side: Perspectives on Deviance.* New York: Free Press, pp. 9–21.
———. 1976. *Everything in Its Path: Destruction of Community in the Buffalo Creek Flood.* New York: Simon & Schuster.
———. 1991. "A New Species of Trouble." In Stephen Robert Couch and J. Stephen Kroll-Smith (Eds.), *Communities at Risk: Collective Responses to Technological Hazards.* New York: Land, pp. 11–29.
———. 1994. *A New Species of Trouble: Explorations in Disaster, Trauma, and Community.* New York: Norton.
Espiritu, Yen Le. 1995. *Filipino American Lives.* Philadelphia: Temple University Press.
Essed, Philomena. 1991. *Understanding Everyday Racism.* Newbury Park, CA: Sage.
Esterberg, Kristin G. 1997. *Lesbian and Bisexual Identities: Constructing Communities, Constructing Self.* Philadelphia: Temple University Press.
Etzioni, Amitai. 1975. *A Comparative Analysis of Complex Organizations: On Power, Involvement, and Their Correlates* (rev. ed.). New York: Free Press.
———. 1994. *The Spirit of Community: The Reinvention of American Society.* New York: Touchstone.
Europa. 2003. "The European Union at a Glance." Retrieved July 26, 2003. Online: http://www.europa.eu.int
Evans, Glen, and Norman L. Farberow. 1988. *The Encyclopedia of Suicide.* New York: Facts on File.
Evans, Peter B., and John D. Stephens. 1988. "Development and the World Economy." In Neil J. Smelser (Ed.), *Handbook of Sociology.* Newbury Park, CA: Sage, pp. 739–773.
Faderman, Lillian (with Ghia Ziong). 1998. *I Begin My Life All Over: The Hmong and the American Immigrant Experience.* Boston: Beacon.
Fagan, Kevin. 2003. "Shame of the City: Homeless Island." *San Francisco Chronicle* (Nov. 30). Retrieved Apr. 11, 2004. Online: http://www.sfgate.com
FAIR (Fairness & Accuracy in Reporting). 2003. "How Many Dead? Major Networks Aren't Counting?" (Dec. 12). Retrieved July 26, 2003. Online: http://www.fair.org/activism/afghanistan-casualties.html
Falk, Patricia. 1989. "Lesbian Mothers: Psychological Assumptions in Family Law." *American Psychologist,* 44: 941–947.
Fallon, Patricia, Melanie A. Katzman, and Susan C. Wooley. 1994. *Feminist Perspectives on Eating Disorders.* New York: Guilford.
Fallows, James. 1997. *Breaking the News: How the Media Undermine American Democracy.* New York: Vintage.
Faludi, Susan. 1999. *Stiffed: The Betrayal of the American Man.* New York: Morrow.
Farb, Peter. 1973. *Word Play: What Happens When People Talk.* New York: Knopf.
Farley, John E. 1995. *Majority–Minority Relations* (3rd ed.). Englewood Cliffs, NJ: Prentice Hall.
Fausto-Sterling, Anne. 1985. *Myths of Gender: Biological Theories About Women and Men.* New York: Basic.
Feagin, Joe R. 1991. "The Continuing Significance of Race: Antiblack Discrimination in Public Places." *American Sociological Review,* 56 (February): 101–116.
Feagin, Joe R., David B. Baker, and Clairece B. Feagin. 2006. *Social Problems: A Critical Power–Conflict Perspective* (6th ed.). Englewood Cliffs. NJ: Prentice Hall.
Feagin, Joe R., and Clairece Booher Feagin. 1994. *Social Problems: A Critical Power–Conflict*

Perspective (4th ed.). Englewood Cliffs, NJ: Prentice Hall.
———. 2003. *Racial and Ethnic Relations* (7th ed.). Upper Saddle River, NJ: Prentice Hall.
———. 2008. *Racial and Ethnic Relations* (8th ed.). Upper Saddle River, NJ: Prentice Hall.
Feagin, Joe R., Anthony M. Orum, and Gideon Sjoberg (Eds.). 1991. *A Case for the Case Study.* Chapel Hill: University of North Carolina Press.
Feagin, Joe R., and Robert Parker. 1990. *Building American Cities: The Urban Real Estate Game* (2nd ed.). Englewood Cliffs, NJ: Prentice Hall.
Feagin, Joe R., and Melvin P. Sikes. 1994. *Living with Racism: The Black Middle-Class Experience.* Boston: Beacon.
Feagin, Joe R., and Hernán Vera. 1995. *White Racism: The Basics.* New York: Routledge.
Feagin, Joe R., Hernán Vera, and Nikitah Imani. 1996. *The Agony of Education: Black Students at White Colleges and Universities.* New York: Routledge.
Fecht, Josh. 2004. "U.S. Cities Cut Civic Services and Staff to Confront Financial Crisis." Retrieved June 12, 2004. Online: http://www.citymayors.com/report/usfiscal_crisis.html
Federal Bureau of Investigation (FBI). 2006. *Crime in the United States: 2005.* Retrieved Mar. 17, 2007. Online: http://www.fbi.gov/ucr/05cius
———. 2007. *Crime in the United States: 2006.* Retrieved Jan. 4, 2008. Online: http://www.fbi.gov/ucr/cius2006
———. 2008. *Crime in the United States: 2007.* Retrieved June 2, 2009. Online: http://www.fbi.gov/ucr/cius2007
Federal Election Commission. 2006. "Congressional Campaigns Spend $966 Million Through Mid October." Retrieved Mar. 24, 2007. Online: http://www.fec.gov/press/press2006/20061102can/20061102can.html
Federal Interagency Forum on Aging-Related Statistics. 2005. "Older Americans 2004: Key Indicators of Well-Being." Retrieved Apr. 16, 2005. Online: http://agingstats.gov/chartbook2004/default.htm
Feifel, H., and W. T. Nagy. 1981. "Another Look at Fear of Death." *Journal of Consulting and Clinical Psychology,* 49: 278–286. Cited in N. Hooyman and H. A. Kiyak, 2002.
Fenstermacher, Gary D. 1994. "The Absence of Democratic and Educational Ideals from Contemporary Educational Reform Initiatives." The Elam Lecture, presented to the Educational Press Association of America, Chicago, June 10.
Fenstermaker, Sarah, and Candace West (Eds.). 2002. *Doing Gender, Doing Difference: Inequality, Power, and Institutional Change.* New York: Routledge.
Ferguson, John. 1977. *War and Peace in the World's Religions.* New York: Oxford University Press.
Ferraro, Gary. 1992. *Cultural Anthropology: An Applied Perspective.* St. Paul, MN: West.
Ferriss, Susan. 2001. "Cold Spell: U.S. Recession Chills Mexico's Economic Hot Spot." *Austin American-Statesman* (Nov. 25): E1, E4 (based on data from the Bank of Mexico).
Fiffer, Steve, and Sharon Sloan Fiffer. 1994. *50 Ways to Help Your Community.* New York: Mainstream/Doubleday.
Findlay, Deborah A., and Leslie J. Miller. 1994. "Through Medical Eyes: The Medicalization of Women's Bodies and Women's Lives." In B. Singh Bolaria and Harley D. Dickinson (Eds.), *Health, Illness, and Health Care in Canada* (2nd ed.). Toronto: Harcourt, pp. 276–306.
Findlay-Kaneko, Beverly. 1997. "In a Breakthrough for Japan, a Woman Takes Over at a National University." *Chronicle of Higher Education* (June 20): A41–A42.
Fine, Michelle, and Lois Weis. 1998. *The Unknown City: The Lives of Poor and Working-Class Young People.* Boston: Beacon.
Fink, Arlene. 1995. *How to Sample in Surveys.* Thousand Oaks, CA: Sage.
Finklea, Kristin M. 2009. "Organized Crime in the United States: Trends and Issues for Congress." *CRS Report for Congress.* Washington, DC: Congressional Research Service. Retrieved May 25, 2009. Online: http://www.fas.org/sgp/crs/misc/R40525.pdf
Finley, M. I. 1980. *Ancient Slavery and Modern Ideology.* New York: Viking.
Finn Paradis, Leonora, and Scott B. Cummings. 1986. "The Evolution of Hospice in America Toward Organizational Homogeneity." *Journal of Health and Social Behavior,* 27: 370–386.
Firestone, Shulamith. 1970. *The Dialectic of Sex.* New York: Morrow.
Fishbein, Diana H., and Susan E. Pease. 1996. *The Dynamics of Drug Abuse.* Boston: Allyn & Bacon.
Fisher-Thompson, Donna. 1990. "Adult Sex-Typing of Children's Toys." *Sex Roles,* 23: 291–303.
Fjellman, Stephen M. 1992. *Vinyl Leaves: Walt Disney World & America.* Boulder, CO: Westview.
Flanagan, William G. 2002. *Urban Sociology: Images and Structures* (4th ed.). Boston: Allyn & Bacon.
Flexner, Abraham. 1910. *Medical Education in the United States and Canada.* New York: Carnegie Foundation.
Florida, Richard, and Martin Kenney. 1991. "Transplanted Organizations: The Transfer of Japanese Industrial Organization to the U.S." *American Sociological Review,* 56 (3): 381–398.
Foderaro, Lisa W. 2007. "Child Wants Cellphone: Reception Is Mixed." *New York Times* (Mar. 29): E1–E2.
Fong-Torres, Ben. 2007. "Hungry Heart." *New York Times Book Review* (Feb. 4): 11.
Forbes. 2007. "The World's Richest People." *Forbes* (Mar. 26): 104–208.
———. 2008. "The World's Billionaires." Retrieved Mar. 19, 2008. Online: http://www.forbes.com/2008/03/05/richest-people-billionaires-billionaires08-cx_lk_0305billie_land.html
Ford, Clyde W. 1994. *We Can All Get Along: 50 Steps You Can Take to Help End Racism.* New York: Dell.
Foucault, Michel. 1979. *Discipline and Punish: The Birth of the Prison.* New York: Vintage.
———. 1988. *Madness and Civilization: A History of Insanity in the Age of Reason.* New York: Vintage (orig. pub. 1961).
———. 1994. *The Birth of the Clinic: An Archeology of Medical Perception.* New York: Vintage (orig. pub. 1963).
Fountain, John W. 2001. "Prayer Warriors Fight Church–State Division." *New York Times* (Nov. 18): A18.
Fox, Julia R., Glory Koloen, and Volkan Sahin. 2007. "No Joke: A Comparison of Substance in *The Daily Show with Jon Stewart* and Broadcast Network Television Coverage of the 2004 Presidential Election Campaign." *Journal of Broadcast & Electronic Media,* 51(2): 213–227.
Frank, Robert H. 1999. *Luxury Fever: Why Money Fails to Satisfy in an Era of Excess.* New York: Free Press.
Frankenberg, Ruth. 1993. *White Women, Race Matters: The Social Construction of Whiteness.* Minneapolis: University of Minnesota Press.
Franklin, John Hope. 1980. *From Slavery to Freedom: A History of Negro Americans.* New York: Vintage.
Freidson, Eliot. 1965. "Disability as Social Deviance." In Marvin B. Sussman (Ed.), *Sociology and Rehabilitation.* Washington, DC: American Sociology Association, pp. 71–99.
———. 1970. *Profession of Medicine.* New York: Dodd, Mead.
———. 1986. *Professional Powers.* Chicago: University of Chicago Press.
French, Howard W. 2008. "Lives of Grinding Poverty, Untouched by China's Boom." *New York Times* (Jan. 13): YT4.
French, Sally. 1999. "The Wind Gets in My Way." In Mairian Corker and Sally French (Eds.), *Disability Discourse.* Buckingham, England: Open University Press, pp. 21–27.
Freud, Sigmund. 1924. *A General Introduction to Psychoanalysis* (2nd ed.). New York: Boni & Liveright.
Freudenberg, Nicholas, and Carl Steinsapir. 1992. "Not in Our Backyards: The Grassroots Environmental Movement." In Riley E. Dunlap and Angela G. Mertig (Eds.), *American Environmentalism: The U.S. Environmental Movement, 1970–1990.* New York: Taylor & Francis, pp. 27–37.
Freudenheim, Milt. 2007. "Showdown Looms in Congress Over Drug Advertising on TV." *New York Times* (Jan. 22): A1.
Friedan, Betty. 1993. *The Fountain of Age.* New York: Simon & Schuster.
Friedman, Debra, and Michael Hechter. 1988. "The Contribution of Rational Choice Theory to Macrosociological Research." *Sociological Theory,* 6: 201–218.
Friedman, Robert I. 2000. *Red Mafiya: How the Russian Mob Has Invaded America.* New York: Little, Brown.
Friedman, Thomas L. 2005a. "It's a Flat World, After All." *New York Times Magazine* (Apr. 3): 33ff.
———. 2005b. *The World Is Flat: A Brief History of the Twenty-First Century.* New York: Farrar, Straus & Giroux.
Friedmann, John. 1995. "The World City Hypothesis." In Paul L. Knox and Peter J. Taylor (Eds.), *World Cities in a World-System.* Cambridge, England: Cambridge University Press, pp. 317–331.
Frisbie, W. Parker, and John D. Kasarda. 1988. "Spatial Processes." In Neil Smelser (Ed.), *The Handbook of Sociology.* Newbury Park, CA: Sage, pp. 629–666.
Funderburg, Lise. 1994. *Black, White, Other: Biracial Americans Talk About Race and Identity.* New York: Morrow.
Futter, Ellen V. 2006. "Failing Science." *New York Times* (Nov. 26): A11.
The Future of Children. 2001. *Caring for Infants and Toddlers.* Retrieved Mar. 27, 2005. Online: http://www.futureofchildren.org/pubs-info2825/pubs-info_show.htm?doc_id579324
Gabriel, Trip. 1996. "High-Tech Pregnancies Test Hope's Limits." *New York Times* (Jan. 7): 1, 10–11.
Galbraith, John Kenneth. 1985. *The New Industrial State* (4th ed.). Boston: Houghton Mifflin.
Gallagher, Charles A. 2003. "Miscounting Race: Explaining Whites' Misperceptions of Racial Group Size." *Sociological Perspectives,* 46 (3): 381–396.
Gambino, Richard. 1975. *Blood of My Blood.* New York: Doubleday/Anchor.
Gamson, William. 1990. *The Strategy of Social Protest* (2nd ed.). Belmont, CA: Wadsworth.
———. 1995. "Constructing Social Protest." In Hank Johnston and Bert Klandermans (Eds.), *Social Movements and Culture.* Minneapolis: University of Minnesota Press, pp. 85–106.

Gandara, Ricardo. 1995. "*Dichos de la Vida:* Homespun Proverbs Link Hispanic Culture's Past with the Present." *Austin American-Statesman* (Jan. 21): E1, E10.

Gans, Herbert. 1974. *Popular Culture and High Culture: An Analysis and Evaluation of Tastes.* New York: Basic.

———. 1982. *The Urban Villagers: Group and Class in the Life of Italian Americans* (updated and expanded ed.; orig. pub. 1962). New York: Free Press.

Garbarino, James. 1989. "The Incidence and Prevalence of Child Maltreatment." In L. Ohlin and M. Tonry (Eds.), *Family Violence.* Chicago: University of Chicago Press, pp. 219–261.

Garber, Judith A., and Robyne S. Turner. 1995. "Introduction." In Judith A. Garber and Robyne S. Turner (Eds.), *Gender in Urban Research.* Thousand Oaks, CA: Sage, pp. x–xxvi.

Garcia Coll, Cynthia T. 1990. "Developmental Outcomes of Minority Infants: A Process-Oriented Look into Our Beginnings." *Child Development,* 61: 270–289.

Gardner, Carol Brooks. 1989. "Analyzing Gender in Public Places: Rethinking Goffman's Vision of Everyday Life." *American Sociologist,* 20 (Spring): 42–56.

Garfinkel, Harold. 1967. *Studies in Ethnomethodology.* Englewood Cliffs, NJ: Prentice Hall.

Garfinkel, Irwin, and Sara S. McLanahan. 1986. *Single Mothers and Their Children: A New American Dilemma.* Washington, DC: Urban Institute Press.

Gargan, Edward A. 1996. "An Indonesian Asset Is Also a Liability." *New York Times* (Mar. 16): 17, 18.

Garreau, Joel. 1991. *Edge City: Life on the New Frontier.* New York: Doubleday.

Garson, Barbara. 1989. *The Electronic Sweatshop: How Computers Are Transforming the Office of the Future into the Factory of the Past.* New York: Penguin.

Gary, Keahn. 2007. "New Report Reveals Top Ten Problems Facing U.S. Students." Retrieved Feb. 11, 2008. Online: http://www.nbc26.com/news/trends/8000952.html

Gatz, Margaret (Ed.). 1995. *Emerging Issues in Mental Health and Aging.* Washington, DC: American Psychological Association.

Gawande, Atul. 2002. *Complications: A Surgeon's Notes on an Imperfect Science.* New York: Picador.

Gaylin, Willard. 1992. *The Male Ego.* New York: Viking/Penguin.

Geertz, Clifford. 1966. "Religion as a Cultural System." In Michael Banton (Ed.), *Anthropological Approaches to the Study of Religion.* London: Tavistock, pp. 1–46.

Gelfand, Donald E. 2003. *Aging and Ethnicity: Knowledge and Services* (2nd ed.). New York: Springer.

Gelles, Richard J., and Murray A. Straus. 1988. *Intimate Violence: The Definitive Study of the Causes and Consequences of Abuse in the American Family.* New York: Simon & Schuster.

"General Facts on Sweden." 1988. *Fact Sheets on Sweden.* Stockholm: Swedish Institute.

General Motors. 2009. "The General Motors Board of Directors." Retrieved July 4, 2009. Online: http://www.gm.com/corporate/investor_information/corp_gov/board.jsp

George, Susan. 1993. "A Fate Worse Than Debt." In William Dan Perdue (Ed.), *Systemic Crisis: Problems in Society, Politics, and World Order.* Fort Worth: Harcourt, pp. 85–96.

Gereffi, Gary. 1994. "The International Economy and Economic Development." In Neil J. Smelser and Richard Swedberg (Eds.), *The Handbook of Economic Sociology.* Princeton, NJ: Princeton University Press, pp. 206–233.

Gerson, Kathleen. 1993. *No Man's Land: Men's Changing Commitment to Family and Work.* New York: Basic.

Gerstel, Naomi, and Harriet Engel Gross. 1995. "Gender and Families in the United States: The Reality of Economic Dependence." In Jo Freeman (Ed.), *Women: A Feminist Perspective* (5th ed.). Mountain View, CA: Mayfield, pp. 92–127.

Gerth, Hans H., and C. Wright Mills. 1946. *From Max Weber: Essays in Sociology.* New York: Oxford University Press.

Gibbs, Lois Marie, as told to Murray Levine. 1982. *Love Canal: My Story.* Albany: SUNY Press.

Gibbs, Nancy. 1994. "Home Sweet School." *Time* (Oct. 31): 62–63.

———. 2003. "Lights Out." *Time* (Aug. 25): 30–39.

Gilbert, Dennis. 2008. *The American Class Structure in an Age of Growing Inequality* (7th ed.). Thousand Oaks, CA: Pine Forge.

Gilkey, Langdon. 1993. "Theories in Science and Religion." In James Huchingson (Ed.), *Religion and the Natural Sciences: The Range of Engagement.* Fort Worth, TX: Harcourt, pp. 61–65.

Gill, Derek. 1994. "A National Health Service: Principles and Practice." In Peter Conrad and Rochelle Kern (Eds.), *The Sociology of Health and Illness* (4th ed.). New York: St. Martin's, pp. 480–494.

Gilligan, Carol. 1982. *In a Different Voice: Psychological Theory and Women's Development.* Cambridge, MA: Harvard University Press.

Gilmore, David D. 1990. *Manhood in the Making: Cultural Concepts of Masculinity.* New Haven, CT: Yale University Press.

Ginorio, Angela, and Michelle Huston. 2000. *¡Sí Puede! Yes, We Can: Latinas in School.* Washington, DC: American Association of University Women.

Glanz, James, and Edward Wong. 2003. "'97 Report Warned of Foam Damaging Tiles." *New York Times* (Feb. 4): A1–A26.

Glaser, Barney, and Anselm Strauss. 1967. *Discovery of Grounded Theory: Strategies for Qualitative Research.* Chicago: Aldine.

———. 1968. *Time for Dying.* Chicago: Aldine.

Glastris, Paul. 1990. "The New Way to Get Rich." *U.S. News & World Report* (May 7): 26–36.

Glenn, Evelyn Nakano. 1986. *Issei, Nisei, War Bride: Three Generations of Japanese American Women in Domestic Service.* Philadelphia: Temple University Press.

Global Policy. 2009. "Comparison of the World's 25 Largest Corporations with the GDP of Selected Countries (2007)." Globalpolicy.org. Retrieved July 4, 2009. Online: http://www.globalpolicy.org/component/content/article/221/47176.html

Goffman, Erving. 1956. "The Nature of Deference and Demeanor." *American Anthropologist,* 58: 473–502.

———. 1959. *The Presentation of Self in Everyday Life.* Garden City, NY: Doubleday.

———. 1961a. *Asylums: Essays on the Social Situation of Mental Patients and Other Inmates.* Chicago: Aldine.

———. 1961b. *Encounters: Two Studies in the Sociology of Interaction.* London: Routledge and Kegan Paul.

———. 1963a. *Behavior in Public Places: Notes on the Social Structure of Gatherings.* New York: Free Press.

———. 1963b. *Stigma: Notes on the Management of Spoiled Identity.* Englewood Cliffs, NJ: Prentice Hall.

———. 1967. *Interaction Ritual: Essays on Face to Face Behavior.* Garden City, NY: Anchor.

———. 1974. *Frame Analysis: An Essay on the Organization of Experience.* Boston: Northeastern University Press.

Gold, Rachel Benson, and Cory L. Richards. 1994. "Securing American Women's Reproductive Health." In Cynthia Costello and Anne J. Stone (Eds.), *The American Woman 1994–95.* New York: Norton, pp. 197–222.

Goldberg, Robert A. 1991. *Grassroots Resistance: Social Movements in Twentieth Century America.* Belmont, CA: Wadsworth.

Gonyea, Judith G. 1994. "The Paradox of the Advantaged Elder and the Feminization of Poverty." *Social Work,* 39 (1): 35–42.

Goode, Erich. 1996. "The Stigma of Obesity." In Erich Goode (Ed.), *Social Deviance.* Boston: Allyn & Bacon, pp. 332–340.

Goode, William J. 1960. "A Theory of Role Strain." *American Sociological Review,* 25: 483–496.

———. 1982. "Why Men Resist." In Barrie Thorne with Marilyn Yalom (Eds.), *Rethinking the Family: Some Feminist Questions.* New York: Longman, pp. 131–150.

Goodman, Mary Ellen. 1964. *Race Awareness in Young Children* (rev. ed.). New York: Collier.

Goodman, Peter S. 1996. "The High Cost of Sneakers." *Austin American-Statesman* (July 7): F1, F6.

Goodnough, Abby, and Jennifer Steinhauer. 2006. "Senate's Failure to Agree on Immigration Plan Angers Workers and Employers Alike." *New York Times* (Apr. 9): A35.

Gordon, David. 1973. "Capitalism, Class, and Crime in America." *Crime and Delinquency,* 19: 163–186.

Gordon, Milton. 1964. *Assimilation in American Life: The Role of Race, Religion, and National Origins.* New York: Oxford University Press.

Gordon, Philip L. 2001. "Federal Judge's Victory Just the First Shot in the Battle Over Workplace Monitoring." Retrieved July 8, 2002. Online: http://www.privacyfoundation.org/workplace/law/law_show.asp?id575&action50

Gotham, Kevin Fox. 1999. "Political Opportunity, Community Identity, and the Emergence of a Local Anti-Expressway Movement." *Social Problems,* 46: 332–354.

Gottdiener, Mark. 1985. *The Social Production of Urban Space.* Austin: University of Texas Press.

———. 1997. *The Theming of America.* Boulder, CO: Westview.

Gouldner, Alvin W. 1960. "The Norm of Reciprocity: A Preliminary Statement." *American Sociological Review,* 25: 161–179.

———. 1970. *The Coming Crisis of Western Sociology.* New York: Basic.

Gray, Paul. 1993. "Camp for Crusaders." *Time* (Apr. 19): 40.

Greeley, Andrew M. 1972. *The Denominational Society.* Glenview, IL: Scott, Foresman.

Green, Donald E. 1977. *The Politics of Indian Removal: Creek Government and Society in Crisis.* Lincoln: University of Nebraska Press.

Greenberg, Edward S., and Benjamin I. Page. 1993. *The Struggle for Democracy.* New York: HarperCollins.

———. 2002. *The Struggle for Democracy* (5th ed.). Boston: Allyn & Bacon.

Greenhouse, Steven. 1997. "Union Membership Drops Worldwide, U.N. Reports." *New York Times* (Nov. 4): A8.

Griswold del Castillo, R. 1984. *La Familia: Chicano Families in the Urban Southwest, 1848 to the Present.* Notre Dame, IN: University of Notre Dame Press.

Gross, Jane. 1991. "More Young Single Men Clinging to Apron Strings." *New York Times* (June 16): A1.

Guha, Ramachandra. 2004. "The Sociology of Suicide." Retrieved Dec. 20, 2007. Online: http://www.indiatogether.org/2004/aug/rgh-suicide.htm

Hadden, Richard W. 1997. *Sociological Theory: An Introduction to the Classical Tradition.* Peterborough, Ontario: Broadview.

Hagan, John. 1989. *Structural Criminology.* New Brunswick, NJ: Rutgers University Press.

Hahn, Harlan. 1987. "Civil Rights for Disabled Americans: The Foundation of a Political Agenda." In Alan Gartner and Tom Joe (Eds.), *Images of the Disabled, Disabling Images.* New York: Praeger, pp. 181–203.

———. 1997. "Advertising the Acceptably Employable Image." In Lennard J. Davis (Ed.), *The Disability Studies Reader.* New York: Routledge, pp. 172–186.

Haines, Valerie A. 1997. "Spencer and His Critics." In Charles Camic (Ed.), *Reclaiming the Sociological Classics: The State of the Scholarship.* Malden, MA: Blackwell, pp. 81–111.

Halberstadt, Amy G., and Martha B. Saitta. 1987. "Gender, Nonverbal Behavior, and Perceived Dominance: A Test of the Theory." *Journal of Personality and Social Psychology,* 53: 257–272.

Hale-Benson, Janice E. 1986. *Black Children: Their Roots, Culture and Learning Styles* (rev. ed.). Provo, UT: Brigham Young University Press.

Hall, Edward. 1966. *The Hidden Dimension.* New York: Anchor/Doubleday.

Hamper, Ben. 1992. *Rivethead: Tales from the Assembly Line.* New York: Warner.

Haraway, Donna. 1994. "A Cyborgo Manifesto: Science, Technology, and Socialist-Feminism in the Late Twentieth Century." In Anne C. Herrmann and Abigail J. Stewart (Eds.), *Theorizing Feminism: Parallel Trends in the Humanities and Social Sciences.* Boulder, CO: Westview, pp. 424–457.

Hardy, Melissa A., and Lawrence E. Hazelrigg. 1993. "The Gender of Poverty in an Aging Population." *Research on Aging,* 15 (3): 243–278.

Harlow, Harry F., and Margaret Kuenne Harlow. 1962. "Social Deprivation in Monkeys." *Scientific American,* 207 (5): 137–146.

———. 1977. "Effects of Various Mother–Infant Relationships on Rhesus Monkey Behaviors." In Brian M. Foss (Ed.), *Determinants of Infant Behavior* (vol. 4). London: Methuen, pp. 15–36.

Harrington, Michael. 1985. *The New American Poverty.* New York: Viking/Penguin.

Harrington Meyer, Madonna. 1990. "Family Status and Poverty Among Older Women: The Gendered Distribution of Retirement Income in the United States." *Social Problems,* 37: 551–563.

———. 1994. "Gender, Race, and the Distribution of Social Assistance: Medicaid Use Among the Frail Elderly." *Gender & Society,* 8 (1): 8–28.

Harris, Anthony, and James W. Shaw. 2000. "Looking for Patterns: Race, Class, and Crime." In Joseph F. Sheley (Ed.), *Criminology: A Contemporary Handbook* (3rd ed.). Belmont, CA: Wadsworth, pp. 128–163.

Harris, Chauncey D., and Edward L. Ullman. 1945. "The Nature of Cities." *Annals of the Academy of Political and Social Sciences* (November): 7–17.

Harris, Marvin. 1974. *Cows, Pigs, Wars, and Witches.* New York: Random House.

———. 1985. *Good to Eat: Riddles of Food and Culture.* New York: Simon & Schuster.

Harrison, Algea O., Melvin N. Wilson, Charles J. Pine, Samuel Q. Chan, and Raymond Buriel. 1990. "Family Ecologies of Ethnic Minority Children." *Child Development,* 61 (2): 347–362.

Hartmann, Heidi. 1976. "Capitalism, Patriarchy, and Job Segregation by Sex." *Signs: Journal of Women in Culture and Society,* 1 (Spring): 137–169.

———. 1981. "The Unhappy Marriage of Marxism and Feminism." In Lydia Sargent (Ed.), *Women and Revolution.* Boston: South End.

Hartung, William D. 1999. "Military–Industrial Complex Revisited: How Weapons Makers Are Shaping U.S. Foreign and Military Policies." Retrieved Sept. 11, 1999. Online: http://www.foreignpolicy-infocus.org/paper/micr/index.html

Haseler, Stephen. 2000. *The Super-Rich: The Unjust New World of Global Capitalism.* New York: St. Martin's.

Hastorf, Albert, and H. Cantril. 1954. "They Saw a Game: A Case Study." *Journal of Abnormal and Social Psychology,* 40 (2): 129–134.

Hattori, James. 2006. "Marathoner Runs Race Against Hunger." MSNBC.com (Jan. 20). Online: http://www.msnbc.msn.com/id/10948443

Hauchler, Ingomar, and Paul M. Kennedy (Eds.). 1994. *Global Trends: The World Almanac of Development and Peace.* New York: Continuum.

Haught, John F. 1995. *Science & Religion: From Conflict to Conversation.* New York: Paulist.

Hauser, Robert M. 1995. "Symposium: The Bell Curve." *Contemporary Sociology: A Journal of Reviews,* 24 (2): 149–153.

Hauser, Robert M., and John Robert Warren. 1996. "Socioeconomic Indexes for Occupations: A Review, Update, and Critique." Retrieved Apr. 6, 2007. Online: http://www.ssc.wisc.edu/cde/cdewp/96-01.pdf

Havighurst, Robert J., Bernice L. Neugarten, and Sheldon S. Tobin. 1968. "Patterns of Aging." In Bernice L. Neugarten (Ed.), *Middle Age and Aging.* Chicago: University of Chicago Press, pp. 161–172.

Haviland, William A. 1993. *Cultural Anthropology* (7th ed.). Orlando, FL: Harcourt.

———. 1999. *Cultural Anthropology* (9th ed.). Orlando, FL: Harcourt.

Hawley, Amos. 1950. *Human Ecology.* New York: Ronald.

———. 1981. *Urban Society* (2nd ed.). New York: Wiley.

Hawton, Keith, Sue Simkin, Jonathan J. Deeks, Susan O'Connor, Allison Keen, Douglas G. Altman, Greg Philo, and Christopher Bulstrode. 1999. "Effects of a Drug Overdose in a Television Drama on Presentations to Hospital for Self Poisoning: Time Series and Questionnaire Study." *British Medical Journal,* 318 (Apr. 10): 972–988.

Haynes, Judie. 1999. "English Language Learners and the 'Hidden Curriculum.'" Retrieved Aug. 7, 2003. Online: http://www.everythingesl.net/inservices/goal3.php

Hays, Laurie. 1996. "Banks' Marketing Blitz Yields Rash of Defaults." *Wall Street Journal* (Sept. 25): B1, B6; cited in Robert H. Frank, *Luxury Fever: Why Money Fails to Satisfy in an Era of Excess.* New York: Free Press, 1999, pp. 47–48.

Headley, Bernard. 1985. "The Atlanta Establishment and the Atlanta Tragedy." *Phylon,* 46(4): 333–340.

Healey, Joseph F. 2002. *Race, Ethnicity, Gender, and Class: The Sociology of Group Conflict and Change* (3rd ed.). Thousand Oaks, CA: Pine Forge.

Hechter, Michael. 1987. *Principles of Group Solidarity.* Berkeley: University of California Press.

Hechter, Michael, and Satoshi Kanazawa. 1997. "Sociological Rational Choice Theory." *Annual Review of Sociology,* 23: 191–215.

Heldrich Center for Workforce Development. 2003. "Work Trends Survey of Employers About People with Disabilities." Retrieved Aug. 23, 2003. Online: http://www.heldrich.rutgers.edu

Henderson, Rick. 1997. "Schools of Thought." *Reasonline* (January). Retrieved July 24, 2002. Online: http://reason.com/9701/fe.rick.shtml

Henley, Nancy. 1977. *Body Politics: Power, Sex, and Nonverbal Communication.* Englewood Cliffs, NJ: Prentice Hall.

Henslin, James M. 1997. *Sociology: A Down-to-Earth Approach* (3rd ed.). Boston: Allyn & Bacon.

Herbert, Bob. 2009. "The Way We Are." *New York Times* (June 13): A17.

Heritage, John. 1984. *Garfinkel and Ethnomethodology.* Cambridge, MA: Polity.

Herrnstein, Richard J., and Charles Murray. 1994. *The Bell Curve: Intelligence and Class Structure in American Life.* New York: Free Press.

Heshka, Stanley, and Yona Nelson. 1972. "Interpersonal Speaking Distances as a Function of Age, Sex, and Relationship." *Sociometry,* 35 (4): 491–498.

Hess, John. 1991. "Geezer-Bashing: Media Attacks on the Elderly." Retrieved Apr. 10, 2005. Online: http://www.fair.org/extra/best-of-extra/geezer-bashing.html

Hesse-Biber, Sharlene, and Gregg Lee Carter. 2000. *Working Women in America: Split Dreams.* New York: Oxford University Press.

Heywood, Leslie, and Shari L. Dworkin. 2003. *Built to Win: The Female Athlete as Cultural Icon.* Minneapolis: University of Minnesota Press.

Hibbard, David R., and Duane Buhrmester. 1998. "The Role of Peers in the Socialization of Gender-Related Social Interaction Styles." *Sex Roles,* 39: 185–203.

Higginbotham, Elizabeth. 1991. "Is Marriage a Priority: Class Differences in Marital Options of Educated Black Women." In Peter Stein (Ed.), *Single Life: Unmarried Adults in Social Context.* New York: St. Martin's.

———. 1994. "Black Professional Women: Job Ceilings and Employment Sectors." In Maxine Baca Zinn and Bonnie Thornton Dill (Eds.), *Women of Color in U.S. Society.* Philadelphia: Temple University Press, pp. 113–131.

Higginbotham, Elizabeth, and Lynn Weber. 1995. "Moving Up with Kin and Community: Upward Social Mobility for Black and White Women." In Margaret L. Andersen and Patricia Hill Collins (Eds.), *Race, Class, and Gender: An Anthology* (2nd ed.). Belmont, CA: Wadsworth, pp. 134–147.

Hight, Bruce. 1994. "A Level Playing Field: Critics Say Minorities Need a Shot Off the Field." *Austin American-Statesman* (June 8): A1, A11.

———. 1999. "Job Hunters: Not Yet Extinct." *Austin American-Statesman* (Sept. 5): J1, J5.

Higley, Stephen Richard. 1997. "Privilege, Power, and Place: The Geography of the American Upper Class." In Diana Kendall (Ed.), *Race, Class, and Gender in a Diverse Society: A Text-Reader.* Boston: Allyn & Bacon, pp. 70–82.

Hill, Paul T. 1997. "Contracting in Public Education." In Diana Ravitch and Joseph P. Viteritti (Eds.), *New Schools for a New Century: The*

Redesign of Urban Education. New Haven, CT: Yale University Press, pp. 61–85.

Hirschi, Travis. 1969. *Causes of Delinquency*. Berkeley: University of California Press.

Hochschild, Arlie Russell. 1983. *The Managed Heart: Commercialization of Human Feeling*. Berkeley: University of California Press.

———. 1997. *The Time Bind: When Work Becomes Home and Home Becomes Work*. New York: Metropolitan.

———. 2003. *The Commercialization of Intimate Life: Notes from Home and Work*. Berkeley: University of California Press.

Hochschild, Arlie Russell, with Ann Machung. 1989. *The Second Shift: Working Parents and the Revolution at Home*. New York: Viking/Penguin.

Hodson, Randy, and Robert E. Parker. 1988. "Work in High Technology Settings: A Review of the Empirical Literature." *Research in the Sociology of Work*, 4: 1–29.

Hodson, Randy, and Teresa A. Sullivan. 2008. *The Social Organization of Work* (4th ed.). Belmont, CA: Wadsworth.

Hoecker-Drysdale, Susan. 1992. *Harriet Martineau: First Woman Sociologist*. Oxford, England: Berg.

Hofferth, Sandra L. 1984. "Kin Networks, Race, and Family Structure." *Journal of Marriage and the Family*, 46: 791–806.

Hoffnung, Michele. 1995. "Motherhood: Contemporary Conflict for Women." In Jo Freeman (Ed.), *Women: A Feminist Perspective* (5th ed.). Mountain View, CA: Mayfield, pp. 162–181.

Holland, Dorothy C., and Margaret A. Eisenhart. 1981. *Women's Peer Groups and Choice of Career*. Final report for the National Institute of Education. ERIC ED 199 328. Washington, DC.

———. 1990. *Educated in Romance: Women, Achievement, and College Culture*. Chicago: University of Chicago Press.

Homans, George. 1958. "Social Behavior as Exchange." *American Journal of Sociology*, 63: 597–606.

———. 1974. *Social Behavior: Its Elementary Forms* (rev. ed.). New York: Harcourt.

Hondagneu-Sotelo, Pierrette. 2001. *Doméstica: Immigrant Workers Cleaning and Caring in the Shadow of Affluence*. Berkeley: University of California Press.

Hoose, Phillip M. 1989. *Necessities: Racial Barriers in American Sports*. New York: Random House.

Hoover, Kenneth R. 1992. *The Elements of Social Scientific Thinking*. New York: St. Martin's.

Hooyman, Nancy R. R., and H. Asuman Kiyak. 2002. *Social Gerontology: A Multidisciplinary Approach* (6th ed.). Boston: Allyn & Bacon.

Horan, Patrick M. 1978. "Is Status Attainment Research Atheoretical?" *American Sociological Review*, 43: 534–541.

Horovitz, Brude. 2006. "More University Students Call for Organic, 'Sustainable' Food." *USA Today* (Sept. 26). Retrieved Feb. 17, 2007. Online: http://www.usatoday.com/money/industries/food/2006-09-26-college-food-usat_x.htm

Horowitz, Allan V. 1982. *Social Control of Mental Illness*. New York: Academic.

Horowitz, R. 1997. "Barriers and Bridges to Class Mobility and Formation: Ethnographies of Stratification." *Sociological Methods and Research*, 25: 495–538.

Hossfeld, Karen. 1992. *Small, Foreign, and Female: Immigrant Women Workers in Silicon Valley*. Berkeley: University of California Press.

———. 1994. "Hiring Immigrant Women: Silicon Valley's 'Simple Formula.'" In Maxine Baca Zinn and Bonnie Thornton Dill (Eds.), *Women of Color in U.S. Society*. Philadelphia: Temple University Press, pp. 65–93.

Hostetler, A. J. 1994. "U.S. Death Rate Falls to Lowest Level Ever Despite Rise in AIDS." *Austin American-Statesman* (Dec. 16): A4.

Howard, Michael E. 1990. "On Fighting a Nuclear War." In Francesca M. Cancian and James William Gibson (Eds.), *Making War, Making Peace: The Social Foundations of Violent Conflict*. Belmont, CA: Wadsworth, pp. 314–322.

Hoyt, Homer. 1939. *The Structure and Growth of Residential Neighborhoods in American Cities*. Washington, DC: Federal Housing Administration.

Hsiung, Ping-Chun. 1996. *Living Rooms as Factories: Class, Gender, and the Satellite Factory System in Taiwan*. Philadelphia: Temple University Press.

Huang, Larke Namhe, and Y. Ying. 1989. "Chinese American Children and Adolescents." In Jenel Taylor Gibbs and Larke Namhe Huang (Eds.), *Children of Color: Psychological Interventions with Minority Youth*. San Francisco: Jossey-Bass, pp. 30–66.

Hudnut-Beumler, James. 1994. *Looking for God in the Suburbs: The Religion of the American Dream and Its Critics, 1945–1965*. New Brunswick, NJ: Rutgers University Press.

Hughes, Everett C. 1945. "Dilemmas and Contradictions of Status." *American Journal of Sociology*, 50: 353–359.

Hull, Gloria T., Patricia Bell-Scott, and Barbara Smith. 1982. *All the Women Are White, All the Blacks Are Men, But Some of Us Are Brave*. Old Westbury, NY: Feminist.

Humphreys, Laud. 1970. *Tearoom Trade: Impersonal Sex in Public Places*. Chicago: Aldine.

Hurst, Charles E. 2007. *Social Inequality: Forms, Causes, and Consequences* (6th ed.). Boston: Allyn & Bacon.

Hurtado, Aida. 1996. *The Color of Privilege: Three Blasphemies on Race and Feminism*. Ann Arbor: University of Michigan Press.

Huston, Aletha C. 1985. "The Development of Sex Typing: Themes from Recent Research." *Developmental Review*, 5: 2–17.

Hwang, S. S., R. Saenz, and B. E. Aguirre. 1995. "The SES-Selectivity of Interracially Married Asians." *International Migration Review*, 29: 469–491.

Hynes, H. Patricia. 1990. *Earth Right: Every Citizen's Guide*. Rocklin, CA: Prima.

Ibrahim, Youseff M. 1990. "Saudi Tradition: Edicts from Koran Produce Curbs on Women." *New York Times* (Nov. 6): A6.

Idaho Association of Soil Conservation Districts. 2004. "Organic Pest Control." Retrieved Feb. 17, 2007. Online: http://www.oneplan.org/Crop/OrganicPestCtrl.shtml

IFAD. 2002. "IFAD in China: The Rural Poor Speak." Rome, Italy: International Fund for Agricultural Development. Retrieved Jan. 22, 2008. Online: http://www.ifad.org/media/success/China.pdf

Inciardi, James A., Ruth Horowitz, and Anne E. Pottieger. 1993. *Street Kids, Street Drugs, Street Crime: An Examination of Drug Use and Serious Delinquency in Miami*. Belmont, CA: Wadsworth.

Inda, Cynthia G. 1997. "Why I Took a Chance on Learning: From Community College to Harvard." *New York Times Education Life Supplement* (Aug. 3): 31, 40, 42.

International Monetary Fund. 1992. *World Economic Outlook*. Washington, DC: International Development Fund.

Ishikawa, Kaoru. 1984. "Quality Control in Japan." In Naoto Sasaki and David Hutchins (Eds.), *The Japanese Approach to Product Quality: Its Applicability to the West*. Oxford: Permagon, pp. 1–5.

Isidore, Chris. 2009. "2 Million Jobs Lost So Far in '09." CNNMoney.com (Apr. 3). Retrieved June 24, 2009. Online: http://www.money.com/2009/04/03/news/economy/jobs_march/index.htm

ITAR/TASS News Agency. 1999a. "Only Three Ingush Men Used Their Right to Polygamy" (Aug. 31). Retrieved Sept. 11, 1999. Online: http://www.comtexnews.com

———. 1999b. "Polygamy Allowed in Southern Russia" (July 21). Retrieved Sept. 11, 1999. Online: http://www.comtexnews.com

Jack, Dana Crowley. 1993. *Silencing the Self: Women and Depression*. New York: HarperPerennial.

Jackson, Kenneth T. 1985. *Crabgrass Frontier: The Suburbanization of the United States*. New York: Oxford University Press.

Jacobs, Emma. 2009. "Lost Job, Lost Identity?" FT.com: Business Life (Apr. 8). Retrieved June 25, 2009. Online: http://www.ft.com/cms/s/f066f606-2461-11de-9a01-00144feabdc0,dwp_uuid5f38b85e4-5

Jacobs, Gloria. 1994. "Where Do We Go from Here? An Interview with Ann Jones." *Ms.* (September/October): 56–63.

Jacquard, Roland. 2001. "The Guidebook of Jihad." *Time* (Oct. 29): 58.

Jameson, Fredric. 1984. "Postmodernism, or, The Cultural Logic of Late Capitalism." *New Left Review*, 146: 59–92.

Janis, Irving. 1972. *Victims of Groupthink*. Boston: Houghton Mifflin.

———. 1989. *Crucial Decisions: Leadership in Policymaking and Crisis Management*. New York: Free Press.

Jankowski, Martin Sanchez. 1991. *Islands in the Street: Gangs and American Urban Society*. Berkeley: University of California Press.

Jaramillo, P. T., and Jesse T. Zapata. 1987. "Roles and Alliances Within Mexican-American and Anglo Families." *Journal of Marriage and the Family*, 49 (November): 727–735.

Jary, David, and Julia Jary. 1991. *The Harper Collins Dictionary of Sociology*. New York: HarperPerennial.

Jensen, Robert. 1995. "Men's Lives and Feminist Theory." *Race, Gender & Class*, 2 (2): 111–125.

Jewell, K. Sue. 1993. *From Mammy to Miss America and Beyond: Cultural Images and the Shaping of US Social Policy*. New York: Routledge.

Johnson, Allan G. 2000. *The Blackwell Dictionary of Sociology* (2nd ed.). Malden, MA: Blackwell.

Johnson, Claudia. 1994. *Stifled Laughter: One Woman's Story About Fighting Censorship*. Golden, CO: Fulcrum.

Johnson, Dirk. 1994. "Equal Loads, Not Pay for Non-union Drivers." *New York Times* (Apr. 10): 10.

Johnson, Earvin "Magic," with William Novak. 1992. *My Life*. New York: Fawcett Crest.

Johnston, David Cay. 2002. "As Salary Grows, So Does a Gender Gap." *New York Times* (May 12): BU8.

Joint Center for Political and Economic Studies. 2003. "Black Elected Officials: A Statistical Summary 2001." Retrieved Mar. 17, 2007. Online: http://www.jointcenter.org/publications1/publication-PDFs/BEO-pdfs/2001-BEO.pdf

Joliet Junior College. 2005. "History of Joliet Junior College." Retrieved Apr. 21, 2005. Online: http://www.jjc.edu/campus/history

Journalism.org. 2004. "The State of the News Media 2004." Retrieved Feb. 23, 2008. Online: http://www.stateofthenewsmedia.org/2007/index.asp

Judson, George. 1995. "Connecticut Wins School Bias Suit." *New York Times* (Apr. 13): A1, A11.

Juergensmeyer, Mark. 1993. *The New Cold War? Religious Nationalism Confronts the Secular State.* Berkeley: University of California Press.

Kabagarama, Daisy. 1993. *Breaking the Ice: A Guide to Understanding People from Other Cultures.* Boston: Allyn & Bacon.

Kahn, Joseph, and Jim Yardley. 2008. "As China Roars, Pollution Reaches Deadly Extremes." *New York Times* (Aug. 26): A1, A6–7.

Kalmijn, Matthijs. 1998. "Intermarriage and Homogamy: Causes, Patterns, Trends." *Annual Review of Sociology,* 24: 395–422.

Kanter, Rosabeth Moss. 1983. *The Change Masters: Innovation and Entrepreneurship in the American Corporation.* New York: Simon & Schuster.

———. 1985. "All That Is Entrepreneural Is Not Gold." *Wall Street Journal* (July 22): 18.

———. 1993. *Men and Women of the Corporation.* New York: Basic (orig. pub. 1977).

Kaplan, David A. 1993. "Dumber Than We Thought." *Newsweek* (Sept. 20): 44–45.

Kaplan, Robert D. 1996. "Cities of Despair." *New York Times* (June 6): A19.

Kaspar, Anne S. 1986. "Consciousness Re-evaluated: Interpretive Theory and Feminist Scholarship." *Sociological Inquiry,* 56 (1): 30–49.

Katz, Michael B. 1989. *The Undeserving Poor: From the War on Poverty to the War on Welfare.* New York: Pantheon.

Katzer, Jeffrey, Kenneth H. Cook, and Wayne W. Crouch. 1991. *Evaluating Information: A Guide for Users of Social Science Research.* New York: McGraw-Hill.

Kaufman, Gayle. 1999. "The Portrayal of Men's Family Roles in Television Commercials." *Sex Roles,* 313: 439–451.

Kaufman, Tracy L. 1996. *Out of Reach: Can America Pay the Rent?* Washington, DC: National Low Income Housing Coalition.

KCTV5.com. 2008. "Viewer Stories: Job Loss Concerns." Retrieved June 25, 2009. Online: http://www.kctv5.com/money/17643714/detail.html

Keller, James. 1994. "'I Treasure Each Moment.'" *Parade* (Sept. 4): 4–5.

Kelley, Tina. 1999. "For That Bowl of Cherries, A Hard Life." *New York Times* (Aug. 11): A11.

Kellner, Thomas. 2004. "Highest-Paid Boards." *Forbes* (May 4). Retrieved May 10, 2005. Online: http://www.forbes.com/2004/05/04/cz_tk_0504directors_print.html

Kelman, Steven. 1991. "Sweden Sour? Downsizing the 'Third Way.'" *New Republic* (July 29): 19–23.

Kemp, Alice Abel. 1994. *Women's Work: Degraded and Devalued.* Englewood Cliffs, NJ: Prentice Hall.

Kendall, Diana. 1980. Square Pegs in Round Holes: Non-Traditional Students in Medical Schools. Unpublished doctoral dissertation, Department of Sociology, the University of Texas at Austin.

———. 2002. *The Power of Good Deeds: Privileged Women and the Social Reproduction of the Upper Class.* Lanham, MD: Rowman & Littlefield.

———. 2004. *Social Problems in a Diverse Society* (3rd ed.). Boston: Allyn & Bacon.

———. 2005. *Framing Class: Media Representations of Wealth and Poverty in America.* Lanham, MD: Rowman & Littlefield.

———. 2008. *Members Only: Elite Clubs and the Process of Exclusion.* Lanham, MD: Rowman & Littlefield.

Kendall, Diana, Rick Linden, and Jane Murray. 2008. *Sociology in Our Times: The Essentials* (4th Canadian edition). Scarborough, Ontario: Nelson Thomson Learning.

Kendall, Diana, Jane Lothian Murray, and Rick Linden. 2004. *Sociology in Our Times* (3rd Canadian edition). Scarborough, Ontario: Nelson Thomson Learning.

Kennedy, Paul. 1987. *The Rise and Fall of the Great Powers.* New York: Random House.

———. 1993. *Preparing for the Twenty-First Century.* New York: Random House.

Kenyon, Kathleen. 1957. *Digging Up Jericho.* London: Benn.

Kerbo, Harold R. 2000. *Social Stratification and Inequality: Class Conflict in Historical, Comparative, and Global Perspective* (4th ed.). New York: McGraw-Hill.

Kershaw, Sarah. 2005. "Crisis of Indian Children Intensifies as Families Fail." *New York Times* (Apr. 5): A13.

Kessler, Ronald C. 1994. "Lifetime and 12-Month Prevalence of DSM-III-R Psychiatric Disorders in the United States: Results of the National Comorbidity Survey." *JAMA, the Journal of the American Medical Association,* 271 (Mar. 2): 654D.

Keyfitz, Nathan. 1994. "The Scientific Debate: Is Population Growth a Problem? An Interview with Nathan Keyfitz." *Harvard International Review* (Fall): 10–11, 74.

Khan, Mahmood Hasan. 2001. "Rural Poverty in Developing Countries: Implications for Public Policy: Economic Issues No. 26." Washington, DC: International Monetary Fund. Retrieved Jan. 21, 2008. Online: http://www.imf.org/external/pubs/ft/issues/issues26/index.htm

Kidron, Michael, and Ronald Segal. 1995. *The State of the World Atlas.* New York: Penguin.

Kilborn, Peter T. 1995. "Women and Minorities Still Face 'Glass Ceiling.'" *New York Times* (Mar. 16): C22.

———. 1997. "Illness Is Turning into Financial Catastrophe for More of the Uninsured." *New York Times* (Aug. 1): A10.

Kilbourne, Jean. 1999. *Deadly Persuasion: The Addictive Power of Advertising.* New York: Simon & Schuster.

———. 2000. *Killing Us Softly 3: Advertising's Image of Women* (Film). Northampton, MA: Media Education Foundation.

Killian, Lewis. 1984. "Organization, Rationality, and Spontaneity in the Civil Rights Movement." *American Sociological Review,* 49: 770–783.

Kim, Ryan. 2006. "The World's a Cell-Phone Stage." *San Francisco Chronicle* (Feb. 27). Retrieved Mar. 30, 2007. Online: http://www.sfgate.com/cgi-bin/article.cgi?file5/chronicle/archive/2006/02/27/BUG2IHECTO1.DTL&type5printable

Kimmell, Michael S., and Michael A. Messner. 2004. *Men's Lives* (6th ed.). Boston: Allyn & Bacon.

King, Gary, Robert O. Keohane, and Sidney Verba. 1994. *Designing Social Inquiry: Scientific Inference in Qualitative Research.* Princeton, NJ: Princeton University Press.

King, Leslie, and Madonna Harrington Meyer. 1997. "The Politics of Reproductive Benefits: U.S. Insurance Coverage of Contraceptive and Infertility Treatments." *Gender and Society,* 11 (1): 8–30.

Kirkpatrick, P. 1994. "Triple Jeopardy: Disability, Race and Poverty in America." *Poverty and Race,* 3: 1–8.

Kitsuse, John I. 1980. "Coming Out All Over: Deviance and the Politics of Social Problems." *Social Problems,* 28: 1–13.

Klein, Alan M. 1993. *Little Big Men: Bodybuilding Subculture and Gender Construction.* Albany: SUNY Press.

Klonoff, Elizabeth A. 1997. *Preventing Misdiagnosis of Women: A Guide to Physical Disorders That Have Psychiatric Symptoms.* Thousand Oaks, CA: Sage.

Kluckhohn, Clyde. 1961. "The Study of Values." In Donald N. Barrett (Ed.), *Values in America.* South Bend, IN: University of Notre Dame Press, pp. 17–46.

Knapp, Caroline. 1996. *Drinking: A Love Story.* New York: Dial.

Knox, Paul L., and Peter J. Taylor (Eds.). 1995. *World Cities in a World-System.* Cambridge, England: Cambridge University Press.

Knudsen, Dean D. 1992. *Child Maltreatment: Emerging Perspectives.* Dix Hills, NY: General Hall.

Kohl, Beth. 2007. "On Indian Surrogates." *The Huffington Post* (Oct. 30). Retrieved Feb. 9, 2008. Online: http://www.huffingtonpost.com/beth-kohl/on-indian-surrogates_b_70425.html

Kohlberg, Lawrence. 1969. "Stage and Sequence: The Cognitive–Developmental Approach to Socialization." In David A. Goslin (Ed.), *Handbook of Socialization Theory and Research.* Chicago: Rand McNally, pp. 347–480.

———. 1981. *The Philosophy of Moral Development: Moral Stages and the Idea of Justice,* vol. 1: *Essays on Moral Development.* San Francisco: Harper & Row.

Kohn, Alfie. 2001. "Five Reasons to Stop Saying 'Good Job!'" Retrieved Jan. 3, 2008. Online: http://www.alfiekohn.org/parenting/gj.htm

Kohn, Melvin L. 1977. *Class and Conformity: A Study in Values* (2nd ed.). Homewood, IL: Dorsey.

Kohn, Melvin L., Atsushi Naoi, Carrie Schoenbach, Carmi Schooler, and Kazimierz M. Slomczynski. 1990. "Position in the Class Structure and Psychological Functioning in the United States, Japan, and Poland." *American Journal of Sociology,* 95: 964–1008.

Kolata, Gina. 1993. "Fear of Fatness: Living Large in a Slimfast World." *Austin American-Statesman* (Jan. 3): C1, C6.

Koopmans, Ruud. 1999. "Political. Opportunity. Structure. Some Splitting to Balance the Lumping." *Sociological Forum* (Mar.): 93–105.

Kornblum, Janet. 2004. "There's a Risk to the Beauty of Surgery." Retrieved Mar. 17, 2005. Online: http://www.usatoday.com/news/health/2004-01-21-plastic-surgery-risks_x.htm

Korsmeyer, Carolyn. 1981. "The Hidden Joke: Generic Uses of Masculine Terminology." In Mary Vetterling-Braggin (Ed.), *Sexist Language: A Modern Philosophical Analysis.* Totowa, NJ: Littlefield, Adams, pp. 116–131.

Korten, David C. 1996. *When Corporations Rule the World.* West Hartford, CT: Kumarian.

Kosmin, Barry A., and Seymour P. Lachman. 1993. *One Nation Under God: Religion in Contemporary American Society.* New York: Crown.

Kovacs, M., and A. T. Beck. 1977. "The Wish to Live and the Wish to Die in Attempted Suicides." *Journal of Clinical Psychology,* 33: 361–365.

Kowinski, William Severini. 2002. *The Malling of America: Travels in the United States of Shopping.* New York: Xlibis.

Kozol, Jonathan. 1986. *Illiterate America.* New York: Anchor/Doubleday.

———. 1991. *Savage Inequalities: Children in America's Schools.* New York: Crown.

Kramnick, Isaac. (Ed.). 1995. *The Portable Enlightenment Reader.* New York: Penguin.

Kranning, Antoinette, and Lee Ehman. 1999. "Help! I'm Lost in Cyberspace." *Social Education,* 63 (3): 152–156. Retrieved Mar. 4, 2007. Online: http://www.indiana.edu/~leeehman/mystery.pdf

Krisberg, Barry. 1975. *Crime and Privilege: Toward a New Criminology.* Englewood Cliffs, NJ: Prentice Hall.

Kristof, Nicholas D. 2006. "Looking for Islam's Luthers." *New York Times* (Oct. 15): A22.

Kristof, Nicholas D., and Sheryl WuDunn. 2000. "Two Cheers for Sweatshops." *New York Times* (Sept. 24), section 6: 70.

Kroloff, Charles A. 1993. *54 Ways You Can Help the Homeless.* Southport, CT: Hugh Lauter Levin Associates; and West Orange, NJ: Behrman.

Kruzel, John J., and Michael J. Carden. 2009. "Defense Department Releases Sexual Assault Statistics." DefenseLink. Retrieved July 9, 2009. Online: http://www.defenselink.mil/news/newsarticle.aspx?id553525

Krysan, Maria, and Reynolds Farley. 1993. "Racial Stereotypes: Are They Alive and Well? Do They Continue to Influence Race Relations?" Paper presented at the annual meeting of the American Sociological Association, Miami Beach, Florida, Aug. 16.

Kübler-Ross, Elisabeth. 1969. *On Death and Dying.* New York: Macmillan.

Kurlantzick, Joshua. 2003. "Charging Ahead." *Washington Monthly* (May 2003). Retrieved Jan. 28, 2007. Online: http://www.washingtonmonthly.com/features/2003/0305.kurlantzick.html

Kurtz, Lester. 1995. *Gods in the Global Village: The World's Religions in Sociological Perspective.* Thousand Oaks, CA: Sage.

Kurz, Demie. 1995. *For Richer, for Poorer: Mothers Confront Divorce.* New York: Routledge.

Kvale, Steinar. 1996. *Interviews: An Introduction to Qualitative Research Interviewing.* Thousand Oaks, CA: Sage.

Lacayo, Richard. 1997. "They'll Vouch for That." *Time* (Oct. 27): 72–74.

Ladd, E. C., Jr. 1966. *Negro Political Leadership in the South.* Ithaca, NY: Cornell University Press.

Lamanna, Marianne, and Agnes Riedmann. 2009. *Marriages and Families: Making Choices and Facing Change* (10th ed.). Belmont, CA: Wadsworth, Cengage Learning.

Lamar, Joe. 2000. "Suicides in Japan Reach a Record High." *British Medical Journal* (Sept. 2). Retrieved Aug. 25, 2001. Online: http://www.findarticles.com/cf_dls/m0999/7260_321/66676910/p1/article.jhtml

Lane, Harlan. 1992. *The Mask of Benevolence: Disabling the Deaf Community.* New York: Vintage.

Lapham, Lewis H. 1988. *Money and Class in America: Notes and Observations on Our Civil Religion.* New York: Weidenfeld & Nicolson.

Lapsley, Daniel K., 1990. "Continuity and Discontinuity in Adolescent Social Cognitive Development." In Raymond Montemayor, Gerald R. Adams, and Thomas P. Gullota (Eds.), *From Childhood to Adolescence: A Transitional Period?* (*Advances in Adolescent Development,* vol. 2). Newbury Park, CA: Sage.

Larimer, Tim. 1999. "The Japan Syndrome." *Time* (Oct. 11): 50–51.

Larson, Magali Sarfatti. 1977. *The Rise of Professionalism: A Sociological Analysis.* Berkeley: University of California Press.

Lasch, Christopher. 1977. *Haven in a Heartless World.* New York: Basic.

Latino Legends in Sports. 2007. "Sports News." Retrieved Mar. 17, 2007. Online: http://www.latinosportslegends.com

Latouche, Serge. 1992. "Standard of Living." In Wolfgang Sachs (Ed.), *The Development Dictionary.* Atlantic Highlands, NJ: Zed, pp. 250–263.

Laumann, Edward O., John H. Gagnon, Robert T. Michael, and Stuart Michaels. 1994. *The Social Organization of Sexuality.* Chicago: University of Chicago Press.

Le Bon, Gustave. 1960. *The Crowd: A Study of the Popular Mind.* New York: Viking (orig. pub. 1895).

Leary, Warren E. 1996. "Even When Covered by Insurance, Black and Poor People Receive Less Health Care." *New York Times* (Sept. 12): A10.

Lee, Felicia R. 1993. "Where Guns and Lives Are Cheap." *New York Times* (Mar. 21): 21.

Lee, Joann Faung Jean. 1992. *Asian Americans: Oral Histories of First to Fourth Generation Americans from China, the Philippines, Japan, India, the Pacific Islands, Vietnam and Cambodia.* New York: New Press.

Leenaars, Antoon A. 1988. *Suicide Notes: Predictive Clues and Patterns.* New York: Human Sciences Press.

Leenaars, Antoon A. (Ed.). 1991. *Life Span Perspectives of Suicide: Time-Lines in the Suicide Process.* New York: Plenum.

Lefrançois, Guy R. 1996. *The Lifespan* (5th ed.). Belmont, CA: Wadsworth.

———. 1999. *The Lifespan* (6th ed.). Belmont, CA: Wadsworth.

Lehmann, Jennifer M. 1994. *Durkheim and Women.* Lincoln: University of Nebraska Press.

Leland, John. 2008. "From the Housing Market to the Maternity Ward." *New York Times* (Feb. 1): A12.

Lemann, Nicholas. 1997. "Let's Guarantee the Key Ingredients." *Time* (Oct. 27): 96.

Lemert, Charles. 1997. *Postmodernism Is Not What You Think.* Malden, MA: Blackwell.

Lemert, Edwin M. 1951. *Social Pathology.* New York: McGraw-Hill.

Lengermann, Patricia Madoo, and Jill Niebrugge-Brantley. 1998. *The Women Founders: Sociology and Social Theory, 1830–1930.* New York: McGraw-Hill.

Lenzer, Gertrud (Ed.). 1998. *The Essential Writings: Auguste Comte and Positivism.* New Brunswick, NJ: Transaction.

Leonard, Andrew. 1999. "We've Got Mail—Always." *Newsweek* (Sept. 20): 58–61.

Leonard, Wilbert M., and Jonathan E. Reyman. 1988. "The Odds of Attaining Professional Athlete Status: Refining the Computations." *Sociology of Sport Journal:* 162–169.

Lerman, Hannah. 1996. *Pigeonholing Women's Misery: A History and Critical Analysis of the Psychodiagnosis of Women in the Twentieth Century.* New York: Basic.

Lerner, Gerda. 1986. *The Creation of Patriarchy.* New York: Oxford University Press.

LeShan, Eda. 1994. *I Want More of Everything.* New York: New Market.

Lester, David. 1988. *Why Women Kill Themselves.* Springfield, IL: Thomas.

———. 1992. *Why People Kill Themselves: A 1990s Summary of Research Findings of Suicidal Behavior* (3rd ed.). Springfield, IL: Thomas.

Lev, Michael A. 1998. "Suicide Imbedded in Japan's Culture." *Chicago Tribune* (Feb. 27). Retrieved Aug. 25, 2001. Online: http://seattletimes.nwsource.com/news/nation-world/html98/altjpan_022798.html

Leventman, Paula Goldman. 1981. *Professionals Out of Work.* New York: Free Press.

Levey, Hilary. 2007. "Here She Is . . . and There She Goes?" *Contexts* (Summer): 70–72.

Levin, William C. 1988. "Age Stereotyping: College Student Evaluations." *Research on Aging,* 10 (1): 134–148.

Levine, Adeline Gordon. 1982. *Love Canal: Science, Politics, and People.* Lexington, MA: Lexington.

Levine, Arthur. 1993. "Student Expectations of College." *Change* (September/October): 4.

Levine, Murray. 1982. "Introduction." In Lois Marie Gibbs, *Love Canal: My Story.* Albany: SUNY Press.

Levine, Nancy E., and Joan B. Silk. 1997. "Why Polyandry Fails: Sources of Instability in Polyandrous Marriages." *Current Anthropology* (June): 375–399.

Levine, Peter. 1992. *Ellis Island to Ebbets Field: Sport and the American Jewish Experience.* New York: Oxford University Press.

Levinthal, Charles F. 2002. *Drugs, Behavior, and Modern Society* (3rd ed.). Boston: Allyn & Bacon.

Levy, Janice C., and Eva Y. Deykin. 1989. "Suicidality, Depression, and Substance Abuse in Adolescence." *American Journal of Psychiatry,* 146 (11): 1462–1468.

Lewin, Tamar, and Jennifer Medina. 2003. "To Cut Failure Rates, Schools Shed Students." *New York Times* (July 31): A1, A22.

Lewis, Paul. 1996. "World Bank Moves to Cut Poorest Nations' Debts." *New York Times* (Mar. 16): 17, 18.

———. 1998. "Marx's Stock Resurges on a 150-Year Tip." *New York Times* (June 27): A17, A19.

Lewis, Tamar. 1997. "School Voucher Program Succeeds in Cleveland." *Austin American-Statesman* (Sept. 21): A31.

Lichter, David T., Felicia B. LeClere, and Diane K. McLaughlin. 1991. "Local Marriage Markets and the Marital Behavior of Black and White Women." *American Journal of Sociology,* 96 (4): 843–867.

Liebow, Elliot. 1967. *Tally's Corner: A Study of Negro Streetcorner Men.* Boston: Little, Brown.

———. 1993. *Tell Them Who I Am: The Lives of Homeless Women.* New York: Free Press.

Lii, Jane H. 1995. "Week in Sweatshop Reveals Grim Conspiracy of the Poor." *New York Times* (Mar. 12): 1, 16.

Lindberg, Richard, and Vesna Markovic. 2001. "Organized Crime Outlook in the New Russia." Search International. Retrieved Apr. 23, 2005. Online: http://www.search-international.com/Articles/crime/russiacrime.htm

Lindblom, Charles. 1977. *Politics and Markets.* New York: Basic.

Linden, Greg. 2005. "Geeking with Greg: Facebook and Building Social Networks" (Oct. 27). Retrieved Jan. 28, 2006. Online: http://gliden.blogspot.com/2005/10/facebook-and-building-social-networks.html

Link, Bruce G., and Bruce P. Dohrenwend. 1989. "The Epidemiology of Mental Disorders." In Howard E. Freeman and Sol Levine (Eds.), *Handbook of Medical Sociology* (4th ed.). Englewood Cliffs, NJ: Prentice Hall, pp. 102–127.

Linnehan, Robert. 2007. "Mayors Get Behind the Wheels to Help Seniors." NJ.com (Mar. 22). Retrieved Mar. 27, 2007. Online: http://www

.nj.com/printer/printer.ssf'?/base/news-2/1174540882150010.xml&coll59

Linton, Ralph. 1936. *The Study of Man.* New York: Appleton-Century-Crofts.

Lips, Hilary M. 2001. *Sex and Gender: An Introduction* (4th ed.). New York: McGraw-Hill.

Lock, Margaret. 1999. "The Politics of Health, Identity, and Culture." In Richard J. Contrada and Richard D. Ashmore (Eds.), *Self, Social Identity, and Physical Health: Interdisciplinary Explorations.* New York: Oxford University Press, pp. 43–68.

Loeb, Paul Rogat. 1994. *Generation at the Crossroads: Apathy and Action on the American Campus.* New Brunswick, NJ: Rutgers University Press.

Lofland, John. 1993. "Collective Behavior: The Elementary Forms." In Russell L. Curtis, Jr., and Benigno E. Aguirre (Eds.), *Collective Behavior and Social Movements.* Boston: Allyn & Bacon, pp. 70–75.

London, Kathryn A. 1991. "Advance Data Number 194: Cohabitation, Marriage, Marital Dissolution, and Remarriage: United States 1988." U.S. Department of Health and Human Services: Vital and Health Statistics of the National Center, Jan. 4.

Longman, Jere. 2009. "Alleging Racism, Soccer Star Seeks 'Moral Compensation.'" *New York Times* (June 14): Y1, Y8.

Lorber, Judith. 1994. *Paradoxes of Gender.* New Haven, CT: Yale University Press.

Lorber, Judith (Ed.). 2005. *Gender Inequality: Feminist Theories and Politics* (3rd ed.). Los Angeles: Roxbury.

Lott, Bernice. 1994. *Women's Lives: Themes and Variations in Gender Learning* (2nd ed.). Pacific Grove, CA: Brooks/Cole.

Low, Setha. 2003. *Behind the Gates: Life, Security, and the Pursuit of Happiness in Fortress America.* New York: Routledge.

Lowe, Maria R. 1998. *Women of Steel: Female Bodybuilders and the Struggle for Self-Definition.* New York: New York University Press.

Lummis, C. Douglas. 1992. "Equality." In Wolfgang Sachs (Ed.), *The Development Dictionary.* Atlantic Highlands, NJ: Zed, pp. 38–52.

Lund, Kristina. 1990. "A Feminist Perspective on Divorce Therapy for Women." *Journal of Divorce,* 13 (3): 57–67.

Lundberg, Ferdinand. 1988. *The Rich and the Super-Rich: A Study in the Power of Money Today.* Secaucus, NJ: Lyle Stuart.

Lupton, Deborah. 1997. "Foucault and the Medicalisation Critique." In Alan Petersen and Robin Bunton (Eds.), *Foucault: Health and Medicine.* London: Routledge, pp. 94–110.

Luttrell, Wendy. 1997. *School-Smart and Mother-Wise: Working-Class Women's Identity and Schooling.* New York: Routledge.

Lynch, Jason, and Todd Gold. 2005. "Happy Drew Year." *People* (Apr. 25): 92–96, 98.

Lynd, Robert S., and Helen M. Lynd. 1929. *Middletown.* New York: Harcourt.

———. 1937. *Middletown in Transition.* New York: Harcourt.

Maag, Christopher. 2007. "When the Bullies Turned Faceless." *New York Times* (Dec. 16). Retrieved Dec. 26, 2007. Online: http://www.nytimes.com/2007/12/16/fashion/16meangirls.html?_r51&oref5slogin

Maccoby, Eleanor E., and Carol Nagy Jacklin. 1987. "Gender Segregation in Childhood." *Advances in Child Development and Behavior,* 20: 239–287.

Mack, Raymond W., and Calvin P. Bradford. 1979. *Transforming America: Patterns of Social Change* (2nd ed.). New York: Random House.

Magid, Larry. 2004. "Talk to Your Kids About Cell Phone Use." Retrieved Mar. 30, 2007. Online: http://www.safekids.com/cellphone.htm

Mahapatra, Rajesh. 2007. "Outsourced Jobs Take Toll on Indians' Health." *Austin American-Statesman* (Dec. 30): H1, H6.

Mahler, Jonathan. 2009. "Slipping Away. G.M, Detroit and the Fall of the Black Middle Class." *New York Times Magazine* (June 28): 30–37, 44–47.

Mahler, Sarah J. 1995. *American Dreaming: Immigrant Life on the Margins.* Princeton, NJ: Princeton University Press.

Males, Mike A. 1996. *The Scapegoat Generation: America's War on Adolescents.* Monroe, ME: Common Courage.

Malinowski, Bronislaw. 1922. *Argonauts of the Western Pacific.* New York: Dutton.

Malthus, Thomas R. 1965. *An Essay on Population.* New York: Augustus Kelley (orig. pub. 1798).

Mangione, Jerre, and Ben Morreale. 1992. *La Storia: Five Centuries of the Italian American Experience.* New York: HarperPerennial.

Mann, Coramae Richey. 1993. *Unequal Justice: A Question of Color.* Bloomington: Indiana University Press.

Mann, Patricia S. 1994. *Micro-Politics: Agency in a Postfeminist Era.* Minneapolis: University of Minnesota Press.

Manning, P. K., and B. Cullum-Swan. 1994. "Narrative, Content, and Semiotic Analysis." In Norman K. Denzin and Y. S. Lincoln (Eds.), *Handbook of Qualitative Research.* Thousand Oaks, CA: Sage.

Manning, Robert D. 1999. *Credit Card Nation.* New York: Basic.

———. 2000. *Credit Card Nation: The Consequences of America's Addiction to Credit.* New York: Basic.

Mansfield, Alan, and Barbara McGinn. 1993. "Pumping Irony: The Muscular and the Feminine." In Sue Scott and David Morgan (Eds.), *Body Matters: Essays on the Sociology of the Body.* London: Falmer, pp. 49–58.

Mantsios, Gregory. 2003. "Media Magic: Making Class Invisible." In Michael S. Kimmel and Abby L. Ferber (Eds.), *Privilege: A Reader.* Boulder, CO: Westview, pp. 99–109.

Marger, Martin N. 1987. *Elites and Masses: An Introduction to Political Sociology* (2nd ed.). Belmont, CA: Wadsworth.

———. 1994. *Race and Ethnic Relations: American and Global Perspectives.* Belmont, CA: Wadsworth.

———. 2003. *Race and Ethnic Relations: American and Global Perspectives* (6th ed.). Belmont, CA: Wadsworth.

———. 2009. *Race and Ethnic Relations: American and Global Perspectives* (8th ed.). Belmont, CA: Wadsworth, Cengage Learning.

Margolis, Richard J. 1990. *Risking Old Age in America.* Boulder, CO: Westview.

Marquand, Robert. 2004. "China's Supersized Mall." *Christian Science Monitor* (Nov. 24). Retrieved Jan. 14, 2006. Online: http://www.csmonitor.com/2004/1124/p01s03-woap.html

Marquart, James W., Sheldon Ekland-Olson, and Jonathan R. Sorensen. 1994. *The Rope, the Chair, and the Needle.* Austin: University of Texas Press.

Marquis, Christopher. 2001. "An American Report Finds the Taliban's Violation of Religious Rights 'Particularly Severe.'" *New York Times* (Oct. 27): B3.

Marsden, Peter V. 1983. "Restricted Access in Networks and Models of Power." *American Journal of Sociology,* 88: 686–717.

Marshall, Gordon. 1998. *A Dictionary of Sociology* (2nd ed.). New York: Oxford University Press.

Martin, Carol L. 1989. "Children's Use of Gender-Related Information in Making Social Judgments." *Developmental Psychology,* 25: 80–88.

Martin, Linda, and Kevin Kinsella. 1994. "Research in the Demography of Aging in Developing Countries." In Linda Martin and Samuel Preston (Eds.), *Demography of Aging.* Washington, DC: National Academic Press.

Martin, Susan Ehrlich, and Nancy C. Jurik. 1996. *Doing Justice, Doing Gender.* Thousand Oaks, CA: Sage.

Martineau, Harriet. 1962. *Society in America* (edited, abridged). Garden City, NY: Doubleday (orig. pub. 1837).

———. 1988. *How to Observe Morals and Manners.* Ed. Michael R. Hill. New Brunswick, NJ: Transaction (orig. pub. 1838).

Marx, Karl. 1967. *Capital: A Critique of Political Economy.* Ed. Friedrich Engels. New York: International (orig. pub. 1867).

Marx, Karl, and Friedrich Engels. 1967. *The Communist Manifesto.* New York: Pantheon (orig. pub. 1848).

———. 1970. *The German Ideology,* Part 1. Ed. C. J. Arthur. New York: International (orig. pub. 1845–1846).

Massey, Douglas J., G. Hugo Arango, A. Kowasuci, A. Pellegrino, and J. E. Taylor. 1993. "Theories of International Migration: A Review and Appraisal." *Population and Development Review,* 19: 431–466.

Matthews, Warren. 2004. *World Religions* (4th ed.). Belmont, CA: Wadsworth.

Maynard, R. A. 1996. *Kids Having Kids: A Robin Hood Foundation Special Report on the Costs of Adolescent Childbearing.* New York: Robin Hood Foundation.

McAdam, Doug. 1982. *Political Process and the Development of Black Insurgency.* Chicago: University of Chicago Press.

———. 1996. "Conceptual Origins, Current Problems, Future Directions." In Doug McAdam, John D. McCarthy, and Meyer N. Zald (Eds.), *Comparative Perspectives on Social Movements.* New York: Cambridge University Press, pp. 23–40.

McAdam, Doug, John D. McCarthy, and Mayer N. Zald. 1988. "Social Movements." In Neil J. Smelser (Ed.), *Handbook of Sociology.* Newbury Park, CA: Sage, pp. 695–737.

McAdoo, Harriet Pipes. 1990. "A Portrait of African American Families in the United States." In S. E. Rix (Ed.), *The American Woman, 1990–91: A Status Report.* New York: Norton.

McCall, George J., and Jerry L. Simmons. 1978. *Identities and Interactions: An Explanation of Human Associations in Everyday Life.* New York: Free Press.

McCall, Nathan. 1994. *Makes Me Wanna Holler: A Young Black Man in America.* New York: Random House.

McCann, Lisa M. 1997. "Patrilocal Co-Residential Units (PCUs) in Al-Barba: Dual Household Structure in a Provincial Town in Jordan." *Journal of Comparative Family Studies* (Summer): 113–136.

McCarthy, John D., and Mayer N. Zald. 1977. "Resource Mobilization and Social Movements: A Partial Theory." *American Journal of Sociology,* 82: 1212–1241.

McDonald, Patrick Range. 1997. "Financial Disaster 101." New Mass Media, Inc.: Advocate: Back to School. Retrieved May 30, 1999. Online: http://www.newhavenadvocate.com/articles/back2school/back2school/html

McDonnell, Janet A. 1991. *The Dispossession of the American Indian, 1887–1934.* Bloomington: Indiana University Press.

McEachern, William A. 2003. *Economics: A Contemporary Introduction.* Mason, OH: Thomson/South-Western.

McGeary, Johanna. 2001. "The Taliban Troubles." *Time* (Oct. 1): 36–43.

McGuire, Meredith R. 2002. *Religion: The Social Context* (5th ed.). Belmont, CA: Wadsworth.

McHale, Susan M., Ann C. Crouter, and C. Jack Tucker. 1999. "Family Context and Gender Role Socialization in Middle Childhood: Comparing Girls to Boys and Sisters to Brothers." *Child Development,* 70: 990–1004.

McKenzie, Roderick D. 1925. "The Ecological Approach to the Study of the Human Community." In Robert Park, Ernest Burgess, and Roderick D. McKenzie, *The City.* Chicago: University of Chicago Press.

McLanahan, Sara, and Karen Booth. 1991. "Mother-Only Families." In Alan Booth (Ed.), *Contemporary Families: Looking Forward, Looking Backward.* Minneapolis: National Council on Family Relations, pp. 405–428.

McLarin, Kimberly J. 1994. "A New Jersey Town Is Troubled by Racial Imbalance Between Classrooms: Would End to Tracking Harm Quality?" *New York Times* (Aug. 11): A12.

McPhail, Clark. 1991. *The Myth of the Maddening Crowd.* New York: Aldine de Gruyter.

McPhail, Clark, and Ronald T. Wohlstein. 1983. "Individual and Collective Behavior Within Gatherings, Demonstrations, and Riots." In Ralph H. Turner and James F. Short, Jr. (Eds.), *Annual Review of Sociology,* vol. 9. Palo Alto, CA: Annual Reviews, pp. 579–600.

McPherson, Barry D., James E. Curtis, and John W. Loy. 1989. *The Social Significance of Sport: An Introduction to the Sociology of Sport.* Champaign, IL: Human Kinetics.

McTague, Jim. 2005. "The Underground Economy." *Wall Street Journal* (classroom ed., Apr.). Retrieved Apr. 8, 2007. Online: http://wsjclassroom.com/archive/05apr/econ_underground.htm

Mead, George Herbert. 1934. *Mind, Self, and Society.* Chicago: University of Chicago Press.

Medicare.gov. 2007. "Nursing Homes." Retrieved Mar. 29, 2007. Online: http://www.medicare.gov/nursing/overview.asp

Mennell, Stephen. 1996. *All Manners of Food: Eating and Taste in England and France from the Middle Ages to the Present.* Urbana: University of Illinois Press.

Mennell, Stephen, Anne Murcott, and Anneke H. van Otterloo. 1993. *The Sociology of Food: Eating, Diet and Culture.* Thousand Oaks, CA: Sage.

Mental Medicine. 1994. "Wealth, Health, and Status." *Mental Medicine Update,* 3 (2): 7.

Merchant, Carolyn. 1983. *The Death of Nature: Women, Ecology, and the Scientific Revolution.* San Francisco: Harper & Row.

———. 1992. *Radical Ecology: The Search for a Livable World.* New York: Routledge.

Merton, Robert King. 1938. "Social Structure and Anomie." *American Sociological Review,* 3 (6): 672–682.

———. 1949. "Discrimination and the American Creed." In Robert M. MacIver (Ed.), *Discrimination and National Welfare.* New York: Harper & Row, pp. 99–126.

———. 1968. *Social Theory and Social Structure* (enlarged ed.). New York: Free Press.

Messerschmidt, James. 1986. *Capitalism, Patriarchy and Crime.* Totowa, NJ: Rowman and Littlefield.

Messner, Michael A. 2002. *Taking the Field: Women, Men and Sports.* Minneapolis: University of Minnesota Press.

Messner, Michael A., Margaret Carlisle Duncan, and Kerry Jensen. 1993. "Separating the Men from the Girls: The Gendered Language of Televised Sports." *Gender & Society,* 7 (1): 121–137.

Meyer, David S., and Debra C. Minkoff. 2004. "Conceptualizing Political Opportunity." *Social Forces* (June): 1457–1492.

Meyer, David S., and Suzanne Staggenborg. 1996. "Movements, Countermovements, and the Structure of Political Opportunity." *American Journal of Sociology,* 101: 1628–1660.

Miall, Charlene. 1986. "The Stigma of Involuntary Childlessness." *Social Problems,* 33 (4): 268–282.

Michael, Robert T., John H. Gagnon, Edward O. Laumann, and Gina Kolata. 1994. *Sex in America.* Boston: Little, Brown.

Michels, Robert. 1949. *Political Parties.* Glencoe, IL: Free Press (orig. pub. 1911).

Mickelson, Roslyn Arlin, and Stephen Samuel Smith. 1995. "Education and the Struggle Against Race, Class, and Gender Inequality." In Margaret L. Andersen and Patricia Hill Collins (Eds.), *Race, Class, and Gender* (2nd ed.). Belmont, CA: Wadsworth, pp. 289–304.

Middleton, Nick. 1999. *The Global Casino: An Introduction to Environmental Issues* (2nd ed.). London: Arnold.

Mies, Maria, and Vandana Shiva. 1993. *Ecofeminism.* Highlands, NJ: Zed.

Miethe, Terance, and Charles Moore. 1987. "Racial Differences in Criminal Processing: The Consequences of Model Selection on Conclusions About Differential Treatment." *Sociological Quarterly,* 27: 217–237.

Milgram, Stanley. 1963. "Behavioral Study of Obedience." *Journal of Abnormal and Social Psychology,* 67: 371–378.

———. 1967. "The Small World Problem." *Psychology Today,* 1: 61–67.

———. 1974. *Obedience to Authority.* New York: Harper & Row.

Miliband, Ralph. 1969. *The State in Capitalist Society.* New York: Basic.

Milkman, Ruth. 1997. *Farewell to the Factory: Auto Workers in the Late Twentieth Century.* Berkeley: University of California Press.

Miller, Casey, and Kate Swift. 1991. *Words and Women: New Language in New Times* (updated ed.). New York: HarperCollins.

Miller, Dan E. 1986. "Milgram Redux: Obedience and Disobedience in Authority Relations." In Norman K. Denzin (Ed.), *Studies in Symbolic Interaction.* Greenwich, CT: JAI, pp. 77–106.

Miller, L. Scott. 1995. *An American Imperative: Accelerating Minority Educational Advancement.* New Haven, CT: Yale University Press.

Miller, Michele. 2003. "Homeschooling: Tuned to the Individual." *St. Petersburg Times Online.* Retrieved Aug. 7, 2003. Online: http://www.sptimes.com/2003/08/03/news_pf/Pasco/Homeschooling_Tuned_.shtml

Mills, C. Wright. 1956. *White Collar.* New York: Oxford University Press.

———. 1959a. *The Power Elite.* Fair Lawn, NJ: Oxford University Press.

———. 1959b. *The Sociological Imagination.* London: Oxford University Press.

———. 1976. *The Causes of World War Three.* Westport, CT: Greenwood.

Mills, Robert J., and Shailesh Bhandari. 2003. "Health Insurance Coverage in the United States: 2002." U.S. Census Bureau, Current Population Reports, P60–223. Washington, DC: U.S. Government Printing Office.

Min, Eungjun (Ed.). 1999. *Reading the Homeless: The Media's Image of Homeless Culture.* Westport, CT: Praeger.

Min, Pyong Gap. 1988. "The Korean American Family." In Charles H. Mindel, Robert W. Habenstein, and Roosevelt Wright, Jr. (Eds.), *Ethnic Families in America: Patterns and Variations* (3rd ed.). New York: Elsevier, pp. 199–229.

Mindel, Charles H., Robert W. Habenstein, and Roosevelt Wright, Jr. (Eds.). 1988. *Ethnic Families in America: Patterns and Variations* (3rd ed.). New York: Elsevier.

mindoh.com. 2007. "I Wish I Knew What to Do?!" Retrieved Feb. 15, 2008. Online: http://www.mindoh.docs/Bullyingbook_excerpt_noCCC.pdf

Mintz, Beth, and Michael Schwartz. 1985. *The Power Structure of American Business.* Chicago: University of Chicago Press.

Mintz, Morton. 2007. "Will Congress Reform Wretched Executive Excess?" Retrieved Mar. 10, 2007. Online: http://www.thenation.com/doc/20070212/mintz

Mirowsky, John. 1996. "Age and the Gender Gap in Depression." *Journal of Health and Social Behavior,* 37 (December): 362–380.

Mirowsky, John, and Catherine E. Ross. 1980. "Minority Status, Ethnic Culture, and Distress: A Comparison of Blacks, Whites, Mexicans, and Mexican Americans." *American Journal of Sociology,* 86, 479–495.

Mishler, Elliot G. 1984. *The Discourse of Medicine: Dialectics of Medical Interviews.* Norwood, NJ: Ablex.

———. 2005. "The Struggle Between the Voice of Medicine and the Voice of the Lifeworld." In Peter Conrad (Ed.), *The Sociology of Health and Illness: Critical Perspectives* (7th ed.). New York: Worth, pp. 319–330.

Miss America. 2008. "Miss America: Key Facts and Figures." Retrieved Jan. 31, 2008. Online: http://www.missamerica.org/organization-info/key-facts-and-figures.asp

Misztal, Barbara A. 1993. "Understanding Political Change in Eastern Europe: A Sociological Perspective." *Sociology,* 27 (3): 451–471.

Mohai, Paul, and Robin Saha. 2007. "Racial Inequality in the Distribution of Hazardous Waste: A National-Level Reassessment." *Social Problems* (August): 343–370.

Monahan, John. 1992. "Mental Disorder and Violent Behavior: Perceptions and Evidence." *American Psychologist,* 47: 511–521.

Moody, Harry R. 2002. *Aging: Concepts and Controversy* (4th ed.). Thousand Oaks, CA: Pine Forge.

Moore, K. A., A. K. Driscoll, and L. D. Lindberg. 1998. *A Statistical Portrait of Adolescent Sex, Contraception, and Childbearing.* Washington, DC: National Campaign to Prevent Teen Pregnancy.

Moore, Patricia, with C. P. Conn. 1985. *Disguised.* Waco, TX: Word.

Moore, R. Laurence. 1995. *Selling God: American Religion in the Marketplace of Culture.* New York: Oxford University Press.

Moore, Robert B. 1992. "Racist Stereotyping in the English Language." In Margaret L. Anderson

and Patricia Hill Collins (Eds.), *Race, Class, and Gender.* Belmont, CA: Wadsworth, pp. 317–329.
Moraga, Cherrié. 1994. "From a Long Line of Vendidas: Chicanas and Feminism." In Anne C. Hermann and Abigail J. Stewart (Eds.), *Theorizing Feminism: Parallel Trends in Humanities and Social Sciences.* Boulder, CO: Westview, pp. 34–48.
Morgan, Leslie, and Suzanne Kunkel. 1998. *Aging: The Social Context.* Thousand Oaks, CA: Pine Forge.
Morrill, C., and Gary A. Fine. 1997. "Ethnographic Contributions to Organizational Sociology." *Sociological Methods and Research,* 25: 424–451.
Morris, Aldon. 1981. "Black Southern Student Sit-In Movement: An Analysis of Internal Organization." *American Sociological Review,* 46: 744–767.
Morse, Jodie. 2001. "Letting God Back In." *Time* (Oct. 22): 71.
Morselli, Henry. 1975. *Suicide: An Essay on Comparative Moral Statistics.* New York: Arno (orig. pub. 1881).
The Motley Fool. 2007. "Scary Debt Stats." Retrieved Feb. 10, 2007. Online: http://www.fool.com/ccc/secrets/secrets.htm
MOWAA. 2007. "Meals on Wheels Association of America." Retrieved Mar. 26, 2007. Online: http://www.mowaa.org/displayContent.asp?mid53¤tid517&type5I
Mowbray, Carol T., Sandra E. Herman, and Kelly L. Hazel. 1992. "Gender and Serious Mental Illness." *Psychology of Women Quarterly,* 16 (March): 107–127.
MSNBC.com. 2007. "Test Scores Lag Behind Rising Grades." (Feb. 22). Retrieved Mar. 3, 2007. Online: http://www.msnbc.msn.com/id/17278393/print.1.displaymode.1098
Mucciolo, Louis. 1992. *Eightysomething: Interviews with Octogenarians Who Stay Involved.* New York: Birch Lane.
Murdock, George P. 1945. "The Common Denominator of Cultures." In Ralph Linton (Ed.), *The Science of Man in the World Crisis.* New York: Columbia University Press, pp. 123–142.
Murphy, Dean E. 2005. "If You Can Plug a Film, Why Not a Budget?" *New York Times* (Feb. 13): ST1, ST2.
Murphy, Robert E., Jessica Scheer, Yolanda Murphy, and Richard Mack. 1988. "Physical Disability and Social Liminality: A Study in the Rituals of Adversity." *Social Science and Medicine,* 26: 235–242.
Musick, Carmen. 2004. "King College Students Tutoring Middle-Schoolers in Reading, Learning." King College Public Relations (Oct. 8). Retrieved Apr. 17, 2005. Online: http://www.king.edu/PublicRelations/100804.htm
Mydans, Seth. 1995. "Part-Time College Teaching Rises, as Do Worries." *New York Times* (Jan. 4): B6.
———. 1997a. "Brutal End for an Architect of Cambodian Brutality." *New York Times* (June 14): 5.
———. 1997b. "Its Mood Dark as the Haze, Southeast Asia Aches." *New York Times* (Oct. 26): 3.
Myerhoff, Barbara. 1994. *Number Our Days: Culture and Community Among Elderly Jews in an American Ghetto.* New York: Meridian.
Myerson, Allen R. 1994. "Jeans Makers Flourish on Border: The Two Faces of an Industry Success Story." *New York Times* (Sept. 29): C1, C13.
Myrdal, Gunnar. 1970. *The Challenge of World Poverty: A World Anti-Poverty Program in Outline.* New York: Pantheon/Random House.
Naffine, Ngaire. 1987. *Female Crime: The Construction of Women in Criminology.* Boston: Allen & Unwin.
NAGIA. 2007. "Graffiti: The Newspaper of the Streets" and "Gang Indicators." National Alliance of Gang Investigators Association. Retrieved Mar. 10, 2007. Online: http://www.nagia.org
Nagourney, Adam. 2002. "The Battleground Shifts." *New York Times* (June 28): A1, A17.
Nasar, Sylvia. 1999. "Jobless Rate in August Again Dipped to a 29-Year Low: Labor Market Was Tight But Wages Barely Rose." *New York Times* (Sept. 4): B1, B3.
National Campaign to Prevent Teen Pregnancy. 1997. *Whatever Happened to Childhood? The Problem of Teen Pregnancy in the United States.* Washington, DC: National Campaign to Prevent Teen Pregnancy.
National Center for Education Statistics. 2005a. "Comparative Indicators of Education in the United States and Other G8 Countries: 2004." National Center for Education Statistics, U.S. Department of Education. Retrieved Mar. 23, 2005. Online: http://nces.ed.gov/pubs2005/2005021.pdf
———. 2005b. "Highlights from the Trends in International Mathematics and Science Study (TIMSS) 2003." Retrieved Apr. 16, 2005. Online: http://nces.ed.gov/pubs2005/timss03/index.asp
———. 2008. "Highlights from TIMSS 2007: Mathematics and Science Achievement of U.S. Fourth- and Eighth-Grade Students in an International Context." Retrieved July 20, 2009. Online: http://nces.ed.gov/pubsearch/pubsinfo.asp?pubid52009001
———. 2009. "Percentage of School-Aged Children Who Were Homeschooled, By Reason Parents Gave." Retrieved Aug. 26, 2009. Online: http://nces.ed.gov/programs/coe/2009/charts/chart06_2.asp?popup5true
National Center for Health Statistics. 2007. "Births: Preliminary Data for 2006." Retrieved Feb. 13, 2008. Online: http://www.cdc.gov/nchs/data/nvsr/nvsr56/nvsr56_07.pdf
National Center for Injury Prevention and Control. 2006. "2004 United States Suicide Injury Deaths and Rates per 100,000." Retrieved Jan. 28, 2007. Online: http://webapp.cdc.gov/sasweb/ncipc/mortrate10_sy.html
National Center on Elder Abuse. 2003. "Types of Elder Abuse in Domestic Settings." Retrieved July 26, 2003. Online: http://www.elderabusecenter.org/pdf/basics/fact1.pdf
National Centers for Disease Control and Prevention. 2001. "43 Percent of First Marriages Break Up Within 15 Years." Retrieved July 14, 2002. Online: http://www.cdc.gov/nchs/releases/01news/firstmarr.htm
———. 2005. "Trend Analysis of the Sex Ratio at Birth in the United States." *National Vital Statistics Report* 53 (20). Retrieved Aug. 25, 2009. Online: http://www.cdc.gov/nchs/data/nvsr/nvsr53/nvsr53_20.pdf
———. 2008. "Understanding School Violence: Fact Sheet." Retrieved Aug. 26, 2009. Online: http://www.cdc.gov/ViolencePrevention/pdf/SchoolViolence_FactSheet-a.pdf
———. 2009. "FastStats." Retrieved July 21, 2009. Online: http://www.cdc.gov/nchs/fastats.htm
National Coalition on Health Care. 2009. "Facts on Health Care Costs." Retrieved July 21, 2009. Online: http://www.nchc.org/documents/Cost%20Fact%20Sheet-2009.pdf
National Council on Crime and Delinquency. 1969. *The Infiltration into Legitimate Business by Organized Crime.* Washington, DC: National Council on Crime and Delinquency.
National Education Association. 2007. "School Safety." Retrieved Apr. 22, 2007. Online: http://www.nea.org/schoolsafety/index.html
National High School Center. 2008. "High School Dropout: A Quick Stats Fact Sheet." Retrieved Aug. 26, 2009. Online: http://www.betterhighschools.org/docs/NHSC_DropoutFactSheet.pdf
National Law Center on Homelessness and Poverty. 2004. "Homelessness and Poverty in America." Retrieved Jan. 21, 2006. Online: http://www.nlchp.org/FA%5FHAPIA
Navarrette, Ruben, Jr. 1997. "A Darker Shade of Crimson." In Diana Kendall (Ed.), *Race, Class, and Gender in a Diverse Society.* Boston: Allyn & Bacon, 1997: 274–279. Reprinted from Ruben Navarrette, Jr., *A Darker Shade of Crimson.* New York: Bantam, 1993.
NBC Nightly News. 2007. "Coaching Themselves to Success." (Mar. 13).
nbc.com. 2007. "The Banker's Blog, February 4, 2007." Retrieved Mar. 17, 2007. Online: http://blog.nbc.com/dealornodeal/2007/02/04-week
NCDC. 2007. "About NCDC: Northwestern Community Development Corps." Retrieved Mar. 31, 2007. Online: http://groups.northwestern.edu/ncdc/about.html
Neergaard, Lauran. 2003. "Elderly Drivers a Concern." *ContraCostaTimes.com.* Retrieved July 20, 2003. Online: http://www.bayarea.com/mld/cctimes/news/6106194.htm
Neimark, Jill. 1997. "On the Front Lines of Alternative Medicine." *Psychology Today* (January/February): 51–68.
Nelson, Margaret K., and Joan Smith. 1999. *Working Hard and Making Do: Surviving in Small Town America.* Berkeley: University of California Press.
Nelson, Mariah Burton. 1994. *The Stronger Women Get, the More Men Love Football: Sexism and the American Culture of Sports.* New York: Harcourt.
Neuborne, Ellen. 1995. "Imagine My Surprise." In Barbara Findlen (Ed.), *Listen Up: Voices from the Next Feminist Generation.* Seattle: Seal, pp. 29–35.
Neuschler, Edward. 1987. *Medicaid Eligibility for the Elder in Need of Long Term Care.* Congressional Research Service Contract No. 86–26 (September). Washington, DC: National Governors Association.
New York Post. 2004. "Death Plunge No. 4: NYU's Grief." (Mar. 10): 1.
New York Times. 2002. "The Landscape on Vouchers." (June 28): A17.
———. 2006. "Far Down the List of Worries." (Apr. 23): WK14.
———. 2007a. "Editorial: Needed Fixes for No Child Left Behind." (Feb. 15). Retrieved Feb. 11, 2008. Online: http://www.nytimes.com/2007/02/15/opinion/15thur3.html
———. 2007b. "Migrating to Poor Countries, Searching for a Better Life." (Dec. 27): A16.
———. 2008. "President Map—Election Results." Online: http://elections.nytimes.com/2008/results/president/map.html
———. 2009. "Obama Takes Oath, and Nation in Crisis Embraces the Moment." (Jan. 21): A1, P6.
Newburger, Eric C. 2001. "Home Computers and Internet Use in the United States, August 2000." Current Population Reports, P23–207. U.S. Census Bureau. Retrieved Sept. 8, 2001. Online: http://www.census.gov/population/www/socdemo/computer.html
Newcott, Bill. 2007. "Movies for Grownups' Awards 2007." *AARP Magazine* (March & April). Retrieved Mar. 25, 2007. Online: http://www.aarpmagazine.org/entertainment/movies/moviesforgrownups_2007awards.html

Newman, Katherine S. 1988. *Falling from Grace: The Experience of Downward Mobility in the American Middle Class.* New York: Free Press.

———. 1993. *Declining Fortunes: The Withering of the American Dream.* New York: Basic.

———. 1999. *No Shame in My Game: The Working Poor in the Inner City.* New York: Knopf and the Russell Sage Foundation.

Newsweek. 1997. "Cult: Now on the Next Level." (Dec. 29/Jan. 5): 17.

———. 1999. "Perils and Promise: Teens by the Numbers." (May 10): 38–39.

Nguyen, Bich Minh. 2007. *Stealing Buddha's Dinner: A Memoir.* New York: Viking.

Niebuhr, H. Richard. 1929. *The Social Sources of Denominationalism.* New York: Meridian.

Nielsen, Joyce McCarl. 1990. *Sex and Gender in Society: Perspectives on Stratification* (2nd ed.). Prospects Heights, IL: Waveland.

Nisbet, Robert. 1979. "Conservativism." In Tom Bottomore and Robert Nisbet (Eds.), *A History of Sociological Analysis.* London: Heinemann, pp. 81–117.

Noble, Barbara Presley. 1995. "A Level Playing Field, for Just $121." *New York Times* (Mar. 5): F21.

Noel, Donald L. 1972. *The Origins of American Slavery and Racism.* Columbus, OH: Merrill.

Nolan, Patrick, and Gerhard E. Lenski. 1999. *Human Societies: An Introduction to Macrosociology* (8th ed.). New York: McGraw-Hill.

Nuland, Sherwin B. 1997. "Heroes of Medicine." *Time* (Fall special edition): 6–10.

Oakes, Jeannie. 1985. *Keeping Track: How High Schools Structure Inequality.* New Haven, CT: Yale University Press.

Oakes, Jeannie, and Martin Lipton. 2003. *Teaching to Change the World* (2nd ed.) New York: McGraw-Hill.

Oberschall, Anthony. 1973. *Social Conflict and Social Movements.* Englewood Cliffs, NJ: Prentice Hall.

Oboler, Suzanne. 1995. *Ethnic Labels, Latino Lives: Identity and the Politics of (Re)presentation in the United States.* Minneapolis: University of Minnesota Press.

O'Connell, Helen. 1994. *Women and the Family. Prepared for the UN-NGO Group on Women and Development.* Atlantic Highlands, NJ: Zed.

O'Connor, James. 1973. *The Fiscal Crisis of the State.* New York: St. Martin's.

Odendahl, Teresa. 1990. *Charity Begins at Home: Generosity and Self-Interest Among the Philanthropic Elite.* New York: Basic.

Ogburn, William F. 1966. *Social Change with Respect to Culture and Original Nature.* New York: Dell (orig. pub. 1922).

O'Hearn, Claudine Chiawei (Ed.). 1998. *Half and Half: Writers on Growing Up Biracial and Bicultural.* New York: Pantheon.

Ohio State University. 2007. "Cultural Diversity: Eating in America." Ohio State University Extension Fact Sheet, Family and Consumer Sciences Series. Retrieved Feb. 11, 2007. Online: http://www.ohioline.ag.ohio-state.edu

Omi, Michael, and Howard Winant. 1994. *Racial Formation in the United States: From the 1960s to the 1990s.* New York: Routledge.

Orenstein, Peggy, in association with the American Association of University Women. 1995. *SchoolGirls: Young Women, Self-Esteem, and the Confidence Gap.* New York: Anchor/Doubleday.

Ortner, Sherry B. 1974. "Is Female to Male as Nature Is to Culture?" In Michelle Rosaldo and Louise Lamphere (Eds.), *Women, Culture, and Society.* Stanford, CA: Stanford University Press.

Orum, Anthony M. 1974. "On Participation in Political Protest Movements." *Journal of Applied Behavioral Science,* 10: 181–207.

Orum, Anthony M., and Amy W. Orum. 1968. "The Class and Status Bases of Negro Student Protest." *Social Science Quarterly,* 49 (December): 521–533.

Orzechowski, Shawna, and Peter Sepielli. 2001. "Net Worth and Asset Ownership of Households: 1998 and 2000." U.S. Census Bureau, Current Population Reports, P70–88. Washington, DC: U.S. Government Printing Office.

Ouchi, William. 1981. *Theory Z: How American Business Can Meet the Japanese Challenge.* Reading, MA: Addison-Wesley.

Outhwaite, William, and Tom Bottomore (Eds.). 1994. *The Blackwell Dictionary of Twentieth-Century Social Thought.* Malden, MA: Blackwell.

Oxendine, Joseph B. 2003. *American Indian Sports Heritage* (rev. ed.). Lincoln: University of Nebraska Press.

Padilla, Felix M. 1993. *The Gang as an American Enterprise.* New Brunswick, NJ: Rutgers University Press.

———. 1997. *The Struggle of Latino/Latina University Students.* New York: Routledge.

Page, Charles H. 1946. "Bureaucracy's Other Face." *Social Forces,* 25 (October): 89–94.

Palen, J. John. 1995. *The Suburbs.* New York: McGraw-Hill.

Palmore, Erdman. 1981. *Social Patterns in Normal Aging: Findings from the Duke Longitudinal Study.* Durham, NC: Duke University Press.

Parenti, Michael. 1994. *Land of Idols: Political Mythology in America.* New York: St. Martin's.

———. 1996. *Democracy for the Few* (5th ed.). New York: St. Martin's.

———. 1998. *America Besieged.* San Francisco: City Lights.

Park, Robert E. 1915. "The City: Suggestions for the Investigation of Human Behavior in the City." *American Journal of Sociology,* 20: 577–612.

———. 1928. "Human Migration and the Marginal Man." *American Journal of Sociology,* 33.

———. 1936. "Human Ecology." *American Journal of Sociology,* 42: 1–15.

Park, Robert E., and Ernest W. Burgess. 1921. *Human Ecology.* Chicago: University of Chicago Press.

Parker, Robert Nash, and Doreen Anderson-Facile. 2000. "Violent Crime Trends." In Joseph F. Sheley (Ed.), *Criminology: A Contemporary Handbook* (3rd ed.). Belmont, CA: Wadsworth, pp. 191–214.

Parrish, Dee Anna. 1990. *Abused: A Guide to Recovery for Adult Survivors of Emotional/Physical Child Abuse.* Barrytown, NY: Station Hill.

Parry, A. 1976. *Terrorism: From Robespierre to Arafat.* New York: Vanguard.

Parsons, Talcott. 1951. *The Social System.* Glencoe, IL: Free Press.

———. 1955. "The American Family: Its Relations to Personality and to the Social Structure." In Talcott Parsons and Robert F. Bales (Eds.), *Family, Socialization and Interaction Process.* Glencoe, IL: Free Press, pp. 3–33.

———. 1960. "Toward a Healthy Maturity." *Journal of Health and Social Behavior,* 1: 163–173.

Passel, Jeffrey S. 2006. "The Size and Characteristics of the Unauthorized Migrant Population in the U.S.: Estimates Based on the March 2005 Current Population Survey." Pew Hispanic Center. Retrieved Mar. 8, 2006. Online: http://pewhispanic.org/reports/print.php?ReportID561

PBS. 1992. "Sex, Power, and the Workplace."

———. 2005a. "The Meaning of Food: Food & Culture." Retrieved Feb. 11, 2007. Online: http://www.pbs.org/opb/meaningoffood

———. 2005b. "Online NewsHour: The Schiavo Case Receives Strong Media Coverage" (Mar. 24). Retrieved Apr. 10, 2005. Online: http://www.pbs.org/newshour/bb/media/jan-june05/schiavo_3-24.html

———. 2008. "Facts About Global Poverty and Microcredit." Retrieved Jan. 19, 2008. Online: http://www.pbs.org/toourcredit/facts_one.htm

Pearce, Diana. 1978. "The Feminization of Poverty: Women, Work, and Welfare." *Urban and Social Change Review,* 11 (1/2): 28–36.

Pearson, Judy C. 1985. *Gender and Communication.* Dubuque, IA: Brown.

People. 2007. "Shopping with Ashley Tisdale." *People* (Feb. 26): 123, 125.

Perrow, Charles. 1986. *Complex Organizations: A Critical Essay* (3rd ed.). New York: Random House.

Perry, David C., and Alfred J. Watkins (Eds.). 1977. *The Rise of the Sunbelt Cities.* Beverly Hills, CA: Sage.

Perry, Steven W. 2004. "American Indians and Crime." U.S. Bureau of Justice Statistics. Retrieved Mar. 19, 2005. Online: http://www.ojp.usdoj.gov/bjs/pub/pdf/aic02.pdf

Petersen, John L. 1994. *The Road to 2015: Profiles of the Future.* Corte Madera, CA: Waite Group.

Peterson, Robert. 1992. *Only the Ball Was White: A History of Legendary Black Players and All-Black Professional Teams.* New York: Oxford University Press (orig. pub. 1970).

Pettigrew, Thomas. 1981. "The Mental Health Impact." In Benjamin Bowser and Raymond G. Hunt (Eds.), *Impacts of Racism on White Americans.* Beverly Hills, CA: Sage, p. 117 (cited in Feagin and Vera, 1995).

Pew Charitable Trusts. 2007. "Economic Mobility: Is the American Dream Alive and Well?" Retrieved Dec. 20, 2007. Online: http://www.economicmobility.org

Pew Forum on Religion & Public Life. 2008. "U.S. Religious Landscape Survey." Retrieved Apr. 23, 2008. Online: http://religions.pewforum.org/reports

Phillips, Jeanne. 2009. "Loss of Job Damages Man's Self-Esteem." *Dear Abby Express.* Retrieved June 25, 2009. Online: http://www.uexpress.com/dearabby/?uc_full_date520090625

Phillips, John C. 1993. *Sociology of Sport.* Boston: Allyn & Bacon.

Phillips, Peter, and Project Censored. 2002. *Censored 2003: The Top 25 Censored Stories.* New York: Seven Stories.

Piaget, Jean. 1932. *The Moral Judgment of the Child.* London: Routledge and Kegan Paul.

———. 1954. *The Construction of Reality in the Child.* Trans. Margaret Cook. New York: Basic.

Pierre-Pierre, Garry. 1997. "Traditional Church's New Life." *New York Times* (Nov. 15): A11.

Pietilä, Hilkka, and Jeanne Vickers. 1994. *Making Women Matter: The Role of the United Nations.* Atlantic Highlands, NJ: Zed.

Pillemer, Karl A. 1985. "The Dangers of Dependency: New Findings on Domestic Violence Against the Elderly." *Social Problems,* 33 (December): 146–158.

Pillemer, Karl A., and David Finkelhor. 1988. "The Prevalence of Elder Abuse: A Random Sample Survey." *Gerontologist,* 28 (1): 51–57.

Pinderhughes, Dianne M. 1986. "Political Choices: A Realignment in Partisanship Among Black Voters?" In James D. Williams (Ed.), *The State of*

Black America 1986. New York: National Urban League, pp. 85–113.

Pinderhughes, Howard. 1997. *Race in the Hood: Conflict and Violence Among Urban Youth.* Minneapolis: University of Minnesota Press.

Pines, Maya. 1981. "The Civilizing of Genie." *Psychology Today,* 15 (September): 28–29, 31–32, 34.

Pitzer, Ronald. 2003. "Rural Children Under Stress." University of Minnesota Extension Service. Retrieved Aug. 9, 2003. Online: http://www.extension.umn.edu/distribution/familydevelopment/components/7269cm.html

Pokin, Steve. 2007. "Pokin Around: A Real Person, a Real Death." *St. Charles Journal.* Retrieved Dec. 26, 2007. Online: http://stcharlesjournal.stltoday.com/articles/2007/11/10/news/sj2tn20071110-1111stc_pokin_1.iil.txt

Polakow, Valerie. 1993. *Lives on the Edge: Single Mothers and Their Children in the Other America.* Chicago: University of Chicago Press.

Polanyi, Karl. 1944. *The Great Transformation: The Political and Economic Origins of Our Time.* New York: Beacon.

Pomice, Eva. 1990. "Madison Avenue's Blind Spot." In Karin Swisher (Ed.), *The Elderly: Opposing Viewpoints.* San Diego: Greenhaven, pp. 42–45.

Population Reference Bureau. 2001. "Human Population: Fundamentals of Growth Patterns of World Urbanization." Retrieved Nov. 22, 2001. Online: http://www.prb.org/Content/NavigationMenu/PRB/E.../Patterns_of_World_Urbanization.html

Porter, Judith D. R. 1971. *Black Child, White Child.* Cambridge, MA: Harvard University Press.

Portes, Alejandro, and Rubén G. Rumbaut. 1996. *Immigrant America: A Portrait* (2nd ed.). Berkeley: University of California Press.

Postman, Neil, and Steve Powers. 1992. *How to Watch TV News.* New York: Penguin.

Postone, Moishe. 1997. "Rethinking Marx (in a Post-Marxist World)." In Charles Camic (Ed.), *Reclaiming the Sociological Classics: The State of the Scholarship.* Malden, MA: Blackwell, pp. 45–80.

Powell, Alvin. 2001. "Partnership Ensures Shelter's Future." *Harvard Gazette* (Jan. 18). Retrieved Feb. 24, 2007. Online: http://www.hno.harvard.edu/gazette/2001/01.18/08-partnership.html

Powell, Brian, and Douglas B. Downey. 1997. "Living in Single-Parent Households: An Investigation of the Same-Sex Hypothesis." *American Sociological Review,* 62 (August): 521–539.

Powell, Michael. 2004. "Evolution Shares a Desk with 'Intelligent Design.'" *Washington Post* (Dec. 26): A1.

Project Censored. 2009. "Top 25 Censored Stories for 2009." projectcensored.org. Retrieved July 9, 2009. Online: http://www.projectcensored.org/top-stories/category/y-2009

Prus, Robert. 1996. *Symbolic Interaction and Ethnographic Research: Intersubjectivity and the Study of Human Lived Experience.* Albany: State University of New York Press.

Pryor, John, and Kathleen McKinney (Eds.). 1991. "Sexual Harassment." *Basic and Applied Social Psychology,* 17 (4). Marketed as a book; Hillsdale, NJ: Erlbaum.

Puette, William J. 1992. *Through Jaundiced Eyes: How the Media View Organized Labor.* Ithaca, NY: ILR.

Puffer, J. Adams. 1912. *The Boy and His Gang.* Boston: Houghton Mifflin.

Quadagno, Jill S. 1984. "Welfare Capitalism and the Social Security Act of 1935." *American Sociological Review,* 49: 632–647.

Quarantelli, E. L., and James R. Hundley, Jr. 1993. "A Test of Some Propositions About Crowd Formation and Behavior." In Russell L. Curtis, Jr., and Benigno E. Aguirre (Eds.), *Collective Behavior and Social Movements.* Boston: Allyn & Bacon, pp. 183–193.

Quart, Alissa. 2003. *Branded: The Buying and Selling of Teenagers.* New York: Basic.

Queen, Stuart A., and David B. Carpenter. 1953. *The American City.* New York: McGraw-Hill.

Quinney, Richard. 1979. *Class, State, and Crime.* New York: McKay.

———. 2001. *Critique of the Legal Order.* Piscataway, NJ: Transaction (orig. pub. 1974).

Qvortrup, Jens. 1990. *Childhood as a Social Phenomenon.* Vienna: European Centre for Social Welfare Policy and Research.

Rabinowitz, Fredric E., and Sam V. Cochran. 1994. *Man Alive: A Primer of Men's Issues.* Pacific Grove, CA: Brooks/Cole.

Radcliffe-Brown, A. R. 1952. *Structure and Function in Primitive Society.* New York: Free Press.

Raffaelli, Marcela, and Lenna L. Ontai. 2004. "Gender Socialization in Latino/a Families: Results from Two Retrospective Studies." *Sex Roles,* 50: 287–299.

Raffalli, Mary. 1994. "Why So Few Women Physicists?" *New York Times Supplement* (January): Sect. 4A, 26–28.

Ramirez, Lisa. 1999. "'Senior Boom' Expected for 100-Year-Olds." *Austin American-Statesman* (Aug. 18): A4.

Ramirez, Marc. 1999. "A Portrait of a Local Muslim Family." *Seattle Times* (Jan. 24). Retrieved Aug. 16, 1999. Online: http://archives.seattletimes.com/cgi-bin/texis.mummy/web/vortex/display?storyID536d4d218

Ratha, Dilip, and William Shaw. 2007. "South–South Migration and Remittances." New York: World Bank Development Prospects Group. Retrieved Jan. 21, 2008. Online: http://siteresources.worldbank.org/INTPROSPECTS/Resources/SouthSouthMigrationandRemittances.pdf

Ravitch, Diana, and Joseph P. Viteritti. 1997. *New Schools for a New Century: The Redesign of Urban Education.* New Haven, CT: Yale University Press.

Reaves, Brian A. 2001. *Felony Defendants in Large Urban Counties, 1998: State Court Processing Statistics.* Washington, DC: U.S. Government Printing Office.

Reckless, Walter C. 1967. *The Crime Problem.* New York: Meredith.

Reich, Robert. 1993. "Why the Rich Are Getting Richer and the Poor Poorer." In Paul J. Baker, Louis E. Anderson, and Dean S. Dorn (Eds.), *Social Problems: A Critical Thinking Approach* (2nd ed.). Belmont, CA: Wadsworth, pp. 145–149. Adapted from *The New Republic,* May 1, 1989.

Reiman, Jeffrey. 1998. *The Rich Get Richer and the Poor Get Prison: Ideology, Class, and Criminal Justice* (5th ed.). Boston: Allyn & Bacon.

Reinharz, Shulamit. 1992. *Feminist Methods in Social Research.* New York: Oxford University Press.

Reinisch, June. 1990. *The Kinsey Institute New Report on Sex: What You Must Know to Be Sexually Literate.* New York: St. Martin's.

Reissman, Catherine. 1991. *Divorce Talk: Women and Men Make Sense of Personal Relationships.* New Brunswick, NJ: Rutgers University Press.

religioustolerance.org. 2005. "Science and Religion." Ontario Consultants on Religious Tolerance. Retrieved Apr. 22, 2005. Online: http://www.religioustolerance.org/sci_rel.htm#menu

Relman, Arnold S. 1992. "Self-Referral—What's at Stake?" *New England Journal of Medicine,* 327 (Nov. 19): 1522–1524.

Reskin, Barbara F., and Irene Padavic. 2002. *Women and Men at Work* (2nd ed.). Thousand Oaks, CA: Pine Forge.

Revkin, Andrew C. 2006. "Yelling 'Fire' On a Hot Planet." *New York Times* (Apr. 23): WK1–WK14.

Richardson, Laurel. 1993. "Inequalities of Power, Property, and Prestige." In Virginia Cyrus (Ed.), *Experiencing Race, Class, and Gender in the United States.* Mountain View, CA: Mayfield, pp. 229–236.

Rigler, David. 1993. "Letters: A Psychologist Portrayed in a Book About an Abused Child Speaks Out for the First Time in 22 Years." *New York Times Book Review* (June 13): 35.

Riley, Matilda White, and John W. Riley, Jr. 1994. "Age Integration and the Lives of Older People." *Gerontologist,* 34 (1): 110–115.

Rios, Delia M. 2004. "Amendments Raise Question: What Is the Constitution For?" *Austin American-Statesman* (Mar. 28): A17.

Ritzer, George. 1995. *Expressing America: A Critique of the Global Credit Card Society.* Thousand Oaks, CA: Pine Forge.

———. 1996. *Sociological Theory* (4th ed.). New York: McGraw-Hill.

———. 1997. *Postmodern Society Theory.* New York: McGraw-Hill.

———. 1998. *The McDonaldization Thesis.* London: Sage.

———. 1999. *Enchanting a Disenchanted World: Revolutionizing the Means of Consumption.* Thousand Oaks, CA: Pine Forge.

———. 2000a. *The McDonaldization of Society.* Thousand Oaks, CA: Pine Forge.

———. 2000b. *Modern Sociological Theory* (5th ed.). New York: McGraw-Hill.

Rizzo, Thomas A., and William A. Corsaro. 1995. "Social Support Processes in Early Childhood Friendships: A Comparative Study of Ecological Congruences in Enacted Support." *American Journal of Community Psychology,* 23: 389–418.

Robbins, Alexandra. 2004. *Pledged: The Secret Life of Sororities.* New York: Hyperion.

Roberts, Keith A. 2004. *Religion in Sociological Perspective* (4th ed.). Belmont, CA: Wadsworth.

Roberts, Sam. 1994. "Black Women Graduates Outpace Male Counterparts." *New York Times* (Oct. 31): A8.

Robinson, Brian E. 1988. *Teenage Fathers.* Lexington, MA: Lexington.

Robson, Ruthann. 1992. *Lesbian (Out)law: Survival Under the Rule of Law.* New York: Firebrand.

Rocca, Mo. 2007. "TV Drug Ads' Side Effects." CBS News *Sunday Morning* (Oct. 14). Retrieved Feb. 24, 2008. Online: http://www.cbsnews.com/stories/2007/10/14/sunday/main3365346.shtml

Rockwell, John. 1994. "The New Colossus: American Culture as Power Export." *New York Times* (Jan. 30): section 2: 1, 30.

Roethlisberger, Fritz J., and William J. Dickson. 1939. *Management and the Worker.* Cambridge, MA: Harvard University Press.

Rogers, Harrell R. 1986. *Poor Women, Poor Families: The Economic Plight of America's Female-Headed Households.* Armonk, NY: Sharpe.

Rollins, Judith. 1985. *Between Women: Domestics and Their Employers.* Philadelphia: Temple University Press.

Romero, Mary. 1992. *Maid in the U.S.A.* New York: Routledge.

———. 1997. "Introduction." In Mary Romero, Pierrette Hondagneu-Sotelo, and Vilma Ortiz (Eds.), *Challenging Fronteras: Structuring Latina and Latino Lives in the U.S.* New York: Routledge, pp. 3–5.

Romo, Harriett D., and Toni Falbo. 1996. *Latino High School Graduation.* Austin: University of Texas Press.

Roob, Nancy, and Ruth McCambridge. 1992. "Private-Sector Funders: Their Role in Homelessness Projects." In Padraig O'Malley (Ed.), *Homelessness: New England and Beyond: New England Journal of Public Policy* (May special issue): 623–646.

Roof, Wade Clark. 1993. *A Generation of Seekers: The Spiritual Journeys of the Baby Boom Generation.* San Francisco: HarperSanFrancisco.

Ropers, Richard H. 1991. *Persistent Poverty: The American Dream Turned Nightmare.* New York: Plenum.

Rose, Jerry D. 1982. *Outbreaks.* New York: Free Press.

Rosenblatt, Robert A. 1994. "Entitlement Seen Taking Up Nearly All Taxes by 2012." *Los Angeles Times* (Aug. 9): 1.

Rosenbloom, Stephanie. 2008. "Putting Your Best Cyberface Forward." *New York Times* (Jan. 3): E1, E6.

Rosengarten, Ellen M. 1995. Communication to author.

Rosenthal, Naomi, Meryl Fingrutd, Michele Ethier, Roberta Karant, and David McDonald. 1985. "Social Movements and Network Analysis: A Case Study of Nineteenth-Century Women's Reform in New York State." *American Journal of Sociology,* 90: 1022–1054.

Rosenthal, Robert, and Lenore Jacobson. 1968. *Pygmalion in the Classroom: Teacher Expectation and Student's Intellectual Development.* New York: Holt, Rinehart, and Winston.

Rosnow, Ralph L., and Gary Alan Fine. 1976. *Rumor and Gossip: The Social Psychology of Hearsay.* New York: Elsevier.

Ross, Dorothy. 1991. *The Origins of American Social Science.* Cambridge, England: Cambridge University Press.

Rossi, Alice S. 1992. "Transition to Parenthood." In Arlene Skolnick and Jerome Skolnick (Eds.), *Family in Transition.* New York: HarperCollins, pp. 453–463.

Rossi, Peter H. 1989. *Down and Out in America: The Origins of Homelessness.* Chicago: University of Chicago Press.

Rossides, Daniel W. 1986. *The American Class System: An Introduction to Social Stratification.* Boston: Houghton Mifflin.

Rostow, Walt W. 1971. *The Stages of Economic Growth: A Non-Communist Manifesto* (2nd ed.). Cambridge: Cambridge University Press (orig. pub. 1960).

———. 1978. *The World Economy: History and Prospect.* Austin: University of Texas Press.

Roth, Guenther. 1988. "Marianne Weber and Her Circle." In Marianne Weber, *Max Weber.* New Brunswick, NJ: Transaction, p. xv.

Rothchild, John. 1995. "Wealth: Static Wages, Except for the Rich." *Time* (Jan. 30): 60–61.

Rotheram, Mary Jane, and Jean S. Phinney. 1987. "Introduction: Definitions and Perspectives in the Study of Children's Ethnic Socialization." In Jean S. Phinney and Mary Jane Rotheram (Eds.), *Children's Ethnic Socialization.* Newbury Park, CA: Sage, pp. 10–28.

Rothman, Robert A. 2005. *Inequality and Stratification: Class, Color, and Gender* (5th ed.). Upper Saddle River, NJ: Prentice Hall.

Rousseau, Ann Marie. 1981. *Shopping Bag Ladies: Homeless Women Speak About Their Lives.* New York: Pilgrim.

Rubin, Lillian B. 1986. "A Feminist Response to Lasch." *Tikkun,* 1 (2): 89–91.

Ruffin, Roy J., and Paul R. Gregory. 2000. *Principles of Economics* (7th ed.). Upper Saddle River, NJ: Pearson Addison-Wesley.

Rural Policy Research Institute. 2004. "Place Matters: Addressing Rural Poverty." Rural Poverty Research Institute. Retrieved Jan. 21, 2008. Online: http://www.rprconline.org/synthesis.pdf

Russo, Nancy Felipe, and Mary A. Jansen. 1988. "Women, Work, and Disability: Opportunities and Challenges." In Michelle Fine and Adrienne Asch (Eds.), *Women with Disabilities: Essays in Psychology, Culture, and Politics.* Philadelphia: Temple University Press.

Rutstein, Nathan. 1993. *Healing in America.* Springfield, MA: Whitcomb.

Rymer, Russ. 1993. *Genie: An Abused Child's Flight from Silence.* New York: HarperCollins.

Sadker, David, and Myra Sadker. 1985. "Is the OK Classroom OK?" *Phi Delta Kappan,* 55: 358–367.

———. 1986. "Sexism in the Classroom: From Grade School to Graduate School." *Phi Delta Kappan,* 68: 512–515.

Sadker, Myra, and David Sadker. 1994. *Failing at Fairness: How America's Schools Cheat Girls.* New York: Scribner.

Safilios-Rothschild, Constantina. 1969. "Family Sociology or Wives' Family Sociology? A Cross-Cultural Examination of Decision-Making." *Journal of Marriage and the Family,* 31 (2): 290–301.

Salt Lake City Sheriff's Department. 2007. "Graffiti: That Writing on the Wall." Retrieved Mar. 10, 2007. Online: http://www.slsheriff.org/html/org/metrogang/graffiti.html

Samovar, Larry A., and Richard E. Porter. 1991a. *Communication Between Cultures.* Belmont, CA: Wadsworth.

———. 1991b. *Intercultural Communication: A Reader* (6th ed.). Belmont, CA: Wadsworth.

Sampson, Robert J. 1986. "Effects of Socioeconomic Context on Official Reaction to Juvenile Delinquency." *American Sociological Review,* 51 (December): 876–885.

———. 1997. "Neighborhoods and Violent Crime: A Multilevel Study of Collective Efficacy." *Science,* 277: 18–25.

Sampson, Robert J., and John Laub. 1993. *Crime in the Making: Pathways and Turning Points Through Life.* Cambridge, MA: Harvard University Press.

Sanchez-Ayendez, Melba. 1995. "Puerto Rican Elderly Women: Shared Meanings and Informal Supportive Networks." In Margaret L. Andersen and Patricia Hill Collins (Eds.), *Race, Class, and Gender: An Anthology.* Belmont, CA: Wadsworth.

Sandals, Leah. 2007. "'Public Space Protection'—But for Which 'Public'?" Retrieved Feb. 24, 2007. Online: http://spacing.ca/wire/?p51466

Sandefur, Gary D., and Arthur Sakamoto. 1988. "American Indian Household Structure and Income." *Demography,* 25 (1): 71–80.

Sapir, Edward. 1961. *Culture, Language and Personality.* Berkeley: University of California Press.

Sargent, Margaret. 1987. *Sociology for Australians* (2nd ed.). Melbourne, Australia: Longman Cheshire.

Sassen, Saskia. 2001. *The Global City: New York, London, Tokyo* (2nd ed.). Princeton, NJ: Princeton University Press.

SAT. 2009. "2008 College-Bound Seniors: Student Background Information and Characteristics." Retrieved July 4, 2009. Online: http://professionals.collegeboard.com/profdownload/Total_Group_Report.pdf

Saulny, Susan. 2009. "In Obama Era, Voices Reflect Rising Sense of Racial Optimism." *New York Times* (May 3): A1, A26.

Savin-Williams, Ritch C. 2004. "Memories of Same-Sex Attractions." In Michael S. Kimmel and Michael A. Messner (Eds.), *Men's Lives* (6th ed.). Boston: Allyn & Bacon, pp. 116–132.

Schaefer, James. 2005. "Reporting Complexity: Science and Religion." In Claire Hoertz Badaracco (Ed.), *Quoting God: How Media Shape Ideas About Religion and Culture.* Waco, TX: Baylor University Press, pp. 211–224.

Schaefer, Richard T., and William W. Zellner. 2007. *Extraordinary Groups: An Examination of Unconventional Lifestyles* (8th ed.). New York: Worth.

Schama, Simon. 1989. *Citizens: A Chronicle of the French Revolution.* New York: Knopf.

Schattschneider, Elmer Eric. 1969. *Two Hundred Americans in Search of a Government.* New York: Holt, Rinehart & Winston.

Schemo, Diana Jean. 1994. "Suburban Taxes Are Higher for Blacks, Analysis Shows." *New York Times* (Aug. 17): A1, A16.

———. 2006. "Most Students in Big Cities Lag Badly in Basic Science." *New York Times* (Nov. 16). Retrieved Mar. 4, 2007. Online: http://www.nytimes.com/2006/11/16/education/16reportcard.html

Schneider, Donna. 1995. *American Childhood: Risks and Realities.* New Brunswick, NJ: Rutgers University Press.

Schneider, Keith. 1993. "The Regulatory Thickets of Environmental Racism." *New York Times* (Dec. 19): E5.

Scholastic Parent & Child. 2007. "How and When to Praise." Retrieved Jan. 3, 2008. Online: http://content.scholastic.com/browse/article.jsp?id52064

Schor, Juliet B. 1999. *The Overspent American: Upscaling, Downshifting, and the New Consumer.* New York: HarperPerennial.

Schubert, Hans-Joachim (Ed.). 1998. "Introduction." In *On Self and Social Organization—Charles Horton Cooley.* Chicago: University of Chicago Press, pp. 1–31.

Schur, Edwin M. 1983. *Labeling Women Deviant: Gender, Stigma, and Social Control.* Philadelphia: Temple University Press.

Schutske, John. 2002. "Keeping Farm Children Safe." University of Minnesota Extension Service. Retrieved Aug. 9, 2003. Online: http://www.extension.umn.edu/distribution/youth development/DA6188.html

Schwartz, John, and John M. Broder. 2003. "Engineer Warned of Consequences of Liftoff Damage." *New York Times* (Feb. 13): A1–A29.

Schwartz, John (with Matthew L. Wald). 2003. "Costs and Risk Clouding Plans to Fix Shuttles." *New York Times* (June 8): A1–A20.

Schwartz, John, and Matthew L. Wald. 2003. "'Groupthink' Is 30 Years Old, and Still Going Strong." *New York Times* (Mar. 9): WK3.

Schwarz, John E., and Thomas J. Volgy. 1992. *The Forgotten Americans.* New York: Norton.

Scott, Alan. 1990. *Ideology and the New Social Movements.* Boston: Unwin & Hyman.

Seccombe, Karen. 1991. "Assessing the Costs and Benefits of Children: Gender Comparisons Among Childfree Husbands and Wives." *Journal of Marriage and the Family,* 53 (1): 191–202.

Seegmiller, B. R., B. Suter, and N. Duviant. 1980. *Personal, Socioeconomic, and Sibling Influences on Sex-Role Differentiation.* Urbana: ERIC Clearinghouse of Elementary and Early Childhood Education, ED 176 895, College of Education, University of Illinois.

Segura, Denise A. 1994. "Inside the Work Worlds of Chicana and Mexican Immigrant Workers." In Maxine Baca Zinn and Bonnie Thornton Dill (Eds.), *Women of Color in U.S. Society.* Philadelphia: Temple University Press, pp. 95–111.

Seid, Roberta P. 1994. "Too 'Close to the Bone': The Historical Context for Women's Obsession with Slenderness." In Patricia Fallon, Melanie A. Katzman, and Susan C. Wooley (Eds.), *Feminist Perspectives on Eating Disorders.* New York: Guilford, pp. 3–16.

Seligman, Martin E. P. 1975. *Helplessness: On Depression, Development and Death.* San Francisco: Freeman.

Sengoku, Tamotsu. 1985. *Willing Workers: The Work Ethic in Japan, England, and the United States.* Westport, CT: Quorum.

Sengupta, Somini. 1997. "At Holidays, Test of Patience of Muslims." *New York Times* (Dec. 25): A12.

Senna, Joseph J., and Larry J. Siegel. 2002. *Introduction to Criminal Justice* (9th ed.). Belmont, CA: Wadsworth.

Seper, Chris. 2003. "Blogging About Your Job? Your Boss Could Take Offense." *Austin American-Statesman* (July 27): J1, J4.

Serbin, Lisa A., Phyllis Zelkowitz, Anna-Beth Doyle, Dolores Gold, and Bill Wheaton. 1990. "The Socialization of Sex-Differentiated Skills and Academic Performance: A Mediational Model." *Sex Roles,* 23: 613–628.

Serrill, Michael S. 1997. "Socialism Dies Again." *Time* (Sept. 22): 44.

Shapin, Steven. 2006. "Paradise Sold: What Are You Buying When You Buy Organic?" *The New Yorker* (May 15). Retrieved Feb. 17, 2007. Online: http://www.newyorker.com/critics/atlarge/articles/060515crat_atlarge

Shapiro, Joseph P. 1993. *No Pity: People with Disabilities Forging a New Civil Rights Movement.* New York: Times/Random House.

Sharma, Arvind. 1995. "Hinduism." In Arvind Sharma (Ed.), *Our Religions.* San Francisco: HarperCollins, pp. 3–67.

Shaw, Randy. 1999. *Reclaiming America: Nike, Clean Air, and the New National Activism.* Berkeley: University of California Press.

Shawver, Lois. 1998. "Notes on Reading Foucault's *The Birth of the Clinic.*" Retrieved Oct. 2, 1999. Online: http://www.california.com/~rathbone/foucbc.htm

Sheen, Fulton J. 1995. *From the Angel's Blackboard: The Best of Fulton J. Sheen.* Ligouri, MO: Triumph.

Sheff, David. 1995. "If It's Tuesday, It Must Be Dad's House." *New York Times Magazine* (Mar. 26): 64–65.

Sheff, Nick. 1999. "My Long-Distance Life." *Newsweek* (Feb. 15): 16.

Shell, Adam. 2007. "Morgan Stanley Settles Sex-Bias Case." *USA Today* (Apr. 24). Retrieved Feb. 2, 2008. Online: http://www.usatoday/com/money/companies/2004-07-12-morgan-stanley-suit_x.htm

Sheppard, Ashley Dawn. 2005. "Facebook Includes Finding Friends, Safety Concerns." *Baylor Lariat* (Jan. 25).

Sherman, Suzanne (Ed.). 1992. "Frances Fuchs and Gayle Remick." In *Lesbian and Gay Marriage: Private Commitments, Public Ceremonies.* Philadelphia: Temple University Press, pp. 189–201.

Shevky, Eshref, and Wendell Bell. 1966. *Social Area Analysis: Theory, Illustrative Application and Computational Procedures.* Westport, CT: Greenwood.

Shorto, Russell. 1997. "Belief by the Numbers." *New York Times Magazine* (Dec. 7): 60.

Shum, Tedd. 1997. "Olympic Gymnast Chow Makes Impact on All Americans." Retrieved Aug. 15, 1999. Online: http://www.dailybruin.ucla.edu/DB/issues/97/05.30/view.shum.html

Shupe, Anson, and Jeffrey K. Hadden. 1989. "Is There Such a Thing as Global Fundamentalism?" In Jeffrey K. Hadden and Anson Shupe (Eds.), *Secularization and Fundamentalism Reconsidered.* New York: Paragon.

Siddiqi, Faraaz, and Harry Anthony Patrinos. 1995. "Child Labor: Issues, Causes and Interventions." World Bank Human Capital Development and Operations Policy working paper 56. Retrieved Mar. 11, 2005. Online: http://www.worldbank.org/html/extdr/hnp/hddflash/workp/wp_00056.html.

Siegel, Larry J. 1998. *Criminology: Theories, Patterns, and Typologies* (6th ed.). Belmont, CA: West/Wadsworth.

———. 2006. *Criminology* (9th ed.). Belmont, CA: Wadsworth.

———. 2007. *Criminology: Theories, Patterns, and Typologies* (9th ed.). Belmont, CA: Wadsworth.

Simmel, Georg. 1950. *The Sociology of Georg Simmel.* Trans. Kurt Wolff. Glencoe, IL: Free Press (orig. written in 1902–1917).

———. 1957. "Fashion." *American Journal of Sociology,* 62 (May 1957): 541–558. Orig. pub. 1904.

———. 1990. *The Philosophy of Money.* Ed. David Frisby. New York: Routledge (orig. pub. 1907).

Simon, David R. 1996. *Elite Deviance* (5th ed.). Boston: Allyn & Bacon.

Simpson, Sally S. 1989. "Feminist Theory, Crime, and Justice." *Criminology,* 27: 605–632.

Singer, Margaret Thaler, with Janja Lalich. 1995. *Cults in Our Midst.* San Francisco: Jossey-Bass.

Sivard, Ruth L. 1991. *World Military and Social Expenditures—1991.* Washington, DC: World Priorities.

———. 1993. *World Military and Social Expenditures—1993.* Washington, DC: World Priorities.

Sjoberg, Gideon. 1965. *The Preindustrial City: Past and Present.* New York: Free Press.

Skocpol, Theda, and Edwin Amenta. 1986. "States and Social Policies." In Ralph H. Turner and James F. Short, Jr. (Eds.), *Annual Review of Sociology,* 12: 131–157.

Sloan, Don. 2005. "The Shame of Child Labor." *Political Affairs* (January). Retrieved Feb. 27, 2005. Online: http://www.politicalaffairs.net/article/view/462

Slugoski, B. F., and G. B. Ginsburg. 1989. "Ego Identity and Explanatory Speech." In John Shotter and Kenneth J. Gergen (Eds.), *Texts of Identity.* London: Sage, pp. 36–55.

Smelser, Neil J. 1963. *Theory of Collective Behavior.* New York: Free Press.

———. 1988. "Social Structure." In Neil J. Smelser (Ed.), *Handbook of Sociology.* Newbury Park, CA: Sage, pp. 103–129.

Smith, Adam. 1976. *An Inquiry into the Nature and Causes of the Wealth of Nations.* Ed. Roy H. Campbell and Andrew S. Skinner. Oxford, England: Clarendon (orig. pub. 1776).

Smith, Allen C., III, and Sheryl Kleinman. 1989. "Managing Emotions in Medical School: Students' Contacts with the Living and the Dead." *Social Science Quarterly,* 52 (1): 56–69.

Smith, Denise. 2003. "The Older Population in the United States: March 2002." U.S. Census Bureau, Current Population Reports, P20-546. Washington, DC: U.S. Government Printing Office.

Smith, Dorothy E. 1999. *Writing the Social: Critique, Theory, and Investigations.* Toronto: University of Toronto Press.

Smith, Douglas, Christy Visher, and Laura Davidson. 1984. "Equity and Discretionary Justice: The Influence of Race on Police Arrest Decisions." *Journal of Criminal Law and Criminology,* 75: 234–249.

Smith, Huston. 1991. *The World's Religions.* San Francisco: HarperSanFrancisco.

Smith, Wallace Charles. 1985. *The Church in the Life of the Black Family.* Valley Forge, PA: Judson.

Smith, Wes. 2001. *Hope Meadows: Real-Life Stories of Healing and Caring from an Inspiring Community.* New York: Berkley.

Smolkin, Rachel. 2007. "What the Mainstream Media Can Learn from Jon Stewart." *American Journalism Review* (June/July 2007). Retrieved Feb. 16, 2008. Online: http://www.ajr.org/Article.asp?id54329

Snow, David A., and Leon Anderson. 1991. "Researching the Homeless: The Characteristic Features and Virtues of the Case Study." In Joe R. Feagin, Anthony M. Orum, and Gideon Sjoberg (Eds.), *A Case for the Case Study.* Chapel Hill: University of North Carolina Press, pp. 148–173.

———. 1993. *Down on Their Luck: A Case Study of Homeless Street People.* Berkeley: University of California Press.

Snow, David A., and Robert Benford. 1988. "Ideology, Frame Resonance, and Participant Mobilization." In Bert Klandermans, Hanspeter Kriesi, and Sidney Tarrow (Eds.), *International Social Movement Research,* Vol. 1, *From Structure to Action.* Greenwich, CT: JAI, pp. 133–155.

Snow, David A., E. Burke Rochford, Jr., Steven K. Worden, and Robert D. Benford. 1986. "Frame Alignment Processes, Micromobilization, and Movement Participation." *American Sociological Review,* 51: 464–481.

Snow, David A., Louis A. Zurcher, and Robert Peters. 1981. "Victory Celebrations as Theater: A Dramaturgical Approach to Crowd Behavior." *Symbolic Interaction,* 4 (1): 21–41.

Snyder, Benson R. 1971. *The Hidden Curriculum.* New York: Knopf.

Solem, Per Erik. 2008. "Age Changes in Subjective Work Ability." *International Journal of Aging and Later Life* 3 (2): 43–70.

Solomon, Jay. 2001. "How Mr. Bambang Markets Big Macs in Muslim Indonesia." *Wall Street Journal* (Oct. 26): A1–A7.

Sommers, Ira, and Deborah R. Baskin. 1993. "The Situational Context of Violent Female Offending." *Journal of Research in Crime and Delinquency,* 30 (2): 136–162.

South, Scott J., Charles M. Bonjean, Judy Corder, and William T. Markham. 1982. "Sex and Power in the Federal Bureaucracy." *Work and Occupations,* 9 (2): 233–254.

Spitzer, Steve. 1975. "Toward a Marxian Theory of Deviance." *Social Problems,* 22: 638–651.

sportsnetwork.com. 2003. "International Soccer: UEFA Seeks to Eradicate Racism." Retrieved July 17, 2003. Online: http://www.sportnetwork.com/?c5sportsnetwork&page5soc-cup/news/CAN2477113.htm

Sreenivasan, Sreenath. 1996. "Blind Users Add Access on the Web." *New York Times* (Dec. 2): C7.

Stack, Steven. 1998. "Gender, Marriage, and Suicide Acceptability: A Comparative Analysis." *Sex Roles,* 38: 501–521.

Stack, Steven, and I. Wasserman. 1995. "The Effect of Marriage, Family, and Religious Ties on

African American Suicide Ideology." *Journal of Marriage and the Family,* 57: 215–222.

Stake, Robert E. 1995. *The Art of Case Study Research.* Thousand Oaks, CA: Sage.

Stanley, Alessandra. 2004. "Old-Time Sexism Suffuses New Season." *New York Times* (Oct. 1): B1–B22.

Stannard, David E. 1992. *American Holocaust: Columbus and the Conquest of the New World.* New York: Oxford University Press.

Staples, Robert. 1994. "The Illusion of Racial Equality: The Black American Dilemma." In Gerald Early (Ed.), *Lure and Loathing: Essays on Race, Identity, and the Ambivalence of Assimilation.* New York: Penguin.

Stapleton-Paff, Katie. 2007. "College Students Prefer 'The Daily Show' to Real News." *The Daily of the University of Washington* (May 21, 2007): 1.

Stark, Rodney, and William Sims Bainbridge. 1981. "American-Born Sects: Initial Findings." *Journal for the Scientific Study of Religion,* 20: 130–149.

Starr, Paul. 1982. *The Social Transformation of Medicine: The Rise of a Sovereign Profession and the Making of a Vast Industry.* New York: Basic.

Steffensmeier, Darrell, and Emilie Allan. 2000. "Looking for Patterns: Gender, Age, and Crime." In Joseph F. Sheley (Ed.), *Criminology: A Contemporary Handbook* (3rd ed.). Belmont, CA: Wadsworth, pp. 85–128.

Stein, Peter J. 1976. *Single.* Englewood Cliffs, NJ: Prentice Hall.

Stein, Peter J. (Ed.). 1981. *Single Life: Unmarried Adults in Social Context.* New York: St. Martin's.

Steinmetz, Erika. 2006. "Americans with Disabilities: 2000." U.S. Census Bureau, Current Population Reports, P70–107. Washington, DC: U.S. Government Printing Office.

Steinmetz, Suzanne K. 1987. "Elderly Victims of Domestic Violence." In Carl D. Chambers, John H. Lindquist, O. Z. White, and Michael T. Harter (Eds.), *The Elderly: Victims and Deviants.* Athens: Ohio University Press, pp. 126–141.

Stevenson, Mary Huff. 1988. "Some Economic Approaches to the Persistence of Wage Differences Between Men and Women." In Ann H. Stromberg and Shirley Harkess (Eds.), *Women Working: Theories and Facts in Perspective* (2nd ed.). Mountain View, CA: Mayfield, pp. 87–100.

Stewart, Abigail J. 1994. "Toward a Feminist Strategy for Studying Women's Lives." In Carol E. Franz and Abigail J. Stewart (Eds.), *Women Creating Lives: Identities, Resilience, and Resistance.* Boulder, CO: Westview, pp. 11–35.

Stickler, Christine. 2004. "One Response to Special Needs in the Classroom: Utilizing College Students as an Untapped Resource." *New Horizons for Learning* (September). Retrieved Apr. 17, 2005. Online: http://www.newhorizons.org/lifelong/higher_ed/stickler.htm

Stiehm, Judith Hicks. 1989. *Arms and the Enlisted Woman.* Philadelphia: Temple University Press.

St. John, Warren. 2007. "A Laboratory for Getting Along." *New York Times* (Dec. 25): A1, A14.

Stoller, Eleanor Palo, and Rose Campbell Gibson. 1997. *Worlds of Difference: Inequalities in the Aging Experience* (2nd ed.). Thousand Oaks, CA: Sage.

StopGlobalWarming.org. 2006. "Marchers." Retrieved May 6, 2006. Online: http://www.stopglobalwarming.org

Stross, Randall. 2006. "Cellphone as Tracker: X Marks Your Doubts." *New York Times* (Nov. 19). Retrieved Mar. 30, 2007. Online: http://select.nytimes.com/search/restricted/article?res5F30F1FF63F5A0C7A8DDDA80994DE404482

Struck, Doug. 2006. "'Rapid Warming' Spreads Havoc in Canada's Forests." Washingtonpost.com (Mar. 1). Retrieved May 2, 2006. Online: http://www.washingtonpost.com/wp-dyn/content/article/2006/02/28/AR2006022801772

Substance Abuse and Mental Health Services Administration. 2000. *National Household Survey on Drug Abuse, 2000.* Retrieved Nov. 20, 2000. Online: http://www.samhsa.gov/oas/NHSDA/2kNHSDA/chapter2.htm

Suhr, Jim. 2007. "Police: Elderly Driver in School Crash Was Bound for Driving School." *Chicago Tribune* (Jan. 30). Retrieved Mar. 25, 2007. Online: http://www.chicagotribune.com/news/local/illinois/chi-ap-il-carhits school,1,3832602.story?coll5chi-newsap_il-hed&ctrack51&cset5true

Sullivan, Thomas J. 2000. *Introduction to Social Problems* (5th ed.). Boston: Allyn & Bacon.

Sumner, William G. 1959. *Folkways.* New York: Dover (orig. pub. 1906).

Sutherland, Edwin H. 1939. *Principles of Criminology.* Philadelphia: Lippincott.

———. 1949. *White Collar Crime.* New York: Dryden.

Swarns, Rachel L. 2006. "The Immigrant Debate: The Overview; Immigrants Rally in Scores of Cities for Legal Status." *New York Times* (Apr. 11): A1.

Swidler, Ann. 1986. "Culture in Action: Symbols and Strategies." *American Sociological Review,* 51 (April): 273–286.

Szasz, Thomas S. 1984. *The Myth of Mental Illness: Foundations of a Theory of Personal Conduct.* New York: HarperCollins.

Tabb, William K., and Larry Sawers. 1984. *Marxism and the Metropolis: New Perspectives in Urban Political Economy* (2nd ed.). New York: Oxford University Press.

Takaki, Ronald. 1989. *Strangers from a Different Shore: A History of Asian Americans.* New York: Penguin.

———. 1993. *A Different Mirror: A History of Multicultural America.* Boston: Little, Brown.

Tanamachi, Cara. 1997. "American Dream Gets a Makeover." *Austin American-Statesman* (July 20): A1, A13.

Tannen, Deborah. 1993. "Commencement Address, State University of New York at Binghamton." Reprinted in *Chronicle of Higher Education* (June 9): B5.

Tarbell, Ida M. 1925. *The History of Standard Oil Company.* New York: Macmillan (orig. pub. 1904).

Tavris, Carol. 1993. *The Mismeasure of Woman.* New York: Touchstone.

Tax Policy Center. 2006. "Historical Number of Households, Average Pretax and After-Tax Income and Shares, and Minimum Income." Retrieved Feb. 26, 2006. Online: http://www.taxpolicycenter.org/TaxFacts/TFDB/TFTemplate.cfm?Docid5461

———. 2009. "Historical Number of Households, Average Pretax and After-Tax Income and Shares, and Minimum Income." Retrieved May 28, 2009. Online: http://www.taxpolicycenter.org/UploadedPDF/901006_taxpolicy.pdf

Taylor, Howard F. 1995. "Symposium: The Bell Curve." *Contemporary Sociology: A Journal of Reviews,* 24 (2): 153–157.

Taylor, Robert Joseph, Linda M. Chatters, and V. Mays. 1988. "Parents, Children, Siblings, In-Laws, and Non-Kin Sources of Emergency Assistance to Black Americans." *Family Relations,* 37: 298–304.

Taylor, Steve. 1982. *Durkheim and the Study of Suicide.* New York: St. Martin's.

Teicher, Stacy A. 2006. "Researchers Say That Middle-School Bullying Could Be Curbed by Showing That It's Not Normal." *Christian Science Monitor* (Aug. 17). Retrieved Feb. 12, 2008. Online: http://www.csmonitor.com/2006/0817/p15s02-legn.html

Tergat, Paul. 2005. "Food That Changed My Life." *The Guardian* (Apr. 15). Retrieved Feb. 24, 2006. Online: http://www.countercurrents.org/tergat150405.htm

Terkel, Studs. 1990. *Working: People Talk About What They Do All Day and How They Feel About What They Do.* New York: Ballantine (orig. pub. 1972).

———. 1996. *Coming of Age: The Story of Our Century by Those Who've Lived It.* New York: St. Martin's Griffin.

Texeira, Erin. 2005. "Multiracial Scenes Now Common in TV Ads But Critics Say Commercials Gloss Over Complicated Racial Realities." MSNBC.com (Feb. 15). Retrieved Mar. 27, 2005. Online: http://www.msnbc.msn.com/id/6975669

That, Sovanny. 2007. "Refugee Women's Alliance." Retrieved Mar. 31, 2007. Online: http://students.washington.edu/sovannyt/communityservice.htm

Thayer, Kay. 2008. Posted on Hockenberry, John, and Adaora Udoji. 2008. "The College Credit-Card Crunch." *The TakeAway* (July 23). Retrieved Feb. 3, 2009. Online: http://www.thetakeaway.org/stories/2008/jul/23/the college-credit-card-crunch

theadventuresofiman.com. 2007. "The Adventures of Iman." Retrieved Mar. 18, 2007. Online: http://www.theadventuresofiman.com/AboutIman.asp

thewritemarket.com. 2003. "Building an Internet Community: What Is an 'Internet Community'?" Retrieved July 5, 2003. Online: http://www.thewritemarket.com/promotion/community.htm

Thomas, Katie. 2009. "Title IX Ruling Could Lead to More Strict Standards." *New York Times* (June 19): B11.

Thompson, Becky W. 1994. *A Hunger So Wide and So Deep: American Women Speak Out on Eating Problems.* Minneapolis: University of Minnesota Press.

Thornberry, Terence P., Marvin D. Krohn, Alan J. Lizotte, and Deborah Chard-Wierschem. 1993. "The Role of Juvenile Gangs in Facilitating Delinquent Behavior." *Journal of Research in Crime and Delinquency,* 30 (1): 55–87.

Thorne, Barrie. 1993. *Gender Play: Girls and Boys in School.* New Brunswick, NJ: Rutgers University Press.

———. 1995. "Girls and Boys Together . . . But Mostly Apart: Gender Arrangements in Elementary Schools." In Michael S. Kimmel and Michael A. Messner (Eds.), *Men's Lives* (3rd ed.). Boston: Allyn & Bacon, pp. 61–73.

Thorne, Barrie, Cheris Kramarae, and Nancy Henley. 1983. *Language, Gender, and Society.* Rowley, MA: Newbury.

Thornton, Arland, and Deborah Freedman. 1983. "The Changing American Family." *Population Bulletin,* 38 (October).

Thornton, Michael C., Linda M. Chatters, Robert Joseph Taylor, and Walter R. Allen. 1990. "Sociodemographic and Environmental Correlates of Racial Socialization by Black Parents." *Child Development,* 61: 401–409.

Thornton, Russell. 1984. "Cherokee Population Losses During the Trail of Tears: A New Perspective and a New Estimate." *Ethnohistory,* 31: 289–300.

Williams, David R., David T. Takeuchi, and Russell K. Adair. 1992. "Socioeconomic Status and Psychiatric Disorders Among Blacks and Whites." *Social Forces,* 71: 179–195.

Williams, Lena. 1995. "A Silk Blouse on the Assembly Lines? (Yes, the Boss's)." *New York Times* (Feb. 5): F7.

Williams, Norma. 1990. *The Mexican American Family: Tradition and Change.* Dix Hills, NY: General Hall.

Williams, Robin M., Jr. 1970. *American Society: A Sociological Interpretation* (3rd ed.). New York: Knopf.

Williamson, Robert C., Alice Duffy Rinehart, and Thomas O. Blank. 1992. *Early Retirement: Promises and Pitfalls.* New York: Plenum.

Willie, Charles V. 1991. *A New Look at Black Families* (4th ed.). Dix Hills, NY: General Hall.

Wilson, David (Ed.). 1997. "Globalization and the Changing U.S. City." *Annals of the American Academy of Political and Social Sciences,* 551 (May special issue).

Wilson, Edward O. 1975. *Sociobiology: A New Synthesis.* Cambridge, MA: Harvard University Press.

Wilson, Elizabeth. 1991. *The Sphinx in the City: Urban Life, the Control of Disorder, and Women.* Berkeley: University of California Press.

Wilson, James Q. 1996. "Foreword." In George L. Kelling and Catherine M. Coles (Eds.), *Fixing Broken Windows: Restoring Order and Reducing Crime in Our Communities.* New York: Touchstone, pp. xiii–xvi.

Wilson, William Julius. 1978. *The Declining Significance of Race: Blacks and Changing American Institutions.* Chicago: University of Chicago Press.

———. 1996. *When Work Disappears: The World of the New Urban Poor.* New York: Knopf.

Winik, Lyric Wallwork. 1997. "Oh Nurse, More Beluga Please." *Forbes FYI: The Good Life* (Winter): 157–166.

Winkleby, Marilyn A., and Catherine Cubbin. 2003. "Influence of Individual and Neighbourhood Socioeconomic Status on Mortality Among Black, Mexican-American, and White Women and Men in the United States." *Journal of Epidemiology and Community Health,* 57: 444–452.

Winn, Maria. 1985. *The Plug-in Drug: Television, Children, and the Family.* New York: Viking.

Wirth, Louis. 1938. "Urbanism as a Way of Life." *American Journal of Sociology,* 40: 1–24.

Wischnowsky, Dave. 2005. "Small Town a Bastion of Bigfoot Belief." *Chicago Tribune* (Oct. 10): 2.

Wiseman, Jacqueline. 1970. *Stations of the Lost: The Treatment of Skid Row Alcoholics.* Chicago: University of Chicago Press.

Wollstonecraft, Mary. 1974. *A Vindication of the Rights of Woman.* New York: Garland (orig. pub. 1797).

Women's Research & Education Institute. 2009. "Breaking Through the Brass Ceiling." Retrieved July 9, 2009. Online: http://www.wrei.org/News.html

Wonders, Nancy. 1996. "Determinate Sentencing: A Feminist and Postmodern Story." *Justice Quarterly,* 13: 610–648.

Wood, Daniel B. 2002. "As Homelessness Grows, Even Havens Toughen Up." *Christian Science Monitor* (Nov. 21). Retrieved July 1, 2003. Online: http://www.csmonitor.com/2002/1121/p01s04-ussc.htm

Wood, Julia T. 1994. *Gendered Lives: Communication, Gender, and Culture.* Belmont, CA: Wadsworth.

———. 1999. *Gendered Lives: Communication, Gender, and Culture* (3rd ed.). Belmont, CA: Wadsworth.

Wooley, Susan C. 1994. "Sexual Abuse and Eating Disorders: The Concealed Debate." In Patricia Fallon, Melanie A. Katzman, and Susan C. Wooley (Eds.), *Feminist Perspectives on Eating Disorders.* New York: Guilford, pp. 171–211.

World Bank. 2003a. "East Asia Navigates Short-Term Shocks for a Stronger Future." Retrieved July 15, 2003. Online: http://web.worldbank.org/WBSITE/EXTERNAL/NEWS/0,,contentMDK:20106953~menuPK:34463-pagePK:34370~piPK:34424~theSitePK:4607,00.html#

———. 2003b. "PovertyNet: Listen to the Voices." Retrieved July 10, 2003. Online: http://www.worldbank.org/poverty/voices/listen-findings.htm

———. 2003c. *World Development Indicators 2003.* Retrieved July 12, 2003. Online: http://www.worldbank.org/data/wdi2003/worldview.pdf

———. 2005. *World Development Indicators 2005.* Retrieved Mar. 4, 2006. Online: http://devdata.worldbank.org/wdi2005/Cover.htm

———. 2007. "Poverty in China: What Do the Numbers Say?" Washington, DC: World Bank. Retrieved Jan. 20, 2008. Online: http://web.worldbank.org

———. 2009. *World Development Indicators 2009.* Retrieved June 6, 2009. Online: http://siteresources.worldbank.org/DATASTATISTICS/Resources/wdi09introch1.pdf

World Health Organization. 2003. *The World Health Report 2003.* Retrieved Jan. 8, 2004. Online: http://www.who.int/whr/2003/en/overview_en.pdf

———. 2004a. "Infant and Under Five Mortality Rates by WHO Region. Year 2000." Retrieved Feb. 28, 2004. Online: http://www.who.int/child-adolescent-health/OVERVIEW/CHILD_HEALTH/Mortality_Rates_00.pdf

———. 2004b. "Suicide Rates and Absolute Numbers of Suicide by Country." Retrieved June 19, 2004. Online: http://www/who.int/mental_health/prevention/suicide/suicideprevent/en

Worster, Donald. 1985. *Natures Economy: A History of Ecological Ideas.* New York: Cambridge University Press.

Wouters, Cas. 1989. "The Sociology of Emotions and Flight Attendants: Hochschild's Managed Heart." *Theory, Culture & Society,* 6: 95–123.

Wright, Erik Olin. 1978. "Race, Class, and Income Inequality." *American Journal of Sociology,* 83 (6): 1397.

———. 1979. *Class Structure and Income Determination.* New York: Academic Press.

———. 1985. *Class.* London: Verso.

———. 1997. *Class Counts: Comparative Studies in Class Analysis.* Cambridge, England: Cambridge University Press.

Wright, Erik Olin, Karen Shire, Shu-Ling Hwang, Maureen Dolan, and Janeen Baxter. 1992. "The Non-Effects of Class on the Gender Division of Labor in the Home: A Comparative Study of Sweden and the U.S." *Gender & Society,* 6 (2): 252–282.

Wright, John W. (Ed.). 1997. *The New York Times 1998 Almanac.* New York: Penguin Reference.

Wuthnow, Robert. 1992. *Rediscovering the Sacred: Perspectives on Religion in Contemporary Society.* Grand Rapids, MI: Eerdmans. Prepared for Religion on Line by William E. Chapman. Retrieved July 12, 2009. Online: http://www.religion-online.org/showchapter.asp?title51509&C51350

Yablonsky, Lewis. 1997. *Gangsters: Fifty Years of Madness, Drugs, and Death on the Streets of America.* New York: New York University Press.

Yelin, Edward H. 1992. *Disability and the Displaced Worker.* New Brunswick, NJ: Rutgers University Press.

Yinger, J. Milton. 1960. "Contraculture and Subculture." *American Sociological Review,* 25 (October): 625–635.

———. 1982. *Countercultures: The Promise and Peril of a World Turned Upside Down.* New York: Free Press.

Young, John. 1990. *Sustaining the Earth: The Story of the Environmental Movement—Its Past Efforts and Future Challenges.* Cambridge, MA: Harvard University Press.

Young, Michael Dunlap. 1994. *The Rise of the Meritocracy.* New Brunswick, NJ: Transaction (orig. pub. 1958).

Zakaria, Fareed. 2005. "The Wealth of Yet More Nations." *New York Times Book Review* (May 1): 10–11.

Zald, Mayer N., and John D. McCarthy (Eds.). 1987. *Social Movements in an Organizational Society.* New Brunswick, NJ: Transaction.

Zavella, Patricia. 1987. *Women's Work and Chicano Families: Cannery Workers of the Santa Clara Valley.* Ithaca, NY: Cornell University Press.

Zelizer, Viviana. 1985. *Pricing the Priceless Child: The Changing Social Value of Children.* New Haven, CT: Yale University Press.

Zellner, William M. 1978. Vehicular Suicide: In Search of Incidence. Unpublished M.A. thesis, Western Illinois University, Macomb. Quoted in Richard T. Schaefer and Robert P. Lamm. 1992. *Sociology* (4th ed.). New York: McGraw-Hill, pp. 54–55.

Zill, N., and C. W. Nord. 1994. *Running in Place.* Washington, DC: Child Trends.

Zipp, John F. 1985. "Perceived Representativeness and Voting: An Assessment of the Impact of 'Choices' vs. 'Echoes.'" *American Political Science Review,* 60 (3): 738–759.

Zuboff, Shoshana. 1988. *In the Age of the Smart Machine: The Future of Work and Power.* New York: Basic.

Photo Credits

This page constitutes an extension of the copyright page. We have made every effort to trace the ownership of all copyrighted material and to secure permission from copyright holders. In the event of any question arising as to the use of any material, we will be pleased to make the necessary corrections in future printings. Thanks are due to the following authors, publishers, and agents for permission to use the material indicated.

Chapter 1. 2: © Plush Studios/Brand X Pictures/Getty Images **7:** Courtesy of Charlene Sullivan **8:** center, © Andrew Ward/Life File/Getty Images **8:** right, © DEA/M. BORCHI/Getty Images **8:** left, © Syracuse Newspapers/John Berry/The Image Works **10:** © AP Images/Eugene Hoshiko **12:** © Hulton Archive/Getty Images **13:** Auguste Comte (1798-1857) (oil on canvas), Etex, Louis Jules (1810-1889)/Temple de la Religion de l'Humanite, Paris, France/The Bridgeman Art Library International **14:** bottom, © Hulton Archive/Getty Images **14:** top, © Spencer Arnold/Getty Images **15:** © Bettmann/CORBIS **16:** © North Wind Picture Archives. All rights reserved. **18:** © Colin Young-Wolff/PhotoEdit **19:** © Hulton Archive/Getty Images **21:** bottom, © Simon Jarratt/CORBIS **21:** top, © The Granger Collection, New York **22:** top, © AP Images **22:** bottom, © Bettmann/CORBIS **24:** bottom, © Bob Daemmrich/The Image Works **24:** top right, © Estate of Robert K. Merton, photo by Sandra Still **24:** top left, © The Granger Collection, New York **26:** bottom, © AP Images/Denis Farrell **26:** top, © Hulton Archive/Getty Images **29:** © Michael Bezjian/WireImage/Getty Images **30:** © Bill Aron/PhotoEdit

Chapter 2. 36: © AP Images/Tom Gannam **39:** © AP Images/Gene Blythe **42:** © Yellow Dog Productions/Getty Images **47:** © Charlie Newham/Alamy **48:** © Junko Kimura/Getty Images **50:** © AP Images/HO **51:** © Michael Newman/PhotoEdit **54:** © Masterfile-RF **59:** © Ariel Skelley/Getty Images **60:** left, Abbreviated credit used for on-page credit - Cover used by permission of W. W. Norton & Company, Inc. Cover photo © Camilo Jose Vergara. **60:** right, © Addison Geary **62:** © Cate Gillon/Getty Images **65:** © AP Images/Dmitry Lovetsky **67:** © Geri Engberg/The Image Works

Chapter 3. 70: © Bob Daemmrich/The Image Works **72:** bottom, bottom, © Opla/Shutterstock **72:** center, center, © Andresr/Shutterstock **72:** top, top, © Gabriela Trojanowska/Shutterstock **75:** top left, © Celia Peterson/Getty Images **75:** top right, © Frans Lemmens/Getty Images **75:** bottom, © Mark Henley/Impact/HIP/The Image Works **76:** top right, © Mark Richards/PhotoEdit **76:** bottom, © Michael Greenlar/The Image Works **76:** top left, © Spencer Grant/PhotoEdit **79:** © Jim West/The Image Works **80:** © AP Images/Ann Johansson **82:** © Getty Images **84:** © Francis Dean/Dean Pictures/The Image Works **85:** © Spencer Grant/PhotoEdit **88:** © AP Images/Tony Dejak **89:** © Victor Englebert/Time Life Pictures/Getty Images **91:** © Sonda Dawes/The Image Works **92:** © Reuters/Landov **95:** bottom, © AP Images/John Raoux **95:** top, © David McNew/Getty Images **97:** © AP Images/ Gary Malerba

Chapter 4. 102: © Frank Trapper/CORBIS **106:** © Martin Rogers/Getty Images **107:** left, © Bill Aron/PhotoEdit **107:** right, © Jose Luis Pelaez, Inc./Getty Images **108:** © Bettmann/CORBIS **111:** top, © Tony Freeman/PhotoEdit **111:** center, © Tony Freeman/PhotoEdit **112:** © PhotoAlto/Alamy **115:** top right, © Daniel Bosler/Getty Images **115:** top left, © Digital Vision/Alamy **115:** center, © Stella/Getty Images **118:** © Michael Newman/PhotoEdit **120:** © Yellow Dog Productions/Getty Images RF **121:** © Michael Newman/PhotoEdit **122:** bottom, © Jeff Greenberg/The Image Works **122:** top, © SW Productions/Getty Images **123:** left, Courtesy of the National Runaway Switchboard **123:** bottom right, © DEAN J. KOEPFLER/Tacoma News Tribune/MCT/Landov **123:** top right, © Tim Boyle/Getty Images **124:** not needed **124:** top, © RAFIQUR RAHMAN/Reuters/Landov **126:** © Image Source RF/ Dinodia **128:** © AP Images **129:** © David Young-Wolff/PhotoEdit **131:** © Sonda Dawes/The Image Works **132:** © Journal Courier/The Image Works **133:** © Lisette LeBon/SuperStock

Chapter 5. 136: © Bob Collins/The Image Works **141:** © Jeff Brass/Getty Images **142:** bottom, © Rachel Epstein/PhotoEdit **142:** top, © Xinhua/Landov **144:** © Kim Eriksen/zefa/CORBIS **146:** left, © Gilles Mingasson **146:** right, © MARIO ANZUONI/Reuters/Landov **149:** © Jason Laure/The Image Works **151:** © Steve Satushek/Getty Images **153:** © Nevada Wier/Corbis **156:** © Mark Ludak/The Image Works **157:** © Bruce Ayres/Getty Images **159:** left, © MANDEL NGAN/AFP/Getty Images; right, © AP Images/Reed Saxon **161:** bottom, © AP Images/Alex Brandon **161:** top, © AP Images/Carolyn Kaster **163:** right, © David Silverman/Getty Images **163:** left, © Tom Prettyman/PhotoEdit **164:** left, © Dana White/PhotoEdit **164:** bottom right, © Krzysztof Mystkowski/AFP/Getty Images **164:** bottom right, © Krzysztof Mystkowski/AFP/Getty Images **164:** top right, © Richard Ross/Getty Images **165:** © David Young-Wolff/PhotoEdit **167:** Courtesy of Gretchen Otto

Chapter 6. 172: © PSL Images/Alamy **178:** © Colin Young-Wolff/PhotoEdit **179:** top, © Michael Newman/PhotoEdit **179:** bottom, © Michael Newman/PhotoEdit **181:** top, © Duncan Hale-Sutton/Alamy **181:** bottom, © Jeff Greenberg/Alamy **182:** top, © Chris Hondros/Getty Images **182:** bottom, © Ilene MacDonald/Alamy **187:** bottom right, NASA Kennedy Space Center (NASA-KSC) **187:** top right, © AP Images/Chris O'Meara **187:** left, © AP Images/Dr. Scott Lieberman **189:** top right, © A. Ramey/PhotoEdit **189:** top left, © AP Images/Vincent Thian **189:** bottom right, © David Grossman/The Image Works **191, left:** left, left, © John Aikins/CORBIS **191, right:** right, right, © Losevsky Pavel/Shutterstock **192:** © Sean Justice/Getty Images **194:** top, © Blend Images/Alamy **194:** bottom, © Mark Richards/PhotoEdit **196:** © TWPhoto/CORBIS

Chapter 7. 200: © A. Ramey/PhotoEdit **208:** © Andrew Holbrooke/Corbis **210:** © AP Images/Luke Palmisano **211:** © Nick Koudis/Getty Images **213:** © LEZLIE STERLING/MCT/Landov **214:** © AP Images/Kevork Djansezian **216:** left, © HBSS/Corbis **216:** right, © Photofusion Picture Library/Alamy **222:** © AP Images/Charles Rex Arbogast **225:** © Sean Cayton/The Image Works **229:** © AP Images/Bebeto Matthews **232:** © Robin Nelson/PhotoEdit **7.02:** © AP Images **F07.076B:** top center, © Image99/Jupiterimages **F07.07A:** top left, left, © Morgan Lane Photography/Shutterstock **F07.07C:** top right, right, © Image Source/Jupiterimages

Chapter 8. 240: © AP Images/Alex Brandon **246:** bottom left, © Alan Sussman/The Image Works **246:** top, © Hulton Archive/Getty Images **246:** © Jupiterimages RF/Getty Images **248:** © Robyn Beck/AFP/Getty Images **249:** © LYNN ISCHAY/The Plain Dealer /Landov **251:** Trae Patton/© NBC/Courtesy: Everett Collection **256:** bottom right, Cover of NICKEL AND DIMED : On (Not) Getting By in America. Cover design copyright 2008 by Henry Holt and Company. Text by Barbara Ehrenreich. Reprinted by permission of Henry Holt and Company, LLC. (Credit Abbrev. for on-page to: Cover design of Nickel and Dimed, copyright 2008 by Henry Holt and Company.) **256:** bottom left, © AP Images/Andrew Shurtleff **258:** top, © AP Images/Adrian Wyld/CP **258:** bottom, © ERIC DRAPER/WHITE HOUSE/UPI/ Landov **259:** bottom, © David Frazier/ PhotoEdit **264:** © AP Images/Michigan Lottery **265:** center, © AP Images/James Branaman/Kitsap Sun **265:** top, © Frances M. Roberts/Alamy **265:** bottom, © Liz Hafalia/San Francisco Chronicle/CORBIS **266:** bottom, not needed by CL contract with ABC **266:** top, © AP Images/Tim Boyd **267:** © Image Source/Getty Images **270:** © Sonda Dawes/The Image Works **272:** © AP Images/Paul Beaty **274:** © Steven Rubin/The Image Works **276:** © Joe Sohm/ The Image Works

Chapter 9. 280: © AP Images/Herbert Knosowski **285:** © Viviane Moos/CORBIS **286:** © Reinhard Krause/Reuters/CORBIS **292:** © Lauren Goodsmith/The Image Works **294:** © Jenny Matthews/Alamy **297:** © Chris Hondros/Getty Images **300:** © James Marshall/The Image Works **302:** © Scott Olson/Getty Images **304:** © AP Images/Sandra Boulanger

Chapter 10. 308: © AP Images/ U.S. Air Force, Master Sgt. Gerold Gamble **313:** © Jeff Greenberg/PhotoEdit **315:** © AP Images **316:** © AP Images/Martial Trezzini/ Keystone **318:** © Michael Greenlar/The Image Works **322:** left, © Bettmann/CORBIS **322:** right, © TIM JOHNSON/Reuters/ Landov **324:** © Shelly Katz/Getty Images **325:** © Bill Aron/PhotoEdit **329:** © AP Images/Mike Yoder/Lawrence Journal-World **332:** © AP Images/Charles Dharapak **334:** © Miramax Films/Photofest **335:** © AP Images **337:** © AP Images/Don Ryan **340:** © Jim West/The Image Works

Chapter 11. 344: © TIM SHAFFER/Reuters/ Landov **350:** left, © Alex Farnsworth/The Image Works **350:** right, © MOHAMMED AMEEN/Reuters/Landov **351:** right, © A. Ramey/PhotoEdit **351:** left, © David Young-Wolff/PhotoEdit **354:** Courtesy of Rima Khorebi **359:** © Barbara Campbell **360:** © Mary Kate Denny/PhotoEdit **361:** © AP Images **362:** © Bill Crump/ Brand X Pictures/Alamy **367:** right, © Spencer Grant/ PhotoEdit **367:** left, © Will Hart/PhotoEdit **369:** © Ron Sachs/Pool/CNP/Corbis **370:** © Image Source Pink/Getty Images **371:** top, © Ariel Skelley/Getty Images **371:** bottom, © Ed Bock/CORBIS **371:** center, © Tim Pannell/Corbis **372:** no credit needed–CL contract with ABC **372:** top, © moodboard/ Alamy **375:** top, © Larry Kolvoord/The Image Works **375:** bottom, © Mauritius/ Photolibrary

Chapter 12. 12.07: © Sonda Dawes/The Image Works **378:** © AP Images/ Sgt. 1st Class Kevin Mcdaniel **380:** © Bonnie Kamin/PhotoEdit **384:** © Jacob Halaska/ Photolibrary **386:** left, © George Shelley/ Masterfile **386:** right, © Olivier Martel/ Corbis **393:** © tom carter/Alamy **395:** "She's Got the Look" photo courtesy of TV Land. ©2009 Hudson Street Productions, Inc. All Rights Reserved. **397:** © AP Images/Marcio Jose Sanchez **402:** right, © Sonda Dawes/ The Image Works **402:** left, © Susan Van Etten/PhotoEdit

Chapter 13. 13.09: © AP Images/Harry Lynch/News & Obsesrver **408:** © AP Images/ Louis Lanzano **413:** right, © AP Images/Alan Marler **413:** left, © Mark Richards/PhotoEdit **414:** © AP Images/ Charles Bennett **418:** © AP Images **419:** © AP Images/Richard Drew **425:** © Alex Segre/Alamy **426:** © AP Images/Victoria Arocho **430:** © Peter Hvizdak/The Image Works **432:** © JLP/Sylvia Torres/CORBIS **433:** © Bill Aron/PhotoEdit **434:** © Kayte M. Deioma/PhotoEdit **435:** © AP Images/Mark Lennihan **436:** © PhotoEdit **438:** © Michael Newman/PhotoEdit **440:** © Mike Blank/ Getty Images

Chapter 14. 14.07: © AP Images/Reed Saxon **444:** © AP Images/Jason DeCrow **449:** bottom, Steve Petteway, Collection of the Supreme Court of the United States **449:** top, © AP Images/Will Burgess/Pool **449:** center, © Tim Graham/Getty Images **452:** © plainpicture GmbH/Alamy **453:** © Samir Hussein/WireImage/Getty Images **457:** © AP Images/Stephen J. Boitano **461:** bottom, © HANS DERYK/Reuters/Landov **461:** top, © Petrified Collection/Getty Images **463:** © AP Images/Lynne Sladky **468:** © AP Images/ Alex Brandon **470:** © AP Images/ Gloria Wright **472:** © AP Images/ Evan Vucci

Chapter 15. 476: © Myrleen Ferguson Cate/Photo Edit **480:** left, © Michael Newman/PhotoEdit **480:** right, © Steven Rubin/The Image Works **483:** © Tony Howarth/Woodfin Camp & Associates **485:** © Thinkstock/Corbis **487:** top left, © Bob Barkany/Getty Images **487:** top right, © Corbis Super RF/Alamy **487:** bottom right, © Michael Newman/PhotoEdit **489:** © AP Images/FarmersOnly.com **490:** left, © AP Images/Hector Mata **490:** right, © AP Images/ Christopher Gannon **492:** © Ariel Skelley/Getty Images **495:** bottom, © Fox Searchlight/Photofest **495:** top, © James Devaney/WireImage/Getty Images **497:** © AP Images/Ajit Solanki **503:** © Ariel Skelley/Blend Images/Corbis **506:** © Davis Barber/PhotoEdit

Chapter 16. 510: © Sky Bonillo/PhotoEdit **515:** © Kayte M. Deioma/PhotoEdit **516:** bottom, © Alan Oddie/PhotoEdit **518:** © Enigma/Alamy **521:** © Radius Images/ Jupiterimages **523:** © Cleve Bryant/ PhotoEdit **524:** © Gabe Palmer/Alamy **525:** © Spencer Grant/PhotoEdit **528:** Wellness/ Recreation Center, University of Northern Iowa. Photo provided by University Marketing & Public Relations **530:** bottom, © Bob Daemmrich/ The Image Works **530:** top, © Justin Sullivan/ Getty Images **532:** © Rob Crandall/Stock Connection/Alamy Images **536:** © AP Images/The News & Advance RVSHR/Chet White **539:** © BRIAN SNYDER/REUTERS/ Landov **542:** © Don Smetzer/PhotoEdit **543:** © Kevin Cooley/Getty Images **545:** © AP Images/Don Ryan

Chapter 17. 550: © AP Images/Jay Laprete **556:** © Jeff Greenberg/Alamy **557:** © David Young-Wolff/PhotoEdit **561:** © AP Images **562:** © Bob Daemmrich/The Image Works **564:** © James Marshall/Getty Images **566:** © Devendra M. Singh/AFP/Getty Images **567:** © Gary Conner/PhotoEdit **568:** © PhotoEdit **570:** © AP Images/Saiful Islam **571:** © Cindy Charles/PhotoEdit **574:** © AP Images **576:** © Rachel Epstein/The Image Works

Chapter 18. 582: © ERIK JACOBS/The New York Times/Redux Pictures **587:** top, © Andrew Holbrooke/The Image Works **587:** bottom, © Lew Lause/SuperStock **588:** © JB REED/Landov **589:** © AP Images/Steve Mitchell **590:** © David M. Grossman/The Image Works **591:** © Michael Newman/ PhotoEdit **595:** © Stanley B. Burns, MD & The Burns Archive N.Y./Photo Researchers, Inc. **596:** © AP Images/Gary Kazanjian **602:** © Royalty-Free/CORBIS **603:** © Sean Justice/Getty Images **605:** © Thinkstock/ Jupiterimages **611:** © Michael Newman/ PhotoEdit

Chapter 19. 618: © A. Ramey/PhotoEdit **624:** top, © Chris Hondros/Getty Images **624:** bottom, © David Young-Wolff/ PhotoEdit **626:** © AP Images/Robert F. Bukaty **627:** top left, © AP Images/Ann Heisenfelt **627:** right, © AP Images/Charles Krupa **627:** bottom left, © Alexander Tamargo/Getty Images **628:** bottom right, no credit needed **628:** top left, © Tony Freeman/ PhotoEdit **635:** © AP Images/George Osodi **636:** © Eyecon Images RF/Alamy **637:** © Comstock RF/Getty Images **640:** © Andrew Holbrooke/The Image Works **642:** © Jeff Greenberg/PhotoEdit **648:** right, © AP Images/Mark Duncan **648:** left, © THOM BAUR/AFP/Getty Images

Chapter 20. 654: © ROGER L. WOLLENBERG/ UPI/Landov **659:** © William Campbell/Sygma/CORBIS **660:** © AP Images/Jason E. Miczek **661:** © AP Images **662:** © David Young-Wolff/ PhotoEdit **664:** © Jonathan Fickies/Getty Images **666:** © AP Images **668:** © Dallas Events Inc./Shutterstock **671:** left, © AP Images/Rich Pedroncelli **671:** right, © Saul Loeb/epa/CORBIS **673:** © AP Images/Wen bao/ICHPL, Imagechina **675:** © Andrew Lichtenstein/The Image Works **677:** right, © AP Images/Bill Haber **680:** © Jeff Greenberg/PhotoEdit

Name Index